Climate Change, Photosynthesis and Advanced Biofuels

Ashwani Kumar • Yuan-Yeu Yau •
Shinjiro Ogita • Renate Scheibe
Editors

Climate Change, Photosynthesis and Advanced Biofuels

The Role of Biotechnology in the Production of Value-added Plant Bio-products

Springer

Editors
Ashwani Kumar
Department of Botany
and P.G. School of Biotechnology
University of Rajasthan
Jaipur, Rajasthan, India

Yuan-Yeu Yau
Department of Natural Sciences
Northeastern State University
Broken Arrow, OK, USA

Shinjiro Ogita
Faculty of Life and Environmental Sciences, Department of Life Sciences
Prefectural University of Hiroshima
Syoubara, Hiroshima, Japan

Renate Scheibe
Faculty of Biology and Chemistry, Plant Physiology
University of Osnabrück
Osnabrück, Niedersachsen, Germany

ISBN 978-981-15-5227-4 ISBN 978-981-15-5228-1 (eBook)
https://doi.org/10.1007/978-981-15-5228-1

This Springer imprint is published by the registered company Springer Nature Singapore Pte Ltd.
The registered company address is: 152 Beach Road, #21-01/04 Gateway East, Singapore 189721, Singapore

Dedication

Govindjee, Mister Photosynthesis
—An extraordinary ambassador of photosynthesis research to the world

A 2019 photograph of Govindjee. Source: College of Liberal Arts & Sciences, University of Illinois at Urbana-Champaign

Since 2019, Govindjee (who had always used one name only) began to use "Govindjee Govindjee" in order to be able to travel with ease around the world. He was born on October 24, 1932, at Allahabad, Uttar Pradesh, India; he married Rajni Varma on October 24, 1957, at Urbana, Illinois, USA; they have two children: Anita and Sanjay; he was naturalized to be a citizen of the USA in 1972. On the academic side, he has been Professor Emeritus of Plant Biology, Biophysics, and Biochemistry, at the University of Illinois at Urbana-Champaign (UIUC) since 1999. Going back in time, he was Professor of Biophysics and Plant Biology for 30 years (1969–1999), Associate Professor of Botany and Biophysics (1965–1969), and Assistant Professor of Botany (1961–1965), all at UIUC. During 1960–1961, he served as United States Public Health (USPH) Service Biophysics Post-Doctoral Fellow, and during his PhD days, 1956–1960, he held a UIUC Graduate Fellowship in Physico-Chemical Biology (Biophysics), as well a research assistantship in Botany earlier, i.e., during 1954–1956, he taught Plant Physiology in the Department of Botany at Allahabad University. He was trained first, during his MSc (1952–1954), in Plant Physiology, by Shri Ranjan, who was a student of Frederick Frost

Blackman (https://en.wikipedia.org/wiki/Frederick_Blackman). During his PhD (1956–1960), he was trained by Robert Emerson (https://en.wikipedia.org/wiki/Robert_Emerson_(scientist)), a student of the 1931 Nobel laureate Otto Warburg (https://www.nobelprize.org/prizes/medicine/1931/warburg/biographical/), and by Eugene Rabinowitch (https://en.wikipedia.org/wiki/Eugene_Rabinowitch), a post-doc student of 1926 Nobel laureate James Franck (https://www.nobelprize.org/prizes/physics/1925/franck/biographical/). After his PhD, and while he was on UIUC faculty, he also did collaborative research in several top photosynthesis laboratories, e.g., that of Bessel Kok (RIAS, Baltimore, MD, USA); C. Stacy French (Carnegie Institution of Science, Stanford, CA, USA); Jean Lavorel & Martin Kamen (CNRS, Gif-sur-Yvette, France); Louis N.M. Duysens (University of Leiden, Leiden, The Netherlands); and Horst Witt (Technical University, Berlin, Germany).

Govindjee has received many scholarships, fellowships, and has been recognized by several societies. The list includes Fulbright Lecture Award; Japanese Society for Promotion of Science (JSPS) Fellow; Fellow of the American Association for the Advancement of Science (AAAS); Fellow and Life Member of the National Academy of Science, India; President, American Society of Photobiology; Honorary President, International Photosynthesis Congress, Montreal, Canada; honored for his "Lifetime Contributions in the Field of Photosynthesis," by the Indian Society of Photobiology; the first Lifetime Achievement Award in Basic Biology, The Rebeiz Foundation; the Communication Award of the International Society of Photosynthesis Research (ISPR); The Professor B.M. Johri Memorial Award for outstanding plant scientists; and (*Pravasi*) Fellow of the National Academy of Agricultural Science, India.

Last but not least is his love for teaching which includes drama (outdoors or indoors) where students play the role of molecules involved in electron (and proton) transport from water to NADP, and in making ATP (see Photosynth Res https://doi.org/10.1007/s11120-016-0317-z, and https://doi.org/10.1007/s11120-014-0034-4). Further, for education around the world, he provides posters linked from his main web page: http://www.life.illinois.edu/govindjee/ (prepared by D. Shevela & coauthors) for teaching photosynthesis.

For further information on Govindjee and his life, see, e.g., http://www.life.illinois.edu/govindjee/recent_papers.html; **https://news.illinois.edu/view/6367/801235;** https://www.youtube.com/watch?v=cOzuL0vxEi0; and https://www.youtube.com/watch?v=OBKusHcjMzw

We, the editors of this volume, which focuses on new developments in biofuels and sustainable energy, based on photosynthesis and related processes, dedicate this book to Professor Emeritus Govindjee, one of the most respected authorities in the field of photosynthesis. He had contributed the Foreword for the previous volume entitled "*Biofuels: Greenhouse Gas Mitigation and Global Warming—Next Generation Biofuels and Role of Biotechnology*." Back in 2017, the editors pointed out that his outstanding contributions in the field of photosynthesis, in particular on the capture of light, energy conversion in the "light reactions," and electron transport (that uniquely requires bicarbonate), are fundamental for the development and distribution of this basic knowledge. This time, we wish to acknowledge his continued interest and engagement and congratulate him on his 88th birthday in the year 2020.

Jaipur, Rajasthan, India	Ashwani Kumar
Broken Arrow, OK, USA	Yuan-Yeu Yau
Syoubara, Hiroshima, Japan	Shinjiro Ogita
Osnabrück, Niedersachsen, Germany	Renate Scheibe

Foreword

Climate change threatens human welfare at a global scale. Depending on the degree of temperature rise, it must be expected that weather extremes in the future become even more frequent and more extreme than recently experienced. Drought resulting in forest fires, soil salinity, and desertification on the one hand and flooding on the other hand are two sides of the coin. The cause in both cases is an increase of thermal energy in the atmosphere due to the greenhouse effect. Among other greenhouse gases such as methane, ozone, and nitrous oxide, carbon dioxide plays a prominent role, contributing about 50% to the global temperature increase. Since the beginning of industrialization and the increasing combustion of fossil fuels, atmospheric carbon dioxide concentration has increased from 0.028% to more than 0.04%. In preindustrial times, the carbon dioxide concentration was in equilibrium, governed by four major processes: respiration and diffusion of carbon dioxide out of oceans releasing about 210 gigatons of carbon per year, whereas diffusion of carbon dioxide into oceans and photosynthesis balanced carbon dioxide release. Human activities increasingly caused higher carbon dioxide release, mainly due to the burning of fossil fuels and deforestation. Although part of the estimated 10 gigatons additional liberation of carbon dioxide is compensated by photosynthesis and diffusion into oceans, there is a net excess of roughly 5 gigatons carbon annually released into the atmosphere.

The net diffusion of carbon dioxide into oceans provokes another problem; the dissolution of carbon dioxide in water forms carbonic acid that decreases pH of the seawater. This in turn may harm sensitive marine life particularly in coral reefs. Thus, photosynthesis is the only significant process that contributes to a sustainable reduction of carbon dioxide concentration in the atmosphere. Presently, 123 gigatons of carbon are annually consumed by photosynthesis. An increase of global photosynthesis by 5% would more than compensate the additional anthropogenic carbon dioxide release. Besides abandoning the consumption of fossil fuels, (re)forestation and the use of renewable resources are key measures to limit further net release of carbon dioxide into the atmosphere. Large areas worldwide are available for sustainable plant production on hitherto barren land.

In the present book "*Climate change, photosynthesis, and advanced biofuels: Role of biotechnology in production of value-added plant products*," scientists from all over the world report their research results how photosynthesis may better contribute to the withdrawal of carbon dioxide from the atmosphere. The plant endeavor to collect carbon dioxide from a very low concentration and divert it to the various products is an extremely efficient process that has been genetically optimized during millions of years of evolution. However, with the development of crop plants, it has also become apparent that survival and propagation are not the main goals when plants are cultivated for food, feed, fiber, and fuel. Research on resistance against abiotic stresses, resource efficiency, as well as source-sink relationships has revealed that plants usually do not conduct photosynthesis at maximum intensity but with a downregulated rate. This helps to avoid the accumulation of metabolites that may compromise metabolism. At least five processes help to compensate excess photosynthesis relative to assimilate consumption in order to avoid the formation of reactive oxygen species that may impair nucleic acids and membranes:

1. Alternative respiration is less efficient in ATP production than conventional respiration and releases thermal energy.
2. Also, uncoupling proteins helps to dissipate some of the excess energy in the form of heat.
3. Photorespiration of C_3 plants consumes excess energy by releasing ammonia and carbon dioxide.
4. Root exudates that mobilize nutrients in the rhizosphere and support the translocation of signals into roots via phloem may reach as much as 30–40% of all photoassimilates.
5. Finally, even in unstressed situations plants may release as much as 10% of photoassimilates via volatilization of terpenes and similar compounds.

Consequently, in crop plants there is a large reservoir of photosynthetic potential that is not utilized in the production of generative storage organs. Even for vegetatively utilized fodder plants, there is evidence that sink activity rather than photosynthetic activity limits yield. Unraveling the bottlenecks during the establishment of sink activity will greatly help to divert more assimilates into storage organs such as grains, tubers, or beets. This, at the same time, will allow to enhance net photosynthesis, water, and nutrient efficiency, as well as crop yields. Sustainable production of food, feed, fiber, and fuel for an ever-growing word population is a continuing but feasible goal that at the same time may contribute to mitigate climate change.

This book provides valuable information, and I hope it will meet the expectations of researchers, students, and decision-makers alike on the subject of climate change and related matters. I highly recommend the book.

Institute of Plant Nutrition, Justus Liebig University Sven Schubert
Giessen, Germany

Prof. **Dr. Sven Schubert** studied agricultural sciences at Justus Liebig University Giessen, Germany. In 1985, he got his Ph.D. in agriculture with a thesis on proton exudation by plant roots under the guidance of Prof. Dr. Drs. h.c. Konrad Mengel. After a postdoc year at the University of California in Davis, USA, with Prof. Dr. Dr. h.c. André Läuchli, Prof. Schubert finalized his habilitation on mechanisms of salt resistance in maize plants in 1991. Following a call to the University of Hohenheim, Germany, in 1992, he spent 5 years as Professor for Plant Nutrition in Stuttgart before he was appointed Chair of the Institute of Plant Nutrition at Justus Liebig University Giessen in 1997. His main research areas are nutrient acquisition of plants, salt and acidity stress, and membrane biochemistry. Besides more than 120 refereed publications and 100 further publications, Prof. Schubert is author of the two German textbooks *Pflanzenernährung* (Plant Nutrition) and *Biochemie* (Biochemistry).

Selected Publications:

Yan F, Zhu Y, Müller C, Zörb C, Schubert S (2002) Adaptation of H^+-pumping and plasma membrane H^+ ATPase activity in proteoid roots of white lupin under phosphate deficiency. Plant Physiol 129:50–63

Qadir M, Oster JD, Schubert S, Noble AD, Sahrawat KL (2007) Phytoremediation of sodic and saline-sodic soils. Adv Agron 96:197–247

Hatzig S, Kumar A, Neubert A, Schubert S (2010) PEP-carboxylase activity: a comparison of its role in a C_4 and a C_3 species under salt stress. J Agron Crop Sci 196:185–192

Hanstein S, Wang X, Qian X, Friedhoff P, Fatima A, Shan Y, Feng K, Schubert S (2011) Changes in cytosolic Mg^{2+} levels can regulate the activity of the plasma membrane H^+-ATPase in maize. Biochem J 435:93–101

Hütsch BW, Osthushenrich T, Faust F, Kumar A, Schubert S (2016) Reduced sink activity in growing shoot tissues of maize (*Zea mays*) under salt stress of the first phase can be compensated by increased PEP-carboxylase activity. J Agron Crop Sci 202:384–393

Faust F, Schubert S (2017) *In vitro* protein synthesis of sugar beet (*Beta vulgaris*) and maize (*Zea mays*) is differentially inhibited when potassium is substituted by sodium. Plant Physiol Biochem 118:228–234

Hütsch BW, Schubert S (2017) Maize harvest index and water use efficiency can be improved by inhibition of gibberellin biosynthesis. J Agron Crop Sci 1–10
Qadir M, Schubert S, Oster JD, Sposito G, Minhas PS, Cheragji SAM, Murtaza G, Mirzabaev A, Saqib M (2018) High-magnesium waters and soils: Emerging environmental and food security constraints. Sci Total Environ 642:1108–1117
Hütsch BW, Jahn D, Schubert S (2019) Grain yield of wheat (*Triticum aestivum* L.) under long-term heat stress is sink-limited with stronger inhibition of kernel setting than grain filling. J Agron Crop Sci 205:22–32

Preface

During the Fifteenth session of Conference of the Parties (COP15), December 2009, almost 190 countries worldwide signed an agreement to keep a global temperature rise below 2 °C in this century. This agreement has a focus on temperature changes due to greenhouse gas emission with a preferable goal of remaining within 1.5 °C above preindustrial level (https://unfccc.int/process-and-meetings/the-paris-agreement/what-is-the-paris-agreement). During COP25 (December 2019), deliberations regarding implementation of policy ended with goals to be achieved by individual nations. Attendees agreed, it is not "climate change"; it is "climate emergency." With increasing global surface temperatures, the possibility of more heat waves, droughts, and increased powerful storms will likely occur (www.usgs.gov). Recent natural disasters such as droughts in California (USA) and South Africa fire in Australia, and vanishing islands in the oceans cannot be ignored. Worldwide increases in sea levels and acidification of seawater are causing loss of flora and fauna. All this negatively influences human health, food supply, and potentially promotes hunger and misery. However, an increasing human population, industrialization, and affluence all drive up the demand for energy. Currently, fossil fuels meet 88% of the demand, resulting in rising carbon dioxide (CO_2) and other greenhouse gas emissions. Carbon dioxide levels are increasing at alarming rates. The past 40 years have tracked an enormous rise in recorded temperature correlated with increasing carbon dioxide levels. Forests act as carbon sinks, but they are diminished worldwide with some vanishing in the Amazonian area with unprecedented loss of biodiversity.

It is common to hear people say, "I believe in the environment. I need clean water and clean air." The problem is energy sources for growth and development, which largely come from fossil fuels. Can we shift to green energy without negatively affecting economic growth? How quickly can we shift to green energy? How to find solutions to create a bright and sustainable future is important and urgent.

In our previous book, "*Biofuels: Greenhouse Gas Mitigation and Global Warming—Next Generation Biofuels and Role of Biotechnology*," it was pointed out that a small increase in global mean temperature results in a profound change in climate. Even a small increase leads to frequent floods, cyclones, droughts, and other natural calamities. Negative impacts of climate change are increasing in frequency and intensity. Much of the world has seen a steep rise in temperature during the last 15 years, including North America and Europe. Additionally, we are also witnessing unprecedented snowfall and cold. Changing weather patterns are reflective of climate change, according to the Intergovernmental Panel for Climate Change (IPCC). Global concerns regarding the use and benefit of biofuel can be addressed appropriately through proper species choice of plant, algae, and bacteria to improve biomass productivity.

The world's leading research groups have contributed valuable articles to current book from a variety of perspectives. Their insights include photosynthesis, source-sink relationships, stress resistance, productivity, nutrient uptake, and recycling. In addition, the role of biotechnology in developing next-generation biofuels and bio-products, and improving biofuels and new alternatives to meet global energy and food demands is also discussed. This book has three major parts. Each highlights one important approach to generating new biofuels/bio-products using plants, algae, or bacteria. Readers will learn about background physiology of plants and algae used as renewable energy sources. Optimally, attempts can be made without impacting food production by competing with agriculture. This can only be achieved by improving grain yield and primary productivity of specially designed crops capable of growing under stress conditions.

Part 1: Photosynthesis and Biomass Production under Changing Conditions

This section describes the present situation and role of new biofuels in a world experiencing dramatic climate change. The basic principles of photosynthetic processes, namely light reactions for energy capture and conversion, fixation and assimilation of oxidized carbon, nitrogen, and sulfur for the synthesis of all organic matter are described. In addition, all related pathways such as photorespiration and respiration, nutrient uptake and recycling, source-sink relationships, and stress resistance influencing productivity are included in this section. Regulatory principles allowing plants to maintain homeostasis under changing conditions are also discussed, which need to be understood for genetically modifying or inserting new pathways in plants to produce technically useful compounds. This section is most relevant for the successful setup of new approaches.

Part 2: Microalgae and Engineered Crops for Production of Biofuels and High-Value Products

Algal cultures are useful options for hydrogen production, and synthetically designed pathways for technologically useful products. An overview of bioproduction based on microalgae species is given. Specifically, the taxonomic distribution of major microalgae species used in industry is described. This section highlights the utility and many recent algae advance.

Part 3: Genetic Resources and Engineering Methods to Improve Crop Plants

Contemporary biotechnology options and potential improvements are presented. Improvements in biofuel production complement an array of value-added products to improve economic viability. Basic principles of the suggested methodologies, with physiological, genetic, and molecular information required for the production of functionally superior plant resources, are explained. Hybrid vigor (heterosis) is economically important for plant breeding. It plays a role in increasing fertility, growth rate, yield, and stress resistance in hybrids. In this regard, heterosis can be exploited to increase biomass production for biofuel crops. In this section, heterosis is discussed by an expert team.

The book provides plentiful resources for biofuel researchers and is designed to provide both general and specific information for students, teachers, academic researchers, industrial teams, as well as laymen who are interested in new developments for the production of biofuels containing value-added properties.

We are thankful to Professor Dr. Sven Schubert of Justus Liebig University in Gießen, Germany, for writing the foreword for our book. We heartily thank all of our coauthors and colleagues who have contributed to this book. We dedicate our book to Professor Dr. Govindjee of the University of Illinois at Urbana Champaign (USA) on his 88th birthday. Dr. Govindjee has contributed immensely to the basic understanding of photosynthesis and is commonly called *Mr. Photosynthesis*. His detailed biodata is enclosed.

Jaipur, Rajasthan, India — Ashwani Kumar
Broken Arrow, OK, USA — Yuan-Yeu Yau
Syoubara, Hiroshima, Japan — Shinjiro Ogita
Osnabrück, Niedersachsen, Germany — Renate Scheibe

Acknowledgments

We acknowledge source material from UNFCC, IPCC, COP, and other agencies of UN quoted in this book. Alexander von Humboldt Foundation Fellowship to one of the editors (AK) and constant encouragement received from Professor Dr. Sven Schubert from Institute for Plant Nutrition, Justus Liebig Universität, Gießen, Germany, Professor Dr. Raghavendra from Hyderabad, Professor Dr. Govindjee from USA, Professor Dr. Jack Widholm from USA, and Dr P.V. Sane from Pune.

We also thank and acknowledge 46 chapter contributors from 8 countries: **Yagut Allahverdiyeva** (Turku, Finland); **Most Tanziman Ara** (Hiroshima, Japan); **Wagner L. Araújo** (Viçosa, Brazil); **Willian Batista-Silva** (Viçosa, Brazil); **James A. Birchler** (Columbia, USA)); **Lindsey Brennan** (Broken Arrow, USA); **Marina Cvetkovska** (Ottawa, Canada); **Paula Da Fonseca-Pereira** (Viçosa, Brazil); **Avishek Dey** (Guwahati, India); **Mona Esterling** (Tulsa, USA); **Norman P.A. Hüner** (London, Canada); **Birgit W. Hütsch** (Gießen, Germany); **Alexander G. Ivanov** (Sofia, Bulgaria); **Osamu Iwata** (Kanagawa, Japan); **Martina Jokel** (Turku, Finland); **Stephan Jung** (Cologne, Germany); **Johannes Knuesting** (Osnabrück, Germany); **Sergey Kosourov** (Turku, Finland); **Ashwani Kumar** (Jaipur, India); **Sanjeev Kumar** (Guwahati, India); **Keiichi Mochida** (Kanagawa, Japan); **Chakravarthi Mohan** (Gainesville, USA); **Sujatha Mulpuri** (Hyderabad, India); **Ashwin Narayan** (Coimbatore, India); **Misaki Nishibe** (Hiroshima, Japan); **Srinivasan Nithiyanantham** (Hyderabad, India); **Greta Nölke** (Aachen, Germany); **Adriano Nunes-Nesi** (Viçosa, Brazil); **Shinjiro Ogita** (Hiroshima, Japan); **Paul Porter** (London, Canada); **Jyoti Porwal** (Dehradun, India); **Suheel Porwal** (Dehradun, India); **Marc Possmayer** (Ottawa, Canada); **Nurhidayah Syahira Muhammad Radzi** (Hiroshima, Japan); **Lingaraj Sahoo** (Guwahati, India); **Renate Scheibe** (Osnabrück, Germany); **Stefan Schillberg** (Aachen, Germany); **Sven Schubert** (Gießen, Germany); **Jennifer Selinski** (Bielefeld, Germany); **Kuldeep Singh** (Mullana, India); **Raman Singh** (Mullana, India); **Marleen Steinbach** (Gießen, Germany); **Beth Szyszka-Mroz** (London, Canada); **Lam-Son Phan Tran** (Kanagawa, Japan); **Yuan-Yeu Yau** (Broken Arrow, USA); **Agustin Zsögön** (Viçosa, Brazil).

We also acknowledge figures reproduced from papers listed below: Espaux, L., Mendez-Perez, D., Li, R. and Keasling, J. D. (2015) Synthetic biology for microbial production of lipid-based biofuels. *Current Opinion in Chemical Biology* 29:58–65.

Figure was reproduced under license no. 4652791194061 (Figs. 2.1 and 2.4); Zhang, Y. P. (2015) Production of biofuels and biochemicals by *in vitro* synthetic biosystems: Opportunities and challenges. *Biotechnology Advances* 33(7):1467–1483. Figure was reproduced under license no. 4652950482642 (Fig. 2.2); Albers, S. C., Berklund, A. M., and Graff, G. D. (2016) The rise and fall of innovation in biofuels. *Nature Biotechnology* 34(8):814–821. Figure was reproduced under license number 4642520128912 (Fig. 2.3a–d); Nielsen, J. and Keasling, J. D. (2016) Engineering Cellular Metabolism. *Cell* 164(6):1185–1197. Figure was reproduced under license no. 4666911059117 (Fig. 2.5); Jagadevan, S., Banerjee, A., Banerjee, C., Guria, C., Tiwari, R. and Baweja, M. (2018) Biotechnology for Biofuels: Recent developments in synthetic biology and metabolic engineering in microalgae towards biofuel production. *Biotechnology for Biofuels* 11:1–21. Used under Creative Commons license (Fig. 2.6); Liao, J. C., Mi, L., Pontrelli, S. and Luo, S. (2016). Fuelling the future: microbial engineering for the production of sustainable biofuels. *Nature Review Microbiology* 14(5):288–304. Figure was reproduced under license no. 4645730007098 from Rights Link (Fig. 2.7); Alper, H. and Stephanopoulos, G. (2009) Engineering for biofuels: exploiting innate microbial capacity or importing biosynthetic potential, *Nature Reviews Microbiology* 7:715–723. Retrieved from https://doi.org/10.1038/nrmicro. Figure was reproduced under license no. 46456400840514 (Figs. 2.8 and 11.1); Peralta-Yahya P.P. et al. (2012) Microbial engineering for the production of advanced biofuels. *Nature* 488:320–328. Reproduced under license no. 4643340791481 (Figs. 2.9 and 2.10); Su, H., Lin, J. and Wang, G. (2016) Metabolic engineering of *Corynebacterium crenatium* for enhancing the production of higher alcohols. *Scientific Reports* 6:39543. Open Access. Used under Creative Commons license (Fig. 2.11); Zhang, Y. P. (2015) Production of biofuels and biochemicals by *in vitro* synthetic biosystems: Opportunities and challenges. *Biotechnology Advances* 33(7):1467–1483. Reproduced under license no. 4652950482642 (Fig. 2.12); Liao, J. C., Mi, L., Pontrelli, S. and Luo, S. (2016) Fuelling the future: microbial engineering for the production of sustainable biofuels. *Nature Review Microbiology* 14(5):288–304. Reproduced under license no. 4645730007098 (Fig. 11.2); Martien J.I. and Amador-Noguez D. (2017) Recent applications of metabolomics to advance microbial biofuel production. *Current Opinion in Biotechnology* 43:118–126. Reproduced under license no. 4666750205840 (Fig. 11.3); Jones, J.A., Ö. Duhan Toparlak and Mattheos AG Koffas (2015) Metabolic pathway balancing and its role in the production of biofuels and chemicals. *Current Opinion in Biotechnology* 33:52–59. Reproduced with permission no. 4671031226483 (Fig. 11.4); Peralta-Yahya, P. P., Ouellet, M., Chan, R., Mukhopadhyay, A., Keasling, J. D. and Lee, T. S. (2011) Identification and microbial production of a terpene-based advanced biofuel. *Nature Communications* 2:483–488. This is an open-access article distributed under the terms of the Creative Commons CC-BY license, which permits unrestricted use, distribution, and reproduction in any medium, provided the original work is properly cited (Figs. 11.5 and 11.6); Ma, T., Shi, B., Ye, Z., Li, X., Liu, M., Chen, Y. and Nielsen, J. (2019) Lipid engineering combined with systematic metabolic engineering of *Saccharomyces cerevisiae* for high-yield production of lycopene. *Metabolic*

Engineering 52:134–142. Reproduced under license no. 4651230668162 (Figs. 11.7 and 11.8); Peralta-Yahya P. P. et al. (2012) Microbial engineering for the production of advanced biofuels. Reproduced under license no. 4643340791481 (Fig. 11.9); Zhang, Y. P. (2015) Production of biofuels and biochemicals by *in vitro* synthetic biosystems: Opportunities and challenges. *Biotechnology Advances* 33(7):1467–1483. Reproduced under license no. 4652950482642 (Fig. 11.10); Jagadevan, S., Banerjee, A., Banerjee, C., Guria, C., Tiwari, R. and Baweja, M. (2018) Biotechnology for Biofuels Recent developments in synthetic biology and metabolic engineering in microalgae towards biofuel production. *Biotechnology for Biofuels* 11:1–21. Used under Creative Commons license (Fig. 11.11); Georgianna, D. R. and Stephen, P. (2012) Exploiting diversity and synthetic biology for the production of algal biofuels. *Nature* 488:330–335. Reproduced under license no. 4646381493445 from Rights Link (Fig. 11.12); Georgianna, D. R. and Stephen, P. (2012) Exploiting diversity and synthetic biology for the production of algal biofuels, *Nature* 488:330–335. Reproduced under license no. 4646381493445 (Fig. 11.13); Georgianna, D. R. and Stephen, P. (2012) Exploiting diversity and synthetic biology for the production of algal biofuels. *Nature* 488:330–335. Reproduced under license no. 4646381493445 (Fig. 11.14); Huo, Y.-X., Cho, K. M., Rivera, J. G. L., Monte, E., Shen, C. R., Yan, Y. and Liao, J. C. (2011) Conversion of proteins into biofuels by engineering nitrogen flux. *Nature Biotechnology* 29(4):346–351. Reproduced with permission under license no. 4646190098001 (Fig. 11.15).

We would like to thank the Springer Nature for publishing the book and the support from Dr. Akansha Tyagi of Springer Nature for guiding us through the process and brings out this book so nicely.

University of Rajasthan, Jaipur, India	Ashwani Kumar
Broken Arrow, OK, USA	Yuan-Yeu Yau
Hiroshima, Japan	Shinjiro Ogita
Osnabrück, Germany	Renate Scheibe
Jan 2020	

Contents

Editors and Contributors

About the Editors

Ashwani Kumar is an Alexander von Humboldt Fellow (Germany) and Professor Emeritus, Department of Botany, University of Rajasthan, Jaipur, India. He initially worked with Professor Dr. K. H. Neumann at the Institute of Plant Nutrition, Justus Liebig University Giessen, Germany, on the photosynthetic apparatus in vitro and in vivo, and subsequently with Professor Dr. Sven Schubert on the physiology and role of enzymes in salinity stress resistance with support from an Alexander von Humboldt Fellowship. A former JSPS visiting professor to Japan, Dr. Kumar has published 220 research papers and 23 books. Currently, he is the President of the Indian Botanical Society.

Yuan-Yeu Yau (aka Frank Yau) received his B.Sc. in Botany from National Taiwan University (NTU) in Taiwan, and received his Master and Ph.D. from the University of Wisconsin, Madison, USA. He subsequently worked at the University of California Berkeley and the Plant Gene Expression Center (USDA-ARS) in Albany (California, USA); the Chinese Academy of Sciences (Full Professor); and Northeastern State University (Oklahoma, USA). His work focuses on developing clean-gene technologies using microbial site-specific recombination (SSR). He has also worked on cotton transformation and the production of anti-stroke drugs using molecular farming. He is an active member of Research Gate, a professional network for scientists and researchers.

Shinjiro Ogita has over two decades of experience in the field of plant biotechnology. In 1992, he began his research career as a master student at the Graduate School of Agriculture, Tokyo University of Agriculture and Technology (TUAT), Japan. In 1997, he received his Ph.D. in Agriculture from the United Graduate School of Agriculture, TUAT, Japan. He is an expert on cell and tissue cultures, and on transformation technologies for higher plants. He is currently a full professor and director of the Field Sciences Center, Prefectural University of Hiroshima (PUH), Japan.

Renate Scheibe held the Chair of Plant Physiology until 2019, teaching on all aspects of plant physiology. Her research chiefly focuses on redox-regulation and energy metabolism; her interests include the molecular mechanisms for maintenance of redox-homeostasis in connection with stress adaptation, which she explores using approaches ranging from protein chemistry and enzymology, to ecophysiology. She has published more than 160 papers, and is now actively contributing to the scientific community with editorial tasks and as a member of the German Research Integrity group. She was elected as a Corresponding Member of the American Society of Plant Biologists and an Honorary Member of the German Botanical Society.

Contributors

Yagut Allahverdiyeva Molecular Plant Biology Unit, Department of Biochemistry, University of Turku, Turku, Finland

Most Tanziman Ara Program in Biological System Science, Graduate School of Comprehensive Scientific Research, Prefectural University of Hiroshima, Hiroshima, Japan

Wagner L. Araújo Departamento de Biologia Vegetal, Universidade Federal de Viçosa, Viçosa, Minas Gerais, Brazil

J. Ashwin Narayan Division of Crop Improvement, ICAR—Sugarcane Breeding Institute, Coimbatore, India

Willian Batista-Silva Departamento de Biologia Vegetal, Universidade Federal de Viçosa, Viçosa, Minas Gerais, Brazil

James A. Birchler Division of Biological Sciences, University of Missouri-Columbia, Columbia, MO, USA

Lindsey Brennan Department of Natural Sciences, Northeastern State University, Broken Arrow, OK, USA

Marina Cvetkovska Department of Biology and the Biotron Centre for Experimental Climate Change Research, University of Western Ontario, London, ON, Canada

Department of Biology, University of Ottawa, Ottawa, ON, Canada

Paula da Fonseca-Pereira Departamento de Biologia Vegetal, Universidade Federal de Viçosa, Viçosa, Minas Gerais, Brazil

Avishek Dey Department of Bioscience and Bioengineering, Indian Institute of Technology Guwahati, Guwahati, Assam, India

Mona Easterling Department of Natural Sciences, Northeastern State University, Broken Arrow, OK, USA

Tulsa Community College, Northeast Campus, Tulsa, OK, USA

Norman P. A. Hüner Department of Biology and The Biotron Experimental Climate Change Research Centre, University of Western Ontario, London, ON, Canada

Birgit W. Hütsch Institute of Plant Nutrition, Justus Liebig University (iFZ), Giessen, Germany

Alexander G. Ivanov Department of Biology and the Biotron Centre for Experimental Climate Change Research, University of Western Ontario, London, ON, Canada

Institute of Biophysics and Biomedical Engineering, Bulgarian Academy of Sciences, Sofia, Bulgaria

Osamu Iwata RIKEN Center for Sustainable Resource Science, Bioproductivity Informatics Research Team RIKEN Cluster for Science, Technology, and Innovation Hub, RIKEN Baton Zone Program, Microalgae Production Control Technology Laboratory, Yokohama, Japan

Martina Jokel Molecular Plant Biology Unit, Department of Biochemistry, University of Turku, Turku, Finland

Stephan Jung Institute of Plant Nutrition, Justus Liebig University (iFZ), Giessen, Germany

Johannes Knuesting Department of Plant Physiology, Faculty of Biology and Chemistry, Osnabrueck University, Osnabrueck, Germany

Sergey Kosourov Molecular Plant Biology Unit, Department of Biochemistry, University of Turku, Turku, Finland

Ashwani Kumar Department of Botany and P.G. School of Biotechnology, University of Rajasthan, Jaipur, India

Sanjeev Kumar Department of Bioscience and Bioengineering, Indian Institute of Technology Guwahati, Guwahati, Assam, India

Keiichi Mochida RIKEN Center for Sustainable Resource Science, Bioproductivity Informatics Research Team RIKEN Cluster for Science, Technology and Innovation Hub, RIKEN Baton Zone Program, Microalgae Production Control Technology Laboratory, Yokohama, Japan

Chakravarthi Mohan, PhD Department of Genetics and Evolution, Federal University of São Carlos, São Carlos, Brazil

Sujatha Mulpuri ICAR-Indian Institute of Oilseeds Research, Hyderabad, India

Misaki Nishibe Program in Biological System Science, Graduate School of Comprehensive Scientific Research, Prefectural University of Hiroshima, Hiroshima, Japan

Srinivasan Nithiyanantham Tierra Seed Science Private Limited, Hyderabad, India

Greta Nölke Department of Plant Biotechnology, Fraunhofer Institute for Molecular Biology and Applied Ecology IME, Aachen, Germany

Adriano Nunes-Nesi Departamento de Biologia Vegetal, Universidade Federal de Viçosa, Viçosa, Minas Gerais, Brazil

Shinjiro Ogita Program in Biological System Science, Graduate School of Comprehensive Scientific Research, Prefectural University of Hiroshima, Hiroshima, Japan

Paul Porter Department of Biology and the Biotron Centre for Experimental Climate Change Research, University of Western Ontario, London, ON, Canada

Jyoti Porwal CSIR-Indian Institute of Petroleum, Dehradun, Uttarakhand, India

Suheel K. Porwal DIT University, Dehradun, Uttarakhand, India

Marc Possmayer Department of Biology and the Biotron Centre for Experimental Climate Change Research, University of Western Ontario, London, ON, Canada

Department of Biology, University of Ottawa, Ottawa, ON, Canada

Nurhidayah Syahira Muhammad Radzi Program in Biological System Science, Graduate School of Comprehensive Scientific Research, Prefectural University of Hiroshima, Hiroshima, Japan

Lingaraj Sahoo Department of Bioscience and Bioengineering, Indian Institute of Technology Guwahati, Guwahati, Assam, India

Renate Scheibe Department of Plant Physiology, Faculty of Biology and Chemistry, Osnabrueck University, Osnabrueck, Germany

Stefan Schillberg Fraunhofer Institute for Molecular Biology and Applied Ecology IME, Aachen, Germany

Phytopathology Department, Institute for Phytopathology and Applied Zoology, Justus-Liebig University Giessen, Giessen, Germany

Sven Schubert Institute of Plant Nutrition, Justus Liebig University (iFZ), Giessen, Germany

Jennifer Selinski Department of Biochemistry and Physiology of Plants, Faculty of Biology, Bielefeld University, Bielefeld, Germany

Kuldeep Singh Maharishi Markandeshwar (Deemed to be University), Mullana, Ambala, Haryana, India

Raman Singh Maharishi Markandeshwar (Deemed to be University), Mullana, Ambala, Haryana, India

Marleen Steinbach Institute of Plant Nutrition, Justus Liebig University (iFZ), Giessen, Germany

Beth Szyszka Department of Biology and the Biotron Centre for Experimental Climate Change Research, University of Western Ontario, London, ON, Canada

Lam-Son Phan Tran RIKEN Center for Sustainable Resource Science, Yokohama, Japan

Yuan-Yeu Yau Department of Natural Sciences, Northeastern State University, Broken Arrow, OK, USA

Agustin Zsögön, PhD Departamento de Biologia Vegetal, Universidade Federal de Viçosa, Viçosa, Minas Gerais, Brazil

Introduction 1

Ashwani Kumar, Yuan-Yeu Yau, Shinjiro Ogita, and Renate Scheibe

Global climate change due to an accumulation of greenhouse gases (GHGs) causes concern regarding the statue of fossil fuels as a primary energy source. Current climate damage and estimates of future risk have grown, bringing previously implausible targets into consideration. One example is a complete carbon phase-out by 2050. However, the potential impact of such rapid reductions raised human rights concerns. A rapid carbon phase-out would be very demanding for all countries, particularly in developing countries. Human rights impacts of climate change have been widely assessed and provide a strong justification for a rapid reduction of emissions. Even greater are the risks of profound climate change, if temperatures rise to more than 2 °C, which is much more likely if mitigation is delayed. A carbon phase-out rapid enough to keep warming less than 2 °C will require extremely ambitious mitigation action in both rich and poor countries.

Furthermore, as has been widely noted, a low-to-zero-carbon future means a majority of world fossil fuel reserves will never be burned. This has potential for countries with fossil resources to forego revenue that could be put to developmental objectives. Along with these "stranded assets" goes a wide range of related infrastructure and human capital. Most directly, an increase in energy costs due to the

A. Kumar (✉)
Department of Botany and P.G. School of Biotechnology, University of Rajasthan, Jaipur, India

Y.-Y. Yau
Department of Natural Sciences, Northeastern State University, Broken Arrow, OK, USA

S. Ogita
Faculty of Life and Environmental Sciences, Department of Life Sciences, Prefectural University of Hiroshima, Hiroshima, Japan
e-mail: ogita@pu-hiroshima.ac.jp

R. Scheibe
Faculty of Biology and Chemistry, Department of Plant Physiology, University of Osnabrück, Osnabrück, Germany
e-mail: rscheibe@uni-osnabrueck.de

A. Kumar et al. (eds.), *Climate Change, Photosynthesis and Advanced Biofuels*,
https://doi.org/10.1007/978-981-15-5228-1_1

closing of fossil fuel-driven industrialization may adversely affect overall development prospects particularly in poorer countries. The anticipated impacts of climate change are often large scale, unpredictable, and irreversible, with extended time lags and outside of human control. In contrast, threats posed by mitigation activities are generally of limited scale, more predictable, and not generally marked by long time lags and remain within human control, as governed primarily by the socioeconomic process. There is reason to believe risks posed by mitigation can be handled, provided there is an ambitious and shared global effort to achieve a rapid carbon phase-out that preserves human rights. A commitment to integrating human rights and equity in all national climate policies is necessary.

Economic analyses suggest a rapid carbon phase-out can be achieved at an aggregate global cost that is affordable and much less than potential costs of climate impacts. There is a strong evidence that a rapid, total (or near-total) carbon phase-out will be technically feasible, both for developed and developing countries. Hydrocarbons produced from biomass using microbial fermentation processes can serve as high-quality liquid transportation fuels and may contribute to a reduction in GHG emissions.

In our previous book (*Biofuels: Greenhouse Gas Mitigation and Climate Change*), we discussed establishing bio-based hydrocarbon production from cheap feedstocks, lowering the cost of developing efficient and robust microbial cell factories, and establishing more efficient routes for biomass hydrolysis to sugars for fermentation. This book provides ample information on overcoming barriers and advancing opportunities for bio-based production of hydrocarbons to produce novel value-added products, in addition to biodiesel and jet biofuels. The chapters are grouped in three major sections: compiling (1) basic facts and attempts concerning metabolic processes and many-faceted adaptations, (2) possibilities for improving or changing metabolism, and (3) examples of successful approaches and genetic techniques for improving plant production of value-added compounds.

1.1 Part I: Photosynthesis and Biomass Production in a Changing World

The increasing rate of population growth, industrialization, and prosperity have led to extensive use of energy. In addition to natural climatic variability, anthropogenic climate change is taking place, due to emissions of greenhouse gases causing environmental damage (IEA 2007; IPCC 2007; Stocker et al. 2013; IPCC 2014; Kumar 2018a, b; Kumar et al. 2019). Fossil fuels produce a major share of greenhouse gases (GHGs) (IPCC 2014). Almost 88% of current energy usage comes from burning of fossil fuels, which contributes to increased CO_2 concentrations. The CO_2 level in 2012 was about 40% higher than nineteenth-century levels. CO_2 is a major contributor to greenhouse gases. Three major non-CO_2 groups of gases—CH_4, N_2O, and fluorinated gases (F-gases), including CF_4, HFCs, and SF_6—also contribute to GHG emissions. Increased levels of greenhouse gas emissions are leading to climate change with adverse effects reportedly causing floods, droughts, forest fires, melting of glaciers at a faster rate, and other natural calamities. During the Conference of the Parties (COP21) Paris Climate Conference (2015), a legally binding and universal agreement on climate change was achieved, with the aim of keeping global warming below 2 °C. Achieving this goal will require drastic emission reductions to stabilize GHG concentration in the atmosphere. Replacement of fossil oil with biofuel derived from plant biomass has the potential to greatly reduce greenhouse gas emissions (see Kumar et al. 2018a; https://doi.org/10.1007/978-81-322-3763-1).

Bioenergy is a renewable energy from biological sources. *Biofuels* are fuels that can be produced from biomass, which are renewable compared to fossil fuels (Kumar 2018a, b). *Biomass* can be defined as the collection of all organic matter composing biological organisms. The main components utilized for biofuel production are sugars (starch, simple sugars, and lignocelluloses) and lipids (Kumar 2001; Kumar 2013; Kumar et al. 2018a, b, 2019). Kumar and Roy (2006) detailed different factors affecting yield of *Jatropha curcas* capable of being increased by experimental manipulations and deposited high-yielding accessions at NBPGR, New Delhi, in a Department of Biotechnology, Govt. of India-funded project. Lactiferous plants growing in arid and semi-arid regions are rich in triterpenoids, which can be converted into biofuel (Kumar 2018a, b). Terpenoids comprise the largest family of natural products and have widespread applications. Extensive studies have been carried out on laticiferous plants growing in arid and semi-arid regions: *Euphorbia antisyphilitica*, *Euphorbia lathyris*, *Euphorbia tirucalli*, *Calotropis procera* (Kumar 2018b), and *Pedilanthus tithymaloides* (Kumar 2013; Kumar et al. 2018a, b).

There is a growing appreciation for solidarity concerning human rights protections to overcome the narrow pursuits of economic nationalism. However, climate inaction poses a risk to human rights, and the climate impacts far outweigh the risks to human rights posed by climate action consistent with meeting the 1.5 °C goal set in the Paris Agreement (see Kumar et al. 2018a; https://doi.org/10.1007/978-81-322-3763-1). Replacement of fossil oil with biofuel derived from plant biomass has the potential to greatly reduce greenhouse gas emissions (Chap. 2).

The impact of climate change and the rising demand for food, feed, and biofuels require an increase in crop productivity without the use of additional land, water, or agrochemicals. Despite recent progress in plant breeding and biotechnology, improving crop productivity beyond existing yield potentials remains one of the greatest challenges in agricultural research. Photosynthesis is the basis of primary productivity on the planet. Crop breeding has sustained steady improvements in yield to keep pace with a growing population. Yet these advances have not resulted in improving the photosynthetic process per se, but rather in altering the way carbon is partitioned within the plant. Given that the pathways of photosynthesis and respiration catalyze partially opposing processes, it follows that their relative activities must be carefully regulated within plant cells. Exciting recent developments regarding interaction between respiration and photosynthesis and the potential mechanisms linking mitochondrial function and photosynthetic efficiency will be reviewed (Chap. 3).

We emphasize challenges and opportunities to further understand the complex interplay between photosynthesis and related metabolic processes. Knowledge of regulatory interconnections is prerequisite to successful manipulations that improve photosynthesis. We summarize recent advances in alternative manipulation and enhancement of photosynthesis and possible applications for crop improvement.

In photosynthetic organisms, energy input is highly variable due to changes in light intensity and other environmental factors. Fast and flexible adjustments for energy distribution and consumption are required to avoid stress (Scheibe 2019). Optimization of photosynthesis and avoidance of wasteful processes have been the focus of many crop improvement studies. Production and consumption of the energy carriers ATP, NADPH, and NADH must be kept in balance since their pool sizes are small and reducing equivalents cannot be transported directly across compartment borders (Selinski and Scheibe 2019). Furthermore, reversible post-translational modifications of target enzymes at regulatory cysteines and fine-tuning fluxes at respective redox switches by small molecules can adjust actual enzyme activities and maintain homeostasis (Knüsting and Scheibe 2018). Finally, sustained stress causes signal transfer via cytosolic redox switches leading to changes in gene expression and induction of salvage pathways required to maintain homeostasis. For any biotechnological approach aiming to produce tailored products and increase yield, consideration of regulatory networks and mechanisms of energy distribution between sinks and sources are of basic importance (Chap. 4).

Several biotechnological approaches have been proposed to increase photosynthetic rate in important C3 crops, including engineered RuBisCO, enhancing Calvin-Benson cycle enzyme activity, introducing CO_2-concentrating mechanisms, and manipulating photorespiration. However, few of these strategies have led to significantly higher crop yields in practice. In Chap. 5, limitations of photosynthesis in C3 plants are briefly discussed, and a focus is placed on current strategies to overcome bottlenecks and achieve higher agricultural productivity. In closing, remaining challenges and perspectives for future development of novel strategies to enhance photosynthetic efficiency are considered.

Cold acclimation and cold adaptation in phototrophs confer the capacity to respond to excess excitation energy. Although modulation of the photosynthetic apparatus redox state is a common feature in sensing excess excitation energy, the response to this redox sensing/signaling mechanism is species-dependent. These concepts are discussed with respect to acclimation and adaptation of green algae, cyanobacteria, and terrestrial plants to extreme environments represented by Antarctic and Arctic ecosystems. We suggest there is an urgent need for more comprehensive research focused on physiology, biochemistry, genomics, and metabolomics of the myriad, yet undiscovered, organisms inhabiting these extreme environments, which may provide novel biotechnological applications to industry, agriculture, and medicine (Chap. 6).

Grain yield and its determinants, i.e., kernel number and single kernel weight, were recorded at maturity and related to physiologically relevant parameters 2 days after controlled pollination. Single kernel weight was unaffected by saline conditions. Decreased number of kernels was not caused by source limitation, because availability of sucrose as a main transport metabolite was consistently higher in developing kernels under salt stress than in control conditions. Acid invertase activity is a key factor for sink activity that was reduced or unchanged under salt stress; hexose concentrations were higher in developing kernels. This points to no limitation. Hexoses are needed for metabolic processes and energy supply in order to enable cell division and extension growth. Plasma membrane H^+-ATPase activity was significantly reduced in the salt-stressed kernels, resulting in a smaller pH gradient. Thus, kernel development and subsequent grain yield performance under salt stress seem to be limited by a transport problem, caused by inhibition of plasma membrane H^+-ATPase (Chap. 7).

1.2 Part II: Microalgae and Engineered Crops for Production of Biofuels and High-Value Products

Microalgae have attracted increasing attention as a renewable energy source and feedstock because of their potential for use in bio-based fuels and material production. Chapter 8 provides an overview of bioproduction based on microalgae species. Specifically, they describe the taxonomic distribution of major industrially exploited microalgae species for biomass use and highlight their utilities and recent advances.

Molecular hydrogen (H_2) is a promising energy carrier for a future sustainable economy. There are a number of different approaches for industrial production of H_2 fuel. However, renewable production of H_2 remains a challenge. Some photosynthetic green algae possess hydrogenase enzyme(s) and naturally photoproduce H_2 gas. In view of the high sensitivity of hydrogenases to O_2 and also to other cellular metabolic hindrances, H_2 photoproduction is not yet efficient enough for industrial applications. Chapter 9 summarizes different protocols developed to date for production of H_2 in algal cultures, including two novel and promising approaches, and discusses advantages and disadvantages of these methods.

Biodiesel and bioethanol are primary biofuels, yet they have limitations in feedstock and production process. Synthetic biofuel can be produced from any type of biomass. Biofuels have a diverse array of feedstocks and pathways available. It may be sustainable, renewable alternative fuel over fossil fuels. It may be a boon in GHG mitigation or reduction of world-level carbon dioxide (CO_2) (Chap. 10).

Recent findings of plant metabolic pathways reconstituted in heterologous hosts, and engineered metabolism of crop plants to improve biofuel production has given new hope for molecular biological approaches in improving food and biofuel production. The de novo engineering of genetic circuits, biological modules, and synthetic pathways is beginning to address these crucial problems and is being used in related practical applications (Chap. 11).

1.3 Part III: Genetic Resources and Engineering Methods to Improve Crop Plants

Hybrid vigor (heterosis) is the phenomenon that hybrids are superior to their parents in biomass and fertility. Heterosis is agronomically important in plant breeding because superior performance can appear as growth rate, yield, and stress tolerance. Heterosis can be extensively exploited for increasing productivity in agriculture (Hochholdinger and Baldauf 2018). Some recent studies of so-called single-gene heterosis and large-scale genomic/phenomic studies have suggested genes typically involved with quantitative traits seem to be major determinants of heterosis rather than complementation of random mutations in a variety of functions. As usual, once a new insight emerges, one can recognize prescient experimental results in classical studies foreshadowing such realizations. In Chap. 12, the history of these prescient results is traced back and related to new ideas concerning evaluation of quantitative traits with a special emphasis on heterosis. Recent advances in analytical platforms and information techniques have enabled us to identify genes or small RNAs (e.g., small interfering RNA or siRNA) which might be involved in plant heterosis. For example, researchers have found increased enzyme activity involved in carbon fixation pathways and net photosynthetic rate in super-hybrid rice LY2186 (Song et al. 2010), indicating photosynthetic capacity is important for heterosis. The global expression profile of siRNA changed in hybrids, usually down-regulated (He et al. 2010). By altering expression levels of these genes and small RNAs, researchers have the potential to increase heterosis, such as biomass heterosis (Ni et al. 2009). This chapter also discusses challenges and possibilities using genetic engineering and gene editing to foster heterosis.

Chapter 13 provides an overview of recent advances in genome resource development in biomass plants. Specifically, this section focuses on grass species such as maize, sugarcane, sorghum, switchgrass, and *Miscanthus* spp. as well as oil crops such as soybean, sunflower, jatropha, and oil palm. The highlights involve genome-based efforts and information resources to improve crop biomass productivity.

Sugarcane is an important worldwide cash crop used for both sugar and ethanol production. Improvement of sugarcane through conventional breeding practices has been limited by its complex polyploid genome. Production of transgenic sugarcane is an alternative method to improve sugarcane traits. RNA interference (RNAi) and recent popular CRISPR (clustered regularly interspaced short palindrome repeats)/Cas9 (CRISPR-associated) genome-editing systems are powerful tools for crop trait improvement (Jinek et al. 2012). RNAi technology enables scientists to modify gene expression, while CRISPR-Cas9 allows genetic elements to be added, deleted, or modified at particular locations in the genome. However, both technologies require a robust transformation method. Efficient sugarcane transformation protocols will be vital in harnessing the potential of this energy crop. High transformation efficiencies are now on the horizon due to improved methods. Chapter 14 discusses the discovery of RNAi, its underlying molecular mechanism, and its applications in gene function study plus crop trait improvement. This section also describes popular RNAi vectors used to induce gene silencing in plants. Finally, the limitations of

RNAi technology are explained. Chapter 15 describes recent advances in sugarcane transformation and highlights novel improvement strategies to enhance target gene expression.

Yield loss in sugarcane due to insect pests ranges from 10 to 30%. Despite the application of insecticides, pesticides, other chemicals, and different integrated pest management (IPM) techniques, a need for improvement remains. Use of chemicals to control pests can potentially cause soil and water contamination and possess a toxic effect on non-target organisms. Transgenesis approach can, however, increase both quality and productivity of crops in an environmentally friendly manner. With genetic engineering advances, considerable success has been achieved in sugarcane genetic improvement. To this end, Chap. 16 focuses on the development of transgenic sugarcane for disease and pest resistance.

A simple, fast, and efficient plant genetic transformation system can facilitate functional genomic studies. Tobacco is a model plant for genetic transformation, with leaf disk transformation being the most commonly used method for its transformation. However, leaves taken from larger tobacco seedlings or plants are usually used. In Chap. 17, we discuss a method using tiny tobacco cotyledons for genetic transformation. This protocol also eliminates submersion of explants in *Agrobacterium* liquid culture and minimizes *Agrobacterium* overgrowth. Maintaining explant fitness for later tissue culturing is one of the many advantages of this method.

Jatropha curcas (jatropha), an oilseed plant with a multitude of uses, is a potential biofuel crop (Kumar and Roy 2006). Most programs are dependent upon germplasm available in undomesticated condition, and the wish list for genetic improvement of this crop is exhaustive. Genetic diversity analysis using molecular markers unarguably confirmed Central American and Mexican regions as treasure troves of *J. curcas* genetic diversity. There is a need to explore varietal development and hybrid breeding programs using this information. With a modest estimate of 6–8 years of concerted efforts, improved germplasm with desired attributes could be available, and improved germplasm could be phased to replace established plantations with unproductive yields (Chap. 18).

Plant-based fuels are generated from renewable sources. Present generation biorefinery focuses on bioengineering of microorganisms to increase target product. Benefits aside, genetically engineered (GE) or genetically modified organisms (GMOs) have been considered a threat to the environment and human health. Therefore, this chapter focused on the combination of metabolic engineering (ME) and plant cell manipulation technology (PCMT) to create alternatives for safer biorefinery production. Satisfactory improvement in metabolic engineering of bamboo and other energy crops has been achieved. Bamboo, the highest biomass producer, or other energy crops can be target organisms for PCMT and ME technologies, substituting GE for a safer biorefinery. Perhaps this technology will create a new generation of biorefinery (Chap. 19).

We hope, in the future, biofuel will become a safe and economical alternative to fossil fuels. More urgently needed are products to replace technically used chemicals presently derived from fossil oil reserves. These need to be replaced by green chemistry. Out of the many promising options, genetic engineering techniques are the basis for success. Therefore, particularly in Europe, a well-defined but not restrictive law concerning GMOs must be agreed upon and accepted. This would allow for application of promising techniques and not hinder research and projects directed to developing solutions for mankind in a favorable environment (see Leopoldina statement 2019). In this respect, this volume is also a useful source of ideas and examples for advisors, policy-makers, and any responsible member of our society. In an ad hoc statement, the National Academy of Sciences in Germany has drawn a current picture of the actual situation and their climate targets to be reached by 2030 (Leopoldina 2019).

This book offers a wide scope of facts and ideas for any scientist or layman or designer and developer of new approaches, strategies, and products. Ample references are provided for these promising and urgently needed directions. We definitely need to minimize climate change by use of any method or approach. Even better would be a combination of available and future techniques, to address global warming and counteract a food and energy shortage. A concerted effort is needed to cope with the upcoming and already threatening environmental, economic, and social problems.

References

He G, Zhu X, Elling AA, Chen L, Wang X, Guo L, Liang M, He H, Zhang H, Chen F et al (2010) Global epigenetic and transcriptional trends among two rice subspecies and their reciprocal hybrids. Plant Cell 22:17–33

Hochholdinger F, Baldauf JA (2018) Heterosis in plants. Curr Biol 28:R1089–R1092

IEA (2007) Energy security and climate policy, assessing. Interactions. http://www.iea.org/textbase/nppdf/free/2007/energy_security_climate_policy.pdf

IPCC (2007) Climate change 2007: synthesis report. In: Core writing team, Pachauri RK, Reisinger A (eds) Contribution of working groups I, II and III to the fourth assessment report of the intergovernmental panel on climate change. IPCC, Geneva, p 104

IPCC (2014) Climate change 2014: synthesis report. In: Core Writing Team, Pachauri RK, Meyer LA (eds) Contribution of working groups i, ii and iii to the fifth assessment report of the intergovernmental panel on climate change. IPCC, Geneva, p 151

Jinek M, Chylinski K, Fonfara I, Hauer M, Doudna JA, Charpentier E (2012) A programmable dual-RNA-guided DNA endonuclease in adaptive bacterial immunity. Science 337:816–821

Knüsting J, Scheibe R (2018) Small molecules govern thiol-redox switches. Trends Plant Sci 23:769–782

Kumar A (2001) Bioengineering of crops for biofuels and bioenergy. In: Bender L, Kumar A (eds) From soil to cell: a broad approach to plant life. Giessen + Electron. Library GEB, 14–29, pp 1–5. http://geb.uni-giessen.de/geb/volltexte/2006/3039/pdf/FestschriftNeumann-2001.pdf

Kumar A (2013) Biofuels utilisation: an attempt to reduce GHG's and mitigate climate change. In: Nautiyal S, Rao K, Kaechele H, Raju K, Schaldach R (eds) Knowledge systems of societies for adaptation and mitigation of impacts of climate change. Environmental science and engineering. Springer, Berlin, pp 199–224

Kumar A (2018a) Alternative biomass from semiarid and arid conditions as a biofuel source: *Calotropis procera* and its genomic characterization. In: Kumar A, Ogita S, Yau YY (eds) Biofuels: greenhouse gas mitigation and global warming next generation biofuels and role of biotechnology. Springer, Heidelberg, pp 241–270

Kumar A (2018b) Global warming, climate change and greenhouse gas mitigation. In: Kumar A, Ogita S, Yau YY (eds) Biofuels: greenhouse gas mitigation and global warming next generation biofuels and role of biotechnology. Springer, Heidelberg, pp 1–16

Kumar A, Roy S (2006) Plant biotechnology and its applications in tissue culture. I.K. International, New Delhi, p 307

Kumar A, Ogita S, Yau YY (eds) (2018a) Biofuels: greenhouse gas mitigation and global warming next generation biofuels and role of biotechnology. Springer, Heidelberg, p 432

Kumar A, Abraham E, Gupta A (2018b) Alternative biomass from saline and semiarid and arid conditions as a source of biofuels: *salicornia*. In: Kumar A, Ogita S, Yau YY (eds) Biofuels: greenhouse gas mitigation and global warming. Springer, New Delhi

Kumar A, Bhansali S, Gupta N, Sharma M (2019) Bioenergy and climate change: greenhouse gas mitigation. In: Rastegari AA, Yadav AN, Gupta A (eds) Prospects of renewable bioprocessing in future energy systems. Biofuel and biorefinery technologies, vol 10. Springer, Cham, pp 269–290

Leopoldina statement (2019). https://www.leopoldina.org/uploads/tx_leopublication/2020_G-Science_Global_Insect_Declines_Statement.pdf

Leopoldina (2019) Climate targets 2030: towards a sustainable reduction of CO_2 emissions. Halle (Saale). https://www.leopoldina.org/uploads/tx_leopublication/2019_Stellungnahme_Klimaziele2030_EN_web.pdf

Ni Z, Kim ED, Ha M, Lackey E, Liu J, Zhang Y, Sun Q, Chen ZJ (2009) Altered circadian rhythms regulate growth vigour in hybrids and allopolyploids. Nature 457:327–331

Scheibe R (2019) Maintaining homeostasis by controlled alternatives for energy distribution in plant cells under changing conditions of supply and demand. Photosynth Res 139:81–91

Selinski J, Scheibe R (2019) Malate valves: old shuttles with new perspectives. Plant Biol 21:21–30

Song GS, Zhai HL, Peng YG, Zhang L, Wei G, Chen XY, Xiao YG, Wang L, Chen YJ, Wu B et al (2010) Comparative transcriptional profiling and preliminary study on heterosis mechanism of super-hybrid rice. Mol Plant 3:1012–1025

Stocker TF, Qin D, Plattner G-K, Tignor M, Allen SK, Boschung J, Nauels A, Xia Y, Bex V and Midgley PM (eds) (2013) Cambridge University Press, Cambridge, p 1535

Ashwani Kumar Alexander von Humboldt Fellow (Germany), Professor *Emeritus,* Department of Botany, University of Rajasthan, Jaipur, India, Studied photosynthetic apparatus in vitro and in vivo initially working with Professor Dr. K H Neumann at Institute of Plant Nutrition at Justus Liebig University Giessen, Germany, and subsequently with Professor Dr. Sven Schubert on physiology role of enzymes in salinity stress resistance with support from Alexander von Humboldt Fellowship. JSPS visiting Professor Japan. Guided 39 students for PhD, published 220 research papers and 23 books. Presently, he is the president of the Indian Botanical Society.

Yuan-Yeu Yau Dr. Yau obtained a BSc in Botany at National Taiwan University (NTU), and received Master and PhD from University of Wisconsin-Madison, USA. He then worked at University of California-Berkeley and Plant Gene Expression Center (USDA-ARS) at Albany (California, USA), Chinese Academy of Sciences (Professor), and Northeastern State University (USA). He works in areas of plant biotechnology to develop clean-gene technology using microbial site-specific recombination system. He also worked on cotton transformation and production of stroke drug using molecular farming. He is also an active member of *ResearchGate*, a professional network for scientists and researchers.

Shinjiro Ogita Dr. Ogita has over two decades of experience in the field of plant biotechnology. Since 1992 he started his research career as a master student at the Graduate School of Agriculture, Tokyo University of Agriculture and Technology (TUAT), Japan. In 1997 he received a PhD in Agriculture (subject; sciences of resources and environment) by the United Graduate School of Agriculture, TUAT, Japan (1997). He is an expert in the field of cell and tissue culture, and transformation technologies for higher plants. He presently works as a full professor and the director of Field Sciences Center, Prefectural University of Hiroshima (PUH), Japan.

Renate Scheibe Dr. Scheibe has held the Chair of Plant Physiology until 2019, teaching the whole field of Plant Physiology. Her research covers topics in the fields of redox regulation and energy metabolism, redox homeostasis in the view of stress adaptation using approaches from protein chemistry, enzymology, and ecophysiology. She published more than 160 papers, and is actively contributing for the scientific community with editorial tasks and as a member of the German Research Integrity group. She was elected as a Corresponding Member of The American Society of Plant Biologists and as Honorary Member of the German Botanical Society.

Part I

Photosynthesis and Biomass Production Under Changing World

Climate Change: Challenges to Reduce Global Warming and Role of Biofuels

2

Ashwani Kumar

Abstract

The increasing level of population growth, industrialization, and prosperity is leading to extensive use of energy. The use of fossil fuels produces a major share of greenhouse gases (GHG). Almost 88% of this energy comes from the burning of fossil fuels. This is contributing to the increase in CO_2 levels. The CO_2 level in 2012 was about 40% higher than it was in the nineteenth century. CO_2 is a major contributor to greenhouse gases. Besides these three major non-CO_2 groups of gases CH_4, N_2O, and fluorinated gases (F-gases), including CF_4, HFCs, and SF_6 also contribute to GHG emissions. The increased levels of greenhouse gas emissions are leading to climate change and its adverse effects are reported to cause floods, droughts, forest fires, and melting of glaciers at a faster rate besides other natural calamities. During the Conference of the Parties (COP21), at the Paris Climate Conference (2015), a legally binding and universal agreement on climate change was achieved, with the aim of keeping global warming below 2 °C. Achieving this goal will require drastic emission reductions to stabilize GHG concentration in the atmosphere. Replacement of fossil oil with biofuel derived from plant biomass has the potential to greatly reduce greenhouse gas emissions.

Keywords

Greenhouse gases · Climate change · Biofuels · Metabolic engineering

A. Kumar (✉)
University of Rajasthan, Jaipur, India

A. Kumar et al. (eds.), *Climate Change, Photosynthesis and Advanced Biofuels*,
https://doi.org/10.1007/978-981-15-5228-1_2

2.1 Introduction

In addition to natural climatic variability, the anthropogenic climate change is taking place due to emissions of greenhouse gases which cause environmental damage to any given area (IEA 2007; IPCC 2007; Stocker et al. 2013; IPCC 2014; Kumar 2018a, b, c; Kumar et al. 2019). Some indicators of climate change are sea-level rise, ocean heat and acidification, and sea-ice and glacier melt continue (Source: https://coyotegulch.blog/2018/11/29/climatechange-signals-and-impacts-continue-in-2018-world-meteorological-organization-actonclimate/. Much has been written in global warming and climate change recently (Kumar et al. 2018a, b; Kumar et al. 2019); the object of this paper is to introduce it to the readers causes of climate change and measures to check it. Since most of the work is already published and the aim of the review is to introduce the theme, the focus is on providing significant developments and what science can provide for the future.

Increasing world population (9.7 billion in 2050 and 11.2 billion in 2100) (United Nations, World Population Prospects (2015 revision) will require increasing levels of energy for economic growth and development, a large portion of which comes from the use of fossil fuels (United Nations: *World Population Prospects:* 2017).

The level of greenhouse gases has been rising continuously and in 2017, greenhouse gas concentrations with CO_2 at 405.5 ± 0.1 parts per million (ppm) increased up to 146%, over the level of the pre-industrial era (Liao et al. 2016). Use of fossil fuel produces the most important greenhouse gas (GHGs) CO_2, and other urban pollutants such as NO_x, CO, CFCs, methane, particulate matters (PMs), unburned hydrocarbons (UHCs), and aromatics keep on accumulating due to anthropogenic activity (IPCC 2007, 2014; Xie et al. 2013).

As per IEA (2007), projections of warming have been made on a credible business-as-usual case extended to 2100. "This case assumes a global annual growth rate of 1.6% in the next 25 years. Under this assumption, CO_2 concentration is projected to increase to 500 ppm in 2050 and 825 ppm by 2100. Such concentrations will yield best-guess average warming, relative to 1990, of 1.5 °C in 2050 and 3.5 °C in 2100. There is still a large range of uncertainty associated with these warming projections; the potential warming in 2100 could be as high as 4.5 °C or as low as 2.1 °C. This warming would be in addition to the 0.4 °C already experienced from 1700 to 1990. Warming would continue into the next century, with equilibrium warming in the 2.3–10.1 °C range, with the best guess at 4.8 °C above 1990 levels" (IEA 2007). The World Health Organization has predicted that between 2030 and 2050, climate change will cause approximately 250,000 additional deaths per year, from malnutrition, malaria, diarrhea, and heat stress (Robinson and Shine 2018).

At the Paris climate conference (COP21) in December 2015, 195 countries adopted the first-ever universal, legally binding global climate deal. This agreement sets out a global action plan to avoid dangerous climate change by limiting global warming to well below 2 °C and pursuing efforts to limit it to 1.5 °C.

In view of the Paris accord of 2015 and IPCC, SR15 report limiting global warming to 1.5 °C above pre-industrial implies reaching net-zero CO_2 emissions globally around 2050 along with deep reductions in emissions of non-CO_2 gases, particularly methane. The commitment made by developed countries in Copenhagen at COP15 and reiterated at COP21 in Paris, to mobilize US$ 100 billion per year of public and private finance by 2020 for climate action in developing countries, may enable developing countries to adopt 1.5 °C-compatible development pathways (Hof et al. 2017).

2.2 Climate Change Mitigation

Different climate models suggest that current atmospheric greenhouse gas concentrations may already be very near those associated with a stable climate at 1.5 °C (Huntingford and Mercado 2016). The world needs reductions of near-term emissions in combination with greenhouse gas removal from the atmosphere (negative emissions) to achieve this lower temperature goal. Besides these transformational changes to energy provision and other sectors including industrial activity and land management, as well as negative emissions will be needed (Hall et al. 1991; Rogelj et al. 2011; Shue 2014; Mollendorf 2013; Liao et al. 2016; Robinson and Shine 2018).

2.2.1 The Market Mechanisms and the Carbon Market

Emissions from developing countries are projected to significantly increase in the near future. Many developing countries are taking mitigation action, including the scaling up of renewables in energy generation or energy efficiency targets.

The Kyoto Protocol introduced three innovative mechanisms, by which Annex I Parties can lower their cost of achieving emission targets (IPCC 2014).

The mechanisms enable countries to access cost-effective opportunities to reduce emissions, or to remove carbon from the atmosphere in other countries:

1. The Clean Development Mechanism (CDM) projects that involve an industrialized country buying carbon credit for a developing country, which uses the payment to produce biofuels or dedicated vehicles, etc., might be a better option to foster biofuels, depending on their greenhouse gas reduction potential (Cécile et al. 2011).
2. Joint Implementation (JI) Funding projects in countries with economies in transition (EITs).
3. Emissions Trading, which allows to trade credits or emission allowances among themselves.

2.2.2 Carbon Capture and Storage Strategies

Photosynthetic activities of living plants can support large-scale carbon dioxide (CO_2) removal from the atmosphere by afforestation/reforestation and avoided deforestation through strategies of Biomass Energy with Carbon Capture and Storage (BECCS) (Smith et al. 2016). The more ambitious mitigation scenarios require an even greater land area for mitigation and/or earlier adoption of CO_2 removal strategies (Harper et al. 2018).

2.3 Renewable Energy Sources

Presently plant biomass provides 10% of global primary energy mainly as ethanol and biodiesel production are both expected to expand to reach, respectively, almost 135 and 39 billion liters by 2024 (5OECD/FAO. Agricultural Outlook, 2015, Available online, https://doi.org/10.1787/data-00736-en) (Smith et al. 2014; Woods et al. 2015). In 2040, the share of biofuels in road transport fuels would range—depending on policies—from 5 to 18% globally, from 11 to 31% in the European Union and from 11 to 29% in the United States (International Energy Agency. World Energy Outlook 2015) (Bopp et al. 2016).

According to Popp et al. (2016a, b), the projected world's primary energy demand by 2050 is expected to be in the range of 600–1000 EJ/year compared to about 500 EJ in 2008. At present, some 55 EJ/year of bioenergy is produced globally. The expert assessment suggests potential deployment levels of bioenergy by 2050 in the range of 100–300 EJ/year. It is expected that annual biomass potential could be of between 200 and 500 EJ/year from energy crops, surplus forest growth, and increased agricultural activity, forestry and agricultural residues other organic wastes (including municipal solid waste).

Energy Agency (IEA 2008) projected that a 27% market penetration of biofuels will be needed by 2050 in order to keep CO_2 emissions below 450 ppm. In the last 35 years, global energy supplies have nearly doubled but the relative contribution from renewables has increased from 13 to 19%, including about 9.3% from traditional biomass and about 9.7% from modern renewables (Sarkar et al. 2012; Espaux et al. 2015).

2.3.1 Bioenergy: Biofuels

Bioenergy is renewable energy from biological sources (see FAO 2008a, b). *Biofuels* are fuels that can be produced from biomass and are renewable as compared to fossil fuels (Kumar 2008, 2011; Cécile et al. 2011). *Biomass* can be defined as the collection of all organic matter composing biological organisms, but the main components utilized for biofuel production are sugars (starch, simple sugars, and lignocelluloses) and lipids (Kumar 2001; Hill et al. 2006; Roy and Kumar 2013; Dugar and Stephanopoulos 2011; Caspeta and Nielsen 2013). The biomass feedstocks include plant oils, starchy materials, sugar crops like sugarcane and sugar beets, cereals, and organic waste (Kumar 2018a, b, c). Switchgrass is a C_4 perennial grass native to the North American great plains that has been developed as a biofuel crop. Sustainable large-scale production of cellulosic biofuels will require the integration of knowledge across many disciplines.

The primary asset of *biofuels* is the convenience that they can be used as blends with conventional fuels in existing vehicles. Biodiesel is defined by ASTM International as a fuel composed of monoalkyl esters of long-chain fatty acids derived from renewable vegetable oils or animal fats meeting the requirements of ASTM D6751 (ASTM 2008).

Since *ethanol* and *biodiesel* both contain oxygen they are better combustibles than the substituted fossil oils, reducing the emission of pollutants such as CO, hydrocarbons (HC), sulfur oxide, and particulates by up to half of these emissions, depending on the biofuel and the blended mix (Murugesan et al. 2009; Lane 2015).

Biofuels reduce emissions because CO_2 produced by fuel combustion is offset by CO_2 captured by growing biomass, which is in turn used to produce more fuel (Espaux et al. 2015) (Fig. 2.1).

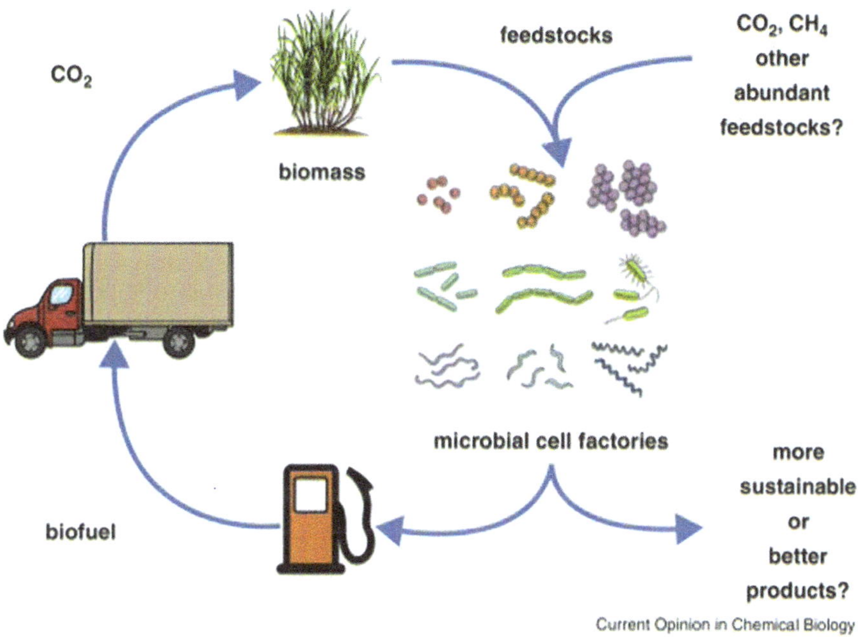

Fig. 2.1 Carbon cycle for a microbial biofuel. Biofuels reduce emissions because CO_2 produced by fuel combustion is offset by CO_2 captured by growing biomass, which is in turn used to produce more fuel. With synthetic biology, it may be possible to produce fuel from various sources of carbon and energy. It may also be possible to produce fuels, or other molecules, with improved properties using the diverse bioconversions observed in living organisms. Additionally, systematic engineering, including host and pathway engineering for terpenoid overproduction, has been reviewed (Bian et al. 2017). Source: Espaux, L., Mendez-Perez, D., Li, R., and Keasling, J. D. (2015). Synthetic biology for microbial production of lipid-based biofuels. *Current Opinion in Chemical Biology 29*: 58–65. https://doi.org/10.1016/j.cbpa.2015.09.009. Reproduced under licence number 4652791194061 from Rights Link

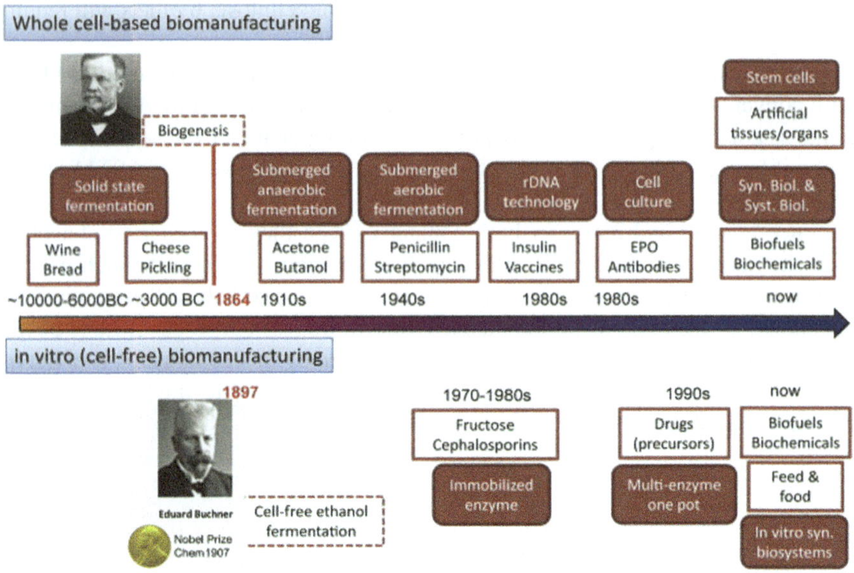

Fig. 2.2 History of biomanufacturing catalyzed by whole cells and in vitro (cell-free) biosystems associated with key milestones. Source: Zhang, Y. P. (2015). Production of biofuels and biochemicals by in vitro synthetic biosystems: Opportunities and challenges. *Biotechnology Advances 33*(7): 1467–1483. https://doi.org/10.1016/j.biotechadv.2014.10.009. Reproduced with licence number 4652950482642 from Rights Link

The history of biomanufacturing catalyzed by whole cells to in vitro is reviewed by Zhang (2015) (Fig. 2.2). With synthetic biology, it may be possible to produce fuel from various sources of carbon and energy (Dale et al. 2014; Espaux et al. 2015).

Albers et al. (2016) compared world fuel price and output, refinery output of petroleum motor fuels and total biofuels and production of fuel ethanol, the main form of biofuel (Figs. 2.2 and 2.3). The share of waste biodiesel feedstocks such as animal fat and used cooking oil increased to 15% in total biodiesel output (Licht 2013).

A correlation among world oil prices, fuel production, fuel ethanol production, and inventions of major fuel types has been established by Albers et al. (2016) in his review article (Fig. 2.3a, b, c, d).

Albers et al. (2016) reported that the second-generation or advanced biofuels—with a promise to environmental, energy security, and economic development promises—have been more difficult to develop or scale up as quickly as had been hoped (Fig. 2.3d).

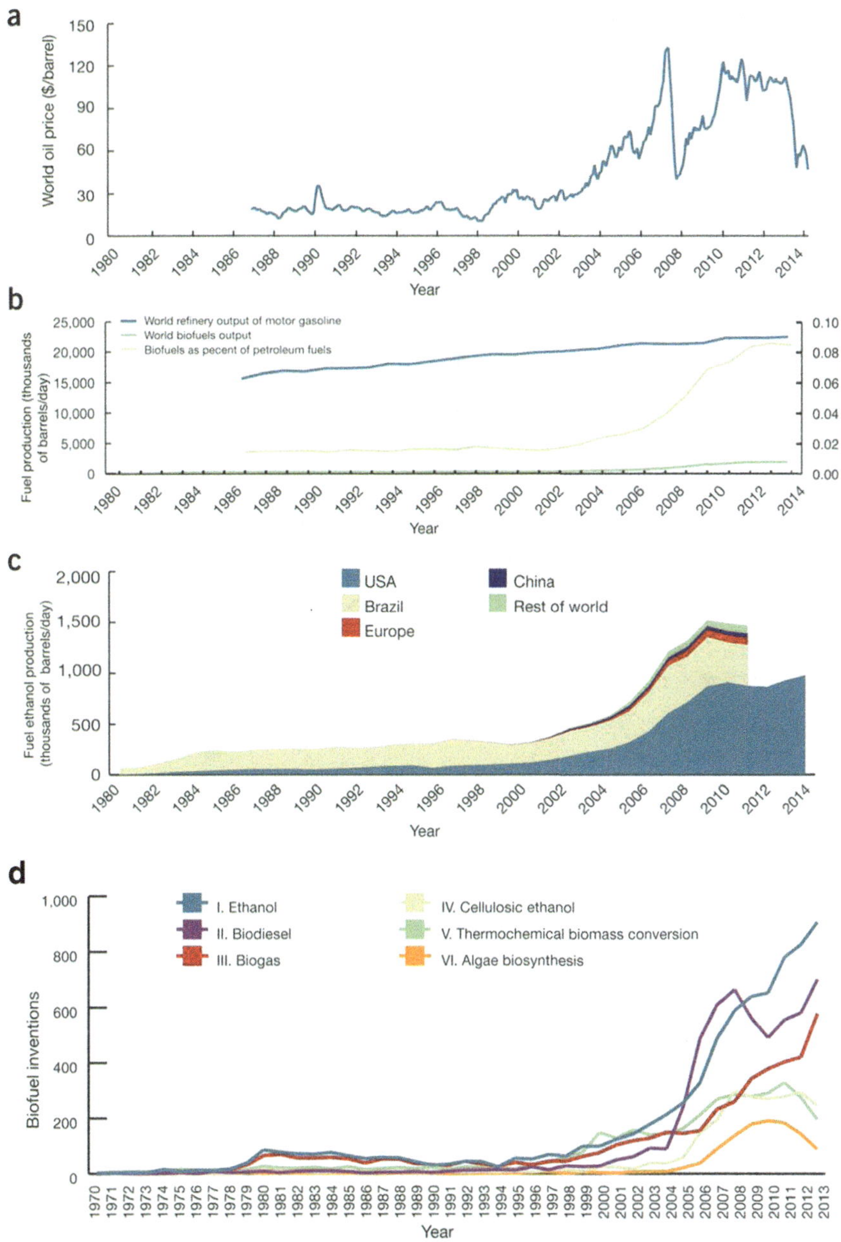

Fig. 2.3 World fuel price and output. (**a**) World oil price (Europe Brent spot price in US dollars per barrel). (**b**) World refinery output of petroleum motor fuels and total biofuels. (**c**) World production of fuel ethanol, the main form of biofuel, broken out by major producing countries. (**d**) Inventions in the six major technical pathways to produce biofuels. Data source: US Department of Energy, Energy Information Agency (EIA) Source: Albers, S. C., Berklund, A. M., and Graff, G. D. (2016). The rise and fall of innovation in biofuels. *Nature Biotechnology 34*(8): 814–821. https://doi.org/10.1038/nbt.3644. Reproduced under licence number 4642520128912

2.4 Biosynthetic Routes for the Production of Natural and Synthetic Fuels from Glucose

Espaux et al. (2015) explained native pathways, heterologous pathways, fatty acid biosynthetic pathway, and isoprenoid pathways (Fig. 2.4).

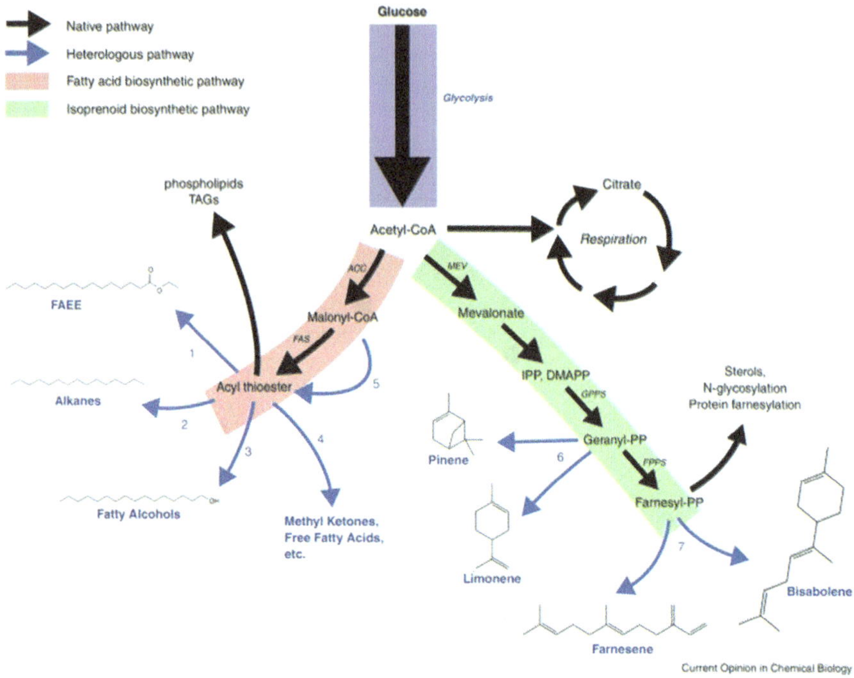

Fig. 2.4 Biosynthetic routes for the production of natural and synthetic fuels from glucose. Fatty acid biosynthesis (pink) naturally produces phospholipids for membrane composition, and TAGs for energy storage. Isoprenoid biosynthesis (green) naturally produces sterols and other compounds. These pathways can be coopted using heterologous genes to produce a number of biofuel molecules. From acyl thioesters: (1) esterification with ethanol by wax synthase to produce FAEE, (2) reduction followed by decarbonylation or PKS-mediated extension-decarboxylation to produce alkanes/enes, (3) reduction either directly or through fatty aldehyde intermediates to produce fatty alcohols, (4) other routes to other products, (5) heterologous FAS pathways, (6) monoterpene synthases can modify C10 geranyl-PP to produce pinene, limonene, or other monoterpenes, (7) sesquiterpene synthases can modify C15 farnesyl-PP to form farnesene, bisabolene, or other sesquiterpenes. Unsaturated lipids can be chemically hydrogenated for biofuel production (e.g., farnesene to farnesane). ACC, acetyl-CoA carboxylase, FAS, fatty acid synthase. The isoprenoid pathway shown is the mevalonate (MEV) pathway. Bacteria employ an alternative route, the DXP pathway, not shown for simplicity. Source: Espaux, L., Mendez-Perez, D., Li, R., and Keasling, J. D. (2015). Synthetic biology for microbial production of lipid-based biofuels. *Current Opinion in Chemical Biology* 29: 58–65. https://doi.org/10.1016/j.cbpa.2015.09.009. Reproduced under licence number 4652791194061 from Rights Link

2.4.1 Biosynthesis of all Building Blocks

Nielsen and Keasling (2016) reported that carbon sources are converted to 12 precursor metabolites that are used for biosynthesis of all secreted metabolites (Fig. 2.5).

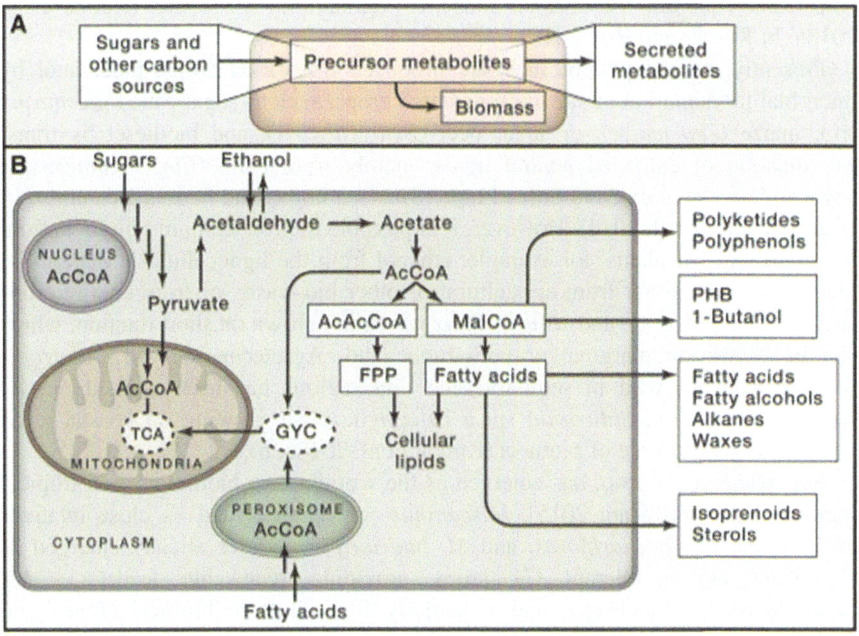

Fig. 2.5 The bow-tie structure of metabolism and acetyl-CoA metabolism in yeast. (**a**) According to the bow-tie structure of metabolism, all carbon sources are converted to 12 precursor metabolites that are used for biosynthesis of all secreted metabolites. The precursor metabolites are also used for the biosynthesis of all building blocks that are needed for synthesizing macromolecules making up the biomass of the cell. The 12 precursor metabolites are: glucose-6-phosphate, fructose-6-phosphate, ribose-5-phosphate, erythrose-4-phosphate, glyceraldehyde-3-phosphate, 3-phosphoglycerate, phosphoenol-pyruvate, pyruvate, acetyl-CoA, 2-oxoglutarate, succinyl-CoA, and oxaloacetate. (**b**) Illustration of how an acetyl-CoA over-producing strain can be used as a platform strain for the production of a range of different molecules. Acetyl-CoA (AcCoA) metabolism in yeast is compartmentalized and there is no direct exchange of this metabolite between the different compartments. AcCoA is formed in the mitochondria from pyruvate and enters the tricarboxylic acid cycle (TCA). AcCoA is also formed in the peroxisome from either fatty acids or acetate and can, via the glyoxylate cycle (GYC), be converted to malate that can be transported to the mitochondria for oxidation. In order to ensure efficient secretion of the product from the cell, it is generally preferred to reconstruct the heterologous pathway in the cytosol, and there is, therefore, a need to ensure efficient provision of cytosolic AcCoA. AcCoA in the cytosol is produced from acetate and is used for the production of acetoacetyl-CoA (AcAcCoA), required for the biosynthesis of sterols via farnesyl pyrophosphate (FPP), and for production of malonyl-CoA (MalCoA), required for fatty acid biosynthesis. AcAcCoA, MalCoA, FPP, and fatty acids can all be converted to commercially interesting products. Source Nielsen, J., & Keasling, J. D. (2016). Engineering Cellular Metabolism. *Cell 164*(6): 1185–1197. https://doi.org/10.1016/j.cell.2016.02.004. Reproduced under licence no 4666911059117

2.5 Energy Crops

Energy crops can be divided into several types: (1) herbaceous energy crops (e.g., rye, switchgrass, grass), (2) short-rotation coppice (SRC) (e.g., poplar, eucalypt, bamboo, and Salix), (3) oilseed crops, (4) hydrocarbon yielding plants, (5) woody biomass—hybrid poplar is considered a promising candidate for a woody energy crop (Labrecque and Teodorescu 2005), (6) agricultural waste, etc. (Kumar et al. 2018a, b; Kumar and Roy 2018).

Presently, most liquid biofuels are produced from food crops: bioethanol by microbial fermentation of sugars from starch crops, such as sugar cane (*Saccharum* sp.), maize (*Zea maize*), or sugar beet (*Beta vulgaris*), and biodiesel by trans-esterification of extracted neutral lipids, mainly from palm (*Elaeis guineensis*), soybean (*Glycine max*), and oilseed rape (*Brassica napus* and *Brassica campestris*) (Kazamia and Smith 2014). However, it is possible to produce liquid biofuels from non-food parts of plants, for example, ethanol from the lignocellulosic material in plant cell walls, either from agricultural or other bio-waste, or from energy crops such as *Miscanthus* sp. and willow (*Salix viminalis*) grown on short rotation, which can be grown on marginal or non-arable land. Agrotechnology of *Calotropis procera* growing wild in semi-arid and arid regions has been worked out by Kumar 2018a, b, c; *Salicornia* sp. a halophytic plant growing on coastal areas could be a good source of biofuels (Kumar et al. 2018a, b).

Sugarcane, a C4 crop, has emerged as the world's best biofuel crop in tropical regions (Sage and Stata 2015). *Miscanthus* × *giganteus* and its close relatives *M. sinensis*, *M. sacchariflorus*, and *M. lutarioriparius* have already emerged as significant chilling-tolerant C4 crops providing renewable feedstocks for bioproducts, for bioenergy, and potentially for cellulosic biofuels (Jones and Walsh 2001; Carroll and Somerville 2009; Heaton et al. 2010; Raghavendra and Sage 2011; Long and Spence 2013). *M.* × *giganteus* is highly efficient for light, nitrogen and water use and is ideally suited for cold climates. A combination of first- and second-generation feedstocks (e.g., corn cobs together with stover) can eliminate bottlenecks and lead to product competitiveness (Paulová et al. 2013). Studies on improving corn productivity have also been reported (Hütsch et al. 2020, this volume).

Perennial plants that use C4 photosynthesis, such as sugarcane, energy cane, elephant grass, switchgrass, and *Miscanthus*, have intrinsically high light, water, and nitrogen use efficiency as compared with that of C_3 species (Boakye-Boaten et al. 2016). There are many potential lignocellulosic feedstocks for 2G bioethanol production in the Mediterranean region. Some of these include feedstocks from processing wastes of cereal crops, tomato and grape, olive solid waste, date palm trunk, and perennial herbaceous lignocellulosic grasses (*Arundo donax*, *Saccharum spontaneum* spp. *aegyptiacum*, and *Miscanthus* × *giganteus*); *Luffa cylindrica* and prickly pear cactus are abundant. Sweet sorghum (*Sorghum bicolor* (L.) Moench), a C4 annual grass (*Poaceae*), has been widely recognized as a promising sugar feedstock crop because it: (1) has low input requirements and wide geographic suitability, (2) has huge breeding potential giving the highest yields of carbohydrates per hectare, and (3) is easily cultivated from seed (Teetor et al. 2011; Li et al. 2014a).

Among these resources, *Stipa tenacissima,* widely present in North Africa, is a promising substrate for the production of 2G bioethanol. Some *Agave* spp., e.g., *A. desertii, A. sisalana,* and *A. salmiana,* use a type of photosynthesis called Crassulacean acid metabolism (CAM) and have a water-use efficiency that may be as much as 6 times greater than that of C_3 species, such as wheat (Borland et al. 2009). They can yield from 7 to 10 dry MT/ha/year in a 6-year cycle.

Boakye-Boaten et al. (2016) reviewed *Miscanthus* × *giganteus* (C4) on bioethanol production from the cool temperate region. The perennial Miscanthus requires relatively fewer fertilizer inputs to sustain growth compared to other annual C3 grass crops (Christian et al. 2008).

2.5.1 Oilseed Crops

Biodiesels are generally produced from a large range of oilseed crops, mainly soybean, rapeseed or canola (*Brassica napus*), *Camelina sativa*, an inedible relative of the mustard plant, oil palm (*Elaeis guineensis*), Chinese pistache (*Pistacia chinensis*), safflower and sunflower, Indian beech (*Pongamia pinnata*), castor bean (*Ricinus communis*), quandong (*Santalum acuminatum*), linseed, peanut, cotton-seed, coconut, restaurant kitchen wastes, *Lesquerella* spp., micro-algae (Cécile et al. 2011), and *Jatropha curcas* in tropical climates (Murugesan et al. 2009; Yong et al. 2010; Kumar 2018a, b, c). *Jatropha curcas* seeds are rich in oil (28–48%), which can be converted to high-quality biodiesel (Kumari et al. 2009; Jongschaap et al. 2009; Devappa et al. 2010; Kumar and Roy 2018). Kumar and Roy (2018) described in detail the effects of different factors affecting the yield of *Jatropha curcas* which could be increased by experimental manipulations and high yielding accessions deposited at NBPGR, New Delhi, in a project funded by the Department of Biotechnology, Govt. of India. Some of the edible plant oils having high oleic acid content used for biodiesel worldwide are rapeseed (84%), sunflower (13%), palm oil (1%), soybean and others (2%) (Atabani et al. 2012).

Jagadevan et al. (2018) reviewed the synergy between enzymes that lead to the formation of lipids (Fig. 2.6).

2.5.1.1 Glycerol Production and Utilization

Glycerol (or glycerin, 1,2,3-propanetriol) is produced in addition to FAAE during transesterification of vegetable oils and animal fats (Lu et al. 2008; Jose and James 2013; Viana et al. 2014; Chen and Liu 2016; Moser 2009). The use of metabolic engineering to improve the performance of industrial strain for converting abundant and low-priced crude glycerol into higher-value products represents a promising route toward glycerol biorefinery, which will significantly increase the economic viability of the biofuels industry (see review Chen and Liu 2016). A broad spectrum of microbes including microalgae, yeasts, and fungi can efficiently utilize crude glycerol for the production of lipids (Tchakouteu et al. 2015; Polburee et al. 2016;

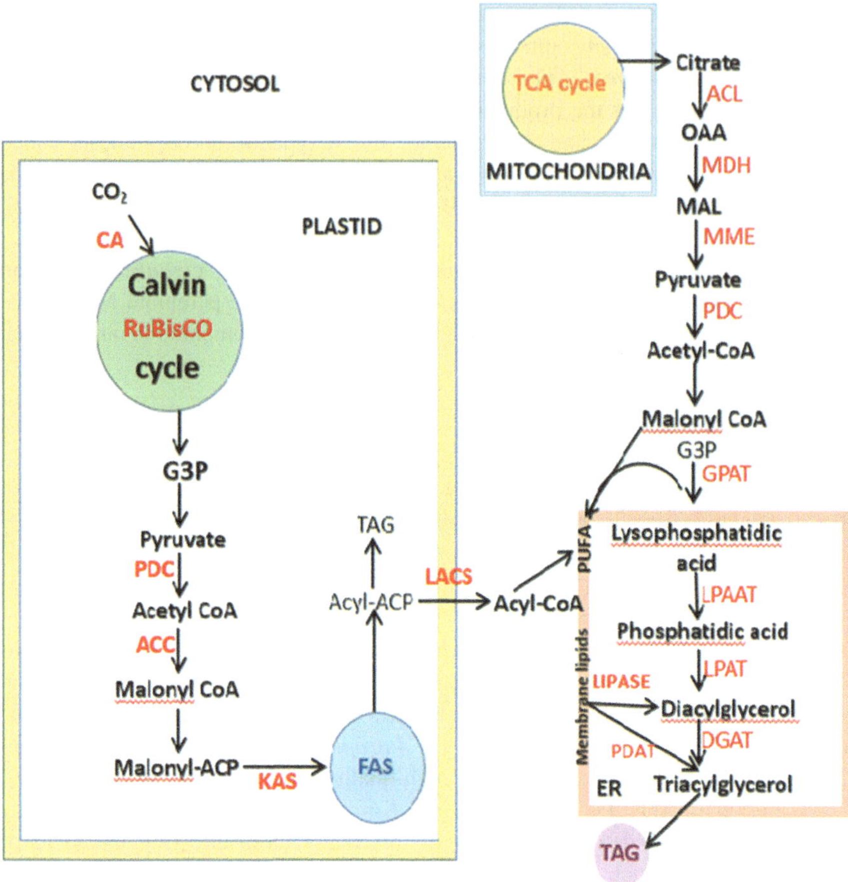

Fig. 2.6 Scheme representing the synergy between enzymes that lead to the formation of lipid (*CA* carbonic anhydrase, *RuBisCO* Ru1,5BP carboxylase/oxygenase, *PDC* pyruvate dehydrogenase complex, *ACC* acetyl-CoA carboxylase, *KAS* 3-ketoacyl-ACP synthase, *ACL* ATP-citrate lyase, *MDH* malate dehydrogenase, *MME* NADP-malic enzyme, *PDC* pyruvate dehydrogenase complex, *GPAT* glycerol-3-phosphate acyltransferase, *LPAAT* lyso-phosphatidic acid acyltransferase, *LPAT* lyso-phosphatidyl choline acyltransferase, *DGAT* diacylglycerol acyltransferase, *PDAT* phospholipid diacylglycerol acyltransferase. Source: Jagadevan, S., Banerjee, A., Banerjee, C., Guria, C., Tiwari, R., & Baweja, M. (2018). Biotechnology for Biofuels: Recent developments in synthetic biology and metabolic engineering in microalgae towards biofuel production. *Biotechnology for Biofuels* 11: 1–21. https://doi.org/10.1186/s13068-018-1181-1. Used under creative commons licence

Karamerou et al. 2016; Xiberras et al. 2019). Viana et al. (2014) reported fermentative H_2 production from residual glycerol.

Saini et al. (2017) engineered *E. coli* for the production of n-butanol from glycerol. Li et al. (2018) investigated biotransformation of glycerol into value-added chemicals with different reduction degrees into pyruvate by an artificial enzymatic reaction cascade composed of alditol oxidase from *Streptomyces coelicolor* A3 (ALDO), dihydroxy acid dehydratase from *Sulfolobus olfataricus* (DHAD), and catalase from *Aspergillus niger* (Gao et al. 2015). Through the rational assembly of thermostable enzymes from various species, they constructed a completely artificially designed in vitro biosystem for the production of valuable chemicals from glycerol (Li et al. 2018).

2.5.2 Hydrocarbon-Yielding Crops

Laticiferous plants growing in arid and semi-arid regions are rich in triterpenoids which can be converted into biofuel (Kumar 2018a, b, c). Terpenoids comprise the largest family of natural products that have widespread applications. Extensive studies have been carried out on laticiferous plants growing in arid and semi-arid regions: *Euphorbia antisyphilitica, Euphorbia lathyris, Euphorbia tirucalli, Calotropis procera* (Kumar 2018a, b), and *Pedilenthus tithymaloides* (see review Kumar 2013, Kumar et al. 2018a, b). Agrotechnology of biofuel production in semi-arid and arid regions has been reviewed by Kumar and Roy (2018). Leavell et al. (2016) reported microbial production using large-scale fermentation of organisms engineered to manufacture terpenoid products, artemisinin, and squalane. There are two major pathways for the biosynthesis of terpenoids: the mevalonate pathway (MVA) (Bloch 1992) and the methylerythritol phosphate (MEP) pathway (Eisenreich et al. 2004). Recently, a modified MVA pathway was also proposed (Chen and Poulter 2010). Both the MVA and MEP pathways are attractive for use in the heterologous microbial production of terpene-based chemicals and fuels (Paddon et al. 2013).

2.5.3 Halophytes for Biofuel

Several halophytic plants are able to grow in saline soils of which *Salicornia* is one major plant in the coastal areas of Gujarat. *Salicornia* species use the C4 pathway to take in carbon dioxide from the surrounding. The Gujarat State Fertilizers and Chemicals Ltd. (GSFC), Baroda, plans to promote the cultivation of *Salicornia*, and CSMCRI, Bhavnagar a CSIR center has carried out experiments on its use (Personal communication). Globally several *Salicornia* spp. has been grown in Middle Eastern countries and Europe for economic benefits using coastal saltwater (Kumar et al. 2018a, b).

2.5.4 Fern *Azolla* as Biofuel

Azolla (mosquito fern, water fern) is a genus with seven species found in ponds, ditches, and wetlands throughout the world, from temperate to tropical regions (Kollah et al. 2016). This aquatic plant is one of the fastest-growing plants capable of doubling its biomass every 5–6 days (Wasiullah et al. 2015; Kollah et al. 2016). *Azolla* can also grow efficiently in nitrogen-depleted media using the nitrogen-fixing capacity of its symbiont, the endophytic cyanobacterium, *Anabaena azollae* Strasburger (*A. azollae*), which grows within its leaf cavities. The chemical composition of Azolla's biomass contains a unique combination of bioenergy molecules found in lignocellulosic, starch- and oil-producing terrestrial bioenergy crops, microalgal and cyanobacterial species. The ability to grow on wastewaters and high growth and productivity rates make *Azolla* species a most attractive feedstock for low cost, low energy-demanding, near-zero maintenance system for production of a wide spectrum of biofuels (see: Miranda et al. 2016).

2.6 Lignocellulosic Feedstocks

Lignocellulosic biomass is composed of three main carbohydrate polymers: (1) cellulose, (2) xylan, and (3) pectin (Wendisch et al. 2016). The intermediate forms of processing include sugar (Lynd et al. 2002), organic acids (Holtzapple and Granda 2009), methane and synthetic gas (syngas) (Rauch et al. 2014; Liew et al. 2012).

Recently, the emphasis is shifting from first-generation biofuel crops to resources such as lignocellulose, algal biomass, and non-food energy crops. Lignocellulose is the most abundant biomass on Earth and consists of about 70% sugars. However, these sugarsrequire chemical, thermal, and biochemical processes before they can be released for microbial fermentation to produce advanced biofuels (Alper et al. 2006; Keasling 2010; Zhang and Keasling 2012; Kumar and Gupta 2018). Atmospheric CO_2 can also be transformed into biofuels through carbon fixation using engineered photosynthetic organisms. Besides this carbon monoxide, from the thermal conversion of lignocellulosic biomass and abundant in steel-mill flue gas, can also be metabolized by microorganisms. According to Nielsen and Keasling (2011), progress in metabolic engineering and synthetic biology combines advanced molecular and system biology techniques with principles of engineering design to produce advanced biofuels with similar properties to petroleum-based fuels (Khalil and Collins 2010). Alper and Stephanopoulos (2009) described engineering for biofuels. In addition to this, the power of using microbial processes for chemical production is twofold: first, renewable carbon sources can serve as substrates, and second, the range and specificity of molecules that can be made biologically surpass that of synthetic chemistry (Chubukov et al. 2016).

Currently, lignocellulose deconstruction to sugars generally starts with size reduction, pretreatment process followed by enzymatic hydrolysis (Houghton et al. 2006; Kumar et al. 2009). Roy and Kumar (2013) reviewed methods for lignocellulosic materials. However, cellulose and hemicellulose, two major sugar polymers of lignocelluloses, have to be depolymerized by hydrolysis to enable more efficient microbial utilization (Acker et al. 2013). An overview of biofuel production from sunlight and atmospheric carbon using microbes has been presented by Liao et al. (2016) (Fig. 2.7).

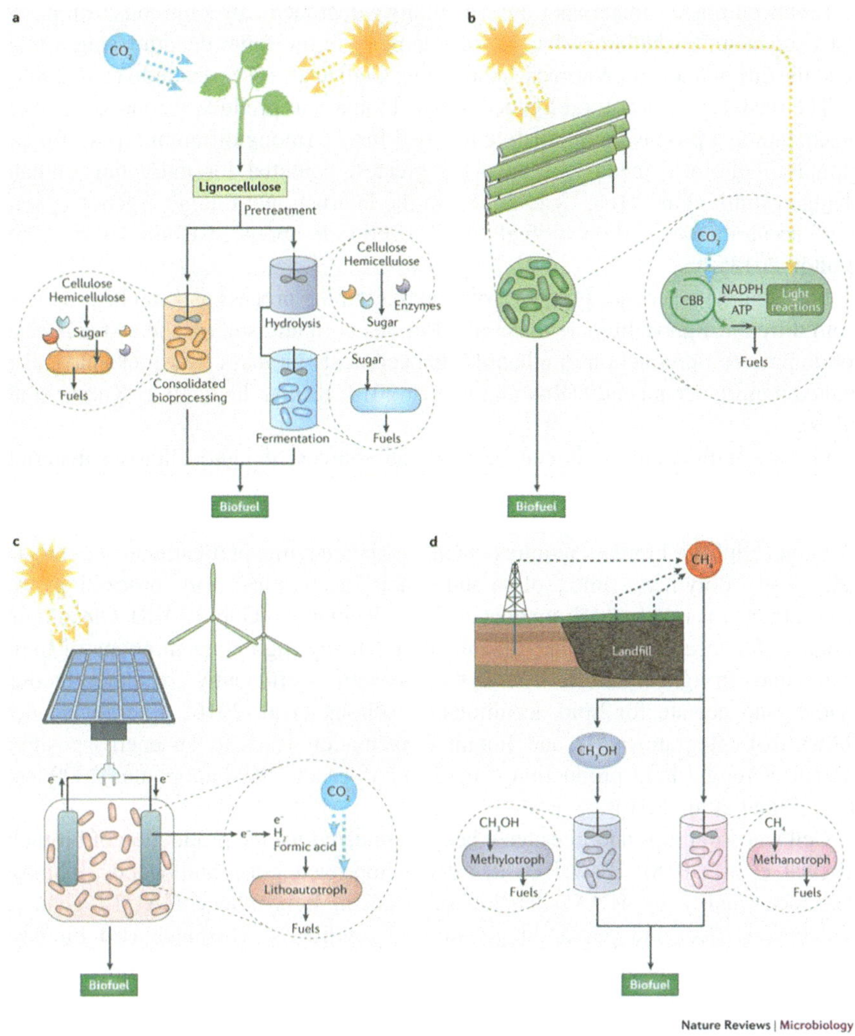

Fig. 2.7 Overview of biofuel production from sunlight and atmospheric carbon. (**a**) The lignocellulose of plant biomass can be converted to fuels through hydrolysis followed by fermentation, or through consolidated bioprocessing, which combines the two processes in one reactor. (**b**) Photosynthetic organisms, such as microalgae and cyanobacteria, can harness energy from sunlight to reduce CO_2 and convert it into liquid fuels. (**c**) A broad range of lithoautotrophs can fix CO_2 to produce fuels with reducing power from electrons or electrochemically generated electron shuttles, such as H_2 and formic acid. (**d**) Low-throughput methane from a landfill or natural gas wells that are otherwise flared can be used directly by methanotrophs to produce fuels, or it can be converted to methanol (CH_3OH) and can then be utilized by methylotrophs for fuel production. CBB, Calvin–Benson–Bassham cycle. (Alkanes Organic molecules with the general formula CnH_{2n+2}). Source: Liao, J. C., Mi, L., Pontrelli, S., and Luo, S. (2016). Fuelling the future: microbial engineering for the production of sustainable biofuels. Nature Review, Microbiology *14*(5): 288–304. https://doi.org/10.1038/nrmicro.2016.32. Reproduced under Licence number 4645730007098 from Rights Link

Lignocellulosic materials are naturally degraded by a consortium of microorganisms, whether in the gut of a termite, in the fungi decomposing a tree, or in the dirt in a forest (Warnecke et al. 2007; Geib et al. 2008; Temudo et al. 2008).

The recovery of non-food lignocellulosic biomass to produce second-generation bioethanol is a promising alternative to fossil fuels. Among different types of plant biomass, cellulosic feedstocks have the greatest potential for mitigating climate change (Smith et al. 2016; Lynd 2017). In the International Energy Agency (Paris) 2 °C scenario, low-carbon biofuels need to provide about 25 exajoules by 2050 (Fulton 2013).

Fuel ethanol is a major product from the biorefining process and can be produced from different lignocellulosic materials. Besides corn and sugar cane as feedstocks for commercial production of ethanol, lignocellulosic biomass feedstocks have also gained importance recently (Roy and Kumar 2013; Kumar 2018a, b, c; Kumar et al. 2019).

Grasses with high yields can be potential sources of lignocellulosic material. Some of the grasses with the highest productivity include Napier Grass (*Pennisetum purpureum*), and *Echinochloa polystachya*.

Plant cellulosic biomass requires chemical and enzymic pretreatments to convert cell wall polymers into oligo-and monosaccharides for processing by microorganisms into fuel (Himmel et al. 2007; Kumar and Gupta 2018). Oleaginous fungus, *Mortierella isabellina*, can tolerate relatively high concentrations of toxic compounds in lignocellulosic hydrolysates as well as efficiently consume glucose, xylose, and acetate for lipid accumulation (Zhong et al. 2016). Energy balance shows that integrating AD and fungal fermentation leads to an energy-positive system of fungal lipid production (Fig. 2.8) (Alper and Stephanopoulos 2009; see also: Zhong et al. 2016).

Cell walls in crops and trees have been engineered for the production of biofuels (Biswal et al. 2018). Biswal et al. (2018) improved grass and woody biomass feedstock quality by RNA silencing of a pectin biosynthetic GAUT4 gene in switchgrass, rice, and poplar in greenhouse conditions. This increased biomass yield and ethanol production from transformed switchgrass lines compared with wild type. Thus transgenic switchgrass with reduced expression GAUT4 gene provided the greatest sugar, absolute ethanol, and biomass yield compared to reduced-recalcitrance feedstock lines modified in lignin and C-1 metabolism (Dumitrache et al. 2017).

Stoichiometric genome-scale metabolic models are now frequently used for considering the entire metabolic network and understanding how alterations in central pathways propagate to the rest of cellular metabolism (King et al. 2015; Chubukov et al. 2016).

Xu and Shanklin (2016) reviewed recent progress in the understanding of triacylglycerol synthesis, turnover, storage, and function in leaves and discussed emerging genetic engineering strategies targeted at enhancing triacylglycerol accumulation in biomass crops. Such plants could potentially be modified to produce oleochemical feedstocks or nutraceuticals.

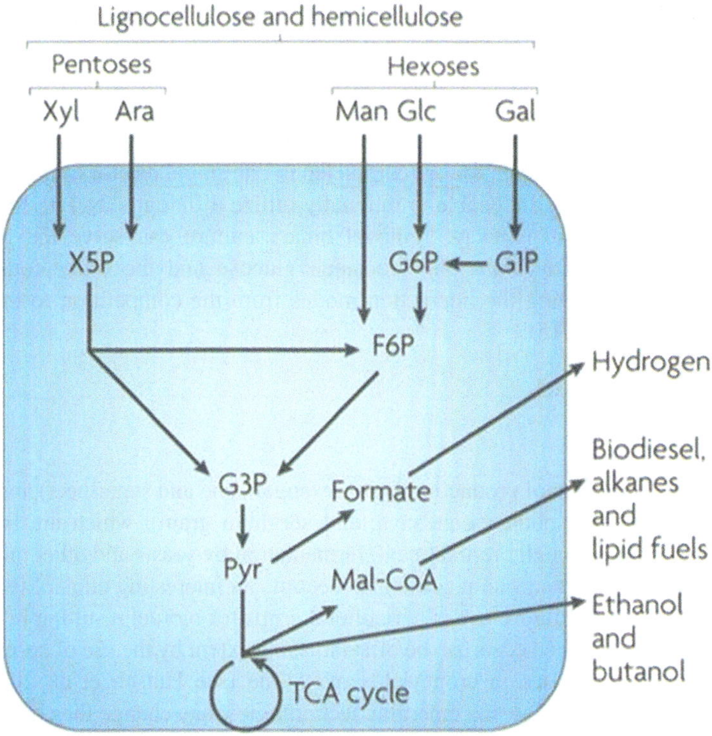

Fig. 2.8 Biofuel production by microorganisms. Five- and six-carbon sugars that are commonly found in lignocellulosic material include xylose (Xyl), arabinose (Ara), glucose (Glc), mannose (Man), and galactose (Gal). These sugars are converted to the phosphorylated forms xylose-5-phosphate (X5P), glucose-6-phosphate (G6P), fructose-6-phosphate (F6P), and glucose-1-phosphate (G1P). These molecules are eventually converted into glyceraldehyde-3-phosphate (G3P), pyruvate (Pyr), and formate. A number of possible biofuels can then be produced. *Mal-CoA* malonyl-CoA. Source: Alper, H., and Stephanopoulos, G. (2009). Engineering for biofuels: exploiting innate microbial capacity or importing biosynthetic potential, *Nature Reviews Microbiology 7*: 715–723. Retrieved from https://doi.org/10.1038/nrmicro. Reproduced under license number 46456400840514 from Rights Link

Advanced CRISPR/Cas9 technology for modification of cell walls in plants used as biomass sources is also highlighted. Case studies are provided, see review (Yau and Easterling 2018).

2.6.1 Cocultivation Systems

In general, two common strategies were developed: one is the incorporation of target product synthesis modules into cellulolytic microbes to achieve product generation from lignocellulose, the other is the introduction of cellulase systems into product-generating microbes (Olson et al. 2012; Yang et al. 2015).

However, the long and complex pathways including cellulase secretion and/or product synthesis would burden the metabolic stress and lead to low amounts of product generated (Shanmugam et al. 2018). On the contrary, microbial consortia offer a simpler and more efficient approach to achieve this goal through the so-called consolidated bioprocessing (CBP), in which enzymes production, substrate hydrolysis, and microbial fermentation are completed in one single reactor (see Jiang et al. 2019). Microbial consortia enable to rationally utilize different substrates based on the specific metabolic pathway. A novel binary culture can solve the problem flexibly, in which one could only consume glucose and the other could only consume xylose, shifting the interaction modes from the competition to the commensalism (Zhang 2015).

2.7 Ethanol

The feedstock for ethanol production is sugar (sugar cane and sugar beet) and starch crops (maize, wheat, potatoes, cassava, and sorghum grain), which are basically equally processed through pre-treatment, fermentation by yeasts and other microbes, and distillation. Sweet sorghum could also become an interesting ethanol feedstock (Li et al. 2014b). The future use of agricultural crops for biofuel resulting in a small increase in livestock feed costs can be offset to some extent by the use of co-products as feed and by increases in crop yields over time (see Hütsch et al. 2020, this volume). A number of new and emerging technologies may change the composition and further improve the nutritional quality and utility of feed co-products. New technologies and practices promise to change the complexion of the ethanol co-products market in the years ahead (Popp et al. 2016a, b).

The techno-economic analysis shows that microbial biofuels provide for significant reductions in CO_2 emissions overusing petroleum fuels (Caspeta and Nielsen 2013; Caspeta et al. 2013; Espaux et al. 2015).

2.7.1 Higher Alcohols

Biosynthesis approaches for the production of higher alcohols as a source of alternative fossil fuels have garnered increasing interest recently. Su et al. (2016) demonstrated that production of higher alcohols including isobutanol, 2-methyl-1-butanol, 3-methyl-1-butanol from glucose and duckweed under simultaneous saccharification and fermentation (SSF) scheme can be improved for biofuel production by the engineering of novel synthetic pathways in microorganisms.

2.7.2 Butanol

Butanol has 84% of the energy content of gasoline, limited miscibility with water, and is completely miscible with gasoline (Lee et al. 2008b) (Fig. 2.9). Natural host of

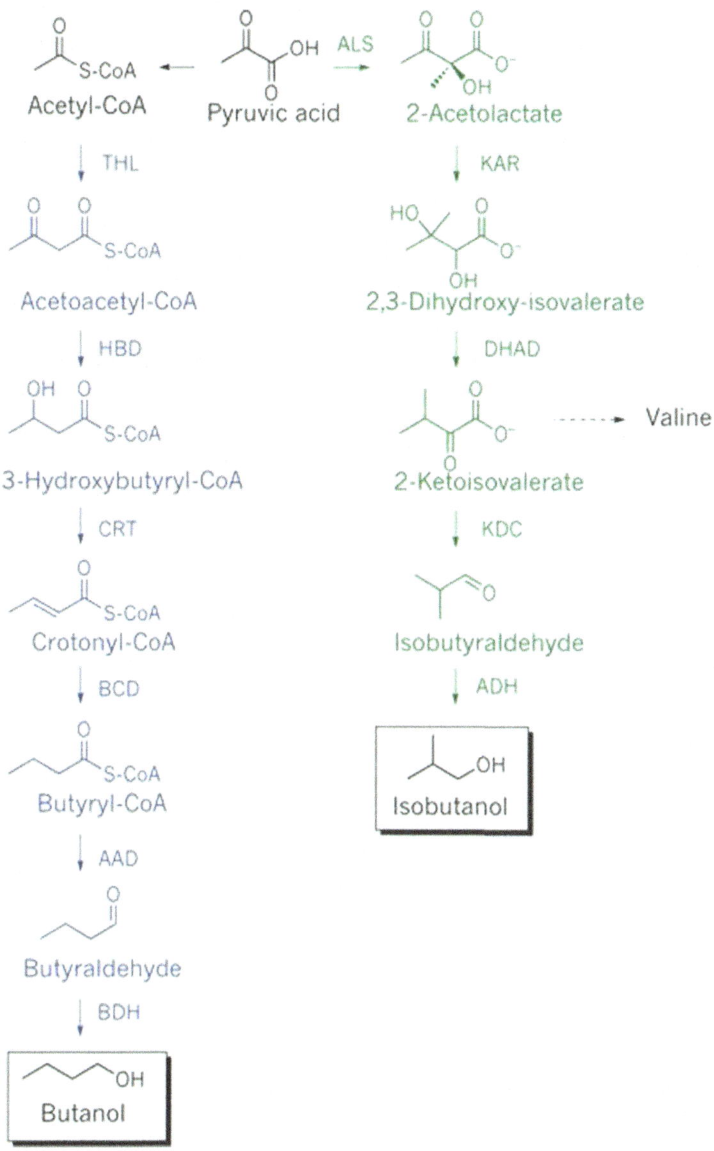

Fig. 2.9 Butanol and isobutanol pathways. In blue, *Clostridium's* butanol pathway converts acetyl-CoA into butanol. *AAD* butyraldehyde dehydrogenase, *BCD* butyryl-CoA dehydrogenase, *BDH* butanol dehydrogenase, *CRT* crotonase, *HBD* 3-hydroxybutyryl-CoA dehydrogenase, *THL* thiolase. In green, the 2-keto acid pathway produces isobutanol from pyruvic acid. *ADH* alcohol dehydrogenase, *ALS* acetolactate synthase, *DHAD* dihydroxy-acid dehydratase, *KAR* ketol-acid reductoisomerase, *KDC* keto-acid decarboxylase. Source: Peralta-Yahya P.P. et al. (2012). Microbial engineering for the production of advanced biofuels. https://doi.org/10.1038/nature 488 320–328. Reproduced with licence no. 4643340791481

butanol production, *Clostridium*, has been engineered to use feedstocks such as glucose (Qureshi and Blaschek 1999), liquefied cornflour (Ezeji et al. 2007) glycerol (a by-product in the production of biodiesel from fats), and even syngas (a mixture of hydrogen and carbon monoxide). Post World War I, *n*-butanol was produced from acetone-butanol-ethanol (ABE) clostridial fermentations (Swidah et al. 2015). Engineered *E. coli* bearing the ABE pathway has been generated in a number of different ways and has been shown to produce high levels of butanol (Bond-Watts et al. 2011; Peralta-Yahya et al. 2012, Fig. 2.2).

Clostridium strains have the capability to utilize diverse carbon sources, including C5 and C6 substrates, which thus allows the production of chemicals from inexpensive and abundant biomass such as corn stover, straw, and woody waste (Cho et al. 2015). *Clostridium tyrobutylicum*, a well-known butyric acid producer, has also been engineered to produce butanol (Cho et al. 2015). *Clostridium* strains have the capability to utilize diverse carbon sources, including C5 and C6 substrates, which thus allows the production of chemicals from inexpensive and abundant biomass such as corn stover, straw, and woody waste (Cho et al. 2015). Recently, several important strategies for the metabolic engineering of *Clostridium* have been developed not only for the enhanced production of these natural products and but also for the production of non-natural isobutanol production. Cho et al. (2015) reviewed the strategies employed for the development of metabolically engineered *Clostridium* strains for the production of such chemicals and provide future perspectives.

2.7.2.1 Isobutanol

There is a growing interest in the use of phototrophic organisms such as cyanobacteria or algae to directly fix carbon dioxide into liquid fuel (Chisti 2007), e.g., *Synechococcus elongatus* PCC7942 has been engineered to produce isobutyraldehyde and isobutanol (Atsumi et al. 2009). Engineered *E. coli* strains can ferment a variety of sugars into ethanol as the predominant product. Although its yield and productivity can be high, other technical barriers need to be overcome in order for this process to become economically viable (Liu and Khosla 2010; Peralta-Yahya et al. 2012) (Fig. 2.9).

2.8 Pathways for Isoprenoid-Derived Fuels

Isoprenoids are a class of compounds widely used as flavors and pharmaceuticals, have the potential to serve as advanced biofuels because of the branches and rings found in their hydrocarbon chain (Lee et al. 2008a, b; Kirby and Keasling 2009; Peralta-Yahya and Keasling 2010) (Fig. 2.3).

Lately, the high capacities of *S. cerevisiae* to form isoprene building blocks and the adducts thereof, i.e., geranyl- and farnesyl-diphosphate, have rendered this particular yeast a favorite commercial production platform for sterols, steroids, and other terpenoids (Ro et al. 2006; Scalcinati et al. 2012; see review Zhang and Keasling 2012).

Isoprenoid production in most bacteria and green algae is carried out by the MEP pathway (Fig. 2.10). Plastid based production of monoterpenes, diterpenes, and carotenoids in plants is also carried out by the MEP pathway (Schwender et al. 1996). The MEP pathway consists of seven steps resulting in the conversion of glyceraldehyde-3-phosphate and pyruvate to IPP and DMAPP (Fig. 2.10).

Terpenoids are important chemicals obtained from hydrocarbon yielding plants like *Calotropis procera* and *Euphorbia* spp. which can be converted into biofuels using the catalytic cracking system (Kumar 2013).

2.9 Biofuels from Protein

In addition to carbohydrates or lipids as raw material for the production of biofuels, the bacteria have also engineered to produce biofuels. Devi and Mohan (2012) suggested that the carbohydrates stored during the growth phase might channel toward the formation of triacylglycerides (TAGs), the efficient ingredients for lipid conversion into biodiesel.

Proteins are the dominant fraction in fast-growing photosynthetic microorganisms (Becker 2007) and industrial fermentation residues. The proteins can be deaminated and converted to fuel or chemicals. The metabolic flow of nitrogen in genetically engineered *Escherichia coli* has been directed to generate the backbone and side chains of amino acids instead of amino acids. Subsequently, the keto acids are then enzymatically converted to two-, four-, and five-carbon alcohol fuels, including ethanol, isobutanol, 2-methyl-1-butanol, and 3-methyl-1-butanol, using a metabolic pathway developed by Atsumi et al. (2008), and see also Huo et al. (2011a, b), Mielenz (2011). Possibly in near future, the biorefining scheme can bypass the need for expensive photobioreactors or the lignocellulose recalcitrance problem by using protein biomass from algal cultures (Sheehan et al. 1998), waste biotreatment, food processing, and the fermentation industry as a long-term, sustainable protein source (Huo et al. 2011a, b).

2.10 Metabolic Engineering for Production of Biofuels

In plants, photosynthetic efficiency in terms of light energy converted to biomass is only ca. 1% and has been identified as one of the most promising targets for improving agricultural productivity (Keasling 2008; Kumar et al. 2014).

Progress in metabolic engineering, and synthetic and systems biology, has allowed the engineering of microbes to produce advanced biofuels with similar properties to petroleum-based fuels (Zhu and Jackson 2015; Keasling 2010; Nielsen and Keasling 2011). Several excellent reviews on systems metabolic engineering and synthetic biology have highlighted the motivation and need for pathway balancing (Völler and Budisa 2017; see also Kumar 2020, this volume).

Genetic modification of plant cell walls has been implemented to reduce lignocellulosic recalcitrance for biofuel production. Genetic modification of plant cell

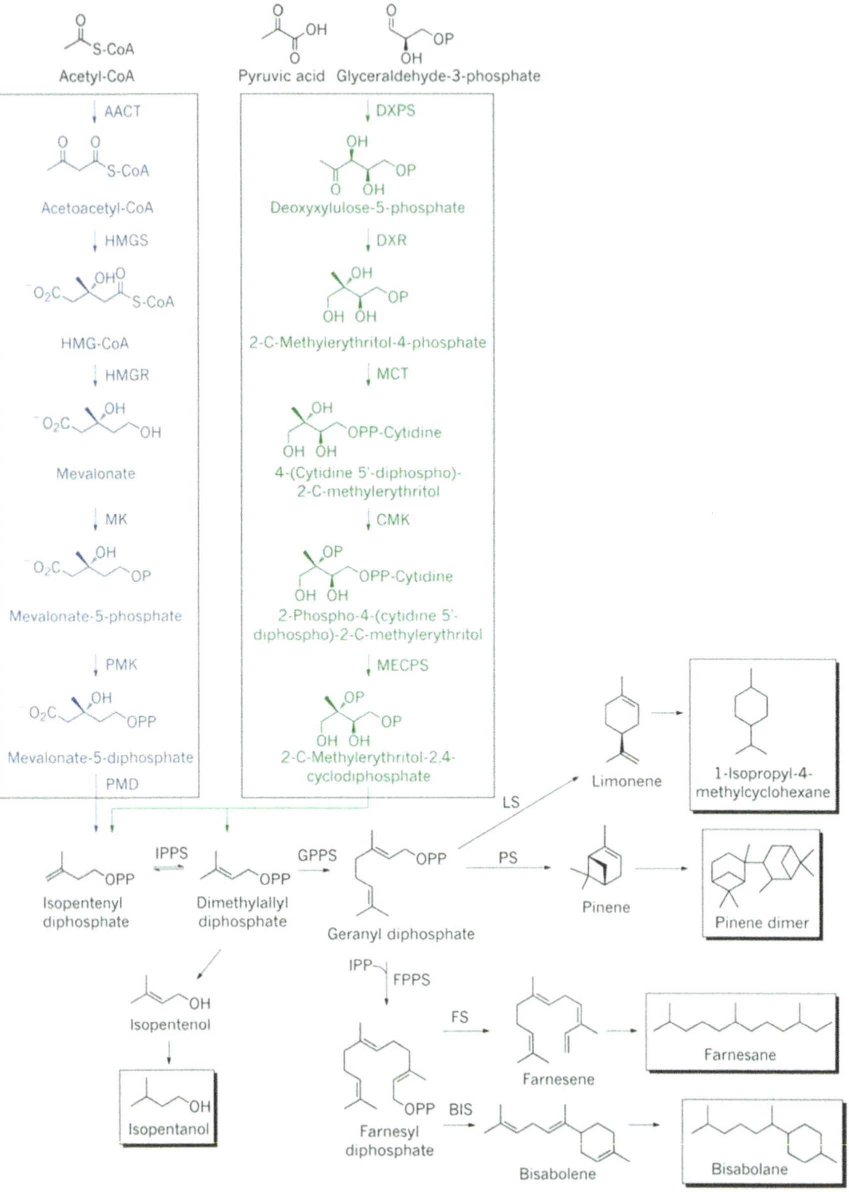

Fig. 2.10 Metabolic pathways used for the production of isoprenoid-based biofuels. In blue is the mevalonate pathway (left) and in green (right) is the deoxyxylulose-5-phosphate (DXP) pathway. *AACT* acetyl-CoA transferase, *BIS* bisabolene synthase, *CMK* 4-(cytidine-5′-diphospho)-2-*C*-methylerythritol kinase, *DXPS* DXP synthase, *DXR* DXP reductoisomerase, *FPPS* farnesyl diphosphate synthase, *FS* farnesene synthase, *GPPS* geranyl diphosphate synthase, *HMGR* HMG-CoA reductase, *HMGS* 3-hydroxy-3-methylglutaryl-CoA (HMG-CoA) synthase, *IPP* isopentenyl diphosphate, *IPPS* IPP isomerase, *LS* limonene synthase, *MCT* 2-*C*-methylerythritol-4-phosphate cytidyltransferase, *MECPS* 2-*C*-methylerythritol-2,4-cyclodiphosphate synthase, *MK* mevalonate

walls has been implicated in the largely enhanced lignocellulose enzymatic saccharification and biofuel production in transgenic crops. Slightly altering cell wall composition and structure and especially improving major wall polymer properties were proposed as feasible approaches for enhanced biomass saccharification and biofuel production (Loqué et al. 2015). Over the past years, attempts have been made to enhance lignocellulose enzymatic hydrolysis by altering hemicellulose features or reducing lignin contents (Li et al. 2014a, b; Eudes et al. 2014; Liu et al. 2014), but most of the transgenic plants displayed defects in growth and strength or limited enhancement of biomass saccharification. Huang et al. (2019) reported that overproduction of native endo β1,4-glucanases leads to largely enhanced biomass saccharification and bioethanol production by specific modification of cellulose features in transgenic rice.

Hybrid processes, combining biochemical and chemical processes, will enhance the competitiveness of bio-based products (Kumar 2010, 2015; Beerthuis et al. 2015; Jones et al. 2015).

As an alternative to lignocellulose, CO_2 can be directly utilized by photosynthetic organisms, such as microalgae and cyanobacteria, or by lithoautotrophic organisms that can use energy that is derived from renewable sources. Finally, methane, which is a more potent greenhouse gas than CO_2 (Yvon-Durocher et al. 2014), can be utilized by methanotrophs or, after activation to methanol, by methylotrophs.

2.11 Algae-based Biofuels

Third-generation biofuels including microalgal biofuels are treated as a technically viable alternative energy solution that overcomes the major drawbacks related to the first and second generations (Das et al. 2011; Lam and Lee 2012; Zhu 2015).

Autotrophic algae have been used to make lipids for conversion into biodiesel, or high-value chemicals such as omega-3 fatty acids (Hossain et al. 2019). Photosynthesis helps to convert 100 billion tons of CO_2 into biomass (Peplow 2014).

There are nine major groups of algae which are cyanobacteria (Cyanophyceae), green algae (Chlorophyceae), diatoms (Bacillariophyceae), yellow-green algae (Xanthophyceae), golden algae (Chrysophyceae), red algae (Rhodophyceae), brown algae (Phaeophyceae), dinoflagellates (Dinophyceae), and "pico-plankton" (Prasinophyceae and Eustigmatophyceae) (Hu et al. 2008).

Oxygenic photoautotrophs have water-splitting oxygen-generating photosystems that can generate reducing power and a proton gradient for the regeneration of ATP. For some oxygenic photoautotrophic cyanobacterial strains of *Synechocystis* spp.

Fig. 2.10 (continued) kinase, *PMD* phosphomevalonate decarboxylase, *PMK* phosphomevalonate kinase, *PS* pinene synthase. Source: Peralta-Yahya P.P. et al. (2012). Microbial engineering for the production of advanced biofuels. https://doi.org/10.1038/nature 488, 320–328. Reproduced with licence no. 4643340791481

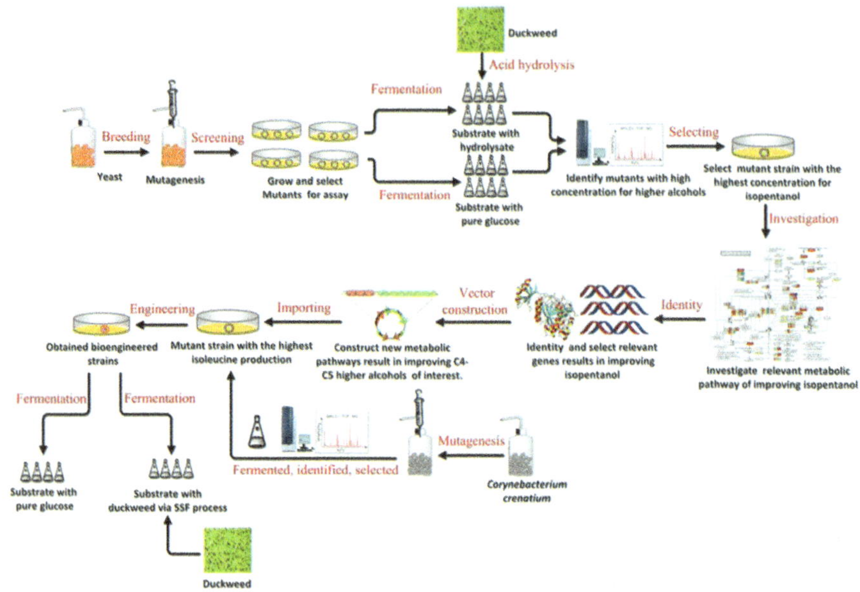

Fig. 2.11 The flowsheet of experiments following the methodology illustrated for screening exogenous mutant enzymes and expression host via undirected whole-cell mutagenesis (UWCM) in vivo and fermentation processes. Source: Su, H., Lin, J., and Wang, G. (2016). Metabolic engineering of *Corynebacterium crenatium* for enhancing the production of higher alcohols. *Nature Publishing Group, Scientific Reports* 6: 39543. DOI: https://doi.org/10.1038/srep39543 (November), 1–20

and *Synechococcus* spp., sets of genetic tools have been developed (Berla et al. 2013; Ramey et al. 2015).

Photosynthetic microorganisms convert CO_2 into biomass by deriving energy from light or inorganic electron donors and have the potential to produce chemicals and biofuels. Modification of autotrophic systems by genetic engineering into heterotrophic model microorganisms has led to increased productivity (see reviews Claassens et al. 2016; Liu et al. 2016; Su et al. 2016) (Fig. 2.11).

Microalgae are already being used to produce lipid-based fuels such as biodiesel. The oil content of *Chlorella* typically ranges between 28 and 32% dry weight (Chisti 2007) but can reach 46% dry weight under stress conditions (Hu et al. 2008) and 55% dry weight when grown heterotrophically (Miao and Wu 2006). Their cell walls are rich in protein rather than cellulose. Besides this releasing protein from algal biomass may be an easier bioconversion process than breaking down lignocellulosic plant matter to fermentable sugars (Pokoo-Aikins et al. 2009; Wijffels and Barbosa 2010; Mielenz 2011; Gimpel et al. 2013; Kawai and Murata 2016; Baritugo et al. 2018).

The batch culture of microalgae is mature and technology-ready, and it has been used in many microalgal species, such as *Chlorella zofingiensis* (Zhu 2015), *Chlorella vulgaris* (Woodworth et al. 2015), *Nannochloropsis salina* (Bellou and Aggelis

2013), *Chlamydomonas polypyrenoideum* (Halfhide et al. 2014), *Cyclotella* sp. (Jeffryes et al. 2013) and many others, ranging from freshwater microalgae to marine microalgae. Developments in several areas, such as genetic and metabolic engineering, are expected to promote microalgal production for biofuel applications still further. Zhu (2015) reviewed microalgal cultivation strategies for the achievement of improved biomass and biofuel productivity.

Cyanobacteria are represented by a diverse group of microorganisms that are marine and freshwater phytoplankton, and significantly contribute to the fixation of atmospheric carbon via photosynthesis. The technology of cyanobacterial biomass conversion to bio-oil and/or cyanodiesel can be further developed and standardized to meet the criteria of the today's demands (Sarsekeyeva et al. 2015; Zhu 2015; Singh et al. 2016; Rizza et al. 2017; Gajraj et al. 2018). Algae, such as *Botryococcus braunii*, produce large quantities of fatty acids. The algal biomass from *Gracilaria verrucosa* from coasts of Orissa and Tamil Nadu, India, has potential for biorefinery development through agar extraction, saccharification of leftover pulp and ethanol fermentation of hydrolysate (Shukla et al. 2016). Industrial processes can also generate useful molecules by direct biological conversion of CO_2.

2.12 Fourth Generation

The fourth-generation biofuels—photobiological solar fuels and electrofuels—are expected to bring fundamental breakthroughs in the field of biofuels. Technology for the production of such solar biofuels is an emerging field and based on direct conversion of solar energy into fuel using raw materials that are inexhaustible, cheap, and widely available. The synthetic biology field is still in its infancy and only a few truly synthetic examples have been published thus far (see Cameron et al. 2014 for a review). Advances in synthetic biology have allowed the transference of metabolic pathways into non-native hosts that are suitable for industrial bioprocesses. Genetic engineering has enabled corresponding improvements in biofuel tolerance (Zhang et al. 2011) and enhanced yields (Ignea et al. 2011). For successful progress, one needs to discover new-to-nature solutions and construct synthetic living factories and designer microorganisms for efficient and direct conversion of solar energy to fuel. Likewise, a combination of photovoltaics or inorganic water-splitting catalysts with metabolically engineered microbial fuel production pathways (electrobiofuels) is a powerful emerging technology for efficient production and storage of liquid fuels.

2.13 Development of In Vitro (Cell-Free) Technologies

Recently, cell-free protein synthesis has been suggested to be the fastest way to make recombinant proteins, even for membrane or complicated proteins (Zhang 2015). In vitro synthetic biosystems emerge as a manufacturing platform by the assembly of numerous enzymes and enzyme complexes from different sources and/or

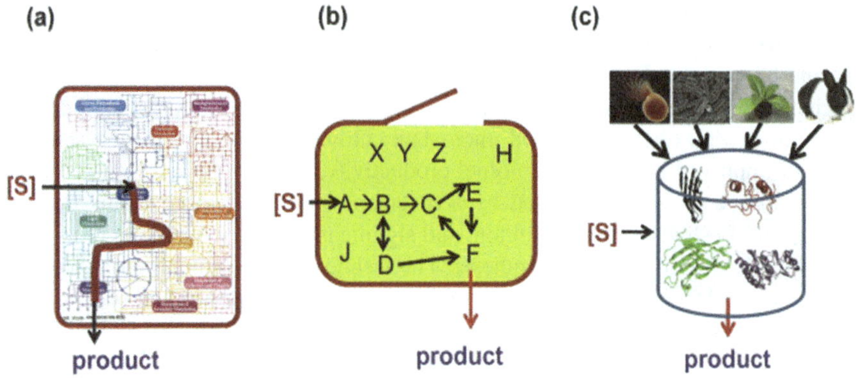

Fig. 2.12 Schemes of biotransformation catalyzed by whole-cell (**a**), cell extract (**b**), and in vitro synthetic biosystem (**c**) Source: Zhang, Y. P. (2015). Production of biofuels and biochemicals by in vitro synthetic biosystems: Opportunities and challenges. *Biotechnology Advances 33*(7): 1467–1483. https://doi.org/10.1016/j.biotechadv.2014.10.009. Reproduced with licence number 4652950482642 from Rights Link

(biomimetic) coenzymes (Zhang 2015, Fig. 2.1c). Such systems could surpass the constraints of whole-cells and cell lysates for implementing some biological reactions that microbe cannot do, for example, high yield production of hydrogen (Martín del Campo et al. 2013), or enzymatic transformation of cellulose to starch (You et al. 2013). Although in vitro synthetic biosystems are on their early stage, they have a great potential to become a disruptive biomanufacturing platform, especially for low-cost production of biofuels and biochemical (Fig. 2.12).

2.14 Direct Photosynthetic Biosynthesis of Fuels and Fuel Precursors (Algae)

The final pathway involves direct biological synthesis by specialized autotrophic photosynthetic microorganisms, using only sunlight as an energy source. This has been perhaps the most ambitious pathway for producing biofuels. Most common within this category are single-celled algae, grown in water within photobioreactors or raceway ponds, and producing a vegetable oil that is converted to fuel (Fig. 2.7). Heterotrophic microorganisms requiring something other than CO_2 as a carbon source were not included in this category. Although a handful of pioneering patents date from the 1970s, inventions in this pathway began to grow significantly in 2005 and 2006 and followed a very similar trajectory to inventions in cellulosic ethanol. Similar to the other two advanced biofuel pathways, inventions involving algae and the direct biosynthesis of biofuels have been in decline since 2010 (Savakis and Hellingwerf 2015; Larom et al. 2010).

2.15 Food vs. Fuel and Environmental Concerns

The major drawback of both first- and second-generation biofuels lies in the fact that the cultivation of these food or non-food crops for biofuel production will compete for limited arable farmlands which will create a conflict of interest between food and fuel prices (Yong et al. 2010; Shaik and Kumar 2014). Given the global scope of the agriculture and energy systems, one approach to gain insights about future land use and inform mitigation strategy design is to examine alternative scenarios with a globally integrated assessment model (IAM) (Angelsen and Kaimowitz 1999; Thomson et al. 2010). Several studies (Wise et al. 2009; Melillo et al. 2009) have shown that, unless appropriate economic incentives are built into a climate mitigation policy, widespread deforestation could result from increasing demands for food and biofuels, in addition to the already existing threats from deforestation (DeFries et al. 2010) and climate change (Malhi et al. 2009; see review Thomson et al. 2010).

Searchinger et al. (2014) assessed the efficiency of changes in land use for mitigating climate change. They reported that land-use changes are critical for climate policy because native vegetation and soils store abundant carbon and their losses from agricultural expansion, together with emissions from agricultural production, contribute about 20–25% of greenhouse gas emissions (Edenhofer et al. 2014).

2.15.1 Peatlands

Globally, peatlands cover an area of 400 million hectares, which is equivalent to 3% of the Earth's land area which is being devastated due to energy plantations (see review Murdiyarso et al. 2010). The main driver of tropical peatlands deforestation is the development of oil-palm and pulpwood plantations which is taking place in Indonesia and Malaysia, accounting for 85% of the world's supply of crude palm oil supply to Chinese, Indian, and European markets (Murdiyarso and Kanninen 2008).

2.16 Policy Aspects of Bio-based Economy

The transition from a fossil fuel-dependent development paradigm toward a development path that takes advantage of bio-based resources and new innovations within biochemistry and the life sciences is prompting the formulation of new strategies and policies (Kircher 2012; Staffas et al. 2013; Khan et al. 2014). With increased research and innovations on bio-based energy forms, chemicals and materials, the use of the terms bioeconomy (BE) and bio-based economy (BBE) has evolved with publication of the Organisation for Economic Cooperation and Development (OECD) document "The Bioeconomy to 2030: Designing a Policy Agenda" (OECD/FAO 2015). China is pursuing a strong position in the bioeconomy with a special focus on biochemistry and life sciences (Fulton 2013; Boterman 2011). Malaysia has a vision for the creation of a bioeconomy (Biotechcorp 2013) as well as a "National Biomass Strategy to 2020", and Brazil issued in 2007 a decree including detailing the development of its bioeconomy (Presidência 2007).

2.17 Climate Action and Human Rights

According to Robinson and Shine (2018), climate actions can have direct and indirect negative impacts on people and their rights (Rights for action 2015). There are already examples from the UN Framework Convention on Climate Change's Clean Development Mechanism and Reduced Emissions from Deforestation and Forest Degradation (REDD+) where climate action has resulted in human rights violations (Schade and Obergassel 2014). This means if climate adaptation and mitigation projects are designed without the participation of local people they can lead to conflict or to the project being rejected by the community (Penz et al. 2011; Hunsberger et al. 2017). Thus climate action including reforestation and afforestation, hydroelectric dams, wind or solar energy installations and biofuel plantations pose risks to human rights including the right to housing and to a livelihood, the right to water and to food, and the right to take part in cultural life (Robinson and Shine 2018). Biofuel programs need to be integrated within a broader context of investment in rural infrastructure and human capital formation (Cécile et al. 2011). There is growing appreciation for the need for solidarity in the protection of human rights to overcome the narrow pursuit of economic nationalism. However, the risks to human rights of climate inaction and of climate impacts far outweigh the risks to human rights posed by climate action consistent with meeting the 1.5 °C goal set in the Paris Agreement.

2.18 Discussion

Most of the scenarios considered in the IPCC Fifth Assessment Report rely upon biomass energy with carbon capture and storage (BECCS) along with afforestation and reforestation to remove CO_2 from the atmosphere (Popp et al. 2016a, b; Harper et al. 2018; Sonntag et al. 2018). More recent studies also find a key role for land-based mitigation in contributing to a 2 °C target (Popp et al. 2017; Griscom et al. 2017). In the Integrated Assessment Model (IAM) scenarios consistent with a 2 °C target, a median of 3.3 GtC per year was removed from the atmosphere through BECCS by 2100, equivalent to one-third of present-day emissions from fossil fuel and industry. This median amount of BECCS would result in cumulative negative emissions of 166 GtC by 2100 (Smith et al. 2016) and would supply ~170 EJ/year of primary energy. The bioenergy crops to deliver such a scale of CO_2 removal could occupy an estimated 380–700 Mha of land (Smith et al. 2016), equivalent to up to ~50% of the present-day cropland area (Klein Goldewijk et al. 2016).

According to Peralta-Yahya et al. (2012) biodiesel produced by the transesterification of vegetable oil or animal fats with methanol, has its own limitations. It has only 91% of the energy content of D2 diesel and, because wax can form in the fuel if the temperature is too low, it is difficult to transport with the current distribution infrastructure, so there are geographical limits to its use (https://doi.org/10.1038/nature 488, 320–328). (OECD/FAO (2015), "OECD-FAO Agricultural Outlook (Edition 2015)", *OECD Agriculture Statistics* (database), https://doi.org/10.1787/data-00736-en (accessed on 11 August 2019)).

Due to the production of biofuels, concerns about their competition for land with food crops have resulted in higher global crop prices (Roberts and Schlenker 2013; Zilberman et al. 2013; Woods et al. 2015). This has led to the conversion of non-cropland to crop production and releasing carbon stored in soils and vegetation resulting in indirect land-use change (ILUC) globally releasing carbon stored in soils and vegetation (Anderson-Teixeira et al. 2012). Efforts to reduce the indirect land-use change (ILUC)-related carbon emissions caused by biofuels have led to the inclusion of an ILUC factor as a part of the carbon intensity of biofuels in a Low Carbon Fuel Standard (see review Wang et al. 2017). When bioenergy crops are placed on pre-existing agricultural land, the biophysical impacts are small. Agricultural practices can result in emissions of both N_2O and CH_4 (Harper et al. 2018).

Land-use change will also impact extreme weather events such as daytime high temperatures (Alkama and Cescatti 2016; Hirsch et al. 2018). Many of these effects require evaluation in a coupled GCM framework, to fully capture local and regional land-atmosphere feedbacks (Fuss et al. 2014).

Chubukov et al. (2016) reviewed the difficulties of taking a microbial production process from conception to commercialization along with the tools that can be used to address some of the challenges and gaps in our knowledge and engineering capabilities. Because of the difficulties of entering new chemical markets, most biological production has focused on molecules with large existing markets (Baeshen et al. 2014). Recently interest has developed in high-volume, low-cost markets such as biofuels, economic considerations have become paramount, and the development of a new project must begin with an analysis of the potential of process commercialization. Adding to these difficulties is the complexity of predicting the extra-economic costs derived from scaling up production and downstream processing (e.g., molecule extraction and purification). According to Alper and Stephanopoulos (2009), ideal microorganism for biofuel production will possess high substrate utilization and processing capacities, fast and deregulated pathways for sugar transport, good tolerance to inhibitors and product, and high metabolic fluxes and produce a single fermentation product. The choice between engineering natural function and importing biosynthetic capacity is affected by current progress in metabolic engineering and synthetic biology (Hutchison et al. 2016). All these factors make target molecule selection the least systematic part of the metabolic engineering process. Even in the cases where biomanufacturing has not achieved economic cost-competitiveness, it can help improve the sustainability of energy sources and other chemical commodity products. However, global trends point to an uncertain future, in particular, for advanced biofuels (Albers et al. 2016).

The combination of synthetic and systems biology is a powerful framework to study fundamental questions in biology and to produce chemicals of practical application such as biofuels, polymers, or therapeutics. However, the engineering of biological systems is more complicated in comparison to physical systems. According to Lee et al. (2008b) and Peralta-Yahya and Keasling (2010) while developing an organism or a pathway to produce an advanced fuel, factors including engine type (spark or compression ignition), energy content, combustion quality or ignition delay, cloud point, volatility, lubricity, viscosity, stability, odor, toxicity, water miscibility, and cost must be considered. Wriessnegger and Pichler (2013) suggested the goals of metabolic engineering: optimization of strains for overproducing recombinant proteins and small molecule chemicals, the extension of substrate range, enhanced productivity and yield, elimination of by-products, improvement of process performance and of cellular properties.

Recent breakthroughs in genomic research and genetic engineering provided the inventory and methods necessary to physically construct and assemble biomolecular parts. Thus the synthetic biology was born with the broad goal of engineering or "wiring" biological circuitry for manifesting logical forms of cellular control.

Some of the potential biofuels which are currently being investigated include biodiesel, butanol, longer-chain alcohols, hydrogen, and synthetic petroleum hydrocarbons. Developments in metabolic engineering and synthetic biology provide alternative methods for producing value-added terpenoids in *Escherichia coli*, *Saccharomyces cerevisiae*, and filamentous fungi with high efficiency (Alper and Stephanopoulos 2009; Bhansali and Kumar 2018).

2.19 Summary

Finding ways to make fuels from renewable sources is among the most active research areas in chemistry. There are several strategies (1) to use biological agents, such as enzymes or whole organisms, (2) to catalyze processes for converting biomass into fuels, or (3) to make fuels by photosynthesis. Biofuel production pathways have been shown by the conversion of syngas, CO_2, algal hydrolysate, and switchgrass into higher alcohols, fatty acids, and isoprenoid-derived biofuels. Some of the challenges are cost-effective to produce advanced biofuels in high yield requiring the engineering of both the substrate use and advanced biofuel-producing pathways. Engineering the fatty-acid pathway has led to the production of several types of biofuel with different physical and combustion properties. Both data-driven and synthetic-biology approaches are powerful tools for troubleshooting and optimizing engineered metabolic pathways. Advanced design tools and experimental techniques will facilitate the tuning of the polyketide-derived fuel properties by manipulating polyketide synthase. Artificial enzymes with new functions can even be created by incorporating unnatural amino acids and computation-based protein design.

Alternatively, the thermochemical approach uses abiological, chemical means to make fuels from biomass. Understanding and engineering of photosynthesis could lead to an improvement in biomass production. In several chapters of this book, an attempt is made in this direction (see this volume).

Acknowledgement The author acknowledges support from the Alexander von Humbodt Foundation (Germany) to work in the laboratory of Late Professor Dr. K-H. Neumann and later on Professor Dr. Sven Schubert at the Institut für Pflanzenernährung, Justus Liebig Unversity, Giessen, Germany, and the JSPS fellowship to work with Professor Shinjiro Ogita during the course of investigations spreading over 20 years. Financial support from the Department of Biotechnology, Govt. of India, New Delhi, and infrastructural support for providing 50 ha land area for experimentation by the University of Rajasthan Syndicate is gratefully acknowledged. I also acknowledge my research scholars and technical staff who have worked in lab and field with me for investigations quoted here. I also thank my wife Professor Vijay R. Kumar for going through the manuscript and making suggestions.

References

Acker RV, Vanholme R, Storme V, Mortimer JC, Dupree P, Boerjan W (2013) Lignin biosynthesis perturbations affect secondary cell wall composition and saccharification yield in *Arabidopsis thaliana*. Biotechnol Biofuels 6:46

Albers SC, Berklund AM, Graff GD (2016) The rise and fall of innovation in biofuels. Nat Biotechnol 34(8):814–821. https://doi.org/10.1038/nbt.3644

Alkama R, Cescatti A (2016) Biophysical climate impacts of recent changes in global forest cover. Science 351:600

Alper H, Stephanopoulos G (2009) Engineering for biofuels: exploiting innate microbial capacity or importing biosynthetic potential? Nat Rev Microbiol 7:715–723

Alper H, Moxley J, Nevoigt E, Fink GR, Stephanopoulos G (2006) Engineering yeast transcription machinery for improved ethanol tolerance and production. Science 314:1565–1568

Anderson-Teixeira KJ et al (2012) Climate-regulation services of natural and agricultural ecoregions of the Americas. Nat Clim Chang 2:177–181

Angelsen A, Kaimowitz D (1999) Rethinking the causes of deforestation: lessons from economic models. World Bank Res Obs 14:73–98

ASTM (2008) Standard specification for biodiesel fuel (B100) blend stock for distillate fuels. In: Annual book of ASTM standards. ASTM International, West Conshohocken, Method D6751–08,

Atabani A, Silitonga A, Badruddin I, Mahlia T, Masjuki H, Mekhilef SA (2012) Comprehensive review on biodiesel as an alternative energy resource and its characteristics. Renew Sustain Energy Rev 16:2070–2093

Atsumi S, Hanai T, Liao JC (2008) Non-fermentative pathways for synthesis of branched-chain higher alcohols as biofuels. Nature 451:86–89. https://doi.org/10.1038/nature06450

Atsumi S, Higashide W, Liao JC (2009) Direct photosynthetic recycling of carbon dioxide to isobutyraldehyde. Nat Biotechnol 27:1177–1180

Baeshen NA et al (2014) Cell factories for insulin production. Microb Cell Factories 13:141

Baritugo KG et al (2018) Recent advances in metabolic engineering of, *Corynebacterium glutamicum* as a potential platform microorganism for biorefinery. Biofuels Bioprod Biorefin 12:899–925

Becker EW (2007) Micro-algae as a source of protein. Biotechnol Adv 25:207–210

Beerthuis R, Rothenber G, Shiju NR (2015) Catalytic routes towards acrylic acid, adipic acid and e-caprolactam starting from biorenewables. Green Chem 17:1341–1361

Bellou S, Aggelis G (2013) Biochemical activities in Chlorella sp. and Nannochloropsis salina during lipid and sugar synthesis in a lab-scale open pond simulating reactor. J Biotechnol 164:318–329

Berla BM et al (2013) Synthetic biology of cyanobacteria: unique challenges and opportunities. Front Microbiol 4:246

Bhansali S, Kumar A (2018) Synthetic and semisynthetic metabolic pathways for biofuel production. In: Kumar A, Ogita S, Yau Y-Y (eds) Biofuels: greenhouse gas mitigation and global warming. Next generation biofuels and role of biotechnology. Springer, Heidelberg, pp 421–432

Bian G, Deng Z, Liu T (2017) Strategies for terpenoid overproduction and new terpenoid discovery. Curr Opin Biotechnol 48:234–241

Biotechcorp (2013) Malaysian biotechnology corporation. http://www.biotechcorp.com.my/. Accessed 4 Apr 2013

Biswal AK et al (2018) Sugar release and growth of biofuel crops are improved by downregulation of pectin biosynthesis. Nat Biotechnol 36(3):249–266

Bloch K (1992) Sterol molecule: structure, biosynthesis, and function. Steroids 57:378–383

Boakye-Boaten NA, Xiu S, Shahbazi A, Wang L, Li R, Mims M, Schimmel K (2016) Effects of fertilizer application and dry/wet processing of Miscanthus x giganteus on bioethanol production. Bioresour Technol 204:98–105

Bond-Watts BB, Bellerose RJ, Chang MC (2011) Enzyme mechanism as a kinetic control element for designing synthetic biofuel pathways. Nat Chem Biol 7:222–227

Bopp et al (2016) Biofuels and their co-products as livestock feed: global economic and environmental implications. Molecules 21:285. https://doi.org/10.3390/molecules21030285

Borland AM, Griffiths H, Hartwell J, Smith JA (2009) J Exp Bot 60:2879

Boterman B (2011) Bio-economie in China. Rathenau Institut and TWA Netwerk Kingdom of the Netherlands, The Hague, p 42

Cameron DE, Bashor CJ, Collins JJ (2014) A brief history of synthetic biology. Nat Rev Microbiol 12:381–390

Carroll A, Somerville C (2009) Cellulosic biofuels. Annu Rev Plant Biol 60:165–182. https://doi.org/10.1146/annurev.arplant.043008.092125

Caspeta L, Nielsen J (2013) Economic and environmental impacts of microbial biodiesel. Nat Biotechnol 31:789–793

Caspeta L, Buijs NAA, Nielsen J (2013) The role of biofuels in the future energy supply. Energy Environ Sci 6:1077

Cécile B, Fabien F, Benoît G, Bruno M (2011) Biofuels, greenhouse gases and climate change. A review. Agron Sustain Dev 31:1–79

Chen Z, Liu D (2016) Biotechnology for biofuels toward glycerol biorefinery: metabolic engineering for the production of biofuels and chemicals from glycerol. Biotechnol Biofuels 9:1–15. https://doi.org/10.1186/s13068-016-0625-8

Chen M, Poulter CD (2010) Characterization of thermophilic archaeal isopentenyl phosphate kinases. Biochemistry 49:207–217

Chisti Y (2007) Biodiesel from microalgae. Biotechnol Adv 25:294–306. https://doi.org/10.1016/j.biotechadv.2007.02.001

Cho C, Jang Y, Moon HG, Lee J, Lee SY (2015) Metabolic engineering of clostridia for the production of chemicals. Biofuels Bioprod Bioref 9:211–225. https://doi.org/10.1002/bbb

Christian DG, Riche AB, Yates NE (2008) Growth, yield and mineral content of *Miscanthus x giganteus* grown as a biofuel for 14 successive harvests. Ind Crop Prod 28:320–327

Chubukov V, Mukhopadhyay A, Petzold CJ, Keasling JD, Martín HG (2016) Synthetic and systems biology for microbial production of commodity chemicals. Syst Biol Appl 2:1–11. https://doi.org/10.1038/npjsba.2016.9. published

Claassens NJ, Sousa DZ, Martins VAP, De Vos WM (2016) Harnessing the power of microbial autotrophy. Nat Rev Microbiol 14(11):692–706. https://doi.org/10.1038/nrmicro.2016.130

Dale BE et al (2014) Take a closer look: biofuels can support environmental, economic and social goals. Environ Sci Technol 48:7200–7203

Das P, Aziz SS, Obbard JP (2011) Two phase microalgae growth in the open system for enhanced lipid productivity. Renew Energy 36:2524–2528

DeFries R et al (2010) Deforestation driven by urban population growth and agricultural trade in the twenty-first century. Nat Geosci 3:178–181

Devappa RK, Maes J, Makkar HPS, Greyt W, Becker K (2010) Quality of biodiesel prepared from phorbol ester extracted *Jatropha curcas* oil. J Am Oil Chem Soc 87(6):697–704. https://doi.org/10.1007/s11746-010-1547-4

Devi MP, Mohan SV (2012) CO_2 supplementation to domestic wastewater enhances microalgae lipid accumulation under mixotrophic microenvironment: effect of sparging period and interval. Bioresour Technol 112:116–123

Dugar D, Stephanopoulos G (2011) Relative potential of biosynthetic pathways for biofuels and bio-based products. Nat Biotechnol 29(12):1074–1078. https://doi.org/10.1038/nbt.2055

Dumitrache A et al (2017) Transgenic switchgrass (*Panicum virgatum* L.) targeted for reduced recalcitrance to bioconversion: a 2-year comparative analysis of field-grown lines modified for target gene or genetic element expression. Plant Biotechnol J 15:688–697

Edenhofer O et al (2014) In Edenhofer O. et al (eds) Climate change 2014: mitigation of climate change. Contribution of working group III to the fifth assessment report of the intergovernmental panel on climate change. Cambridge University Press, Cambridge, pp. 1–30

Eisenreich W, Bacher A, Arigoni D, Rohdich F (2004) Biosynthesis of isoprenoids via the non-mevalonate pathway. Cell Mol Life Sci 61:1401–1426

Espaux L, Mendez-perez D, Li R, Keasling JD (2015) Science direct synthetic biology for microbial production of lipid-based biofuels. Curr Opin Chem Biol 29:58–65. https://doi.org/10.1016/j.cbpa.2015.09.009

Eudes A, Liang Y, Mitra P, Loqué D (2014) Lignin bioengineering. Curr Opin Biotechnol 26:189–198

Ezeji TC, Qureshi N, Blaschek HP (2007) Production of acetone butanol (AB) from liquefied corn starch, a commercial substrate, using *Clostridium beijerinckii* coupled with product recovery by gas stripping. J Ind Microbiol Biotechnol 34:771–777

FAO (2008a) The state of food and agriculture. Biofuels: prospects risks and opportunities. FAO, Rome

FAO (2008b) Bioenergy, food security and sustainability—towards an international framework. FAO. http://www.fao.org/fileadmin/user_upload/foodclimate/HLCdocs/HLC08-inf-3-E.pdf. Accessed 2 Mar 2011

Fulton M (2013) 12th five year plan—Chinese leadership towards a low carbon economy. DB Climate Change Advisors, Deutsche Bank Group, Frankfurt am Main, p 16. Sustainability, 5 2767

Fuss S et al (2014) Betting on negative emissions. Nat Clim Chang 4:850–853

Gajraj R, Singh GP, Kumar A (2018) Third-generation biofuel: algal biofuels as a sustainable energy source. In: Kumar A, Ogita S, Yau Y-Y (eds) Biofuels: greenhouse gas mitigation and global warming next generation biofuels and role of biotechnology. Springer, Heidelberg, pp 307–326

Gao C et al (2015) An artificial enzymatic reaction cascade for a cell-free bio-system based on glycerol. Green Chem 17:804–807

Geib SM et al (2008) Lignin degradation in wood-feeding insects. Proc Natl Acad Sci U S A 105:12932–12937

Gimpel JA, Specht EA, Georgianna DR, Mayfield SP (2013) Advances in microalgae engineering and synthetic biology applications for biofuel production. Curr Opin Chem Biol 17:489–495

Griscom BW et al (2017) Natural climate solutions. Proc Natl Acad Sci U S A 114:11645–11650

Halfhide T, Åkerstrøm A, Lekang OI, Gislerød HR, Ergas SJ (2014) Production of algal biomass, chlorophyll, starch and lipids using aquaculture wastewater under axenic and non-axenic conditions. Algal Res 6:152–159

Hall DO, Mynick HE, Williams RH (1991) Cooling the greenhouse with bioenergy. Nature 353:11

Harper AB, Powell T, Cox PM, House J, Huntingford C, Lenton TM, Shu S (2018) Land-use emissions play a critical role in land-based mitigation for Paris climate targets. Nat Commun 9 (1):2938. https://doi.org/10.1038/s41467-018-05340-z

Heaton EA, Dohleman FG, Miguez AF, Juvik JA, Lozovaya V, Widholm J, Zabotina OA, McIsaac GF, David MB, Voigt TB, Boersma NN, Long SP (2010) *Miscanthus:* a promising biomass crop. In: Kader JC, Delseny M (eds) Advances in botanical research, vol 56. Elsevier, Amsterdam, pp 75–137

Hill J, Nelson E, Tilman D, Polasky S, Tiffany D (2006) Environmental, economic and energetic costs and benefits of biodiesel and ethanol biofuels. Proc Natl Acad Sci U S A 103:11206–11210. https://doi.org/10.1073/pnas.0604600103

Himmel ME et al (2007) Biomass recalcitrance: engineering plants and enzymes for biofuels production. Science 315:804–807

Hirsch AL et al (2018) Biogeophysical impacts of land-use change on climate extremes in low-emission scenarios: results from HAPPI-Land. Earth's Future 6. https://doi.org/10.1002/2017EF000744

Hof AF et al (2017) Global and regional abatement costs of nationally determined contributions (NDCs) and of enhanced action to levels well below 2 °C and 1.5 °C. Environ Sci Policy 71:30–40

Holtzapple M, Granda C (2009) Carboxylate platform: the MixAlco process part 1: comparison of three biomass conversion platforms. Appl Biochem Biotechnol 156:95–106

Hossain N, Mahlia TMI, Saidur R (2019) Biotechnology for biofuels latest development in microalgae biofuel production with nano additives. Biotechnol Biofuels 12:1–16. https://doi.org/10.1186/s13068-019-1465-0

Houghton J, Weatherwax S, Ferrell J (2006) Breaking the biological barriers to cellulosic ethanol: a joint research agenda. US Dep. Energy, 7–9 Dec 2005, Rockville

Hu Q, Sommerfeld M, Jarvis E, Ghirardi M, Posewitz M, Seibert M, Darzins A (2008) Microalgal triacylglycerols as feedstocks for biofuel production: perspectives and advances. Plant J 54:621–639

Huang J, Xia T, Li G, Li X, Li Y, Wang Y, Wang L (2019) Biotechnology for biofuels overproduction of native endo- β-1,4-glucanases leads to largely enhanced biomass saccharification and bioethanol production by specific modification of cellulose features in transgenic rice. Biotechnol Biofuels 12:1–15. https://doi.org/10.1186/s13068-018-1351-1

Hütsch BW, Jung S, Steinbach M, Schubert S (2020) What is the limiting factor? The key question for grain yield of maize as a renewable resource under salt stress. In: Kumar A, Yau YY, Ogita S, Scheibe R (eds) Climate change, photosynthesis and advanced biofuels: role of biotechnology in production of value-added plant products. Springer, Singapore

Hunsberger C et al (2017) Climate change mitigation, land grabbing and conflict: towards a landscape-based and collaborative action research agenda. Can J Dev Stud 38:305–324

Huntingford C, Mercado L (2016) High chance that current atmospheric greenhouse concentrations commit to warmings greater than 1.5 °C over land. Sci Rep 6. https://doi.org/10.1038/srep30294

Huo Y et al (2011a) Conversion of proteins into biofuels by engineering nitrogen flux. Nat Biotechnol 29:346–351

Huo Y-X, Cho KM, Rivera JGL, Monte E, Shen CR, Yan Y, Liao JC (2011b) Conversion of proteins into biofuels by engineering nitrogen flux. Nat Biotechnol 29(4):346–351. https://doi.org/10.1038/nbt.1789

Hutchison CA, Chuang RY, Noskov VN, Assad-Garcia N, Deerinck TJ, Ellisman MH, Gill J, Kannan K, Karas BJ, Ma L et al (2016) Design and synthesis of a minimal bacterial genome. Science 351:aad6253

IEA (2007) Key world energy statistics. OECD/IEA, IEA Publications, Paris, 82p

Ignea C, Cvetkovic I, Loupassaki S, Kefalas P, Johnson CB et al (2011) Improving yeast strains using recyclable integration cassettes, for the production of plant terpenoids. Microb Cell Factories 10:4

International Energy Agency (IEA) (2008) Energy technology perspectives 2008. Organization for Economic Cooperation and Development, IEA, Paris

IPCC (2007) In: CW Team, Pachauri RK, Reisinger A (eds) Climate change 2007: synthesis report. IPCC, Geneva

IPCC (2014) In: Core Writing Team, Pachauri RK, Meyer LA (eds) Climate change 2014: synthesis report. Contribution of working groups I, II and III to the fifth assessment report of the intergovernmental panel on climate change. IPCC, Geneva, 151p
Jagadevan S, Banerjee A, Banerjee C, Guria C, Tiwari R, Baweja M (2018) Biotechnology for biofuels: recent developments in synthetic biology and metabolic engineering in microalgae towards biofuel production. Biotechnol Biofuels 11:1–21. https://doi.org/10.1186/s13068-018-1181-1. Used under creative commons licence
Jeffryes C, Rosenberger J, Rorrer GL (2013) Fed-batch cultivation and bioprocess modeling of Cyclotella sp. for enhanced fatty acid production by controlled silicon limitation. Algal Res 2:16–27
Jiang Y, Wu R, Zhou J, He A, Xu J, Xin F, Zhang W (2019) Biotechnology for biofuels recent advances of biofuels and biochemicals production from sustainable resources using co-cultivation systems. Biotechnol Biofuels 12:1–12. https://doi.org/10.1186/s13068-019-1495-7
Jones MB, Walsh M (2001) Miscanthus for energy and fibre. James and James, London
Jones JA, Toparlak ÖD, Koffas MAG (2015) Metabolic pathway balancing and its role in the production of biofuels and chemicals. Curr Opin Biotechnol 33:52–59. https://doi.org/10.1016/j.copbio.2014.11.013
Jongschaap REE, Blesgraaf RAR, Bogaard TA, van Loo EN, Savenije HHG (2009) The water footprint of bioenergy from *Jatropha curcas* L. Proc Natl Acad Sci U S A 106:E92–E92
Jose A, James M (2013) Fermentation of glycerol and production of valuable chemical and biofuel molecules. Biotechnol Lett 35:831–842
Karamerou EE, Theodoropoulos C, Webb C (2016) A biorefinery approach to microbial oil production from glycerol by *Rhodotorula glutinis*. Biomass Bioenergy 89:113–122
Kawai S, Murata K (2016) Biofuel production based on carbohydrates from both brown and red macroalgae: recent developments in key biotechnologies. Int J Mol Sci 17:1–17
Kazamia E, Smith AG (2014) Assessing the environmental sustainability of biofuels. Trends Plant Sci 19(10):615–618. https://doi.org/10.1016/j.tplants.2014.08.001
Keasling JD (2008) Synthetic biology for synthetic chemistry. ACS Chem Biol 3:64–76
Keasling JD (2010) Manufacturing molecules through metabolic engineering. Science 330:1355–1358
Khalil AS, Collins JJ (2010) Synthetic biology: applications come of age. Nat Rev Genet 11 (5):367–379. https://doi.org/10.1038/nrg2775
Khan NE, Myers JA, Tuerk AL, Curtis WRA (2014) Process economic assessment of hydrocarbon biofuels production using chemoautotrophic organisms. Bioresour Technol 172:201–211
King ZA, Lloyd CJ, Feist AM, Palsson BO (2015) Next-generation genome-scale models for metabolic engineering. Curr Opin Biotechnol 35:23–29
Kirby J, Keasling JD (2009) Biosynthesis of plant isoprenoids: perspectives for microbial engineering. Annu Rev Plant Biol 60:335–355
Kircher M (2012) The transition to a bio-economy: national perspectives. Biofuels Bioprod Biorefin 6:240–245
Klein Goldewijk K, Beusen A, Doelman J, Stehfest E (2016) New anthropogenic land use estimates for the Holocene, HYDE 3.2. Earth Syst Sci Data Discuss 2016:1–40
Kollah B, Patra AK, Mohanty SR (2016) Aquatic microphylla Azolla: a perspective paradigm for sustainable agriculture, environment and global climate change. Environ Sci Pollut R 23 (5):4358–4369
Kumar A (2001) Bioengineering of crops for biofuels and bioenergy. In: Bender L, Kumar A (eds) From soil to cell: a broad approach to plant life. Giessen + Electron. Library GEB, pp 14–29 . http://geb.uni-giessen.de/geb/volltexte/2006/3039/pdf/FestschriftNeumann-2001.pdf
Kumar A (2008) Bioengineering of crops for biofuels and bioenergy. In: Kumar A, Sopory S (eds) Recent advances in plant biotechnology. I.K. International, New Delhi, pp 346–360
Kumar A (2010) Plant genetic transformation and molecular markers. Pointer Publishers, Jaipur. 288p

Kumar A (2011) Biofuel resources for greenhouse gas mitigation and environment protection. In: Trivedi PC (ed) Agriculture biotechnology. Avishkar Publishers, Jaipur, pp 221–246
Kumar A (2013) Biofuels utilisation: an attempt to reduce GHG's and mitigate climate change. In: Nautiyal S, Rao K, Kaechele H, Raju K, Schaldach R (eds) Knowledge systems of societies for adaptation and mitigation of impacts of climate change. Environmental science and engineering. Springer, Berlin, pp 199–224
Kumar A (2015) Metabolic engineering in plants. In: Bahadur B, Venkat Rajam M, Sahijram L, Krishnamurthy KV (eds) Plant biology and biotechnology. II. Plant genomics and biotechnology. Springer, New Delhi, pp 517–526
Kumar A (2018a) A review on first- and second-generation biofuel productions. In: Kumar A, Ogita S, Yau Y-Y (eds) Biofuels: greenhouse gas mitigation and global warming. Next generation biofuels and role of biotechnology. Springer, Heidelberg, pp 141–154
Kumar A (2018b) Alternative biomass from semiarid and arid conditions as a biofuel source: *Calotropis procera* and its genomic characterization. In: Kumar A, Ogita S, Yau Y-Y (eds) Biofuels: greenhouse gas mitigation and global warming. Next generation biofuels and role of biotechnology. Springer, Heidelberg, pp 241–270
Kumar A (2018c) Global warming, climate change and greenhouse gas mitigation. In: Kumar A, Ogita S, Yau Y-Y (eds) Biofuels: greenhouse gas mitigation and global warming. Next generation biofuels and role of biotechnology. Springer, Heidelberg, pp 1–16
Kumar A (2020) Synthetic biology and future production of biofuels and high value products. In: Kumar A, Yau YY, Ogita S, Scheibe R (eds) Climate change, photosynthesis and advanced biofuels: role of biotechnology in production of value-added plant products. Springer, Singapore
Kumar A, Gupta N (2018) Potential of lignocellulosic materials for production of ethanol. In: Kumar A, Ogita S, Yau Y-Y (eds) Biofuels: greenhouse gas mitigation and global warming. Next generation biofuels and role of biotechnology. Springer, Heidelberg, pp 271–290
Kumar A, Roy S (2018) Agrotechnology, production, and demonstration of high-quality planting material for biofuels in arid and semiarid regions. In: Kumar A, Ogita S, Yau Y-Y (eds) Biofuels: greenhouse gas mitigation and global warming. Next generation biofuels and role of biotechnology. Springer, Heidelberg, pp 205–228
Kumar P, Barrett DM, Delwiche MJ, Stroeve P (2009) Methods for pretreatment of lignocellulosic biomass for efficient hydrolysis and biofuel production. Ind Eng Chem Res 48:3713–3729
Kumar A, Sharma M, Basu SK, Asif M, Li X, Chen X (2014) Plant molecular breeding: perspectives from plant biotechnology and marker assisted selection. Am J Soc Humanit 4:177–189
Kumar A, Ogita S, Yau Y-Y (eds) (2018a) Biofuels: greenhouse gas mitigation and global warming. Next generation biofuels and role of biotechnology. Springer, Heidelberg, 432p
Kumar A, Abraham E, Gupta A (2018b) Alternative biomass from saline and semiarid and arid conditions as a source of biofuels: *Salicornia*. In: Kumar A, Ogita S, Yau Y-Y (eds) Biofuels: greenhouse gas mitigation and global warming. Next generation biofuels and role of biotechnology. Springer, Heidelberg, pp 229–240
Kumar A, Bhansali S, Gupta N, Sharma M (2019) Bioenergy and climate change: greenhouse gas mitigation. In: Rastegari AA, Yadav AN, Gupta A (eds) Prospects of renewable bioprocessing in future energy systems. Biofuel and biorefinery technologies, vol 10. Springer, Heidelberg, pp 269–290
Kumari A, Mahapatra P, Garlapati VK, Banerjee R (2009) Enzymatic transesterification of Jatropha oil. Biotechnol Biofuels 2(1):1. https://doi.org/10.1186/1754-6834-2-1
Labrecque M, Teodorescu TI (2005) Field performance and biomass production of 12 willow and poplar clones in short-rotation coppice in southern Quebec (Canada). Biomass Bioenergy 29:1–9
Lam MK, Lee KT (2012) Microalgae biofuels: a critical review of issues, problems and the way forward. Biotechnol Adv 30(3):673–690. https://doi.org/10.1016/j.biotechadv.2011.11.008

Lane J (2015) Biofuels mandates around the world biofuels dig. http://www.biofuelsdigest.com/bdigest/2014/12/31/biofuels-mandates-around-the-world-2015/
Larom S, Salama F, Schuster G, Adir N (2010) Engineering of an alternative electron transfer path in photosystem II. Proc Natl Acad Sci 107:9650–9655. https://doi.org/10.1073/pnas.1000187107
Leavell MD, Mcphee DJ, Paddon CJ (2016) Developing fermentative terpenoid production for commercial usage. Curr Opin Biotechnol 37:114–119. https://doi.org/10.1016/j.copbio.2015.10.007
Lee SK et al (2008a) Metabolic engineering of microorganisms for biofuels production: from bugs to synthetic biology to fuels. Curr Opin Biotechnol 19:556–563
Lee SY et al (2008b) Fermentative butanol production by clostridia. Biotechnol Bioeng 101:209–228
Li M, Feng S, Wu L, Li Y, Fan C, Zhang R, Zou W, Tu Y, Jing HC, Li S, Peng L (2014a) Sugar-rich sweet sorghum is distinctively affected by wall polymer features for biomass digestibility and ethanol fermentation in bagasse. Bioresour Technol 167:14–23
Li Q, Song J, Peng S, Wang JP, Qu G-Z, Sederoff RR, Chiang VL (2014b) Plant biotechnology for lignocellulosic biofuel production. Plant Biotechnol J 12:1174–1192
Li Z, Yan J, Sun J, Xu P, Ma C, Gao C (2018) Production of value-added chemicals from glycerol using in vitro enzymatic cascades. Commun Chem, pp 1–7. https://doi.org/10.1038/s42004-018-0070-7
Liao JC, Mi L, Pontrelli S, Luo S (2016) Fuelling the future: microbial engineering for the production of sustainable biofuels. Nat Rev Microbiol 14:288. https://doi.org/10.1038/nrmicro.2016.32. JOUR
Licht FO (2013) World ethanol and biofuels report. Agra Informa, London
Liew LN, Shi J, Li Y (2012) Methane production from solid state anaerobic digestion of lignocellulosic biomass. Biomass Bioenergy 46:125–132
Liu T, Khosla C (2010) Genetic engineering of *Escherichia coli* for biofuel production. Annu Rev Genet 44:53–69. https://doi.org/10.1146/annurev-genet-102209-163440
Liu C-J, Cai Y, Zhang X, Gou M, Yang H (2014) Tailoring lignin biosynthesis for efficient and sustainable biofuel production. Plant Biotechnol J 12:1154–1162
Liu C, Colón BC, Ziesack M, Silver PA, Nocera DG (2016) Water splitting-biosynthetic system with CO_2 reduction efficiencies exceeding photosynthesis. Science 352:1210–1213
Long SP, Spence AK (2013) Toward cool C(4) crops. Annu Rev Plant Biol 64:701–722. https://doi.org/10.1146/annurev-arplant-050312-120033
Loqué D, Scheller HV, Pauly M (2015) Engineering of plant cell walls for enhanced biofuel production. Curr Opin Plant Biol 25:151–161
Lu X, Vora H, Khosla C (2008) Overproduction of free fatty acids in E. coli: implications for biodiesel production. Metab Eng 10:333–339
Lynd LR (2017) C O M M E N TA R Y: the grand challenge of cellulosic biofuels. Nat Publ Group 35(10):912–915. https://doi.org/10.1038/nbt.3976
Lynd LR, Weimer PJ, van Zyl WH, Pretorius IS (2002) Microbial cellulose utilization: fundamentals and biotechnology. Microbiol Mol Biol Rev 66:506–577
Malhi Y et al (2009) Exploring the likelihood and mechanism of a climate-change- induced dieback of the Amazon rainforest. Proc Natl Acad Sci U S A 106:20610–20615
Martín del Campo JS, Rollin J, Myung S, Chun Y, Chandrayan S, Patiño R et al (2013) High-yield production of dihydrogen from xylose by using a synthetic enzyme cascade in a cell-free system. Angew Chem Int Ed 52(17):4587–4590
Melillo JM et al (2009) Indirect emissions from biofuels: how important? Science 326:1397–1399
Miao X, Wu Q (2006) Biodiesel production from heterotrophic microalgal oil. Bioresour Technol 97:841–846
Mielenz JR (2011) Biofuels from protein A new look at membrane protein binding. Nat Biotechnol 29(4):327–328. https://doi.org/10.1038/nbt0411-327

Miranda AF, Biswas B, Ramkumar N, Singh R, Kumar J, James A et al (2016) Biotechnology for biofuels aquatic plant Azolla as the universal feedstock for biofuel production. Biotechnol Biofuels 9:1–17. https://doi.org/10.1186/s13068-016-0628-5

Mollendorf D (2013) Moral challenge of dangerous climate change: values, poverty and policy. Cambridge University Press, New York, 263pp. https://doi.org/10.1017/CBO9781139083652

Moser BR (2009) Biodiesel production, properties, and feedstocks. In Vitro Cell Dev Biol Plant 45 (3):229–266

Murdiyarso D, Kanninen M (2008) Forests and climate change: an outlook of Asian forests in the new climate regime. In: Loh C, Stevenson A, Tay S (eds) Climate change negotiations: can Asia change the game? Hong Kong Civic Exchange–Singapore Institute of International Affairs, Singapore, pp 74–87

Murdiyarso D, Hergoualc'h K, Verchot LV (2010) Opportunities for reducing greenhouse gas emissions in tropical peatlands. Proc Natl Acad Sci U S A 107(46):19655–19660. https://doi.org/10.1073/pnas.0911966107

Murugesan A, Umarani C, Subramanian R, Nedunchezhian N (2009) Bio-diesel as an alternative fuel for diesel engines—a review. Renew Sust Energ Rev 13:653–662

Nielsen J, Keasling JD (2011) Synergies between synthetic biology and metabolic engineering. Nat Biotechnol 29:693–695

Nielsen J, Keasling JD (2016) Engineering cellular metabolism. Cell 164(6):1185–1197. https://doi.org/10.1016/j.cell.2016.02.004

OECD/FAO (2015) Agricultural outlook. https://doi.org/10.1787/data-00736-en. Assessed 10 Dec 2015

Olson DG, McBride JE, Shaw AJ, Lynd LR (2012) Recent progress in consolidated bioprocessing. Curr Opin Biotechnol 23(3):396–405

Paddon CJ, Westfall PJ, Pitera DJ, Benjamin K, Fisher K, McPhee D, Leavell MD, Tai A, Main A, Eng D et al (2013) High-level semi-synthetic production of the potent antimalarial artemisinin. Nature 496:528–532

Paulová L, Melzoch K, Rychtera M, Patáková P (2013) Production of 2nd generation of liquid biofuels. In: Fang Z (ed) Liquid, gaseous and solid biofuels: conversion techniques. InTech Open Access Publisher, Rijeka, pp 47–78

Penz P, Drydyk J, Bose PS (2011) Displacement by development: ethics, rights and responsibilities. Cambridge University Press, Cambridge. 344pp

Peplow M (2014) Cellulosic ethanol fights for life. Nature 507(7491):152–153

Peralta-Yahya PP, Keasling JD (2010) Advanced biofuel production in microbes. Biotechnol J 5:147–162

Peralta-Yahya PP, Fuzhong Z, del Cardayre SB, Keasling JD (2012) Microbial engineering for the production of advanced biofuels. Nature 488:320. https://doi.org/10.1038/nature11478

Pokoo-Aikins G, Nadim A, El-Halwagi MM, Mahalec V (2009) Design and analysis of biodiesel production from algae grown through carbon sequestration. Clean Techn Environ Policy 12 (3):239–254

Polburee P, Yongmanitchai W, Honda K, Ohashi T (2016) Lipid production from biodiesel-derived crude glycerol by *Rhodosporidium fluviale* DMKU-RK253 using temperature shift with high cell density. Biochem Eng J 112:208–218

Popp J et al (2016a) Biofuels and their co-products as livestock feed: global economic and environmental implications. Molecules 21:1–16. https://doi.org/10.3390/molecules21030285

Popp et al (2016b) Biofuels and their co-products as livestock feed: global economic and environmental implications. Molecules 21:1–26. https://doi.org/10.3390/molecules21030285

Popp A et al (2017) Land-use futures in the shared socio-economic pathways. Glob Environ Chang 42:331–345

Presidência da República (2007) Decreto no 6041, Anexo Política de Desenvolvimento da Biotecnologia, DECRETO N ° 6.041, DE 8 DE FEVEREIRO DE 2007. Presidência da República, Brasil, Brasilia, p 35

Qureshi N, Blaschek HP (1999) Production of acetone butanol ethanol (ABE) by a hyper-producing mutant strain of *Clostridium beijerinckii* BA101 and recovery by pervaporation. Biotechnol Prog 15:594–602

Raghavendra AS, Sage RF (2011) C_4 photosynthesis and related CO_2 concentrating mechanisms. Springer, Dordrecht. 412pp

Ramey CJ, Barón-Sola Á, Aucoin HR, Boyle NR (2015) Genome engineering in cyanobacteria: where we are and where we need to go. ACS Synth Biol 4:1186–1196

Rauch R, Hrbek J, Hofbauer H (2014) Biomass gasification for synthesis gas production and applications of the syngas. WIREs Energy Environ 3:343–362

Rights for Action (2015) Putting people at the centre of action on climate change (Mary Robinson Foundation—Climate Justice). https://go.nature.com/2LkU4hc

Rizza LS, Smachetti MES, Nascimento MD, Salerno GL, Curatti L (2017) Bioprospecting for native microalgae as an alternative source of sugars for the production of bioethanol. Algal Res 22:140–147

Ro DK, Paradise EM, Ouellet M, Fisher KJ, Newman KL, Ndungu JM et al (2006) Production of the antimalarial drug precursor artemisinic acid in engineered yeast. Nature 440:940–943

Roberts MJ, Schlenker W (2013) Identifying supply and demand elasticities of agricultural commodities: implications for the US ethanol mandate. Am Econ Rev 103:2265–2295

Robinson M, Shine T (2018) Achieving a climate justice pathway to 1.5 °C. Nat Clim Chang, 8 July. https://doi.org/10.1038/s41558-018-0189-7

Rogelj J, Hare W, Lowe J, van Vuuren DP, Riahi K, Matthews B, Meinshausen M (2011) Emission pathways consistent with a 2 °C global temperature limit. Nat Clim Chang 1(8):413–418. https://doi.org/10.1038/nclimate1258

Roy A, Kumar A (2013) Pretreatment methods of lignocellulosic materials for biofuel production: a review. J Emerg Trends Eng Appl Sci 4(2):181–193

Sage RF, Stata M (2015) Photosynthetic diversity meets biodiversity: the C4 plant example. J Plant Physiol 172:104–119

Saini M, Wang ZW, Chiang CJ, Chao YP (2017) Biotechnology for biofuels metabolic engineering of *Escherichia coli* for production of *n*-butanol from crude glycerol. Biotechnol Biofuels 10:1–8. https://doi.org/10.1186/s13068-017-0857-2

Sarkar N et al (2012) Bioethanol production from agricultural wastes: an overview. Renew Energy 37(1):19–27

Sarsekeyeva F, Zayadan BK, Usserbaeva A, Bedbenov VS, Sinetova MA, Los DA (2015) Cyanofuels: biofuels from cyanobacteria. Reality and perspectives. Photosynth Res 125 (1–2):329–340. https://doi.org/10.1007/s11120-015-0103-3

Savakis P, Hellingwerf KJ (2015) Engineering cyanobacteria for direct biofuel production from CO2. Curr Opin Biotechnol 33:8–14

Scalcinati G, Partow S, Siewers V, Schalk M, Daviet L, Nielsen J (2012) Combined metabolic engineering of precursor and co-factor supply to increase alpha-santalene production by *Saccharomyces cerevisiae*. Microb Cell Factories 11:117

Schade J, Obergassel W (2014) Human rights and the clean development mechanism. Camb Rev Int Aff 27:717–735

Schwender J, Seemann M, Lichtenthaler HK, Rohmer M (1996) Biosynthesis of isoprenoids (carotenoids, sterols, prenyl side-chains of chlorophylls and plastoquinone) via a novel pyruvate/glyceraldehyde 3-phosphate non-mevalonate pathway in the green alga *Scenedesmus obliquus*. Biochem J 316(Pt 1):73–80

Searchinger T et al (2014) Creating a sustainable food future. A menu of solutions to sustainably feed more than 9 billion people by 2050. World Resources Institute, Washington

Shaik N, Kumar A (2014) Energy crops for bio fuel and food security. J Pharm Sci Innov 3 (6):507–515. https://doi.org/10.7897/2277-4572.036206

Shanmugam S, Sun C, Zeng X, Wu YR (2018) High-efficient production of biobutanol by a novel *Clostridium* sp. strain WST with uncontrolled pH strategy. Bioresour Technol 256:543–547

Sheehan J, Dunahay T, Benemann J, Roessler PA (1998) Look back at the US Department of Energy's aquatic species program: biodiesel from algae, close-out report. National Renewable Energy Laboratory, Golden. http://www.nrel.gov/biomass/pdfs/24190.pdf

Shue H (2014) Climate justice: vulnerability and protection. Oxford University Press, Oxford, pp 27–46

Shukla R, Kumar M, Chakraborty S, Gupta R, Kumar S, Sahoo D, Kuhad RC (2016) Process development for the production of bioethanol from waste algal biomass of *Gracilaria verrucosa*. Bioresour Technol 220:584–589. https://doi.org/10.1016/j.biortech.2016.08.096

Singh RSA, Pandey E, Gnansounou E (2016) Biofuels. In: Production and future perspective. CRC Press, Boca Raton. 558p

Smith P et al (2014) In: Edenhofer O et al (eds) Climate change: mitigation of climate change. Contribution of working group III to the fifth assessment report of the intergovernmental panel on climate change. Cambridge University Press, Cambridge/New York

Smith P et al (2016) Biophysical and economic limits to negative CO_2 emissions. Nat Clim Chang 6:42–50

Sonntag S et al (2018) Quantifying and comparing effects of climate engineering methods on the earth system. Earth's Future 6:149–168

Staffas L, Gustavsson M, Mccormick K (2013) Strategies and policies for the bioeconomy and bio-based economy: an analysis of official national approaches. Sustainability 5:2751–2769. https://doi.org/10.3390/su5062751

Stocker TF, Qin D, Plattner G-K, Tignor M, Allen SK, Boschung J, Nauels A, Xia Y, Bex V, Midgley PM (eds.) (2013) Climate change 2013: the physical science basis. Contribution of working group I to the fifth assessment report of the intergovernmental panel on climate change. Cambridge University Press, Cambridge/New York, 1535p

Su H, Lin J, Wang G (2016) Metabolic engineering of *Corynebacterium crenatium* for enhancing production of higher alcohols. Sci Rep 6:39543:1–20. https://doi.org/10.1038/srep39543

Swidah R, Wang H, Reid PJ, Ahmed HZ, Pisanelli AM, Persaud KC, Ashe MP (2015) Butanol production in *S. cerevisiae* via a synthetic ABE pathway is enhanced by specific metabolic engineering and butanol resistance. Biotechnol Biofuels 8:1–9. https://doi.org/10.1186/s13068-015-0281-4

Tchakouteu S, Kalantzi O, Gardeli C, Koutinas A, Aggelis G, Papanikolaou S (2015) Lipid production by yeasts growing on biodiesel-derived crude glycerol: strain selection and impact of substrate concentration on the fermentation efficiency. J Appl Microbiol 118:911–927

Teetor VH, Duclos DV, Wittenber ETG, Young KM, Chawhuaymak J, Riley MR, Ray DT (2011) Effects of planting date on sugar and ethanol yield of sweet sorghum grown in Arizona. Ind Crop Prod 34:1293–1300

Temudo M, Muyzer G, Kleerebezem R, van Loosdrecht M (2008) Diversity of microbial communities in open mixed culture fermentations: impact of the pH and carbon source. Appl Microbiol Biotechnol 80:1121–1130

Thomson AM, Calvin KV, Chini LP, Hurtt G, Edmonds JA, Bond-Lamberty B, Janetos AC (2010) Climate mitigation and the future of tropical landscapes. Proc Natl Acad Sci U S A 107 (46):19633–19638. https://doi.org/10.1073/pnas.0910467107

United Nations (2017) World population prospects: the 2017 revision, key findings and advance tables. United Nations Department of Economic and Social Affairs/Population Division, New York

Viana QM, Viana MB, Vasconcelos EA, Santaella ST, Leitao RC (2014) Fermentative H_2 production from residual glycerol: a review. Biotechnol Lett 36:1381–1390

Völler J-S, Budisa N (2017) Coupling genetic code expansion and metabolic engineering for synthetic cells. Curr Opin Biotechnol 48:1–7. https://doi.org/10.1016/j.copbio.2017.02.002

Wang W, Hudiburg TW, Delucia EH, Khanna M (2017) The social inefficiency of regulating indirect land use change due to biofuels. Nat Commun 8:1–9. https://doi.org/10.1038/ncomms15513

Warnecke F et al (2007) Metagenomic and functional analysis of hindgut microbiota of a wood-feeding higher termite. Nature 450:560–565

Wasiullah DR, Malaviya D, Pandiyan K, Singh UB, Sahu A, Shukla R, Singh BP, Rai JP, Sharma PK et al (2015) Bioremediation of heavy metals from soil and aquatic environment: an overview of principles and criteria of fundamental processes. Sustainability 7(2):2189–2212

Wendisch VF, Brito LF, Gil Lopez M, Hennig G, Pfeifenschneider J, Sgobba E et al (2016) The flexible feedstock concept in industrial biotechnology: metabolic engineering of *Escherichia coli, Corynebacterium glutamicum, Pseudomonas, Bacillus* and yeast strains for access to alternative carbon sources. J Biotechnol 234:139–157

Wijffels RH, Barbosa MJ (2010) An outlook on microalgal biofuels. Science 329:796–799

Wise M et al (2009) Implications of limiting CO_2 concentrations for land use and energy. Science 324:1183–1186

Woods J et al (2015) In: Souza GM, Victoria R, Joly C, Verdade L (eds) Bioenergy and sustainability: bridging the gaps. SCOPE, Paris, pp 258–300

Woodworth BD, Mead RL, Nichols CN, Kolling DRJ (2015) Photosynthetic light reactions increase total lipid accumulation in carbon-supplemented batch cultures of *Chlorella vulgaris*. Bioresour Technol 179:159–164

Wriessnegger T, Pichler H (2013) Progress in lipid research yeast metabolic engineering—targeting sterol metabolism and terpenoid formation. Prog Lipid Res 52(3):277–293. https://doi.org/10.1016/j.plipres.2013.03.001

Xiberras J, Klein M, Nevoigt E (2019) Glycerol as a substrate for *Saccharomyces cerevisiae* based bioprocesses—knowledge gaps regarding the central carbon catabolism of this 'non-fermentable' carbon source. Biotechnol Adv 37(6):107378. https://doi.org/10.1016/j.biotechadv.2019.03.017

Xie T, Liu J, Du K, Liang B, Zhang Y (2013) Enhanced biofuel production from high-concentration bioethanol wastewater by a newly isolated heterotrophic microalga, *Chlorella vulgaris* LAM-Q. J Microbiol Biotechnol 23:1460–1471

Xu C, Shanklin J (2016) Triacylglycerol metabolism, function, and accumulation in plant vegetative tissues. Annu Rev Plant Biol 67:179–206. https://doi.org/10.1146/annurev-arplant-043015-111641

Yang X, Xu M, Yang ST (2015) Metabolic and process engineering of *Clostridium cellulovorans* for biofuel production from cellulose. Metab Eng 32:39–48

Yau Y-Y, Easterling M (2018) Lignocellulosic feedstock improvement for biofuel production through conventional breeding and biotechnology. In: Kumar A, Ogita S, Yau Y-Y (eds) Biofuels: greenhouse gas mitigation and global warming next generation biofuels and role of biotechnology. Springer, Heidelberg, p 1. https://doi.org/10.1007/978-81-322-3763-1_7

Yong JWH, Ng YF, Tan SN, Chew AYL (2010) Effect of fertilizer application on photosynthesis and oil yield of *Jatropha curcas* L. Photosynthetica 48(2):208–218

You C, Chen H, Myung S, Sathitsuksanoh N, Ma H, Zhang X-Z et al (2013) Enzymatic transformation of nonfood biomass to starch. Proc Natl Acad Sci U S A 110:7182–7187

Yvon-Durocher G et al (2014) Methane fluxes show consistent temperature dependence across microbial to ecosystem scales. Nature 507:488–491

Zhang YP (2015) Production of biofuels and biochemicals by in vitro synthetic biosystems: opportunities and challenges. Biotechnol Adv 33(7):1467–1483. https://doi.org/10.1016/j.biotechadv.2014.10.009

Zhang F, Keasling JD (2012) Microbial engineering for the production of advanced biofuels. Nature 488:320–328. https://doi.org/10.1038/nature11478

Zhang YHP, Myung S, You C, Zhu Z, Rollin JA (2011) Toward low-cost biomanufacturing through in vitro synthetic biology: bottom-up design. J Mater Chem 21:18877–18886

Zhong Y, Liu Z, Isaguirre C, Liu Y, Liao W (2016) Biotechnology for biofuels fungal fermentation on anaerobic digestate for lipid-based biofuel production. Biotechnol Biofuels 9:1–11. https://doi.org/10.1186/s13068-016-0654-3

Zhu L (2015) Microalgal culture strategies for biofuel production: a review. Biofuels Bioprod Biorefin 9:801–814. https://doi.org/10.1002/bbb

Zhu Q, Jackson EN (2015) Metabolic engineering of *Yarrowia lipolytica* for industrial applications. Curr Opin Biotechnol 36:65–72. https://doi.org/10.1016/j.copbio.2015.08.010

Zilberman D, Hochman G, Rajagopal D, Sexton S, Timilsina G (2013) The impact of biofuels on commodity food prices: assessment of findings. Am J Agric Econ 95:275–281

Ashwani Kumar von Humboldt Fellow (Germany), Professor *Emeritus,* Department of Botany, University of Rajasthan, Jaipur, India, Studied photosynthetic apparatus in vitro and in vivo initially working with Professor Dr. K H Neumann at Institute of Plant Nutrition at Justus Liebig University Giessen, Germany, and subsequently with Professor Dr. Sven Schubert on physiology role of enzymes in salinity stress resistance with support from Alexander von Humboldt Fellowship. JSPS visiting Professor Japan. Guided 39 students for PhD, published 220 research papers and 23 books. Presently, he is the president of the Indian Botanical Society.

The Multifaceted Connections Between Photosynthesis and Respiratory Metabolism

3

Paula da Fonseca-Pereira, Willian Batista-Silva, Adriano Nunes-Nesi, Agustin Zsögön, and Wagner L. Araújo

Abstract

Photosynthesis is the basis of primary productivity on the planet. Crop breeding has sustained steady improvements in yield to keep pace with population growth increases. Yet these advances have not resulted from improving the photosynthetic process per se, but rather from altering the way carbon is partitioned within the plant. Given that the pathways of photosynthesis and respiration catalyze partially opposing processes, it follows that their relative activities must be carefully regulated within plant cells. Exciting recent developments in efforts to understanding the interaction between respiration and photosynthesis during the last decades and the potential mechanisms linking mitochondrial function and photosynthetic efficiency will be reviewed. We additionally emphasize the challenges and opportunities to further understand the complex interplay between photosynthesis and related metabolic processes that has limited success in the manipulation and improvement of photosynthesis. We will also summarize recent advances of alternative approaches for the manipulation and enhancement of photosynthesis and their possible application for crop improvement.

Keywords

Respiration · Photosynthesis · Productivity · Targeted manipulation · Mitochondria · Crop yield

Paula da Fonseca-Pereira and Willian Batista-Silva authors contributed equally to this work.

P. da Fonseca-Pereira · W. Batista-Silva · A. Nunes-Nesi · A. Zsögön · W. L. Araújo (✉)
Departamento de Biologia Vegetal, Universidade Federal de Viçosa, Viçosa, Minas Gerais, Brazil
e-mail: nunesnesi@ufv.br; zsogon@ufv.br; wlaraujo@ufv.br

A. Kumar et al. (eds.), *Climate Change, Photosynthesis and Advanced Biofuels*,
https://doi.org/10.1007/978-981-15-5228-1_3

Abbreviations

2-OGDH	2-Oxoglutarate dehydrogenase
2-PG	2-phosphoglycolate
3-PGA	3-Phosphoglyceric acid
ACO	Aconitase
AOX	Alternative oxidase
CBC	Calvin-Benson cycle
CCM	CO_2-concentrating mechanisms
CMSII	Cytoplasmic male sterile II
CUE	Carbon-use efficiency
ENO	Enolase
$FADH_2$	Flavin adenine dinucleotide reduced
GAP	Glyceraldehyde-3-phosphate
GAPC1	Phosphorylating glyceraldehyde-3-phosphate dehydrogenase 1
GDC	Glycine decarboxylase complex
GLDH	L-galactono-1,4-lactone dehydrogenase
IMS	Inner membrane space
IPGAM	2,3-biphosphoglycerate-independent phosphoglycerate mutase
MDH	Malate dehydrogenase
*m*ETC	Mitochondrial electron transport chain
MPC	Mitochondrial pyruvate carrier
N	Nitrogen
NADH	Nicotinamide adenine dinucleotide phosphate
NADPH	Nicotinamide adenine dinucleotide phosphate, reduced
ND_{ex}	External type II NAD(P)H dehydrogenase
ND_{in}	Internal type II NAD(P)H dehydrogenase
NDs	Type II NAD(P)H dehydrogenases
NMS1	Nuclear male sterile 1
NP-GAPDH	Non-phosphorylating glyceraldehyde 3-phosphate dehydrogenase
NPP	Net primary productivity
OAA	Oxaloacetate
OPPP	Oxidative pentose phosphate pathway
PAR	Photosynthetically active radiation
PGAM	Phosphoglycerate mutase
P_i	Inorganic phosphate
PSII	Photosystem II
R_d	Dark respiration
R_L	Light respiration
ROS	Reactive oxygen species
RT-PCR	Reverse transcription PCR
RubisCO	Ribulose-1,5-bisphosphate carboxylase/oxygenase
RuBP	Ribulose-1,5-bisphosphate
SBPase	Sedoheptulose-1,7-bisphosphatase

SDH	Succinate dehydrogenase
SHAM	Salicylhydroxamic acid
TCA	Tricarboxylic acid
UCP	Uncoupling protein
WT	Wild type

3.1 Introduction

Like in all eukaryotic cells, mitochondria are vital organelles in plant cells, representing the primary site of energy transduction and ATP generation through respiration (Noctor et al. 2007; Araújo et al. 2014b; O'Leary et al. 2018). In addition, mitochondria are responsible for many other important cellular processes such as the synthesis of reducing equivalents and metabolic intermediates for use in biosynthesis elsewhere in the cell (for a review see Araújo et al. 2012a) and are today recognized to play an important role in optimizing photosynthesis. Given that photosynthesis and respiration catalyze partially opposing processes, sharing carbon dioxide (CO_2) and oxygen (O_2) as substrate and product or as product and substrate, respectively (Padmasree et al. 2002), it follows that their relative activities must be carefully regulated within plant cells, being intimately linked to each other (Nunes-Nesi et al. 2011).

The impact of chloroplasts on respiration has been extensively demonstrated. While mitochondrial reactions are supported by substrates directly provided by the chloroplastidial reactions in the illuminated leaf, the same substrates are provided indirectly from photosynthesis via its storage pools in heterotrophic tissues and in the darkened leaf (Nunes-Nesi et al. 2011). Conversely, the impact of mitochondria on photosynthesis has only been more recently demonstrated. Mitochondrial metabolism, particularly oxidative electron transport and phosphorylation, is essential for the proper maintenance of intracellular redox gradients, to allow considerable rates of photorespiration and in turn efficient photosynthesis (Araújo et al. 2014a, b; Raghavendra and Padmasree 2003). Further, mitochondria protect the chloroplast photosynthetic machinery against photoinhibition through not only the oxidative electron transport/oxidative phosphorylation but also through the key photorespiratory reactions under suboptimal conditions (Padmasree et al. 2002).

Collectively, the data discussed above indicate that the photosynthetic function can be benefited from mitochondrial metabolism in different ways. Moreover, this chapter attempts to demonstrate that the examples provided above are by no means the only ones that illustrate the multiple modes by which the function of the chloroplast is optimized by the complementary nature of mitochondrial metabolism. Thus, here we critically assess and emphasize the challenges and opportunities to further understand the connections between photosynthesis and respiration in the light of the research gained during the last decades and the potential mechanisms linking mitochondrial function and photosynthetic efficiency.

We further provide a perspective on our current understanding of the complex interplay between photosynthesis and related metabolic processes that has limited success toward delivering more productive crops. We further summarize the current state of the ongoing efforts in molecular engineering that have been recently identified for the manipulation and improvement of photosynthetic efficiency and their potential application for plant growth and yield.

3.2 The Balance Between Respiration and Photosynthesis Determining Plant Biomass Accumulation

Increasing plant productivity is of unprecedented importance and one of the major challenges that our society faces nowadays (Bar-Even 2018; Éva et al. 2018; Nowicka et al. 2018). The accelerated growth of the global population, the shortage of agricultural areas, and the associated effects of climate change are the main reasons behind the urgent demand for improving plant productivity (Éva et al. 2018; Fernie and Yan 2019). Therefore, plant productivity should thus be improved in the context of climate change and limited natural resources (Nunes-Nesi et al. 2016). Given that photosynthesis is the main driving force for plant growth and biomass production, much attention has been given to the improvement of photosynthetic efficiency as a strategy for the optimization of crop productivity (Nowicka et al. 2018).

The approaches used to increase plant biomass and yield by altering photosynthetically related processes are diverse and have received considerable attention (Nowicka et al. 2018; Heyneke and Fernie 2018). Accordingly, a growing body of evidence has been accumulated supporting a major role for mitochondria in the modulation of photosynthetic metabolism (Raghavendra and Padmasree 2003; Noguchi and Yoshida 2008; Araújo et al. 2014b; Dahal and Vanlerberghe 2018). From these studies, it has become evident that mitochondrial reactions, particularly oxidative electron transport and phosphorylation, are essential for sustaining photosynthetic carbon assimilation in land plants (Raghavendra and Padmasree 2003).

Photosynthesis and respiration are the major pathways of energy production which are largely confined to the plastid and mitochondria, respectively (Nunes-Nesi et al. 2008). Given that their pathways catalyze partially opposing processes, it implies that they complement and interact with each other (Padmasree et al. 2002). Moreover, while photosynthesis involves the enzyme-catalyzed reduction of atmospheric CO_2 (CO_2 assimilation), the reactions of respiration are the reversal of the photosynthesis and involve the oxidation of carbon compounds with the simultaneous release of CO_2 (CO_2 efflux). As the consequence of the complementary nature of these two processes, much of the carbon fixed by photosynthesis is subsequently lost by respiration and, therefore, plant respiration is directly associated to biomass and yield (Amthor et al. 2019).

In light of the aforementioned information, a number of strategies are currently pursued to increase photosynthesis by minimizing respiratory carbon-loss (Ort et al. 2015; Amthor et al. 2019). Remarkably, several efforts have been made to identify a set of engineering strategies that lead to lower respiratory carbon-loss (for a review *see* Amthor et al. 2019). It is important to highlight that plant respiration can be conceptually divided into "growth" and "maintenance" fractions (Amthor 2000). Both growth and maintenance respiration must be considered to assess potential plant productivity and to understand plant responses to environmental factors. Briefly, the strategies to enhance crop productivity by reducing respiratory carbon-loss aim to increase the proportion of carbon that stays in biomass by shrinking the slices for growth or maintenance respiration. Given that part of the biomass is allocated to harvested organs, these strategies have thus the potential to optimize productivity. Notably, higher biomass is expected to be achieved by the combination of both respiratory carbon-loss and carbon-gain strategies rather than using each strategy alone (Amthor et al. 2019). Therefore, carbon-loss approaches allow new possibilities in the current efforts to drive the main strategies to enhance crop productivity.

3.3 On the Operation of Plant Mitochondrial Metabolism During Photosynthesis

Mitochondrial metabolism is directly (and indirectly) involved in the photosynthetic metabolism in several ways. In addition to their crucial roles in respiration and photorespiration, mitochondria are of fundamental importance in processes such as nitrogen (N) metabolism, redox regulation, and signaling and for the provision of ATP for the cytosolic synthesis of sucrose (Kromer 1995; Noguchi and Yoshida 2008; Tcherkez et al. 2008). Moreover, mitochondrial metabolism is essential for the supply of a wide range of intermediates for biosynthetic reactions elsewhere in the cell (Araújo et al. 2014b).

Leaf day respiration (non-photorespiratory CO_2 evolution in the light) is an essential metabolic pathway that accompanies photosynthetic CO_2 assimilation and photorespiration (Tcherkez et al. 2008). It is widely accepted that leaf respiration is inhibited in the light (Tcherkez et al. 2008, 2012, 2017; Gauthier et al. 2010; Foyer et al. 2011; Lothier et al. 2019). Indeed, the activity of the tricarboxylic acid (TCA) cycle as well as the TCA-cycle-related enzyme, the mitochondrial pyruvate dehydrogenase (PDH), is partly compromised in the illuminated leaf, resulting in a lower CO_2 evolution rate in the light respiration (R_L) relative to the dark respiration (R_d) (Tcherkez et al. 2005). In consequence, there is a diurnal metabolic control whereby the TCA cycle does not fully operate as a cycle and mostly synthesizes malate (and/or fumarate) and 2-oxoglutarate for N assimilation in the light, and returns to a cycle generating organic acids during the dark period (Lothier et al. 2019; Sweetlove et al. 2010; Tcherkez et al. 2012).

Our understanding of the precise details of the different flux modes of plant respiration is still incomplete as well as the degree of inhibition of the TCA cycle in the light remains somewhat controversial (Nunes-Nesi et al. 2011; Zhang and Fernie 2018). The divergence between both in vitro measurements and flux profiles with the results from transgenic plants are the main reasons behind this conflict (Zhang and

Fernie 2018). These findings can be reconciled if the operation of different flux modes in the light and in the dark (Foyer et al. 2011; Zhang and Fernie 2018) is actually considered. Thus, although partly inhibited in illuminated leaves, light respiration is seemingly essential to sustain photosynthesis and redox homeostasis (Raghavendra and Padmasree 2003; Lothier et al. 2019).

Additional evidence from the involvement of mitochondrial metabolism on photosynthesis has been obtained in several independent studies with either mitochondrial mutants defective in respiration or by using specific inhibitors. Briefly, the aforementioned studies suggest that changes in levels of redox-active metabolites communicated between mitochondria and chloroplasts can, in fact, affect photosynthesis (Noctor et al. 2007). Compelling evidence has demonstrated that NAD(P)H oxidation by the respiratory chain may regulate cellular redox balance and maintain optimal photosynthesis in illuminated leaves (Noguchi and Yoshida 2008). In the presence of antimycin A, a specific inhibitor of the cellular respiration at the level of the mitochondrial Complex III, the triose-P/3-PGA ratio was increased rather than the Mal/OAA ratio, whereas the Mal/OAA ratio was strongly increased in the presence of Salicylhydroxamic acid (SHAM), an inhibitor of the Alternative Oxidase (AOX) (Padmasree and Raghavendra 1999b). The usage of the inhibitors antimycin A and SHAM has suggested that the cytochrome c pathway is able to maintain the export of triose phosphate and the supply of ATP to the cytosolic synthesis of sucrose, whereas the AOX activity maintains the oxidation of malate in the light (Padmasree and Raghavendra 1999b; Noguchi and Yoshida 2008). In good agreement, the AOX pathway plays an important role in Photosystem II (PSII) photoprotection by maintaining photorespiration to detoxify glycolate and via the indirect export of excess reducing equivalents from chloroplasts by the malate/OAA shuttle in C3 plants (Scheibe 2004; Zhang et al. 2017; Selinski and Scheibe 2019). Collectively, these data provided compelling evidence that respiration and photosynthesis pathways must act in a coordinated manner to optimize energy metabolism in the illuminated leaf.

3.4 Examples of Mitochondrial Manipulation that May Affect Photosynthesis: A Perspective on Current Knowledge and Future Trends

Respiration is the fundamental energy-conserving process common to all living organisms (Millar et al. 2011). In plants, the main substrates for respiration are sucrose and starch (Plaxton 1996; Plaxton and Podestá 2006). Thus, a large amount of free energy is released from the controlled oxidation of these highly reduced carbohydrates in a series of sequential reactions that involves several distinct steps in different cellular compartments (Millar et al. 2011). Here, our main goal was to highlight specific targets of respiratory metabolism that can be associated with photosynthesis, and as such we will first provide an overview of plant respiration and further, we focus, in the context of current models and future trends, on the aspects of plant respiration that were previously demonstrated to exert influence on photosynthesis.

Generally, plant respiration is grouped into four major processes: glycolysis, the oxidative pentose phosphate pathway (OPPP), the TCA cycle, and the mitochondrial electron transport chain (*m*ETC) (for details see Fig. 3.1). Altogether, these processes work in conjunction and play specific and important roles, not only for energy provision but also for a wide range of other physiological functions (Fernie et al. 2004). Glycolysis prepares the substrates for subsequent oxidation in mitochondria and produces a relatively small amount of chemical energy in the form of ATP and nicotinamide adenine dinucleotide reduced (NADH) (Plaxton and Podestá 2006). Further, the pyruvate generated in the cytosol during glycolysis is transported through the inner mitochondrial membrane via the mitochondrial pyruvate carrier (MPC) (Herzig et al. 2012) and the respiratory process continues within the mitochondrion. Once inside the mitochondrial matrix, pyruvate is decarboxylated by the large complex pyruvate dehydrogenase yielding NADH, CO_2, and acetyl-CoA as products. Subsequently, the acetyl group of acetyl-CoA derived from pyruvate is condensed by citrate synthase, the first enzyme in the TCA cycle, with a four-carbon dicarboxylic acid (oxaloacetate) to generate a six-carbon TCA intermediate (citrate). Citrate then goes through a series of chemical transformations carried out by seven more enzymes (aconitase, isocitrate dehydrogenase, 2-oxoglutarate dehydrogenase, succinyl-CoA ligase, succinate dehydrogenase, fumarase, and malate dehydrogenase) where it is completely oxidized to CO_2 (Plaxton and Podestá 2006). In the end, the complete oxidation of pyruvate generates the major amount of reducing power (16 NADH + 4 $FADH_2$ per sucrose) and energy by the substrate-level phosphorylation of ADP to ATP.

Apart from glycolysis, the OPPP, dually located in both cytosol and plastids, also serves as an available route for the oxidation of sugars in plant cells. OPPP plays an essential role in plant metabolism by producing reducing power conserved in the form of NADPH. Further, the electrons from NADH (produced by both glycolysis and TCA cycle), $FADH_2$ (from TCA cycle), and NADPH (from OPPP) are transferred to oxygen in the oxidative phosphorylation along with electron transport proteins bound to the mitochondrial membrane that collectively forms the *m*ETC. This electron transfer releases a large amount of free energy in the form of ATP, which is synthesized from ADP and inorganic phosphate (Pi) by the ATP synthase. In addition to the so-called classical *m*ETC, comprising four large protein complexes (Complexes I, II, III, and IV), plant mitochondria possess alternative electron transport pathways that do not pump protons such as NAD(P)H dehydrogenases on different sides of the inner membrane and the AOX (Rasmusson et al. 2004, 2008). Nevertheless, plant mitochondria also contain a set of carriers and channels that provide the substrates and cofactors for these respiratory processes from the cytosol and that facilitate the release of the products of respiration to the rest of the cell, assuming, thus, a critical influence on the functions of plant mitochondria in the metabolism as a whole (Laloi 1999; Millar et al. 2011).

Nearly all enzymes and proteins that participate in respiratory pathways have been investigated over a relatively long time period (Fernie et al. 2004; van Dongen

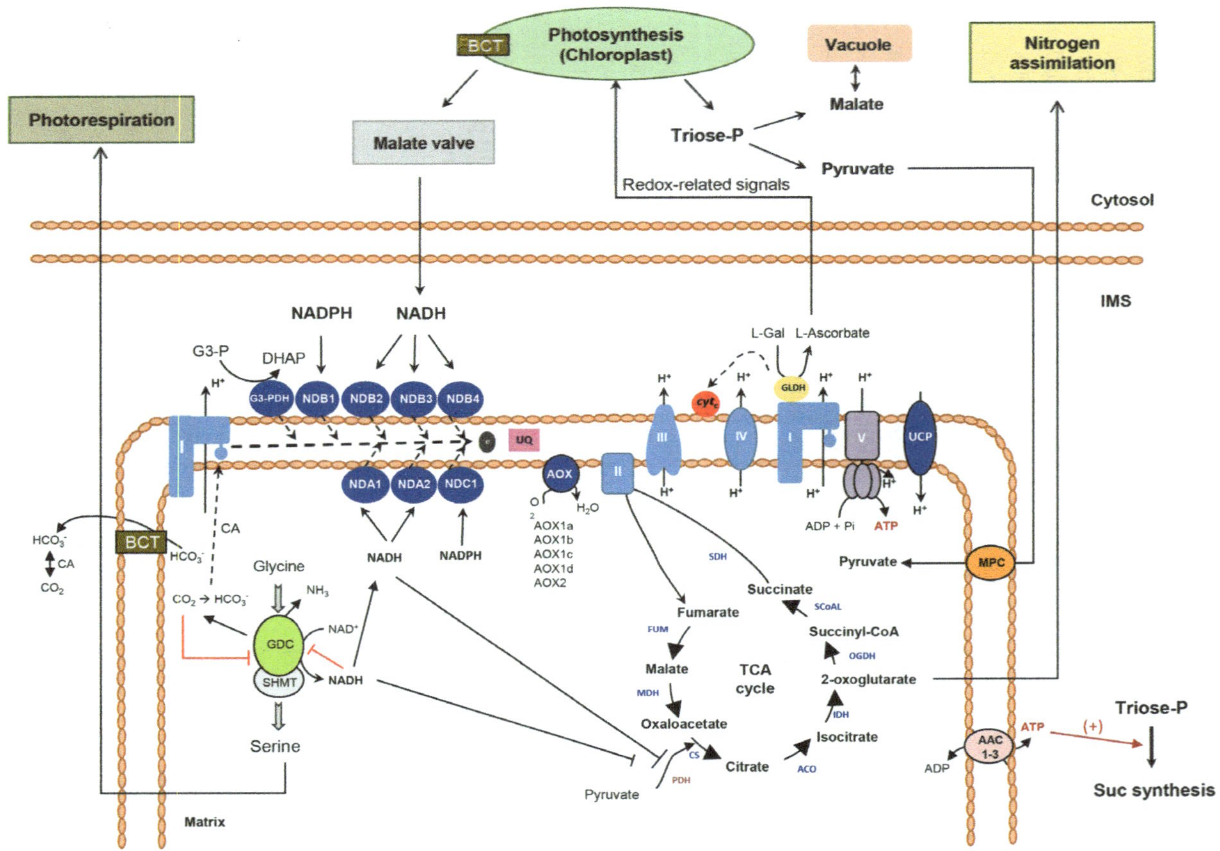

Fig. 3.1 Illustrative model of major interactions between mitochondrial activity, photosynthesis, and photorespiration. The *m*ETC contains four standard protein complexes (I-IV) (light blue) and the alternative electron transport pathways (dark blue) that do not pump protons such as NAD(P)H dehydrogenases on

different sides of the inner membrane and the AOX. ND_{in} participates in the recycling of NAD^+ from NADH produced by the photorespiratory GDC, reflecting thus a critical role of ND_{in} in the maintenance of photorespiratory metabolism. AOX promotes dynamic redox adjustments in dissipating excess chloroplastic reducing equivalents to optimize and protect photosynthesis from photoinhibition or prevent photooxidative stress in the light. Together, NDs, AOX, and UCP help in the maintenance of the redox poise of the *m*ETC to facilitate photosynthetic metabolism. The ATP generated by ATP synthase (purple) in the mitochondrial matrix is exported to the cytosol via the ATP/ADP carrier (AAC) (light pink) and supplied for the conversion of triose-P to sucrose. GLDH (yellow), which catalyzes the last step of the main biosynthetic pathway of ascorbate, is an assembly factor of Complex I in Arabidopsis. The conversion of L-galactono-1,4-lactone to ascorbate coupled to the *m*ETC leads to an upregulation of photosynthesis by a mechanism involving redox regulation. CAs compose an extra spherical domain directly attached to the membrane arm of Complex I. CAs are proposed to help in the maintenance of GDC function and thus photorespiratory fluxes by diffusion of CO_2 across organellar membranes or by active transport of HCO_3 by bicarbonate transporters (BCTs, brown). The malate produced from triose-P transported from chloroplasts in the cytosol can be stored in the vacuole or transported to mitochondria (not shown) for mitochondrial oxidation by the TCA cycle. Abbreviation: *AOX* alternative oxidase, *CAs* carbonic anhydrases, *GDC* glycine decarboxylase, *GLDH* L-galactono-1,4-lactone dehydrogenase, *NDin* internal type II NAD(P)H dehydrogenase, *mETC* mitochondrial electron transport chain, *UCP* uncoupling protein. For further information concerning the enzymes see Table 3.1

Table 3.1 Selected genetic modifications in main enzymes of glycolysis, TCA cycle, and *m*ETC that resulted in impacts in either maximum photosynthesis or biomass accumulation

Changed enzyme	Genetic manipulation background	Species	Phenotype	Reference
Glycolysis				
Glyceraldehyde-3-P dehydrogenase (NP-GAPDH)	T-DNA loss-of-function mutant lines	*Arabidopsis thaliana*	Plants lacking NP-GAPDH exhibit delayed growth and decreased CO_2 fixation associated with the downregulation of several photosynthetic genes	Rius et al. (2006); Taniguchi and Miyake (2012)
Glyceraldehyde-3-P dehydrogenase NAD-dependent (GAPC-1)	T-DNA loss-of-function mutant lines	*Arabidopsis thaliana*	Deficiency in the cytosolic GAPC results in modifications of carbon flux and mitochondrial dysfunction, leading to an alteration of plant and embryo development with a decreased number of seeds, reducing productivity	Rius et al. (2008)
Phosphoglycerate mutase (PGAM)	T-DNA loss-of-function mutant lines	*Arabidopsis thaliana*	Loss-of-function in PGAM affects stomatal movement, vegetative growth, and pollen production	Zhao and Assmann (2011)
Phosphoglycerate mutase (PGAM)	Antisense orientation	*Solanum tuberosum*	Transgenic potato lines increase 3-PGA levels, impacting the photosynthesis and growth without changes in starch content	
Enolase	Antisense orientation	*Nicotiana tabacum*	Decreased enolase activity affected photosynthetic rates, retarding growth	Voll et al. (2009)
Phosphoenolpyruvate/ carboxylase (PEPC)	T-DNA loss-of-function mutant lines	*Arabidopsis thaliana*	Disruption of PPC results in profound changes in C/N metabolism and plant growth	Shi et al. (2015)
TCA cycle enzymes				
Aconitase (ACO)	Reduced expression of aconitase in Aco1	*Solanum lycopersicum*	Reduced expression of aconitase enhanced ~150% assimilation rate (photosynthesis) with significative increases in fruit yield (~ 600%) in comparison with control	Carrari et al. (2003); Heyneke and Fernie (2018)

Succinate dehydrogenase (SDH)	SDH cloned in antisense orientation	*Solanum lycopersicum*	Reduced expression of *SDH* gene enhanced ~25% assimilation of CO_2 rate (photosynthesis) mediated by stomatal movements, followed by significant increases (~30%) in fruit yield	Araújo et al. (2011b); Heyneke and Fernie (2018)
	Mutation of SDH with downregulation by RNAi	*Arabidopsis thaliana*	A deficiency in the *AtSDH* gene improved photosynthesis mediated by stomatal pore aperture and mitochondrial ROS production leading to increased plant growth	Fuentes et al. (2011); Jardim-Messeder et al. (2015)
2-Oxoglutarate dehydrogenase (2-OGDH)	Mitochondrial OGDH was cloned in antisense orientation	*Solanum lycopersicum*	Transgenic antisense in *Sl2-OGDH* affects plant respiration and fruit maturity without changes in fruit yield	Araújo et al. (2012b)
Malate dehydrogenase (MDH)	Mitochondrial MDH cloned in antisense orientation	*Solanum lycopersicum*	Reduction in mitochondrial MDH increases photosynthesis (120% increase ~150%), increasing substantially in fruit yield	Nunes-Nesi et al. (2005a); Van Der Merwe et al. (2009)
Fumarase (Fum)	Antisense orientation of Fum	*Solanum lycopersicum*	Antisense lines in fumarase affect photosynthesis (reduced ~50%) in comparison with wild-type mainly by modulation in stomatal aperture	Nunes-Nesi et al. (2007); Van Der Merwe et al. (2009)
Mitochondrial electron transport chain				
Complex I				
Cytoplasmic male sterile II (CMSII)	Protoplast culture mutant	*Nicotiana sylvestris*	Plants lacking a functional complex I (CMSII) exhibits reduced growth by increases in photorespiration by increases in photorespiration	Pineau et al. (2005); Priault et al. (2006); Lothier et al. (2019)
Nuclear male sterile 1 (NMS1)		*Nicotiana sylvestris*	The functional lack of NMS 1 leads to decreased photosynthesis, in correlation with lower leaf conductance	Lothier et al. (2019)
Carbonic anhydrase (CA2)	T-DNA knock-out insertion mutants	*Arabidopsis thaliana*	Reduced growth rate and reduced respiration rate of a suspension cell culture of CA2 mutant lines	Wang et al. (2012); Senkler et al. (2017)

(continued)

Table 3.1 (continued)

Changed enzyme	Genetic manipulation background	Species	Phenotype	Reference
Carbonic anhydrase-like (CL1::CL2)	*cal1* T-DNA knock-out *cal2* RNAi downregulated	*Arabidopsis thaliana*	Significative reduction in complex I lead to reduced plant growth with alteration in the central mitochondrial metabolism	Fromm et al. (2016b); Senkler et al. (2017)
Carbonic anhydrase; carbonic anhydrase-like (CA2::CL1)	*ca2::cl1* T-DNA knock-out insertion mutants	*Arabidopsis thaliana*	Reduced plant growth and photorespiration by increases in glycine levels as well as increases in ROS content was observed in double KO	Soto et al. (2015); Fromm et al. (2016b)
Carbonic anhydrase (Ca1:: CA2)	*ca1::ca2* T-DNA knock-out insertion mutants	*Arabidopsis thaliana*	Double mutant in CA1 and CA2 drastically decreased plant development, with the complete absence of complex I and affecting the light capture by changes in photosynthetic apparatus	Villarreal et al. (2009); Fromm et al. (2016b)
	Natural occurrence (absence of complex I)	*Viscum album*	Complete absence of complex I in European mistletoe alters mETC with increases in alternative dehydrogenases (NDA, NDB), AOX1,2, decreases in SDH, Cyt, COX and ATP synthase	da Fonseca-Pereira et al. (2018b); Maclean et al. (2018); Senkler et al. (2018)
Uncoupling protein				
	SAIL collection	*Arabidopsis thaliana*	Absence of UCP1 results in photosynthetic inefficiency with a decrease in the rate of oxidation of photorespiratory glycine in the mitochondrion.	Sessions et al. (2002); Sweetlove et al. (2006)
Alternative oxidase				
Alternative oxidase 1	*aox1a* knock-out mutants T-DNA lines	*Arabidopsis thaliana*	The functional lack of aox1a regulates cellular redox and ROS homeostasis to optimize photosynthesis	Strodtkötter et al. (2009); Xu et al. (2011); Vishwakarma et al. (2015)

Type II NAD(P)H dehydrogenases				
NDA-type mitochondrial NAD(P)H dehydrogenase	*nda1* and *nda2* RNAi downregulation	*Arabidopsis thaliana*	RNAi mutant lines were characterized by impacts on the NADPH/$NADP^+$ ratio, growth and development. Notably, such changes were not directly associated with photosynthetic processes, but the results suggest an important role in the reoxidation of photorespiratory NADH	Wallström et al. (2014a)

et al. 2011). Moreover, biochemical studies and reverse genetic characterization of mutant plants with alterations in enzymes related to respiratory metabolism have been, in recent years, fundamental for a far greater understanding of the exact contribution of respiration for photosynthetic performance. In the following sections, we discuss these recent developments in the light of the analysis of plant mutants defective in specific components of respiratory pathways and the different modes by which these alterations were shown to somehow affect photosynthesis-related parameters.

3.4.1 Glycolysis

Glycolysis reactions produce ATP, serving as a source of energy for cellular metabolism. Notably, two cytosolic enzymes for the oxidation of glyceraldehyde-3-phosphate (GAP) to 3-phosphoglycerate (3PGA) are present in plants (Rius et al. 2006, 2008; Piattoni et al. 2013). The parallel occurrence of both routes generates an important difference for cell energy metabolism, since different amounts of energy (ATP) and/or reducing power (NADPH) will be provided, depending on the relative activity levels of each enzyme in the cytosol (Piattoni et al. 2013). This can occur either via the couple GAPDH (known as GAPC) plus phosphoglycerate kinase, generating NADH and ATP or in a single step catalyzed by the non-phosphorylating glyceraldehyde-3-P dehydrogenase (NP-GAPDH) and generating NADPH (but not ATP) (Rius et al. 2008; Taniguchi and Miyake 2012). It has been proposed that NP-GAPDH participates in a shuttle of triose-P/phosphate that indirectly transfers photosynthetically reduced $NADP^+$ from chloroplast to cytosol during photosynthesis (Kelly and Gibbs 1973; Rius et al. 2008). Accordingly, an Arabidopsis mutant lacking NP-GAPDH exhibits delayed growth and decreased CO_2 fixation associated with the downregulation of several photosynthetic genes (Rius et al. 2006; Taniguchi and Miyake 2012). In addition, microarray and RT-PCR results from Arabidopsis plants deficient in GAPC1 activity suggest modifications of carbon flux and photosynthetic metabolism, which seems to induce a stress-like situation that could account for the phenotype exhibited by these mutants (Rius et al. 2008). Furthermore, GAPC-deficient plants displayed higher stomatal conductance and a higher rate of photosynthesis than the wild type (WT), probably allowing more CO_2 uptake and increased nutrient transport than the WT (Guo et al. 2012). Moreover, physiological analysis of double mutants for the two Arabidopsis genes encoding phosphoglycerate mutase (PGAM), glycolytic enzymes that catalyze the interconversion of 3-PGA to 2-PGA, also indicates a critical role of glycolysis in stomatal movement, vegetative growth, and pollen production (Zhao and Assmann 2011). Given that transgenic inhibition of the glycolytic enzyme enolase *eno-1* in tobacco (Voll et al. 2009) and *ipgam* (2,3-bisphosphoglycerate-independent phosphoglycerate mutase) antisense in potato (Westram et al. 2002) were both characterized by reduced photosynthetic rates, it most likely explains the retarded growth phenotype of *ipgam* double mutants of Arabidopsis (Zhao and Assmann 2011). Unfortunately, *ipgam* double mutants of Arabidopsis are extremely small, precluding gas exchange technology assessment of photosynthesis.

3.4.2 TCA Cycle

Several independent studies have demonstrated that photosynthetic performance can be improved by modifications of the activities of the TCA cycle (Heyneke and Fernie 2018; Sweetlove et al. 2010; Zhang and Fernie 2018). This fact aside, these studies have also revealed a surprising complexity in this response (Sweetlove et al. 2010). While suppression of some enzymes leads to increased photosynthesis (Carrari et al. 2003; Nunes-Nesi et al. 2005a; Araújo et al. 2011b), others lead to decreased photosynthesis (Nunes-Nesi et al. 2007) and yet others have virtually no effect (Studart-Guimaraes et al. 2007; Sienkiewicz-Porzucek et al. 2008). Interestingly, from the selected genetic modifications that have yielded increased maximum photosynthesis and more biomass accumulated, overall significantly higher values have been achieved by suppression of TCA cycle enzymes than even by enzymes of photosynthetic pathways (Heyneke and Fernie 2018). As we detail in the next sections, a number of factors might explain the great contribution of mitochondrial metabolism and, in particular, of TCA cycle reactions, for photosynthetic performance. Among other factors, regulation of metabolite distribution as a means to balance cellular redox status, modulation of organic acids in order to keep guard cell function and buffering of metabolism by photorespiration can explain, at least partially, why the inhibition of TCA cycle activity led increases in both photosynthesis and growth.

Metabolic changes observed in transgenic lines with modified expression of photorespiratory or respiratory enzymes provide compelling evidence for the close relationship between these two pathways (for a review see Obata et al. 2016). Several TCA cycle intermediates were demonstrated to be significantly modified in T-DNA mutants disrupted in photorespiratory enzymes, which are proposed to mainly result from a higher or lower rate of the utilization of these intermediates (for a review see Obata et al. 2016). Conversely, transgenic studies of the enzymes of the TCA cycle in tomato plants revealed a number of changes in the levels of glycine, serine, and glycerate, three intermediates of the photorespiratory pathway (Obata et al. 2016). Altogether, these data support the strong influence that TCA cycle manipulation has on photorespiratory metabolism and, in turn, on photosynthesis and plant growth. Thus, it seems reasonable to assume that modification of TCA cycle enzymes is a very promising alternative for improving photosynthetic performance.

3.4.3 Aconitase

Aconitase catalyzes the reversible isomerization of citrate into isocitrate and two isoforms of aconitase have been detected in plants (Carrari et al. 2003). The mitochondrial isoform is involved in the TCA cycle (Carrari et al. 2003), whereas the cytosolic one participates in the glyoxylate cycle as well as in citrate metabolism (Hayashi et al. 1995). Genetic lesion in the gene encoding aconitase in tomato (Aco-1; aconitate hydratase) resulted in lower expression of the Aco-1 transcript and lower levels of both cytosolic and mitochondrial aconitase protein and activity (Carrari et al. 2003), suggesting that, at least in tomato, this gene product is dual-targeted (Cavalcanti et al. 2014). Interestingly, the same mutation affecting both mitochondrial and cytosolic ACO activities resulted in much higher rates of photosynthesis (~150%) coupled with an increase in fruit yield (~600%) (Carrari et al. 2003; Heyneke and Fernie 2018). However, the reasons behind increased plant performance (more details described in Sect. 3.6) in the aconitase lines are still unclear (Zhang and Fernie 2018).

3.4.4 Complex II (Succinate Dehydrogenase)

Component of both the TCA cycle and the *m*ETC, SDH is a flavoprotein subunit encoded by two nuclear genes (*SUCCINATE DEHYDROGENASE, SDH1-1,* and *SDH1-2*) in Arabidopsis (Hagerhall 1997; Figueroa et al. 2001; Fuentes et al. 2011). Transgenic tomato plants deficient in the expression of the iron-sulfur subunit of SDH displayed increased rates of net photosynthesis (25%) and growth under normal greenhouse conditions as well as enhanced rates of net photosynthesis under suboptimal carbon dioxide concentrations via an organic acid-mediated effect on stomatal aperture (Araújo et al. 2011b). Moreover, Arabidopsis plants with reduced SDH activity displayed significantly higher CO_2 assimilation rates and enhanced growth together with increased stomatal aperture and density (Fuentes et al. 2011).

3.4.5 Oxoglutarate Dehydrogenase

The 2-oxoglutarate dehydrogenase (2-OGDH) plays an important role in the metabolism by controlling the levels of 2-oxoglutarate and several other important organic acids in plant cells (Araújo et al. 2014a, b). OGDH catalyzes the oxidative decarboxylation of 2-oxoglutarate to succinyl-CoA and also generates the reduced coenzyme NADH. 2-OGDH tomato antisense lines with significantly higher levels of 2-oxoglutarate were characterized by clear reductions in respiration and considerable shifts in the metabolism (Araújo et al. 2012b). In spite of the observation that little difference was found in the photosynthetic capacity of young plants, these plants displayed reduced photosynthetic capacity at later stages of development due to the early onset of senescence (Araújo et al. 2012b). Therefore, these data further corroborate the implication of OGDH in N remobilization during leaf senescence (Taylor et al. 2010; Araújo et al. 2012b). Moreover, based on cyanobacterial investigations, it seems that, in addition to its essential role as a carbon skeleton for N assimilation, 2-oxoglutarate also serves as a signal of N limitation (for a review see Zhang et al. 2018). Similarly, the photorespiratory intermediate 2-phosphoglycolate (2-PG) was shown to act as a signal of inorganic carbon limitation (Haimovich-Dayan et al. 2015; Klähn et al. 2015; Zhang et al. 2018). The signaling nature of 2-oxoglutarate and 2-PG is most likely a conserved feature in photosynthetic organisms. However, it deserves further investigation to comprehend to what extent the signaling mechanisms identified from cyanobacteria are a common feature in eukaryotic algae and land plants (Zhang et al. 2018).

The well-characterized involvement of 2-oxoglutarate in providing C skeletons for N assimilation (Hodges 2002) together with the possibility that 2-oxoglutarate contributes to the C/N balance in plants (Zhang et al. 2018) clearly implies the participation of the 2-OGDH in photosynthetic and respiratory metabolism as well as in programs of plant development connected to CN interactions. OGDH antisense lines demonstrated that steady-state levels of photorespiratory intermediates, namely glycerate, and glycine, were reduced, coupled with a significant reduction in the label redistribution to glycine and serine (Araújo et al. 2012b). In accordance, downregulation of the TCA cycle activity found in 2-OGDH lines is most likely associated with an upregulation in the flux through the photorespiration metabolism as part of reprogramming to maintain either mitochondrial NADH homeostasis and/or the glutamate pool size (Araújo et al. 2014b). Altogether, these results demonstrated that 2-oxoglutarate metabolism has greater impacts on plant respiration and its connections than previously expected.

3.4.6 Malate Dehydrogenase

The mitochondrial MDH catalyzes the reversible oxidation of malate to oxaloacetate, producing another molecule of NADH. Antisense lines of mitochondrial MDH showed enhanced photosynthetic activity (up to 11%) and increased shoot, leaf and fruit biomass (up to 19%), but a reduced root biomass under optimal growth conditions (Nunes-Nesi et al. 2005a; Zhang and Fernie 2018). These lines were also characterized by a large increase in the synthesis of ascorbate, which was likely due to an elevated utilization by the *m*ETC of L-galactono-lactone, the terminal precursor of ascorbate biosynthesis that works as an alternative electron donor, given the reduced availability of electrons from the TCA cycle (Carrari et al. 2005; Nunes-Nesi et al. 2005a; Zhang and Fernie 2018). Experiments in which ascorbate was fed to isolated leaf discs from WT plants also resulted in increased rates of photosynthesis (~34%) providing a strong indication for an ascorbate-mediated link between the energy-generating processes of respiration and photosynthesis (Nunes-Nesi et al. 2005a). This led to the proposition that ascorbate acts as a signal that allows the coordination of plant energy metabolism between the mitochondria and the chloroplast (Nunes-Nesi et al. 2005a; Zhang and Fernie 2018). Intriguingly, *Aco1* mutant also displayed a dramatically elevated ascorbate level, suggesting that higher ascorbate, following a restriction of flux through the TCA cycle, could possibly explain the elevated photosynthesis in *Aco-1* mutants per se (Nunes-Nesi et al. 2005b, 2007; Urbanczyk-Wochniak et al. 2006). However, since the increase in the photosynthetic rate was greater in the *Aco1* mutant than that found in the MDH antisense plants, it is thus suggested that factors other than those mediated by ascorbate also mediate the enhancement of photosynthesis in *Aco-1* mutants (Urbanczyk-Wochniak et al. 2006)

(Fig. 3.1). In summary, the precise mechanism responsible for these findings remains to be elucidated and clearly deserving further investigation.

It was also further demonstrated that a knockout mutant of mMDH1, the isoform that accounts for ~60% of the total mMDH activity, shows a strong growth impairment when grown under low CO_2 (Lindén et al. 2016). By using ^{13}C labeling, it was shown that, under low CO_2 conditions, the glycine/serine ratio increased with a concomitant altered glutamine/glutamate/2-oxoglutarate relation clearly indicating that proper mMDH activity is essential to shuttle reductants out of the mitochondria to support photorespiratory flux (Lindén et al. 2016). It has been also demonstrated that MDHs from different cellular locations act in combination with malate/OAA shuttles as key enzymes connecting distinct cell compartments (Selinski and Scheibe 2019). In addition, these malate valves seem to participate in the interplay between N-assimilation and energy metabolism which enables the adaptation to both short- and long-term metabolic and environmental changes by optimally balancing the ATP/NADPH ratio in each case (Selinski and Scheibe 2014).

3.4.7 Fumarase

Also known as fumarate hydratase, fumarase catalyzes the reversible hydration of fumarate to malate. Reduction in mitochondrial fumarase activity in tomato was demonstrated to have a negative impact on photosynthetic performance (Nunes-Nesi et al. 2007), reflecting the potential importance of malate metabolism in C3 photosynthesis (Martinoia and Rentsch 1995). A more detailed characterization of the mutants for mitochondrial fumarase in tomato revealed major changes in stomatal function (Nunes-Nesi et al. 2007). Moreover, several reactions of the TCA cycle can be bypassed by steps in the cytosol, the only exceptions being the reactions catalyzed by citrate synthase and SDH, providing complications inherent in studying the TCA cycle in plants (Sweetlove et al. 2010). Thus, at least in Arabidopsis, the conversion of malate to fumarate can also be catalyzed by a cytosolic isoform (FUM2) previously identified (Pracharoenwattana et al. 2010). In contrast to the mitochondrial fumarase (FUM1), FUM2 is not required for plant growth (Pracharoenwattana et al. 2010). However, FUM2 is required for the massive accumulation of carbon in the form of fumarate that occurs in Arabidopsis leaves during the day which is, in turn, apparently required for rapid N assimilation and growth on high N (Pracharoenwattana et al. 2010). These facts are consistent with the proposed role of fumarate in pH regulation during nitrate assimilation (Tschoep et al. 2009; Araújo et al. 2011a). Additionally, fumarate accumulation provided by FUM2 was demonstrated to have a function in sensing low temperatures possibly by modulating metabolic or redox signals (Dyson et al. 2016). Briefly, *fum2* mutants accumulated higher concentrations of phosphorylated sugar intermediates, starch, and malate in response to cold conditions (Dyson et al. 2016). Different from WT plants, *fum2* displayed clear downregulation of transcripts for proteins involved in photosynthesis which demonstrate the significance of fumarate for the ability to acclimate photosynthesis to low temperature.

3.4.8 Oxidative Pentose-Phosphate Pathway (OPPP)

Together with glycolysis and TCA cycle, the OPPP is one of the major pathways involved in plant carbohydrate metabolism (Rius et al. 2006). Given that ferredoxin is reduced directly by components of Photosystem I, the OPPP needs to be inactivated in illuminated chloroplasts in order to potentially avoid futile interactions with the Calvin-Benson cycle (CBC) (Kruger and Von Schaewen 2003). In contrast, there is a need to maintain the flux through the plastidic OPPP in non-photosynthetic tissues, even in the face of the high stromal NADPH/NADP$^+$ levels that are required to drive ferredoxin-dependent reactions such as nitrite assimilation (Kruger and Von Schaewen 2003). Furthermore, OPPP provides the reducing power for nitrite reductase and glutamine-2-oxoglutarate aminotransferase (GOGAT) in roots (Oji et al. 1985; Bowsher et al. 1989, 1992), being also potentially interesting for its role in N metabolism (Lejay et al. 2008). Interestingly, genes encoding root ion carriers in Arabidopsis, namely, NRT1.1 (NO_3^- transporter, formerly CHL1) and NRT2.1 (NO_3^- transporter), among others, were demonstrated to be controlled by a signal originated from the OPPP. Given that the control over root uptake systems has often been attributed to the regulatory action of sugars produced by photosynthesis and transported downward to roots (Lejay et al. 2008), the existence of an OPPP-dependent sugar signaling pathway provides additional support for the important link between OPPP and photosynthesis.

3.4.9 Mitochondrial Electron Transport Chain (*m*ETC)

The direct assessment of the in vivo role of the *m*ETC in photosynthesis has been obtained by following both the biochemical and physiological characterization of available mutants with specific genetic lesions in key components of the *m*ETC. Members of the *m*ETC and additional components that serve as non-phosphorylating by-passes have been demonstrated to cause significant effects on the rate of photosynthesis and will be briefly discussed below.

3.4.9.1 Complex I

The major electron entry point into the *m*ETC, Complex I is a multimeric enzyme with 49 subunits in Arabidopsis (Braun et al. 2014). The Complex I is responsible for the electron transfer from matrix-located NADH to ubiquinone and translocate protons to the inner membrane space (IMS) during this process, accounting for ~40% of mitochondrial ATP production (Petersen et al. 2015). Thus, it seems reasonable to anticipate that the deficiency of a functional Complex I culminates with a remarkable reduction in energetic efficiency, which in turn impacts photosynthesis (Petersen et al. 2015). For instance, Arabidopsis mutants lacking Complex I display increased glycolytic fluxes to produce ATP (Kühn et al. 2015a; Fromm et al. 2016c). Similarly, the hemiparasitic European mistletoe (*Viscum album*), the unique recognized multicellular eukaryote able to cope with the complete absence of mitochondrial Complex I, displays a large rearrangement of its metabolism to generate ATP through glycolysis rather than mitochondrial respiration (da Fonseca-Pereira et al. 2018b; Maclean et al. 2018; Senkler et al. 2018). Interestingly, the high levels of glycolytic substrates required by *Viscum* can either be supplied from the host or by its own photosynthetic capacity (Maclean et al. 2018).

Direct evidence showing that Complex I plays an important role in photosynthesis comes from various experimental findings (reviewed in Braun et al. 2014). The best-characterized respiratory chain mutant in plants is a *Nicotiana sylvestris* mitochondrial mutant, *cytoplasmic male sterile II* (*cmsii*), lacking a functional Complex I (Gutierres et al. 2002; Lothier et al. 2019), which suffers from a decreased efficiency of NADH oxidation (Gutierres et al. 2002) and has a lower photosynthetic activity than WT plants under photorespiratory conditions (Priault et al. 2006). The *cmsii* mutant is characterized by a re-adjustment of whole-cell redox homeostasis, gene expression, as well as metabolic pathways that use pyridine nucleotides (Noctor et al. 2004). The nuclear male sterile 1 (NMS1), previously isolated in *N. sylvestris* and associated with the nuclear-encoded NAD4 subunit of Complex I (part of the H^+-translocation module) (De Paepe et al. 1990), is also extensively studied as a mitochondrial mutation with impacts in photosynthesis (Sabar et al. 2000; Lothier et al. 2019). Both CMSII and NMS1 exhibit decreased photosynthesis and lower leaf conductance, highlighting a mitochondrial control on photosynthesis (Sabar et al. 2000). Interestingly, both CMSII and NMS1 tobacco mutants have higher R_L and R_d (Lothier et al. 2019). In consequence of the negative correlation between C use efficiency (CUE) and respiration [intrinsic leaf CUE = $(A - R_n)/(A + R_d)$; after Gifford 2003], both mutants have a higher relative respiratory loss at the leaf level when expressed with respect to net photosynthesis as the CUE (Lothier et al. 2019).

Another important connection between Complex I and photosynthesis originates with the mitochondrial L-galactono-1,4-lactone dehydrogenase (GLDH), which catalyzes the last step of the main biosynthetic pathway of ascorbate and is an assembly factor of Complex I in Arabidopsis (Schimmeyer et al. 2016). GLDH donates electrons to cytochrome c in the *m*ETC (Bartoli et al. 2000; Millar et al. 2003) and, accordingly, cytochrome c is a substrate for GLDH (Ntagkas et al. 2018). Therefore, the activity of GLDH may compete with Complex III to access oxidized cytochrome c (Rasmusson and Møller 2010). Furthermore, as shown for intact potato mitochondria (Bartoli et al. 2000), the availability of oxidized cytochrome c favors higher ascorbate biosynthesis (Ntagkas et al. 2018). As previously discussed, ascorbate levels in plants are linked to respiration and photosynthesis (Fig. 3.1). Of special interest in this respect is the observation that light is an important factor modulating ascorbate levels in plants (Ntagkas et al. 2018, 2019), supporting previous interactions between the *m*ETC, photosynthesis, and light signaling (Rasmusson and Møller 2010). The further identification of a plastid ascorbate transporter (Miyaji et al. 2015) has provided insights into how ascorbate acts as a signal affecting photosynthesis, although the exact nature of its transduction still requires considerable further research (Zhang and Fernie 2018).

In addition to GLDH, mitochondrial γ-carbonic anhydrases are required for Complex I assembly and compose an extra spherical domain directly attached to the membrane arm of Complex I in plants (Fromm et al. 2016a). The proposed role of mitochondrial γ-carbonic anhydrase subunits in recycling photorespiratory CO_2 and sustaining efficient photosynthesis under ambient conditions (Braun and Zabaleta 2007; Zabaleta et al. 2012; Soto et al. 2015) offers additional evidence for the role of *m*ETC on photosynthetic performance. Moreover, Arabidopsis mutants defective in two mitochondrial carbonic anhydrase subunits displayed low Complex I levels as well as reduced levels of proteins associated with photorespiration, as well as affected central mitochondrial metabolism (Fromm et al. 2016b; Hodges et al. 2016), demonstrating the significance of carbonic anhydrases for (photo)respiratory metabolism (Fig. 3.1). Thus, carbonic anhydrases, together with mitochondrial MDH would help to maintain glycine decarboxylase (GDC) function and thus photorespiratory fluxes by removing CO_2 and NADH that are inhibitory to GDC activity (Bykova et al. 2014).

3.4.9.2 Uncoupling Protein

Integral to the inner mitochondrial membrane, uncoupling protein (UCP) functions by dissipating the mitochondrial proton gradient as heat (Ricquier and Boillaud 2000). The absence of UCP results in a dramatic reduction in the rates of CO_2 assimilation linked to a reduced rate of photorespiratory glycine oxidation inside the mitochondrion (Sweetlove et al. 2006). Collectively, these results suggest that the main physiological role of UCP1 in Arabidopsis leaves is related to maintaining the redox poise of the *m*ETC to facilitate photosynthetic metabolism. Moreover, a number of studies have demonstrated that UCP1 may decrease reactive oxygen species (ROS) production under stress conditions by uncoupling the electrochemical gradient from ATP synthesis. For instance, it was shown that overexpression of *AtUCP1* in tobacco induces a better performance under conditions of drought and salt (Begcy et al. 2011) and hypoxia (Barreto et al. 2016). In addition, UCP1-overexpressing lines exhibit increased mitochondrial biogenesis and amplification of a broad stress response, which includes the upregulation of hundreds of stress-responsive genes and reduced ROS accumulation (Barreto et al. 2014). Therefore, it seems reasonable to assume that UCP1 overexpression can potentially be used to develop crops that are more tolerant to abiotic stress conditions.

3.4.9.3 Alternative Oxidase (AOX)

Differently from UCPs and arguably the most extensively studied component of the plant *m*ETC (Kühn et al. 2015b), AOX does not directly affect H^+ transport and thereby is not directly linked to ATP production. In addition, together with internal type II NAD(P)H dehydrogenase (ND_{in}), AOX is seemingly important for the oxidation of photorespiratory NADH (Igamberdiev et al. 1997). Studied in a wide variety of plant species, AOX catalyzes the cyanide-insensitive respiration. In Arabidopsis, five genes encode AOX (*AtAOX1a-1d* and *AtAOX2*), and *AtAOX1a* is mainly expressed in green leaves (Clifton et al. 2006; Noguchi and Yoshida 2008). Numerous studies have investigated the functional link between cyanide-insensitive AOX pathways and chloroplast metabolism (Strodtkötter et al. 2009; Zhang et al. 2010; Florez-Sarasa et al. 2011; Vishwakarma et al. 2014; Dinakar et al. 2016). Overall, impairments of AOX by specific inhibitors or genetic approaches negatively impact photosynthesis (Padmasree and Raghavendra 1999a). From these studies, it has become clear that AOX is important in dissipating excess of chloroplastic reducing equivalents to optimize and protect photosynthesis from photoinhibition or by preventing photooxidative stress in the light (Vishwakarma et al. 2014). Together with the malate valve, AOX can respond directly to and also be seen as sensors of energy status for maintenance of homeostasis in mitochondria and chloroplasts, respectively (Scheibe 2019). Moreover, AOX function optimizes biomass accumulation when the cytochrome pathway is restricted (Selinski et al. 2018) and it is also suggested to play a key role in maintaining the cellular redox balance by dissipating excess reducing equivalents produced by the photochemical reactions and the photorespiratory glycine oxidation (Yoshida et al. 2006; Strodtkötter et al. 2009). Consequently, AOX probably works in conjunction with UCP in extreme environments, such as high light, in order to counteract higher photorespiration (Rasmusson and Møller 2010).

It is important to mention that the main isoforms of AOX protein have likely evolved from a common ancestor di-iron carboxylate protein present in both proteobacteria and archaebacteria (for a review see: Selinski et al. 2018). This early evolution of AOX has been proposed as a result of the ability of di-iron proteins to reduce oxygen to water (Selinski et al. 2018), and this has been of pivotal significance to deal with the transition from an anaerobic to an aerobic world and not being inhibited by sulfide. It seems reasonable to suggest that this, at least partially, can explain the wide distribution of AOX across kingdoms and why AOX likely exerts a protective role toward oxidative stress and hydrogen sulfide-mediated inhibition of cytochrome c oxidase.

3.4.9.4 Type II NAD(P)H Dehydrogenases (NDs)

In addition to AOX and UCP, alternative type II NAD(P)H dehydrogenases (NDs) exist as non-phosphorylating pathways in the *m*ETC. NDs bypass Complex I and play an important role in dissipating excess reductants in the chloroplast mostly under high light (Noguchi and Yoshida 2008). External NDs (ND_{ex}) are found in the inner membrane facing the IMS while internal NDs (ND_{in}) exist in the matrix surface of the inner mitochondrial membrane (Fernie et al. 2004; Rasmusson et al. 2004). Thus, ND_{in} and ND_{ex} oxidize matrix and cytosolic NAD(P)H, respectively (Liu et al. 2008). Both NDs do not transport H^+, and as such do not directly contribute to ATP synthesis (Liu et al. 2008), being also insensitive to the Complex I inhibitor rotenone (Melo et al. 1996; Rasmusson et al. 1999). Both ND_{in} and ND_{ex} are thought to function only when high NADH concentrations are found in the matrix. This is based on the higher K_m (NADH) for the rotenone-insensitive NADH oxidation compared with Complex I-mediated activity (Rasmusson et al. 2004; Noguchi and Yoshida 2008; Wallström et al. 2014a). The Arabidopsis genome contains seven putative genes encoding NDs (Michalecka et al. 2003; Moore et al. 2003; Elhafez et al. 2006; Millar et al. 2011). Based on in vitro import assays, gene products of these seven putative genes fall into three subgroups in Arabidopsis: *AtNDA* (1, 2) and *AtNDC* are classified as internal, while *AtNDB* (1–4) are external (Elhafez et al. 2006; Millar et al. 2011; Smith et al. 2011). Both NDA and NDB proteins are derived from the proteobacterial endosymbiont, whereas the NDC type is of cyanobacterial origin (Michalecka et al. 2003; Hao and Rasmusson 2016), suggesting that this gene entered the eukaryotic cell via the chloroplast progenitor (Michalecka et al. 2003, 2004).

Since NDs have been studied less intensely as compared with AOX (Liu et al. 2008), relatively less is known about their connection with chloroplast metabolism. This fact aside, transgenic modification of NDB1 gene expression in *A. thaliana* (Wallström et al. 2014b) and *Nicotiana sylvestris* (Liu et al. 2008, 2009) affected $NADPH/NADP^+$ ratios, growth, and development although these changes were not directly associated with photosynthetic processes (Hao and Rasmusson 2016). Moreover, gene expression of *At*NDB1 was not associated with light metabolism (Escobar et al. 2004; Liu et al. 2009; Hao and Rasmusson 2016). These observations are consistent with the ancient eukaryotic origin of the non-acidic-motif NDB1 (NADPH, non-acidic type), which indicates that external NADPH oxidation has an original cellular function that is likely not connected to photosynthesis (Hao and Rasmusson 2016). However, transcript levels of each ND appear to be regulated in a tissue, development or stress-inducible manner (Elhafez et al. 2006). In this context, the strong light enhanced expression of ND_{in} genes *NDA1*, *NDC1* in Arabidopsis (Escobar et al. 2004; Elhafez et al. 2006) and *NDA1* in potato (Svensson and Rasmusson 2001) and the higher oxidation of NADH by ND_{in} in mature leaves in potato (Svensson and Rasmusson 2001) suggest a significant role for this pathway in the photosynthetically associated mitochondrial metabolism (Svensson and Rasmusson 2001; Michalecka et al. 2003; Raghavendra and Padmasree 2003; Noguchi and Yoshida 2008). Collectively, the induction of alternative NDs in the light relative to dark (Escobar et al. 2004) suggests a role for them in the photorespiratory pathway (Escobar et al. 2004; Lothier et al. 2019).

Taking into consideration that photorespiration occurs and generates NADH by GDC in the mitochondrial matrix in the light, it is reasonable to suggest that a corresponding increase in the redox capacity of the *m*ETC is most likely required to maintain photorespiration without causing overreduction of the basal *m*ETC and any resulting ROS production (Maxwell et al. 1999; Møller 2001; Escobar et al. 2004; Rasmusson et al. 2004). Given the light induction of NDs, as previously mentioned, as well the fact that glycine oxidation of leaf mitochondria was rotenone-insensitive, ND_{in} are thought to participate in the recycling of NAD^+ from NADH produced by the photorespiratory GDC (Igamberdiev et al. 1997; Gardeström et al. 2002; Escobar et al. 2004; Noguchi and Yoshida 2008; Obata et al. 2016). Thus, it is possible that light regulation of NDA1 and NDC1 reflects a critical role in the maintenance of photorespiratory metabolism (Escobar et al. 2004), which could act in conjunction with AOX to promote dynamic, short-term redox adjustments that could maintain multiprotein complexes I and III, and ubiquinone, in a relatively oxidized state (Rasmusson et al. 1998; Escobar et al. 2004). Although the metabolic relationship with photorespiration is less clear in NDs than that in both UCP and AOX, NDA suppression lines also showed increased levels of glycine and serine under high light conditions and serine was reduced in the NDB line (Wallström et al. 2014a, b). Altogether, these results clearly indicate the close interaction of *m*ETC and photorespiration, most likely due to the re-oxidation of photorespiratory NADH, which is necessary to support high photorespiratory flux in mitochondria (Sweetlove et al. 2006; Strodtkötter et al. 2009; Bykova et al. 2014; Obata et al. 2016). Nevertheless, since there is no consistent photorespiratory pattern in Complex I mutants, and thus no consistent effect on the glycine-oxidizing capacity (Obata et al. 2016; Lothier et al. 2019), caution must be taken when interpreting such results. For example, photorespiration is increased in CMSII and *ndufs1*, but unchanged in *ndufs8*, whereas, despite the considerable variation, most photorespiratory mutants show also alterations in N metabolism (Obata et al. 2016; Lothier et al. 2019). Remarkably, elucidation of the precise function of NDs is further complicated by the fact that several NDs have a dual localization and are also present in subcellular compartments other than mitochondria (Carrie et al. 2008; Millar et al. 2011; Smith et al. 2011; Xu et al. 2013), suggesting a far complex physiological role for these enzymes in the plant. In addition to mitochondria, NDB1 and NDA1–2 are present in plant peroxisomes, while NDC1 is also found in the chloroplast (Smith et al. 2011). Thus, the interpretation of knockout studies for these genes is highly complex (Millar et al. 2011), demonstrating that the functions of NDs within plant cells and their connection to photosynthesis have to be re-evaluated in light of this distinct subcellular targeting information.

3.5 Mitochondrial Metabolite Transporters

Metabolic interactions between chloroplasts and mitochondria require the existence of an efficient system for the import of respiratory substrates and the export of products of respiration for the provision of light-dependent processes, such as photosynthesis and photorespiration. The mitochondrial carriers (MCs) constitute a large family of nuclear-encoded proteins known as the mitochondrial carrier family (MCF) (Picault et al. 2004; Palmieri et al. 2011). MCs have an essential function in the maintenance of the metabolic communication of the mitochondrion with the cytosol (Picault et al. 2004; Palmieri et al. 2011). For instance, the export of ATP synthesized via oxidative phosphorylation relies on adenine nucleotides carriers belonging to the MCF and located in the mitochondrion (da Fonseca-Pereira et al. 2018a). The in silico survey of the expression pattern of adenylate carriers demonstrated its higher induction under stresses, which might be a consequence of the use of ATP for energy-consuming processes under these conditions (da Fonseca-Pereira et al. 2018a). Moreover, the number of genes co-expressed with adenylate carriers was significantly higher under stress conditions, suggesting that these transporters display far more complex expression patterns across different tissue types under adverse environmental conditions (da Fonseca-Pereira et al. 2018a).

3.6 Advances on Photosynthetic Performance: Why Is it so Difficult to Improve?

Photosynthesis is the basis of primary productivity and undoubtedly one of the most well-known biological processes over the world. Photosynthesis is an endergonic process which uses the solar energy to reduce the atmospheric CO_2 to the level of sugar and must be accompanied by oxidation of water to yield oxygen (O_2) (Paul and Pellny 2003; Heyneke and Fernie 2018). In addition, it is a fundamental process governing biomass production and is directly influenced by several environmental cues (Heyneke and Fernie 2018). Although it has been extensively studied over the years, our knowledge about the earliest origins of photosynthesis remains obscure. According to De Marais (2000), there is suggestive evidence that photosynthetic organisms were present approximately 3.2–3.5 billion years ago in a form of stromatolites, exhibiting layered structures similar to forms that are produced by some modern cyanobacteria. Additionally, there are also evidences that the first photosynthetic organism appeared at least 2.5 billion years ago (Archean Eon) (Niklas 2016). Noteworthy, it is still highly controversial and has engendered a great deal of spirited discussion in the present literature as previously raised (Buick 2008). In spite of an ongoing debate related to the exact moment of photosynthesis appearance, there is no doubt that the photosynthesis was originally an aquatic-based process occurring in a strongly reducing atmosphere. Evidence indicates that free O_2 began to accumulate by 2.4 billion years ago in the atmosphere, although the ability to perform oxygenic photosynthesis probably began somewhat earlier (Buick 2008). Consequently, the transition to a terrestrial environment with an increased O_2 atmosphere shaped the photosynthetic pathway into its current form (Hohmann-Marriott and Blankenship 2011). From an evolutionary point of view, land plant cells contain chloroplasts, which are the site of photosynthesis and have originated from endosymbiosis of a cyanobacteria-like organism (Raven and Allen 2003; Timmis et al. 2004). Notably, chloroplasts host numerous essential metabolic pathways including photosynthesis, which makes chloroplasts the primary source of chemical energy on earth (Zoschke and Bock 2018). This characteristic is supported by a high number of proteins (~3000), the vast majority of which are nucleus-encoded and post-translationally imported into the organelle (Goldschmidt-Clermont 1997; Zoschke and Bock 2018).

In the chloroplast membranes, the sunlight is directly captured by chlorophyll and other accessory pigments and used to energize electrons derived from a water molecule in the thylakoid membrane of the chloroplast (Fig. 3.2a).

These high-energy electrons are then transferred to carrier molecules, which can donate them for the reduction of CO_2 to triose-phosphates in the chloroplast stroma (Martin et al. 2006). Remarkably, only around 40% of the incident solar energy (from 400 to 740 nm) is used for photosynthesis (Leister 2019). Therefore, the oxygenic photosynthesis is somehow limited in an energetic yield point of view and the light reactions in plants provide ample scope for their improvement (Leister 2012, 2019; Moses 2019). This fact apart, with the advances in genetic engineering,

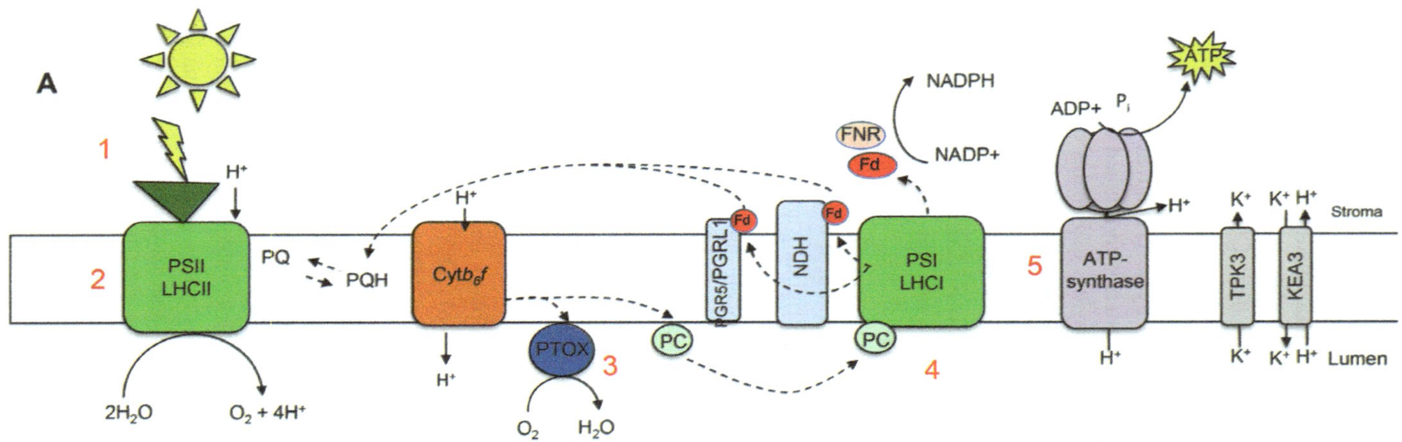

B

PSII-LHCII supercomplex	Cytochrome *b6f*		PSI-LHCI supercomplex	ATP-synthase
(1) Light harvesting	(2) PSII core	(3) Alternative Electrons flow	(4) PSI core	(5) ATP synthase
• Alter size of antenna • Synthesize pigments • Introduce new antenna complex • Increase range of PPFD	• Exchange of conserved modules (e.g. PsBA/D1, PsB/CP47, PsbC/CP43, PsbH, PsaA or the PSII core • PSII repair proteins	• Introduce electron escape valves (Flv, PTOX); • Control the amount of cyclic and pseudo cyclic electron flow	• Express a red-shifted PSI-like centre (e.g., bacteriochlorophyll with an absorption maximum at 1100 nm); • PSI coupled to platinum or bio-photovoltaics • PSI-450 hybrids; • PSI-hydrogenase hybrids; • PSI bio-nanohybrids	• Modify the ATP synthase; • Alterations in the number of *c* subunits

Fig. 3.2 Scheme of the light reactions of photosynthesis and its potential targets for improvement light reactions and photoprotection. (**a**) Linear electron flow from water oxidation in the PSII-LHCII supercomplex (light green) to $NADP^+$ reduction by ferredoxin (red) and ferredoxin:$NADP^+$ reductase (light orange)

releasing protons into the lumen, which is then used to drive the ATP biosynthesis by ATP synthase (purple). In addition, there is an alternative electron transfer pathway that mediates cyclic electron flow such as PGR5/PGRL1- and NDH (light blue) or PTOX (circle dark blue). These pathways mediate the water-to-water cycle, conferring dynamic protection and preventing the production of reactive oxygen species (ROS). Noteworthy, ion channels can be modulated by light fluctuations, regulating the proton motive force (*pmf*) in the chloroplast which can be an important regulator of ATP biosynthesis such as TPK3 and KEA3 (light gray) where are exchanging K^+ and H^+ between stroma and lumen. All these points have become important targets to optimize and improve photosynthetic efficiency. The reaction presented here is not balanced. Dashed arrows indicated the electron flux. (**b**) Recent advances based on photosynthesis reaction and its potential main targets for manipulation aiming to enhance light reactions and energy conversion. Five different main targets are described in B and schematically represented in A. For details, see the main text and Table 3.1. Abbreviations: *APX* ascorbate peroxidase, *Fd* ferredoxin, *FNR* ferredoxin: $NADP^+$ reductase, *KEA3* potassium efflux antiporter 3, *LHC* light-harvesting chlorophyll-*a/b*-binding complex, *NDH* NADH dehydrogenase-like complex, *PC* plastocyanin, *PQ* plastoquinone, PQH_2 plastoquinol, *PGR5* proton gradient regulation 5, *PGRL1* PGR5-like protein 1, *PSI* Photosystem I, *PSII* Photosystem II, *PTOX* plastid terminal oxidase, *SOD* superoxide dismutase, *TPK3* two-pore potassium channel 3

the current state of the ongoing efforts in molecular engineering to improve photosynthesis, plant growth, and yield has been recently reviewed (Foyer et al. 2017).

The photosynthesis is limited by the slowest step in the process, the so-called *limiting factor* under any particular condition (Blackman 1905). Following this idea, the photosynthesis can be limited by both, light and CO_2 concentration (Harbinson 2012; Orr et al. 2017). It is important to mention that ideally the light absorbed by the leaf would be used to fix CO_2 with a constant quantum efficiency across the naturally occurring light intensities, then this process would result in a linear relationship between light intensity and the rate of CO_2 fixation (Harbinson 2012). In intact leaves, three major metabolic properties are important for optimal photosynthetic performance namely: ribulose-1,5-biphosphate carboxylase/oxygenase (RubisCO) activity (Parry et al. 2013), regeneration of ribulose bisphosphate (RuBP) (Rosenthal et al. 2011), and metabolism of triose phosphate (Yang et al. 2016). Altogether, these important steps are the main targets to be improved nowadays.

The enzyme responsible for the first step in CO_2 fixation is the RubisCO in the Calvin Benson cycle (CBC). During carbon assimilation, RubisCO must first be carbamylated by an activator CO_2, in addition, the substrate CO_2, and must bind Mg^{2+} before binding the five-carbon substrate RuBP resulting in two molecules of 3-PGA that are integrated into the CBC ultimately to form sugars (Lorimer 1981). Accordingly, the oxygenation of RubisCO produces 2-PG, which is later converted back to 3-PGA in the photorespiratory cycle (Hagemann and Bauwe 2016). The photorespiratory process requires energy (ATP) and reducing equivalents such as NAD(P)H used during CO_2 and ammonia (NH_3) re-fixation (Bergman et al. 1981; Maurino and Peterhänsel 2010). This energy-dependence and reducing equivalents and part of the previous CO_2 fixed is again released as CO_2. These are the main reasons why photorespiration is often viewed as a wasteful process (Hagemann and Bauwe 2016). Additionally, the CO_2 losses in the photorespiration process and the low availability of environmental CO_2 led to the evolution of plants that developed CO_2-concentrating mechanisms such as those found in aquatic organisms and C4 and CAM higher plants (Martin et al. 2006). Thus, the introduction of the C4 trait into other plant species that may lack the ability to naturally evolve it is a challenging task (Denton et al. 2013; Schuler et al. 2016; Wang et al. 2016), a point that will be discussed.

Given that photosynthesis is a crucial process responsible to produce biomass, approaches aiming at increasing biomass and gain of yield have received considerable attention as recently reviewed (Heyneke and Fernie 2018). Recent years have witnessed a growing body of evidence that provided a detailed picture of the intricacies of the photosynthetic process and further suggested potential avenues for its improvement (Orr et al. 2017; Eisenhut and Weber 2019; Kubis and Bar-Even 2019). Here, we have focused on covering the most recent approaches that regulate the rate of photosynthesis and influence biomass yield, by further addressing also aspects connecting photosynthesis, photorespiration, and mitochondrial metabolism (Evans 2013; Cardona et al. 2018; Heyneke and Fernie 2018; Kubis and Bar-Even 2019).

We start by exploring the advances regarding optimized response to changes in light-use efficiency (Fig. 3.2). Light can be described as a wave of particles known as photons and the light-use efficiency can be defined as the ratio of net primary productivity (NPP) to absorbed photosynthetically active radiation (PAR) (Medlyn 1998). Accordingly, the maximum conversion efficiency of solar energy to biomass in plants is estimated to be 4.6% for C3 photosynthesis and 6% for C4 photosynthesis at 30 °C and 380 ppm atmospheric CO_2 (Zhu et al. 2008). Notably, the major losses of energy conversion during plant biomass formation occur during light absorption and the photochemical reactions (Amthor 2010; Stitt 2013; Amthor et al. 2019). In fact, several independent studies have attempted to maximize the capacity to process the influx of energy by the photosynthetic apparatus, and it seems reasonable to anticipate that this is still a challenge that must be pursued to ameliorate energy losses during plant photosynthesis. Indeed, there are several reasons to be very optimistic about enhancing photosynthetic efficiency, but still many appealing ideas are on the drawing board (Cardona et al. 2018). Thus, in the next paragraphs, we have compiled a number of recent pieces of evidence and promising results that have somehow increased light-use efficiency, culminating in significant increases in total biomass production.

Recent efforts aiming to optimize RubisCO performance are subsequently summarized. RubisCO is an ancient enzyme that evolved in a CO_2-rich atmosphere devoid of O_2, made of eight copies each of a large subunit (RbcL) encoded by the chloroplast genome and of a small subunit (RbcS) encoded by the nuclear genome (Pottier et al. 2018). RubisCO, the key enzyme of the CBC, is probably the most abundant protein in the biosphere and is responsible for assimilating the vast majority of inorganic carbon (Raven 2013). However, despite its abundance and biochemical roles in plants, RubisCO is considerably slower compared with most enzymes in central metabolism (Bar-Even et al. 2011). Furthermore, the enzyme is not completely specific for CO_2 but also accepts O_2, producing a toxic product that should be further assimilated by a wasteful process called photorespiration (Kubis and Bar-Even 2019; Maurino and Peterhänsel 2010; Peterhansel et al. 2010). Given that RubisCO is often considered as the limiting step in photosynthesis it is not surprising that it has been a target for metabolic engineering in several different species. Not only RubisCO but also the sedoheptulose-1,7-bisphosphatase (SBPase) is another enzyme of importance on the control of carbon assimilation, tightly linked with carbon flux in the CBC, by regenerating RuBP (Driever et al. 2017; Zhu et al. 2007, 2008). SBPase is an important enzyme that may limit the rate of RuBP regeneration and is another way to increase the rate of photosynthesis (Raines et al. 2001; Driever et al. 2017). We also discuss the current knowledge on recent genetic manipulation of photorespiration aiming to increase plant productivity. To this end, recent results involving synthetic bypass and photorespiratory engineering approaches aiming to improve plant productivity by reducing photorespiratory losses are summarized. Finally, we also summarized recent progress on the manipulation of carbon concentrating mechanisms (CCMs), which are evolutionary solutions to counter RubisCO inefficiency and a new research perspective about synthetic biology introducing the C4 cycle in C3 crops.

3.7 Advances in Plant Light-Use Efficiency

The photochemical reactions of photosynthesis use solar energy to reduce equivalents further fixing CO_2 to biochemical energy during the biochemical reaction in the CBC reactions using redox power generated in the photochemical reaction (Martin et al. 2006; Amthor 2010; Hohmann-Marriott and Blankenship 2011). In fact, however, only a small fraction of solar energy that reaches the earth's surface is actually fixed (Moses 2019). The main reason behind this fact is that, from the total amount of solar spectrum that reaches the earth's surface, land plants use only a fraction of the light spectrum, mainly 48% as photosynthetically active light. In addition, a significant part of this light is lost during absorption and photochemical reactions (Stitt 2013). The maximum light-efficiency used to convert solar energy to plant carbon biomass is estimated to be around 4.6% in C3 plants and 6% for C4 plants at environmental CO_2 of 380 ppm at 30 °C, which means a very low efficiency (Leister 2019; Zhu et al. 2007). Two main ways to maximize the sunlight use efficiency by plants have been suggested (Leister 2012, 2019). First, by expanding the spectral band used for photosynthesis and shifting the saturation of the process to higher light intensities, even minor enhancements the efficiency or stress resistance of light reactions should somehow positively impact plant biomass production, and ultimately crop yield (Cardona et al. 2018; Leister 2012, 2019; Long et al. 2015). Therefore, it has been possible to engineer plants for improved photosynthetic efficiency as evidenced by several studies in different species (Kromdijk et al. 2016; Foyer et al. 2017; Orr et al. 2017; Cardona et al. 2018; Kubis and Bar-Even 2019).

In plants, each photosystem is associated with large antenna systems with large complexes of proteins, chlorophyll, and other cofactors to convert light into chemical energy (Qin et al. 2015; Mazor et al. 2017; Su et al. 2017). Decreasing the light-harvesting antenna size of the photosystems in tobacco helps to increase the photosynthetic productivity, accumulating 25% more stem and leaf biomass in comparison with the wild-type counterpart (Kirst et al. 2017). Additionally, downregulation of the *CpSRP43* gene confers a truncated light-harvesting antenna (TLA) and enhances biomass in tobacco (Kirst et al. 2018). The light capture can be also improved by alteration in chlorophyll content. Thus, photosynthesis increases were associated with N availability once lower amounts of leaf chlorophyll could result in 9% savings in leaf N without penalties in the rates of photosynthesis (Walker et al. 2018; Evans and Clarke 2018).

Considering the sunlight proprieties and that plants do not use a wide range of sunlight, another approach that should be mentioned is the possibility of engineering crops with light-harvesting systems that do not naturally occur in plants (Cardona et al. 2018). Although there is a complex system of organism evolution, it has been already hypothesized that the extension of photosynthetically usable light to 750 nm may result in an increased number (~19%) of available photons (Chen and Blankenship 2011). The recent introduction of proteins responsible for chlorophyll *f* biosynthesis into crop plants could potentially expand the range of wavelengths that such plants use during photosynthesis and thereby increase their growth efficiency

(Ho et al. 2016). Therefore, the avenue to improve the light-harvesting system is definitively open (Croce and van Amerongen 2014). In a complementary manner, improvement of the photosynthetic process may consist of re-engineering the photosystems where the scientists are trying to express a red-shifted PSI-like reaction center from anoxygenic photosynthetic bacteria containing bacteriochlorophyll with an absorption maximum at 1100 nm (Blankenship et al. 2011). Considering that, recently Ort et al. (2015) raised a pertinent question: Is it still possible to get a more efficient arrangement of light energy conversion and harvesting antenna complex by plants? Indeed, as an interesting approach, it has been suggested that to replace photosystem I (PSI) with a reaction center and its associated cyclic electron transport machinery from a purple photosynthetic bacterium that uses bacteriochlorophyll *b* instead of the chlorophyll *a* used by oxygenic organisms. It is equally important to mention that these more radical approaches are still on the drawing board as recently reviewed (Cardona et al. 2018).

Not only the approaches related to the structure of the antenna complex but also studies improving photoprotection in a fluctuating environment have been performed (Orr et al. 2017; Cardona et al. 2018). In the field, the intensity of light (light quantity) or increased photochemistry to use the absorbed energy and the spectral profile (light quality) changes through the day, and as such several photoprotective mechanisms are induced to protect the photosynthetic antenna complexes from overexcitation (Long et al. 1994; Külheim et al. 2002; Li et al. 2009). Moreover, plants have developed mechanisms to support and mitigate cell damage by spatially and temporally operating different excitation and electron transfer processes that maximize efficient light use (Roach and Krieger-Liszkay 2014; Cardona et al. 2018). These processes include alterations in the PSIIs cofactors preventing dangerous back-reactions (Brinkert et al. 2016), non-photochemical quenching (NPQ) mechanisms to dissipate excess energies and against photooxidation (Szabó et al. 2005; Dall'Osto et al. 2017), as well as regulation of cyclic and linear electron flow at the level of the cytochrome *b6f* complex (Joliot and Johnson 2011; Shikanai 2014). Finally, the plastid terminal oxidase (PTOX) is involved in an alternative electron transport pathway that mediates electron flow from plastoquinol to O_2 (Fig. 3.2; McDonald et al. 2011; Alric and Johnson 2017).

It has been suggested that NPQ is an area where there is room for improvement by positively influencing the photosynthesis (Davison et al. 2002; Murchie and Niyogi 2011). Remarkably, overexpression of two xanthophyll cycle enzymes (de-epoxidase and zeaxanthin epoxidase) improved the NPQ performance in tobacco plants (Kromdijk et al. 2016). In addition, these alterations allowed the plant to bounce back from a dissipative heat-producing state into a productive light-harvesting state at a faster rate than the WT counterpart, resulting in an increase in dry-weight biomass of up to 20% in glasshouse trials and around 15% in duplicated field trials (Kromdijk et al. 2016). Changes in the energy dissipation capacity (NPQ) also positively influenced photosynthesis in tobacco plants (Kromdijk et al. 2016). The conservation of NPQ across plants suggests that this approach may also serve to improve the growth of other crops (Orr et al. 2017).

Different aspects of NPQ have emerged, and recent approaches have raised the motion that improved photosynthesis may be based on NPQ alterations. However, it should be born in mind that more studies should be developed given that the temptation to increase the capacity to process the influx of energy by the photosynthetic apparatus is a challenge that must be pursued to ameliorate energy losses during plant photosynthesis (Chen and Blankenship 2011; Orr et al. 2017). Therefore, the use of genetic engineering in light reactions of photosynthesis could lead to potential improvements in crop yields although it seems clear that many proposals to improve photosynthesis in plants are still at a very early stage of conception (Cardona et al. 2018).

The recent progress and prospects in synthetic biology (SynBio) to the light reaction of photosynthesis studies have been recently explored (Moses 2019; Leister 2019). It has been further suggested that both, to enhance the efficiency of light use and to couple the light reactions in novel ways to enzymes or non-biological components that use the reducing power from the light reactions, are promising approaches. Additionally, Boehm and Bock (2019) explored the plastome as a platform for engineering chloroplastidic metabolism whereas SynBio was further used to raise crop yields by increasing the CO_2 concentration around RubisCO (Weber and Bar-Even 2019). The idea is to reduce photorespiration and explore the possible by-pass or even new pathways that can switch carbon losses by the photorespiratory process into carbon-gaining processes. All of these studies were discussed elsewhere (Hanson et al. 2019).

3.8 The Advances in RubisCO Engineering

RubisCO is considerably slower than most enzymes in central metabolism, and therefore, to compensate for the low K_{cat} leaves contain large amounts of RubisCO, often 30% of total leaf protein (Farquhar et al. 2001; Bar-Even et al. 2011). On top of that, the bi-specificity of RubisCO for O_2 and CO_2 has called the attention of researchers to further improve photosynthesis efficiency and increase crop productivity by altering the RubisCO specificity and the photorespiration process (Whitney et al. 2011; Parry et al. 2013). The great fundament is grounded by the underlying idea that CO_2 enrichment can increase crop yield and photosynthesis (Long et al. 2006; Zhu et al. 2010b; Kubis and Bar-Even 2019). As observed in C4 plants metabolism, greater photosynthetic rates lead to higher biomass production for a given amount of sunlight in comparison with C3 crops; therefore, such findings are somehow key strategies to improve photosynthetic capacity in C3 plants, improving the CO_2 fixation capacity consequently, improving crop yield (Parry et al. 2013). Indeed, during the last decades, several studies have attempted to find and produce the "super RubisCO" by optimizing its catalytic potential (von Caemmerer and Evans 2010; Evans 2013; Parry et al. 2013; Pottier et al. 2018; Kubis and Bar-Even 2019). The concept of improving photosynthesis to raise crop yield has been expertly reviewed elsewhere (Evans 2013; Filatov et al. 2010; Heyneke and Fernie 2018; Leister 2019; Long et al. 2006; Orr et al. 2017). Five important targets to improve photosynthesis are listed in these reviews, namely: (1) improving RubisCO kinetic properties (Whitney et al. 2011; Parry et al. 2013), (2) introduction of C4 metabolism into C3 plants (Schuler et al. 2016; Leister 2019), (3) relaxation of photoprotection (Murchie and Niyogi 2011), (4) manipulation in SBPase activity (Raines et al. 2001; Rosenthal et al. 2011; Zhu et al. 2007), and finally, (5) improved canopy architecture (Evans 2013).

Modification in RubisCO kinetic properties by increasing the CO_2 specificity should decrease photorespiration, consequently improving photosynthesis mainly when both, light and CO_2 are not limiting (Carmo-Silva et al. 2015; Evans 2013; Peterhänsel et al. 2008; Whitney et al. 2011). In nature, some organisms with improved RubisCO specificity such as red algae are already known (Uemura et al. 1997), while C4 plants possess RubisCO with superior catalytic turnover rates, converting sunlight into biomass with a greater efficiency than C3 crops (Long et al. 2006; Pottier et al. 2018; Sheehy et al. 2007). Substantial increases in canopy photosynthesis could follow from incorporating a "better RubisCO" into C3 crop species (von Caemmerer and Evans 2010). Noteworthy, evolutionary analysis of RubisCO in C4 plants from the genus Flaveria (Asteraceae) has demonstrated the presence of two main residues in the RubisCO large subunit that could explain, at least partially, the different catalytic properties of this enzyme (Filatov et al. 2010). In addition, catalytic improvements have been found to be transposable to RubisCO in tobacco (Zhu et al. 2010a, b). It was later experimentally verified in tobacco that a single amino acid mutation acts as a catalytic "switch" to convert RubisCO from different species from a "C3 style" into a "C4 style" and vice versa (Evans 2013; Orr et al. 2017; Whitney et al. 2011). In addition, RubisCO mutants in *C. reinhardtii* and cyanobacteria showed improvements in their catalytic process and CO_2 affinity (Zhu et al. 2010a; Morel et al. 2016; Eason-Hubbard et al. 2017).

Recent developments have been also observed with the manipulation in the small RubisCO subunit increasing catalytic turnover rate of RubisCO in different species such as rice (Ishikawa et al. 2011; Ogawa et al. 2012), Arabidopsis (Makino et al. 2012), and *Chlamydomonas* (Genkov et al. 2010). Although subjected to several technical limitations, an alternative approach to alter the extant RubisCO in a crop species is to replace it with better-performing natural variants (Orr et al. 2017). Therefore, the introduction of foreign RubisCO was exemplified by the introduction of the cyanobacterial CCM into land plants where the RubisCO from *Synechococcus elongatus* PCC7942 has been successfully introduced into tobacco plants, which also supported higher plant growth under elevated CO_2 concentrations (Lin et al. 2014; Occhialini et al. 2016). In plants, the genetic engineering of RubisCO is hindered by the disparate location of rbcL and RbcS in different genomes (Whitney et al. 2011). The plastome transformation of rbcL has successfully demonstrated the feasibility of replacing land plant RubisCO with phylogenetically distinct bacterial *Rhodospirillum rubrum* (L_2) and archeal *Methanococcoides burtonii* (L_{10}) RubisCO (Whitney et al. 2011). Remarkably, in these transplastomic plants, both growth and photosynthetic properties have corresponded with the content and catalytic properties of the recombinant RubisCO, confirming the accuracy of the models used to predict photosynthetic carbon assimilation (Whitney et al. 2011; Parry et al. 2013).

3.9 Calvin-Benson Cycle Optimization

In order to produce plants with increased yield research efforts have been extensively placed to identify limiting points in the photosynthetic process (Raines 2003). As described above, several attempts have been performed to improve photosynthetic carbon fixation mostly focused on altering the catalytic properties of RubisCO (Uemura et al. 1997; Zhu et al. 2010a; Makino et al. 2012; Evans 2013; Eason-Hubbard et al. 2017). Since the CBC comprises 11 different enzymes, catalyzing 13 reactions, it presents itself as viable targets to accelerate plant carbon fixation (Rosenthal et al. 2011; Raines 2011; Orr et al. 2017). For instance, an important enzyme in the CBC is the SBPase, which possesses the second greatest control coefficient for carbon assimilation (Heyneke and Fernie 2018). SBPase is involved in either the rate of RuBP regeneration or in the production of sucrose or starch, therefore it has been considered an important point to increase the rate of photosynthesis (Raines et al. 2001; Ding et al. 2016; Driever et al. 2017). In fact, increases in SBPase activity in tobacco resulted in elevated sucrose levels and starch accumulation accompanied by an increased leaf area index and increased biomass (Zhu et al. 2010a; Ding et al. 2016). In agreement, increases in SBPase activity stimulate photosynthesis and growth from the early developmental stage under normal conditions (Lefebvre et al. 2005) as well as under field elevated [CO_2] (Rosenthal et al. 2011) in transgenic tobacco plants. Furthermore, SBPase overexpression in rice plants enhances photosynthesis under both higher temperatures (Feng et al. 2007b) and salt stress (Feng et al. 2007a). Recently, transgenic wheat plants expressing SBPase from *Brachypodium distachyon* were characterized by enhanced photosynthesis, increased total biomass, and dry seed yield (Driever et al. 2017). Changes in the activity of SBPase altered photosynthetic capacity, growth, and tolerance to chilling stress in transgenic tomato plants (Ding et al. 2016). All of these efforts have shown that the effects of increased SBPase activity on photosynthesis are rather positive, but effects on plant growth or yield may vary, particularly in important food crop species (Driever et al. 2017). Altogether, these results provide mounting evidence to suggest that increasing the activity of SBPase has the potential to improve photosynthesis. Furthermore, overexpression in SBPase and fructose 1,6-bisphosphate aldolase (FBPA) in tobacco plants resulted in a cumulative by increased biomass (Simkin et al. 2015). Later, alterations in the expression levels of three different enzymes involved in the CBC and photorespiratory pathway, namely SBPase, FBPA, and the photorespiratory glycine decarboxylase-H (GDH-H), lead to improvements in vegetative biomass production and seed yield (Simkin et al. 2017). Remarkably, recent approaches have focused on reducing the CO_2 energy costs of photorespiration by the usage of synthetic biology namely the "photorespiratory bypass" pathways in the chloroplast (Kebeish et al. 2007; Nölke et al. 2014; Dalal et al. 2015). The power of this remarkable approach has been demonstrated, and it has been expertly reviewed elsewhere (Maurino and Peterhänsel 2010; Eisenhut and Weber 2019). In fact, ongoing investigations suggest that expanded efforts could ultimately optimize this exciting biotechnological process.

3.10 Synthetic Photorespiration Bypass

Together, the CBC and the photorespiratory pathway, are responsible for nearly all biological CO_2 fixation in the earth. Notably, the occurrence of the photorespiratory pathway allows the CBC to operate even in the presence of molecular oxygen (Husic et al. 1987; Timm et al. 2016). Briefly, the oxygenation by RubisCO leads to the formation of the toxic product 2-phosphoglycolate (Erb and Zarzycki 2018). The photorespiratory pathway is essential to maintain photosynthesis under O_2-containing atmosphere and further metabolize 2-phosphoglycolate. This fact aside, photorespiration is an essential process that cannot be avoided as previously demonstrated in many independent studies that have attempted to abolish or reduce the activity of photorespiratory enzymes (Timm and Bauwe 2013; Timm et al. 2016). For these reasons, genetic engineering of this pathway has gained attention since it offers the potential to enhance photosynthesis capacity and crop productivity (Heyneke and Fernie 2018; Eisenhut and Weber 2019; Kubis and Bar-Even 2019). In fact, these efforts are mostly inspired by natural mechanisms present in organisms such as cyanobacteria and algae (Zarzycki et al. 2013; Bar-Even 2018). In addition, synthetic metabolic routes to redirect the canonical pathway of CO_2 assimilation and photorespiration are also available currently (Zarzycki et al. 2013; Eisenhut and Weber 2019). Therefore, replacing photorespiration with a synthetic alternative could tackle one or more of these drawbacks (Bar-Even 2018; Kubis and Bar-Even 2019). Indeed, during the last decade, several CO_2-releasing photorespiration bypass routes have been suggested and, at least partially, already implemented (Kebeish et al. 2007; Carvalho et al. 2011; Maier et al. 2012; Fahad et al. 2019; Shen et al. 2019).

The enzymes required at different stages of the photorespiratory pathway are well known (Peterhansel et al. 2010; Peterhänsel et al. 2012; Hagemann and Bauwe 2016). Briefly, this entire pathway occurs to metabolize the glycolate produced by RuBP oxygenation, while minimizing the losses of carbon, N, and energy, and ultimately to avoid the accumulation of photorespiratory intermediates (Shen et al. 2019). It is important to mention that plants engineered with photorespiration bypasses that ultimately reduced photorespiratory activity culminated with significant increases in photosynthesis and yield (Fig. 3.3) (Carvalho et al. 2011; Dalal et al. 2015; Fahad et al. 2019; Kebeish et al. 2007; Maier et al. 2012; Shen et al. 2019). Briefly, during bypass 1, glycolate is diverted into glycerate within the chloroplast, shifting the release of CO_2 from mitochondria to chloroplasts, and reducing ammonia release. During bypass 2, two *Escherichia coli* enzymes were implemented into the peroxisome to catalyze the conversion of glyoxylate into hydroxypyruvate and CO_2 in a two-step process. Finally, in bypass 3, glycolate is completely oxidized into CO_2 inside chloroplasts by both newly introduced and

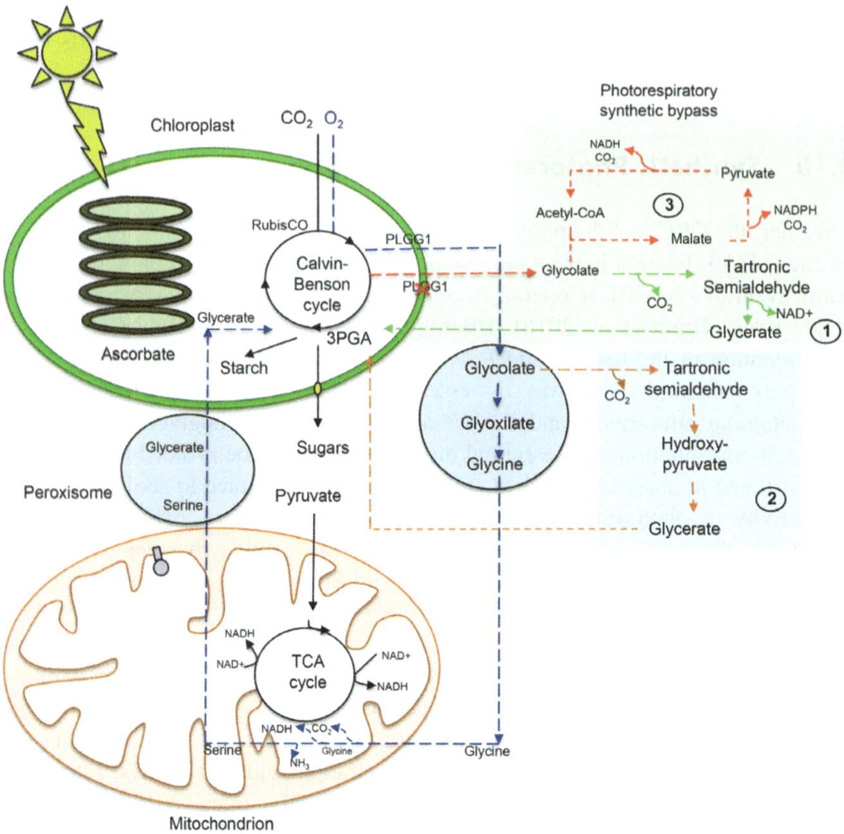

Fig. 3.3 The multicompartment photorespiratory pathway and new advances in the photorespiration bypasses. The Ribulose-1,5-biphosphate carboxylase/oxygenase (RubisCO) catalyze CO_2 and O_2 fixation. The product of CO_2 fixation is the 3-phosphoglycerate that enters in the Calvin-Benson cycle and can be directed to starch biosynthesis and/or sugars in the cytosol converted into pyruvate which is then completely oxidized in the mitochondria during respiration (Black arrow). In addition, the oxygenation drives the production of phosphoglycolate which is metabolized in the C2 photorespiratory metabolism (Blue arrows). Photorespiration, which involves at least three organelles namely chloroplast, peroxisome and mitochondria, is associated with carbon losses. Recently, three new synthetic photorespiratory bypasses have been proposed in an attempt to improve the carbon assimilation and reduce the photorespiration in C3 plants. (**1**) In bypass 1, glycolate is diverted into glycerate within the chloroplast, shifting the release of CO_2 from mitochondria to chloroplasts, and reducing ammonia release (dashed light green arrows) (For details see Kebeish et al. 2007); (**2**) Bypass 2 is a peroxisomal pathway, catalyzed by two *Escherichia coli* enzymes which convert glyoxylate into hydroxypyruvate and CO_2 in a two-step process (dashed orange arrows) (for details see Peterhänsel et al. 2012); (**3**) Lastly, bypass 3 is considered as a non-real bypass since the glycolate is completely oxidized into CO_2 inside chloroplasts by both newly introduced and native enzymes (dashed red arrows) (For details see Peterhänsel et al. 2012). Abbreviations: *3-PGA* the 3-phosphoglycerate, *PLGG1* plastidial glycolate/glycerate transporter 1

native enzymes (Carvalho et al. 2011; Peterhänsel et al. 2013). Further details are shown in Fig. 3.3.

The metabolic engineering of photorespiratory bypasses promoted significant changes in total photosynthesis yield and the main benefit of these bypasses is the fact that ammonia release in the mitochondrion is omitted, thereby saving ammonia re-assimilation costs (Maier et al. 2012; Peterhänsel et al. 2013). Accordingly, the main goal of this approach is to increase the CO_2/O_2 ratio inside the chloroplast, avoiding RuBP oxygenation as previously argued (Peterhänsel et al. 2013). In this context, 75% of glycolate fed into bypass 1 was returned to the CBC enhancing the final biomass production by up to 30% in Arabidopsis (Kebeish et al. 2007; Maier et al. 2012). Furthermore, the expression of a recombinant glycolate dehydrogenase polyprotein in plastids of potato (*Solanum tuberosum*) strongly enhances photosynthesis and tuber yield (Nölke et al. 2014). In addition, the re-engineering of bypass lines in *Camelina sativa* led to increased vegetative biomass with faster development, affecting also seed final yield (Dalal et al. 2015). Interestingly, the advantages of this bypass were only detectable under short-day conditions where the energy efficiency of carbon fixation is more limiting for growth than under long-day conditions (Peterhänsel et al. 2013). Although we have witnessed significant advances during the last years and several strategies have been suggested to make the metabolic engineering feasible, considerable barriers await the researcher who will try to implement them in plants (Bar-Even 2018; Weber and Bar-Even 2019). Accordingly, it has been experimentally demonstrated that manipulation of central metabolism is rather difficult even in simpler systems (e.g., microbes). Due to multiple layers of regulation that operate maintaining metabolic steady state and the highly connected nature of central metabolism can at least partially explain this difficulty (Sweetlove et al. 2017, see also Chap. 4, Knuesting et al.).

3.11 Introducing the C4 Cycle in C3 Crops

The photosynthetic C4 metabolism is one of the most convergent evolutionary phenomena in the whole biological system and it has evolved independently of C3 photosynthesis in several angiosperm families during the last 25 million years in at least 66 independent events (Sage et al. 2011; Edwards 2012). Accordingly, this morphophysiological multiple parallel evolution appears to have occurred as an adaptive response to low atmospheric CO_2 concentrations and high temperature (Sage 2004, 2016). The first evidence of intermediate C3–C4 forms was reported in the 1970s (Kennedy and Laetsch 1974; Sage 2016), leading to intensive efforts to understand the mechanistic basis of the transition from C3 to C4. It seems clear therefore that the transition from C3 to C4 plants requires the evolution of both morphological and physiological traits making the evolution of such a complex trait system in one single step highly unlikely (Schlüter and Weber 2016).

To reduce the rate of the RubisCO oxygenation reaction and thereby the inefficiencies associated with photorespiration, it would be desirable to engineer such carbon concentration mechanisms (CCM) into crops to increase their yield potential (Evans 2013; Heyneke and Fernie 2018; Weber and Bar-Even 2019). The introduction of the C4 trait into plant species that may lack the ability to naturally evolve is nevertheless a challenging task (Denton et al. 2013; Schuler et al. 2016; Wang et al. 2016). Indeed, using complementary approaches including genome and transcriptome analyses, the international C4 Rice Consortium is working toward introducing the C4 mechanism into rice (von Caemmerer et al. 2012). Although this research has already generated exciting results including the identification of metabolite transporters and transcription factors potentially useful for engineering C4 rice (von Caemmerer and Furbank 2016; Wang et al. 2016), further investigation is clearly required.

The C4 photosynthesis is highly complex requiring the differentiation of photosynthetically active vascular bundle sheath cells, modification in the biochemistry of several enzymes, and increased intercellular and intracellular transport of metabolites. This clearly makes the introduction of CCM in C3 plants (e.g., rice, wheat, barley, and Arabidopsis) which may lack the genetic prerequisites to evolve the corresponding leaf anatomy (Christin et al. 2013; Schuler et al. 2016) rather difficult. It is important to mention that the alteration of C4 acid metabolism is likely not beneficial without the appropriate leaf morphology or physiology (Maurino and Weber 2013), and as such, the identification of the relevant genetic regulators has been of paramount importance. Therefore, several efforts to engineer C4 photosynthesis into C3 plant species were hampered by an incomplete list of genes and gene functions required to support the trait (Weber and Bar-Even 2019).

In addition, recent efforts have expanded to different genera beyond Flaveria species, including Cleome and Moricandia, two close relatives of the C3 model Arabidopsis which contain C3 species as well as C3-C4 intermediates and also true C4 species (Kurz et al. 2016). Noteworthy, this provides a powerful opportunity to accelerate advances through comparison with the large amount of data already available for Arabidopsis in order to find the minimal genetic basis of C4 photosynthesis (Orr et al. 2017). Recently, the SynBio approach has suggested that instead of modulating individual components the entire processes could be redesigned to overcome key barriers that cannot be easily solved with currently existents biological systems (Weber and Bar-Even 2019; Hanson et al. 2019).

Research efforts are also current devoted to better understand the key elements required for CAM photosynthesis (Yang et al. 2015). Once CAM plants are typically characterized by high water use efficiency (WUE), it seems reasonable to anticipate that improving crop WUE and expanding the land area capable of supporting agriculture may occur by the rational engineering of CAM into food or bioenergy crops (Borland et al. 2014). It is worth to mention that CAM species may also serve as a suitable source of high-temperature-adapted enzymes of potential application for the engineering of photosynthetic metabolism.

Two fundamental points need to be considered when developing such C4 and CAM projects: (1) alterations of leaf morphology (Kranz anatomy) or physiology (stomatal closure during the day), and (2) leaf metabolism engineered in such a way that the C4 acid carboxylation and decarboxylation cycle characteristic of both C4 and CAM operates at high flux and with appropriate spatial or temporal patterning. Although there are still several open questions, it seems clear that these types of studies are likely to be key to understanding the genetic triggers needed to reorganize leaf anatomy, gene expression, and biochemistry within C4 and CAM plants, possibly paving the way toward engineering C3-C4 or even CAM plants (Orr et al. 2017).

3.12 Conclusion and Future Prospects

The accelerated increase of the global population and the uncertainties derived from climate change make the ultimate optimization of strategies toward increasing plant yield of paramount importance. An impressive number of genetic manipulation approaches in both, respiratory and photosynthetic metabolism have been generated over the last years, which has clearly provided a much deeper understanding of the contribution of individual genes, transcripts, proteins, and metabolites for improved productivity. However, most of the approaches adopted so far have been limited to manipulations at the single-gene level (Sonnewald and Fernie 2018; Zhang and Fernie 2018). We must now turn our attention to emerging metabolic engineering strategies toward multigene transformation (Kopka and Fernie 2018; Sonnewald and Fernie 2018; Zhang and Fernie 2018). We provide current evidence that synthetic biology is largely increasing our knowledge not only in the manipulation of respiratory and CBC elements but also in engineering alternate CCM mechanisms into land plants.

Undoubtedly, expanded research efforts are required to build upon many of the technologies described above by, for example, enhancing our understanding of CAM metabolism and introducing alternative non-plant CCMs into C3 plants. Beyond traditional approaches, metabolic engineering and synthetic biology tools hold promise for further progress toward the improvement of breeding strategies. Future advances in engineering and developing CAM and C4 plants will need to involve integrated analysis together with a further comprehension of the related gene regulatory networks. In addition, another genetic engineering approach with significative importance is the plastid transformation which may deliver valuable traits to increase agricultural production and sustainability (Meyers et al. 2010; Bock 2014; Lu et al. 2017; Piatek et al. 2018; Fuentes et al. 2018). Indeed, transplastomic plants can be used to produce the so-called "green chemicals" and due to the high-level of recombinant protein expression and multigene engineering, this approach holds great promise for plant biotechnology purposes (for details see: Bock 2014). We are anticipating that the success of such projects will lead to an improved balance between photosynthetic and respiratory metabolism, ultimately enhancing crop productivity as well as to produce alternative compounds that could assist in feeding a growing population in the future.

Acknowledgements This work was made possible through financial support from the Max Planck Society, the CNPq (National Council for Scientific and Technological Development, Brazil) [grant No. 402511/2016-6]; the FAPEMIG (Foundation for Research Assistance of the Minas Gerais State, Brazil) [grant Nos. APQ-01357-14, APQ-01078-15, and RED-00053-16]. We also thank the scholarships granted by the Brazilian Federal Agency for Support and Evaluation of Graduate Education (CAPES-Brazil) to PF-P and WB-S. Research fellowships granted by CNPq-Brazil to ANN and WLA are also gratefully acknowledged.

References

Alric J, Johnson X (2017) Alternative electron transport pathways in photosynthesis: a confluence of regulation. Curr Opin Plant Biol 37:78–86

Amthor JS (2000) The McCree-de wit-penning de Vries-Thornley respiration paradigms: 30 years later. Ann Bot 86:1–20

Amthor JS (2010) From sunlight to phytomass: on the potential efficiency of converting solar radiation to phyto-energy. New Phytol 188:939–959

Amthor JS, Bar-even A, Hanson AD et al (2019) Engineering strategies to boost crop productivity by cutting respiratory carbon loss. Plant Cell 31:297–314

Araújo WL, Nunes-Nesi A, Fernie AR (2011a) Fumarate: multiple functions of a simple metabolite. Phytochemistry 72:838–843

Araújo WL, Nunes-Nesi A, Osorio S et al (2011b) Antisense inhibition of the iron-sulphur subunit of succinate dehydrogenase enhances photosynthesis and growth in tomato via an organic acid–mediated effect on stomatal aperture. Plant Cell 23:600–627

Araújo WL, Nunes-Nesi A, Nikoloski Z et al (2012a) Metabolic control and regulation of the tricarboxylic acid cycle in photosynthetic and heterotrophic plant tissues. Plant Cell Environ 35:1–21

Araújo WL, Tohge T, Osorio S et al (2012b) Antisense inhibition of the 2-oxoglutarate dehydrogenase complex in tomato demonstrates its importance for plant respiration and during leaf senescence and fruit maturation. Plant Cell 24:2328–2351

Araújo WL, Martins AO, Fernie AR, Tohge T (2014a) 2-Oxoglutarate: linking TCA cycle function with amino acid, glucosinolate, flavonoid, alkaloid, and gibberellin biosynthesis. Front Plant Sci 5:1–6

Araújo WL, Nunes-Nesi A, Fernie AR (2014b) On the role of plant mitochondrial metabolism and its impact on photosynthesis in both optimal and sub-optimal growth conditions. Photosynth Res 119:141–156

Bar-Even A (2018) Daring metabolic designs for enhanced plant carbon fixation. Plant Sci 273:71–83

Bar-Even A, Noor E, Savir Y et al (2011) The moderately efficient enzyme: evolutionary and physicochemical trends shaping enzyme parameters. Biochemist 50:4402–4410

Barreto P, Okura VK, Neshich IAP et al (2014) Overexpression of UCP1 in tobacco induces mitochondrial biogenesis and amplifies a broad stress response. BMC Plant Biol 14:1–15

Barreto P, Okura V, Pena IA et al (2016) Overexpression of mitochondrial uncoupling protein 1 (UCP1) induces a hypoxic response in *Nicotiana tabacum* leaves. J Exp Bot 67:301–313

Bartoli CG, Pastori GM, Foyer CH (2000) Ascorbate biosynthesis in mitochondria is linked to the electron transport chain between complexes III and IV. Plant Physiol 123:335–344

Begcy K, Mariano ED, Mattiello L et al (2011) An Arabidopsis mitochondrial uncoupling protein confers tolerance to drought and salt stress in transgenic tobacco plants. PLoS One 6(8):e23776

Bergman A, Gardeström P, Ericson I (1981) Release and refixation of ammonia during photorespiration. Physiol Plant 53:528–532

Blackman FF (1905) Optima and limiting factors. Ann Bot 19:281–295

Blankenship RE, Tiede DM, Barber J et al (2011) Comparing photosynthetic and photovoltaic efficiencies and recognizing the potential for improvement. Science 332:805–809

Bock R (2014) Genetic engineering of the chloroplast: novel tools and new applications. Curr Opin Biotechnol 26:7–13
Boehm CR, Bock R (2019) Recent advances and current challenges in synthetic biology of the plastid genetic system and metabolism. Plant Physiol 179:794–802
Borland AM, Hartwell J, Weston DJ et al (2014) Engineering crassulacean acid metabolism to improve water-use efficiency. Trends Plant Sci 19:327–338
Bowsher CG, Hucklesby DP, Emes MJ (1989) Nitrite reduction and carbohydrate metabolism in plastids purified from roots of *Pisum sativum* L. Planta 177:359–366
Bowsher CG, Boulton EL, Rose J et al (1992) Reductant for glutamate synthase in generated by the oxidative pentose phosphate pathway in non-photosynthetic root plastids. Plant J 2:893–898
Braun HP, Zabaleta E (2007) Carbonic anhydrase subunits of the mitochondrial NADH dehydrogenase complex (complex I) in plants. Physiol Plant 129:114–122
Braun HP, Binder S, Brennicke A et al (2014) The life of plant mitochondrial complex I. Mitochondrion 19:295–313
Brinkert K, De Causmaecker S, Krieger-Liszkay A et al (2016) Bicarbonate-induced redox tuning in photosystem II for regulation and protection. Proc Natl Acad Sci U S A 113:12144–12149
Buick R (2008) When did oxygenic photosynthesis evolve? Philos Trans R Soc 363:2731–2743
Bykova NV, Møller IM, Gardeström P, Igamberdiev AU (2014) The function of glycine decarboxylase complex is optimized to maintain high photorespiratory flux via buffering of its reaction products. Mitochondrion 19:357–364
Cardona T, Shao S, Nixon PJ (2018) Enhancing photosynthesis in plants: the light reactions. Essays Biochem 62:85–94. https://doi.org/10.1042/EBC20170015
Carmo-Silva E, Scales JC, Madgwick PJ, Parry MAJ (2015) Optimizing Rubisco and its regulation for greater resource use efficiency. Plant Cell Environ 38:1817–1832
Carrari F, Nunes-Nesi A, Gibon Y et al (2003) Reduced expression of aconitase results in an enhanced rate of photosynthesis and marked shifts in carbon partitioning in illuminated leaves of wild species tomato. Plant Physiol 133:1322–1335
Carrari F, Coll-Garcia D, Schauer N et al (2005) Deficiency of a plastidial adenylate kinase in Arabidopsis results in elevated photosynthetic amino acid biosynthesis and enhanced growth. Plant Physiol 137:70–82
Carrie C, Murcha MW, Kuehn K et al (2008) Type II NAD(P)H dehydrogenases are targeted to mitochondria and chloroplasts or peroxisomes in *Arabidopsis thaliana*. FEBS Lett 582:3073–3079
Carvalho J de FC, Madgwick PJ, Powers SJ, et al (2011) An engineered pathway for glyoxylate metabolism in tobacco plants aimed to avoid the release of ammonia in photorespiration. BMC Biotechnol 11:111
Cavalcanti JHF, Esteves-Ferreira AA, Quinhones CGS et al (2014) Evolution and functional implications of the tricarboxylic acid cycle as revealed by phylogenetic analysis. Genome Biol Evol 6:2830–2848
Chen M, Blankenship RE (2011) Expanding the solar spectrum used by photosynthesis. Trends Plant Sci 16:427–431
Christin P-A, Osborne CP, Chatelet DS et al (2013) Anatomical enablers and the evolution of C4 photosynthesis in grasses. Proc Natl Acad Sci U S A 110:1381–1386
Clifton R, Millar AH, Whelan J (2006) Alternative oxidases in Arabidopsis: a comparative analysis of differential expression in the gene family provides new insights into function of non-phosphorylating bypasses. Biochim Biophys Acta 1757:730–741
Croce R, van Amerongen H (2014) Natural strategies for photosynthetic light harvesting. Nat Chem Biol 10:492–501
da Fonseca-Pereira P, Neri-Silva R, Cavalcanti JHF et al (2018a) Data-mining bioinformatics: connecting adenylate transport and metabolic responses to stress. Trends Plant Sci 23:961–974
da Fonseca-Pereira P, Silva WB, Araújo WL, Nunes-Nesi A (2018b) How does European mistletoe survive without complex I? Trends Plant Sci 23:847–850
Dahal K, Vanlerberghe GC (2018) Growth at elevated CO_2 requires acclimation of the respiratory chain to support photosynthesis. Plant Physiol 178:82–100

Dalal J, Lopez H, Vasani NB et al (2015) A photorespiratory bypass increases plant growth and seed yield in biofuel crop *Camelina sativa*. Biotechnol Biofuels 8:1–22

Dall'Osto L, Cazzaniga S, Bressan M et al (2017) Two mechanisms for dissipation of excess light in monomeric and trimeric light-harvesting complexes. Nat Plants 3:1–9

Davison PA, Hunter CN, Horton P (2002) Overexpression of β-carotene hydroxylase enhances stress tolerance in Arabidopsis. Nature 418:203–206

De Marais DJ (2000) When did photosynthesis emerge on earth? Science 289:1703–1705

De Paepe R, Ch P, Vitart V et al (1990) Several nuclear genes control both male sterility and mitochondrial protein synthesis in *Nicotiana sylvestris* protoclones. Mol Gen Genet 222:206–210

Denton AK, Simon R, Weber APM (2013) C4 photosynthesis: from evolutionary analyses to strategies for synthetic reconstruction of the trait. Curr Opin Plant Biol 16:315–321

Dinakar C, Vishwakarma A, Raghavendra AS, Padmasree K (2016) Alternative oxidase pathway optimizes photosynthesis during osmotic and temperature stress by regulating cellular ROS, malate valve and antioxidative systems. Front Plant Sci 7:1–17

Ding F, Wang M, Zhang S, Ai X (2016) Changes in SBPase activity influence photosynthetic capacity, growth, and tolerance to chilling stress in transgenic tomato plants. Sci Rep 6:32741

Driever SM, Simkin AJ, Alotaibi S et al (2017) Increased SBPase activity improves photosynthesis and grain yield in wheat grown in greenhouse conditions. Philos Trans R Soc 370(1730):1–9

Dyson BC, Miller MAE, Feil R et al (2016) FUM2, a cytosolic fumarase, is essential for acclimation to low temperature in *Arabidopsis thaliana*. Plant Physiol 172:118–127

Eason-Hubbard MR, Rickaby REM, Heureux AMC et al (2017) The role of Rubisco kinetics and pyrenoid morphology in shaping the CCM of haptophyte microalgae. J Exp Bot 68:3959–3969

Edwards EJ (2012) New grass phylogeny resolves deep evolutionary relationships and discovers C4 origins. New Phytol 193:304–312

Eisenhut M, Weber APM (2019) Improving crop yield. Science 363:32–33

Elhafez D, Murcha MW, Clifton R et al (2006) Characterization of mitochondrial alternative NAD (P)H dehydrogenases in Arabidopsis: Intraorganelle location and expression. Plant Cell Physiol 47:43–54

Erb TJ, Zarzycki J (2018) A short history of Rubisco: the rise and fall (?) of Nature's predominant CO_2 fixing enzyme. Curr Opin Biotechnol 49:100–107

Escobar MA, Franklin KA, Svensson ÅS et al (2004) Light regulation of the Arabidopsis respiratory chain. Multiple discrete photoreceptor responses contribute to induction of type II NAD(P) H dehydrogenase genes. Plant Physiol 136:2710–2721

Éva C, Oszvald M, Tamás L (2018) Current and possible approaches for improving photosynthetic efficiency. Plant Sci 280:433–440

Evans JR (2013) Improving photosynthesis. Plant Physiol 162:1780–1793

Evans JR, Clarke VC (2018) The nitrogen cost of photosynthesis. J Exp Bot 70:7–15

Fahad S, Khan FA, Pandupuspitasari N et al (2019) Suppressing photorespiration for the improvement in photosynthesis and crop yields: a review on the role of S-allantoin as a nitrogen source. J Environ Manag 237:644–651

Farquhar GD, von Caemmerer S, Berry JA (2001) Models of photosynthesis. Plant Physiol 125:42–45

Feng L, Han Y, Liu G et al (2007a) Overexpression of sedoheptulose-1,7-bisphosphatase enhances photosynthesis and growth under salt stress in transgenic rice plants. Funct Plant Biol 34:822–834

Feng L, Wang K, Li Y et al (2007b) Overexpression of SBPase enhances photosynthesis against high temperature stress in transgenic rice plants. Plant Cell Rep 26:1635–1646

Fernie AR, Yan J (2019) De novo domestication: an alternative route toward new crops for the future. Mol Plant 12:615–631

Fernie AR, Carrari F, Sweetlove LJ (2004) Respiratory metabolism: glycolysis, the TCA cycle and mitochondrial electron transport. Curr Opin Plant Biol 7:254–261

Figueroa P, León G, Elorza A et al (2001) Three different genes encode the iron-sulfur subunit of succinate dehydrogenase in *Arabidopsis thaliana*. Plant Mol Biol 46:241–250

Filatov DA, Kapralov MV, Kubien DS, Andersson I (2010) Changes in rubisco kinetics during the evolution of C4 photosynthesis in flaveria (Asteraceae) are associated with positive selection on genes encoding the enzyme. Mol Biol Evol 28:1491–1503

Florez-Sarasa I, Flexas J, Rasmusson AG et al (2011) In vivo cytochrome and alternative pathway respiration in leaves of *Arabidopsis thaliana* plants with altered alternative oxidase under different light conditions. Plant Cell Environ 34:1373–1383

Foyer CH, Noctor G, Hodges M (2011) Respiration and nitrogen assimilation: targeting mitochondria-associated metabolism as a means to enhance nitrogen use efficiency. J Exp Bot 62:1467–1482. https://doi.org/10.1093/jxb/erq453

Foyer CH, Ruban AV, Nixon PJ (2017) Photosynthesis solutions to enhance productivity. Philos Trans R Soc 372:20160374

Fromm S, Braun H, Peterhansel C (2016a) Mitochondrial gamma carbonic anhydrases are required for complex I assembly and plant reproductive development. New Phytol 211:194–207

Fromm S, Göing J, Lorenz C et al (2016b) Depletion of the "gamma-type carbonic anhydrase-like" subunits of complex I affects central mitochondrial metabolism in *Arabidopsis thaliana*. Biochim Biophys Acta 1857:60–71

Fromm S, Senkler J, Eubel H et al (2016c) Life without complex I: proteome analyses of an Arabidopsis mutant lacking the mitochondrial NADH dehydrogenase complex. J Exp Bot 67:3079–3093

Fuentes D, Meneses M, Nunes-Nesi A et al (2011) A deficiency in the flavoprotein of Arabidopsis mitochondrial complex II results in elevated photosynthesis and better growth in nitrogen-limiting conditions. Plant Physiol 157:1114–1127

Fuentes P, Armarego-Marriott T, Bock R (2018) Plastid transformation and its application in metabolic engineering. Curr Opin Biotechnol 49:10–15

Gardeström P, Igamberdiev AU, Raghavendra AS (2002) Mitochondrial functions in the light and significance to carbon-nitrogen interactions. In: Foyer CH, Noctor G (eds) Photosynthetic nitrogen assimilation and associated carbon and respiratory metabolism. Advances in photosynthesis and respiration, vol 12. Springer, Dordrecht

Gauthier PPG, Bligny R, Gout E et al (2010) In folio isotopic tracing demonstrates that nitrogen assimilation into glutamate is mostly independent from current CO_2 assimilation in illuminated leaves of *Brassica napus*. New Phytol 185:988–999

Genkov T, Meyer M, Griffiths H, Spreitzer RJ (2010) Functional hybrid rubisco enzymes with plant small subunits and algal large subunits engineered rbcS cDNA for expression in Chlamydomonas. J Biol Chem 285:19833–19841

Gifford RM (2003) Plant respiration in productivity models : conceptualisation, representation and issues for global terrestrial carbon-cycle research. Funct Plant Biol 30:171–186

Goldschmidt-Clermont M (1997) Coordination of nuclear and chloroplast gene expression in plant cells. Int Rev Cytol 177:115–180

Guo L, Devaiah SP, Narasimhan R et al (2012) Cytosolic glyceraldehyde-3-phosphate dehydrogenases interact with phospholipase Dδ to transduce hydrogen peroxide signals in the Arabidopsis response to stress. Plant Cell 24:2200–2212

Gutierres S, Sabar M, Lelandais C et al (2002) Lack of mitochondrial and nuclear-encoded subunits of complex I and alteration of the respiratory chain in *Nicotiana sylvestris* mitochondrial deletion mutants. Proc Natl Acad Sci U S A 94:3436–3441

Hagemann M, Bauwe H (2016) Photorespiration. In: Encyclopedia of applied plant sciences. American Society of Plant Biologists, pp 86–89

Hagerhall C (1997) Succinate: quinone oxidoreductases variations on a conserved theme. Biochim Biophys Acta 1320:107–141

Haimovich-Dayan M, Lieman-Hurwitz J, Orf I et al (2015) Does 2-phosphoglycolate serve as an internal signal molecule of inorganic carbon deprivation in the cyanobacterium *Synechocystis* sp. PCC 6803? Environ Microbiol 17:1794–1804

Hanson AD, Hibberd JM, Koffas MAG et al (2019) Focus issue editorial: synthetic biology. Plant Physiol 179:772–774

Hao MS, Rasmusson AG (2016) The evolution of substrate specificity-associated residues and Ca^{2+}-binding motifs in EF-hand-containing type II NAD(P)H dehydrogenases. Physiol Plant 157:338–351

Harbinson J (2012) Modeling the protection of photosynthesis. Proc Natl Acad Sci U S A 109:15533–15534

Hayashi M, Debellis L, Alpi A, Nishimura M (1995) Cytosolic aconitase participates in the glyoxylate cycle in etiolated pumpkin cotyledons. Plant Cell Physiol 36:669–680

Herzig S, Raemy E, Montessuit S et al (2012) Identification and functional expression of the mitochondrial pyruvate carrier. Science 337:93–96

Heyneke E, Fernie AR (2018) Metabolic regulation of photosynthesis. Biochem Soc Trans 46:321–328

Ho M-Y, Shen G, Canniffe DP et al (2016) Light-dependent chlorophyll f synthase is a highly divergent paralog of PsbA of photosystem II. Science 353

Hodges M (2002) Enzyme redundancy and the importance of 2-oxoglutarate in plant ammonium assimilation. Agronomie 53:905–916

Hodges M, Dellero Y, Keech O et al (2016) Perspectives for a better understanding of the metabolic integration of photorespiration within a complex plant primary metabolism network. J Exp Bot 67:3015–3026

Hohmann-Marriott MF, Blankenship RE (2011) Evolution of photosynthesis. Annu Rev Plant Biol 62:515–548

Husic DW, Husic HD, Tolbert NE, Black CC (1987) The oxidative photosynthetic carbon cycle or C2 cycle. CRC Crit Rev Plant Sci 5:45–100

Igamberdiev AU, Bykova NV, Gardeström P (1997) Involvement of cyanide-resistant and rotenone-insensitive pathways of mitochondrial electron transport during oxidation of glycine in higher plants. FEBS Lett 412:265–269

Ishikawa C, Hatanaka T, Misoo S et al (2011) Functional incorporation of sorghum small subunit increases the catalytic turnover rate of rubisco in transgenic rice. Plant Physiol 156:1604–1611

Jardim-Messeder D, Caverzan A, Rauber R et al (2015) Succinate dehydrogenase (mitochondrial complex II) is a source of reactive oxygen species in plants and regulates development and stress responses. New Phytol 208:776–789

Joliot P, Johnson GN (2011) Regulation of cyclic and linear electron flow in higher plants. Proc Natl Acad Sci U S A 108:13317–13322

Kebeish R, Niessen M, Thiruveedhi K et al (2007) Chloroplastic photorespiratory bypass increases photosynthesis and biomass production in *Arabidopsis thaliana*. Nat Biotechnol 25:593–599

Kelly GJ, Gibbs M (1973) Nonreversible D-glyceraldehyde 3-phosphate dehydrogenase of plant tissues. Plant Physiol 52:111–118

Kennedy RA, Laetsch WM (1974) Plant species intermediate for C3, C4 photosynthesis. Science 184:1087–1089

Kirst H, Gabilly ST, Niyogi KK et al (2017) Photosynthetic antenna engineering to improve crop yields. Planta 245:1009–1020

Kirst H, Shen Y, Vamvaka E et al (2018) Downregulation of the CpSRP43 gene expression confers a truncated light-harvesting antenna (TLA) and enhances biomass and leaf-to-stem ratio in *Nicotiana tabacum* canopies. Planta 248:139–154

Klähn S, Orf I, Schwarz D et al (2015) Integrated transcriptomic and metabolomic characterization of the low-carbon response using an ndhR mutant of *Synechocystis* sp. PCC 6803. Plant Physiol 169:1540–1556

Kopka J, Fernie AR (2018) Editorial overview: plant synthetic and systems biology. Curr Opin Biotechnol 49:viii–xi

Kromdijk J, Głowacka K, Leonelli L et al (2016) Improving photosynthesis and crop productivity by accelerating recovery from photoprotection. Science 354

Kromer S (1995) Respiration during photosynthesis. Annu Rev Plant Physiol Plant Mol Biol 46:45–70

Kruger NJ, Von Schaewen A (2003) The oxidative pentose phosphate pathway: structure and organisation. Curr Opin Plant Biol 6:236–246

Kubis A, Bar-Even A (2019) Synthetic biology approaches for improving photosynthesis. J Exp Bot 70:1425–1433

Kühn K, Obata T, Feher K et al (2015a) Complete mitochondrial complex I deficiency induces an up-regulation of respiratory fluxes that is abolished by traces of functional complex I. Plant Physiol 168:1537–1549

Kühn K, Yin G, Duncan O et al (2015b) Decreasing electron flux through the cytochrome and/or alternative respiratory pathways triggers common and distinct cellular responses dependent on growth conditions. Plant Physiol 167:228–250

Külheim C, Ågren J, Jansson S (2002) Rapid regulation of light harvesting and plant fitness in the field. Science 297:91–93

Kurz S, Mettler-Altmann T, Schlüter U et al (2016) Photosynthesis in C3–C4 intermediate *Moricandia* species. J Exp Bot 68:191–206

Laloi M (1999) Plant mitochondrial carriers: an overview. Cell Mol Life Sci 56:918–944

Lefebvre S, Lawson T, Fryer M et al (2005) Increased sedoheptulose-1,7-bisphosphatase activity in transgenic tobacco plants stimulates photosynthesis and growth from an early stage in development. Plant Physiol 138:451–460

Leister D (2012) How can the light reactions of photosynthesis be improved in plants? Front Plant Sci 3:1–3

Leister D (2019) Genetic engineering, synthetic biology and the light reactions of photosynthesis. Plant Physiol 179:778–793

Lejay L, Wirth J, Pervent M et al (2008) Oxidative pentose phosphate pathway-dependent sugar sensing as a mechanism for regulation of root ion transporters by photosynthesis. Plant Physiol 146:2036–2053

Li Z, Wakao S, Fischer BB, Niyogi KK (2009) Sensing and responding to excess light. Annu Rev Plant Biol 60:239–260

Lin MT, Occhialini A, Andralojc PJ et al (2014) A faster Rubisco with potential to increase photosynthesis in crops. Nature 513:547–550

Lindén P, Keech O, Stenlund H et al (2016) Reduced mitochondrial malate dehydrogenase activity has a strong effect on photorespiratory metabolism as revealed by ^{13}C labelling. J Exp Bot 67:3123–3135

Liu Y, Norberg F, Szilágyi A et al (2008) The mitochondrial external NADPH dehydrogenase modulates the leaf NADPH/$NADP^+$ ratio in transgenic Nicotiana sylvestris. Plant Cell Physiol 49:251–263

Liu Y-J, Nunes-Nesi A, Wallström SV et al (2009) A redox-mediated modulation of stem bolting in transgenic Nicotiana sylvestris differentially expressing the external mitochondrial NADPH dehydrogenase. Plant Physiol 150:1248–1259

Long SP, Humphries S, Falkowski PG (1994) Photoinhibition of photosynthesis in nature. Annu Rev Plant Physiol Plant Mol Biol 45:633–662

Long SP, Zhu X-G, Naidu SL, Ort DR (2006) Can improvement in photosynthesis increase crop yields? Plant Cell Environ 29:315–330

Long SP, Marshall-Colon A, Zhu XG (2015) Meeting the global food demand of the future by engineering crop photosynthesis and yield potential. Cell 161:56–66

Lorimer GH (1981) The carboxylation and oxygenation of ribulose 1,5-bisphosphate: the primary events in photosynthesis and photorespiration. Annu Rev Plant Physiol 32:349–382

Lothier J, De Paepe R, Tcherkez G (2019) Mitochondrial complex I dysfunction increases CO_2 efflux and reconfigures metabolic fluxes of day respiration in tobacco leaves. New Phytol 221:750–763

Lu Y, Stegemann S, Agrawal S et al (2017) Horizontal transfer of a synthetic metabolic pathway between plant species. Curr Biol 27:3034–3041.e3

Maclean AE, Hertle AP, Ligas J et al (2018) Absence of complex I is associated with diminished respiratory chain function in European mistletoe. Current Biol 28:1614–1619.e3

Maier A, Fahnenstich H, von Caemmerer S et al (2012) Transgenic introduction of a glycolate oxidative cycle into *A. thaliana* chloroplasts leads to growth improvement. Front Plant Sci 3:38

Makino A, Tsunoda H, Izumi M et al (2012) RBCS1A and RBCS3B, two major members within the Arabidopsis RBCS multigene family, function to yield sufficient Rubisco content for leaf photosynthetic capacity. J Exp Bot 63:2159–2170
Martin W, Scheibe R, Schnarrenberger C (2006) The Calvin cycle and its regulation. Springer, Netherlands
Martinoia E, Rentsch D (1995) Malate compartmentation-responses to a complex metabolism. Annu Rev Plant Physiol 45:447–467
Maurino VG, Peterhänsel C (2010) Photorespiration: current status and approaches for metabolic engineering. Curr Opin Plant Biol 13:248–255
Maurino VG, Weber APM (2013) Engineering photosynthesis in plants and synthetic microorganisms. J Exp Bot 64:743–751
Maxwell DP, Wang Y, McIntosh L (1999) The alternative oxidase lowers mitochondrial reactive oxygen production in plant cells. Proc Natl Acad Sci U S A 96:8271–8276
Mazor Y, Borovikova A, Caspy I, Nelson N (2017) Structure of the plant photosystem I supercomplex at 2.6 Å resolution. Nat Plants 3:17014
McDonald AE, Ivanov AG, Bode R et al (2011) Flexibility in photosynthetic electron transport: the physiological role of plastoquinol terminal oxidase (PTOX). Biochim Biophys Acta 1807:954–967
Medlyn BE (1998) Physiological basis of the light use efficiency model. Tree Physiol 18:167–176
Melo AMP, Roberts TH, Moiler M (1996) Evidence for the presence of two rotenone-insensitive NAD(P)H dehydrogenases on the inner surface of the inner membrane of potato tuber mitochondria. Biochim Biophys Acta 2728:133–139
Meyers B, Zaltsman A, Lacroix B et al (2010) Nuclear and plastid genetic engineering of plants: comparison of opportunities and challenges. Biotechnol Adv 28:747–756
Michalecka AM, Svensson ÅS, Johansson FI et al (2003) Arabidopsis genes encoding mitochondrial type II NAD(P)H dehydrogenases have different evolutionary origin and show distinct responses to light. Plant Physiol 133:642–652
Michalecka AM, Agius SC, Mùller IM, Rasmusson AG (2004) Identication of a mitochondrial external NADPH dehydrogenase by overexpression in transgenic Nicotiana sylvestris. Plant J 37:415–425
Millar AH, Mittova V, Kiddle G et al (2003) Control of ascorbate synthesis by respiration and its implications for stress responses. Plant Physiol 133:443–447
Millar AH, Whelan J, Soole KL, Day DA (2011) Organization and regulation of mitochondrial respiration in plants. Annu Rev Plant Biol 62:79–104
Miyaji T, Kuromori T, Takeuchi Y et al (2015) AtPHT4;4 is a chloroplast-localized ascorbate transporter in Arabidopsis. Nat Commun 6:1–11
Møller IM (2001) Plant mitochondria and oxidative stress: electron transport, NADPH turnover, and metabolism of reactive oxygen species. Annu Rev Plant Physiol Plant Mol Biol 52:561–591
Moore CS, Cook-Johnson RJ, Rudhe C et al (2003) Identification of AtNDI1, an internal non-phosphorylating NAD(P)H dehydrogenase in Arabidopsis mitochondria. Plant Physiol 133:1968–1978
Morel FMM, Young JN, Heureux AMC et al (2016) Large variation in the Rubisco kinetics of diatoms reveals diversity among their carbon-concentrating mechanisms. J Exp Bot 67:3445–3456
Moses T (2019) Shedding light on the power of light. Plant Physiol 179:775–777
Murchie EH, Niyogi KK (2011) Manipulation of photoprotection to improve plant photosynthesis. Plant Physiol 155:86–92
Niklas KJ (2016) Plant evolution: an introduction to the history of life. University of Chicago Press, Chicago
Noctor G, Dutilleul C, De Paepe R et al (2004) Use of mitochondrial electron transport mutants to evaluate the effects of redox state on photosynthesis, stress tolerance and the integration of carbon/nitrogen metabolism. J Exp Bot 55:49–57
Noctor G, De Paepe R, Foyer CH (2007) Mitochondrial redox biology and homeostasis in plants. Trends Plant Sci 12:125–134

Noguchi K, Yoshida K (2008) Interaction between photosynthesis and respiration in illuminated leaves. Mitochondrion 8:87–99
Nölke G, Houdelet M, Kreuzaler F et al (2014) The expression of a recombinant glycolate dehydrogenase polyprotein in potato (*Solanum tuberosum*) plastids strongly enhances photosynthesis and tuber yield. Plant Biotechnol J 12:734–742
Nowicka B, Ciura J, Szymańska R, Kruk J (2018) Improving photosynthesis, plant productivity and abiotic stress tolerance—current trends and future perspectives. J Plant Physiol 231:415–433
Ntagkas N, Woltering EJ, Marcelis LFM (2018) Light regulates ascorbate in plants: an integrated view on physiology and biochemistry. Environ Exp Bot 147:271–280
Ntagkas N, Woltering E, Nicole C et al (2019) Light regulation of vitamin C in tomato fruit is mediated through photosynthesis. Environ Exp Bot 158:180–188
Nunes-Nesi A, Carrari F, Lytovchenko A et al (2005a) Enhanced photosynthetic performance and growth as a consequence of decreasing mitochondrial malate dehydrogenase activity in transgenic tomato plants. Plant Physiol 137:611–622
Nunes-Nesi A, Carrari F, Lytovchenko A, Fernie AR (2005b) Enhancing crop yield in Solanaceous species through the genetic manipulation of energy metabolism. Biochem Soc Trans 33:1430–1434
Nunes-Nesi A, Carrari F, Gibon Y et al (2007) Deficiency of mitochondrial fumarase activity in tomato plants impairs photosynthesis via an effect on stomatal function. Plant J 50:1093–1106
Nunes-Nesi A, Sulpice R, Gibon Y, Fernie AR (2008) The enigmatic contribution of mitochondrial function in photosynthesis. J Exp Bot 59:1675–1684
Nunes-Nesi A, Araujo WL, Fernie AR (2011) Targeting mitochondrial metabolism and machinery as a means to enhance photosynthesis. Plant Physiol 155:101–107
Nunes-Nesi A, Nascimento VDL, Magnum F et al (2016) Natural genetic variation for morphological and molecular determinants of plant growth and yield. J Exp Bot 67:2989–3001
O'Leary BM, Asao S, Millar AH, Atkin OK (2018) Core principles which explain variation in respiration across biological scales. New Phytol:670–686
Obata T, Florian A, Timm S et al (2016) On the metabolic interactions of (photo)respiration. J Exp Bot 67:3003–3014
Occhialini A, Lin MT, Andralojc PJ et al (2016) Transgenic tobacco plants with improved cyanobacterial Rubisco expression but no extra assembly factors grow at near wild-type rates if provided with elevated CO_2. Plant J 85:148–160
Ogawa S, Suzuki Y, Yoshizawa R et al (2012) Effect of individual suppression of RBCS multigene family on Rubisco contents in rice leaves. Plant Cell Environ 35:546–553
Oji Y, Watanabe M, Wakiuchi N, Okamoto S (1985) Nitrite reduction in barley-root plastids: dependence on NADPH coupled with glucose-6-phosphate and 6-phosphogluconate dehydrogenases, and possible involvement of an electron carrier and a diaphorase. Planta 165:85–90
Orr DJ, Pereira AM, da Fonseca PP et al (2017) Engineering photosynthesis: progress and perspectives. F1000 Res 6:1891
Ort DR, Merchant SS, Alric J et al (2015) Redesigning photosynthesis to sustainably meet global food and bioenergy demand. Proc Natl Acad Sci U S A 112:8529–8536
Padmasree K, Raghavendra AS (1999a) Importance of oxidative electron transport over oxidative phosphorylation in optimizing photosynthesis in mesophyll protoplasts of pea (*Pisum sativum* L.). Physiol Plant 105:546–553
Padmasree K, Raghavendra AS (1999b) Response of photosynthetic carbon assimilation in mesophyll protoplasts to restriction on mitochondrial oxidative metabolism: metabolites related to the redox status and sucrose biosynthesis. Photosynth Res 62:231–239
Padmasree K, Padmavathi L, Raghavendra AS (2002) Essentiality of mitochondrial oxidative metabolism for photosynthesis: optimization of carbon assimilation and protection against photoinhibition. Crit Rev Biochem Mol Biol 37:71–119
Palmieri F, Pierri CL, De Grassi A et al (2011) Evolution, structure and function of mitochondrial carriers: a review with new insights. Plant J 66:161–181
Parry MAJ, Andralojc PJ, Scales JC et al (2013) Rubisco activity and regulation as targets for crop improvement. J Exp Bot 64:717–730

Paul MJ, Pellny TK (2003) Carbon metabolite feedback regulation of leaf photosynthesis and development. J Exp Bot 54:539–547
Peterhansel C, Horst I, Niessen M et al (2010) Photorespiration. Arabidopsis Book 8:e0130. https://doi.org/10.1199/tab.0130
Peterhänsel C, Niessen M, Kebeish RM (2008) Metabolic engineering towards the enhancement of photosynthesis. Photochem Photobiol 84:1317–1323
Peterhänsel C, Blume C, Offermann S (2012) Photorespiratory bypasses: how can they work? J Exp Bot 64:709–715
Peterhänsel C, Krause K, Braun HP et al (2013) Engineering photorespiration: current state and future possibilities. Plant Biol 15:754–758
Petersen G, Cuenca A, Møller IM, Seberg O (2015) Massive gene loss in mistletoe (*Viscum*, Viscaceae) mitochondria. Sci Rep 5:1–7
Piatek AA, Lenaghan SC, Neal Stewart C (2018) Advanced editing of the nuclear and plastid genomes in plants. Plant Sci 273:42–49
Piattoni CV, Guerrero SA, Iglesias AA (2013) A differential redox regulation of the pathways metabolizing glyceraldehyde-3-phosphate tunes the production of reducing power in the cytosol of plant cells. Int J Mol Sci 14:8073–8092
Picault N, Hodges M, Palmieri L, Palmieri F (2004) The growing family of mitochondrial carriers in Arabidopsis. Trends Plant Sci 9:138–146
Pineau B, Mathieu C, Gérard-Hirne C et al (2005) Targeting the NAD7 subunit to mitochondria restores a functional complex I and a wild type phenotype in the *Nicotiana sylvestris* CMS II mutant lacking nad7. J Biol Chem 280:25994–26001
Plaxton WC (1996) The organization and regulation of plant glycolysis. Annu Rev Plant Physiol Plant Mol Biol 47:185–214
Plaxton WC, Podestá FE (2006) The functional organization and control of plant respiration. CRC Crit Rev Plant Sci 25:159–198
Pottier M, Gilis D, Boutry M (2018) The hidden face of Rubisco. Trends Plant Sci 23:382–392
Pracharoenwattana I, Zhou W, Keech O et al (2010) Arabidopsis has a cytosolic fumarase required for the massive allocation of photosynthate into fumaric acid and for rapid plant growth on high nitrogen. Plant J 62:785–795
Priault P, Tcherkez G, Cornic G et al (2006) The lack of mitochondrial complex I in a CMSII mutant of *Nicotiana sylvestris* increases photorespiration through an increased internal resistance to CO_2 diffusion. J Exp Bot 57:3195–3207
Qin X, Suga M, Kuang T, Shen J-R (2015) Structural basis for energy transfer pathways in the plant PSI-LHCI supercomplex. Science 348:989–995
Raghavendra AS, Padmasree K (2003) Beneficial interactions of mitochondrial metabolism with photosynthetic carbon assimilation. Trends Plant Sci 8:546–553
Raines CA (2003) The Calvin cycle revisited. Photosynth Res 75:1–10
Raines CA (2011) Increasing photosynthetic carbon assimilation in C3 plants to improve crop yield: current and future strategies. Plant Physiol 155:36–42
Raines CA, Harrison EP, Olcer H et al (2001) Small decreases in SBPase cause a linear decline in the apparent RuBP regeneration rate, but do not affect Rubisco carboxylation capacity. J Exp Bot 52:1779–1784
Rasmusson AG, Møller IM (2010) Mitochondrial electron transport and plant stress. In: Plant mitochondria. Springer, New York, pp 357–382
Rasmusson AG, Heiser V, Zabaleta E, Brennicke A (1998) Physiological, biochemical and molecular aspects of mitochondrial complex I in plants. Biochim Biophys Acta 1364:101–111
Rasmusson AG, Svensson ÅS, Knoop V et al (1999) Homologues of yeast and bacterial rotenone-insensitive NADH dehydrogenases in higher eukaryotes: two enzymes are present in potato mitochondria. Plant J 20:79–87
Rasmusson AG, Soole KL, Elthon TE (2004) Alternative NAD(P)H dehydrogenases of plant mitochondria. Annu Rev Plant Biol 55:23–39
Rasmusson AG, Geisler DA, Møller IM (2008) The multiplicity of dehydrogenases in the electron transport chain of plant mitochondria. Mitochondrion 8:47–60

Raven JA (2013) Rubisco: still the most abundant protein of earth? Rubisco content of algae and plants. New Phytol 198:1–3
Raven JA, Allen JF (2003) Genomics and chloroplast evolution: what did cyanobacteria do for plants? Genome Biol 4:209
Ricquier D, Boillaud F (2000) Mitochondrial uncoupling proteins: from mitochondria to the regulation of energy balance. J Physiol:3–10
Rius P, Casati P, Iglesias AA, Gomez-Casati DF (2006) Characterization of an *Arabidopsis thaliana* mutant lacking a cytosolic non-phosphorylating glyceraldehyde-3-phosphate dehydrogenase. Plant Mol Biol 61:945–957
Rius P, Casati P, Iglesias AA, Gomez-casati DF (2008) Characterization of Arabidopsis lines deficient in GAPC-1, a cytosolic NAD-dependent gluceraldeyde-3-phosphate dehydrogenase. Plant Physiol 148:1655–1667
Roach T, Krieger-Liszkay A (2014) Regulation of photosynthetic electron transport and photoinhibition. Curr Protein Pept Sci 15:351–362
Rosenthal DM, Locke AM, Khozaei M et al (2011) Over-expressing the C3 photosynthesis cycle enzyme sedoheptulose-1-7 bisphosphatase improves photosynthetic carbon gain and yield under fully open air CO_2 fumigation (FACE). BMC Plant Biol 11:123
Sabar M, De Paepe R, de Kouchkovsky Y (2000) Complex I impairment, respiratory compensations, and photosynthetic decrease in nuclear and mitochondrial male sterile mutants of *Nicotiana sylvestris*. Plant Physiol 124:1239–1250
Sage RF (2004) The evolution of C4 photosynthesis. New Phytol 161:341–370
Sage RF (2016) A portrait of the C4 photosynthetic family on the 50th anniversary of its discovery: species number, evolutionary lineages, and hall of fame. J Exp Bot 67:4039–4056
Sage RF, Edwards EJ, Christin P-A (2011) The C4 plant lineages of planet earth. J Exp Bot 62:3155–3169
Scheibe R (2004) Malate valves to balance cellular energy supply. Physiol Plant 120:21–26
Scheibe R (2019) Maintaining homeostasis by controlled alternatives for energy distribution in plant cells under changing conditions of supply and demand. Photosynth Res 139:81–91
Schimmeyer J, Bock R, Meyer EH (2016) L-Galactono-1,4-lactone dehydrogenase is an assembly factor of the membrane arm of mitochondrial complex I in Arabidopsis. Plant Mol Biol 90:117–126
Schlüter U, Weber AP (2016) The road to C4 photosynthesis: evolution of a complex trait via intermediary states. Plant Cell Physiol 57:881–889
Schuler ML, Mantegazza O, Weber APM (2016) Engineering C4 photosynthesis into C3 chassis in the synthetic biology age. Plant J 87:51–65
Selinski J, Scheibe R (2014) Lack of malate valve capacities lead to improved N-assimilation and growth in transgenic *A. thaliana* plants. Plant Signal Behav 9:e29057
Selinski J, Scheibe R (2019) Malate valves: old shuttles with new perspectives. Plant Biol 21:21–30
Selinski J, Scheibe R, Day DA, Whelan J (2018) Alternative oxidase is positive for plant performance. Trends Plant Sci 23:588–597
Senkler J, Senkler M, Braun H-P (2017) Structure and function of complex I in animals and plants—a comparative view. Physiol Plant 161:6–15
Senkler J, Rugen N, Eubel H et al (2018) Absence of complex I implicates rearrangement of the respiratory chain in european mistletoe. Current Biol 28:1606–1613.e4
Sessions A, Burke E, Presting G et al (2002) A high-throughput Arabidopsis reverse genetics system. Plant Cell 14:2985–2994
Sheehy JE, Ferrer AB, Mitchell PL et al (2007) How the rice crop works and why it needs a new engine. In: Charting new pathways to C4 rice. International Rice Research Institute. World Scientific, Los Banos, pp 3–26
Shen B-R, Wang L-M, Lin X-L et al (2019) Engineering a new chloroplastic photorespiratory bypass to increase photosynthetic efficiency and productivity in rice. Mol Plant 12:199–214
Shi J, Yi K, Liu Y et al (2015) Phosphoenolpyruvate carboxylase in arabidopsis leaves plays a crucial role in carbon and nitrogen metabolism. Plant Physiol 167:671–681
Shikanai T (2014) Central role of cyclic electron transport around photosystem I in the regulation of photosynthesis. Curr Opin Biotechnol 26:25–30

Sienkiewicz-Porzucek A, Nunes-Nesi A, Sulpice R et al (2008) Mild reductions in mitochondrial citrate synthase activity result in a compromised nitrate assimilation and reduced leaf pigmentation but have no effect on photosynthetic performance or growth. Plant Physiol 147:115–127
Simkin AJ, McAusland L, Headland LR et al (2015) Multigene manipulation of photosynthetic carbon assimilation increases CO_2 fixation and biomass yield in tobacco. J Exp Bot 66:4075–4090
Simkin AJ, Lopez-Calcagno PE, Davey PA et al (2017) Simultaneous stimulation of sedoheptulose 1,7-bisphosphatase, fructose 1,6-bisphosphate aldolase and the photorespiratory glycine decarboxylase-H protein increases CO_2 assimilation, vegetative biomass and seed yield in Arabidopsis. Plant Biotechnol J 15:805–816
Smith C, Barthet M, Melino V et al (2011) Alterations in the mitochondrial alternative NAD(P)H dehydrogenase NDB4 lead to changes in mitochondrial electron transport chain composition, plant growth and response to oxidative stress. Plant Cell Physiol 52:1222–1237
Sonnewald U, Fernie AR (2018) Next-generation strategies for understanding and influencing source—sink relations in crop plants. Curr Opin Plant Biol 43:63–70
Soto D, Córdoba JP, Villarreal F et al (2015) Functional characterization of mutants affected in the carbonic anhydrase domain of the respiratory complex I in *Arabidopsis thaliana*. Plant J 83:831–844
Stitt M (2013) Progress in understanding and engineering primary plant metabolism. Curr Opin Biotechnol 24:229–238
Strodtkötter I, Padmasree K, Challabathula D et al (2009) Induction of the AOX1D isoform of alternative oxidase in *A. thaliana* T-DNA insertion lines lacking isoform AOX1A is insufficient to optimize photosynthesis when treated with antimycin A. Mol Plant 2:284–287
Studart-Guimaraes C, Fait A, Nunes-Nesi A et al (2007) Reduced expression of succinyl-coenzyme A ligase can be compensated for by up-regulation of the -aminobutyrate shunt in illuminated tomato leaves. Plant Physiol 145:626–639
Su X, Ma J, Wei X et al (2017) Structure and assembly mechanism of plant C2S2M2-type PSII-LHCII supercomplex. Science 357:815–820
Svensson ÊS, Rasmusson AG (2001) Light-dependent gene expression for proteins in the respiratory chain of potato leaves. Plant J 28
Sweetlove LJ, Lytovchenko A, Morgan M et al (2006) Mitochondrial uncoupling protein is required for efficient photosynthesis. Proc Natl Acad Sci U S A 103:19587–19592
Sweetlove LJ, Nielsen J, Fernie AR (2017) Engineering central metabolism - a grand challenge for plant biologists. Plant J 90:749–763
Sweetlove LJ, Beard KFM, Nunes-Nesi A et al (2010) Not just a circle: flux modes in the plant TCA cycle. Trends Plant Sci 15:462–470
Szabó I, Bergantino E, Giacometti GM (2005) Light and oxygenic photosynthesis: energy dissipation as a protection mechanism against photo-oxidation. EMBO Rep 6:629–634
Taniguchi M, Miyake H (2012) Redox-shuttling between chloroplast and cytosol : integration of intra-chloroplast and extra-chloroplast metabolism. Curr Opin Plant Biol 15:252–260
Taylor L, Nunes-Nesi A, Parsley K et al (2010) Cytosolic pyruvate, orthophosphate dikinase functions in nitrogen remobilization during leaf senescence and limits individual seed growth and nitrogen content. Plant J 62:641–652
Tcherkez G, Cornic G, Bligny R et al (2005) *In vivo* respiratory metabolism of illuminated leaves. Plant Physiol 138:1596–1606
Tcherkez G, Bligny R, Gout E et al (2008) Respiratory metabolism of illuminated leaves depends on CO_2 and O_2 conditions. Proc Natl Acad Sci U S A 105:797–802
Tcherkez G, Boex-Fontvieille E, Mahé A, Hodges M (2012) Respiratory carbon fluxes in leaves. Curr Opin Plant Biol 15:308–314
Tcherkez G, Gauthier P, Buckley TN et al (2017) Leaf day respiration: low CO_2 flux but high significance for metabolism and carbon balance. New Phytol 216
Timm S, Bauwe H (2013) The variety of photorespiratory phenotypes—employing the current status for future research directions on photorespiration. Plant Biol 15:737–747
Timm S, Florian A, Fernie AR, Bauwe H (2016) The regulatory interplay between photorespiration and photosynthesis. J Exp Bot 67:2923–2929

Timmis JN, Ayliffe MA, Huang CY, Martin W (2004) Endosymbiotic gene transfer: organelle genomes forge eukaryotic chromosomes. Nat Rev Genet 5:123

Tschoep H, Gibon Y, Carillo P et al (2009) Adjustment of growth and central metabolism to a mild but sustained nitrogen-limitation in Arabidopsis. Plant Cell Environ 32:300–318

Uemura K, Anwaruzzaman MS, Yokota A (1997) Ribulose-1,5-bisphosphate carboxylase/oxygenase from thermophilic red algae with a strong specificity for CO_2 fixation. Biochem Biophys Res Commun 233:568–571

Urbanczyk-Wochniak E, Usadel B, Thimm O et al (2006) Conversion of MapMan to allow the analysis of transcript data from Solanaceous species: effects of genetic and environmental alterations in energy metabolism in the leaf. Plant Mol Biol 60:773–792

Van Der Merwe MJ, Osorio S, Moritz T et al (2009) Decreased mitochondrial activities of malate dehydrogenase and fumarase in tomato lead to altered root growth and architecture via diverse mechanisms. Plant Physiol 149:653–669

van Dongen JT, Gupta KJ, Ramírez-Aguilar SJ et al (2011) Regulation of respiration in plants: a role for alternative metabolic pathways. J Plant Physiol 168:1434–1443

Villarreal F, Martín V, Colaneri A et al (2009) Ectopic expression of mitochondrial gamma carbonic anhydrase 2 causes male sterility by anther indehiscence. Plant Mol Biol 70:471–485

Vishwakarma A, Bashyam L, Senthilkumaran B et al (2014) Physiological role of AOX1a in photosynthesis and maintenance of cellular redox homeostasis under high light in *Arabidopsis thaliana*. Plant Physiol Biochem 81:44–53

Vishwakarma A, Tetali SD, Selinski J et al (2015) Importance of the alternative oxidase (AOX) pathway in regulating cellular redox and ROS homeostasis to optimize photosynthesis during restriction of the cytochrome oxidase pathway in *Arabidopsis thaliana*. Ann Bot 116:555–569

Voll LM, Hajirezaei MR, Czogalla-Peter C et al (2009) Antisense inhibition of enolase strongly limits the metabolism of aromatic amino acids, but has only minor effects on respiration in leaves of transgenic tobacco plants. New Phytol 184:607–618

von Caemmerer S, Evans JR (2010) Enhancing C3 photosynthesis. Plant Physiol 154:589–592

von Caemmerer S, Furbank RT (2016) Strategies for improving C4 photosynthesis. Curr Opin Plant Biol 31:125–134

von Caemmerer S, Quick WP, Furbank RT (2012) The development of C4 rice: current progress and future challenges. Science 336:1671–1672

Walker BJ, Drewry DT, Slattery RA et al (2018) Chlorophyll can be reduced in crop canopies with little penalty to photosynthesis. Plant Physiol 176:1215–1232

Wallström SV, Florez-Sarasa I, Araújo WL et al (2014a) Suppression of NDA-type alternative mitochondrial NAD(P)H dehydrogenases in *Arabidopsis thaliana* modifies growth and metabolism, but not high light stimulation of mitochondrial electron transport. Plant Cell Physiol 55:881–896

Wallström SV, Florez-Sarasa I, Araújo WL et al (2014b) Suppression of the external mitochondrial NADPH dehydrogenase, NDB1, in *Arabidopsis thaliana* affects central metabolism and vegetative growth. Mol Plant 7:356–368

Wang Q, Fristedt R, Yu X et al (2012) The γ-carbonic anhydrase subcomplex of mitochondrial complex I is essential for development and important for photomorphogenesis of Arabidopsis. Plant Physiol 160:1373–1383

Wang P, Vlad D, Langdale JA (2016) Finding the genes to build C4 rice. Curr Opin Plant Biol 31:44–50

Weber APM, Bar-Even A (2019) Update on improved carbon fixation update: improving the efficiency of photosynthetic carbon reactions. Plant Physiol 179:803–812

Westram A, Lloyd JR, Roessner U et al (2002) Increases of 3-phosphoglyceric acid in potato plants through antisense reduction of cytoplasmic phosphoglycerate mutase impairs photosynthesis and growth, but does not increase starch contents. Plant Cell Environ 25:1133–1143

Whitney SM, Houtz RL, Alonso H (2011) Advancing our understanding and capacity to engineer nature's CO_2-sequestering enzyme, Rubisco. Plant Physiol 155:27–35

Xu F, Yuan S, Lin HH (2011) Response of mitochondrial alternative oxidase (AOX) to light signals. Plant Signal Behav 6:55–58

Xu L, Law SR, Murcha MW et al (2013) The dual targeting ability of type II NAD(P)H dehydrogenases arose early in land plant evolution. BMC Plant Biol 13

Yang X, Cushman JC, Borland AM et al (2015) A roadmap for research on crassulacean acid metabolism (CAM) to enhance sustainable food and bioenergy production in a hotter, drier world. New Phytol 207:491–504

Yang JT, Preiser AL, Li Z et al (2016) Triose phosphate use limitation of photosynthesis: short-term and long-term effects. Planta 243:687–698

Yoshida K, Terashima I, Noguchi K (2006) Distinct roles of the cytochrome pathway and alternative oxidase in leaf photosynthesis. Plant Cell Physiol 47:22–31

Zabaleta E, Martin MV, Braun HP (2012) A basal carbon concentrating mechanism in plants? Plant Sci 187:97–104

Zarzycki J, Axen SD, Kinney JN, Kerfeld CA (2013) Cyanobacterial-based approaches to improving photosynthesis in plants. J Exp Bot 64:787–798

Zhang Y, Fernie AR (2018) On the role of the tricarboxylic acid cycle in plant productivity. J Integr Plant Biol 60:1199–1216

Zhang DW, Xu F, Zhang ZW et al (2010) Effects of light on cyanide-resistant respiration and alternative oxidase function in Arabidopsis seedlings. Plant Cell Environ 33:2121–2131

Zhang ZS, Liu MJ, Scheibe R et al (2017) Contribution of the alternative respiratory pathway to PSII photoprotection in C3 and C4 plants. Mol Plant 10:131–142

Zhang CC, Zhou CZ, Burnap RL, Peng L (2018) Carbon/nitrogen metabolic balance: lessons from cyanobacteria. Trends Plant Sci 23:1116–1130

Zhao Z, Assmann SM (2011) The glycolytic enzyme, phosphoglycerate mutase, has critical roles in stomatal movement, vegetative growth, and pollen production in *Arabidopsis thaliana*. J Exp Bot 62:5179–5189

Zhu X-G, de Sturler E, Long SP (2007) Optimizing the distribution of resources between enzymes of carbon metabolism can dramatically increase photosynthetic rate: a numerical simulation using an evolutionary algorithm. Plant Physiol 145:513–526

Zhu X-G, Long SP, Ort DR (2008) What is the maximum efficiency with which photosynthesis can convert solar energy into biomass? Curr Opin Biotechnol 19:153–159

Zhu G, Kurek I, Liu L (2010a) Engineering photosynthetic enzymes involved in CO_2–assimilation by gene shuffling. In: Rebeiz CA, Benning C, Bohnert HJ et al (eds) Advances in photosynthesis and respiration. Springer, Dordrecht, pp 307–322

Zhu X-G, Long SP, Ort DR (2010b) Improving photosynthetic efficiency for greater yield. Annu Rev Plant Biol 61:235–261

Zoschke R, Bock R (2018) Chloroplast translation: structural and functional organization, operational control, and regulation. Plant Cell 30:745–770

Paula Da Fonseca-Pereira is currently a postdoctoral researcher fellow at the Universidade Federal de Viçosa, Brazil. She received her PhD in Plant Physiology in 2017. She also spent 1 year (2015–2016) as guest PhD student at Max Planck Institute of Molecular Plant Physiology, Potsdam-Golm. Her research interests include the regulation of respiration mediated by thioredoxins and related processes in plants.

Willian Batista-Silva is a postdoctoral researcher fellow at the Plant Biology Department of the Universidade Federal de Viçosa, Viçosa, Brazil. He received his PhD in Plant Physiology from the same University in 2018. He also has worked in the Institut für Pflanzengenetik in Hannover, Germany. His research focuses on plant hormones and its interaction with central metabolism (e.g., photosynthesis and respiration).

Adriano Nunes-Nesi is an agronomist and received his doctorate in Plant Physiology in 2004 from the Universidade Federal de Viçosa, Brazil, where he is currently an Associate Professor in the Plant Biology Department. Dr. Nunes-Nesi worked for 6 years as a researcher at the Max Planck Institute of Molecular Plant Physiology (Potsdam-Golm, Germany), where he studied molecular plant physiology, with a focus on primary metabolism. His current interests include the interactions of mitochondrial metabolism with photosynthesis in plants and microalgae under stress conditions.

Agustin Zsögön is a geneticist and received his PhD in Plant Sciences from the Australian National University. He is an adjunct professor of Plant Physiology at the Universidade Federal de Viçosa (Minas Gerais, Brazil). His research group studies the relationship between form and function in plants using genetic and physiological approaches in Solanaceae species. Special focus is given to the genetic basis of relevant physiological traits harnessed over the course of crop domestication and improvement. Main research interests: natural genetic variation; genetic manipulation of plant growth habit; genetic basis of plant resistance to abiotic stress; relationship between leaf structure and function; plant transformation and genome engineering; water-use efficiency in crops; coordination and regulation of gas exchange in leaves.

Wagner L. Araújo obtained his PhD at the Max Planck Institute of Molecular Plant Physiology, Potsdam-Golm, 2011. Assistant professor in the Plant Biology Department at Universidade Federal de Viçosa, Viçosa, Brazil. His research is currently devoted to increasing our understanding of the regulation of respiration and the alternative donation of electrons to the mitochondrial electron transfer chain as well as the linkage between hormone metabolism and primary metabolism as a whole.

Regulatory Principles of Energy Fluxes and Their Impact on Custom-Designed Plant Productivity

4

Johannes Knuesting, Renate Scheibe, and Jennifer Selinski

Abstract

In organisms performing oxygenic photosynthesis, energy input is highly variable due to changes in light intensities and other environmental factors. Therefore, fast and flexible adjustments for energy distribution and consumption are required to avoid any stress. Optimization of photosynthesis and avoidance of wasteful processes have been in the focus of many studies in the field of crop improvement. Production and consumption of the energy carriers ATP, NADPH, and NADH need to be kept in balance since their pool sizes are small and reducing equivalents cannot be transported directly across compartment borders. Furthermore, reversible redox-modification at regulatory cysteines of the target enzymes and the adjustment of fluxes at the respective redox switches by small molecules adjust the actual enzyme activities to the metabolic situation to maintain homeostasis. Finally, upon sustained imbalances, signal transfer via cytosolic redox-switches can lead to changes in gene expression for maintenance of homeostasis. For any biotechnological approach aiming for the production of tailored products and increased yield, consideration of the regulatory networks and mechanisms of energy distribution between sinks and sources are of basic importance.

J. Knuesting · R. Scheibe
Department of Plant Physiology, Faculty of Biology and Chemistry, Osnabrueck University, Osnabrueck, Germany
e-mail: scheibe@biologie.uni-osnabrueck.de

J. Selinski (✉)
Department of Biochemistry and Physiology of Plants, Faculty of Biology, Bielefeld University, Bielefeld, Germany
e-mail: jennifer.selinski@uni-bielefeld.de

A. Kumar et al. (eds.), *Climate Change, Photosynthesis and Advanced Biofuels*,
https://doi.org/10.1007/978-981-15-5228-1_4

Keywords

Energy metabolism · Metabolite transporters · Regulation · Redox-homeostasis · Improvement of photosynthesis · Production of specialty compounds

4.1 Importance of Plants for Mankind

Life on earth appeared about 3.5 billion years ago. During evolution, the first photosynthetically active life developed in the Archean, about 2.4 billion years ago. As primary producers, photosynthetic organisms play an essential role in the circle of life as we know it today. The ability to convert the light energy of the sun into chemical energy enables the assimilation and synthesis of valuable biomass which is the basis for bioenergy and nutrition and technical use. As soon as the human society became sessile, and agriculture started with the domestication of wild relatives of our nowadays crops, yield and quality of the harvest was continuously improved by selection over centuries. With the growing world population, more and more areas were converted to arable land in order to satisfy the ever-growing demand for food and renewable energy. The Green Revolution could once more increase yield per area significantly due to the selection of mutants with short shoots that were able to carry large ears with many grains without falling over ("High-yield varieties"). However, even with perfect fertilization and crop protection schemes, a limit seems to be reached. But the human population is still exponentially growing requiring increased production on restricted availability of area.

The discovery of the pathway for CO_2 assimilation by Andrew A. Benson, Melvin Calvin, and James Bassham in the sixties of the last century and the analysis of photosynthesis as the primary process to generate organic matter from inorganic precursors provide the necessary research to aim for improvement of primary production. The knowledge of the whole genomes of a growing number of plants and other photosynthetic organisms is one essential part that opens the field up for many approaches to increase yield and to manipulate plants for valuable bio-products. The importance of renewable resources is ever increasing as the ancient energy reserves are close to being depleted, and food, feed, and energy compounds are urgently needed as efficient and specifically tailored biomass for all human needs. Human activities, at the same time, lead to a shortage of essential resources such as water and are causative for global climate change. Therefore, improved properties of plants need to be generated in order to prevent further deterioration of the ecological status of the globe. In particular, together with the danger of climate extremes that should be kept at a minimum, the availability of adapted plants with resistance to abiotic and biotic stressors should be aimed for in future attempts to engineer plants.

After many years of research to develop high-energy plants for fuel production it became clear that the use of valuable land for the products of biomass that is only converted into fuel is not cost-effective since the conversion of biomass into fuel is connected with even more energy input. It is also ecologically of an adverse outcome if we consider the loss of biodiversity due to large natural areas being destroyed and occupied by monocultures. Therefore, the development of better plants in the fourth generation is now aiming for high-value specialty products that can replace conventionally produced materials for industrial or for medical purposes. Such high-value specialties would require only smaller areas of valuable land and produce renewable value products not further consuming the oil reserves.

For improvement of amount and kind of biomass production, the knowledge of the biosynthetic steps for its synthesis is required as a basis for their manipulation. Here, not only subcellular localization but also the distribution of intermediates and storage of products across the whole plant, its growth and development as well as its stress-resistance come into the play. At the same time, regulatory aspects need to be considered and are studied, e.g., in the model plant *Arabidopsis thaliana (A. thaliana)*, in order to increase basic knowledge. Finally, as suggested in a recent opinion paper (Jansson et al. 2018), the obvious demand for higher crop yields should be also seen as a task to generate climate-smart crops at the same time so that emissions are kept low.

When the complex network of metabolism interlinked with growth and development in time and space is affected in approaches to engineer plants for the production of specialty products, it is of utmost importance to also consider energy distribution between compartments in differently differentiated cells. Knowledge of the regulatory mechanisms at all levels as well as of the shuttle systems for the delivery of the proper kind and amount of energy (NAD(P)H or ATP) at each time and in each compartment is needed (Scheibe 2019). If any imbalance is coming up, redox-homeostasis will be disturbed, and high-energy carriers without the immediate presence of the proper target will lead to the formation of radicals, oxidative stress, and cell damage if the redox-buffering devices such as ascorbate and glutathione (GSH) are exhausted (Foyer and Noctor 2011). Light absorption and efficiency of energy conversions for usage, distribution of electrons, flexible pathways of electron flow for production of reduced ferredoxin, NADPH and ATP, their distribution for the assimilatory processes, as well as the removal of any excess, all need to act in concert for effective biomass production. Thermal dissipation of energy in the thylakoids, as well as the alternative oxidases in mitochondria, are also important parts of such a system.

In this chapter, we will provide examples for the importance of a flexible machinery to allow for redox-balancing and the simultaneous adjustment of metabolic fluxes under changed conditions or in the presence of different substrates to be reductively assimilated for the production of the desired product (Fig. 4.1). Fine-tuning of the natural and the engineered biochemical pathways in concert with the various pathways of energy production and consumption is required. Initially, a short overview is given of approaches and current attempts to improve productivity. Better usage of light and CO_2, as well as the removal of wasteful processes, are long-lasting goals to achieve the basis for better yield. Also, the higher stability of metabolic processes under stressful conditions is aimed at in many ways.

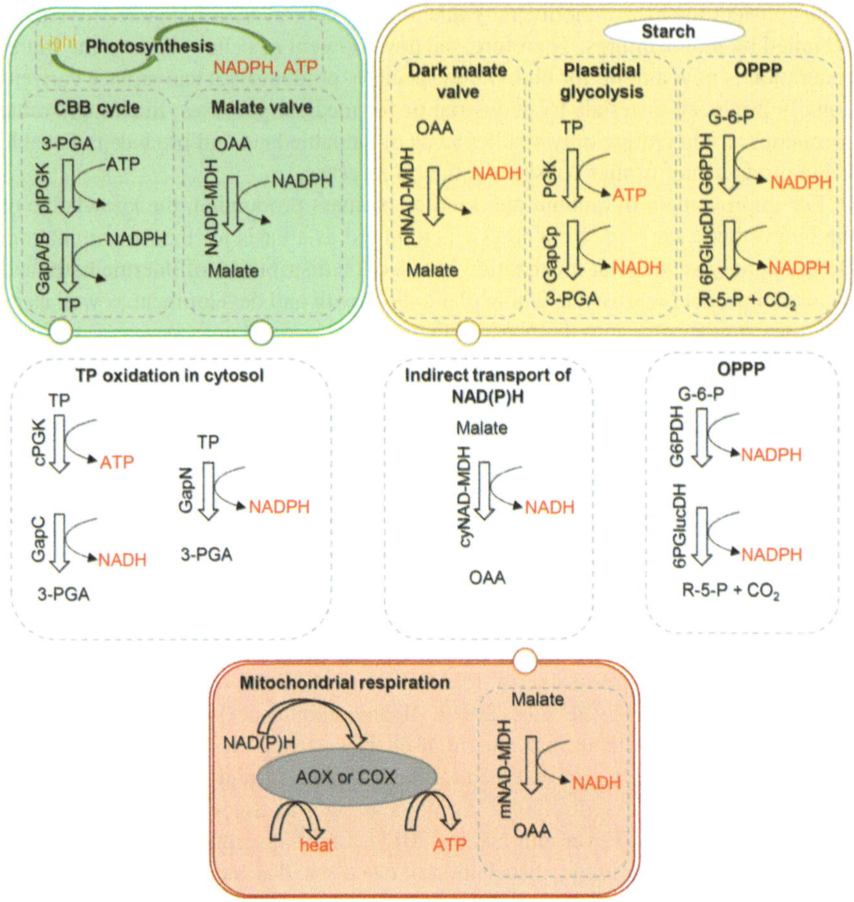

Fig. 4.1 Distribution and interconversion of energy and reducing equivalents between cell compartments. In illuminated green tissues, ATP and NADPH are generated during photosynthesis. A large portion of these energy carriers is consumed for CO_2 assimilation in the Calvin-Benson-Bassham (CBB) cycle (*plPGK*, chloroplast phosphoglycerate kinase; *GapA/B*, chloroplast light-activated isoform of glyceraldehyde-3-phosphate dehydrogenase). Any excess of reducing equivalents is converted into malate by the NADP-dependent light-activated malate dehydrogenase (NADP-MDH) which is part of the malate valve in the light. Malate acts as a carrier for reducing equivalents for removal of excess NADPH (or NADH in the dark) as it is converted back to oxaloacetate (OAA) by the cytosolic NAD-dependent malate dehydrogenase (cyNAD-MDH). Triosephosphates (TP) from the CBB cycle can be exported into the cytosol via translocators. TP can be oxidized either in the glycolytic steps of 3-phosphoglycerate kinase (cyPGK) and NAD-dependent glyceraldehyde 3-phosphate dehydrogenase (GapC) leading to the generation of ATP and NADH or by the non-phosphorylating 3-phosphoglycerate dehydrogenase (GapN) which is an irreversibly functioning aldehyde reductase-type enzyme. In darkness or in plastids of non-green tissues, starch degradation yields hexose-phosphate or triose phosphate for carbohydrate oxidation, which provides ATP and reducing equivalents in the initial glycolytic steps (substrate phosphorylation). Or it can be oxidized in the oxidative pentose phosphate (OPP) pathway both, in the plastids or in the cytosol yielding NADPH. When triose phosphates are oxidized by the plastidial glycolytic enzymes plPGK and NAD-GAPDH (GapCp), both ATP and NADH are generated. Usually, for anabolic processes, only ATP plus NADPH from OPP pathway are required. Then, NADH from plastidial glycolysis is indirectly exported as malate via the "dark malate valve"

4.2 Improvement of Light Usage and Assimilatory Processes

Genes coding for transcription factors, phytohormone synthesis, protein synthesis, cell division, and expansion have been shown to affect growth positively. An overview of genes that have been introduced into Arabidopsis to increase growth and leaf size is given in a review (Gonzalez et al. 2009). However, a general picture of their relative importance is still lacking since experimental conditions and genetic background varied from lab to lab.

Melvin Calvin has suggested photosynthesis as a resource for energy and materials already in the seventies of the last century. He has proposed (and experimentally proven) the culture of fast-growing plants for hydrocarbon production or algae for hydrogen or even synthetic systems functioning according to the principle of photosynthesis (Calvin 1976). From then on, multiple, continued attempts are made to improve these fundamental processes of light-dependent assimilation. In particular, the improvement of ribulose-1,5-bisphosphate carboxylase/oxygenase (RubisCO) efficiency and removal of photorespiratory losses are in the scope of such research for yield improvement (Evans 2013; Kebeish et al. 2007; Murchie et al. 2009; Ort et al. 2015) (Nölke and Schillberg, this book).

Energy conversion and distribution in the chloroplast, primary and secondary reactions, as well as shuttling of energy equivalents and intermediates of metabolism between compartments, should be considered as possible sites for changes and improvements in terms of yield of desired products (Simkin et al. 2019). We will shortly summarize some more recent developments in basic research that are unveiling more and more complex structures of the regulatory system which enables high variability and flexibility.

4.2.1 Improvement of Energy Capture and Extension of the Usable Light Spectrum

Photosynthesis starts with the absorption of light energy at the light-harvesting complexes (LHCs). Carotenoids and chlorophylls increase the absorption spectra of LHCs and pass the light energy to the reaction centers in photosystem I and photosystem II. Various factors were shown to limit the efficiency of light utilization to generate biomass. For instance, chlorophyll and carotenoids are only capable of

Fig. 4.1 (continued) consisting of the plastidial NAD-MDH (plNAD-MDH) and the corresponding cyNAD-MDH. Remaining excess of reducing equivalents (NADPH and NADH or malate) which are not consumed for anabolism are transferred into the mitochondria. Electron transfer in the respiratory chain leads to the generation of either ATP via oxidative phosphorylation (COX, cytochrome oxidase) or alternatively excess energy can be dissipated via alternative oxidase (AOX) as heat. Any challenge or additional synthetic pathways would require adjustment of these steps in order to maintain homeostasis of the energy carriers in each compartment according to the demand

absorbing visible light between 400 and 700 nm to drive photosynthesis, but the major fraction of sunlight lies in the far-red region that is beyond 700 nm. Cyanobacteria possess chlorophyll *d* and *f* that exhibit a broader absorption of up to 750 nm (Allakhverdiev et al. 2016).

A Eustigmatophyte alga can absorb far-red light using a chlorophyll *a*-containing antenna complex and is even able to grow when cultivated under far-red light solely (Wolf et al. 2017). Using these naturally occurring pigments together with a reduction of the antenna size, especially in the upper canopy leaves, will shift the photosynthetic light-use efficiency to higher intensities (Blankenship and Chen 2013; Jin et al. 2016; Ort and Melis 2011). Engineering plants possessing fewer light-harvesting pigments per photosystem and fewer photosystems in their uppermost leaves might lead to a greater proportion of absorbed photons that can be converted to biomass and enhance plant productivity.

Besides engineering pigment composition and antenna sizes in plants, researchers aim to redesign the two photosystems and optimize components of the electron transport chain and downstream acceptors to increase photosynthetic efficiency. To avoid competition on available light and split electron flow, each photosystem should work independently of the other. This could be reached by replacing photosystem I with a reaction center and its associated cyclic electron transport machinery by bacteriochlorophyll *b* (Ort et al. 2015). Furthermore, plants expressing variable amounts of the Rieske iron-sulfur protein as a component of the cytochrome b_6f complex were shown to enhance leaf photosynthesis and increase grain yield in *Oryza sativa* (Yamori 2016) and Arabidopsis (Simkin et al. 2017a). Photosynthetic efficiency has also been increased by introducing the cytochrome c_6 gene (similar to plastocyanin) from algae in *A. thaliana* (Chida et al. 2007) and overexpression of chloroplastic NAD kinase in rice (Takahara et al. 2010).

Plants experience dramatic fluctuations in light intensities in their natural environment. Under high-light conditions or rapid increases in light intensity as the transition from shade to light, several photoprotective mechanisms are induced to protect the photosynthetic complexes from over-reduction. Excess light can be dissipated as heat in the antenna complexes of photosystem II. This process is called non-photochemical quenching of chlorophyll fluorescence (NPQ). Although NPQ can be fast, it does not happen immediately due to the conformational changes that are necessary. Furthermore, the recovery rate of NPQ relaxation is slower than the rate of induction and impaired by repeated exposure to excess light (Perez-Bueno et al. 2008). The slow induction rate of photosynthesis during shade-to-sun shifts has been estimated to cost about 21% of productivity in *Triticum aestivum* (Taylor and Long 2017). The engineering of genotypes capable of more rapid recovery from the photoprotected state was estimated to increase carbon uptake (Zhu et al. 2004). In this context, acceleration of the xanthophyll cycle and overexpression of the photosystem II subunit S has been shown to lead to a faster restoration of the maximum efficiency of CO_2 assimilation under rapidly changing light intensities that in turn leads to increased plant productivity in *Nicotiana tabacum* (*N. tabacum*) (Kromdijk et al. 2016).

4.2.2 Creating an Optimal Environment for RubisCO

RubisCO is one of the most abundant enzymes on earth and the central enzyme in CO_2 fixation. CO_2 capture by this enzyme is in the focus of research as a step for further improvement (Orr et al. 2017; Parry et al. 2013). CO_2-concentrating mechanisms such as algal carboxysomes, expression of cyanobacterial bicarbonate transporters, as well as the implementation of the advantages of the C4-pathway or CAM, might lead to higher production of sugars since then the CO_2 concentration at the RubisCO active site is increased, and oxygenation leading to carbon loss is minimized (Borland et al. 2011; Häusler et al. 2002; Leegood 2002; McGrath and Long 2014; Price et al. 2011; Rolland et al. 2016).

The majority of plants and crops pursue C3 photosynthesis, and 3-phosphoglycerate (3-PGA) is produced as the first organic carbon compound. Under high temperature and high light, RubisCO has a high affinity for oxygen leading to an increase in its oxygenation reaction and concomitant reduction of photosynthetic and water-use efficiency due to photorespiration. In contrast to C3 plants, C4 plants minimize photorespiration by separating initial CO_2 fixation (in mesophyll cells) and the CBB cycle (in bundle-sheath cells) in space while CAM plants minimize photorespiration and save water by separating these two phases in time. In CAM and C4 plants, the enzyme PEP carboxylase (PEPC) that has no affinity for O_2 catalyzes the formation of the four-carbon compound oxalo-acetate (OAA). Subsequently, OAA is converted to malate (or aspartate) that can be transported into the bundle-sheath cells (in case of C4 plants) or stored in the vacuole in the form of malic acid (in case of CAM plants). Inside the bundle-sheath cells (C4 plants) or during the day (CAM plants) malate is cleaved by the malic enzyme into pyruvate and CO_2 that is in turn fixed by RubisCO and enters the CBB cycle. Via this mechanism, C4 and CAM plants increase CO_2 concentrations around RubisCO and circumvent or reduce the occurrence of photorespiration and the concomitant loss of carbon and energy (Driever and Kromdijk 2013; Schuler et al. 2016; von Caemmerer et al. 2012).

The following two types of CO_2-concentrating mechanism could be engineered in C3 crops: On the one hand, single-cell CO_2-concentrating mechanisms depend on the presence of carboxysomes in the chloroplast, and inorganic carbon transporters that import bicarbonate into the chloroplast (Hanson et al. 2016; Price et al. 2013). Inside the carboxysome, carbonic anhydrase converts bicarbonate to CO_2 and water leading to an increase of CO_2 concentration. Engineering single-cell C4 photosynthesis into C3 plants would not require the induction of the Kranz anatomy that would be necessary for the two-cell CO_2-concentrating mechanism. Comparative transcriptome studies have led to the generation of a model containing putative genetic regulators for the induction of the Kranz anatomy (Fouracre et al. 2014; Wang et al. 2013), but this model remains to be tested.

Engineering the CAM pathway in C3 crops aims to increase plant water-use efficiency. Distinctive features of CAM are (1) CO_2 assimilation and fixation at night, (2) an inverse stomatal behavior, in which stomata are closed during the day and open at night, and (3) the decarboxylation of stored organic acids leading to the release of CO_2 during the day. The released CO_2 is then refixed by RubisCO to produce carbohydrates via the CBB cycle. To introduce the CAM pathway in C3 crops, comparative analyses of the evolution, genomic features, and regulatory mechanisms of CAM are necessary (Borland et al. 2014; Yang et al. 2015). Although engineering C4 and CAM photosynthesis are both still in the initial steps of application, these approaches offer the potential to sustain plant productivity in hotter and drier climates triggered by climate changes in temperate regions. Besides the approaches targeting C4 and CAM photosynthesis, overexpression of aquaporins has been shown to increase CO_2 conductance (Flexas et al. 2006; Hanba et al. 2004; Kawase et al. 2013).

RubisCO requires post-translational activation to perform its carboxylation reaction. On the one hand, a conserved lysine residue in its active site needs to be carbamylated by CO_2 in order to form a complex with a Mg^{2+} ion that is in turn required for activity. On the other hand, the average turnover of RubisCO catalysis is between 1 and 10 s^{-1} that is limiting for photosynthetic CO_2 fixation under optimal conditions. Therefore, various approaches aim at the neofunctionalization of RubisCO by generating RubisCO variants possessing enhanced carboxylation rates, lowered oxygenation rates, and an increased specificity toward CO_2 (Mueller-Cajar and Whitney 2008; Satagopan et al. 2017; Shih et al. 2016; Whitney et al. 2011). For instance, engineering a RubisCO that is characterized by a lower binding affinity for its inhibitors or increasing the thermal stability of RubisCO activase has been shown to increase photosynthesis and growth (Kumar et al. 2009; Kurek et al. 2007; Yamori et al. 2012).

4.2.3 Avoidance of Photorespiration

Besides the carboxylation of ribulose 1,5-bisphosphate (RuBP) yielding two molecules of 3-PGA, RubisCO also catalyzes its oxygenation leading to no gain of fixed carbon, but only one molecule of 3-PGA and one molecule of 2-phosphoglycolate (2-PG). In C3 plants, RubisCO is characterized by an oxygenation rate of more than 20% of its carboxylation rate that can even increase to up to more than 40% at high temperatures or low intracellular CO_2. Thus, a substantial amount of 2-PG is formed during photosynthesis that needs to be recycled to regain three carbon atoms from two molecules of 2-PG in an energy-demanding process called photorespiration (Walker et al. 2016; Zhu et al. 2010).

The introduction of an *Escherichia coli* (*E. coli*) bypass for photorespiration that is a glycolate catabolic pathway was shown to increase net photosynthesis and biomass production in *A. thaliana* and *N. tabacum* (Carvalho et al. 2011; Kebeish et al. 2007; Maier et al. 2012). In respect to bio-engineering photorespiration, another approach also led to the increase in biomass and seed yield in Arabidopsis. Facilitating photorespiratory carbon flow through overexpression of subunits of the mitochondrial glycine decarboxylase system in combination with the overexpression of CBB cycle enzymes has been shown to increase biomass and yield (Simkin et al.

2017b). These approaches represent a proof of concept to target primary metabolism by bio-engineering to improve biomass and resilience in plants.

In a more recent study, it was shown that synthetic biology approaches could increase growth and biomass production of tobacco plants. To avoid energy loss during photorespiration, the authors developed three different alternative photorespiratory pathways and introduced them into the chloroplast. On the one hand, five genes of the glycolate oxidation pathway from *E. coli* were introduced, which increased the biomass by 13% in the greenhouse. On the other hand, glycolate oxidase from Arabidopsis, malate synthase from pumpkin, and catalase from *E. coli* were transformed into tobacco with no significant effect on yield. Interestingly, the introduction of malate synthase from pumpkin and of algal glycolate dehydrogenase increased biomass by 24% when grown in the greenhouse, and in the field even by >40% (South et al. 2019). As a further positive example, an increase in yield could be achieved by introducing a photorespiratory bypass in rice (Shen et al. 2019). Circumventing photorespiration through a bypass has improved biofuel yield also in *Camelina sativa* (Dalal et al. 2015). Synthetic photorespiration allows for surplus CO_2 fixation compensating the oxygenation reaction of RubisCO and its concomitant loss of carbon in cyanobacteria (Shih et al. 2014).

Changing photosynthetic carbon metabolism cannot only be realized by manipulating RubisCO but also by substituting the CBB cycle by a natural or synthetic pathway. Furthermore, introducing RubisCO-independent CO_2 fixation pathways such as the Malonyl-CoA-Oxaloacetate-Glyoxylate (MOG) or the CETCH cycle (Bar-Even et al. 2010; Schwander et al. 2016) are promissing approaches in this context. In the MOG cycle, a single carboxylating enzyme such as the PEPC plays a key role in the binding of CO_2, and, therefore, represents an alternative entry point of inorganic carbon. In contrast, the CETCH cycle is based on the highly efficient and versatile class of enoyl-CoA carboxylases/reductases (Erb et al. 2007; Peter et al. 2015) that are insensitive to O_2 and have been shown to lead to higher carbon fixation rates than the CBB cycle in vitro (Rosenthal et al. 2014, 2017; Schwander et al. 2016). However, both cycles generate glyoxylate. Due to the fact that the assimilation of glyoxylate involves a decarboxylation reaction, these cycles seem rather inefficient in planta. In general, transplanting pathways either from foreign organisms or the design of enzymes with new functions are possible approaches. The latter can be achieved in two ways by (1) rational protein engineering or (2) directed protein evolution, both approaches being described as future options to reach the goal (Rao 2008).

In order to reduce the loss of assimilated carbon, a further approach would be to engineer respiration to reduce costs (Amthor et al. 2019).

4.3 Flexible Distribution of Energy Across Compartment Borders

Energy for the required biosynthesis is available as ATP and NADPH. The indirect transport across membranes between subcellular compartments is enabled by the various shuttle-systems, mainly in the form of malate or as triose-P (Scheibe 2019; Selinski and Scheibe 2019). These organic intermediates of primary metabolism, as well as other intermediates of central metabolism, are easily converted into further compounds, such as sugar phosphates and dicarboxylic acids that can be used not only as sources for energy in the different compartments but also as building blocks for the synthesis of the desired organic compounds. When such intermediates are removed to enter into new pathways, a network of catabolic and anaplerotic reactions and transport between compartments is acting in concert. Key points of the branched pathways need to be strictly regulated in such a way that energy and precursor pools are kept homeostatically balanced even when fluxes are changed.

Plant cells are highly compartmentalized and possess a complex metabolic network (Lunn 2007; Szecowka et al. 2013; Tegeder and Weber 2006). Plant metabolism is driven by the energy-transducing reactions of chloroplasts and mitochondria, respectively. In general, metabolic pathways are interconnected across several compartments and depend on the supply of metabolic precursors from other parts of the cell. In addition, detoxification of unfavorable byproducts of plant metabolism as for instance 2-PG that is formed during photorespiration comprises several compartments, namely chloroplasts, mitochondria, and peroxisomes. Metabolite transport between cell compartments enables plants to coordinate metabolic pathways and facilitates the transmission of information. Furthermore, metabolic requirements of different developmental stages and defense change dynamically with time and according to the actual environmental condition.

4.3.1 Metabolite Exchange Across the Inner Chloroplast Membrane

In chloroplasts, TPs are generated via the CBB cycle and are subsequently incorporated into transport sugars for export to various heterotrophic tissues (sinks) or the fixed carbon can be used for storage as transitory starch within the chloroplast (Hoefnagel et al. 1998). This differential distribution of photosynthates generating transitory starch in the chloroplast or sucrose in the cytosol is termed partitioning. For sucrose synthesis, assimilated carbon is exported in the form of TPs from the chloroplast to the cytosol. This counter-exchange of TPs and/or 3-PGA with inorganic phosphate (P_i) is catalyzed by the TP/phosphate translocator (TPT) (Fig. 4.2) (Fliege et al. 1978; Flügge 1999). The lack of TPT in *Solanum lycopersicum*, *N. tabacum*, and *A. thaliana* results in an increased steady-state level of transitory starch indicating that altered metabolism can compensate for TPT deficiency via redistribution of assimilated carbon (Riesmeier et al. 1993; Schneider et al. 2002; Walters et al. 2004).

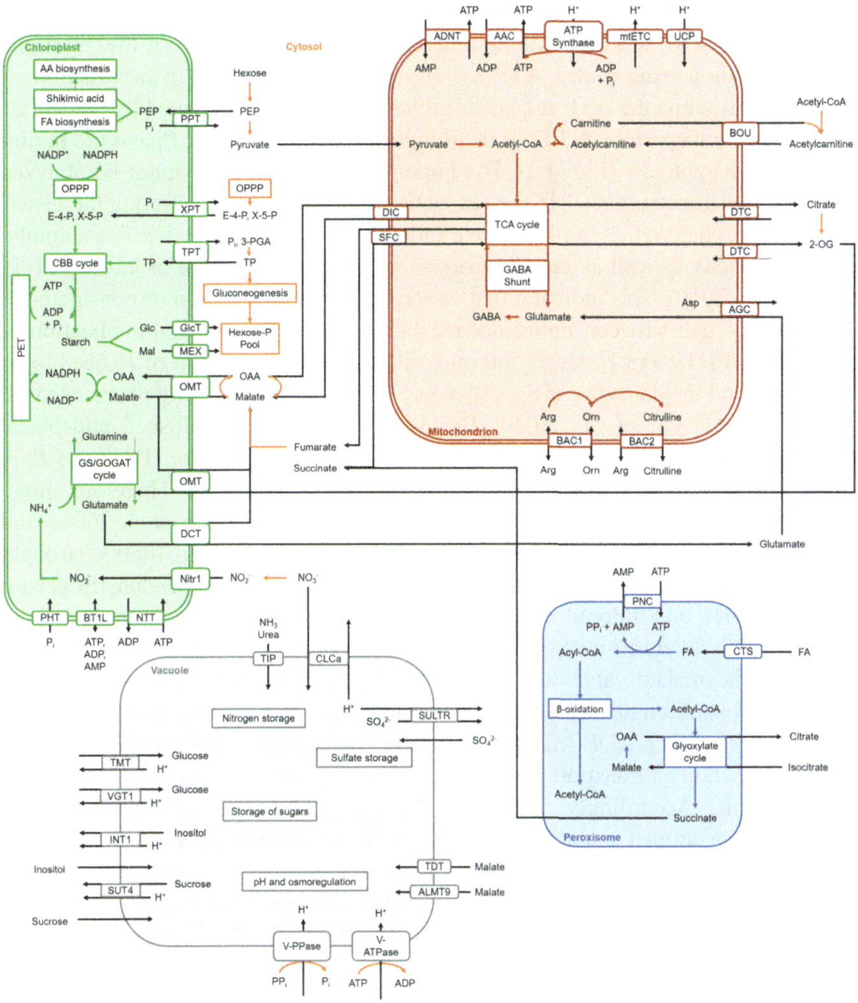

Fig. 4.2 Metabolite shuttling between subcellular compartments in plant cells. Transport processes of metabolites such as carbon and nitrogen as well as energy compounds across membranes of subcellular compartments in plant cells are depicted. Aiming for the generation of plants exhibiting increased resilience and yield necessitates the consideration of metabolite transport between subcellular compartments and sinks and sources. Metabolic fluxes and metabolic pathways are strictly regulated at various steps, for instance, feedback inhibition of metabolic intermediates. Therefore, combinatorial approaches of interacting metabolic pathways and transporters need to be designed for successful application as for instance shown by South et al. (2019) who introduced three different alternative photorespiratory pathways combined with the suppression of the glycolate/glycerate transporter in the chloroplast preventing the transport of glycolate and concomitant inhibition of the native photorespiratory pathway ultimately leading to an increased biomass of >40%. *2-OG* 2-oxoglutarate, *3-PGA* 3-phosphoglycerate, *AA* amino acid, *AAC* ATP/ADP carrier, *ADNT* adenine nucleotide carrier, *AGC* aspartate/glutamate transporter, *ALMT9* aluminum-activated malate transporter 9, *Arg* arginine, *Asp* aspartate, *BAC* basic amino acid carrier, *BOU* carnitine carrier, *BT1L* BT1-like transporter, *CBB cycle* Calvin-Benson-Bassham cycle, *CLCa* nitrate/proton antiporter, *CoA* coenzyme A, *CTS* fatty acid importer, *DCT* glutamate/malate translocator, *DIC* dicarboxylate carrier, *DTC* dicarboxylate/tricarboxylate carrier, *E-4-P*

The maintenance of P_i homeostasis in chloroplasts is essential for photosynthetic ATP synthesis and carbon partitioning between starch and sucrose. A low plastidial P_i concentration exacerbates ATP synthesis and leads to the over-reduction of photosystems while the ATP import in exchange with ADP mediated by nucleotide triphosphate transporters (NTT) leads to higher concentrations of P_i and the inhibition of starch synthesis (Fig. 4.2). The import of P_i into the chloroplast is catalyzed by phosphate transporters (PHT). Approaches in plant breeding targeting the generation of phosphate efficient plants have failed so far due to excessive P_i accumulation and toxicity as well as growth arrest in plants overexpressing or lacking PHTs (Wu and Xu 2010). This indicates that various other factors need to be considered to minimize the gap between molecular mechanisms and the practical application of engineering PHTs. For instance, internal utilization and allocation of P_i need to be optimized, and the phosphorus use efficiency of PHT overexpressing plants needs to be determined (Gu et al. 2016). Besides PHTs, TPT, xylulose 5-phosphate/phosphate translocator (XPT), and the PEP/phosphate translocator (PPT) use P_i in counter-exchange with its corresponding substrates (Fig. 4.2). Therefore, these transporters contribute to the poising of plastidic P_i levels (Chen et al. 2008; Guo et al. 2008). Due to the fact that PHT2.1 is mainly expressed in shoots, strongly induced by light and co-regulated with the NADP-dependent malate dehydrogenase (NADP-MDH) and thioredoxin, a cross-talk of P_i and redox homeostasis in chloroplasts is likely to occur (Rausch et al. 2004). CO_2 assimilation in the CBB cycle and the malate valve with NADP-MDH as a key enzyme are essential in $NADP^+$ regeneration allowing continued ATP production via photosynthetic electron transport. The NADP-MDH uses excess NADPH to convert OAA to malate enabling the indirect transport of reducing equivalents between different subcellular compartments. Accordingly, malate/OAA shuttles (OMT) with MDHs as key enzymes (also termed malate valves) act as powerful systems for balancing the ATP/NAD(P)H ratio and maintaining energy homeostasis in plant cells (Scheibe 1987, 2004; Selinski and Scheibe 2019). OMT catalyzes the import of 2-OG in counter exchange with malate. Besides its high affinity for 2-OG and malate, OMT displays a high affinity for OAA and has been proposed as the OAA/malate valve

Fig. 4.2 (continued) erythrose-4-phosphate, *FA* fatty acid, *GABA* γ-aminobutyric acid, *Glc* glucose, *GlcT* glucose transporter, *GOGAT* glutamine-2-oxoglutarate aminotransferase, *GS* glutamine synthetase, *Hexose-P* hexose phosphate, *INT1* inositol transporter 1, *Mal* maltose, *MEX* maltose exporter, *mtETC* mitochondrial electron transport chain, *Nitr1* nitrite transporter, *NTT* nucleotide triphosphate transporter, *OAA* oxaloacetate, *OMT* 2-oxoglutarate/malate translocator, malate/oxaloacetate translocator, *OPPP* oxidative pentose phosphate pathway, *Orn* ornithine, *PEP* phosphoenol pyruvate, *PET* photosynthetic electron transport, *PHT* phosphate transporter, *PPT* phosphoenol pyruvate/phosphate translocator, *Pi* inorganic phosphate, *PNC* peroxisomal adenine nucleotide carrier, *PPi* pyrophosphate, *SFC* succinate/fumarate carrier, *SULTR* sulfate/proton symporter, *SUT4* sucrose transporter 4, *TCA cycle* tricarboxylic acid cycle, *TDT* tonoplast dicarboxylate transporter, *TIP* urea transporter, *TMT* tonoplast monosaccharide transporter, *TP* triose phosphate, *TPT* triose phosphate/phosphate translocator, *UCP* uncoupling protein, *VGT1* vacuolar glucose transporter 1, *XPT* xylulose-5-phosphate/phosphate translocator, *X-5-P* xylulose-5-phosphate

(Taniguchi et al. 2002). However, inside the chloroplast, 2-OG serves as a precursor of nitrate assimilation generating glutamate via the glutamine-2-OG aminotransferase (GOGAT). The resulting glutamate can then either be exported in exchange for malate via the glutamate/malate translocator (DCT) (Renné et al. 2003; Riebeseel et al. 2010; Schneidereit et al. 2006; Weber and Flügge 2002; Weber et al. 1995) or it can re-enter the glutamine synthetase (GS)/GOGAT cycle. OMT and DCT have both been shown to mediate carbon/nitrogen metabolism and their crucial role in photorespiratory nitrogen/ammonia re-assimilation (Kinoshita et al. 2011; Renné et al. 2003; Taniguchi et al. 2002).

Transitory starch serves as the major energy source during the night, and its degradation drives plastidial glycolysis as well as the synthesis of sucrose in the cytosol (Geiger and Servaites 1994). Starch is a complex polysaccharide composed of linear amylose and highly branched amylopectin. These branched chains have to be phosphorylated in order to be accessible for degrading enzymes. The incorporation of phosphoryl groups to the polysaccharide is catalyzed by the glucan-water dikinase and phosphoglucan-water dikinase. After phosphorylation, β-amylase and a debranching enzyme catalyze the hydrolysis of starch granules producing maltose and glucose. Maltose and glucose can be transported into the cytosol acting as precursors for sucrose synthesis, or glucose directly enters plastidial glycolysis (Fig. 4.2).

The OPPP represents an alternative biochemical pathway in parallel to glycolysis for oxidizing carbohydrates in the dark. This oxidation is coupled with NADPH synthesis, the generation of pentoses, and the maintenance of the redox potential necessary to protect plants against oxidative stress (Juhnke et al. 1996; Neuhaus and Emes 2000). The OPPP can be divided into two distinct phases: the irreversible, oxidative phase and the reversible, non-oxidative phase. In the former, glucose 6-phosphate is oxidized to ribulose 5-phosphate (Ru-5-P). This step represents the source of reducing equivalents in the shape of NADPH. The subsequent non-oxidative phase consists of a series of transaldolase and transketolase reactions resulting in fructose 6-phosphate and 3-PGA. Depending on the environmental conditions and metabolic needs, 3-PGA and fructose 6-phosphate can be converted to glucose 6-phosphate for re-entry into the oxidative phase of the OPPP, maximizing the formation of NADPH or both molecules may enter glycolysis synthesizing ATP. In addition, the intermediates ribose 5-phosphate and erythrose 4-phosphate (E-4-P) can be withdrawn from the OPPP for nucleotide synthesis and phenylpropanoid production via the shikimic acid pathway (Kruger and von Schaewen 2003) (Fig. 4.2). For most organisms, the OPPP takes place in the cytosol. However, it was shown in spinach that the complete set of OPPP enzymes could be found in plastids, whereas only enzymes covering the oxidative phase of the pathway could be identified in the cytosol (Schnarrenberger et al. 1995). The OPPP localized in plastids and partially in the cytosol can interact through XPT that is present in the inner plastid envelope and catalyzes the counter-exchange of TPs, X-5-P, Ru-5-P, and E-4-P for P_i (Eicks et al. 2002; Kruger and von Schaewen 2003; Weber 2004; Weber et al. 2004, 2005).

4.3.2 Metabolite Exchange Across the Inner Mitochondrial Membrane

Besides chloroplasts, mitochondria represent energy-transducing organelles within the cell. The conversion of glucose 6-phosphate to pyruvate via glycolysis in the cytosol has a relatively low energy contribution to the cellular ATP pool upon aerobic conditions. However, the oxidation of pyruvate during mitochondrial respiration yields a much higher supply of energy. Respiration comprises cytosolic glycolysis, the mitochondrial electron transport chain (mtETC), and the mitochondrial tricarboxylic acid (TCA) cycle. The TCA cycle is a fundamental metabolic pathway that provides reducing equivalents for other cell compartments and carbon skeletons for many metabolic pathways, such as synthesis of nucleotides, amino acids, lipids, and vitamins and therefore, a specific exchange of molecules between the cytosol and mitochondria is required (Haferkamp 2007; Laloi 1999; Picault et al. 2004). Following plastidial and cytosolic glycolysis, pyruvate enters mitochondria, either directly from the cytosol or it is generated within the mitochondrial matrix via malic enzyme and is subsequently interconverted to acetyl-CoA. Acetyl-CoA derives not only from pyruvate but also from the oxidation of fatty acids (-β-oxidation) that takes place in peroxisomes (Fig. 4.2).

The TCA cycle in mitochondria is linked to various metabolic processes such as gluconeogenesis and nucleic acid synthesis in the cytosol, the glyoxylate cycle in peroxisomes, and the biosynthesis of amino acids in chloroplasts. Thus, TCA cycle intermediates, such as di- and tricarboxylates, are transported across the inner mitochondrial membrane via various transporters such as the dicarboxylate/tricarboxylate carrier (DTC) that mediates transport of the dicarboxylates 2-OG, OAA, malate and succinate, and the tricarboxylates citrate, isocitrate, and aconitate as well as the succinate/fumarate carrier (SFC) that facilitates the counter-exchange of succinate and fumarate and the dicarboxylate carrier (DIC) (Fig. 4.2) (Haferkamp 2007; Laloi 1999; Picault et al. 2004, 2002). DTC as well as DIC exhibits a broad substrate spectrum. However, DTC is ubiquitously expressed in plants and has been proposed to fulfill a housekeeping role in plant metabolism that requires a flux of organic acids to or from mitochondria (Linka and Weber 2010). Furthermore, DTC catalyzes the export of 2-OG from mitochondria that is used for nitrate assimilation in chloroplasts (Picault et al. 2002). DICs have been discussed to function as a malate/OAA shuttle to enable the indirect transfer of reducing equivalents to other subcellular compartments (Palmieri et al. 2008). In contrast to DTC and DIC, SFC does not exhibit a broad substrate spectrum. This carrier mediates the import of succinate that is generated in the glyoxylate cycle in peroxisomes in counter-exchange with fumarate that can enter gluconeogenesis in the cytosol (Catoni et al. 2003). Therefore, SFC might link the glyoxylate cycle in peroxisomes, the TCA cycle in mitochondria, and gluconeogenesis in the cytosol (Catoni et al. 2003; Palmieri et al. 1997).

Di- and tricarboxylates that are imported into mitochondria can enter the TCA cycle leading to increased electron transport in the mtETC and concomitant ATP production. The ADP/ATP carrier (AAC), as well as the adenine nucleotide carrier

(ADNT), enable the supply of other subcellular compartments with ATP (Klingenberg 2008). ATP export via AAC is regulated by the establishment of a proton motive force across the inner mitochondrial membrane that is established during mtETC and at the same time drives ATP synthesis via ATP synthase. Mitochondrial electron transfer is highly branched, both at the primary input oxidation of substrates (alternative NAD(P)H dehydrogenases) and at the final reduction of O_2 to H_2O (alternative oxidase, AOX), and also uncoupling proteins (UCPs) to avoid the build-up of a proton motive force. These components of the respiratory chain in plants mediate bypasses around complex I (catalyzed by alternative NAD(P)H dehydrogenases), complexes III and IV (catalyzed by AOX), and the ATP synthase (catalyzed by UCP) (Fernie et al. 2004; Millenaar and Lambers 2003; Møller 2001; Rasmusson et al. 2008, 2004; Saha et al. 2016; Selinski et al. 2018; Siedow and Umbach 1995; Sweetlove et al. 2006; Vanlerberghe 2013; Vanlerberghe and McIntosh 1997; Vercesi et al. 2006).

Electron transport through alternative NAD(P)H dehydrogenases and AOX occurs without proton translocation and ATP synthesis. In contrast, UCP allows for proton translocation but without ATP synthesis. Consequently, these alternative electron transport pathways mediate respiration that is uncoupled from energy conservation, and most of the energy is dissipated as heat. In this way, metabolic flexibility is achieved, which is required for acclimation to changing environmental conditions experienced in the natural environment, and also in plant growth and development.

Almost all amino acids are synthesized in chloroplasts. Therefore, the import of amino acids into plant mitochondria is necessary for protein biosynthesis (Haferkamp 2007; Laloi 1999; Picault et al. 2004). Two mitochondrial transporters have been identified in Arabidopsis that mediate basic amino acid transport (BAC1 and BAC2) across the inner mitochondrial membrane (Catoni et al. 2003; Hoyos et al. 2003). Furthermore, in plant mitochondria, an aspartate/glutamate carrier (AGC) can be found that exports aspartate to the cytosol in exchange for glutamate. The AGC is involved in various cellular functions such as the control of mitochondrial respiration, calcium signaling, redox homeostasis, and defense (Amoedo et al. 2016). The export of aspartate from mitochondria to the cytosol is used to regenerate GSH (Contreras and Satrustegui 2009) whereas glutamate that is imported in counter-exchange with aspartate can enter the γ-aminobutyric acid (GABA) shunt (Fig. 4.2). GABA is a non-protein amino acid that accumulates dramatically in response to abiotic and biotic stresses, and its concentration can even exceed that of amino acids involved in protein synthesis (Michaeli and Fromm 2015; Ramesh et al. 2017). However, although GABA has been hypothesized to play an important role in carbon and nitrogen metabolism as well as signaling, experimental proof is lacking so far.

4.3.3 Metabolite Exchange Between Peroxisomes and Cytosol

Peroxisomes play an important role in a variety of metabolic processes such as storage oil mobilization, photorespiration, generation and removal of reactive oxygen species (ROS), biosynthesis of phytohormones, breakdown of purine and branched-chain amino acids, and pathogen defense (Charton et al. 2019; Hayashi and Nishimura 2006; Pracharoenwattana and Smith 2008; Reumann and Weber 2006). Various metabolites need to be imported into and exported from the peroxisomal matrix. A porin-like channel mediates the transfer of metabolites across the peroxisomal membrane that is specific for glycolate, glycerate, malate, OAA, succinate, glutamate, and 2-OG (Reumann 2000; Reumann et al. 1997, 1995, 1998).

Due to the fact that no ATP-generating systems exist in plant peroxisomes (Arai et al. 2008; Linka et al. 2008), adenine nucleotide carriers (PNC) facilitating ATP import in counter-exchange with ATP, ADP, or AMP are inevitable (Fig. 4.2) (Arai et al. 2008; Linka et al. 2008). The import of cytosolic ATP into peroxisomes is required for storage oil mobilization and the activation of fatty acids (FA) that subsequently enter β-oxidation to support early seedling growth (Arai et al. 2008; Linka et al. 2008). FA transport across the peroxisomal membrane is mediated by an ATP binding cassette (ABC) transporter named COMATOSE (CTS) (Fig. 4.2) (Footitt et al. 2002; Hayashi et al. 2002; Zolman et al. 2000).

4.3.4 Metabolite Exchange Across the Tonoplast

The vacuole plays an important role in long- and short-term storage of primary metabolites as for instance mono- and disaccharides, oligo- and polysaccharides, sugar alcohols, dicarboxylic acids, amino acids, and nutrients (Martinoia et al. 2007; Neuhaus 2007). Metabolite and ion uptake into the vacuole are driven by a proton or an electrochemical gradient that is generated by the activity of a vacuolar H^+-transporting ATPase (V-ATPase) and a pyrophosphatase (V-PPase) (Fig. 4.2) (Maeshima 2001). The V-ATPase has been shown to be important in maintaining cytosolic ion homeostasis and cellular metabolism. Furthermore, the V-ATPase functions as a stress-responsive enzyme undergoing changes in conformation and in the expression of its subunits under adverse growth conditions (Ratajczak 2000). The V-PPase is the second vacuolar proton pump and uses pyrophosphate (PP_i) that is generated during DNA and RNA synthesis, sucrose, and cellulose synthesis or the conversion of pyruvate to PEP as its energy source. Arabidopsis plants overexpressing V-PPase exhibit increased drought and salt tolerance that is likely due to the increased proton gradient across the tonoplast (Gaxiola et al. 2001; Graus et al. 2018).

The vacuole is the main cellular storage pool for excess mono- and disaccharides that are generated during photosynthesis. In Arabidopsis, a vacuolar sucrose transporter (SUT4) has been identified that facilitates the export of sucrose (Endler et al. 2006). The tonoplast monosaccharide transporter (TMT) and the vacuolar glucose transporter (VGT1) mediate a glucose/proton antiport (Fig. 4.2) (Aluri and Büttner

2007; Neuhaus 2007; Wormit et al. 2006). Interestingly, Arabidopsis mutants exhibiting increased TMT activity have been shown to produce higher seed biomass that is of high importance for plant breeding approaches (Wingenter et al. 2010). The lack of VGT1 in Arabidopsis leads to reduced seed germination and late flowering with retarded growth of the shoot. Therefore, VGT1 might play an important role in vacuolar hexose accumulation necessary for the formation and maintenance of the cell turgor that drives cell expansion (Aluri and Büttner 2007).

Malate is a versatile compound in plant metabolism that is important as a photosynthate in C4, and CAM plants (Fig. 4.2), an intermediate of the TCA cycle, a pH regulator and it can easily be transported across subcellular membranes enabling the indirect transport of reducing equivalents. Furthermore, malate is a component of root exudates and a regulatory osmolyte mediating stomatal function (Fernie and Martinoia 2009). Excess malate can be stored within the vacuole and is imported via the tonoplast dicarboxylate transporter (TDT) and the aluminum-activated malate transporter (ALMT) (Fig. 4.2). While *tdt* mutants exhibit significantly decreased levels of malate in leaves, *amlt9* mutants exhibit only a slightly reduced malate content (Emmerlich et al. 2003; Kovermann et al. 2007). Thus, ALMT9 function can be partially compensated by TDT or another so far unknown vacuolar malate transporter (Kovermann et al. 2007). However, the severely compromised cellular acidification in *tdt* mutants indicates that TDT activity is critical for the regulation of pH homeostasis in plant cells (Hurth et al. 2005).

The major nitrogen source within plant cells is the pool of free nitrate stored inside the vacuole. Plant vacuoles possess voltage-dependent chloride channels (CLCa) that mediate proton-coupled nitrate import and are involved in nitrate homeostasis (Bergsdorf et al. 2009; De Angeli et al. 2006; Harada et al. 2004). Essential storage forms of nitrogen are ammonia and urea that are imported via tonoplast intrinsic proteins (TIPs) and accumulate at high concentrations in the vacuole to detoxify the cytosol (Dynowski et al. 2008; Jahn et al. 2004; Loqué et al. 2005; Wudick et al. 2009).

In contrast to nitrogen import, the identification of sulfate transporters mediating the import of sulfate has failed so far. However, it has been shown that the import of excess sulfate into the vacuole depends on an electrochemical gradient (Massonneau et al. 2000). In contrast, the existence of sulfate export systems (SULTR) has been reported (Kataoka et al. 2004).

Although there has been great progress in understanding metabolic fluxes within plant cells, there are still various unknowns that need to be identified and analyzed in more detail. Furthermore, the understanding of sink-source relationships and limiting steps of metabolic fluxes have to be investigated for successful application in plant breeding research that aims to increase biomass production, yield, and resilience in plants.

4.4 Protection from Oxidative Stress: Reinforcement of Antioxidants

Plants are constantly subjected to changing environmental conditions, causing them to adjust their metabolism in order to maintain the balance between energy generation and consumption (Scheibe 2019). Various abiotic and biotic stress conditions such as high light, drought, salinity, chilling, and pathogen attack can lead to enhanced generation and accumulation of ROS causing oxidative stress, due to metabolic imbalances in plant cells (Suzuki et al. 2012).

The production of ROS such as superoxide radicals, hydrogen peroxide, and singlet oxygen is an unavoidable consequence of aerobic metabolism. At high concentrations, ROS are extremely harmful, since they cause irreversible oxidative damage to cellular components (Mittler 2002). However, despite their potential to cause oxidative damage, ROS can additionally act as second messengers involved in the control of plant growth and development as well as in the activation of signaling pathways by ROS-responsive regulatory genes (Foyer and Shigeoka 2011; Schwarzländer and Finkemeier 2013; Sharma et al. 2012). Whether ROS result in oxidative damage or function as signaling molecules depends on the equilibrium between ROS generation and scavenging by antioxidative systems, thereby regulating redox homeostasis. Due to the multiple functions of ROS, a controlled dose-dependent response of plant cells is essential to avoid oxidative damage, thereby remaining a basal level of ROS for signaling purposes (Veal et al. 2007).

Almost every adverse condition leads to the production of ROS due to electron transfer to molecular oxygen when no electron acceptor is present. Any over-reduction in electron transport chains in chloroplasts or mitochondria has the potential of ROS production. Prior to such a situation, therefore, the release of excess energy by thermal dissipation prevents oxidative stress. Dissipation of energy as heat can be detected as part of NPQ in the thylakoids. In plant mitochondria, finally, any excess energy arriving there increases the pool of reduced ubiquinone and is released as heat via the various isoforms of AOX of the alternative respiratory pathway which is not energy-conserving.

Scavenging of excess ROS can be achieved by antioxidative systems that are present in various cell compartments including enzymatic and non-enzymatic antioxidants. Non-enzymatic antioxidants include the major redox buffers ascorbate and GSH, as well as proline, carotenoids, phenolics, and tocopherols. Enzymatic antioxidants comprise enzymes of the ascorbate-GSH cycle such as ascorbate peroxidase, monodehydroascorbate reductase, dehydroascorbate reductase, and GSH reductase as well as superoxide dismutase, catalase, and peroxiredoxins. The multiple attempts to generate stress-tolerant crops by manipulating ROS-dependent pathways have been summarized in a recent review (Gómez et al. 2019). As part of the antioxidant defense system in plants, peroxiredoxins function as thiol-dependent enzymes that catalyze the reduction of chemically diverse peroxides such as hydrogen peroxide, alkyl hydroperoxides, and peroxynitrite. However, peroxiredoxins have also been shown to play important roles in redox signaling during plant development and adaptation (Dietz 2011; Dietz and Pfannschmidt 2011; Liebthal et al. 2018).

Besides scavenging of excess ROS, poising mechanisms that contribute to the balancing of the ATP/NAD(P)H ratio and the avoidance of generating excess ROS due to limited electron acceptors in electron transport chains play an essential role in plant metabolism. In chloroplasts, the avoidance of oxidative stress is achieved by D1 turnover in the reaction center of photosystem II, state transitions, NPQ, the xanthophyll cycle, chlororespiration, the Mehler reaction and cyclic electron transport (Aro et al. 1993; Shikanai 2007). Besides the cyclic electron transport and other mechanisms, the redox-regulated NADP-MDH as part of the malate valve in illuminated chloroplasts uses excess NADPH generated via photosynthetic electron transport to convert OAA to malate, regenerating the electron acceptor $NADP^+$ thereby counteracting the generation of excess ROS (Hebbelmann et al. 2012; Scheibe 1987, 2004; Scheibe et al. 2005; Selinski and Scheibe 2019; Voss et al. 2013). In addition, the malate valve enables the indirect transport of reducing equivalents to other subcellular compartments that leads to the decrease of electron pressure in the photosynthetic electron transport chain and a concomitant decrease in ROS generation in chloroplasts. However, the indirect transport of reducing equivalents, in turn, leads to an increase of electron pressure in the mitochondrial electron transport chain when those are transported into plant mitochondria.

The electron transport chain in plant mitochondria is highly branched, both at the primary input in the oxidation of substrates (alternative NAD(P)H dehydrogenases) and the final reduction of O_2 to H_2O (Alternative Oxidase, AOX). These components mediate bypasses around complex I (catalyzed by alternative NAD(P)H dehydrogenases) and complexes III and IV (catalyzed by AOX) (Fernie et al. 2004; Millenaar and Lambers 2003; Møller 2001; Rasmusson et al. 2008, 2004; Saha et al. 2016; Selinski et al. 2018; Siedow and Umbach 1995; Vanlerberghe 2013; Vanlerberghe and McIntosh 1997). Therefore, electron transport through alternative NAD(P)H dehydrogenases and AOX occurs without proton translocation and ATP synthesis. Consequently, these alternative electron transport pathways mediate respiration that is uncoupled from energy conservation, and most of the energy is dissipated as heat. Based on this, these enzymes are seen as "energy wasteful" proteins. However, alternative NAD(P)H dehydrogenases and AOX are beneficial to plants due to their contribution to the maintenance of redox homeostasis and the concomitant avoidance of oxidative stress. Under adverse growth conditions, alternative electron transport pathways act as safety valves oxidizing surplus reducing equivalents in a two-electron step, thereby preventing an over-reduction of the mitochondrial electron transport chain, excess ROS formation, and feedback inhibition of metabolism (Millenaar and Lambers 2003). Besides their role in the avoidance of oxidative stress, alternative electron transport pathways in plant mitochondria were shown to be involved in various developmental processes.

Fruit ripening is accompanied by a peak in respiration and a concomitant burst of ethylene. In *Mangifera indica* and *Solanum lycopersicum*, AOX transcript and protein levels both peak at the ripe stage and AOX-RNAi tomato plants showed retarded fruit ripening, reduced carotenoid amounts, respiration, and ethylene production (Considine et al. 2001; Xu et al. 2012). Furthermore, AOX was shown to play a central role in reproductive development and fecundity in *Glycine max* (soybean). In GmAOX2b antisense plants, seed set was reduced, and ovule abortion increased compared to the wild type. This was accompanied by increased rates of pollen abortion in vivo and reduced rates of pollen germination in vitro (Chai et al. 2010). In addition, RNAi of single alternative NAD(P)H dehydrogenases has been shown to affect plant growth and development (Podgórska et al. 2018; Wallström et al. 2014a, b). In general, various scavenging and poising mechanisms have been demonstrated to be important hubs linking stress tolerance and development in plants. Therefore, components of the antioxidant defense system are promising targets for plant breeding research. However, these components underly a rather complex regulatory network that is not fully understood yet and necessitates further investigation.

4.5 Regulatory Steps (Checkpoints) in Complex Networks

The energy that is generated in the light as reducing equivalents and ATP should be optimally distributed into all endergonic reactions, and any imbalance or oxidative damage should be avoided. Such a functional network requires regulation at all levels to allow for the specific responses that are needed under changing and stressful conditions. To drive the regulatory mechanisms, additional reducing equivalents and ATP are consumed (Fig. 4.3).

A genome-wide association-mapping approach demonstrates the diversity of the systems available for the coordination of metabolism, growth, and development (Fusari et al. 2017). Also, the checkpoints that partition carbon into starch and sucrose, as well as C/N crosstalk, require highly sensitive regulatory mechanisms at all checkpoints. In nature, both N-sources, nitrate and ammonia, are generally present, and their assimilation is characterized by compartment-specific reductive steps. A well-studied example is a shift of the N-source from nitrate to ammonia assimilation when altered energy distribution is needed (Podgórska et al. 2015). In another example, for Chlamydomonas suitable for culturing in fermenters, availability of ATP and reductant at sufficient rates and extent for lipid in place of starch production could only be achieved accordingly by introducing additional compartment-specific energy-generating pathways into the plastids (Johnson and Alric 2013).

Genetic engineering of plant metabolism for the synthesis of useful products, therefore, requires considering such regulatory networks. Due to the high flexibility and adaptability, plants can adjust to major changes without any problem. Analyzing work with antisense and knock-out lines affected in metabolic pathways has led to the conclusion that alternative pathways and bypasses are then used to maintain a functional metabolism (Hanke et al. 2009). Under certain conditions, not only maintenance but even improved growth has been observed, e.g., when plastidial malate dehydrogenase (NAD- and NADP-dependent isoforms, respectively) has

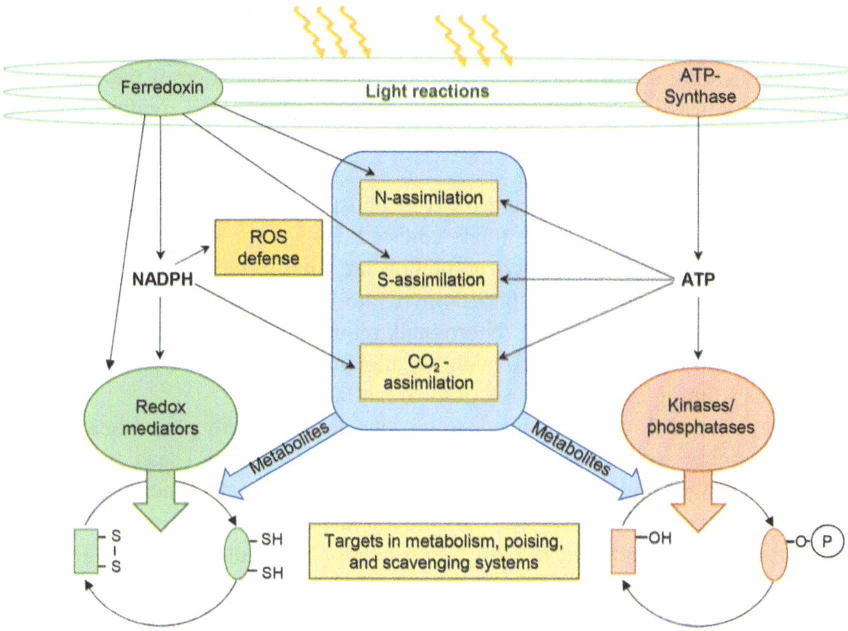

Fig. 4.3 Distribution of photosynthetic electrons and ATP for assimilation, regulation, poising, and avoidance of oxidative stress. The assimilation of C, N, and S requires electrons from the photosynthetic electron transport in the light in the form of reduced ferredoxin or as NADPH (via FNR) as well as ATP from photophosphorylation. NADPH and ATP are consumed in the assimilatory processes (CO_2, nitrate (N) and sulfate (S)) or translocated via "valve systems" across compartmental boarders (poising systems) or used to detoxify ROS (ROS scavenging) thus maintaining ROS protection. NADPH and ATP are additionally consumed for regulatory proteins that are either reversibly redox-modified by redox mediators or regulated by protein phosphorylation/dephosphorylation (kinases/phosphatases). The large number of Redox mediators is indicated by the circle named comprising numerous isoforms of thioredoxins (Trx), glutaredoxins (Grx), peroxiredoxins (Prx), and related "redoxins" consuming reductant for regulation. The protein kinases/phosphatases continuously consume ATP for the sake of regulation. The metabolites from central metabolism act as effectors on these regulatory switches thus maintaining fluxes under control according to the actual demand

been knocked out leading to enhanced alternative pathways so that growth on NH_4^+ or NO_3^- exclusively is even enhanced (Selinski and Scheibe 2014). In these cases, any redox-imbalances have been coped with by expression of alternative enzymes, i.e., regulation has occurred at the level of gene transcription. In addition to these longer-term adaptations, plants respond quickly to imbalances by adjusting the activation state at various control points by covalently or non-covalently modifying enzymes thus changing their activation states or affecting their function in other ways.

Examples for reversible redox-modifications are various chloroplast enzymes located in the different sections of the CBB cycle, in the malate valve, or in the plastidial OPPP. Here, regulation of the continuous redox-cycle is achieved specifically at each of the control steps by small molecules, namely substrates or products of the respective enzymes of the central metabolism (Fig. 4.3). They act as effectors or inhibitors of reductive and/or oxidative modification at the regulatory cysteine residues shifting the steady-state between reduced and oxidized enzyme forms and adjusting the fluxes according to the demand (Knuesting and Scheibe 2018).

Similarly, enzyme activities controlled by phosphorylation/dephosphorylation at regulatory serine residues can be finely adjusted depending on the metabolic situation. The main examples of such regulation are the cytosolic enzymes sucrose-phosphate synthase (SPS), phosphoenolpyruvate carboxylase (PEPC), and nitrate reductase (NR), all involved in C/N crosstalk required for coordinated assimilation of CO_2 and nitrate. Lastly, non-covalent regulation of enzymes is a potent means to adjust fluxes through the various sections of pathways.

Post-translational regulation needs to be considered, in particular, when external changes require fast responses or when manipulated pathways lead to changing demand as the generation of energy carriers and precursors are concerned. Also, in genetically engineered systems, not only the implementation of the immediate target (s) will determine effectiveness, but readjustment of the whole systems might either help to obtain the desired product or interfere with its synthesis. Therefore, in next-generation biotech crops, the whole plant exposed to its growth environment needs to be analyzed also when negative effects on productivity and (in case of food) safety are in focus (Martino-Catt and Sachs 2008). In chickpea, the introduction of a gene coding for an ABC transporter was found to positively affect yield also of cultivars with desired traits (Basu et al. 2019).

As the cellular environment is characterized by low content of free water and, therefore, by molecular crowding, protein-protein interactions are likely to be enhanced, and channeling should be considered as well. Optimal metabolite concentrations at the active sites of the enzymes and avoidance of the escape of toxic intermediates are resulting from microcompartmentation. To engineer such metabolons for new pathways by introducing artificial channels is a promising approach (Pröschel et al. 2015).

4.6 Plants for the Production of Tailored Products

Metabolic engineering of plant metabolism is a large field of research with the potential application for the production of tailored pharmaceuticals or nutriceuticals (Dixon 2005). For some valuable compounds, model systems have been established as a basis for a better understanding of the pathways and potential ways for their manipulation. Alkaloid biosynthesis is extensively studied in opium poppy (*Papaver somniferum*) and Madagascar periwinkle (*Catharanthus roseus*), but we are still far from understanding the coordination of all the steps concerning the regulatory mechanisms at all levels (Facchini and De Luca 2008). To understand the principles and regulation of phenylpropanoid and polyketide synthesis is another challenge and might allow producing tailored hydrocarbon chains in the future (Yu and Jez 2008).

Promising research is also progressing in the fields of aroma and pigment production that can be used for multiple purposes (Schwab et al. 2008; Tanaka et al. 2008). In addition to their importance in biotic interactions of plants with insects and animals as pollinators or for seed dispersal, and in defense, the physiological properties of such compounds are also important nutritional or medical plant-derived products of high value.

In all these cases, metabolic pathways with potentially toxic intermediates leading to physiologically active compounds should be separated from cellular metabolism of the producing cells to avoid damage. To this end, channeling creates protected subcompartments, and end products are excreted into the apoplast or stored in the vacuole. Individual transporters are present for these steps such as the ABC transporters that can even transport new metabolites as long as they are linked to GSH or glucose (Hwang et al. 2016).

Since individual products obtained from engineered plants are aimed for in current research, new enzymes and whole pathways from other organisms open up a broad range of applications. Molecular farming for the production of valuable proteins, e.g., in tobacco, can be performed in the field or cell cultures. Plants in nature produce an immense variety of chemicals for their defense. These compounds with significant effects on other organisms are traditionally already used in medicine and for other purposes. Coordinated manipulation of these pathways is now feasible using genetic engineering (see also Special Issue of The Plant Journal 54, 2008) (Dixon 2005). Also, starch and fatty acids or oils with specifically designed properties for industrial use or food processing are developed, although such genetically modified products for food is not accepted at present. Less critical is the production of cellulose or lignin for technical purposes or of hydrogen or hydrocarbons or oils as renewable resources. Such attempts have been made frequently by cultivating algae or cyanobacteria in large fermenters (Kosourov et al. 2017; Pinnola et al. 2017; Potters et al. 2010; Schenk et al. 2008). Vegetable oils are energy-rich and versatile plant products used for food, and fuel, and as precursors for technically used chemicals. In particular, the oilseed plant *Camelina sativa* is promising as it can be used for redesigning lipid metabolism to obtain the desired products (Bansal and Durrett 2016; Haslam et al. 2016).

In parallel to the attempts to modify plant metabolism for the synthesis of specialty products, it is necessary to think about long-term effects on the ongoing climate change provoked by gaseous emissions. Agriculture, with respect to the production of plant products, has the potential to influence the budget of emitted gasses if appropriate means are taken and considered. Plant-microbe interactions possess a high potential to be engineered for sustainable agriculture with less CH_4 and N_2O emissions (Philippot and Hallin 2011). Furthermore, the potential of succulent plants to use CO_2 also during the night phase by switching to CAM could be made use of for a better CO_2 balance, in particular under potentially limiting conditions of water availability (Borland et al. 2011). By introducing genes that provide resistance against drought, salinity, flooding, and other climatic challenges, the extension of arable land for the cultivation of specialty crops will be advanced. A large pool of options realized in algae and plants that are exposed to extremely cold conditions might provide the knowledge and the tools to introduce such adaptations into organisms designed for the production of specialty products (see Chap. 6 Huner et al., this book).

Alterations of carbon partitioning often lead to an increase of yield, but a limit will be reached rather soon. Instead, the manipulations of plant metabolism in every step, starting from energy capture in the photosynthetic apparatus to addition or removal of pathways to channel more energy into the desired product, are likely to improve plants as a renewable resource (see also Chap. 3 Araujo).

Biotechnological methods and genetically modified plants are needed and should become acceptable in our society if future generations are to be provided with sufficient food and energy. Science education is a prerequisite for an unbiased development and acceptance of the techniques required to engineer plants for our future survival (Borlaug 2000; Khush 2005; Ort et al. 2015).

References

Allakhverdiev SI, Kreslavski VD, Zharmukhamedov SK, Voloshin RA, Korol'kova DV, Tomo T, Shen JR (2016) Chlorophylls *d* and *f* and their role in primary photosynthetic processes of cyanobacteria. Biochem Mosc 81:201–212

Aluri S, Büttner M (2007) Identification and functional expression of the *Arabidopsis thaliana* vacuolar glucose transporter 1 and its role in seed germination and flowering. Proc Natl Acad Sci U S A 104:2537–2542

Amoedo ND, Punzi G, Obre E, Lacombe D, De Grassi A, Pierri CL, Rossignol R (2016) AGC1/2, the mitochondrial aspartate-glutamate carriers. Biochim Biophys Acta 1863:2394–2412

Amthor JS, Bar-Even A, Hanson AD, Millar AH, Stitt M, Sweetlove LJ, Tyerman SD (2019) Engineering strategies to boost crop productivity by cutting respiratory carbon loss. Plant Cell 31:297–314

Arai Y, Hayashi M, Nishimura M (2008) Proteomic identification and characterization of a novel peroxisomal adenine nucleotide transporter supplying ATP for fatty acid β-oxidation in soybean and Arabidopsis. Plant Cell 20:3227–3240

Aro EM, Virgin I, Andersson B (1993) Photoinhibition of photosystem II. Inactivation, protein damage and turnover. Biochim Biophys Acta 1143(2):113–114

Bansal S, Durrett TP (2016) *Camelina sativa*: an ideal platform for the metabolic engineering and field production of industrial lipids. Biochimie 120:9–16

Bar-Even A, Noor E, Lewis NE, Milo R (2010) Design and analysis of synthetic carbon fixation pathways. Proc Natl Acad Sci U S A 107:8889–8894

Basu U et al (2019) ABC transporter-mediated transport of glutathione conjugates enhances seed yield and quality in chickpea. Plant Physiol 180:253–275

Bergsdorf EY, Zdebik AA, Jentsch TJ (2009) Residues important for nitrate/proton coupling in plant and mammalian CLC transporters. J Biol Chem 284:11184–11193

Blankenship RE, Chen M (2013) Spectral expansion and antenna reduction can enhance photosynthesis for energy production. Curr Opin Chem Biol 17:457–461

Borland AM, Barrera Zambrano VA, Ceusters J, Shorrock K (2011) The photosynthetic plasticity of Crassulacean acid metabolism: an evolutionary innovation for sustainable productivity in a changing world. New Phytol 191:619–633

Borland AM et al (2014) Engineering Crassulacean acid metabolism to improve water-use efficiency. Trends Plant Sci 19:327–338

Borlaug (2000) Ending world hunger: the promise of biotechnology and the threat of antiscience zealotry. Plant Physiol 124:487–490

Calvin M (1976) Photosynthesis as a resource for energy and materials. Photochem Photobiol 23:425–444

Carvalho J, Madgwick PJ, Powers SJ, Keys AJ, Lea PJ, Parry MAJ (2011) An engineered pathway for glyoxylate metabolism in tobacco plants aimed to avoid the release of ammonia in photorespiration. BMC Biotechnol 11:1–17

Catoni E et al (2003) Identification of an *Arabidopsis* mitochondrial succinate-fumarate translocator. FEBS Lett 534:87–92

Chai TT, Simmonds D, Day DA, Colmer TD, Finnegan PM (2010) Photosynthetic performance and fertility are repressed in GmAOX2b antisense soybean. Plant Physiol 152:1638–1649

Charton L, Plett A, Linka N (2019) Plant peroxisomal solute transporter proteins. J Integr Plant Biol 61(7):817–835. https://doi.org/10.1111/jipb.12790

Chen YF, Wang Y, Wu WH (2008) Membrane transporters for nitrogen, phosphate and potassium uptake in plants. J Integr Plant Biol 50:835–848

Chida H et al (2007) Expression of the algal cytochrome c_6 gene in *Arabidopsis* enhances photosynthesis and growth. Plant Cell Physiol 48:948–957

Considine MJ, Daley DO, Whelan J (2001) The expression of alternative oxidase and uncoupling protein during fruit ripening in mango. Plant Physiol 126:1619–1629

Contreras L, Satrustegui J (2009) Calcium signaling in brain mitochondria: interplay of malate aspartate NADH shuttle and calcium uniporter/mitochondrial dehydrogenase pathways. J Biol Chem 284:7091–7099

Dalal J et al (2015) A photorespiratory bypass increases plant growth and seed yield in biofuel crop *Camelina sativa*. Biotechnol Biofuels 8:175–197

De Angeli A, Monachello D, Ephritikhine G, Frachisse JM, Thomine S, Gambale F, Barbier-Brygoo H (2006) The nitrate/proton antiporter AtCLCa mediates nitrate accumulation in plant vacuoles. Nature 442:939–942

Dietz KJ (2011) Peroxiredoxins in plants and cyanobacteria. Antioxid Redox Signal 15:1129–1159

Dietz KJ, Pfannschmidt T (2011) Novel regulators in photosynthetic redox control of plant metabolism and gene expression. Plant Physiol 155:1477–1485

Dixon RA (2005) Engineering of plant natural product pathways. Curr Opin Plant Biol 8:329–336

Driever SM, Kromdijk J (2013) Will C_3 crops enhanced with the C_4 CO_2-concentrating mechanism live up to their full potential (yield)? J Exp Bot 64:3925–3935

Dynowski M, Mayer M, Moran O, Ludewig U (2008) Molecular determinants of ammonia and urea conductance in plant aquaporin homologs. FEBS Lett 582:2458–2462

Eicks M, Maurino V, Knappe S, Flügge UI, Fischer K (2002) The plastidic pentose phosphate translocator represents a link between the cytosolic and the plastidic pentose phosphate pathways in plants. Plant Physiol 128:512–522

Emmerlich V, Linka N, Reinhold T, Hurth MA, Traub M, Martinoia E, Neuhaus HE (2003) The plant homolog to the human sodium/dicarboxylic cotransporter is the vacuolar malate carrier. Proc Natl Acad Sci U S A 100:11122–11126

Endler A et al (2006) Identification of a vacuolar sucrose transporter in barley and Arabidopsis mesophyll cells by a tonoplast proteomic approach. Plant Physiol 141:196–207

Erb TJ, Berg IA, Brecht V, Muller M, Fuchs G, Alber BE (2007) Synthesis of C5-dicarboxylic acids from C2-units involving crotonyl-CoA carboxylase/reductase: the ethylmalonyl-CoA pathway. Proc Natl Acad Sci U S A 104:10631–10636

Evans JR (2013) Improving photosynthesis. Plant Physiol 162:1780–1793

Facchini PJ, De Luca V (2008) Opium poppy and Madagascar periwinkle: model non-model systems to investigate alkaloid biosynthesis in plants. Plant J 54:763–784

Fernie AR, Martinoia E (2009) Malate. Jack of all trades or master of a few? Phytochemistry 70:828–832

Fernie AR, Carrari F, Sweetlove LJ (2004) Respiratory metabolism: glycolysis, the TCA cycle and mitochondrial electron transport. Curr Opin Plant Biol 7:254–261

Flexas J et al (2006) Tobacco aquaporin NtAQP1 is involved in mesophyll conductance to CO_2 *in vivo*. Plant J 48:427–439

Fliege R, Flügge UI, Werdan K, Heldt HW (1978) Specific transport of inorganic phosphate, 3-phosphoglycerate and triosephosphates across the inner membrane of the envelope in spinach chloroplasts. Biochim Biophys Acta 502:232–247

Flügge UI (1999) Phosphate translocators in plastids. Annu Rev Plant Physiol Plant Mol Biol 50:27–45

Footitt S et al (2002) Control of germination and lipid mobilization by *COMATOSE*, the *Arabidopsis* homologue of human ALDP. EMBO J 21:2912–2922

Fouracre JP, Ando S, Langdale JA (2014) Cracking the Kranz enigma with systems biology. J Exp Bot 65:3327–3339

Foyer CH, Noctor G (2011) Ascorbate and glutathione: the heart of the redox hub. Plant Physiol 155:2–18

Foyer CH, Shigeoka S (2011) Understanding oxidative stress and antioxidant functions to enhance photosynthesis. Plant Physiol 155:93–100

Fusari CM et al (2017) Genome-wide association mapping reveals that specific and pleiotropic regulatory mechanisms fine-tune central metabolism and growth in Arabidopsis. Plant Cell 29:2349–2373

Gaxiola RA, Li J, Undurraga S, Dang LM, Allen GJ, Alper SL, Fink GR (2001) Drought- and salt-tolerant plants result from overexpression of the AVP1 H^+-pump. Proc Natl Acad Sci U S A 98:11444–11449

Geiger DR, Servaites JC (1994) Diurnal regulation of photosynthetic carbon metabolism in C_3 plants. Annu Rev Plant Physiol Plant Mol Biol 45:235–256

Gómez R, Vicino P, Carrillo N, Lodeyro AF (2019) Manipulation of oxidative stress responses as a strategy to generate stress-tolerant crops. From damage to signaling to tolerance. Crit Rev Biotechnol 39:693–708

Gonzalez N et al (2009) David and Goliath: what can the tiny weed Arabidopsis teach us to improve biomass production in crops? Curr Opin Plant Biol 12:157–164

Graus D et al (2018) High V-PPase activity is beneficial under high salt loads, but detrimental without salinity. New Phytol 219:1421–1432

Gu M, Chen A, Sun S, Xu G (2016) Complex regulation of plant phosphate transporters and the gap between molecular mechanisms and practical application: what is missing? Mol Plant 9:396–416

Guo B, Jin Y, Wussler C, Blancaflor EB, Motes CM, Versaw WK (2008) Functional analysis of the Arabidopsis PHT4 family of intracellular phosphate transporters. New Phytol 177:889–898

Haferkamp I (2007) The diverse members of the mitochondrial carrier family in plants. FEBS Lett 581:2375–2379

Hanba YT, Shibasaka M, Hayashi Y, Hayakawa T, Kasamo K, Terashima I, Katsuhara M (2004) Overexpression of the barley aquaporin HvPIP2;1 increases internal CO_2 conductance and CO_2 assimilation in the leaves of transgenic rice plants. Plant Cell Physiol 45:521–529

Hanke GT, Holtgrefe S, König N, Strodtkötter I, Voss I, Scheibe R (2009) Use of transgenic plants to uncover strategies for maintenance of redox homeostasis during photosynthesis. In: Jacquot J-P (ed) Advances in botanical research, vol 52. Academic, Burlington, pp 207–251

Hanson MR, Lin MT, Carmo-Silva AE, Parry MA (2016) Towards engineering carboxysomes into C3 plants. Plant J 87:38–50

Harada H, Kuromori T, Hirayama T, Shinozaki K, Leigh RA (2004) Quantitative trait loci analysis of nitrate storage in Arabidopsis leading to an investigation of the contribution of the anion channel gene, AtCLC-c, to variation in nitrate levels. J Exp Bot 55:2005–2014

Haslam RP, Sayanova O, Kim HJ, Cahoon EB, Napier JA (2016) Synthetic redesign of plant lipid metabolism. Plant J 87:76–86

Häusler RE, Hirsch A, Kreuzaler F, Peterhänsel C (2002) Overexpression of C4-cycle enzymes in transgenic C3 plants: a biotechnological approach to improve C3-photosynthesis. J Exp Biol 53:591–607

Hayashi M, Nishimura M (2006) *Arabidopsis thaliana* - a model organism to study plant peroxisomes. Biochim Biophys Acta 1763:1382–1391

Hayashi M et al (2002) Ped3p is a peroxisomal ATP-binding cassette transporter that might supply substrates for fatty acid β-oxidation. Plant Cell Physiol 43:1–11

Hebbelmann I et al (2012) Multiple strategies to prevent oxidative stress in Arabidopsis plants lacking the malate valve enzyme NADP-malate dehydrogenase. J Exp Bot 63:1445–1459

Hoefnagel MHN, Atkin OK, Wiskich JT (1998) Interdependence between chloroplasts and mitochondria in the light and the dark. Biochim Biophys Acta 1366:235–255

Hoyos ME, Luigi P, Wertin T, Arrigoni R, Polacco JC, Palmieri F (2003) Identification of a mitochondrial transporter for basic amino acids in *Arabidopsis thaliana* by functional reconstitution into liposomes and complementation in yeast. Plant J 33:1027–1035

Hurth MA et al (2005) Impaired pH homeostasis in Arabidopsis lacking the vacuolar dicarboxylate transporter and analysis of carboxylic acid transport across the tonoplast. Plant Physiol 137:901–910

Hwang JU et al (2016) Plant ABC transporters enable many unique aspects of a terrestrial plant's lifestyle. Mol Plant 9:338–355

Jahn TP et al (2004) Aquaporin homologues in plants and mammals transport ammonia. FEBS Lett 574:31–36

Jansson C, Vogel J, Hazen S, Brutnell T, Mockler T (2018) Climate-smart crops with enhanced photosynthesis. J Exp Bot 69(16):3801–3809

Jin H et al (2016) Optimization of light-harvesting pigment improves photosynthetic efficiency. Plant Physiol 172:1720–1731

Johnson X, Alric J (2013) Central carbon metabolism and electron transport in *Chlamydomonas reinhardtii*: metabolic constraints for carbon partitioning between oil and starch. Eukaryot Cell 12:776–793

Juhnke H, Krems B, Kötter P, Entian KD (1996) Mutants that show increased sensitivity to hydrogen peroxide reveal an important role for the pentose phosphate pathway in protection of yeast against oxidative stress. Mol Gen Genet 252:456–464

Kataoka T et al (2004) Vacuolar sulfate transporters are essential determinants controlling internal distribution of sulfate in Arabidopsis. Plant Cell 16:2693–2704

Kawase M, Hanba YT, Katsuhara M (2013) The photosynthetic response of tobacco plants overexpressing ice plant aquaporin McMIPB to a soil water deficit and high vapor pressure deficit. J Plant Res 126:517–527

Kebeish R et al (2007) Chloroplastic photorespiratory bypass increases photosynthesis and biomass production in *Arabidopsis thaliana*. Nat Biotechnol 25:593–599

Khush GS (2005) What it will take to feed 5.0 billion rice consumers in 2030. Plant Mol Biol 59:1–6

Kinoshita H et al (2011) The chloroplastic 2-oxoglutarate/malate transporter has dual function as the malate valve and in carbon/nitrogen metabolism. Plant J 65:15–26

Klingenberg M (2008) The ADP and ATP transport in mitochondria and its carrier. Biochim Biophys Acta 1778:1978–2021

Knuesting J, Scheibe R (2018) Small molecules govern thiol redox switches. Trends Plant Sci 23:769–782

Kosourov S, Murukesan G, Seibert M, Allahverdiyeva Y (2017) Evaluation of light energy to H_2 energy conversion efficiency in thin films of cyanobacteria and green alga under photoautotrophic conditions. Algal Res 28:253–263

Kovermann P et al (2007) The Arabidopsis vacuolar malate channel is a member of the ALMT family. Plant J 52:1169–1180

Kromdijk J, Glowacka K, Leonelli L, Gabilly ST, Iwai M, Niyogi KK, Long SP (2016) Improving photosynthesis and crop productivity by accelerating recovery from photoprotection. Science 354:857–861

Kruger NJ, von Schaewen A (2003) The oxidative pentose phosphate pathway: structure and organisation. Curr Opin Plant Biol 6:236–246

Kumar A, Li C, Portis AR Jr (2009) *Arabidopsis thaliana* expressing a thermostable chimeric Rubisco activase exhibits enhanced growth and higher rates of photosynthesis at moderately high temperatures. Photosynth Res 100:143–153

Kurek I, Chang TK, Bertain SM, Madrigal A, Liu L, Lassner MW, Zhu G (2007) Enhanced Thermostability of *Arabidopsis* Rubisco activase improves photosynthesis and growth rates under moderate heat stress. Plant Cell 19:3230–3241

Laloi M (1999) Plant mitochondrial carriers: an overview. Cell Mol Life Sci 56:918–944

Leegood RC (2002) C_4 photosynthesis: principles of CO_2 concentration and prospects for its introduction into C_3 plants. J Exp Bot 53:581–590

Liebthal M, Maynard D, Dietz KJ (2018) Peroxiredoxins and redox signaling in plants. Antioxid Redox Signal 28:609–624

Linka N, Weber APM (2010) Intracellular metabolite transporters in plants. Mol Plant 3:21–53

Linka N, Theodoulou FL, Haslam RP, Linka M, Napier JA, Neuhaus HE, Weber AP (2008) Peroxisomal ATP import is essential for seedling development in *Arabidopsis thaliana*. Plant Cell 20:3241–3257

Loqué D, Ludewig U, Yuan L, von Wiren N (2005) Tonoplast intrinsic proteins AtTIP2;1 and AtTIP2;3 facilitate NH_3 transport into the vacuole. Plant Physiol 137:671–680

Lunn JE (2007) Compartmentation in plant metabolism. J Exp bot 58:35-47

Maeshima M (2001) Tonoplast transporters: organization and function. Annu Rev Plant Physiol Plant Mol Biol 52:469–497

Maier A, Fahnenstich H, von Caemmerer S, Engqvist MK, Weber AP, Flügge UI, Maurino VG (2012) Transgenic introduction of a glycolate oxidative cycle into *A. thaliana* chloroplasts leads to growth improvement. Front Plant Sci 3:38

Martino-Catt SJ, Sachs ES (2008) Editor's choice series: the next generation of biotech crops. Plant Physiol 147:3–5

Martinoia E, Maeshima M, Neuhaus HE (2007) Vacuolar transporters and their essential role in plant metabolism. J Exp Bot 58:83–102

Massonneau A, Martinoia E, Dietz KJ, Mimura T (2000) Phosphate uptake across the tonoplast of intact vacuoles isolated from suspension-cultured cells of *Catharanthus roseus* (L.) G. Don. Planta 211:390–395

McGrath JM, Long SP (2014) Can the cyanobacterial carbon-concentrating mechanism increase photosynthesis in crop species? A theoretical analysis. Plant Physiol 164:2247–2261

Michaeli S, Fromm H (2015) Closing the loop on the GABA shunt in plants: are GABA metabolism and signaling entwined? Front Plant Sci 6:419

Millenaar FF, Lambers H (2003) The alternative oxidase: in vivo regulation and function. Plant Biol 5:2–15

Mittler R (2002) Oxidative stress, antioxidants and stress tolerance. Trends Plant Sci 7:405–410

Møller IM (2001) Plant mitochondria, and oxidative stress: electron transport, NADPH turnover, and metabolism of reactive oxygen species. Annu Rev Plant Physiol Plant Mol Biol 52:561–591

Mueller-Cajar O, Whitney SM (2008) Evolving improved Synechococcus Rubisco functional expression in *Escherichia coli*. Biochem J 414:205–214

Murchie EH, Pinto M, Horton P (2009) Agriculture and the new challenges for photosynthesis research. New Phytol 181:532–552

Neuhaus HE (2007) Transport of primary metabolites across the plant vacuolar membrane. FEBS Lett 581:2223–2226

Neuhaus HE, Emes MJ (2000) Nonphotosynthetic metabolism in plastids. Annu Rev Plant Physiol Plant Mol Biol 51:111–140

Orr DJ, Pereira AM, da Fonseca PP, Pereira-Lima IA, Zsögön A, Araújo WL (2017) Engineering photosynthesis: progress and perspectives. F1000 Res 6:1891

Ort DR, Melis A (2011) Optimizing antenna size to maximize photosynthetic efficiency. Plant Physiol 155:79–85

Ort DR et al (2015) Redesigning photosynthesis to sustainably meet global food and bioenergy demand. Proc Natl Acad Sci U S A 112:8529–8536

Palmieri L, Lasorsa FM, De Palma A, Palmieri F, Runswick MJ, Walker JE (1997) Identification of the yeast ACR1 gene product as a succinate-fumarate transporter essential for growth on ethanol or acetate. FEBS Lett 417:114–118

Palmieri L, Picault N, Arrigoni R, Besin E, Palmieri F, Hodges M (2008) Molecular identification of three *Arabidopsis thaliana* mitochondrial dicarboxylate carrier isoforms: organ distribution, bacterial expression, reconstitution into liposomes and functional characterization. Biochem J 410:621–629

Parry MAJ, Andralojc PJ, Scales JC, Salvucci ME, Carmo-Silva AE, Alonso H, Whitney SM (2013) Rubisco activity and regulation as targets for crop improvement. J Exp Biol 64:717–730

Perez-Bueno ML, Johnson MP, Zia A, Ruban AV, Horton P (2008) The Lhcb protein and xanthophyll composition of the light harvesting antenna controls the ΔpH-dependency of non-photochemical quenching in *Arabidopsis thaliana*. FEBS Lett 582:1477–1482

Peter DM, von Borzyskowski LS, Kiefer P, Christen P, Vorholt JA, Erb TJ (2015) Screening and engineering the synthetic potential of carboxylating reductases from central metabolism and polyketide biosynthesis. Angew Chem Int Ed Engl 54:13457–13461

Philippot L, Hallin S (2011) Towards food, feed and energy crops mitigating climate change. Trends Plant Sci 16:476–480

Picault N, Palmieri L, Pisano I, Hodges M, Palmieri F (2002) Identification of a novel transporter for dicarboxylates and tricarboxylates in plant mitochondria: bacterial expression, reconstitution, functional characterization, and tissue distribution. J Biol Chem 277:24204–24211

Picault N, Hodges M, Palmieri L, Palmieri F (2004) The growing family of mitochondrial carriers in Arabidopsis. Trends Plant Sci 9:138–146

Pinnola A, Formighieri G, Bassi R (2017) Algae: a new biomass resource. In: Meyers R (ed) Encyclopedia of sustainability science and technology. Springer, New York, pp 1–33

Podgórska A, Ostaszewska M, Gardeström P, Rasmusson AG, Szal B (2015) In comparison with nitrate nutrition, ammonium nutrition increases growth of the *frostbite1 Arabidopsis* mutant. Plant Cell Environ 38:224–237

Podgórska A, Ostaszewska-Bugajska M, Borysiuk K, Tarnowska A, Jakubiak M, Burian M, Rasmusson AG, Szal B (2018) Suppression of external NADPH dehydrogenase-NDB1 in *Arabidopsis thaliana* confers improved tolerance to ammonium toxicity via efficient glutathione/redox metabolism. Int J Mol Sci 19:1412

Potters G, van Goethem D, Schutte F (2010) Promising biofuel resources: lignocellulose and algae. Nat Edu 3:14–19

Pracharoenwattana I, Smith SM (2008) When is a peroxisome not a peroxisome? Trends Plant Sci 13:522–525

Price GD, Badger MR, von Caemmerer S (2011) The prospect of using cyanobacterial bicarbonate transporters to improve leaf photosynthesis in C3 crop plants. Plant Physiol 155:20–26

Price GD et al (2013) The cyanobacterial CCM as a source of genes for improving photosynthetic CO_2 fixation in crop species. J Exp Bot 64:753–768

Pröschel M, Detsch R, Boccaccini AR, Sonnewald U (2015) Engineering of metabolic pathways by artificial enzyme channels. Front Bioeng Biotechnol 3:168

Ramesh SA, Tyerman SD, Gilliham M, Xu B (2017) γ-Aminobutyric acid (GABA) signalling in plants. Cell Mol Life Sci 74:1577–1603

Rao AG (2008) The outlook for protein engineering in crop improvement. Plant Physiol 147:6–12

Rasmusson AG, Soole KL, Elthon TE (2004) Alternative NAD(P)H dehydrogenases of plant mitochondria. Annu Rev Plant Biol 55:23–39

Rasmusson AG, Geisler DA, Møller IM (2008) The multiplicity of dehydrogenases in the electron transport chain of plant mitochondria. Mitochondrion 8:47–60

Ratajczak R (2000) Structure, function and regulation of the plant vacuolar H^+-translocating ATPase. Biochim Biophys Acta 1465:17–36

Rausch C, Zimmermann P, Amrhein N, Bucher M (2004) Expression analysis suggests novel roles for the plastidic phosphate transporter Pht2;1 in auto- and heterotrophic tissues in potato and *Arabidopsis*. Plant J 39:13–28

Renné P, Dreáen U, Hebbeker U, Hille D, Flügge UI, Westhoff P, Weber APM (2003) The *Arabidopsis* mutant *dct* is deficient in the plastidic glutamate/malate translocator DiT2. Plant J 35:316–331

Reumann S (2000) The structural properties of plant peroxisomes and their metabolic significance. Biol Chem 381:639–648

Reumann S, Weber APM (2006) Plant peroxisomes respire in the light: some gaps of the photorespiratory C2 cycle have become filled—others remain. Biochim Biophys Acta 1763:1496–1510

Reumann S, Maier E, Benz R, Heldt HW (1995) The membrane of leaf peroxisomes contains a porin-like channel. J Biol Chem 29:17559–17565

Reumann S, Bettermann M, Benz R, Heldt HW (1997) Evidence for the presence of a porin in the membrane of glyoxysomes of castor bean. Plant Physiol 115:891–899

Reumann S, Maier E, Heldt HW, Benz R (1998) Permeability properties of the porin of spinach leaf peroxisomes. Eur J Biochem 251:359–366

Riebeseel E et al (2010) The 2-oxoglutarate/malate translocator mediates amino acid and storage protein biosynthesis in pea embryos. Plant J 61:350–363

Riesmeier JW, Flügge UI, Schulz B, Heineke D, Heldt HW, Willmitzer L, Frommer WB (1993) Antisense repression of the chloroplast triose phosphate translocator affects carbon partitioning in transgenic potato plants. Proc Natl Acad Sci U S A 90:6160–6164

Rolland V, Badger MR, Price GD (2016) Redirecting the cyanobacterial bicarbonate transporters BicA and SbtA to the chloroplast envelope: soluble and membrane cargos need different chloroplast targeting signals in plants. Front Plant Sci 7:185

Rosenthal RG, Ebert MO, Kiefer P, Peter DM, Vorholt JA, Erb TJ (2014) Direct evidence for a covalent ene adduct intermediate in NAD(P)H-dependent enzymes. Nat Chem Biol 10(1):50–55

Rosenthal RG, Vogeli B, Wagner T, Shima S, Erb TJ (2017) A conserved threonine prevents self-intoxication of enoyl-thioester reductases. Nat Chem Biol 13:745–749

Saha B, Borovskii G, Panda SK (2016) Alternative oxidase and plant stress tolerance. Plant Signal Behav 11:e1256530

Satagopan S, Sun Y, Parquette JR, Tabita FR (2017) Synthetic CO_2-fixation enzyme cascades immobilized on self-assembled nanostructures that enhance CO_2/O_2 selectivity of RubisCO. Biotechnol Biofuels 10:175

Scheibe R (1987) $NADP^+$-malate dehydrogenase in C3-plants: regulation and role of a light-activated enzyme. Physiol Plant 71:393–400

Scheibe R (2004) Malate valves to balance cellular energy supply. Physiol Plant 120:21–26

Scheibe R (2019) Maintaining homeostasis by controlled alternatives for energy distribution in plant cells under changing conditions of supply and demand. Photosynth Res 139:81–91

Scheibe R, Backhausen JE, Emmerlich V, Holtgrefe S (2005) Strategies to maintain redox homeostasis during photosynthesis under changing conditions. J Exp Bot 56:1481–1489

Schenk PM et al (2008) Second generation biofuels: high efficiency microalgae for biodiesel production. Bioenergy Res 1:20–43

Schnarrenberger C, Flechner A, Martin W (1995) Enzymatic evidence for a complete oxidative pentose phosphate pathway in chloroplasts and an incomplete pathway in the cytosol of spinach leaves. Plant Physiol 108:609–614

Schneider A et al (2002) An *Arabidopsis thaliana* knock-out mutant of the chloroplast triose phosphate/phosphate translocator is severely compromised only when starch synthesis, but not starch mobilisation is abolished. Plant J 32:685–699

Schneidereit J, Häusler RE, Fiene G, Kaiser WM, Weber AP (2006) Antisense repression reveals a crucial role of the plastidic 2-oxoglutarate/malate translocator DiT1 at the interface between carbon and nitrogen metabolism. Plant J 45:206–224

Schuler ML, Mantegazza O, Weber AP (2016) Engineering C4 photosynthesis into C3 chassis in the synthetic biology age. Plant J 87:51–65

Schwab W, Davidovich-Rikanati R, Lewinsohn E (2008) Biosynthesis of plant-derived flavor compounds. Plant J 54:712–732

Schwander T, Schada von Borzyskowski L, Burgener S, Cortina NS, Erb TJ (2016) A synthetic pathway for the fixation of carbon dioxide in vitro. Science 354:900–904

Schwarzländer M, Finkemeier I (2013) Mitochondrial energy and redox signaling in plants. Antioxid Redox Signal 18:2122–2144

Selinski J, Scheibe R (2014) Lack of malate valve capacities lead to improved N-assimilation and growth in transgenic *A. thaliana* plants. Plant Signal Behav 29057:1–5

Selinski J, Scheibe R (2019) Malate valves: old shuttles with new perspectives. Plant Biol 21:21–30

Selinski J, Scheibe R, Day DA, Whelan J (2018) Alternative oxidase is positive for plant performance. Trends Plant Sci 23:588–597

Sharma P, Jha AB, Dubey RS, Pessarakli M (2012) Reactive oxygen species, oxidative damage, and antioxidative defense mechanism in plants under stressful conditions. J Bot 2012:1–26

Shen BR et al (2019) Engineering a new chloroplastic photorespiratory bypass to increase photosynthetic efficiency and productivity in rice. Mol Plant 12:199–214

Shih PM, Zarzycki J, Niyogi KK, Kerfeld CA (2014) Introduction of a synthetic CO_2-fixing photorespiratory bypass into a cyanobacterium. J Biol Chem 289:9493–9500

Shih PM, Occhialini A, Cameron JC, Andralojc PJ, Parry MA, Kerfeld CA (2016) Biochemical characterization of predicted Precambrian RuBisCO. Nat Commun 7:10382. https://doi.org/10.1038/ncomms10382

Shikanai T (2007) Cyclic electron transport around photosystem I: genetic approaches. Annu Rev Plant Biol 58:199–217

Siedow JN, Umbach AL (1995) Plant mitochondrial electron transfer and molecular biology. Plant Cell 7:821–831

Simkin AJ, McAusland L, Lawson T, Raines CA (2017a) Overexpression of the RieskeFeS protein increases electron transport rates and biomass yield. Plant Physiol 175:134–145

Simkin AJ et al (2017b) Simultaneous stimulation of sedoheptulose 1,7-bisphosphatase, fructose 1,6-bisphosphate aldolase and the photorespiratory glycine decarboxylase-H protein increases CO_2 assimilation, vegetative biomass and seed yield in Arabidopsis. Plant Biotechnol J 15:805–816

Simkin AJ, Lopez-Calcagno PE, Raines CA (2019) Feeding the world: improving photosynthetic efficiency for sustainable crop production. J Exp Bot 70:1119–1140

South PF, Cavanagh AP, Liu HW, Ort DR (2019) Synthetic glycolate metabolism pathways stimulate crop growth and productivity in the field. Science 363:1–9

Suzuki N, Koussevitzky S, Mittler R, Miller G (2012) ROS and redox signalling in the response of plants to abiotic stress. Plant Cell Environ 35:259–270

Sweetlove LJ et al (2006) Mitochondrial uncoupling protein is required for efficient photosynthesis. Proc Natl Acad Sci U S A 103:19587–19592

Szecowka M et al (2013) Metabolic fluxes in an illuminated Arabidopsis rosette. Plant Cell 25:694–714

Takahara K et al (2010) Metabolome and photochemical analysis of rice plants overexpressing Arabidopsis NAD kinase gene. Plant Physiol 152:1863–1873

Tanaka Y, Sasaki N, Ohmiya A (2008) Biosynthesis of plant pigments: anthocyanins; betalains and carotenoids. Plant J 54:733–749

Taniguchi M et al (2002) Identifying and characterizing plastidic 2-oxoglutarate/malate and dicarboxylate transporters in *Arabidopsis thaliana*. Plant Cell Physiol 43(7):706–717

Taylor SH, Long SP (2017) Slow induction of photosynthesis on shade to sun transitions in wheat may cost at least 21% of productivity. Phil Trans R Soc Lond B 372: 20160543

Tegeder M, Weber APM (2006) Metabolite transporters in the control of plant primary metabolism. In: Plaxton WC, McManus MT (eds) Control of primary metabolism in plants, vol 4. Blackwell Publishing, Oxford, pp 85–120

Vanlerberghe GC (2013) Alternative oxidase: a mitochondrial respiratory pathway to maintain metabolic and signaling homeostasis during abiotic and biotic stress in plants. Int J Mol Sci 14:6805–6847

Vanlerberghe GC, McIntosh L (1997) Alternative oxidase: from gene to function. Annu Rev Plant Physiol Plant Mol Biol 48:703–734
Veal EA, Day AM, Morgan BA (2007) Hydrogen peroxide sensing and signaling. Mol Cell 26:1–14
Vercesi AE, Borecký J, de Godoy MI, Arruda P, Cuccovia IM, Chaimovich H (2006) Plant uncoupling mitochondrial proteins. Annu Rev Plant Biol 57:383–404
von Caemmerer S, Quick WP, Furbank RT (2012) The development of C4 rice: current progress and future challenges. Science 336:1671–1672
Voss I, Sunil B, Scheibe R, Raghavendra AS (2013) Emerging concept for the role of photorespiration as an important part of abiotic stress response. Plant Biol 15:713–722
Walker BJ, VanLoocke A, Bernacchi CJ, Ort DR (2016) The costs of photorespiration to food production now and in the future. Annu Rev Plant Biol 67:107–129
Wallström SV et al (2014a) Suppression of the external mitochondrial NADPH dehydrogenase, NDB1, in *Arabidopsis thaliana* affects central metabolism and vegetative growth. Mol Plant 7:356–368
Wallström SV et al (2014b) Suppression of NDA-type alternative mitochondrial NAD(P)H dehydrogenases in *Arabidopsis thaliana* modifies growth and metabolism, but not high light stimulation of mitochondrial electron transport. Plant Cell Physiol 55:881–896
Walters RG, Ibrahim DG, Horton P, Kruger NJ (2004) A mutant of Arabidopsis lacking the triose-phosphate/phosphate translocator reveals metabolic regulation of starch breakdown in the light. Plant Physiol 135:891–906
Wang P, Kelly S, Fouracre JP, Langdale JA (2013) Genome-wide transcript analysis of early maize leaf development reveals gene cohorts associated with the differentiation of C4 Kranz anatomy. Plant J 75:656–670
Weber APM (2004) Solute transporters as connecting elements between cytosol and plastid stroma. Curr Opin Plant Biol 7:247–253
Weber A, Flügge UI (2002) Interaction of cytosolic and plastidic nitrogen metabolism in plants. J Exp Bot 53:865–874
Weber A, Menzlaff E, Arbinger B, Gutensohn M, Eckerskorn C, Flügge UI (1995) The 2-oxoglutarate/malate translocator of chloroplast envelope membranes: molecular cloning of a transporter containing a 12-helix motif and expression of the functional protein in yeast cells. Biochemistry 34:2621–2627
Weber APM, Schneidereit J, Voll LM (2004) Using mutants to probe the in vivo function of plastid envelope membrane metabolite transporters. J Exp Bot 55:1231–1244
Weber APM, Schwacke R, Flügge UI (2005) Solute transporters of the plastid envelope membrane. Annu Rev Plant Biol 56:133–164
Whitney SM, Houtz RL, Alonso H (2011) Advancing our understanding and capacity to engineer nature's CO_2-sequestering enzyme, Rubisco. Plant Physiol 155:27–35
Wingenter K et al (2010) Increased activity of the vacuolar monosaccharide transporter TMT1 alters cellular sugar partitioning sugar signaling, and seed yield in Arabidopsis. Plant Physiol 154:665–677
Wolf BM, Niedzwiedzki DM, Magdaong NCM, Roth R, Goodenough U, Blankenship RE (2017) Characterization of a newly isolated freshwater Eustigmatophyte alga capable of utilizing far-red light as its sole light source. Photosynth Res 135:177–189
Wormit A et al (2006) Molecular identification and physiological characterization of a novel monosaccharide transporter from Arabidopsis involved in vacuolar sugar transport. Plant Cell 18:3476–3490
Wu P, Xu J (2010) Does OsPHR2 central P_i signaling regulator regulate some unknown factors crucial for plant growth? Plant Signal Behav 5:712–714
Wudick MM, Luu DT, Maurel C (2009) A look inside: localization patterns and functions of intracellular plant aquaporins. New Phytol 184:289–302
Xu F, Yuan S, Zhang DW, Lv X, Lin HH (2012) The role of alternative oxidase in tomato fruit ripening and its regulatory interaction with ethylene. J Exp Bot 63:5705–5716
Yamori W (2016) Photosynthetic response to fluctuating environments and photoprotective strategies under abiotic stress. J Plant Res 129(3):379–395

Yamori W, Masumoto C, Fukayama H, Makino A (2012) Rubisco activase is a key regulator of non-steady-state photosynthesis at any leaf temperature and, to a lesser extent, of steady-state photosynthesis at high temperature. Plant J 78:871–880

Yang X et al (2015) A roadmap for research on Crassulacean acid metabolism (CAM) to enhance sustainable food and bioenergy production in a hotter, drier world. New Phytol 207:491–504

Yu O, Jez JM (2008) Nature's assembly line: biosynthesis of simple phenylpropanoids and polyketides. Plant J 54:750–762

Zhu J et al (2004) An Arabidopsis homeodomain transcription factor gene, *HOS9*, mediates cold tolerance through a CBF-independent pathway. Proc Natl Acad Sci U S A 101:9873–9878

Zhu XG, Long SP, Ort DR (2010) Improving photosynthetic efficiency for greater yield. Annu Rev Plant Biol 61:235–261

Zolman BK, Yoder A, Bartel B (2000) Genetic analysis of indole-3-butyric acid responses in *Arabidopsis thaliana* reveals four mutant classes. Genetics 156:1323–1337

Johannes Knuesting received his PhD in the laboratory of Plant Physiology headed by Prof. Dr. Renate Scheibe, and continued as a postdoctoral researcher in the same lab. He now works in a start-up company in Canada. His major research interests are redox-dependent cellular processes and their physiological impact on metabolism, growth, and development.

Renate Scheibe has held the Chair of Plant Physiology until 2019, teaching the whole field of Plant Physiology. Her research covers topics in the fields of redox regulation and energy metabolism, redox homeostasis in the view of stress adaptation using approaches from protein chemistry, enzymology, and ecophysiology. She published more than 160 papers, and is actively contributing for the scientific community with editorial tasks and as a member of the German Research Integrity group. She was elected as a Corresponding Member of The American Society of Plant Biologists and as Honorary Member of the German Botanical Society.

Jennifer Selinski received her PhD in Plant Physiology at Osnabrück University, Germany, supervised under Professor Dr. Renate Scheibe. Afterwards, she held a postdoctoral research position at La Trobe University, Melbourne, Australia, followed by a Feodor-Lynen Research Fellowship from the Alexander von Humboldt Foundation to continue her research overseas. During her time in Australia, she was associated to the Australian Research Council Centre of Excellence in Plant Energy Biology Currently, she is working at Bielefeld University in the Department of Biochemistry and Physiology of Plants headed by Prof. Dr. Karl-Josef Dietz as a Feodor-Lynen Return Fellow. Her research focuses on redox homeostasis and redox signaling as well as energy metabolism in plants.

5 Strategies to Enhance Photosynthesis for the Improvement of Crop Yields

Greta Nölke and Stefan Schillberg

Abstract

The impact of climate change and the rising demand for food, feed, and biofuels requires an increase in crop productivity without the use of additional land, water, or agrochemicals. Despite recent progress in plant breeding and biotechnology, improving crop productivity beyond existing yield potentials remains one of the greatest challenges in agricultural research. Photosynthesis is a critical process that underlies plant growth and agronomic performance, so the improvement of photosynthetic efficiency is necessary to achieve higher crop yields. Several biotechnological approaches have been proposed to increase the rate of photosynthesis in important C3 crops, including the engineering of RuBisCO, enhancing the activity of Calvin cycle enzymes, introducing CO_2-concentration mechanisms and manipulating photorespiration. However, few of these strategies have led to significantly higher crop yields in practice. In this review, we will briefly discuss the limitations of photosynthesis in C3 plants before focusing on current strategies to overcome the bottlenecks and achieve higher agricultural productivity. Finally, we consider the remaining challenges and perspectives for the future development of novel strategies to enhance the efficiency of photosynthesis.

G. Nölke
Fraunhofer Institute for Molecular Biology and Applied Ecology IME, Aachen, Germany
e-mail: greta.noelke@ime.fraunhofer.de

S. Schillberg (✉)
Fraunhofer Institute for Molecular Biology and Applied Ecology IME, Aachen, Germany

Institute for Phytopathology, Justus-Liebig University Giessen, Giessen, Germany
e-mail: stefan.schillberg@ime.fraunhofer.de

A. Kumar et al. (eds.), *Climate Change, Photosynthesis and Advanced Biofuels*,
https://doi.org/10.1007/978-981-15-5228-1_5

Keywords

Biomass · Carbon concentration Mechanism · C3 plants · Photorespiration · RuBisCo

5.1 Introduction

Over the next three decades, global crop production must double to avoid catastrophic food shortages caused by the growing population and the resulting competitive demands on land use for urbanization, biofuel and biomass production (Foley et al. 2011; Tilman et al. 2011; http://esa.un.org/unpd/wpp/index.htm). Moreover, resource scarcity and CO_2 emissions that accelerate the effects of climate change, including the frequency of unforeseen droughts, flooding, and soil erosion (IPCC 2018; Asadieh et al. 2016), pose a severe and urgent challenge in terms of increasing crop productivity. Substantial increases in the yield of staple crops such as wheat and rice were achieved half a century ago by the Green Revolution through a combination of breeding strategies and greater inputs of fertilizers, pesticides, and water, but this will not be enough to meet the projected demand for food and feed in 2030 and beyond. Recent crop-production trends are showing significant stagnation and even a decline in yield improvement, suggesting that crop yields are approaching the ceiling of maximum yield potential (Tilman et al. 2002, 2011; Zhu et al. 2008; Reynolds et al. 2012; Ray et al. 2013). This trend is already apparent across 24–39% of the growing areas of four key crops—maize, rice, wheat, and soybean—which currently represent 64% of agricultural calorie production (Ray et al. 2012). To meet the challenges associated with the growing demand for yield improvement, next-generation biotechnological solutions include increasing the efficiency of photosynthesis as the basis of primary plant productivity (Zhu et al. 2010; Long et al. 2015; Orr et al. 2017; Simkin et al. 2019).

Photosynthesis involves a series of biophysical and biochemical processes in which sunlight is captured and transformed into the energy and reducing equivalents needed to convert CO_2 into sugars and, in turn, storage carbohydrates. More than 90% of the CO_2 converted into biomass is fixed by the enzyme ribulose-1,5-bisphosphate carboxylase/oxygenase (RuBisCO) which accounts for 30–50% of the soluble protein in plant leaves. Indeed, RuBisCO is one of nature's most abundant proteins, produced by bacteria, archaea, and eukaryotes in sufficient quantities to provide 5 kg for every person on the planet (Phillips and Milo 2009). Despite its hegemonic role, RuBisCO is actually an incredibly inefficient enzyme, with a turnover frequency (for an "average" RuBisCO) of only 0.1–1 s^{-1} (http://brenda-enzymes.org), making it a limiting factor in photosynthetic CO_2 fixation under optimal conditions. Moreover, RuBisCO has the capacity to catalyze both the carboxylation and the oxygenation of ribulose 1,5-bisphosphate (RuBP). The carboxylation reaction produces two molecules of 3-phosphoglycerate (3PG) that enter the Calvin-Benson cycle to regenerate RuBP and ultimately produce glucose, sucrose, and starch, but in C3 plants the oxidative reaction produces single molecules of 3PG and 2-phosphoglycolate (2-PG). The latter is toxic to plants and is metabolized by photorespiration, causing a significant waste of resources (Peterhänsel and Maurino 2011).

The efficiency of photosynthesis in C3 plants is far from its theoretical maximum and is limited not only by the slow catalytic rate of RuBisCO and the use of O_2 for the competing process of photorespiration, but also by other factors such as the low specificity of RuBisCO for CO_2 rather than O_2 (especially at higher temperatures), the restricted availability of CO_2 at the site of CO_2 fixation, and the inability of crops to intercept solar radiation effectively. Looking at the complex process of photosynthesis in C3 plants, both the light-dependent and light-independent reactions involve proteins (enzymes, transporters, and accessory proteins) that can, in principle, be targeted for improvement. Therefore, a broad range of targets to improve photosynthesis has been identified over the last three decades, and several biotechnological approaches have been proposed and evaluated to overcome the bottlenecks in photosynthesis (Raines 2011; Long et al. 2015; Ort et al. 2015; Betti et al. 2016; Sharwood et al. 2016; Orr et al. 2017; Simkin et al. 2019; Weber and Bar-Even 2019). These include more active versions of RuBisCO and the Calvin-Benson cycle enzymes, the introduction of CO_2-concentration mechanisms (CCMs), the optimization of the light reaction, and the manipulation of photorespiration. For example, recent progress in the optimization of light collection and utilization appears very promising (Orr et al. 2017; Cardona et al. 2018). Transgenic tobacco plants with smaller light-harvesting antennae proteins accumulated 25% more stem and leaf biomass (Song et al. 2017). Rice plants with a 50% lower chlorophyll content showed a 40% increase in the rate of photosynthesis and grew faster, achieving similar yields to wild-type plants in less time (Gu et al. 2017). Finally, accelerated relaxation of non-photochemical quenching (NPQ) in tobacco plants increased biomass production under both greenhouse (20%) and field (15%) conditions (Kromdijk et al. 2016).

Other strategies aim to improve the source-sink interaction by boosting source and sink capacities simultaneously. This has been achieved by the overexpression of two plastidial translocators in potato, increasing the tuber yield by 89% under near-field conditions (Jonik et al. 2012; Sonnewald and Fernie 2018). Furthermore, the overexpression of Calvin-Benson enzymes such as sedoheptulose 1,7-bisphosphatase or fructose 1,6-bisphosphate aldolase, either alone or in combination with the photorespiratory enzymes or cyanobacterial carbon transporter B (*ictB*), has significantly enhanced carbon assimilation and biomass production in several plant species (Simkin et al. 2019). Here we briefly review current efforts to increase crop yields, focusing on the manipulation of photorespiration and the integration of CCMs into the chloroplasts of C3 plants.

5.2 Manipulating Photorespiration

Photorespiration is the light-dependent consumption of O_2 initiated by the oxygenase activity of RuBisCO, which occurs simultaneously with photosynthetic CO_2 uptake and O_2 release. The carboxylase/oxygenase activity of RuBisCO is common to all photosynthetic organisms, but whereas cyanobacteria, algae, and C4 plants have evolved strategies to reduce the impact of the oxygenase activity, in C3 plants every third to fourth molecule of RuBP is oxygenated rather than carboxylated at present-day air CO_2/O_2 ratios (Sharkey 1988; Tcherkez 2013). The situation is even worse at high temperatures as a consequence of the lower affinity of RuBisCO for CO_2 and the higher availability of O_2 within leaves (Jordan and Ogren 1984; Peterhänsel and Maurino 2011). Photorespiration is important for plants because it recovers 75% of the fixed carbon (Leegood et al. 1995; Tolbert 1997) and efficiently removes toxic 2-PG, which inhibits the Calvin-Benson cycle enzymes, RuBisCO activation, chloroplast functions, and RuBP regeneration (Anderson 1971; Kelly and Latzko 1976; Artus et al. 1986; Häusler et al. 1996; Gonzalez-Moro et al. 1997; Flügel et al. 2017).

Recently, photorespiration in C3 plants has been re-evaluated and may be more efficient than previously anticipated by stimulating the production of malate in the chloroplast, allowing energy-demanding reactions such as those involved in nitrate assimilation (Bloom and Lancaster 2018). However, photorespiration also consumes energy and reducing equivalents (3.5 ATP and 2 NADPH per RuBP oxygenated and regenerated) and 25% of the already-fixed carbon is again released as CO_2 in the mitochondria (Peterhänsel et al. 2010; Peterhänsel and Maurino 2011; Betti et al. 2016; Walker et al. 2016a, 2016b). Photorespiration is therefore considered a futile cycle (Walker et al. 2016b; Betti et al. 2016) and has been an attractive target for crop improvement ever since the energy losses via this pathway were identified in the 1970s (Peterhänsel and Maurino 2011; Betti et al. 2016; Huma et al. 2018; South et al. 2018).

Initial approaches to improve photorespiration focused on screening for highly productive genotypes with naturally high photosynthetic rates combined with naturally low photorespiration rates (Medrano et al. 1995; Betti et al. 2016). But after 40 years of fruitless searching, the modification of photorespiration flux by genetic engineering now appears as the most promising approach to increase productivity. Supported by modeling studies indicating that even a modest reduction in the abundance of photorespiratory proteins could achieve better nitrogen distribution and higher CO_2 assimilation (Zhu et al. 2007), antisense strategies were designed to reduce the activity of individual photorespiratory enzymes. The first attempts to increase yields by inhibiting photorespiration were unsuccessful because the photorespiration mutant phenotype was lethal under ambient CO_2 conditions (Somerville 1984, 2001). Antisense suppression of the key photorespiratory enzymes also has a negative effect on photosynthesis, growth rates, and productivity in C3 plants such as Arabidopsis, potato, and rice (Betti et al. 2016; Simkin et al. 2019). In contrast, the overexpression of glycine cleavage system components such as the H-protein and L-protein boosted photosynthesis and achieved a significant biomass increase in Arabidopsis (Timm et al. 2012, 2015, 2016; Simkin et al. 2015, 2017). Similarly, mesophyll-specific overexpression of the H-protein in tobacco increased biomass by up to 47% in the field, suggesting that a higher photorespiratory capacity can minimize the accumulation of toxic intermediates and reduce their negative feedback on the Calvin-Benson cycle (Lopez-Calcagno et al. 2018).

Another approach used to improve crop yields is the re-engineering of photorespiration by introducing energetically favorable photorespiratory bypass systems that optimizes the recycling of 2-PG. Several alternative pathways (APs) have been reported in C3 plants. Kebeish et al. (2007) first demonstrated the feasibility of such an approach by introducing the *Escherichia coli* glycolate pathway (AP1) into Arabidopsis chloroplasts to convert glycolate into glycerate and release CO_2 in the vicinity of RuBisCO. The plants were subsequently transformed with genes encoding three subunits (D, E, and F) of glycolate dehydrogenase (GlcDH), as well as glyoxylate carboxyligase and tartronic semialdehyde reductase. In the transgenic plants, there was less flux through the native photorespiratory pathway, a higher rate of photosynthesis, higher carbohydrate levels, and up to 30% more shoot and root biomass under short-day conditions (Kebeish et al. 2007). Similarly, when the bacterial glycolate pathway was expressed in the chloroplasts of the oilseed crop *Camelina sativa*, the seed yield increased by 57–73% (Dalal et al. 2015). Remarkably, most of these effects were also achieved when only the genes encoding the three subunits of bacterial GlcDH, the first enzyme in the pathway, were introduced (Kebeish et al. 2007; Dalal et al. 2015) providing evidence for a glycolate oxidation pathway in the plastids of wild-type plants (Peterhänsel and Maurino 2011).

Blume et al. (2013) provided additional data showing that some of the glyoxylate produced in the chloroplast by GlcDH activity may be completely oxidized to CO_2 by an endogenous pyruvate dehydrogenase. However, the installation of GlcDH activity requires the overexpression of all three subunits (D, E, and F) of the enzyme. In *E. coli* and Arabidopsis, the subunits are expressed at different levels, with the relative scarcity of the F subunit limiting the assembly of the complete DEF complex (Pellicer et al. 1996; Kebeish et al. 2007). To avoid this happening *in planta*, the three subunits have been fused into a polyprotein by joining the three corresponding bacterial genes (*glc*D, *glc*E, and *glc*F) with flexible linkers to ensure stoichiometric expression (Nölke et al. 2014). The resulting recombinant polyprotein (DEFp) retained the activity of the native GlcDH when expressed in *E. coli*, confirming the strategy was viable. Accordingly, transgenic potato lines expressing DEFp showed reduced photorespiration and more efficient CO_2 uptake with a significant impact on carbon metabolism. The higher carbohydrate levels produced in the leaves were utilized by the strong sink capacity of the tubers, increasing the tuber yield by 2.3-fold in the greenhouse (Nölke et al. 2014) and by 29% in semi-field trials performed in collaboration with KWS (Fig. 5.1). Comparable phenotypic effects were also observed when tobacco plants overexpressing DEFp were grown under normal and nitrogen-restricted conditions (Fig. 5.2). In addition to the higher shoot and root biomass, the transgenic lines also suffered less chlorosis under nitrogen-restricted conditions indicating that bypassing photorespiration resulted in higher nutrient use efficiency, in agreement with the prediction that plants expressing a photorespiratory bypass may gain a 15% increase in nitrogen use efficiency (Long et al. 2015; Xin et al. 2015).

The second photorespiratory bypass approach (AP2) was to introduce a pathway for the complete oxidation of glycolate to CO_2 into Arabidopsis chloroplasts

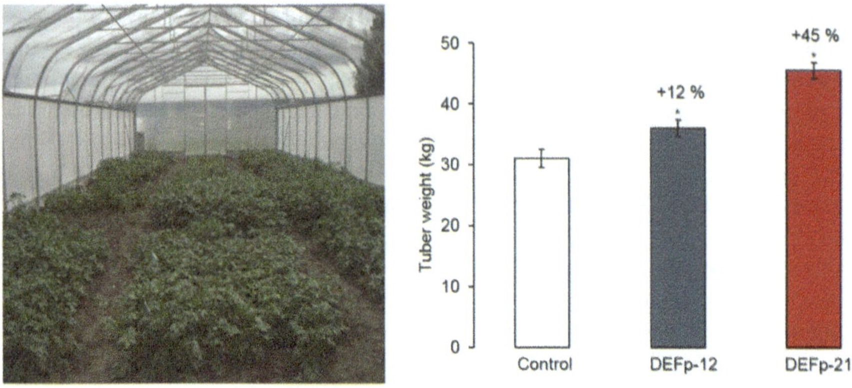

Fig. 5.1 Semi-field trails of transgenic potato plants expressing engineered glycolate dehydrogenase polyprotein (DEFp). Left: Representative image of semi-field trials performed from May to October 2013 at KWS Saat (Einbeck, Germany). Right: tuber yield in semi-field-grown wild-type and DEFp transgenic lines 12 and 21. Data are means ± standard deviations ($n = 36$; $*p < 0.05$)

Fig. 5.2 Phenotype of 6-week-old representative wild-type and transgenic DEFp tobacco grown for 3 weeks in hydroponic culture under restricted-nitrogen conditions. Plants expressing DEFp produced more shoot and root biomass and suffered less chlorosis than wild-type control plants

(Fahnenstich et al. 2007; Maier et al. 2012). Despite the higher energetic costs (Peterhänsel et al. 2013) and the predicted low rate of RuBP re-supply suggesting potentially slower rates of photosynthesis (Xin et al. 2015), this approach also significantly reduced photorespiratory flux, increased the rate of photosynthesis and improved plant growth (Maier et al. 2012).

According to Peterhänsel and Maurino (2011), both AP1 and AP2 offer several advantages compared to the native photorespiratory pathway in C3 plants, including the release of CO_2 in the chloroplast to build up the CO_2 concentration around RuBisCO, the prevention of ammonia release and related energy costs for re-fixation, and the production of additional reducing equivalents in the chloroplast. However, the impact of these theoretical benefits has yet to be investigated in detail.

To avoid the release of ammonia during photorespiration, Carvalho et al. (2011) engineered an alternative photorespiratory bypass pathway similar to that found in cyanobacteria (Eisenhut et al. 2008), which is based on the expression of glyoxylate carboxyligase and hydroxypyruvate isomerase in the peroxisome. However, the transgenic tobacco plants produced glyoxylate carboxyligase but not hydroxypyruvate isomerase, and chlorotic lesions appeared close to the leaf veins when the plants grew in normal air (Carvalho et al. 2011). The results also indicated that photorespiration was not completely bypassed and some glyoxylate was diverted from glycine into a deleterious short-circuit of the photorespiratory nitrogen cycle.

South et al. (2019) compared the performance of three photorespiratory bypass pathways in tobacco plants grown in the greenhouse and field. AP1 was the *E. coli* glycolate pathway (Kebeish et al. 2007), AP2 was the glycolate oxidase, malate synthase and catalase pathway (Maier et al. 2012), and AP3 was the malate synthase and green algal glycolate dehydrogenase pathway. To maximize flux through these pathways, RNA interference (RNAi) was used to silence the native chloroplast glycolate transporter in the photorespiratory pathway. In greenhouse experiments, AP2 showed no significant improvement compared to wild-type plants, but AP1 increased the biomass by almost 13% and AP3 by 18% without RNAi and by 24% with RNAi. Furthermore, the AP1 and AP3 transgenic plants performed better than wild-type plants in the field, with 16% and >25% increases in biomass, respectively. The combination of AP3 and RNAi produced tobacco plants with >40% higher productivity in the field (South et al. 2019).

Most recently, a novel photorespiratory bypass was established in rice (Shen et al. 2019), which resembles the AP2 strategy because it involves the complete oxidation of glycolate to CO_2 (Maier et al. 2012). This so-called GOC pathway (named after the three endogenous rice enzymes glycolate oxidase, oxalate oxidase, and catalase) differs from the AP2 strategy in that it produces no additional reducing equivalents but instead yields H_2O_2. Based on energy demand calculations (Peterhänsel et al. 2013) the GOC bypass is energetically the most wasteful pathway, but GOC transgenic rice plants nevertheless showed greater photosynthetic efficiency, faster growth, and productivity in the field increased by more than 25% (Shen et al. 2019). The authors suggested that these improvements resulted mainly from a photosynthetic CO_2-concentrating effect in the chloroplast, rather than an improved energy balance.

Given the promising results achieved by engineering photorespiratory bypass pathways and the recent advances in synthetic biology, more ambitious photorespiratory bypasses could be designed in the future to successfully mitigate the negative effects of the native photorespiratory flux in C3 plants.

5.3 Integrating CO2-Concentrating Mechanisms into the Chloroplasts of C3 Plants

Another approach to increase the CO_2 concentration near RuBisCO and thus to mitigate the effect of RuBisCO oxygenation activity is the integration of a CO_2-concentration mechanism (CCM) into the chloroplasts of C3 plants. Cyanobacteria, algae and non-C3 higher plants (CAM and C4 plants) have evolved efficient biochemical and biophysical CCMs as an adaptive response to low atmospheric CO_2 concentrations, higher photorespiratory pressure, and high temperatures (Sage et al. 2012; Rae et al. 2017; South et al. 2018). Inspired by these naturally occurring CCMs, researchers have attempted to transfer such mechanisms into C3 plants (South et al. 2018). For example, the C4 Rice Consortium is currently looking at ways to introduce the C4 pathway into rice (http://c4rice.irri.org; Langdale 2011; von Caemmerer et al. 2012), whereas the RIPE consortium (https://ripe.illinois.edu/) is investigating multiple CCM strategies to increase the photosynthetic efficiency and yields of staple food crops such as soybean, rice, cassava, and cowpea in Sub-Saharan Africa. CO_2 fixation in C4 plants is dependent on interactions between leaf mesophyll cells and photosynthetic bundle sheath cells, so the transfer of C4-like photosynthesis into C3 plants is a long-term goal whose feasibility has yet to be demonstrated (Sedelnikova et al. 2018; Weber and Bar-Even 2019).

Cyanobacteria and unicellular green algae such as *Chlamydomonas reinhardtii* have evolved highly efficient CCMs that allow cells to accumulate intracellular carbon up to 1000-fold from low-CO_2 environments (Wang et al. 2015). Algal and cyanobacterial CCMs share three major functional similarities enabling them to cope not only with the low solubility and diffusion of CO_2 through water but also with the highly variable supply of inorganic carbon (Ci) (Mangan et al. 2016; Rae et al. 2017). First, both CCMs involve active energy-dependent Ci uptake via bicarbonate transporters to increase the cellular bicarbonate concentration.

Second, the bicarbonate is dehydrated to CO_2 by carbonic anhydrases located near RuBisCO and sequestered in subcellular micro-compartments (cyanobacterial carboxysomes) or regions (algal pyrenoids). Finally, both CCMs include strategies to minimize CO_2 diffusion from the site of carboxylation (Kerfeld and Melnicki 2016; Meyer et al. 2016).

Theoretical considerations based on CO_2/HCO_3^- diffusion-reaction kinetic models indicated that introducing cyanobacterial or algal CCMs into C3 plants would enhance photosynthetic efficiency and crop yields (Price et al. 2011, 2013; McGrath and Long 2014). This would require three key modifications: first, the expression of active Ci transporters in leaf cells to increase CO_2 levels in the vicinity of RuBisCO (Price et al. 2013); second, the formation of compartments resembling carboxysomes (Price et al. 2011; Zarzycki et al. 2013) or pyrenoids (Badger and Price 1994; Raven 2010; McGrath and Long 2014); and third, the expression of a heterologous and/or engineered RuBisCO with the ability to assemble into CCM-like structures that accelerate CO_2 fixation (Lin et al. 2014a).

The first promising steps toward these goals were the assembly of the shell proteins of β-carboxysomes in higher plants (Lin et al. 2014b), followed by the expression of cyanobacterial RuBisCO in the stroma of tobacco chloroplasts alone or in combination with either the assembly chaperone RbcX or the internal carboxysomal protein CcmM35 (Lin et al. 2014a; Occhialini et al. 2016). The engineered RuBisCO showed higher rates of CO_2 fixation per unit of enzyme, but the transplastomic tobacco lines were able to survive only in a CO_2-enriched atmosphere (Occhialini et al. 2016). To ensure the incorporation of RuBisCO into the pyrenoid-like structures, Atkinson et al. (2017) re-engineered Arabidopsis RuBisCO to incorporate either the algal RuBisCO small subunit or surface-exposed algal small subunit α-helices that are considered essential for the recruitment of RuBisCO to the pyrenoid. Recently, tobacco plants have been engineered to replace endogenous RuBisCO with a cyanobacterial counterpart in carboxysomes by expressing the cyanobacterial Form-1A RuBisCO large and small subunit genes (Long et al. 2018). Although the transgenic plants grew more slowly than wild-type controls, they could be used as background lines for the addition of further CCM components to C3 plants.

Modeling has predicted that the integration of one or two active bicarbonate transporters could lead to a 5–10% increase in biomass on a daily basis (Price et al. 2011; McGrath and Long 2014). Even a single cyanobacterial bicarbonate transporter (BicA, BCT1 or SbtA) could increase light-saturated CO_2 assimilation rates by 9%, whereas incorporating all three together could achieve gains of up to 16% under light-saturated conditions (McGrath and Long 2014). However, the expression of BicA alone in tobacco chloroplasts did not improve CO_2 assimilation rates or growth compared to wild-type control plants (Pengelly et al. 2014) suggesting that BicA was inactive (or at least less active) in C3 chloroplasts, as also reported in *E. coli* (Du et al. 2014). The expression of additional cyanobacterial bicarbonate transporters has been hampered because it was not clear how to target nuclear-encoded transmembrane proteins from organisms lacking plastids to the chloroplast envelope in plants. However, Rolland et al. (2016) successfully targeted BicA and SbtA to the chloroplast envelope of *Nicotiana benthamiana* cells by transient expression using the N-terminus of Arabidopsis inner-envelope proteins containing a cleavable chloroplast transit peptide and a membrane protein leader. The effect of BicA and SbtA on photosynthesis and plant growth has yet to be evaluated.

In *C. reinhardtii*, at least 14 genes are needed to maintain a fully functional CCM under ambient or low CO_2 concentrations (Wang et al. 2015). Recent efforts to understand the algal CCM and to identify key components have resulted in the discovery of various inorganic bicarbonate transporters and carbonic anhydrases among the core proteins, similar to the cyanobacterial system (Meyer et al. 2016).

Several algal CCM components, including carbonic anhydrases and putative bicarbonate transporters, have been successfully incorporated into higher plant chloroplasts (Meyer et al. 2016; Atkinson 2016; Nölke et al. 2019). Recently, the efficient transport of *C. reinhardtii* bicarbonate transporter LCIA to the inner membrane of tobacco chloroplasts was confirmed by electron microscopy (Nölke et al. 2019). Although the expression of individual components of the *C. reinhardtii* CCM did not enhance the growth of Arabidopsis plants, constitutive LCIA expression in the inner membrane of tobacco chloroplasts enhanced the rate of photosynthesis by 9%, consistent with theoretical models (Price et al. 2011). The transgenic tobacco plants grew more rapidly, accumulated more carbohydrates, and produced 48% more biomass than wild-type controls. Similarly, the expression of the algal α-type carbonic anhydrase CAH3 in the thylakoid lumen boosted the efficiency of photosynthetic CO_2 assimilation and resulted in the accumulation of more biomass (Nölke et al. 2019). The substantial improvements achieved using single CCM components suggest that the introduction of other cyanobacterial and/or algal CCM components may accomplish even greater increases in photosynthetic performance and biomass accumulation in C3 plants.

5.4 Challenges and Future Prospects

The modification of photosynthesis to achieve higher crop yields is a promising strategy, but individual approaches are insufficient to meet the challenges for food and feed production over the next 30 years. Plant growth is a complex and multifactorial process, and combinatorial approaches are therefore needed to achieve the next breakthrough in agricultural productivity, addressing not only photosynthesis but also water and nitrogen use efficiency, source-sink interactions as well as plant defenses against biotic and abiotic stress. Recent advances in plant transformation, genome editing, and synthetic biology will allow the simultaneous introduction of many transgenes combined with the targeted editing of endogenous genes, thus allowing the implementation of more sophisticated strategies for the improvement of crop yields.

Acknowledgments The authors gratefully acknowledge Dr. Wolfgang Koch and Dr. Markus Niessen (KWS Saat, Einbeck, Germany) for the design and management of semi-field studies. We thank Dr. Richard M. Twyman for manuscript editing.

References

Anderson LE (1971) Chloroplast and cytoplasmic enzymes. II. Pea leaf triose phosphate isomerases. Biochim Biophys Acta 235:237–244

Artus NN, Somerville SC, Somerville CR (1986) The biochemistry and cell biology of photorespiration. CRC Crit Rev Plant Sci 4:121–147

Asadieh B, Krakauer NY, Fekete BM (2016) Historical trends in mean and extreme runoff and streamflow based on observations and climate models. Water 8:189. https://doi.org/10.3390/

w8050189. Atkinson N, Feike D, Mackinder LC, Meyer MT, Griffiths H, Jonikas MC, Smith AM, McCormick

Atkinson J (2016) Introducing an algal carbon-concentrating mechanism into higher plants: location and incorporation of key components. Plant Biotechnol J 14:1302–1315

Atkinson N, Leitao N, Orr DJ, Meyer MT, Carmo-Silva E, Griffiths H, Smith AM, McCormick AJ (2017) Rubisco small subunits from the unicellular green alga Chlamydomonas complement Rubiscodeficient mutants of Arabidopsis. New Phytol 214:655–667

Badger MR, Price GD (1994) The role of carbonic anhydrase in photosynthesis. Annu Rev Plant Physiol Plant Mol Biol 45:369–392

Betti M, Bauwe H, Busch FA, Fernie AR, Keech O, Leveys M, Ort DO, Parry AAJ, Sage R, Timm S, Walker B, Weber APM (2016) Manipulating photorespiration to increase plant productivity: recent advances and perspectives for crop improvement. J Exp Bot 67:2977–2988

Bloom AJ, Lancaster KM (2018) Manganese binding to Rubisco could drive a photorespiratory pathway that increases the energy efficiency of photosynthesis. Nat Plants 4:414–422

Blume C, Behrens C, Eubel H, Braun H-P, Peterhänsel C (2013) A possible role for the chloroplast pyruvate dehydrogenase complex in plant glycolate and glyoxylate metabolism. Phytochemistry 95:168–176

Cardona T, Shao S, Nixon PJ (2018) Enhancing photosynthesis in plants: the light reactions. Essays Biochem 62:85–94

Carvalho JF, Madgwick PJ, Powers SJ, Keys AJ, Lea PJ, Parry MAJ (2011) An engineered pathway for glyoxylate metabolism in tobacco plants aimed to avoid the release of ammonia in photorespiration. BMC Biotechnol 11:111. https://doi.org/10.1186/s13068-015-0357-1

Dalal J, Lopez H, Vasani NB, Hu ZH, Swift JE, Yalamanchili R, Dvora M, Lin XL, Xie DY, Qu RD, Sederoff HW (2015) A photorespiratory bypass increases plant growth and seed yield in biofuel crop *Camelina sativa*. Biotechnol Biofuels 8:175. https://doi.org/10.1186/s13068-015-0357-1

Du J, Förster B, Rourke L, Howitt SM, Price DG (2014) Characterisation of cyanobacterial bicarbonate transporters in *E. coli* shows that SbtA homologs are functional in this heterologous expression system. PLoS One 9:e115905. https://doi.org/10.1371/journal.pone.0115905

Eisenhut M, Ruth W, Haimovich M, Bauwe H, Kaplan A, Hagemann M (2008) The photorespiratory glycolate metabolism is essential for cyanobacteria and might have been conveyed endo-symbiontically to plants. Proc Natl Acad Sci U S A 105:17199–17204

Fahnenstich H, Saigo M, Niessen M, Zanor MI, Andreo CS, Ferni AR, Drincovich MF, Flügge U-I, Maurino VG (2007) Alteration of organic acid metabolism in *Arabidopsis thaliana* overexpressing the maize C4- NADP-malic enzyme causes accelerated senescence during extended darkness. Plant Physiol 145:640–652

Flügel F, Timm S, Arrivault S, Florian A, Stitt M, Fernie AR, Bauwe H (2017) The photorespiratory metabolite 2-phosphoglycolate regulates photosynthesis and starch accumulation in Arabidopsis. Plant Cell 29:2537–2551

Foley JA, Ramankutty N, Brauman KA, Cassidy ES, Gerber JS, Johnston M, Mueller ND, O'Connell C, Ray DK, West PC et al (2011) Solutions for a cultivated planet. Nature 478:337–342

Gonzalez-Moro B, Lacuesta M, Becerril JM, Gonzalez-Murua C, Munoz-Rueda A (1997) Glycolate accumulation causes a decrease of photosynthesis by inhibiting RUBISCO activity in maize. J Plant Physiol 150:388–394

Gu JF, Zhou ZX, Li ZK, Chen Y, Wang ZQ, Zhang H (2017) Rice (Oryza sativa L.) with reduced chlorophyll content exhibit higher photosynthetic rate and efficiency, improved canopy light distribution, and greater yields than normally pigmented plants. Field Crop Res 200:58–70

Häusler RE, Bailey KJ, Lea PJ, Leegood RC (1996) Control of photosynthesis in barley mutants with reduced activities of glutamine synthetase and glutamate synthase. 3. Aspects of glyoxylate metabolism and effects of glyoxylate on the activation state of ribulose-1,5-bisphosphate carboxylase-oxygenase. Planta 200:388–396

Huma B, Kundu S, Poolman MG, Kruger NJ, Fell DA (2018) Stoichiometric analysis of the energetics and metabolic impact of photorespiration in C3 plants. Plant J 96:1228–1241
IPCC (2018) Global warming of 1.5 °C special report. October 2018. https://www.ipcc.ch/sr15/
Jonik C, Sonnewald U, Hajirezaei MR, Flügge UI, Ludewig F (2012) Simultaneous boosting of source and sink capacities doubles tuber starch yield of potato plants. Plant Biotechnol J 10:1088–1098
Jordan DB, Ogren WL (1984) The CO2/O2 specificity of ribulose 1,5-bisphosphate carboxylase/oxygenase. Planta 161:308–313
Kebeish R, Niessen M, Thiruveedhi K, Bari R, Hirsch HJ, Rosenkranz R, Stäbler N, Schönfeld B, Kreuzaler F, Peterhänsel C (2007) Chloroplastic photorespiratory bypass increases photosynthesis and biomass production in *Arabidopsis thaliana*. Nat Biotechnol 25:593–599
Kelly GJ, Latzko E (1976) Inhibition of spinach-leaf phosphofructokinase by 2-phosphoglycollate. FEBS Lett 68:55–58
Kerfeld CA, Melnicki MR (2016) Assembly, function and evolution of cyanobacterial carboxysomes. Curr Opin Plant Biol 31:66–75
Kromdijk J, Głowacka K, Leonelli L, Gabilly ST, Iwai M, Niyogi KK, Long SP (2016) Improving photosynthesis and crop productivity by accelerating recovery from photoprotection. Science 354:857–861
Langdale JA (2011) C4 cycles: past, present, and future research on C4 photosynthesis. Plant Cell 23:3879–3892
Leegood R, Lea PJ, Adcock MD, Häusler RE (1995) The regulation and control of photorespiration. J Exp Bot 46:1397–1414
Lin MT, Occhialini A, Andralojc PJ, Parry MAJ, Hanson MR (2014a) A faster Rubisco with potential to increase photosynthesis in crops. Nature 513:547–550
Lin MT, Occhialini A, Andralojc PJ, Devonshire J, Hines KM, Parry MA, Hanson MR (2014b) β-Carboxysomal proteins assemble into highly organized structures in Nicotiana chloroplasts. Plant J 79:1–12
Long SP, Marshallcolon A, Zhu XG (2015) Meeting the global food demand of the future by engineering crop photosynthesis and yield potential. Cell 161:56–66
Long BM, Hee WY, Sharwood RE, Rae BD, Kaines S, Lim YL, Nguyen ND, Massey B, Bala S, von Caemmerer S, Badger MR, Price GD (2018) Carboxysome encapsulation of the CO2-fixing enzyme Rubisco in tobacco chloroplasts. Nat Commun 9:3570. https://doi.org/10.1038/s41467-018-06044-0
Lopez-Calcagno PE, Fisk S, Brown KL, Bull SE, South PF, Raines CA (2018) Overexpressing the H-protein of the glycine cleavage system increases biomass yield in glasshouse and field grown transgenic tobacco plants. Plant Biotechnol J 16:1–11
Maier A, Fahnenstich H, von Caemmer S, Engqvist MKM, Weber APM, Flügge U-I, Maurino VG (2012) Transgenic introduction of a glycolate oxidative cycle into *A. thaliana* chloroplasts leads to growth improvement. Front Plant Sci 3:1–12
Mangan NM, Flamholz A, Hood RD, Milo R, Savage DF (2016) pH determines the energetic efficiency of the cyanobacterial CO2 concentrating mechanism. Proc Natl Acad Sci U S A 113: E5354–E5362. https://doi.org/10.1073/pnas.1525145113
McGrath JM, Long SP (2014) Can the cyanobacterial carbon-concentrating mechanism increase photosynthesis in crop species? A theoretical analysis. Plant Physiol 164:2247–2261
Medrano H, Keys AJ, Lawlor DW, Parry MAJ, Azcon-Bieto J, Delgado E (1995) Improving plant production by selection for survival at low CO2 concentrations. J Exp Bot 46:1389–1396
Meyer MT, McCormick AJ, Griffiths H (2016) Will an algal CO2-concentrating mechanism work in higher plants? Curr Opin Plant Biol 31:181–188
Nölke G, Houdelet M, Kreuzaler F, Peterhänsel C, Schillberg S (2014) The expression of a recombinant glycolate dehydrogenase polyprotein in potato (*Solanum tuberosum*) plastids strongly enhances photosynthesis and tuber yield. Plant Biotechnol J 12:734–742

Nölke G, Barsoum M, Houdelet M, Arcalis E, Kreuzaler F, Fischer R, Schillberg S (2019) The integration of algal carbon concentration mechanism components into tobacco chloroplasts increases photosynthetic efficiency and biomass. Biotechnol J 14:e1800170

Occhialini A, Lin MT, Andralojc PJ, Hanson MR, Parry MAJ (2016) Transgenic tobacco plants with improved cyanobacterial Rubisco expression but no extra assembly factors grow at near wild-type rates if provided with elevated CO2. Plant J 85:148–160

Orr DJ, Pereira AM, Pereira PF, Pereira-Lima IA, Zsögön A, Araujo WL (2017) Engineering photosynthesis: progress and perspectives. F1000 Res 6:1891. https://doi.org/10.12688/f1000re-search.12181.1

Ort DR, Merchant SS, Alric J, Barkan A, Blankenship RE, Bock R, Croce R et al (2015) Redesigning photosynthesis to sustainability meet global food and energy demand. Proc Natl Acad Sci U S A 112:8529–8536

Pellicer MT, Badia J, Aguilar J, Baldoma L (1996) Glc locus of *Escherichia coli*: characterization of genes encoding the subunits of glycolate oxidase and the glc regulator protein. J Bacteriol 178:2051–2059

Pengelly JJ, Forster B, von Caemmerer S, Badger M, Price GD, Whitney SM (2014) Transplastomic integration of cyanobacteria bicarbonate transporter into tobacco chloroplasts. J Exp Bot 12:3071–3080

Peterhänsel C, Maurino VG (2011) Photorespiration redesigned. Plant Physiol 155:49–55

Peterhänsel C, Horst I, Niessen M, Blume C, Kebeish R, Kürkcüoglu S, Kreuzaler F (2010) Photorespiration. Arabidopsis Book 8:e0130

Peterhänsel C, Blume C, Offermann S (2013) Photorespiratory bypasses: how can they work? J Exp Bot 64:709–715

Phillips R, Milo R (2009) A feeling for the numbers in biology. Proc Natl Acad Sci U S A 106:21465–21471

Price GD, Badger MR, von Caemmerer S (2011) The prospect of using cyanobacterial bicarbonate transporters to improve leaf photosynthesis in C3 crop plants. Plant Physiol 155:20–26

Price GD, Pengelly JJL, Forster B, Du J, Whitney SM, von Caemmerer S, Badger MR, Howitt SM, Evans JR (2013) The cyanobacterial CCM as a source of genes for improving photosynthetic CO2 fixation in crop species. J Exp Bot 3:753–768

Rae BD, Long BM, Föster B, Nguyen ND, Velanis CN, Atkinson N, Hee WY, Mukherjee B, Price D, McCormick AJ (2017) Progress and challenges of engineering a biophysical CO2-concentrating mechanism into higher plants. J Exp Bot 68:3717–3737

Raines CA (2011) Increasing photosynthetic carbon assimilation in C3 plants to improve crop yield: current and future strategies. Plant Physiol 155:36–42

Raven JA (2010) Inorganic carbon acquisition by eukaryotic algae: four current questions. Photosynth Res 106:123–134

Ray DK, Ramankutty N, Mueller ND, West PC, Foley JA (2012) Recent patterns of crop yield growth and stagnation. Nat Commun 3:1293. https://doi.org/10.1038/ncomms2296

Ray DK, Mueller ND, West PC, Foley JA (2013) Yield trends are insufficient to double global crop production by 2050. PLoS One 8:e66428

Reynolds M, Foulkes J, Furbank R, Griffiths S, King J, Murchie E, Parry M, Slafer G (2012) Achieving yield gains in wheat. Plant Cell Environ 35:1799–1823

Rolland V, Badger MR, Price GD (2016) Redirecting the cyanobacterial bicarbonate transporters BicA and SbtA to the chloroplast envelope: soluble and membrane cargos need different chloroplast targeting signals in plants. Front Plant Sci 7:185

Sage RF, Sage TL, Kocacinar F (2012) Photorespiration and the evolution of C4 photosynthesis. Annu Rev Plant Biol 63:19–47

Sedelnikova OV, Hughes TE, Langdale JA (2018) Understanding the genetic basis of C4 Kranz anatomy with a view to engineering C3 crops. Annu Rev Genet 52:249–270

Sharkey TD (1988) Estimating the rate of photorespiration in leaves. Physiol Plant 73:147–152

Sharwood RE, Sonawane BV, Ghannoum O, Whitney SM (2016) Improved analysis of C4 and C3 photosynthesis via refined in vitro assays of their carbon fixation biochemistry. J Exp Bot 67:3137–3148

Shen BR, Wang LM, Lin XL, Yao Z, Xu HW, Zhu CH, Teng HY, Cui LL, Liu EE, Zhang JJ, He ZH, Peng XX (2019) Engineering a new chloroplastic photorespiratory bypass to increase photosynthetic efficiency and productivity in rice. Mol Plants 12(2):199–214. https://doi.org/10.1016/j.molp.2018.11.013

Simkin AJ, McAusland L, Headland LR, Lawson T, Raines CA (2015) Multigene manipulation of photosynthetic carbon assimilation increases CO2 fixation and biomass yield in tobacco. J Exp Bot 66:4075–4090

Simkin AJ, Lopez-Calcagno PE, Davey PA, Headland LR, Lawson T, Timm S, Bauwe H, Raines CA (2017) Simultaneous stimulation of sedoheptulose 1,7-bisphosphatase, fructose 1,6-bisphosphate aldolase and the photorespiratory glycine decarboxylase-H protein increases CO2 assimilation, vegetative biomass and seed yield in Arabidopsis. Plant Biotechnol J 15:805–816

Simkin AJ, Lopez-Calcagno PE, Raines CA (2019) Feeding the world: improving photosynthetic efficiency for sustainable crop production. J Exp Bot 70:1119–1140

Somerville CR (1984) The analysis of photosynthetic carbon dioxide fixation and photorespiration by mutant selection. Oxf Surv Plant Mol Cell Biol 1:103–131

Somerville CR (2001) An early Arabidopsis demonstration. Resolving a few issues concerning photorespiration. Plant Physiol 125:20–24

Song Q, Wang Y, Qu M, Ort DR, Zhu XG (2017) The impact of modifying photosystem antenna size on canopy photosynthetic efficiency: development of a new canopy photosynthesis model scaling from metabolism to canopy level processes. Plant Cell Environ 40:2946–2957

Sonnewald U, Fernie AR (2018) Next-generation strategies for understanding and influencing source-sink relations in crop plants. Curr Opin Plant Biol 43:63–70

South PF, Cavanagh A, Lopez-Calcagno PE, Raines CA, Ort DR (2018) Optimizing photorespiration for improved crop productivity. J Integr Plant Biol 60:1217–1230

South PF, Cavanagh A, Liu HW, Ort DR (2019) Synthetic glycolate metabolism pathways stimulate crop growth and productivity in the field. Science 363. https://doi.org/10.1126/science.aat9077

Tcherkez G (2013) Modelling the reaction mechanism of ribulose 1,5 bisphosphate carboxylase/oxygenase and consequences for kinetic parameters. Plant Cell Environ 36:7246–7251

Tilman D, Cassman KG, Matson PA, Naylor R, Polasky S (2002) Agricultural sustainability and intensive production practices. Nature 418:671–677

Tilman D, Balzer C, Hill J, Befort BL (2011) Global food demand and the sustainable intensification of agriculture. Proc Natl Acad Sci U S A 108:20260–20264

Timm S, Florian A, Arrivault S, Stitt M, Fernie AR, Bauwe H (2012) Glycine decarboxylase controls photosynthesis and plant growth. FEBS Lett 586:3692–3697

Timm S, Wittmiss M, Gamlien S, Ewald R, Florian A, Frank M, Wirtz M, Hell R, Fernie AR, Bauwe H (2015) Mitochondrial dihydrolipoyl dehydrogenase activity shapes photosynthesis and photorespiration of *Arabidopsis thaliana*. Plant Cell 27:1968–1984

Timm S, Florian A, Fernie AR, Bauwe H (2016) The regulatory interplay between photorespiration and photosynthesis. J Exp Bot 67:2923–2929

Tolbert NE (1997) The C2 oxidative photosynthetic carbon cycle. Annu Rev Plant Physiol Plant Mol Biol 48:1–25

von Caemmerer S, Quick WP, Furbank RT (2012) The development of C4 rice: current progress and future challenges. Science 336:1671–1672

Walker BJ, South PF, Ort DR (2016a) Physiological evidence for plasticity in glycolate/glycerate transport during photorespiration. Photosynth Res 129:93–103

Walker BJ, VanLoocke A, Bernacchi CJ, Ort DR (2016b) The costs of photorespiration to food production now and in the future. Annu Rev Plant Biol 67:107–129

Wang Y, Stessman DJ, Spalding MH (2015) The CO2-concentrating mechanism and photosynthetic carbon assimilation in limiting CO2: how Chlamydomonas works against the gradient. Plant J 82:429–448
Weber APM, Bar-Even A (2019) Update: improving the efficiency of photosynthetic carbon reactions. Plant Physiol 179:803–812
Xin CP, Tholen D, Devloo V, Zhu XG (2015) The benefits of photorespiratory bypasses: how can they work? Plant Physiol 167:574–585
Zarzycki J, Axen SD, Kinney JN, Kerfeld CA (2013) Cyanobacteria-based approaches to improving photosynthesis in plants. J Exp Bot 64:787–798
Zhu XG, de Sturler E, Long SP (2007) Optimizing the distribution of resources between enzymes of carbon metabolism can dramatically increase photosynthetic rate: a numerical simulation using an evolutionary algorithm. Plant Physiol 145:513–526
Zhu XG, Long SP, Ort DR (2008) What is the maximum efficiency with which photosynthesis can convert solar energy into biomass? Curr Opin Biotechnol 19:153–159
Zhu XG, Long SP, Ort DR (2010) Improving photosynthetic efficiency for greater yield. Annu Rev Plant Biol 61:235–261

Greta Nölke is leader of the research group of Metabolic Engineering. Conducting applied research with focus on improvement of resource-use efficiency and productivity in crop plants, metabolic engineering, and development of novel assays for biosensing and pathogen detection.

Stefan Schillberg is member of the Institute Management Team (acting) and lecturer at the Justus-Liebig University Giessen (Germany). Conducting applied research with a focus on recombinant protein production and downstream processing, cell-free expression, metabolic engineering, and improved crop traits including engineering photosynthesis to increase plant biomass.

Photosynthetic Acclimation and Adaptation to Cold Ecosystems

6

Norman P. A. Hüner, Alexander G. Ivanov, Marina Cvetkovska, Beth Szyszka, Marc Possmayer, and Paul Porter

Abstract

Cold-tolerant, photosynthetic organisms are able to either acclimate or, alternatively, adapt to low temperatures. The former are designated as psychrotolerant phototrophs, whereas the latter are defined as psychrophilic phototrophs. Central to cold acclimation and cold adaptation in phototrophs is the capacity to respond to excess excitation energy. This requires the integration of both low temperature sensing/signal transduction pathways and light sensing/signal transduction mechanisms to maintain photostasis, that is, cellular energy balance. The generation of excess excitation energy by high light is mimicked by exposure to low temperatures. Although modulation of the redox state of the photosynthetic apparatus is a common feature in sensing excess excitation energy, the response to this redox sensing/signalling mechanism is species dependent. These concepts are discussed with respect to acclimation and adaptation of green algae, cyanobacteria, and terrestrial plants to the extreme environments represented by

N. P. A. Hüner (✉) · B. Szyszka · P. Porter
Department of Biology and the Biotron Centre for Experimental Climate Change Research, University of Western Ontario, London, ON, Canada
e-mail: nhuner@uwo.ca; pporter@uwo.ca

A. G. Ivanov
Department of Biology and the Biotron Centre for Experimental Climate Change Research, University of Western Ontario, London, ON, Canada

Institute of Biophysics and Biomedical Engineering, Bulgarian Academy of Sciences, Sofia, Bulgaria
e-mail: aivanov@uwo.ca

M. Cvetkovska · M. Possmayer
Department of Biology and the Biotron Centre for Experimental Climate Change Research, University of Western Ontario, London, ON, Canada

Department of Biology, University of Ottawa, Ottawa, ON, Canada
e-mail: mcvetkov@uottawa.ca; mpossmay@uottawa.ca

A. Kumar et al. (eds.), *Climate Change, Photosynthesis and Advanced Biofuels*,
https://doi.org/10.1007/978-981-15-5228-1_6

the Antarctic and Arctic ecosystems. Our comparative discussions illustrate that phototrophs have evolved different strategies to deal with excess excitation energy which results in an impressive array of phenotypes in response to acclimation and adaptation to cold environments. Consequently, we conclude that the photosynthetic apparatus is not only a critical energy transformer for all phototrophs, but it is also a crucial sensor of the abiotic environment. Furthermore, our comparative analyses of green algae, cyanobacteria, and terrestrial plants native to Antarctic and Arctic ecosystems clearly indicate that psychrophily is not essential for survival in these extreme cold environments. We suggest that there is an urgent need for more comprehensive research focussed on the physiology, biochemistry, genomics, and metabolomics of the myriad, yet undiscovered, organisms that inhabit these extreme environments, which may provide novel biotechnological applications to industry, agriculture, and medicine.

Keywords

Psychrophiles · Photostasis · Excess excitation energy · Redox · Energy partitioning · Sensing · Signalling · Low temperature stress · Salt stress · Plants · Green algae · Cyanobacteria · Arctic · Antarctic · Biotechnology

6.1 Introduction

Although terrestrial and aquatic photosynthetic organisms are exposed to and must adjust to myriad changes in their environment, this review is focussed on temperature, light, and the interactions of these two important environmental cues. Changes in light environment can occur over a range of temporal scales (from seconds to hours to days to seasonal changes) as well as spatial scales from deep shade and sun flecks to full sunlight (Demmig-Adams et al. 2012; Adams et al. 2013; Verhoeven 2014). Furthermore, when considering light one must not only include changes in photon flux density (PFD) but also light quality as assessed by changes in wavelength. In contrast to light, ambient environmental temperature extremes generally are considered to fluctuate on longer timescales (hours to days to seasonal) under natural conditions. However, although natural changes in PFD and light quality versus changes in temperature appear to occur on different timescales, such fluctuations in these important environmental cues do not change independently of each other. For example, in the northern hemisphere, terrestrial plants are exposed to changes in light quality on a daily basis (dawn to dusk) as well as on a seasonal basis (spring-summer vs autumn-winter) which are typically associated with concomitant alterations in the temperature regime (warm spring-summer vs cold autumn-winter). Consequently, photosynthetic organisms exhibit mechanisms to sense changes in these environmental cues and then track them over time and subsequently integrate and couple the sensing mechanisms to the activation of appropriate signal transduction pathways that convert this information into appropriate responses at the levels of transcription and translation which, in turn, govern the alterations in cellular metabolism, growth, and development.

Determination of maximum growth rates as a function of any environmental parameter such as temperature, PFD, water availability, or nutrient levels typically defines the optimal conditions for the growth and reproduction of any living organism. Under such optimal growth conditions, an organism may be considered to be in a physiologically optimal steady state for energy conversion and metabolism needed for its normal growth and development. Any sudden change in an organism's ambient environment usually disrupts this optimal steady-state condition which is usually reflected in an inhibition in either its growth rate or in the rate of some physiological processes such as photosynthesis and respiration. This inhibition signifies a *stress response* to an environmental change (Hopkins and Huner 2009). Such a stress-induced disruption of the optimal, cellular steady state typically leads to the generation of reactive oxygen species (ROS) due to an imbalance in cellular energy budget (Tang and Vincent 1999; Mittler et al. 2011; Baxter et al. 2014; Dietz et al. 2016a, b). Stress-intolerant plants and algae are less able to adjust and recover from a stress event and consequently succumb to the stress. In contrast, stress-tolerant species are able to recover, grow, and reproduce after exposure to a stress event due to their inherent ability to establish a new physiological steady state over time called the *acclimated state* (Hopkins and Huner 2009). Although stress-induced ROS levels decreased significantly once the acclimated state was established in *Arabidopsis thaliana* (Bode et al. 2016), ROS have been implicated in the stimulation of signal transduction networks (Mittler et al. 2004; Bailey-Serres and Mittler 2006; Dietz et al. 2016a, b; Noctor and Foyer 2016) which contribute to the phenotypic plasticity associated with stress-tolerant species and the establishment of the new, acclimated state. These definitions of stress versus acclimation are consistent with the analyses of cold stress in wheat, barley, and rice (Janmohammadi et al. 2015), cyanobacteria (Tang and Vincent 1999; Miskiewicz et al. 2000, 2002; Zakhia et al. 2008), as well as *Arabidopsis thaliana* (Badawi et al. 2007; Bode et al. 2016). Unlike the process of acclimation to a stress event which is usually associated with significant and specific changes in transcript abundance (Yamaguchi-Shinozaki and Shinozaki 2006; Bode et al. 2016), *adaptation* to the environment requires a heritable genomic change (Vincent 2000; Hopkins and Huner 2009).

The fact that the Earth is a cold place is a surprise to many people. If one considers that the world oceans are, on average, 5 °C or less and combined with cold terrestrial habitats of the northern and southern hemispheres including alpine environments, it is estimated that approximately 80% of the Earth's ecosystems exist in extreme habitats which include acidic (Quatrini and Johnson 2016), high temperature (Li 2015), high salt (Papke and Oren 2014), and cold temperatures (Russell 1984, 1990; Priscu et al. 1998; Vincent 2000, 2007; Morgan-Kiss et al. 2006; Mock and Thomas 2008; Yumoto 2013; Quatrini and Johnson 2016). The study of the myriad species that occupy these extreme environments including the terrestrial and aquatic habitats in polar regions provides an unprecedented opportunity to explore the basis of phenotypic plasticity associated with adaptation to extreme environments and the mechanisms by which evolution has triumphed over the constraints imposed by life at the edge (Priscu et al. 1998; Vincent 2000, 2007; Morgan-Kiss et al. 2006; Zakhia et al. 2008; Bielewicz et al. 2011; Bakermans 2012; Dolhi et al. 2013; Yumoto 2013; Chown et al. 2015). Most studies on adaptation to extreme environments are focussed on heterotrophic Archaea and Bacteria. Hence, in our opinion, the scientific literature on adaptation to polar environments is biased toward heterotrophic organisms and microorganisms (Feller and Gerday 1997, 2003; Hochachka and Somero 2002; Collins et al. 2008). However, this review is focussed on a comparison of the mechanisms by which psychrotolerant and psychrophilic polar algae, cyanobacteria, and terrestrial plants alter the structure and function of their photosynthetic apparatus in response to the interactive effects of low temperature and light through acclimation and/or adaptation. Adaptation of phototrophic organisms to cold, polar environments is more complex than that of heterotrophic microorganisms since the former require the capacity to integrate both light-dependent and temperature-dependent sensing/signalling pathways. Since we are not able to cover all aspects of acclimation and adaptation to polar environments, we direct the reader to several excellent books (Whitton and Potts 2000; Seckbach 2007; Margesin et al. 2008) and reviews (Priscu et al. 1998; Vincent 2000, 2007; Morgan-Kiss et al. 2006; Mock and Thomas 2008; Zakhia et al. 2008; Dolhi et al. 2013) that have been published on microbial acclimation and adaptation of phototrophic microbes to polar environments.

6.2 Photostasis and Acclimation to Light and Low Temperature

Photoacclimation customizes the structure and composition of the photosynthetic apparatus to suit any new light environment to which phototrophs may be exposed and results in impressive phenotypic plasticity. In terrestrial plants and green algae, this restructuring occurs through retrograde signal transduction pathways between the chloroplast and the nucleus (Pfannschmidt 2003; Nott et al. 2006; Woodson and Chory 2008; Jung and Chory 2010; Pfannschmidt and Yang 2012). In cyanobacteria, changes in the redox status of the photosynthetic electron transport chain (PETC) govern, in part, the remodelling of the photosynthetic apparatus in response to changes in their light and nutrient environment (Fujita et al. 1994; Bhaya et al. 2000; Vincent 2007; Zakhia et al. 2008; Grossman et al. 2010). It is incumbent upon all photosynthetic organisms to balance the energy trapped through the rapid, temperature-insensitive, photobiophysical, and photochemical processes that occur within the reaction centers of photosystem I (PSI) and photosystem II (PSII) that act as energy sources with energy utilized at much slower rates through temperature-dependent, enzyme-catalyzed, metabolic reactions involved in the reduction of C, N, and S and growth which represent sinks for the photosynthetically generated electrons. The establishment of cellular energy balance in phototrophs is called photostasis (Melis 1998; Huner et al. 2003; Hollis and Huner 2014).

Comparison of photoreceptor mutants of *Arabidopsis thaliana* (L.) Heynh. cv. Landsberg *erecta* with wild type indicated that all mutants retained the capacity for photoacclimation as assessed by their ability to modulate the structure and function of their photosynthetic apparatus in response to changes in growth PFD (Walters et al. 1999). This was interpreted to indicate that photoacclimation of mature chloroplasts to changes in PFD occurs independent of photoreceptors such as phytochrome. However, more recent evidence supports the thesis that photoacclimation involves an integration of photomorphogenic, photoreceptor-mediated pathways (Bhaya et al. 2000; Pogson et al. 2008) and redox signalling pathways in terrestrial plants (Ruckle et al. 2007, 2012; Larkin and Ruckle 2008), cyanobacteria (Bhaya et al. 2000; Grossman et al. 2010), as well as algae (Hollis et al. 2019). These results are consistent with the recent model of Guadagno et al. (2018) which proposes that photoacclimation is the result of a complex integration of environmental modulation of the circadian clock with chloroplast redox status.

Just as light is an essential environmental cue to which all photosynthetic organisms sense and respond, changes in temperature are another crucial environmental cue for development (Casal et al. 2004). All organisms, photosynthetic as well as nonphotosynthetic, must sense and respond to changes in environmental temperature (Hochachka and Somero 2002; Feller and Gerday 2003; Siddiqui and Cavicchioli 2006). However, in the case of phototrophs, they must exhibit the capacity to sense and respond to these environmental cues in a coincident manner since they tend to fluctuate in parallel on a daily as well as a seasonal basis. Consequently, photosynthetic organisms, unlike heterotrophs, must integrate these two apparently disparate environmental cues (Casal et al. 2004). However, a common feature of light and temperature is that both represent energy: light, the energy of electromagnetic radiation (E) that is defined by its wavelength (λ) ($E = \mathrm{hc} / \lambda$), and temperature which is a reflection of the kinetic energy of molecules. The Arrhenius equation ($k = \mathrm{Ae}^{\text{-Ea/RT}}$) provides the relationship between the rate constant of a chemical or biochemical reaction (k) and changes in the absolute temperature (T) from which one can calculate the activation energy (Ea) for the chemical or biochemical reaction (Feller and Gerday 1997; Hochachka and Somero 2002; Siddiqui and Cavicchioli 2006; Hopkins and Huner 2009). Thus, phototrophs respond to changes to their abiotic environment through the integration of light and temperature sensing/signalling pathways (Huner et al. 1998, 2013; Oquist and Huner 2003; Ensminger et al. 2006; Kurepin et al. 2013).

Whereas there is consensus that variation in light quality is perceived by the combination of photoreceptors including phytochromes, cryptochromes, and phototropins (Quail et al. 1995; Casal et al. 2004; Kianianmomeni and Hallmann 2014), the identification of specific temperature sensors has remained more elusive. Plant developmental changes collectively associated with high temperature but below the temperatures to induce heat stress are termed thermomorphogenesis (Quint et al. 2016). By exploiting the model plant, *Arabidopsis thaliana*, modulation of the transcription factor, PIF4 involved in regulating phytochrome levels (Jung et al. 2016; Legris et al. 2016), is a critical player in thermomorphogenesis which is the result of the integration of various light sensing/signalling pathways with the circadian clock as well as epigenetic modifications to the genome (Jung et al. 2016; Legris et al. 2016; Quint et al. 2016).

Cold acclimation may proceed over several days to weeks which generates a new, homeostatic state called the *cold-acclimated state* (Wilson et al. 2006a, b; Hopkins and Huner 2009; Kurepin et al. 2013) characterized by increased plant freeze tolerance (Levitt 1980; Steponkus 1984; Guy 1990; Houde and Dhindsa 1992; Gray et al. 1997; Thomashow 1999; Mahfoozi et al. 2001a, b; Pocock et al. 2001; Wilson et al. 2006a, b; Thomashow 2010; Fowler 2012; Gusta and Wisniewski 2013). However, plant cold acclimation is also dependent on the interaction of photoperiod and low temperature. Steponkus and Lanphear (1968) were one of the first to assess the role of photoperiod on plant freezing tolerance and suggested that cold acclimation of *Hedera helix* appeared to exhibit a photoperiod-dependent component. More recently, the molecular basis for the photoperiod dependence for cold acclimation and freezing tolerance has been elucidated. Light quality and low temperature appear to interact to govern freezing tolerance in *Arabidopsis thaliana* as a consequence of the differential temperature-dependent expression of the phytochrome gene family combined with the expression of the *Cor* regulon (Halliday and Whitelam 2003; Franklin and Whitelam 2007; Franklin 2009; Patel and Franklin 2009; Lee and Thomashow 2012; Franklin et al. 2014).

Vernalization is the developmental process by which certain plant species require exposure to a combination of low temperature and short photoperiod to induce flowering (Fowler et al. 1996, 2014; Sung and Amasino 2005; Galiba et al. 2009; Oliver et al. 2009; Trevaskis 2010, 2015; Winfield et al. 2010; Zhu et al. 2014). This low temperature-photoperiod requirement varies within species. For example, spring wheat cultivars do not require an exposure to low temperature to flower, whereas winter wheat cultivars do which is generally true for all winter versus spring plant varieties (Trevaskis et al. 2007; Trevaskis 2010, 2015). This difference between spring and winter cultivars is a consequence of differences in the VRN1 locus which is located on chromosome 5A of wheat. Since VRN1 is contiguous with FR2, the frost tolerance locus (Limin and Fowler 2006; Galiba et al. 2009; Winfield et al. 2009, 2010), vernalization and cold acclimation are linked. Vernalization ensures that winter varieties slowly transition, during exposure to decreasing temperatures and photoperiod in autumn and winter, from the vegetative state to the reproductive state in the spring when conditions are suitable for flowering and seed production.

In addition to photoperiod, changes in PFD also affect acclimation to low temperature (Gray et al. 1997; Huner et al. 1998, 2012; Ensminger et al. 2006; Wilson et al. 2006a, b; Hollis and Huner 2014). In nature photosynthetic organisms are frequently exposed to PFD that exceeds the capacity for the metabolic sinks to utilize this absorbed energy. Thus, to attain photostasis during exposure to excessive excitation energy (EEE) (Karpinski et al. 1997, 1999), photosynthetic organisms evolved photoprotective, non-photochemical quenching (NPQ) processes to dissipate EEE which protects the photosynthetic apparatus from photodamage (Demmig-Adams and Adams 1992; Horton et al. 1996; Melis 1998, 1999; Huner et al. 2003;

Takahashi and Murata 2008; Murchie et al. 2009; Demmig-Adams et al. 2012; Foyer et al. 2012; Horton 2012). Changes in excitation pressure, measured as the relative redox state of $Q_A[Q_{Ared}/Q_{Ared} + Q_{Aox}]$, the first, stable quinone electron acceptor in the reaction centers of PSII (Dietz et al. 1985; Huner et al. 1996, 1998; Ensminger et al. 2006; Wilson et al. 2006a, b), provide an estimate of the cellular energy imbalance. Due to the differential sensitivity to temperature of the photobiophysical and photochemical processes involved in the conversion of photons into electrons by PSII and PSI (the source) versus the biochemical reactions that constitute the metabolism and growth (the sinks), exposure to low temperature at a constant irradiance mimics the effects of high light with respect to the extent of the light-dependent closure of PSII reaction centers and, consequently, comparable high excitation pressure. Consequently, exposure to low temperature generates a comparable high excitation pressure phenotype as exposure to high light in green algae, cyanobacteria, and winter rye through to chloroplast redox retrograde regulation (Fig. 6.1) (Maxwell et al. 1994; 1995a, b; Gray et al. 1997; Huner et al. 1998; Miskiewicz et al. 2000, 2002). Thus, plants, algae, and cyanobacteria can sense temperature changes through alterations in the redox status of the chloroplast. Thus, it has been suggested that photosynthesis has a dual role. Not only is it the primary energy transformer for the biosphere, but it is also an important environmental redox sensor for phototrophs (Huner et al. 1998, 2013; Ensminger et al. 2006; Murchie et al. 2009; Pfannschmidt and Yang 2012, 2016).

Low temperature-induced changes in cell membrane viscosity appear to be involved in sensing and acclimation to low temperatures (Wada et al. 1993; Nishida and Murata 1996; Murata and Los 1997; Hochachka and Somero 2002). Low

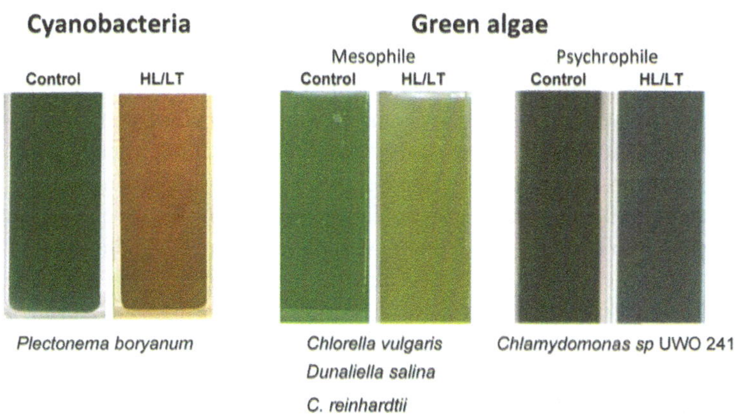

Fig. 6.1 Phenotypic plasticity in cyanobacteria and green algae in response to acclimation to either high light or low temperature. *HL* high light, *LT* low temperature. The lack of apparent phenotypic response in the psychrophile, UWO241 differs markedly from the phenotypic responses of the psychrotrophs *Plectonema boryanum*, *Chlorella vulgaris*, *Dunaliella salina*, and *Chlamydomonas reinhardtii*

temperature-induced increase in cyanobacterial cell membrane viscosity activates a two-component histidine kinase cascade which affects the expression profile of specific fatty acid desaturases involved in the enhanced membrane fluidity in response to low growth temperature (Los and Murata 2002; Murata and Los 2006; Los et al. 2013). In addition, activation of cell membrane Ca^{2+} channels and rapid modulation of cytosolic Ca^{2+} levels has also been shown to be a component of low temperature as well as drought and salinity signalling pathways in *Arabidopsis thaliana* (Knight et al. 1997, 1998; Knight and Knight 2000). Cytosolic Ca^{2+} accumulation activates a protein kinase required for the phosphorylation of the transcription factor, ICE1, which regulates the expression of the *CBF* (cold binding factor) family of transcription factors which, in turn, govern the expression of the *COR* genes necessary for cold acclimation (Monroy et al. 1993; Sarhan et al. 1997; Thomashow 1999, 2010; Zarka et al. 2003; Badawi et al. 2007; Rapacz et al. 2008).

Although plants, algae, and cyanobacteria sense environmentally induced changes in chloroplast redox state as assessed by changes in excitation pressure, the phenotypic, physiological, and molecular responses to this chloroplast redox signal are species dependent (Huner et al. 2003; Kurepin et al. 2013; Hüner et al. 2016). Green algae, cyanobacteria, as well as spring cereals decrease photosynthetic efficiency due to a limited ability to adjust sink capacity which is associated with an increase in the capacity to dissipate EEE through NPQ (Leonardos et al. 2003; Huner and Grodzinski 2011; Huner et al. 2014, 2016). This maximizes photoprotection and allows the organisms to survive but minimizes carbon gain. In contrast, cold-tolerant cereals such as winter rye and winter wheat are able to adjust growth and sink capacity by coupling increased CO_2 assimilation rates and enhanced rates of sucrose and fructan metabolism coordinated with increased rates of leaf carbon export to the active sinks (Leonardos et al. 2003). Consequently, winter cereals have a minimal dependence on NPQ for photoprotection which results in the ability to maximize light conversion efficiency. This is translated into increased vegetative biomass production as well as increased seed yield (Dahal et al. 2012a, b, 2014). The enhanced photosynthetic performance and seed yield of winter compared to spring cereals must be a reflection, at least in part, of the vernalization dependence of winter cereals (Fowler et al. 1996; Mahfoozi et al. 2001a, b; Limin and Fowler 2006; Trevaskis et al. 2007; Trevaskis 2015; Hüner et al. 2016). However, elucidation of the molecular basis linking vernalization with enhanced photosynthetic performance at low temperatures remains to be elucidated.

Based on the discussion above, we suggest that plants, algae, and cyanobacteria may not exhibit a single low temperature sensor, but rather, they appear to integrate information regarding changes in temperature through changes in the redox state of the photosynthetic electron transport chain and stimulation of photoreceptors such as phytochrome, as well as specific cell membrane and low temperature sensors that govern observed phenotypic plasticity and the establishment of the cold-acclimated state (Fig. 6.2) (Kurepin et al. 2013, 2015).

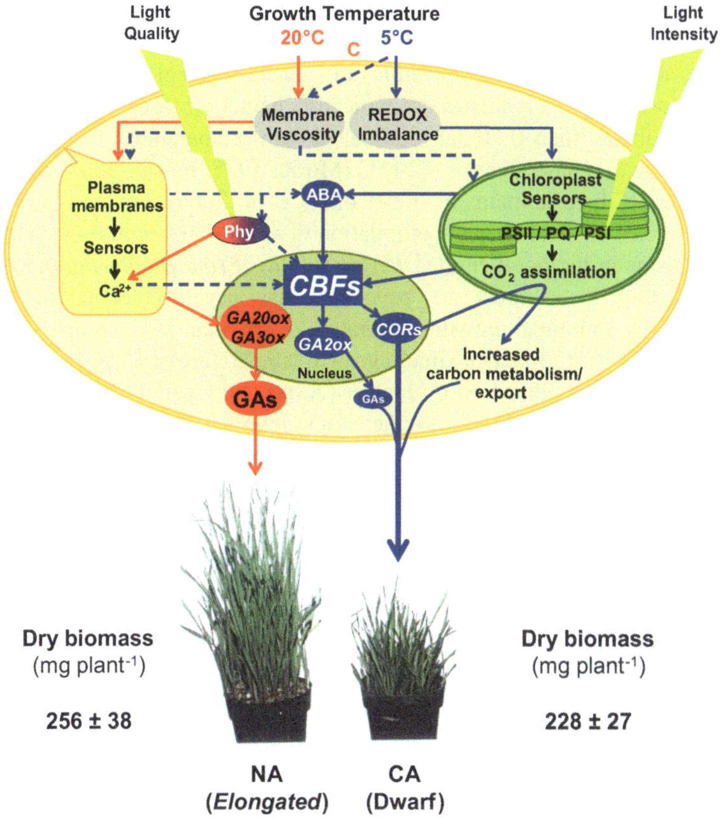

Fig. 6.2 CBFs as integrators of cold acclimation: a schematic model of plant responses to growth temperature of 20 and 5 °C. Growth of winter rye plants at 20 °C ("red" pathway) is likely regulated by photoreceptors such as phytochrome as well as the plasma membrane and translated in the downstream mode with Ca^{2+} as second messenger which ultimately leads to the appearance of a typical elongated phenotype. This is dependent upon the upregulation of *GA20ox* and *GA3ox* genes which result in high levels of growth-active GAs. In contrast, growth of winter rye plants at a cold acclimation temperature of 5 °C (the "blue" pathway) to a comparable developmental stage as 20 °C-grown plants is sensed by changes in plasma membrane viscosity (dashed blue line pathway) and phytochrome and chloroplast redox imbalance (solid blue line pathway). The increase in plasma membrane viscosity increases Ca^{2+} levels which eventually leads to upregulation of ABA biosynthesis. ABA biosynthesis may also be induced by a redox imbalance in chloroplast which causes changes in the redox state of the photosynthetic electron transport chain (PSII/PQ/PSI). The increase in ABA biosynthesis activates *CBF* gene expression which can also be activated independently of ABA, likely directly by chloroplast redox imbalance. Upregulation of *CBF* genes causes an increase in the expression of *COR* and *GA2ox* genes. Upregulation of *GA2ox* genes reduces the amount of growth-active GA genes which is integral in the generation of the dwarf phenotype. Upregulation of *COR* genes not only increases cold and freezing tolerance but also enhances photosynthetic performance via its effect on increased CO_2 assimilation, photosynthetic carbon metabolism, and carbon export. This results in plants which exhibit a dwarf phenotype at 5 °C and a dry biomass comparable to plants grown at 20 °C. The comparable dry mass between the elongated and dwarf phenotype is accounted for by increased leaf thickness, decreased cellular water content,

6.3 Adaptations to Low Temperature

Organisms adapted to low temperature are generally labelled as psychrophiles. Obligate psychrophilic organisms are typically characterized by a minimal growth temperature of less than 0 °C, an optimal growth temperature of ≤15 °C, but a maximum growth temperature of ≤20 °C (Morita 1975; Finster 2008) and include species from all three domains of life (Collins et al. 2008). For an overview, the reader is directed to the excellent compilation of specific research on psychrophilic microorganisms in the book entitled *Psychrophiles: From Biodiversity to Biotechnology* (2008). In contrast to psychrophiles, psychrotolerant species are characterized by a minimal growth temperature of less than 7 °C, an optimal growth temperature of ≤20 °C, but a maximum growth temperature of 35 °C or less (Finster 2008). However, as emphasized by Finster (2008) and Cvetkovska et al. (2017), a lack of agreement with respect to the strict definition of these terms remains confusing in the scientific literature which leads to some difficulty with respect to classification of species in these categories. In this review we will employ the following definition for psychrophilic eukaryotic algae: any naturally occurring phototrophic Eukarya "that are metabolically active and able to reproduce at temperatures permanently close to the freezing point of water and that cannot tolerate more moderate (mesophilic) temperatures (≥20 °C)" (Cvetkovska et al. 2017). Because obligate psychrophiles are adapted to such restricted growth temperature range, they are considered stenothermic (Vincent 2000).

Sensitivity to environmental temperatures between 0 and 40 °C is a characteristic of most biological organisms (Collins et al. 2008). Depending upon the species, organisms can exhibit temperature acclimation which results in an upward or downward shift in the thermal optimum of specific physiological processes (Way and Yamori 2014). This is characteristic of most eurythermal organisms. Due to thermodynamic constraints imposed by low temperature on biochemical reaction rates, low temperature typically results in inhibition of physiological process such as growth, transcription, translation, as well as energy metabolism associated with photosynthesis and respiration in mesophilic organisms (Hochachka and Somero 2002; Hopkins and Huner 2009). In contrast to mesophiles, many psychrophiles retain the capacity to maintain unusually high rates of metabolism at low temperature (Collins et al. 2008). Much research on microbial psychrophiles has attempted to elucidate the biochemical and molecular basis for this apparent anomalous thermodynamic behavior. Since rates of metabolism are governed by the structure and function of enzymes within any metabolic pathway, efforts have focussed on the comparative structural and functional biochemistry of homologous enzymes from

---

Fig. 6.2 (continued) and increased size of crown tissue in cold-acclimated versus non-acclimated plants. *ABA* abscisic acid, *CA* cold-acclimated, *CBFs* C-repeat binding factors, *COR* cold regulated, *Gas* growth-active gibberellins, *GA2ox* GA2 oxidase, *GA3ox* GA3 oxidase, *GA20ox* GA20 oxidase, *NA* non-acclimated, *Phy* phytochromes, *PQ* plastoquinone, *PSI or PSII* photosystem I or II (Kurepin et al. 2013)

psychrophiles and mesophiles (Feller and Gerday 1997, 2003; Siddiqui and Cavicchioli 2006; Collins et al. 2008; Aquist et al. 2017). Electrostatic interactions in proteins tend to be stabilized at low temperatures, whereas hydrophobic interactions tend to be destabilized by low temperature (Collins et al. 2008). Consequently, adaptation to low temperature at the protein level in psychrophiles is assumed to be the result of alterations in amino acid sequences that alter the contributions of electrostatic versus hydrophobic interactions to stabilize protein structure and function at low temperature. Such adaptive changes in amino acid sequence that enhance stability of proteins at low temperature may be one reason that accounts for the higher than expected enzyme activity at low temperature in psychrophilic versus homologous mesophilic enzymes. Conversely, this also accounts for decreased stability of psychrophilic enzymes at warm temperatures (Feller and Gerday 1997; D'Amico et al. 2003; Siddiqui and Cavicchioli 2006; Collins et al. 2008; Cvetkovska et al. 2018). Thus, it appears that the higher activity at low temperature is associated with a decreased structural stability at higher temperatures. It is proposed that the apparent conundrum between reaction rates and enzyme stability is the result of increased flexibility of the enzyme structure that allows for enhanced molecular motion required for enzyme activity at low temperature (Feller and Gerday 2003; Siddiqui and Cavicchioli 2006; Collins et al. 2008; Aquist et al. 2017). As of 2008, crystal structures of 22 enzymes from psychrophilic organisms had been reported in the literature (Collins et al. 2008). A startling conclusion from these studies is that the native conformations of the same enzyme crystallized from organisms grown at different temperature regimes are surprisingly similar with high residue conservation associated with the catalytic sites (Collins et al. 2008). A limitation of such x-ray crystallographic studies is that the study of enzyme crystals reflects a static analysis of the enzyme structure "frozen in time." Protein molecular dynamics may be a promising computational approach to overcome this limitation allowing one to follow the modulation of protein conformation over a very rapid, nanosecond time frame with the enzyme in solution (Aquist et al. 2017; Possmayer 2018). In addition to modulation of enzyme activity at low temperature, adaptation to low temperature may also be associated with higher cellular concentrations of specific enzymes such as the photosynthetic, CO_2-fixing enzyme, Rubisco (Morgan-Kiss et al. 2006; Dolhi et al. 2013), and ferredoxin (Cvetkovska et al. 2018) in algal psychrophiles than mesophilic algae which can be combined with different isoforms of the same enzyme (Hochachka and Somero 2002; Cvetkovska et al. 2018). However, the ability to increase cellular enzyme concentrations in response to low temperature is energetically costly and is not a trait specific to psychrophily since eurythermal, cold-tolerant winter cereals such as rye, wheat, and canola, which are not psychrophiles, double their Rubisco levels upon growth at low temperature (Dahal et al. 2012a, b).

Membrane integrity is crucial to the maintenance of cellular homeostasis and adaptation to low temperatures in all three domains of life (Russell and Fukunaga 1990; Hazel 1995; Nishida and Murata 1996; Murata and Los 1997; Hochachka and Somero 2002; Chintalapati et al. 2004; Russell 2008; Ernst et al. 2016; Siliakus et al. 2017). Modulation of membrane lipid and fatty acid composition in response to temperature is called homeoviscous adaptation (Hazel 1995) which enhances membrane fluidity to ensure optimal function not only of cell membranes but also those of mitochondria and chloroplasts. Low temperature-induced increase in cyanobacterial cell membrane viscosity activates a two-component histidine kinase cascade which affects the expression profile of specific fatty acid desaturases involved in the enhanced membrane fluidity in response to low growth temperature (Los and Murata 2002; Murata and Los 2006; Los et al. 2013). This two-component cascade includes a Hik33 kinase present in the cell membrane which senses changes in membrane viscosity and governs the transcription of downstream desaturase genes (Susuki et al. 2000). Based on the discussion above, psychrophiles do not appear to exhibit specific cold-adapted membrane lipid and fatty acid compositions but rather appear to adjust their fatty acid compositions in response to low temperature in a manner very similar to that of mesophiles (Morgan-Kiss et al. 2002; Russell 2008). Consequently, homeoviscous adaptation does not appear to distinguish psychrophiles from mesophiles.

6.4 Photosynthetic Adaptations to Cold Ecosystems

6.4.1 Aquatic Ecosystems

Psychrophiles can be excellent biological systems to elucidate the physiological, biochemical, and molecular bases for adaptation to extreme environments (Siddiqui and Cavicchioli 2006; Siddiqui et al. 2013). Furthermore, it has been suggested that psychrophiles from extreme Antarctic environments such as the McMurdo Dry Valleys (Wynn-Williams 2000; Morgan-Kiss et al. 2006) may also provide insights into exobiology, that is, the possibilities for life on other planets (Priscu et al. 1998; Wynn-Williams 2000). Photoautotrophs that flourish in the Earth's cold environments fix a significant proportion of the total CO_2 in our biosphere and, consequently, contribute substantially to the mediation of climate change (Lyon and Mock 2014).

6.4.1.1 *Chlamydomonas* sp. UWO241: A Model Green Algal System

Chlamydomonas sp. UWO241 was isolated from Lake Bonney, Antarctica, where it exists 17 m below its permanently ice-covered surface (Neale and Priscu 1995; Priscu et al. 1998) at low but constant temperatures (4–6 °C) combined with high salt concentrations (700 mM) (Morgan-Kiss 2006; Morgan-Kiss et al. 2006; Dolhi et al. 2013). It was originally named *Chlamydomonas subcaudata* (Morgan et al. 1998) which was subsequently changed to *Chlamydomonas raudensis* UWO241 based on comparative DNA sequencing of the internal transcribed spacer (ITS) 1 and ITS2 regions (including the 5.8S) of the ribosomal operon (Pocock et al. 2004). Surprisingly, in collaboration with the SAG Culture Collection, the sequencing data for UWO241 were identical to those of the mesophile *Chlamydomonas raudensis* Ettl (SAG 49.72). Thus, it was concluded that the psychrophile, UWO241, was not *Chlamydomonas subcaudata* but, in fact, a strain of the mesophile, *Chlamydomonas raudensis* Ettl (SAG 49.72) (Pocock et al. 2004). Hence, UWO241 was renamed

Chlamydomonas raudensis UWO241 (Gudynaite-Savitch et al. 2006, 2007; Pocock et al. 2007; Szyszka et al. 2007; Possmayer et al. 2011; Dolhi et al. 2013). However, further independent comparative genome profiling of UWO241 with that of the *Chlamydomonas raudensis* SAG49.72 strain and various strains of *C. reinhardtii* using RAPD analyses were incongruent with the conclusion that UWO241 was a strain of *Chlamydomonas raudensis* Ettl (SAG 49.72) (Gupta 2013). The RAPD analyses were confirmed by our independent re-sequencing of nuclear rDNA (18S and 28S) and the plastid-encoded large subunit of Rubisco (rbcL), which indicated that UWO241 is a distinct species from SAG 49.72 positioned in the Moewusinia clade of the Chlamydomonadales. Henceforth, the Antarctic psychrophile is renamed *Chlamydomonas* sp. UWO241 (Possmayer et al. 2016).

The structure and function of the photosynthetic apparatus of UWO241 has been examined in detail since 1995, and its genome has been sequenced (Cvetkovska et al. 2018, 2019). Thus, UWO241 represents an excellent candidate to become a model algal system to study psychrophily and photosynthetic adaptation to cold temperatures (Morgan-Kiss et al. 2006; Dolhi et al. 2013; Cvetkovska et al. 2017). Although the natural habitat of UWO 241 is one of high salt, its growth rates are maximum at low salt (10 mM) and low temperature but dies at growth temperatures above 18 °C which classifies UWO241 as a halotolerant, obligate psychrophile (Lizotte and Priscu 1992; Morgan et al. 1998; Morgan-Kiss et al. 2006; Pocock et al. 2007; Takizawa et al. 2009; Possmayer et al. 2011; Dolhi et al. 2013; Possmayer 2018). UWO241 is found at the lake's lowest trophic zone, which is characterized by low photon flux density (PFD) (<50 μmol photons m^{-2} s^{-1}) enriched in the blue-green region of the visible spectrum (450–550 nm) during the 6 months of austral summer. Nevertheless, maximum growth rates for UWO 241 are attained at light levels (250 μmol photons m^{-2} s^{-1}) that are at least fivefold higher than its natural growth light (Morgan-Kiss et al. 2006). Furthermore, during the austral winter, UWO241 must survive total darkness. As a consequence of adaptation to low temperature, high salt, low light, and an extreme photoperiod, UWO241 is considered a polyextremophile.

Structurally, UWO241 cells are present either as motile, biflagellate single cells or as immobile structures called palmelloids that are surrounded by a limiting membrane and comprised of up to 16 individual flagellated cells each containing a single chloroplast (Pocock et al. 2004; Possmayer et al. 2016). UWO241 exhibits the normal complement of photosynthetic pigments and an active xanthophyll cycle involved in non-photochemical quenching (NPQ) (Morgan et al. 1998; Pocock et al. 2007; Szyszka et al. 2007). However, UWO241 chloroplasts exhibit a PSI:PSII ratio of about 0.5, contributing to an unusually low chlorophyll a/b ratio (~1.8–2.2) for intact cells and isolated thylakoids (Morgan et al. 1998; Szyszka et al. 2007). UWO241 exhibits high levels of lipid unsaturation in the major chloroplast galactolipids (MGDG, DGDG, SQDG) and the phospholipid, PG, which result in a 30% higher unsaturation index coupled with a lower stability of the PSII supercomplex than the mesophile, *C. reinhardtii*, upon exposure to high temperature stress (Morgan-Kiss et al. 2002). The lipid and fatty acid compositions for UWO241 are consistent with adaptation and acclimation of plants, algae, and cyanobacteria to low temperature (Wada et al. 1993; Nishida and Murata 1996; Murata and Los 1997; Moellering et al. 2010). This is consistent with the conclusion that homeoviscous adaptation (Russell 2008) does not appear to distinguish psychrophiles from mesophiles.

Functionally, UWO241 exhibits maximum light-saturated rates of photosynthesis near its optimal growth temperature of 8 °C which are equivalent to maximum rates of photosynthesis in *C. reinhardtii* measured at its optimal growth temperature of 28 °C (Pocock et al. 2007). Consequently, the light response curve for excitation pressure, measured as 1-qL, in UWO241 at its optimal growth temperature (8 °C) is comparable to that of the mesophile, *Chlamydomonas raudensis* SAG49.72 at 28 °C (Szyszka et al. 2007). Thus, the psychrophile maintains a comparable energy balance as that of the mesophile indicating that both the polyextremophile and this mesophile maintain a comparable redox status of their intersystem photosynthetic electron transport chain (PETC) in response to increasing PFD at their respective optimum growth temperatures. However, in contrast to other green algae such as *Chlorella vulgaris*, *Dunaliella salina*, and the cyanobacterium, *Plectonema boryanum*, UWO241 does not appear to photoacclimate since it does not alter its pigmentation, Chla/b ratio, or the level of LHCII polypeptides in response to growth at high light (Morgan-Kiss et al. 2006) (Fig. 6.1).

Although UWO241 has retained the capacity for long-term adjustment of energy distribution between PSI and PSII by modulating photosystem stoichiometry in response to light quality (Morgan-Kiss et al. 2005), this psychrophile was previously reported to exhibit an aberrant capacity to regulate energy distribution between PSII and PSI through traditional state transitions as assessed by 77 K fluorescence emission spectra and LHCII phosphorylation based on immunodetection of P-Thr residues while maintaining high rates of photosynthesis under its normal growth regime (Morgan-Kiss et al. 2002; Takizawa et al. 2009). This is very unusual since the regulation state transitions appear to be essential for all known green algae and terrestrial plants (Nelson and Ben-Shem 2004; Eberhard et al. 2008; Rochaix 2011, 2014). The molecular basis for the regulation of state transitions is thought to include a reversible, LHCII phosphorylation-dependent mechanism in plants and green algae (Fork and Satoh 1986; Rochaix 2011, 2013, 2014; Ünlü et al. 2014; Ueno et al. 2016) and is considered to be regulated by the thylakoid protein kinases, Stt7 and Stl1, in *C. reinhardtii* and their orthologues, STN7 and STN8, in *Arabidopsis thaliana* (Pesaresi et al. 2011; Wunder et al. 2013; Rochaix 2014). The protein kinase activities are sensitive to the reduction state of the intersystem photosynthetic electron transport chain (Allen et al. 1981; Oxborough et al. 1987; Bennett 1991; Allen 1992; Zer and Ohad 2003) and phosphorylate a specific mobile population of LHCII which induces disengagement from PSII, migration in the plane of the membrane to become associated with PSI.

Despite the apparent deficiency in state transitions in UWO241 as discussed above, recent examination of the UWO241 genome (Cvetkovska et al. 2018) indicated the presence of thylakoid protein kinases, *Stt7* and *Stl1*, and protein levels comparable to those observed for *C. reinhardtii* (Szyszka-Mroz et al. 2019). In contrast with *C. reinhardtii*, thylakoid polypeptide radiolabelling with ^{33}P-ATP in

UWO241 indicated that the kinases were active at its low, permissive growth temperature (5 °C) but inhibited at 25 °C. Furthermore, contrary to our previous reports (Morgan-Kiss et al. 2002; Takizawa et al. 2009), our most recent data show that UWO241 is able to phosphorylate LHCII and undergo state transitions in response to the redox state of the PETC. However, due to the reorganization of PSI and PSII units within thylakoid membranes of UWO241 based on digitonin fraction (Szyszka-Mroz et al. 2019), the 77 K emission spectra are unique, andconsequently, modulation of PSI/PSII energy distribution in UWO241 is more difficult to detect and quantify due to a marked blue shift in the emission maximum of PSI. Similar results for the modulation of state transitions have been reported for the psychrophilic *Chlorella* sp. strain BI isolated from a transitory pond near Bratina Island, Antarctica (Morgan-Kiss et al. 2008), as well as the Arctic psychrophile, *Chlamydomonas* sp. RCC2488 (Fig. 6.3). The unique 77 K fluorescence emission spectra of UWO241 may be due, in part, to either a reduction or absence of 7 of the 11 Lhca polypeptides normally associated with PSI (Morgan et al. 1998). Unlike *C. reinhardtii*, quantification of state transitions in UWO241 requires deconvolution of the 77 K fluorescence emission spectra to detect the major PSI emission band (Szyszka-Mroz et al. 2019). Consequently, UWO241 exhibits an atypical modulation of state transitions which remains qualitatively distinct from the traditional state transition response associated with *C. reinhardtii*. Furthermore, the kinase domain of the Stt7 kinase from UWO241 that regulates state transitions exhibited significant structural alterations which we suggest may predispose it to function maximally at low temperature and may contribute to its relative insensitivity to the protein kinase inhibitor, staurosporine, compared to that of *C. reinhardtii*. Thus, the Stt7/Stl1 kinases in UWO241 appear to be examples of cold-adapted, membrane-bound enzymes that function optimally at low temperature and exhibit a decrease in activity at moderate to high temperatures (Aquist et al. 2017). The reorganization of the thylakoid membranes of UWO241 with respect to PSII/PSI distribution coupled with its strong quenching capacity (Szyszka et al. 2007) may reflect a state transition phenomenon based on PSII-PSI spillover mechanism similar to that reported by Slavov et al. (Slavov et al. 2013, 2016). This conclusion remains equivocal since further experimentation is required to differentiate the potential contribution of the LHCII phosphorylation-dependent mechanism for state transitions from the PSII-PSI spillover mechanism.

Photosynthetic ferredoxins (Fds) are crucial control sites for the distribution of photosynthetically generated electrons from PSI to various essential metabolic reactions in cyanobacteria, green algae, and plants (Knaff 1996; McKay et al. 1999; Boehm et al. 2015; Schorsch et al. 2018). Recently, a global interaction network was established for *C. reinhardtii*, providing putative roles for Fds in redox metabolism, carbohydrate modification, fatty acid biosynthesis, hydrogen production, nitrogen and sulfur metabolism, state transitions, and dark anoxia (Peden et al. 2013). The best-characterized isoform of the family of photosynthetic ferredoxins is PETF or Fd-1 which represents about 98% of all transcribed Fd genes in *C. reinhardtii* (Terauchi et al. 2009). Recently, we reported that ferredoxin from UWO241 is specifically adapted to function at low temperatures. The purified

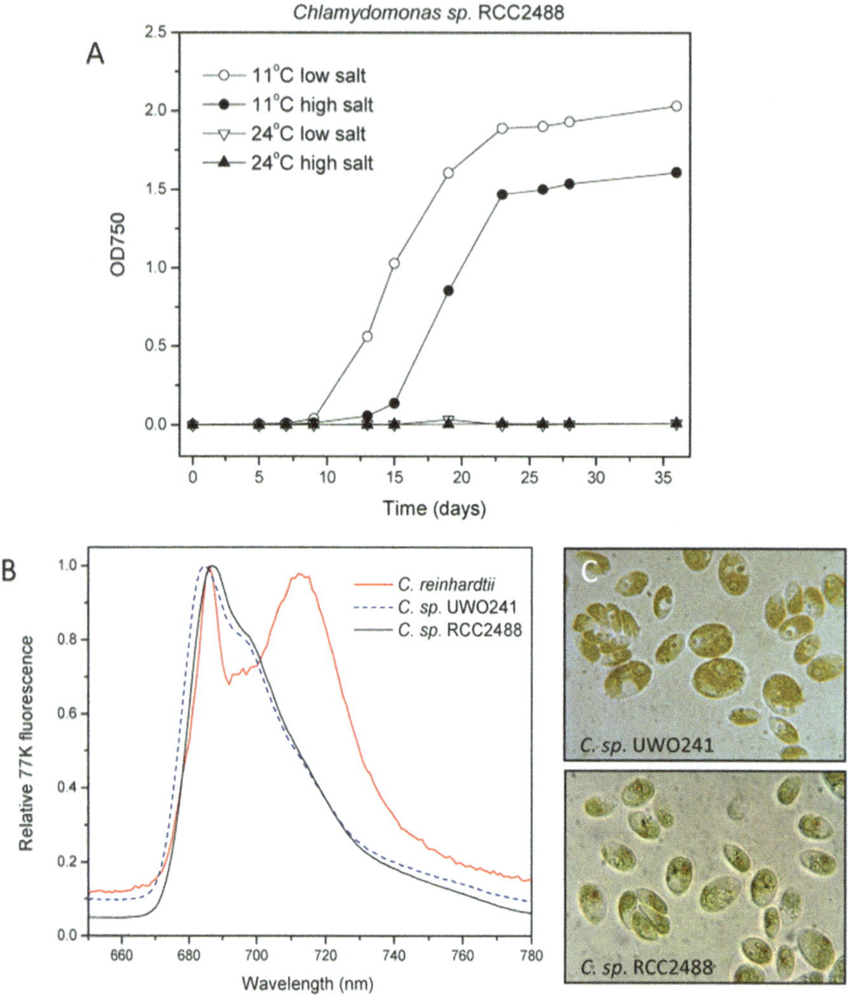

Fig. 6.3 Comparisons of the Arctic *Chlamydomonas* sp. RCC2488 and Antarctic *Chlamydomonas* sp. UWO241. (**a**) Growth curves for *Chlamydomonas* sp. RCC2488 based on light scattering at OD_{750} indicate that it is psychrophilic. (**b**) The 77 K fluorescence emission spectra of RCC2488 (black solid line) and UWO241 (blue broken line) are distinct from that of the model green alga, *Chlamydomonas reinhardtii* (red solid line). (**c**) Light microscopic images of mid-log phase cultures of the Antarctic psychrophile, *Chlamydomonas* sp. UWO241, are similar to that of the Arctic psychrophile, *Chlamydomonas* sp. RCC2488

enzyme exhibited highest structural stability and activity at 10 °C but is more labile at 60 °C than that of *C. reinhardtii*. However, an unusual feature of the psychrophilic Fd is its ability to maintain high activity even at moderate temperatures (40 °C) comparable to that of *C. reinhardtii* (Cvetkovska et al. 2018). Most psychrophilic enzymes are inhibited at such moderate temperatures (Feller and Gerday 1997, 2003;

Collins et al. 2008). Genomic comparisons of UWO241 Fd-1 with 21 other Fds from green algae, plants, diatoms, and cyanobacteria indicated that all Fds were 94–96 amino acids long and were highly conserved at the primary protein sequence (90.4–62.8% identity). Predicted amino acid sequence indicated that the mesophilic *C. reinhardtii* and the psychrophilic UWO241 Fd differed by only 11 amino acids which occur in regions distant from the active site and involved in Fe binding and protein-protein interactions. Nevertheless, such relatively minor differences in predicted amino acid sequences appear to lead to significant increases in cold stability and activity. These observations for Fd-1A and Fd-1B in UWO241 are consistent with those of other psychrophilic enzymes examined to date (Feller and Gerday 2003; D'Amico et al. 2006; Aquist et al. 2017).

Our transcriptome and biochemical analyses confirmed the expression and accumulation of two isoforms of Fd in UWO241, Fd1A and Fd1B. Furthermore, these transcriptome and protein analyses indicated that UWO241 Fd-1 accumulated to levels twice that of Fd-1 in *C. reinhardtii* (Cvetkovska et al. 2018). Therefore, we suggest that UWO241 Fd-1 is unique among psychrophilic proteins since it appears to combine two important strategies. Not only does the Fd-1 exhibit enhanced activity and structural stability at low temperature in UWO241, but it accumulates to levels twice that observed in mesophilic *C. reinhardtii*. The combination of these two characteristics would provide higher capacity for photosynthetic electron transport under its natural cold conditions. We suggest that the presence of two forms of Fd-1in UWO241 is a consequence of a gene duplication event that may be part of a mechanism for adaptation to an extreme environment (Cvetkovska et al. 2017, 2018).

UWO241 is not only adapted to low temperature but is also adapted to an extreme photoperiod with approximately 6 months of sunlight during the austral summer combined with 6 months of darkness during the austral winter. What adaptations allow UWO241 to persist to through such a prolonged dark period? To address this question, an exquisite field experiment was performed by Morgan-Kiss and colleagues whereby a laboratory culture of UWO241 contained in dialysis bags containing water from Lake Bonney was transferred back into Lake Bonney (McMurdo Dry Valleys, Antarctica) and suspended at the 17 m depth to allow the psychrophile to respond to the natural light, temperature, and dissolved ions of Lake Bonney. Chemical, physiological, and environmental molecular analyses of the transplanted UWO241 cultures were compared with the natural phytoplankton community of Lake Bonney over a 6-week transition period from the 24-h light period of austral summer to the 24-h darkness associated with austral winter (Morgan-Kiss et al. 2015). As PFD decreased, the natural communities ceased CO_2 fixation which was accompanied by a downregulation of expression of genes involved in carbon fixation (*rbcL*) as well as PSII photochemistry (psbA) which was matched in the transplanted UWO241 monocultures. Transplanted UWO241 shifted from light-adapted photochemistry to a shade-adapted state in response to the transition to polar night which is consistent with the previously proposed model (Morgan-Kiss et al. 2006, 2015; Mock and Thomas 2008). The mechanism(s) for reactivation of photosynthetic electron transport and the induction of photosynthetic CO_2 assimilation upon exposure to austral summer light conditions remains unknown.

The shade-adapted state exhibited by UWO241 during the summer-to-winter photoperiod transition indicates that the psychrophile is able to maintain Chl levels in the dark. Plants and algae exhibit two distinct enzymes for reducing protochlorophyllide to chlorophyllide, a rate-limiting step in the chlorophyll biosynthetic pathway. LPOR is the light-dependent and DPOR is the light-independent protochlorophyllide oxidoreductase (Reinbothe and Reinbothe 1996). Plants and most algae etiolate in response to darkness (Nemhauser and Chory 2009). Despite being adapted to seasonal, prolonged darkness, UWO241 is able to maintain high levels of Chla and Chlb during this prolonged dark period (Morgan-Kiss et al. 2015). However, sequencing, assembly, and annotation of the UWO241 plastome indicated that the three genes (*chlL*, *chlN*, and *chlB*) that encode DPOR are not only absent from the plastome but also not present in either the UWO241 nuclear genome or its mitochondrial genome (Cvetkovska et al. 2019). However, two other duplicated genes, *GUN4* and *CAO*, were detected in the genome of UWO241. The former is a subunit of the magnesium chelatase which governs Chl biosynthesis as well as retrograde signalling, and the latter regulates Chlb biosynthesis (Chory and Wu 2001; Nott et al. 2006; Tanaka and Tanaka 2007). Thus, UWO241 does have a functional Chl biosynthetic pathway that is totally dependent on LPOR but has lost DPOR even though this alga is exposed to prolonged seasonal darkness. How can this astonishing conundrum be rationalized? Lake Bonney at 17 m below the ice where UWO241 was isolated exhibits extremely high O_2 concentrations (Morgan-Kiss et al. 2006). DPOR is very sensitive to O_2 concentrations due the presence of an essential Fe-S cluster which is absent in LPOR. Consequently, we presume that there would be no deleterious effect of a mutation(s) that eliminated DPOR from UWO241. Such a nonadaptive strategy may explain the absence of *chlL*, *chlN*, or *chlB* genes in UWO241. To confirm this hypothesis with respect O_2 concentrations and the absence of DPOR, more phototrophs from Lake Bonney must be examined. Furthermore, what is the mechanism by which UWO241 is able to maintain its Chl levels and stabilize its photosynthetic apparatus during prolonged darkness? Morgan-Kiss et al. (2006) have suggested that the photosynthetic apparatus of UWO241 is converted into a highly quenched state during austral winter which we suggest may be similar to the sustained quenching mode observed in overwintering evergreen conifers (see below). The answer to these important questions remains a challenge for the future.

6.4.1.2 Cyanobacteria

The photosynthetic apparatus of chloroplast of plants and green algae is characterized by a light-harvesting pigment-protein complex that is an integral thylakoid membrane complex (Melis 1991; Nelson and Ben-Shem 2004; Eberhard et al. 2008). In contrast, cyanobacteria utilize phycobilisomes consisting of protein-bound phycoerythrin, phycocyanin, and allophycocyanin that are arranged on the surfaces of their thylakoid membranes to harvest light energy (Grossman et al. 1993, 1994; Gantt 1994; Sidler 1994). Alteration in pigmentation of cyanobacteria represents a photoprotective response to the abiotic environment and typically occurs as a consequence of changes in the structure and composition of the

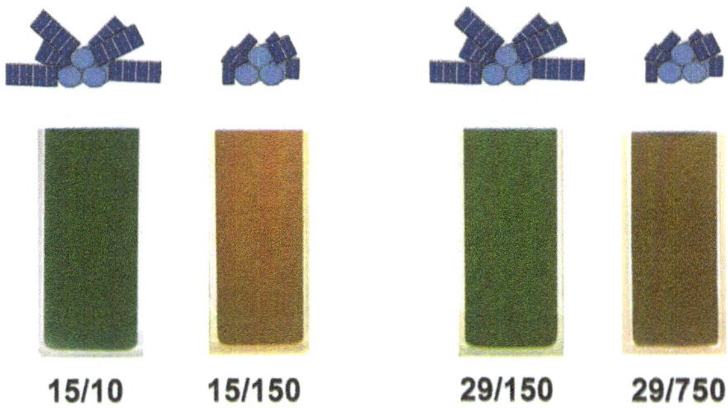

Fig. 6.4 Acclimation to low temperature mimics acclimation to high light in the cyanobacterium, *Plectonema boryanum*. The cyanobacterium was grown at low temperature (15 °C) at either low (10 μmol m^{-2} s^{-1}) or higher PFD (150 μmol m^{-2} s^{-1}). Alternatively, *Plectonema boryanum* was grown at optimal temperatures (29 °C) at either moderate (150 μmol m^{-2} s^{-1}) or high PFD (750 μmol m^{-2} s^{-1}). The blue phycobilisome structures associated are illustrated above each culture. High excitation pressure generated by modulating the PFD at either low temperature (15 °C) or high temperature (29 °C) results in comparable phenotype due to changes in the phycobilisome structure and composition as well as the accumulation of the carotenoid, myxoxanthophyll

phycobilisomes which reduces the light-harvesting efficiency of cyanobacteria in response to EEE (Grossman et al. 2003; Bailey and Grossman 2008). The filamentous cyanobacterium, *Plectonema boryanum*, reduces the size and composition of its phycobilisomes in response to growth at high PFD which is mimicked by growth at low temperature (Miskiewicz et al. 2000, 2002) (Fig. 6.4). This was interpreted to represent a photosynthetic redox response to comparable EEE generated either by exposure to high light or low temperature. The decrease in light-harvesting efficiency was accompanied by the accumulation of a major cyanobacterial carotenoid, myxoxanthophyll, localized in the cell wall/cell membrane of *Plectonema boryanum* which presumably screens the photosynthetic apparatus from EEE.

This protects the *P. boryanum* from photoinhibition (Miskiewicz et al. 2000, 2002; Huner et al. 2005).

Many phycobilisome-containing aquatic cyanobacteria are characterized by the presence of an orange carotenoid protein (OCP) to quench the excess energy absorbed by the phycobilisome under EEE in order to increase energy dissipation in the form of heat (Kirilovsky 2007, 2015; Wilson et al. 2008). This protects the photosynthetic apparatus of cyanobacteria by decreasing the energy transfer efficiency between the phycobilisome and the reaction centers (Wilson et al. 2006a, b). Recently, it was reported that desiccated field samples of the terrestrial cyanobacterium, *N. flagelliforme*, accumulated red proteins that are orthologs of OCP (Yang et al. 2019). However, these red proteins were absent in *N. flagelliforme* grown in a standard liquid medium which is consistent with the conclusion that the induction of the red proteins in *N. flagelliforme* is a response to EEE during desiccation stress (Yang et al. 2019). Thus, mesophilic cyanobacteria exhibit myriad mechanisms to protect the photosynthetic apparatus from EEE due to various environmental stresses (Fig. 6.5).

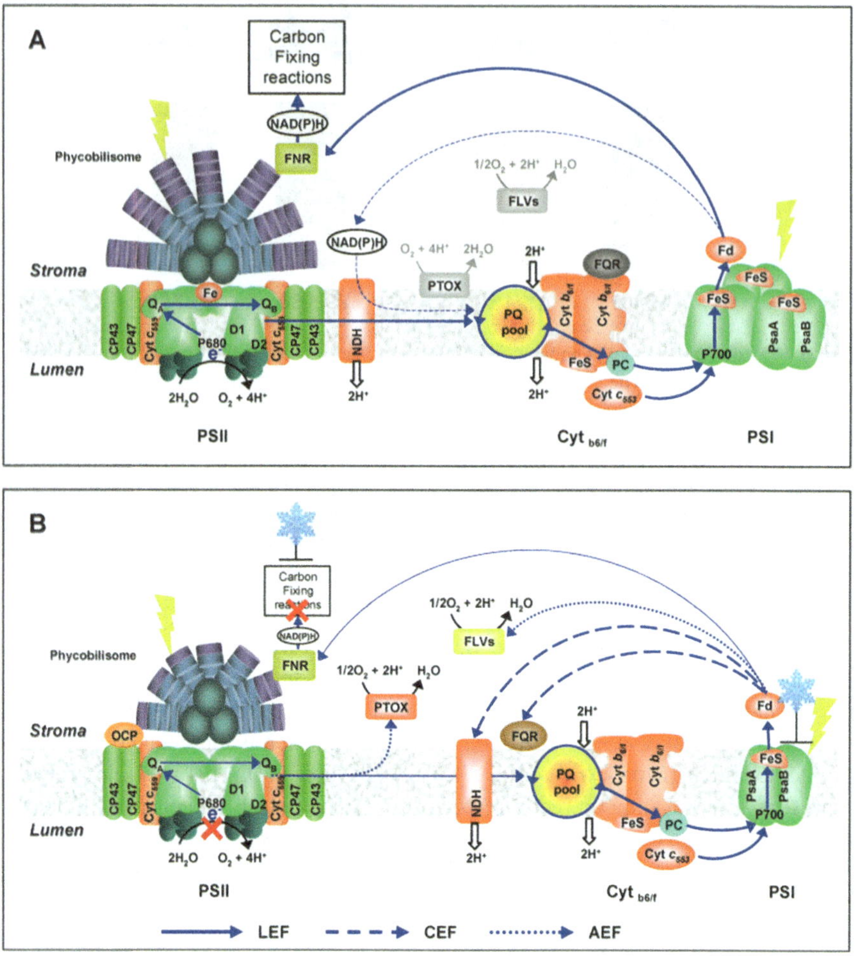

Fig. 6.5 Electron transport pathways (dark blue arrows) within the thylakoid membranes of cyanobacteria acclimated to cold environmental conditions. (**a**) During growth and development under optimal conditions, the PQ pool remains preferentially oxidized because the rate of consumption of photosynthetic electrons through metabolic sinks such as carbon-fixing reactions, as well as N and S reduction, keeps pace with the rate at which PSII undergoes charge separation to reduce the PQ pool. Under these conditions the linear photosynthetic electron flow (LEF, dark blue solid arrows) from PSII (water splitting) to PSI ($NADP^+$ generation) dominates, although additional alternative electron transport pathways (gray dashed arrows) might also be present. (**b**) Acclimation to cold imposing acute limitations at the acceptor site of PSI, the excitation pressure over PSII increases, and the PQ pool becomes predominantly reduced. This is a prerequisite for PSII photoinhibition and leads to severely restricted LEF. The excess electrons not utilized by carbon metabolism can be diverted from the linear electron flow and utilized through alternative oxygen-dependent electron sinks and/or recirculated by the NDH- and/or FQR-dependent PSI-dependent CET pathways. This could be achieved through upregulation of PTOX-dependent and/or flavodiiron proteins (FLVs)-dependent electron donation to O_2. Alternative oxygen-dependent electron sinks can accept electrons downstream from PSII and can directly oxidize the PQ pool (dark blue dotted arrows), thus avoiding its over-reduction. The latter would facilitate the upregulation of PSI-dependent CET pathways

Cyanobacteria not only dominate the freshwater lakes and streams of the Arctic and Antarctic as consortia in the form of microbial mats but are also the dominant phototroph found in glacial cryoconite holes (Vincent 2000; Zakhia et al. 2008). In contrast to these ecosystems, cyanobacteria are comparatively less evident in cold open oceans (Vincent 2000). Since cyanobacteria are so dominant in cold polar environments, this has led to some confusion regarding the nature of their cold tolerance. For example, D'Amico and co-workers suggest that psychrophily and psychrotolerance are interchangeable and that all microorganisms that grow well around the freezing point of water can be considered psychrophiles (D'Amico et al. 2006). Tang and co-workers examined the effect of growth temperature of 27 cyanobacterial isolates from the Arctic as well as the Antarctic (Tang et al. 1997). The growth temperature optima for these 27 isolates ranged from 15 to 35 °C from which they concluded that cyanobacteria generally are psychrotolerant and not psychrophilic. This is consistent with the results of Nadeau and Castenholz (2000) except that two of the oscillatorian cyanobacteria isolated from the meltwater of the McMurdo ice shelf in Antarctica are indeed psychrophilic (Nadeau and Castenholz 2000). Thus, it appears that psychrotolerance may be a more prevalent survival strategy than psychrophily in cyanobacteria. In fact, growth rates of extant polar cyanobacterial species do not exhibit any exceptional performance at low temperatures (Vincent 2000, 2007). The success of cyanobacteria in polar regions is suggested to be in part a consequence of slow but steady growth combined with minimal grazing which results in the seasonal accumulation of a significant inoculum which stimulates growth when temperatures are more favorable for photosynthesis and maximizing growth rates (Vincent 2007).

A detailed study of two cyanobacterial species, *Phormidium subfuscum* isolated from a microbial mat in an Antarctic lake on the McMurdo ice shelf and *Phormidium tenue* isolated from the rock surface in an Alaskan riverbed, indicated considerable plasticity with respect to their growth and photosynthetic response to temperature. *Phormidium tenue* grew over a wider range of temperatures (10–40 °C) than *Phormidium subfuscum* (5–20 °C) which was reflected in a greater capacity to adjust photosynthetic performance over a wider temperature range than *Phormidium subfuscum* even though both species originated from a polar environment. It is suggested that such a significant difference in photosynthetic plasticity is due to the fact that the natural temperature and PFD to which *Phormidium tenue* is exposed is more variable compared to that of *Phormidium subfuscum.* Thus, the former is an example of a eurythermal species, whereas the latter is stenothermal (Tang and Vincent 1999). Thus, as suggested by Vincent (2000), a eurythermal acclimation strategy may provide an advantage in environments that fluctuate significantly on a daily basis, whereas a stenothermal strategy may represent a significant benefit in environments that are relatively constant with respect to seasonal temperatures such as exhibited in perennially ice-covered lakes of Antarctica (Morgan-Kiss et al. 2006).

Phormidium tenue and *Phormidium subfuscum* also differed phenotypically in their response to temperature. Eurythermal *P. tenue* decreased its Chla content in response to decreased growth temperature which decreased its light-harvesting capacity and light use efficiency, whereas *P. subfuscum* exhibited minimal changes in pigmentation in response to low temperature (Tang and Vincent 1999). The phenotypic and photosynthetic response of *P. tenue* to low temperature is consistent with the responses of the eurythermal green algae, *Chlorella vulgaris* and *Dunaliella salina* (Maxwell et al. 1994, 1995a, b; Krol et al. 1997), as well as the filamentous cyanobacterium, *Plectonema boryanum* (Miskiewicz et al. 2000, 2002), to changes in excitation pressure (Huner et al. 1998, 2013; Ensminger et al. 2006). Furthermore, the minimal flexibility in phenotype exhibited by *P. subfuscum* is similar to the absence of a phenotypic response to low temperature exhibited by the psychrophile, UWO241 (Morgan-Kiss et al. 2006), which is correlated with the fact that both of these species are stenothermal.

In addition to a slow-growth strategy, the capacity to protect against EEE at low temperature may be another reason for the success and dominance of polar cyanobacteria. Isolates of Antarctic cyanobacteria exhibit a high capacity for photoacclimation in response to PFD (Vincent 2000) as well as low temperature (Tang et al. 1997; Tang and Vincent 1999). However, the photoacclimatory response is species dependent as discussed above. A common response involved in photoacclimation in cyanobacteria is an adjustment in light-harvesting efficiency by altering the structure and composition of its phycobilisomes (Figs. 6.4 and 6.5).

Isolates of Antarctic cyanobacteria also exhibit similar changes in light-harvesting efficiency and carotenoid accumulation in response to light and low growth temperature as mesophilic *Plectonema boryanum* (Tang et al. 1997; Tang and Vincent 1999). Like *P. boryanum*, these Antarctic isolates concomitantly accumulate the UV-absorbing compounds, scytonemin and mycosporine-like amino acids. Consequently, not only do these changes in the structure and composition of the photosynthetic apparatus protect against photoinhibition, but the accumulation of carotenoids combined with the UV-absorbing compounds decreases the susceptibility of the Antarctic cyanobacteria to damage due to UV radiation (Tang and Vincent 1999; Vincent 2000, 2007; Zakhia et al. 2008). These photoprotective mechanisms help to explain the dominance of cyanobacteria in these extreme Antarctic and Arctic aquatic ecosystems. It has been suggested that the dominance of cyanobacteria in Arctic and Antarctic lakes and streams is due to their slow but consistent growth rates over many seasons combined with minimum loss of biomass due to grazing (Tang and Vincent 1999; Nadeau and Castenholz 2000; Vincent 2000).

6.4.2 Terrestrial Plants

6.4.2.1 Evergreens

The high-latitude boreal forests of North America and Eurasia cover approximately 1.3 billion hectares and store up to 33% of all terrestrial carbon on Earth. Consequently, this boreal ecosystem is crucial in governing the global carbon cycle in response to climate change (FAO 2001). Boreal forests of the northern hemisphere consist of both deciduous and evergreen species which exhibit quite distinct strategies as discussed in detail elsewhere (Oquist and Huner 2003). Due to the seasonal fluctuations in temperature from summer to winter, needles of boreal evergreens are exposed to EEE because they continue to absorb light during the winter even though this absorbed energy cannot be used productively due to the low temperature-induced inhibition of photosynthesis (Öquist 1983; Öquist and Martin 1986; Öquist and Huner 1991; Krivosheeva et al. 1996; Ivanov et al. 2001; Oquist and Huner 2003; Ensminger et al. 2004; Sveshnikov et al. 2006; Way and Sage 2008; Fréchette et al. 2015; Stinziano et al. 2015). Boreal evergreens develop frost hardiness or freezing tolerance which enables them to survive extreme winter conditions by exploiting two primary environmental cues. The shortened photoperiod in late summer and the autumn induces dormancy which terminates active growth and is coupled to low and freezing temperatures in the fall and winter that induce maximum frost hardiness within the genetically determined limitations of individual species (Oquist and Huner 2003; Fréchette et al. 2016).

As discussed above, the dynamic regulation of NPQ appears to occur in all higher plants and algae and is relatively well understood when considered over a rather short timescale of minutes to hours (Horton et al. 2008; Li et al. 2009; Murchie et al. 2009; Demmig-Adams et al. 2012, 2014; Derks et al. 2015; Duffy and Ruban 2015; Park et al. 2019). However, the mechanism(s) of sustained quenching as observed in evergreens over an entire winter season remains to be elucidated. This phenomenon appears to be associated not only with the xanthophyll cycle (Adams et al. 1995; Verhoeven et al. 1999; Demmig-Adams et al. 2012; Fréchette et al. 2015) but is also correlated with the aggregation of components of the photosynthetic apparatus in *Pinus sylvestris* which is reversed upon warming in the spring to produce a fully functional photosynthetic apparatus with no requirement for de novo chlorophyll biosynthesis (Ottander et al. 1995). The sustained quenching phenomenon observed in *Pinus sylvestris* appears to be associated with a reorganization of the photosynthetic apparatus to maximize quenching of absorbed light and its safe dissipation as heat during the winter months.

PSI is known to quench excitation energy (Butler 1978) especially under stress conditions (Butler 1978; Slavov et al. 2013, 2016). Winter needles of *Pinus sylvestris* are more metabolically active upon thawing than previously assumed because PSI functions in the cyclic mode (Ivanov et al. 2001). The presence of PSI in the winter aggregates of *Pinus sylvestris* (Ottander et al. 1995) indicates that PSI may act as an important component of the sustained quenching mode of overwintering needles of *Pinus sylvestris*. Such a photoprotective role for PSI has also been suggested for *Geum montanum*, an alpine plant (Manuel et al. 1999).

Thus, it appears that overwintering evergreens such as conifers can shift between two modes: a dynamic quenching mode governed by the xanthophyll cycle during active growth and photosynthesis in the spring and summer and a sustained quenching mode induced by cold temperatures in the autumn and winter to protect the photosynthetic apparatus from photodamage. The sustained winter quenching prevents photooxidative damage to the photosynthetic pigments, which contributes to the evergreen phenotype of most conifers.

6.4.2.2 Herbaceous Plants

Eutrema salsugineum, previously named *Thellungiella salsuginea*, is a cold-tolerant halophyte native to the Arctic, and its Yukon ecotype can be found growing in the Takhini salt flats (60°51.292 N 135°43.042 W) near Whitehorse, Yukon Territory, Canada (Griffith et al. 2007). Since this species is a close relative of *Arabidopsis thaliana*, the Yukon (Griffith et al. 2007) and Shandong ecotypes (Stepien and Johnson 2009) of *Eutrema salsugineum* are considered an excellent model system for comparative analyses to elucidate the genetic, molecular, and biochemical basis of plant stress tolerance (Griffith et al. 2007; Kazachkova et al. 2018). *Eutrema* is generally more tolerant than *Arabidopsis* to a number of abiotic stresses including freezing (Griffith et al. 2007), salt (Inan et al. 2004), N deficiency (Kant et al. 2008), heat, and phosphate stress (Velasco et al. 2016).

The photosynthetic performance of the Shandong ecotype of *E. salsugineum* was compared to WT *Arabidopsis thaliana* with respect to salt stress (Stepien and Johnson 2009). Similar to other plant species exposed to a stress condition, *Arabidopsis* exhibited an inhibition of photosynthetic CO_2 assimilation with a concomitant decrease in rates of photosynthetic linear electron flow (LEF) due to feedback inhibition of photosynthesis as a consequence of lower sink activity under high salt. This was compensated by enhanced cyclic electron flow (CEF) around PSI combined with stimulation of NPQ to dissipate absorbed energy and protect the photosynthetic apparatus from EEE experienced during high salt stress (Stepien and Johnson 2009). In contrast, exposure of the halophyte, *Eutrema salsugineum*, to high salt resulted in minimal inhibition of photosynthetic carbon assimilation and, consequently, minimal induction of either CEF or NPQ due to its ability to continue to consume photosynthetically generated electrons. This was associated with elevated levels of the plastid terminal oxidase (PTOX) which keeps the PQ pool oxidized by reducing O_2 to water and minimizing the accumulation of ROS (Aluru and Rodermel 2004; Streb et al. 2005; McDonald et al. 2011; Nawrocki et al. 2015; Johnson and Stepien 2016). Thus, utilization of O_2 rather than CO_2 as a terminal electron acceptor may be important in establishing resilience to either low temperature, in the case of *D. antarctica* (Perez-Torres et al. 2007), or high salt in the case of *Eutrema salsugineum* (Stepien and Johnson 2009; Johnson and Stepien 2016).

On the Antarctic continent, there are only two angiosperms, *Deschampsia antarctica* Desv., a monocot, and *Colobanthus quitensis* (Kunth) Bartl., a dicot, that are found in the maritime Antarctic peninsula, 68°42’ S. Both species grow during austral summer when the mean temperature is 3 °C but remain under the snow and presumably dormant during austral winter (Bravo et al. 2001; Bravo and Griffith 2005; Bascunan-Godoy et al. 2006). The light-dependent reduction of O_2 by the PETC has been implicated as an alternative pathway for the consumption of photosynthetically generated electrons especially upon exposure to EEE (Asada 1994a, b; Fryer et al. 1998; Ort 2001; Stepien and Johnson 2009; Ivanov et al. 2012; Queval and Foyer 2012; Johnson and Stepien 2016). One consequence of this is the generation of reactive oxygen species (ROS) which potentially contribute oxidative damage and therefore must be accompanied by the accumulation of antioxidants to mitigate the potential damage from ROS (Asada 1994a, b). However, the photoreduction of O_2 may play an important role in protection of leaves from EEE (Ort and Baker 2002). Furthermore, ROS have also been shown to be important components of signalling networks induced by myriad environmental stresses (Karpinski et al. 1999; Baxter et al. 2014; Dietz et al. 2016a, b; Foyer et al. 2017). Photosynthetic electron transport in *C. quitensis* was reported to be insensitive to changes in oxygen concentration under non-photorespiratory conditions, whereas in *D. antarctica*, approximately 30% of its PETC activity was linked to O_2 (Perez-Torres et al. 2007). However, it is interesting to note that cold-acclimated *C. quitensis* did not accumulate increased levels of the O_2-scavenging enzymes superoxide dismutase, ascorbate peroxidase (APx), or glutathione reductase (Perez-Torres et al. 2004). These results were interpreted to indicate that O_2 was not exploited as a photosynthetic electron acceptor in *C. quitensis* which was associated with a greater dependence on NPQ for photoprotection than was *D. antarctica*. Thus, it appears that these two Antarctic terrestrial species may be dependent on different photoprotective mechanisms to manage and survive the harsh Antarctic environment.

We note with interest that although *Deschampsia antarctica* Desv., *Colobanthus quitensis* (Kunth) Bartl, and *Eutrema salsugineum* are considered extremophiles, we find no published experimental evidence to indicate that these plant species or any terrestrial plant species, for that matter, have been designated as psychrophilic despite their apparent adaptation to natural cold habitats. We suggest that similar to polar cyanobacteria (Tang and Vincent 1999; Vincent 2007), these Antarctic and Arctic terrestrial plant species are psychrotolerant rather than psychrophilic.

6.5 Biotechnology

The diversity of psychrophiles and psychrotolerant microorganisms isolated from cold environments includes species from all three domains of life. Exploitation of this microbial diversity in adaptations to cold environments for industrial, agricultural, and medicinal purposes is in its infancy. Since so few species from the three domains of life have been isolated from these cold environments, a major challenge remains the generation and maintenance of culture collections of these psychrophiles and psychrotolerant microbes which is essential for screening and future biotechnological applications (Cavicchioli et al. 2002; Huston 2008). Since psychrophilic algae accumulate higher levels of lipids relative to algal mesophiles, it has been suggested that photopsychrophiles may be exploited for commercial biofuel production through advanced biotechnological approaches (Chaffin et al. 2012; Griffiths et al. 2012; Mou et al. 2012). One limitation for the use of microalgae for the generation of biofuels is biomass production that is sufficiently cost-effective. A novel approach to this problem of algal biomass production was the use of external static magnetic fields in attempts to enhance the biomass production of the microalga, *Chlorella kessleri*, using a small-scale raceway pond (Small et al. 2011). The entire volume of the raceway pond, but only 1% (v/v) at any specific instant, was periodically exposed to static magnetic fields (Small et al. 2011). An exposure to a 10 mT static magnetic field resulted in a doubling in growth rates, threefold increase in rates of respiration, a doubling in light-saturated rates of photosynthesis, and a fourfold increase in daily biomass production (Small et al. 2011). It was suggested that exposure of algal cultures to a 10 mT static magnetic field to enhance algal biomass production may contribute to making algal biofuel cost competitive. However, the effect of scaling up the small raceway pond to an industrial scale for biomass production has not been tested and may represent a major obstacle for successful industrial application. In addition to exploitation of psychrophiles and psychrotolerant microalgae for biofuels (Chaffin et al. 2012; Griffiths et al. 2012; Mou et al. 2012), cold- adapted microbes are being exploited for the potential biodegradation of petroleum in cold marine environments (Brakstad 2008), bioremediation of polychlorophenols (Langwaldt et al. 2008), as well as treatment of acid mine drainage (Kaksonen et al. 2008).

Acclimation of *Plectonema boryanum* to EEE not only increased its resistance to photoinhibition but also decreased its susceptibility to UV damage (Ivanov et al. 2000). This was correlated with the redox-regulated accumulation of the carotenoid myxoxanthophyll and concomitant accumulation of the UV-absorbing compounds, scytonemin and mycosporine amino acids. These compounds appear to act synergistically as a natural sunscreen to protect the photosynthetic apparatus from UV radiation (Tang and Vincent 1999; Ivanov et al. 2000; Vincent 2000, 2007; Zakhia et al. 2008). These results were subsequently exploited to produce a natural sunscreen for the protection of human skin (Huner et al. 2004). However, the efficacy of the cyanobacterial-derived sunscreen skin cream has yet to be tested on humans.

6.6 General Summary

The concept of photostasis in response to exposure to excessive excitation energy (EEE) is central to acclimation in eukaryotic terrestrial plants and green algae (Fig. 6.5) and cyanobacteria (Fig. 6.6) as well as adaptation of psychrophiles to cold environments.

The role of chloroplast redox state in sensing EEE is central in relation to acclimation in psychrotrophic phototrophs.

Psychrophily is not essential for survival in cold environments. Different strategies have evolved in different terrestrial plant, eukaryotic, and prokaryotic phototrophic species.

The proposed model system, *Chlamydomonas* sp. UWO241, exhibits a reorganization of its genome as indicated by apparent gene duplication events.

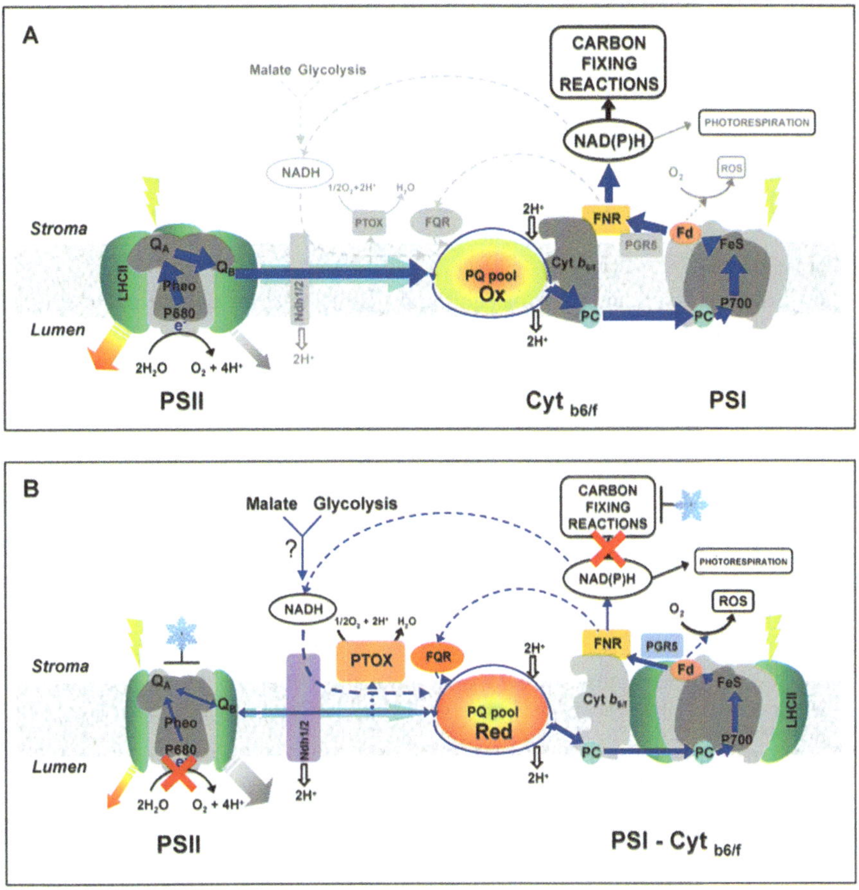

Fig. 6.6 Simplified overview of possible electron transport pathways (dark blue arrows) and energy partitioning of absorbed light energy into fractions utilized by PSII photochemistry (Φ_{PSII}, light blue arrows), thermally dissipated via ΔpH- and xanthophyll-dependent energy quenching (Φ_{NPQ}, red arrows), and nonregulated ΔpH-independent energy quenching (Φ_{NO}, gray arrows), within the thylakoid membranes of higher plants/green algae acclimated to cold environmental conditions. (**a**) During growth and development under optimal conditions, the PQ pool remains preferentially oxidized because the rate of consumption of photosynthetic electrons through metabolic sinks such as carbon-fixing reactions, as well as N and S reduction, keeps pace with the rate at which PSII undergoes charge separation to reduce the PQ pool. Under these conditions the linear photosynthetic electron flow (LEF, dark blue solid arrows) from PSII (water splitting) to PSI ($NADP^+$ generation) dominates, although additional alternative electron transport pathways (gray dashed arrows) might also be present. The absorbed light energy is preferentially utilized by PSII photochemistry (Φ_{PSII}, light blue arrow), and a small fraction is thermally dissipated via ΔpH- and xanthophyll-dependent NPQ (Φ_{NPQ}, red arrow). (**b**) Acclimation to cold imposing acute limitations at the acceptor site of PSI, the excitation pressure over PSII increases, and the PQ pool becomes predominantly reduced. This is a prerequisite for PSII photoinhibition and leads to severely restricted LEF. Since under cold stress conditions only a small fraction of the absorbed light energy can be utilized by PSII photochemistry (Φ_{PSII}, light blue arrow) and the regulated NPQ (Φ_{NPQ}, red arrow) is thermodynamically restricted, the excess energy is dissipated mostly through

6.7 Future Directions

There is a paucity of sequenced genomes from psychrophilic algae (Mock et al. 2017; Cvetkovska et al. 2018, 2019). More genomic data for psychrophilic and psychrotolerant phototrophs is urgently needed not only to better understand the molecular basis of cold adaptation in the Eukarya but also to provide a global approach to the identification of potential novel metabolites and enzymes associated with adaptation to cold environments that could be utilized through the application of biotechnology.

Coordinated international expansion and maintenance of culture collections of polar psychrophilic and psychrotolerant phototrophs represents an important, long-term requirement.

Coordinated genome sequencing of psychrophilic and psychrotolerant phototrophs needs continued expansion and support.

An outstanding unanswered question that requires attention is the elucidation of the mechanism(s) by which Antarctic and Arctic psychrophilic and psychrotolerant phototrophs survive the prolonged darkness.

References

Adams WW III, Demmig-Adams B, Verhoeven AS, Barker DH (1995) 'Photoinhibition' during winter stress: involvement of sustained xanthophyll cycle-dependent energy dissipation. Aust J Plant Physiol 22:261–276

Adams W III, Muller O, Cohu C, Demmig-Adams B (2013) May photoinhibition be a consequence, rather than a cause, of limited plant productivity? Photosynth Res 117:31–44

Allen JF (1992) Protein phosphorylation in regulation of photosynthesis. Biochim Biophys Acta 1098:275–335

Allen JF, Bennett J, Steinback KE, Arntzen CJ (1981) Chloroplast protein phosphorylation couples plastoquinone redox state to distribution of excitation energy between photosystems. Nature 291:21–25

Aluru MR, Rodermel SR (2004) Control of chloroplast redox by the IMMUTANS terminal oxidase. Physiol Plant 120:4–11

Aquist J, Isaksen GV, Brandsdal BO (2017) Computation of enzyme cold adaptation. Nat Rev Chem 1:1–14

Fig. 6.6 (continued) nonregulated constitutive energy quenching (Φ_{PSII}, gray arrow) possibly located within the reaction center of PSII. The excess electrons not utilized by carbon metabolism can be diverted from the linear electron flow and utilized through alternative oxygen-dependent electron sinks and/or recirculated by the NDH- and/or FQR-dependent PSI-dependent CET pathways. This could be achieved through upregulation of PTOX-dependent electron donation to O_2. The higher abundance of PTOX proteins in cold-acclimated plants supports this suggestion. Alternative oxygen-dependent electron sinks can accept electrons downstream from PSII and can directly oxidize the PQ pool (dark blue dotted arrows), thus avoiding its over-reduction. The latter would facilitate the upregulation of PSI-dependent CET pathways. The employment of PTOX-mediated alternative electron pathways as safety valves may play a critical role in balancing/regulating the linear/cyclic photosynthetic electron flows when the acceptor side of PSI is limited by low temperatures

Asada K (1994a) Mechanisms for scavenging reactive molecules generated in chloroplasts under light stress. In: Baker NR, Bowyer JR (eds) Photoinhibition of photosynthesis. From molecular mechanisms to the field. Bios Scientific, Oxford, pp 129–142

Asada K (1994b) In: Foyer CH, Mullineaux PM (eds) Production and action of active oxygen species in photosynthetic tissues. CRC Press, Boca Raton, pp 77–104

Badawi M, Danyluk J, Boucho B, Houde M, Sarhan F (2007) The CBF gene family in hexaploid wheat and its relationship to the phylogenetic complexity of cereal *CBFs*. Mol Gen Genet 277:533–554

Bailey S, Grossman A (2008) Photoprotection in cyanobacteria: regulation of light harvesting. Photochem Photobiol 84:1410–1420

Bailey-Serres J, Mittler R (2006) The roles of reactive oxygen species in plant cells. Plant Physiol 141:311

Bakermans C (2012) Psychrophiles: life in the cold. In: Aniton RP (ed) Extremophiles: microbiology and biotechnology. Caister Academic, Norfolk, pp 53–76

Bascunan-Godoy L, Uribe E, Zuniga-Feest A, Corcuera L, Bravo L (2006) Low temperature regulates sucrose-phosphate synthase activity in *Colobanthus quitensis* (Kunth) Bartl. by decreasing its sensitivity to Pi and increased activation by glucose-6-phosphate. Polar Biol 29:1011–1017

Baxter A, Mittler R, Suzuki N (2014) ROS as key players in plant stress signalling. J Exp Bot 65:1229–1240

Bennett J (1991) Protein phosphorylation in green plant chloroplasts. Annu Rev Plant Physiol Plant Mol Biol 42:281–311

Bhaya D, Schwarz R, Grossman AR (2000) Molecular responses to environmental stress. In: Whitton BA, Potts M (eds) The ecology of cyanobacteria. Kluwer Academic, Dordrecht, pp 397–442

Bielewicz S, Bell E, Kong W, Friedberg I, Priscu JC, Morgan-Kiss RM (2011) Protist diversity in a permanently ice-covered Antarctic Lake during the polar night transition. ISME J 5:1559–1564

Bode R, Ivanov AG, Hüner NPA (2016) Global transcriptome analyses provide evidence that chloroplast redox state contributes to intracellular as well as long-distance signalling in response to stress and acclimation in Arabidopsis. Photosynth Res 128:287–312

Boehm M, Alahuhta M, Mulder DW, Peden EA, Long H, Brunecky R, Lunin VV, King PW, Ghirardi ML, Dubini A (2015) Crystal structure and biochemical characterization of Chlamydomonas FDX2 reveal two residues that, when mutated, partially confer FDX2 the redox potential and catalytic properties of FDX1. Photosynth Res 128:45–57

Brakstad OG (2008) Natural and stimulated biodegradation of petroleum in cold marine environments. In: Margesin R, Schinner F, Marx J-C, Gerday C (eds) Psychrophiles. From biodiversity to biotechnology. Springer, Berlin, pp 389–407

Bravo LA, Griffith M (2005) Characterization of antifreeze activity in Antarctic plants. J Exp Bot 56:1189–1196

Bravo LA, Ulloa N, Zuniga GE, Casanova A, Corcuera LJ, Alberdi M (2001) Cold resistance in Antarctic angiosperms. Physiol Plant 111:55–65

Butler WL (1978) Energy distribution in the photochemical apparatus of photosynthesis. Annu Rev Plant Physiol 29:345–378

Casal JJ, Fankhauser C, Coupland G, Blazquez MA (2004) Signalling for developmental plasticity. Trends Plant Sci 9:309–314

Cavicchioli R, Siddiqui KS, Andrews D, Sowers KR (2002) Low-temperature extremophiles and their applications. Curr Opin Biotechnol 13:253–261

Chaffin J, Mishra S, Kuhaneck R, Heckathorn S, Bridgeman T (2012) Environmental controls on growth and lipid content for the freshwater diatom, Fragilaria capucina: a candidate for biofuel production. J Appl Phycol 24:1045–1051

Chintalapati S, Kiran MD, Shivaji S (2004) Role of membrane lipid fatty acids in cold adaptation. Cell Mol Biol 50:631–642

Chory J, Wu D (2001) Weaving the complex web of signal transduction. Plant Physiol 125:77–80

Chown SL, Clarke A, Fraser CI, Cary SC, Moon KL, McGeoch MA (2015) The changing form of Antarctic biodiversity. Nature 522:431–438
Collins T, Roulling F, Piette F, Marx J-C, Feller G, Gerday C, D'Amico S (2008) Fundamentals of cold-adapted enzymes. In: Margesin R, Schinner F, Marx J-C, Gerday C (eds) Psychrophiles. From biodiversity to biotechnology. Springer, Berlin, pp 211–227
Cvetkovska M, Hüner NPA, Smith DR (2017) Chilling out: the evolution and diversification of psychrophilic algae with a focus on Chlamydomonadales. Polar Biol 40:1169–1184
Cvetkovska M, Szyszka-Mroz B, Possmayer M, Pittock P, Lajoie G, Smith DR, Hüner NPA (2018) Characterization of photosynthetic ferredoxin from the Antarctic alga Chlamydomonas sp. UWO241 reveals novel features of cold adaptation. New Phytol 219:588–604
Cvetkovska M, Orgnero S, Hüner NPA, Smith DR (2019) The enigmatic loss of light-independent chlorophyll biosynthesis from an Antarctic green alga in a light-limited environment. New Phytol 222(2):651–656
D'Amico S, Marx J-C, Gerday C, Feller G (2003) Activity-stability relationships in extremophilic enzymes. J Biol Chem 278:7891–7896
D'Amico S, Marx J-C, Feller G, Gerday C (2006) Psychrophilic microorganisms: challenges for life. EMBO Rep 7:385–389
Dahal K, Gadapati W, Savitch L, Singh J, Huner N (2012a) Cold acclimation and *BnCBF17*-over-expression enhance photosynthetic performance and energy conversion efficiency during long-term growth of *Brassica napus* under elevated CO_2 conditions. Planta 236:1639–1652
Dahal K, Kane K, Gadapati W, Webb E, Savitch LV, Singh J, Sharma P, Sarhan F, Longstaffe FJ, Grodzinski B, Huner NPA (2012b) The effects of phenotypic plasticity on photosynthetic performance in winter rye, winter wheat and Brassica napus. Physiol Plant 144:169–188
Dahal K, Weraduwage SM, Kane K, Rauf SA, Leonardos ED, Gadapati W, Savitch L, Singh J, Marillia E-F, Taylor DC, Micallef MC, Knowles V, Plaxton W, Barron J, Sarhan F, Hüner N, Grodzinski B, Micallef BJ (2014) Enhancing biomass production and yield by maintaining enhanced capacity for CO_2 uptake in response to elevated CO_2. Can J Plant Sci 94:1075–1083
Demmig-Adams B, Adams WW III (1992) Photoprotection and other responses of plants to high light stress. Annu Rev Plant Physiol Plant Mol Biol 43:599–626
Demmig-Adams B, Cohu C, Muller O, Adams W (2012) Modulation of photosynthetic energy conversion efficiency in nature: from seconds to seasons. Photosynth Res 113:75–88
Demmig-Adams B, Stewart JJ, Burch TA, Adams WW III (2014) Insights from placing photosynthetic light harvesting into context. J Phys Chem Lett 5:2880–2889
Derks A, Schaven K, Bruce D (2015) Diverse mechanisms for photoprotection in photosynthesis. Dynamic regulation of photosystem II excitation in response to rapid environmental change. Biochim Biophys Acta 1847:468–485
Dietz K-J, Schreiber U, Heber U (1985) The relationship between the redox state of Q_A and photosynthesis in leaves at various carbon-dioxide, oxygen and light regimes. Planta 166:219–226
Dietz K-J, Mittler R, Noctor G (2016a) Recent progress in understanding the role of reactive oxygen species in plant cell signaling. Plant Physiol 171:1535–1539
Dietz K-J, Turkan I, Krieger-Liszkay A (2016b) Redox- and reactive oxygen species-dependent signaling into and out of the photosynthesizing chloroplast. Plant Physiol 171:1541–1550
Dolhi JM, Maxwell DP, Morgan-Kiss RM (2013) Review: the Antarctic *Chlamydomonas raudensis*: an emerging model for cold adaptation of photosynthesis. Extremophiles 17:711–722
Duffy CDP, Ruban AV (2015) Dissipative pathways in the photosystem-II antenna in plants. J Photochem Photobiol B Biol 152:215–226
Eberhard S, Finazzi G, Wollman F-A (2008) The dynamics of photosynthesis. Annu Rev Genet 42:463–515
Ensminger I, Sveshnikov D, Campbell DA, Funk C, Jansson S, Lloyd J, Shibistova O, Oquist G (2004) Intermittent low temperatures constrain spring recovery of photosynthesis in boreal Scots pine forests. Glob Chang Biol 10:995–1008

Ensminger I, Busch F, Huner NPA (2006) Photostasis and cold acclimation: sensing low temperature through photosynthesis. Physiol Plant 126:28–44
Ernst R, Ejsing CS, Antony B (2016) Homeoviscous adaptation and the regulation of membrane lipids. J Mol Biol 428:4776–4791
FAO (2001) Climate change and forests. In: State of the world's forests. FAO, Rome
Feller G, Gerday C (1997) Psychrophilic enzymes: molecular basis of cold adaptation. Cell Mol Life Sci 53:830–841
Feller G, Gerday C (2003) Psychrophilic enzymes: hot topics in cold adaptation. Nat Rev Microbiol 1:200–208
Finster K (2008) Anaerobic bacteria and Archaea in cold environments. In: Margesin R, Schinner F, Marx J-C, Gerday C (eds) Psychrophiles. From biodiversity to biotechnology. Springer, Berlin, pp 103–119
Fork DC, Satoh K (1986) The control by state transitions of the distribution of excitation energy in photosynthesis. Annu Rev Plant Physiol 37:335–361
Fowler DB (2012) Wheat production in the high winter stress climate of the Great Plains of North America. An experiment in crop adaptation. Crop Sci 52:11–20
Fowler DB, Limin AE, Wang SY, Ward RW (1996) Relationship between low-temperature tolerance and vernalization response in wheat and rye. Can J Plant Sci 76:37–42
Fowler DB, Byrns BM, Greer KJ (2014) Overwinter low-temperature responses of cereals: analyses and simulation. Crop Sci 54:2395–2405
Foyer CH, Neukermans J, Queval G, Noctor G, Harbinson J (2012) Photosynthetic control of electron transport and the regulation of gene expression. J Exp Bot 63:1637–1661
Foyer CH, Ruban AV, Noctor G (2017) Viewing oxidative stress through the lens of oxidative signalling rather than damage. Biochem J 474:877–883
Franklin KA (2009) Light and temperature signal crosstalk in plant development. Curr Opin Plant Biol 12:63–68
Franklin KA, Whitelam GC (2007) Light-quality regulation of freezing tolerance in *Arabidopsis thaliana*. Nat Genet 39:1410–1413
Franklin KA, Toledo-Ortiz G, Pyott DE, Halliday KJ (2014) Interaction of light and temperature signalling. J Exp Bot 65:2859–2871
Fréchette E, Wong CYS, Junker LV, Chang CY-Y, Ensminger I (2015) Zeaxanthin-independent energy quenching and alternative electron sinks cause a decoupling of the relationship between the photochemical reflectance index (PRI) and photosynthesis in an evergreen conifer during spring. J Exp Bot 66:7309–7323
Fréchette E, Chang CY-Y, Ensminger I (2016) Photoperiod and temperature constraints on the relationship between the photochemical reflectance index and the light use efficiency of photosynthesis in *Pinus strobus*. Tree Physiol 36:311–324
Fryer MJ, Andrews JR, Oxborough K, Blowers DA, Baker NR (1998) Relationship between CO_2 assimilation, photosynthetic electron transport, and active O_2 metabolism in leaves of maize during periods of low temperature. Plant Physiol 116:571–580
Fujita Y, Murakami A, Aizawa K, Ohki K (1994) Short-term and long-term adaptation of the photosynthetic apparatus: homeostatic properties of thylakoids. In: Bryant DA (ed) Advances in photosynthesis. The molecular biology of cyanobacteria, vol 1. Kluwer Academic, Dordrecht, pp 677–692
Galiba G, Vágújfalvi A, Li C, Soltész A, Dubcovsky J (2009) Regulatory genes involved in the determination of frost tolerance in temperate cereals. Plant Sci 176:12–19
Gantt E (1994) Supramolecular membrane organization. In: Bryant DA (ed) Advances in photosynthesis. Molecular biology of cyanobacteria, vol 1. Kluwer Academic, Dordrecht, pp 119–138
Gray GR, Chauvin L-P, Sarhan F, Huner NPA (1997) Cold acclimation and freezing tolerance. A complex interaction of light and temperature. Plant Physiol 114:467–474

Griffith M, Timonin M, Wong ACE, Gray GR, Akhter SR, Saldanha M, Rogers MA, Weretilnyk EA, Moffatt B (2007) Thellungiella: an Arabidopsis-related model plant adapted to cold temperatures. Plant Cell Environ 30:529–538

Griffiths M, Hille R, Harrison S (2012) Lipid productivity, settling potential and fatty acid profile of 11 microalgal species grown under nitrogen replete and limited conditions. J Appl Phycol 24:989–1001

Grossman AR, Schaefer MR, Chiang GG, Collier JL (1993) The phycobilisome, a light-harvesting complex responsive to environmental conditions. Microbiol Rev 57:725–749

Grossman AR, Schaefer MR, Chiang GG, Collier JL (1994) The responses of cyanobacteria to environmental conditions: light and nutrients. In: Bryant DA (ed) Advances in photosynthesis. The molecular biology of cyanobacteria, vol 1. Kluwer Academic, Dordrecht, pp 641–675

Grossman AR, Waasenbergen LG, Kehoe DM (2003) Environmental regulation of phycobilisome biosynthesis. In: Green BG, Parson W (eds) Advances in photosynthesis and respiration. Light-harvesting antennas in photosynthesis, vol 13. Kluwer Academic, Dordrecht, pp 471–493

Grossman AR, Mackey KRM, Bailey S (2010) A perspective on photosynthesis in the oligotrophic oceans: hypothesis concerning alternate routes of electron flow. J Phycol 46:629–634

Guadagno CR, Ewers BE, Weinig C (2018) Circadian rhythms and redox state in plants: till stress do us part. Front Plant Sci 9:247

Gudynaite-Savitch L, Gretes M, Morgan-Kiss R, Savitch L, Simmonds J, Kohalmi S, Huner N (2006) Cytochrome *f* from the Antarctic psychrophile, *Chlamydomonas raudensis* UWO 241: structure, sequence, and complementation in the mesophile, *Chlamydomonas reinhardtii*. Mol Gen Genomics 275:387–398

Gudynaite-Savitch L, Loiselay C, Savitch LV, Simmonds J, Kohalmi S, Choquet Y, Huner NPA (2007) The small domain of cytochrome f from the psychrophile, *Chlamydomonas raudensis* UWO 241, modulates the apparent molecular mass and decreases the accumulation of cytochrome f in the mesophile, *Chlamydomonas reinhardtii*. Biochem Cell Biol 85:616–627

Gupta R (2013) Genome profiling of two strains of the green alga *Chlamydomonas raudensis*. MSc Thesis, University of Western Ontario

Gusta LV, Wisniewski M (2013) Understanding plant cold hardiness: an opinion. Physiol Plant 147:4–14

Guy CL (1990) Cold acclimation and freezing tolerance: role of protein metabolism. Annu Rev Plant Physiol Plant Mol Biol 41:187–223

Halliday KJ, Whitelam GC (2003) Changes in photoperiod or temperature alter the functional relationships between phytochromes and reveal roles for phyD and phyE. Plant Physiol 131:1913–1920

Hazel JR (1995) Thermal adaptation in biological membranes: is homeoviscous adaptation the explanation? Annu Rev Physiol 57:19–42

Hochachka PW, Somero GN (2002) Biochemical adaptation. Mechanisms and process in physiological evolution. Oxford University Press, Oxford

Hollis L, Huner NPA (2014) Retrograde operational sensing and signalling pathways maintain photostasis in green algae, cyanobacteria and terrestrial plants. Trends Photochem Photobiol 16:47–61

Hollis L, Ivanov AG, Hüner NPA (2019) Chlorella vulgaris integrates photoperiod and chloroplast redox signals in response to growth at high light. Planta 249:1189–1205

Hopkins WG, Huner NPA (2009) Introduction to plant physiology, 4th edn. Wiley and Sons, Hoboken

Horton P (2012) Optimization of light harvesting and photoprotection: molecular mechanisms and physiological consequences. Philos Trans R Soc B 367:3455–3465

Horton P, Ruban AV, Walters RG (1996) Regulation of light harvesting in green plants. Annu Rev Plant Physiol Plant Mol Biol 47:655–684

Horton P, Johnson MP, Perez-Bueno ML, Kiss AZ, Ruban AV (2008) Photosynthetic acclimation: does the dynamic structure and macro-organisation of photosystem II in higher plant grana membranes regulate light harvesting states? FEBS J 275:1069–1079

Houde M, Dhindsa RS (1992) A molecular marker to select for freezing tolerance in Gramineae. Mol Gen Genet 234:43–48

Huner NPA, Grodzinski B (2011) Photosynthesis and photoautotrophy. In: Moo-Young M (ed) Comprehensive biotechnology, vol 1, 2nd edn. Elsevier, Amsterdam, pp 315–322

Huner NPA, Maxwell DP, Gray GR, Savitch LV, Krol M, Ivanov AG, Falk S (1996) Sensing environmental temperature change through imbalances between energy supply and energy consumption: redox state of photosystem II. Physiol Plant 98:358–364

Huner N, Oquist G, Sarhan F (1998) Energy balance and acclimation to light and cold. Trends Plant Sci 3:224–230

Huner NPA, Oquist G, Melis A (2003) Photostasis in plants, green algae and cyanobacteria: the role of light harvesting antenna complexes. Advances in photosynthesis and respiration. In: Green BR, Parson WW (eds) Light harvesting antennas in photosynthesis, vol 13. Kluwer Academic, Dordrecht, pp 401–421

Huner NPA, Krol M, Ivanov AG, Sarhan F (2004) Solar radiation protection. US Patent No. 6787147 B1

Huner NPA, Wilson KE, Miskiewicz E, Maxwell DP, Gray GR, Krol M, Ivanov AG (2005) Regulation of light harvesting in photosystem II of plants, green algae and cyanobacteria. In: Andrews DI (ed) Energy harvesting materials. World Scientific, London, pp 97–142

Huner N, Dahal K, Hollis L, Bode R, Rosso D, Krol M, Ivanov AG (2012) Chloroplast redox imbalance governs phenotypic plasticity: the "grand design of photosynthesis" revisited. Front Plant Physiol 3:255

Huner NPA, Bode R, Dahal K, Busch FA, Possmayer M, Szyszka B, Rosso D, Ensminger I, Krol M, Ivanov AG, Maxwell DP (2013) Shedding some light on cold acclimation, cold adaptation, and phenotypic plasticity. Botany 91:127–136

Huner NPA, Dahal K, Kurepin LV, Savitch L, Singh J, Ivanov AG, Kane K, Sarhan F (2014) Potential for increased photosynthetic performance and crop productivity in response to climate change: role of CBFs and gibberellic acid. Front Chem 2:18

Hüner NPA, Dahal K, Bode R, Kurepin LV, Ivanov AG (2016) Photosynthetic acclimation, vernalization, crop productivity and 'the grand design of photosynthesis'. J Plant Physiol 203:29–43

Huston AL (2008) Biotechnological aspects of cold-adapted enzymes. In: Margesin R, Schinner F, Marx J-C, Gerday C (eds) Psychrophiles. From biodiversity to biotechnology. Springer-Verlag, Berlin, pp 347–363

Inan G, Zhang Q, Li P, Wang Z, Cao Z, Zhang H, Zhang C, Quist TM, Goodwin SM, Zhu J, Shi H, Damsz B, Charbaji T, Gong Q, Ma S, Fredricksen M, Galbraith DW, Jenks MA, Rhodes D, Hasegawa PM, Bohnert HJ, Joly RJ, Bressan RA, Zhu J-K (2004) Salt cress: A halophyte and cryophyte Arabidopsis relative model system and its applicability to molecular genetic analyses of growth and development of extremophiles. Plant Physiol 135:1718–1737

Ivanov AG, Miskiewicz E, Clarke AK, Greenberg BM, Huner NPA (2000) Protection of photosystem II against UV-A and UV-B radiation in the cyanobacterium *Plectonema boryanum*: the role of growth temperature and growth irradiance. Photochem Photobiol 72:772–779

Ivanov AG, Sane PV, Zeinalov Y, Malmberg G, Gardestrom P, Huner NPA, Oquist G (2001) Photosynthetic electron transport adjustments in overwintering Scots pine (*Pinus sylvestris* L). Planta 213:575–585

Ivanov AG, Rosso D, Savitch LV, Stachula P, Rosembert M, Oquist G, Hurry V, Hüner NPA (2012) Implications of alternative electron sinks in increased resistance of PSII and PSI photochemistry to high light stress in cold-acclimated *Arabidopsis thaliana*. Photosynth Res 113:191–206

Janmohammadi M, Zolla L, Rinalducci S (2015) Low temperature tolerance in plants: changes at the protein level. Phytochemistry 117:76–89

Johnson GN, Stepien P (2016) Plastid terminal oxidase as a route to improving plant stress tolerance: known knowns and known unknowns. Plant Cell Physiol 57:1387–1396

Jung H-S, Chory J (2010) Signaling between chloroplasts and the nucleus: can a systems biology approach bring clarity to a complex and highly regulated pathway? Plant Physiol 152:453–459

Jung J-H, Domijan M, Klose C, Biswas S, Ezer D, Gao M, Khattak AK, Box MS, Charoensawan V, Cortijo S, Kumar M, Grant A, Locke JCW, Schäfer E, Jaeger KE, Wigge PA (2016) Phytochromes function as thermosensors in Arabidopsis. Science 354:886

Kaksonen AH, Dopson M, Karnachuk O, Tuovinen OH, Puhakka JA (2008) Biological iron oxidation in the treatment of acid mine drainage at low temperatures. In: Margesin R, Schinner F, Marx J-C, Gerday C (eds) Psychrophiles. From biodiversity to biotechnology. Springer, Berlin, pp 429–454

Kant S, Bi Y-M, Weretilnyk E, Barak S, Rothstein SJ (2008) The Arabidopsis halophytic relative *Thellungiella halophila* tolerates nitrogen-limiting conditions by maintaining growth, nitrogen uptake, and assimilation. Plant Physiol 147:1168–1180

Karpinski S, Escobar C, Karpinska B, Creissen G, Mullineaux PM (1997) Photosynthetic electron transport regulates the expression of cytosolic ascorbate peroxidase genes in Arabidopsis during excess light stress. Plant Cell 9:627–640

Karpinski S, Reynolds H, Karpinska B, Wingsle G, Creissen G, Mullineaux P (1999) Systemic signaling and acclimation in response to excess excitation energy in Arabidopsis. Science 284:654–657

Kazachkova Y, Eshel G, Pantha P, Cheeseman JM, Dassanayake M, Barak S (2018) Halophytism: what have we learnt from *Arabidopsis thaliana* relative model systems? Plant Physiol 178:972–988

Kianianmomeni A, Hallmann A (2014) Algal photoreceptors: in vivo functions and potential applications. Planta 239:1–26

Kirilovsky D (2007) Photoprotection in cyanobacteria: the orange carotenoid protein (OCP)-related non-photochemical-quenching mechanism. Photosynth Res 93:7–16

Kirilovsky D (2015) Modulating energy arriving at photochemical reaction centers: orange carotenoid protein-related photoprotection and state transitions. Photosynth Res 126:3–17

Knaff DB (1996) Ferredoxin and ferredoxin-dependent enzymes. In: Ort DR, Yocum CF (eds) Advances in photosynthesis. Oxygenic photosynthesis: the light reactions, vol 4. Kluwer Academic, Dordrecht, pp 333–361

Knight H, Knight MR (2000) Imaging spatial and cellular characteristics of low temperature calcium signature after cold acclimation in Arabidopsis. J Exp Bot 51:1679–1686

Knight H, Trewavas AJ, Knight MR (1997) Calcium signalling in *Arabidopsis thaliana* responding to salinity. Plant J 12:1067–1078

Knight H, Brandt S, Knight MR (1998) A history of stress alters drought calcium signalling pathways in Arabidopsis. Plant J 16:681–687

Krivosheeva A, Tao D-L, Ottander C, Wingsle G, Dube SL, Oquist G (1996) Cold acclimation and photoinhibition of photosynthesis in Scots pine. Planta 200:296–305

Krol M, Maxwell DP, Huner NPA (1997) Exposure of *Dunaliella salina* to low temperature mimics the high light-induced accumulation of carotenoids and the carotenoid binding protein (Cbr). Plant Cell Physiol 38:213–216

Kurepin L, Dahal K, Savitch L, Singh J, Bode R, Ivanov AG, Hurry V, Huner NPA (2013) Role of CBFs as integrators of chloroplast redox, phytochrome and plant hormone signaling during cold acclimation. Int J Mol Sci 14:12729–12763

Kurepin L, Park J, Lazarovits G, Hüner N (2015) Involvement of plant stress hormones in Burkholderia phytofirmans-induced shoot and root growth promotion. Plant Growth Regul 77:179–187

Langwaldt JH, Tiirola M, Puhakka JA (2008) Microbial adaptation to boreal saturated subsurface: implications for bioremediation of polychlorophenols. In: Margesin R, Schinner F, Marx J-C, Gerday C (eds) Psychrophiles. From biodiversity to biotechnology. Springer, Berlin, pp 409–427

Larkin RM, Ruckle ME (2008) Integration of light and plastid signals. Curr Opin Plant Biol 11:593–599

Lee C-M, Thomashow MF (2012) Photoperiodic regulation of the C-repeat binding factor (CBF) cold acclimation pathway and freezing tolerance in Arabidopsis thaliana. Proc Natl Acad Sci U S A 109:15054–15059

Legris M, Klose C, Burgie ES, Rojas CCR, Neme M, Hiltbrunner A, Wigge PA, Schäfer E, Vierstra RD, Casal JJ (2016) Phytochrome B integrates light and temperature signals in *Arabidopsis*. Science 354:897–900

Leonardos ED, Savitch LV, Huner NPA, Oquist G, Grodzinski B (2003) Daily photosynthetic and C-export patterns in winter wheat leaves during cold stress and acclimation. Physiol Plant 117:521–531

Levitt J (1980) Responses of plants to environmental stresses. In: Chilling, freezing, and high temperature stresses, vol I. Academic, New York, p 497

Li F-L (2015) Thermophilic microorganisms. Caister Academic, Poole, p 254

Li Z, Wakao S, Fischer BB, Niyogi KK (2009) Sensing and responding to excess light. Annu Rev Plant Biol 60:239–260

Limin AE, Fowler DB (2006) Low-temperature tolerance and genetic potential in wheat (*Triticum aestivum* L.): response to photoperiod, vernalization, and plant development. Planta 224:360–366

Lizotte MP, Priscu JC (1992) Spectral irradiance and bio-optical properties in perennially ice-covered lakes of the dry valleys (McMurdo Sound, Antarctica). Ant Res Ser 57:1–14

Los DA, Murata N (2002) Sensing and responses to low temperature in cyanobacteria. In: Storey KB, Storey JM (eds) Sensing, signalling and cell adaptation. Elsevier Science BV, Amsterdam, pp 139–153

Los D, Mironov K, Allakhverdiev S (2013) Regulatory role of membrane fluidity in gene expression and physiological functions. Photosynth Res 116:489–509

Lyon BR, Mock T (2014) Polar microalgae: new approaches to an extreme and changing environment. Biology 3:56–80

Mahfoozi S, Limin AE, Fowler DB (2001a) Developmental regulation of low-temperature tolerance in winter wheat. Ann Bot 87:751–757

Mahfoozi S, Limin AE, Fowler DB (2001b) Influence of vernalization and photoperiod responses on cold hardiness in winter cereals. Crop Sci 41:1006–1011

Manuel N, Cornic G, Aubert S, Choler P, Bligny R, Heber U (1999) Protection against photoinhibition in the alpine plant Geum montanum. Oecologia 119:149–158

Margesin R, Schinner F, Marx J-C, Gerday C (2008) Psychrophiles: from biodiversity to biotechnology. Springer, Berlin

Maxwell DP, Falk S, Trick CG, Huner NPA (1994) Growth at low temperature mimics high-light acclimation in *Chlorella vulgaris*. Plant Physiol 105:535–543

Maxwell DP, Falk S, Huner NPA (1995a) Photosystem II excitation pressure and development of resistance to photoinhibition I. LHCII abundance and zeaxanthin content in Chlorella vulgaris. Plant Physiol 107:687–694

Maxwell DP, Laudenbach DE, Huner NPA (1995b) Redox regulation of light-harvesting complex II and *cab* mRNA abundance in *Dunaliella salina*. Plant Physiol 109:787–795

McDonald AE, Ivanov AG, Bode R, Maxwell DP, Rodermel SR, Hüner NPA (2011) Flexibility in photosynthetic electron transport: the physiological role of plastoquinol terminal oxidase (PTOX). Biochim Biophys Acta 1807:954–967

McKay RML, La Roche J, Yakunin AF, Durnford DG, Geider RJ (1999) Accumulation of ferredoxin and flavodoxin in a marine diatom in response to Fe. J Phycol 35:510–519

Melis A (1991) Dynamics of photosynthetic membrane composition and function. Biochim Biophys Acta 1058:87–106

Melis A (1998) Photostasis in plants. In: Williams, Thistle (eds) Photostasis and related phenomena. Plenum Press, New York, pp 207–220

Melis A (1999) Photosystem-II damage and repair cycle in chloroplasts: what modulates the rate of photodamage in vivo? Trends Plant Sci 4:130–135

Miskiewicz E, Ivanov AG, Williams JP, Khan MU, Falk S, Huner NPA (2000) Photosynthetic acclimation of the filamentous cyanobacterium, *Plectonema boryanum* UTEX 485, to temperature and light. Plant Cell Physiol 41:767–775

Miskiewicz E, Ivanov AG, Huner NPA (2002) Stoichiometry of the photosynthetic apparatus and phycobilisome structure of the cyanobacterium *Plectonema boryanum* UTEX 485 are regulated by both light and temperature. Plant Physiol 130:1414–1425

Mittler R, Vanderauwera S, Gollery M, Van Breusegem F (2004) Reactive oxygen gene network of plants. Trends Plant Sci 9(10):490–498

Mittler R, Vanderauwera S, Suzuki N, Miller G, Tognetti VB, Vandepoele K, Gollery M, Shulaev V, van Breusegem F (2011) ROS signalling: the new wave? Trends Plant Sci 16:300–308

Mock T, Thomas DN (2008) Microalgae in polar regions: linking functional genomics and physiology with environmental conditions. In: Margesin R, Schinner F, Marx J-C, Gerday C (eds) Psychrophiles. From biodiversity to biotechnology. Springer, Berlin, pp 285–312

Mock T, Otillar RP, Strauss J, McMullan M, Paajanen P, Schmutz J, Salamov A, Sanges R et al (2017) Evolutionary genomics of the cold-adapted diatom *Fragilariopsis cylindrus*. Nature 541:536–540

Moellering ER, Muthan B, Benning C (2010) Freezing tolerance in plants requires lipid remodeling at the outer chloroplast membrane. Science 330:226–228

Monroy AF, Sarhan F, Dhindsa RS (1993) Cold-induced changes in freezing tolerance, protein phosphorylation, and gene expression. Plant Physiol 102:1227–1235

Morgan RM, Ivanov AG, Priscu JC, Maxwell DP, Huner NPA (1998) Structure and composition of the photochemical apparatus of Antarctic green alga, *Chlamydomonas subcaudata*. Photosynth Res 56:303–314

Morgan-Kiss R (2006) Photosynthesis on the edge: phytoplankton life in the ice-covered lakes of the McMurdo Dry Valleys. Can Antarct Res Netw 21:17–18

Morgan-Kiss R, Ivanov AG, Williams J, Mobashsher K, Huner NPA (2002) Differential thermal effects on the energy distribution between photosystem II and photosystem I in thylakoid membranes of a psychrophilic and a mesophilic alga. Biochim Biophys Acta 1561:251–265

Morgan-Kiss RM, Ivanov AG, Pocock T, Krol M, Gudynaite-Savitch L, Huner NPA (2005) The Antarctic psychrophile, Chlamydomonas raudensis ETTL (UWO241) (CHLOROPHYCEAE, CHLOROPHYTA), exhibits a limited capacity to photoacclimate to red light. J Phycol 41:791–800

Morgan-Kiss RM, Priscu JC, Pocock T, Gudynaite-Savitch L, Huner NPA (2006) Adaptation and acclimation of photosynthetic microorganisms to permanently cold environments. Microbiol Mol Biol Rev 70:222–252

Morgan-Kiss R, Ivanov AG, Modla S, Czymmek K, Huner NPA, Priscu JC, Lisle JT, Hanson TE (2008) Identity and physiology of a new psychrophilic eukaryotic green alga, *Chlorella* sp., strain BI, isolated from a transitory pond near Bratina Island, Antarctica. Extremophiles 12:701–711

Morgan-Kiss R, Lizotte MP, Kong W, Priscu JC (2015) Photoadaptation to the polar night by phytoplankton in a permanently ice-covered Antarctic lake. Limnol Oceanogr 61:3–13

Morita RY (1975) Psychrophilic bacteria. Bacteriol Rev 39:144–167

Mou S, Xu D, Ye N, Zhang X, Liang Q, Zhen Z, Zhuang Z, Miao J (2012) Rapid estimation of lipid content in an Antarctic ice alga (*Chlamydomonas* sp.) using the lipophilic fluorescent dye, BODIPY505/515. J Appl Phycol 24:1169–1176

Murata N, Los DA (1997) Membrane fluidity and temperature perception. Plant Physiol 115:875–879

Murata N, Los DA (2006) Histidine kinase Hik33 is an important participant in cold-signal transduction in cyanobacteria. Physiol Plant 126:17–27

Murchie EH, Pinto M, Horton P (2009) Agriculture and the new challenges for photosynthesis research. New Phytol 181:532–552

Nadeau TL, Castenholz RW (2000) Characterization of psychrophilic oscillatorians (cyanobacteria) from Antarctic meltwater ponds. J Phycol 36:914–923

Nawrocki WJ, Tourasse NJ, Taly A, Rappaport F, Wollman F-A (2015) The plastid terminal oxidase: its elusive function points to multiple contributions to plastid physiology. Annu Rev Plant Biol 66:49–74

Neale PJ, Priscu JC (1995) The photosynthetic apparatus of phytoplankton from a perennially ice-covered Antarctic lake: acclimation to an extreme shade environment. Plant Cell Physiol 36:253–263
Nelson N, Ben-Shem A (2004) The complex architecture of oxygenic photosynthesis. Nat Rev Mol Cell Biol 5:971–982
Nemhauser JL, Chory J (2009) Photomorphogenesis. In: The Arabidopsis book. The American Society of Plant Biologists, Rockville, pp 1–12
Nishida I, Murata N (1996) Chilling sensitivity in plants and cyanobacteria: the crucial contribution of membrane lipids. Annu Rev Plant Physiol Plant Mol Biol 47:541–568
Noctor G, Foyer CH (2016) Intracellular redox Compartmentation and ROS-related communication in regulation and signaling. Plant Physiol 171:1581–1592
Nott A, Jung H-S, Koussevitzky S, Chory J (2006) Plastid-to-nucleus retrograde signaling. Annu Rev Plant Biol 57:739–759
Oliver SN, Finnegan EJ, Dennis ES, Peacock WJ, Trevaskis B (2009) Vernalization-induced flowering in cereals is associated with changes in histone methylation at the VERNALIZATION1 gene. Proc Natl Acad Sci U S A 106:8386–8391
Öquist G (1983) Effects of low temperature on photosynthesis: review. Plant Cell Environ 6:281–301
Öquist G, Huner NPA (1991) Effects of cold acclimation on the susceptibility of photosynthesis to photoinhibition in Scots pine and in winter and spring cereals: a fluorescence analysis. Funct Ecol 5:91–100
Oquist G, Huner NPA (2003) Photosynthesis of overwintering evergreen plants. Annu Rev Plant Biol 54:329–355
Öquist G, Martin B (1986) Cold climates. In: Baker NR, Long SP (eds) Photosynthesis in contrasting environments, vol 7. Elsevier, New York, pp 237–293
Ort DR (2001) When there is too much light. Plant Physiol 125:29–32
Ort D, Baker NR (2002) A photoprotective role for O_2 as an alternative electron sink in photosynthesis? Curr Opin Plant Biol 5:193–198
Ottander C, Campbell D, Öquist G (1995) Seasonal changes in photosystem II organization and pigment composition in *Pinus sylvestris*. Planta 197:176–183
Oxborough K, Lee P, Horton P (1987) Regulation of thylakoid protein phosphorylation by high-energy-state quenching. FEBS Lett 221:211–214
Papke RT, Oren A (2014) Halophiles: genetics and genomes. Caister Academic, Poole, p 196
Park S, Steen CJ, Lyska D, Fischer AL, Endelman B, Iwai M, Niyogi KK, Fleming GR (2019) Chlorophyll–carotenoid excitation energy transfer and charge transfer in *Nannochloropsis oceanica* for the regulation of photosynthesis. Proc Natl Acad Sci U S A 116:3385
Patel D, Franklin KA (2009) Temperature-regulation of plant architecture. Plant Signal Behav 4:577–579
Peden EA, Boehm M, Mulder DW, Davis R, Old WM, King PW, Ghirardi ML, Dubini A (2013) Identification of global ferredoxin interaction networks in *Chlamydomonas reinhardtii*. J Biol Chem 288:35192–35209
Perez-Torres E, Dinamarca J, Bravo L, Corcuera L (2004) Responses of *Colobanthus quitensis* (Kunth) Bartl. to high light and low temperature. Polar Biol 27:183–189
Perez-Torres E, Bravo LA, Corcuera LJ, Johnson GN (2007) Is electron transport to oxygen an important mechanism in photoprotection? Contrasting responses from Antarctic vascular plants. Physiol Plant 130:185–194
Pesaresi P, Pribil M, Wunder T, Leister D (2011) Dynamics of reversible protein phosphorylation in thylakoids of flowering plants: the roles of STN7, STN8 and TAP38. Biochim Biophys Acta 1807:887–896
Pfannschmidt T (2003) Chloroplast redox signals: how photosynthesis controls its own genes. Trends Plant Sci 8(1):33–41

Pfannschmidt T, Yang C (2012) The hidden function of photosynthesis: a sensing system for environmental conditions that regulates plant acclimation responses. Protoplasma 249:125–136. Springer Wien

Pocock TH, Hurry V, Savitch LV, Huner NPA (2001) Susceptibility to low-temperature photoinhibition and the acquisition of freezing tolerance in winter and spring wheat: the role of growth temperature and irradiance. Physiol Plant 113:499–506

Pocock T, Lachance M-A, Proschold T, Priscu JC, Kim SS, Huner NPA (2004) Identification of a psychrophilic green alga from lake Bonney Antarctica: *Chlamydomonas raudensis* ETTL. (UWO 241) *Chlorophyceae*. J Phycol 40:1138–1148

Pocock TH, Koziak A, Rosso D, Falk S, Hüner NPA (2007) *Chlamydomonas raudensis* (UWO 241), Chlorophyceae, exhibits the capacity for rapid D1 repair in response to chronic photoinhibition at low temperature. J Phycol 43:924–936

Pogson BJ, Woo NS, Förster B, Small ID (2008) Plastid signalling to the nucleus and beyond. Trends Plant Sci 13:602–609

Possmayer M (2018) Phylogeny, heat-stress and enzymatic heat sensitivity in the Antarctic psychrophile, *Chlamydomonas* UWO241. PhD Thesis, Western University

Possmayer M, Berardi G, Beall BFN, Trick CG, Huner NPA, Maxwell DP (2011) Plasticity of the psychrophilic green alga *Chlamydomonas raudensis* (UWO 241) (Chlorophyta) to supraoptimal temperature stress. J Phycol 47:1098–1109

Possmayer M, Gupta RK, Szyszka-Mroz B, Maxwell DP, Lachance MA, Hüner NPA, Smith DR (2016) Resolving the phylogenetic relationship between *Chlamydomonas* sp. UWO 241 and *Chlamydomonas raudensis* SAG 49.72 (Chlorophyceae) with nuclear and plastid DNA sequences. J Phycol 52:305–310

Priscu JC, Fritsen CH, Adams EE, Giovannoni J, Paerl HW, McKay CP, Doran PT, Gordon DA, Lanoil BD, Pinckney JL (1998) Perennial Antarctic lake ice: an oasis for life in a polar desert. Science 280:2095–2098

Quail PH, Boylan MT, Parks BM, Short TW, Xu Y, Wagner D (1995) Phytochromes: photosensory perception and signal transduction. Science 268:675–680

Quatrini R, Johnson DB (2016) Acidophiles: life in extremely acidic environments. Caister Academic, Poole, p 310

Queval G, Foyer CH (2012) Redox regulation of photosynthetic gene expression. Philos Trans R Soc B 367:3475–3485

Quint M, Delker C, Franklin KA, Wigge PA, Halliday KJ, van Zanten M (2016) Molecular and genetic control of plant thermomorphogenesis. Nat Plants 2:15190

Rapacz M, Wolanin B, Hura K (2008) The effects of cold acclimation on the photosynthetic apparatus and the expression of COR14b in four genotypes of barley (*Hordeum vulgare*) contrasting in their tolerance to freezing and high light treatment in cold conditions. Ann Bot 101:689–699

Reinbothe S, Reinbothe C (1996) Regulation of chlorophyll biosynthesis in angiosperms. Plant Physiol 111:1–7

Rochaix J-D (2011) Regulation of photosynthetic electron transport. Biochim Biophys Acta 1807:878–886

Rochaix J-D (2013) Fine-tuning photosynthesis. Science 342:50–51

Rochaix J-D (2014) Regulation and dynamics of the light-harvesting system. Annu Rev Plant Biol 65:287–309

Ruckle ME, DeMarco SM, Larkin RM (2007) Plastid signals remodel light signaling networks and are essential for efficient chloroplast biogenesis in Arabidopsis. Plant Cell 19:3944–3960

Ruckle ME, Burgoon LD, Lawrence LA, Sinkler CA, Larkin RM (2012) Plastids are major regulators of light signaling in Arabidopsis. Plant Physiol 159:366–390

Russell NJ (1984) Mechanisms of thermal adaptation in bacteria: blueprints for survival. Trends Biochem Sci 9:108–112

Russell NJ (1990) Cold adaptations of microorganisms. Philos Trans R Soc Lond B Biol Sci 326:595–608

Russell NJ (2008) Membrane components and cold sensing. In: Psychrophiles. From biodiversity to biotechnology (Margesin R, Schinner F, Marx J-C, Gerday C, eds) Springer, Berlin: 177–190

Russell NJ, Fukunaga N (1990) A comparison of thermal adaptation of membrane lipids in psychrophilic and thermophilic bacteria. FEMS Microbiol Rev 75:171–182

Sarhan F, Ouellet F, Vazquez-Tello A (1997) The wheat *wcs*120 gene family. A useful model to understand the molecular genetics of freezing tolerance in cereals. Physiol Plant 101:439–445

Schorsch M, Kramer M, Goss T, Eisenhut M, Robinson N, Osman D, Wilde A, Sadaf S, Brückler H, Walder L, Scheibe R, Hase T, Hanke GT (2018) A unique ferredoxin acts as a player in the low-iron response of photosynthetic organisms. Proc Natl Acad Sci U S A 115: E12111

Seckbach J (2007) Algae and cyanobacteria in extreme environments. Springer, Dordrecht

Siddiqui KS, Cavicchioli R (2006) Cold-adapted enzymes. Annu Rev Biochem 75:403–433

Siddiqui KS, Williams TJ, Wilkins D, Yau S, Allen MA, Brown MV, Lauro FM, Cavicchioli R (2013) Psychrophiles. Annu Rev Earth Planet Sci 41:87–115

Sidler WA (1994) Phycobilisome and phycobiliprotein structures. In: Bryant DA (ed) Advances in photosynthesis. Molecular biology of cyanobacteria, vol 1. Kluwer Academic, Dordrecht, pp 139–216

Siliakus MF, van der Oost J, Kengen SWM (2017) Adaptations of archaeal and bacterial membranes to variations in temperature, pH and pressure. Extremophiles 21:651–670

Slavov C, Reus M, Holzwarth AR (2013) Two different mechanisms cooperate in the desiccation-induced excited state quenching in Parmelia lichen. J Phys Chem B 117:11326–11336

Slavov C, Schrameyer V, Reus M, Ralph PJ, Hill R, Büchel C, Larkum AWD, Holzwarth AR (2016) "Super-quenching" state protects *Symbiodinium* from thermal stress—implications for coral bleaching. Biochim Biophys Acta 1857:840–847

Small D, Huner NPA, Wan W (2011) Effect of static magnetic fields on the growth, photosynthesis and ultrastructure of *Chlorella kessleri* microalgae. Bioelectromagnetics 33:298–308

Stepien P, Johnson GN (2009) Contrasting responses of photosynthesis to salt stress in the glycophyte *Arabidopsis* and the halophyte *Thellungiella*: role of the plastid terminal oxidase as an alternative electron sink. Plant Physiol 149:1154–1165

Steponkus PL (1984) Role of the plasma membrane in freezing injury and cold acclimation. Annu Rev Plant Physiol 35:543–584

Steponkus PL, Lanphear FO (1968) The role of light in cold acclimation of *Hedera helix* L. var. Thorndale. Plant Physiol 43:151–156

Stinziano JR, Hüner NPA, Way DA (2015) Warming delays autumn declines in photosynthetic capacity in a boreal conifer, Norway spruce (*Picea abies*). Tree Physiol 35:1303–1313

Streb P, Josse E-M, Gallouet E, Baptist F, Kuntz M, Cornic G (2005) Evidence for alternative electron sinks to photosynthetic carbon assimilation in the high mountain plant species *Ranunculus glacialis*. Plant Cell Environ 28:1123–1135

Sung S, Amasino RM (2005) Remembering winter: toward a molecular understanding of vernalization. Annu Rev Plant Biol 56:491–508

Susuki I, Los DA, Kanesaki Y, Mikami K, Murata N (2000) The pathway for perception and transduction of low-temperature signals in *Synechocystis*. EMBO J 19:1327–1334

Sveshnikov D, Ensminger I, Ivanov AG, Campbell DA, Lloyd J, Funk C, Huner NPA, Oquist G (2006) Excitation energy partitioning and quenching during cold acclimation in Scots pine. Tree Physiol 26:325–336

Szyszka B, Ivanov AG, Huner NPA (2007) Psychrophily is associated with differential energy partitioning, photosystem stoichiometry and polypeptide phosphorylation in *Chlamydomonas raudensis*. Biochim Biophys Acta 1767:789–800

Szyszka-Mroz B, Cvetkovska M, Ivanov AG, Smith DR, Possmayer M, Maxwell DP, Hüner NPA (2019) Protein kinases of the Antarctic polyextremophile, *Chlamydomonas* sp. UWO241, confer a distinct protein phosphorylation pattern associated with a remodelling of thylakoid membrane architecture and light energy distribution between photosystem I and photosystem II. Plant Physiol. (in press)

Takahashi S, Murata N (2008) How do environmental stresses accelerate photoinhibition? Trends Plant Sci 13:178–182

Takizawa K, Takahashi S, Huner NPA, Minagawa J (2009) Salinity affects the photoacclimation of *Chlamydomonas raudensis* Ettl UWO241. Photosynth Res 99:195–203

Tanaka R, Tanaka A (2007) Tetrapyrrole biosynthesis in higher plants. Annu Rev Plant Biol 58:321–346

Tang EPY, Vincent WF (1999) Strategies of thermal adaptation by high-latitude cyanobacteria. New Phytol 142:315–323

Tang EPY, Tremblay R, Vincent WF (1997) Cyanobacterial dominance of polar freshwater ecosystems—high-latitude mat-formers adapted to low temperature. J Phycol 33:171–181

Terauchi AM, Lu S-F, Zaffagnini M, Tappa S, Hirasawa M, Tripathy JN, Knaff DB, Farmer PJ, Lemaire SD, Hase T, Merchant SS (2009) Pattern of expression and substrate specificity of chloroplast ferredoxins from *Chlamydomonas reinhardtii*. J Biol Chem 284:25867–25878

Thomashow MF (1999) Plant cold acclimation: freezing tolerance genes and regulatory mechanisms. Annu Rev Plant Biol 50:571–599

Thomashow MF (2010) Molecular basis of plant cold acclimation: insights gained from studying the CBF cold response pathway. Plant Physiol 154:571–577

Trevaskis B (2010) The central role of the vernalization1 gene in the vernalization response of cereals. Funct Plant Biol 37:479–487

Trevaskis B (2015) Wheat gene for all seasons. Proc Natl Acad Sci U S A 112:11991–11992

Trevaskis B, Hemming MN, Dennis ES, Peacock WJ (2007) The molecular basis of vernalization-induced flowering in cereals. Trends Plant Sci 12:352–357

Ueno Y, Aikawa S, Kondo A, Akimoto S (2016) Energy transfer in cyanobacteria and red algae: confirmation of spillover in intact megacomplexes of phycobilisome and both photosystems. J Phys Chem Lett 7:3567–3571

Ünlü C, Drop B, Croce R, van Amerongen H (2014) State transitions in *Chlamydomonas reinhardtii* strongly modulate the functional size of photosystem II but not of photosystem I. Proc Natl Acad Sci U S A 111:3460–3465

Velasco VME, Mansbridge J, Bremner S, Carruthers K, Summers PS, Sung WWL, Champigny MJ, Weretilnyk EA (2016) Acclimation of the crucifer *Eutrema salsugineum* to phosphate limitation is associated with constitutively high expression of phosphate-starvation genes. Plant Cell Environ 39:1818–1834

Verhoeven AS (2014) Sustained energy dissipation in winter evergreens. New Phytol 201:57–65

Verhoeven AS, Adams WW III, Demmig-Adams B, Croce R, Bassi R (1999) Xanthophyll cycle pigment localization and dynamics during exposure to low temperatures and light stress in *Vinca major*. Plant Physiol 120:727–738

Vincent W (2000) Cyanobacterial dominance in the polar regions. In: Whitton BA, Potts M (eds) The ecology of cyanobacteria. Kluwer Academic, Dordrecht, pp 321–340

Vincent WF (2007) Cold tolerance in cyanobacteria and life in the cryosphere. In: Seckbach J (ed) Algae and cyanobacteria in extreme environments. Springer, Dordrecht

Wada H, Gombos Z, Sakamoto T, Murata N (1993) Role of lipids in low temperature adaptation. In: photosynthetic responses to the environment (Yamomoto HY, Smith CM, eds). Am Soc Plant Physiol 8:78–87

Walters RG, Rogers JJM, Shephard F, Horton P (1999) Acclimation of *Arabidopsis thaliana* to the light environment: the role of photoreceptors. Planta 209:517–527

Way DA, Sage RF (2008) Thermal acclimation of photosynthesis in black spruce *Picea mariana* (Mill.) B.S.P. Plant Cell Environ 31:1250–1262

Way DA, Yamori W (2014) Thermal acclimation of photosynthesis: on the importance of adjusting our definitions and accounting for thermal acclimation of respiration. Photosynth Res 119:89–100

Whitton BA, Potts M (2000) The ecology of cyanobacteria: their diversity in time and space. Kluwer Academic, Dordrecht

Wilson A, Ajlani G, Verbavatz J-M, Vass I, Kerfeld CA, Kirilovsky D (2006a) A soluble carotenoid protein involved in phycobilisome-related energy dissipation in cyanobacteria. Plant Cell 18:992–1007
Wilson KE, Ivanov AG, Oquist G, Grodzinski B, Sarhan F, Huner NPA (2006b) Energy balance, organellar redox status and acclimation to environmental stress. Can J Bot 84:1355–1370
Wilson A, Punginelli C, Gall A, Bonetti C, Alexandre M, Routaboul J-M, Kerfeld CA, van Grondelle R, Robert B, Kennis JTM, Kirilovsky D (2008) A photoactive carotenoid protein acting as light intensity sensor. Proc Natl Acad Sci U S A 105:12075–12080
Winfield MO, Lu CG, Wilson ID, Coghill JA, Edwards KJ (2009) Cold- and light-induced changes in the transcriptome of wheat leading to phase transition from vegetative to reproductive growth. BMC Plant Biol 9:55
Winfield MO, Lu C, Wilson ID, Coghill JA, Edwards KJ (2010) Plant responses to cold: transcriptome analysis of wheat. Plant Biotechnol J 8:749–771
Woodson JD, Chory J (2008) Coordination of gene expression between organellar and nuclear genomes. Nat Rev Genet 9:383–395
Wunder T, Xu W, Liu Q, Wanner G, Leister D, Pribil M (2013) The major thylakoid protein kinases STN7 and STN8 revisited: effects of altered STN8 levels and regulatory specificities of the STN kinases. Front Plant Physiol 4:417
Wynn-Williams DD (2000) Cyanobacteria in deserts- life at the limit? In: Whitton BA, Potts M (eds) The ecology of cyanobacteria. Kluwer Academic, Dordrecht, pp 341–366
Yamaguchi-Shinozaki K, Shinozaki K (2006) Transcriptional regulatory networks in cellular responses and tolerance to dehydration and cold stresses. Annu Rev Plant Biol 57:781–803
Yang Y-W, Yin Y-C, Li Z-K, Huang D, Shang J-L, Chen M, Qiu B-S (2019) Orange and red carotenoid proteins are involved in the adaptation of the terrestrial cyanobacterium Nostoc flagelliforme to desiccation. Photosynth Res 140:103–113
Yumoto I (2013) Cold adapted microorganisms. Caister Academic, Norfolk
Zakhia F, Jungblut A-D, Taton A, Vincent W, Wilmotte A (2008) Cyanobacteria in cold ecosystems. In: Margesin R, Schinner F, Marx J-C, Gerday C (eds) Psychrophiles. From biodiversity to biotechnology. Springer, Berlin, pp 121–135
Zarka DG, Vogel JT, Cook D, Thomashow MF (2003) Cold induction of Arabidopsis CBF genes involves multiple ICE (inducer of CBF expression) promoter elements and a cold-regulatory circuit that is desensitized by low temperature. Plant Physiol 133:910–918
Zer H, Ohad I (2003) Light, redox state, thylakoid-protein phosphorylation and signaling gene expression. Trends Biochem Sci 28:467–470
Zhu J, Pearce S, Burke A, See D, Skinner D, Dubcovsky J, Garland-Campbell K (2014) Copy number and haplotype variation at the VRN-A1 and central FR-A2 loci are associated with frost tolerance in hexaploid wheat. Theor Appl Genet 127:1183–1197

Norman P.A. Hüner Professor *Emeritus*, Dept. of Biology, University of Western Ontario and a Tier 1 Canada Research Chair in Environmental Stress Biology until his retirement in 2017. His research program continues to focus on photosynthetic acclimation and adaptation to extreme environments for which he received the Gold Medal of the Canadian Society of Plant Physiologists. He is an elected Fellow of the Academy of Science, Royal Society of Canada.

Alexander G. Ivanov is Professor of Biophysics at the Institute of Biophysics and Biomedical Engineering, Bulgarian Academy of Sciences, and a Senior Research Associate in the Dept. of Biology, University of Western Ontario. He received his PhD degree in Biophysics from the Bulgarian Academy of Sciences. His research interests include the biophysics of photosynthesis, bioenergetics, membrane dynamics, and plant stress physiology.

Marina Cvetkovska is an assistant professor in the Department of Biology, University of Ottawa. She received her PhD from the University of Toronto (2006–2012), followed by an NSERC-funded postdoctoral position at the University of Western Ontario (2014–2018). Her research interests are focused on the mechanisms that govern the responses of photosynthetic organisms to environmental stress. Through an NSERC-funded research program, the Cvetkovska lab is elucidating the mechanisms by which Antarctic extremophilic algae adapt to the permanent cold and how these organisms respond to elevated temperatures and heat stress.

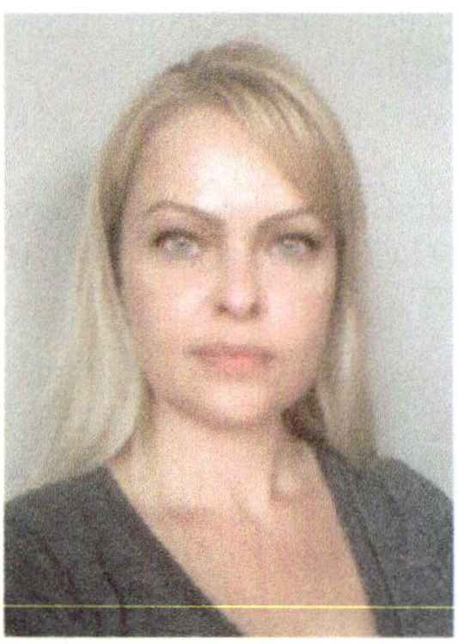

Beth Szyszka research interests are focused on the physiological, biochemical, biophysical, and molecular characterization of the photosynthetic apparatus of the Antarctic psychrophile, *Chlamydomonas sp.* UWO241. She completed her PhD at the University of Western Ontario, during which she was an NSERC Postgraduate Scholar. Presently, she continues her research interests in photosynthesis and psychrophily as a Research Associate in the Dept. of Biology, University of Western Ontario.

Marc Possmayer obtained a BSc in Biology and Environmental Science at McGill University, Montreal, before undertaking studies of the molecular physiology of an Antarctic alga for his MSc and PhD at the University of Western Ontario. He is currently a postdoctoral fellow in the Department of Biology at the University of Ottawa.

Paul Porter is a second-year student in the Faculty of Health Sciences, University of Western Ontario. He is from Six Nations of the Grand River Mohawk Bear clan. Paul was awarded a prestigious scholarship from Western's Head and Heart Programme which provided a summer stipend in support of his research project on psychrophilic algae in the Dept. of Biology, University of Western Ontario.

7 What Is the Limiting Factor? The Key Question for Grain Yield of Maize as a Renewable Resource Under Salt Stress

Birgit W. Hütsch, Stephan Jung, Marleen Steinbach, and Sven Schubert

Abstract

During 4 years, maize (*Zea mays* L.) experiments were conducted to investigate the effects of salt stress on grain yield performance and to identify limiting factors. The plants were grown in large containers, filled with 145 kg soil to a depth of 0.9 m, in a vegetation hall with controlled water supply. Grain yield and its determinants, i.e., kernel number and single kernel weight, were recorded at maturity and related to physiologically relevant parameters 2 days after controlled pollination. The time around pollination is decisive for kernel setting and thus kernel number at maturity, as this parameter was almost exclusively responsible for grain yield reductions under salt stress. Single kernel weight was unaffected by saline conditions. The decreased number of kernels was not caused by source limitation, because the availability of sucrose as main transport metabolite was always higher in developing kernels under salt stress than under control conditions. Although acid invertase activity as a key factor for sink activity was reduced or unchanged under salt stress, the hexose concentrations were always higher in the developing kernels pointing to no limitation. Another key enzyme in the plasma membrane of the sink cells is H^+-ATPase, which provides the pH gradient necessary for H^+-cotransport of hexoses from the apoplast into the cytosol. The hexoses are needed for metabolic processes and energy supply in order to enable cell division and extension growth. Plasma membrane H^+-ATPase activity was significantly reduced in the salt-stressed kernels, resulting in a smaller pH gradient. Thus, kernel development and hence grain yield performance under salt stress seem to be limited by a transport problem, caused by inhibition of plasma membrane H^+-ATPase.

B. W. Hütsch (✉) · S. Jung · M. Steinbach · S. Schubert
Institute of Plant Nutrition, Justus Liebig University (iFZ), Giessen, Germany
e-mail: Birgit.W.Huetsch@ernaehrung.uni-giessen.de; Stephan.Jung@lwk.nrw.de; sven.schubert@ernaehrung.uni-giessen.de

A. Kumar et al. (eds.), *Climate Change, Photosynthesis and Advanced Biofuels*,
https://doi.org/10.1007/978-981-15-5228-1_7

Keywords

Kernel setting · Source-sink relations · Acid invertase · Plasma membrane H^+-ATPase · pH gradient · H^+/hexose cotransport

7.1 Introduction

Salinity is one of the major reasons for yield losses of crops in areas where evapotranspiration exceeds precipitation. This natural occurrence is worsened by irrigation practices that are not adapted to the environmental conditions in arid and semiarid areas (Qadir et al. 2006). Besides increasing efforts to avoid anthropogenic salinization of soils and active attempts of soil desalinization, salt-resistant crops could be grown to diminish salt stress effects. In order to develop those crops, the physiological processes which lead to yield losses by salt stress have to be understood (Munns 2011). Described effects of salinity on plant development mainly occur in the vegetative growth stage. However, they have also consequences for the generative stage. It is well known that abiotic stress such as salt stress reduces the number of maize kernels reaching maturity (Schubert et al. 2009; Henry et al. 2015; Hütsch et al. 2015; Jung et al. 2017).

Maize is grown and utilized not only as food and feed but also as feedstocks for generation of renewable fuel ethanol. Maize kernels make a good biofuel feedstock due to their high starch content and the comparatively easy conversion to ethanol. Three main steps are necessary to process maize kernels for ethanol production: (1) conversion of starch into fermentable sugars, (2) fermentation with yeast to convert sugars into ethanol, and (3) purification to remove byproducts. On a global scale, maize is considered as the primary source of fuel ethanol, the USA being the top ethanol producer in the world with a contribution of 58% in 2017 (Mohanty and Swain 2019). In 2016, almost 95% of the total fuel ethanol production in the USA came from maize starch feedstock (Mohanty and Swain 2019). Global fuel ethanol production has increased significantly in recent years, not only in the USA but also in China and other countries, and this trend will presumably continue (Oladosu et al. 2011). Thus, in the future maize will increasingly be grown on marginal sites, where plants, e.g., face salinity or drought conditions. The osmotic effects of salinity on maize growth and kernel development are comparable to effects of water shortage; thus the results we present here on salt stress can be partly transferred to drought conditions (Hütsch et al. 2015).

The grain yield potential of maize, which generally produces just one cob per plant, is determined by two parameters: kernel number per cob and single kernel weight. Changes in one or both of these determinants have profound effects on the final grain yield at maturity. The kernel number per cob depends on kernel setting at or shortly after anthesis. The individual kernel weight is determined by rate and duration of grain filling. Schubert et al. (2009) demonstrated that the grain yield decrease of various maize hybrids under salt stress was mainly caused by reduced kernel setting and not by reduced grain filling.

Reductions in maize grain yield under salt stress might be caused by a restricted availability of assimilates. Decreases in vegetative shoot growth diminish the photosynthetic capacity resulting in reduced delivery of assimilates. Already fertilized kernels can abort because of insufficient supply with carbohydrates (Abbate et al. 1995). However, to date it is unclear whether these probable negative effects of salt stress on photosynthesis and thus assimilate supply are a limiting factor for grain yield production of maize.

Another important parameter for grain yield performance is sink strength. Assimilates are transported via phloem mainly as sucrose from the source tissues to the sinks (e.g., developing kernels). This transport is driven by the sink strength, which consists of two components, sink capacity and sink activity (Ho 1988). The sink capacity is determined by the number of set kernels and by the kernel size, whereas the sink activity relies on the metabolic activity in the sink organs. The transport of assimilates from the maternal tissues to the daughter cells must take the apoplastic pathway, as there are no symplastic connections between the embryonic and maternal tissues of a developing seed (Ho and Gifford 1984; Thorne 1985; Wang and Fisher 1995). An important enzyme which affects sink activity in the apoplast is acid invertase. Shortly after anthesis acid invertase activity was inhibited in maize kernels by salt stress (Hütsch et al. 2014, 2015; Jung et al. 2017). It has been demonstrated that restrictions in the activity of this enzyme have deleterious effects particularly on kernel set (Chourey et al. 2006). Another key enzyme for kernel setting is plasma membrane H^+-ATPase, which establishes the pH gradient necessary to transport hexoses into the developing kernels via H^+-cotransport (Hütsch and Schubert 2017; Jung et al. 2017). Under salt stress, the activity of plasma membrane H^+-ATPase was also inhibited in the kernel tissue (Jung et al. 2017). On the other hand, grain filling is mainly determined by the activity of starch-synthesizing enzymes. However, only set kernels can be filled; thus the processes around pollination are the most important for grain yield performance. High grain yields can only be achieved with large kernel numbers. Even though acid invertase and plasma membrane H^+-ATPase are inhibited by salt stress, it is unclear whether these enzymes cause limitation of maize grain yield.

In the present study, the long-term effects of salt stress (during the entire vegetation period) on maize were investigated in container technique under near-field conditions in 4 years, focusing on the relation between metabolic changes around the time of pollination and the impact on yield determinants at maturity. The objective was to test whether one yield-limiting factor can be identified for all experiments or if the limiting factor varies with the year of conductance. The relatively salt-resistant maize hybrid Pioneer 3906 was investigated. The electrical conductivity in the soil of the salt stress treatment was set to 11 dS m^{-1}.

7.2 Experimental Approach to Determine Yield-Limiting Factors of Grain Maize Under Salt Stress

In studies of Hütsch et al. (2015) and Jung et al. (2017), the experimental setup for determination of salt stress effects on grain yield performance of maize is described in detail. In the following chapter, a brief overview of the experimental approach and the used methods is given.

7.2.1 Plant Cultivation

Maize (*Zea mays* L. cv. Pioneer 3906) was grown in soil culture in large plastic containers, placed in a vegetation hall (Fig. 7.1a, b, c). The maize cultivar Pioneer 3906 has a relatively strong capability to exclude Na^+ from the shoot, which is achieved by efficient Na^+ exclusion at the root surface and by restricted Na^+ translocation from root to shoot (Schubert et al. 2009). Thus, Pioneer 3906 is regarded as relatively salt-resistant (Sümer et al. 2004). Two different treatments were set up: control and salt stress. The experiments were conducted in the years 2011, 2012, 2014, and 2017.

Fig. 7.1 Maize plants, Pioneer 3906, growing in container technique in the vegetation hall (**a**), early vegetative growth (**b**), late vegetative growth (**c**), cobs at maturity (**d**)

Large plastic containers (120 L) were filled with 145 kg of an air-dry Brown Earth subsoil (loamy sand) in four increments: three layers with 30 kg soil each and a topsoil layer (approx. 0–30 cm), which was fertilized with a compound fertilizer. In the salt treatment, 2.1 g NaCl kg^{-1} soil were applied to each soil layer in order to obtain an electrical conductivity of 11 dS m^{-1} throughout the whole soil depth of 0.9 m. Thus the plants were facing salinity right from germination until maturity. Each soil layer was moistened immediately after filling.

Between middle of May and beginning of June (depending on the year), maize was sown with nine seeds per container, and 10 days later the number of plants was reduced to four per container, and water content was adjusted to 60% maximum water-holding capacity (WHC). During the whole vegetation period, water supply was recorded for each container and adjusted at least daily to the desired WHC. Thus, water consumption was determined for the entire growth period, and water-use efficiency was calculated (WUE_{grain} = grain dry matter/total water consumption). The plants grew in the vegetation hall of the experimental station of the Institute of Plant Nutrition in Giessen under natural light conditions. The containers were set up in a completely randomized design, and their position was changed at least once a week. During the vegetation period, additional compound fertilizer was applied when appropriate.

Controlled pollination took place 5 days after silk emergence, which is considered to be the time with best receptivity (Cárcova et al. 2000). Fresh pollen from additionally cultivated donor plants, grown under control conditions, was used. The time from sowing until pollination varied with treatment and year and is inserted in Fig. 7.2.

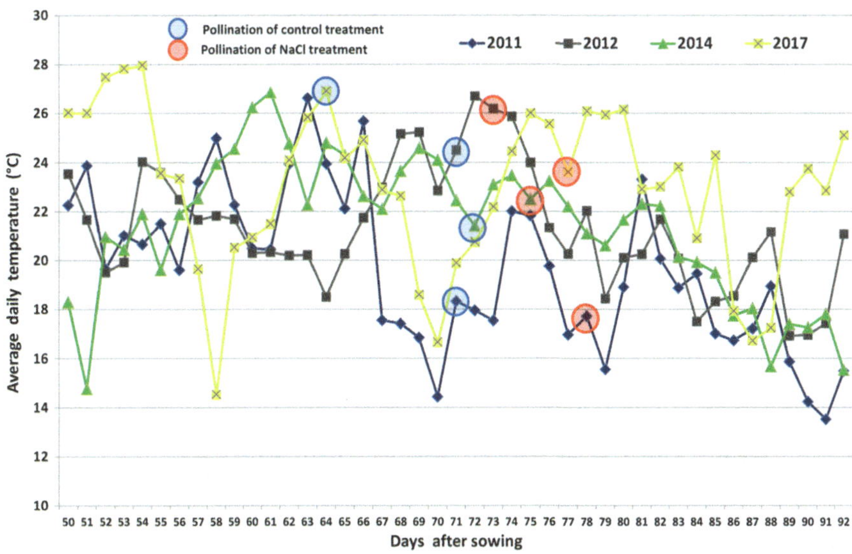

Fig. 7.2 Average daily temperature in the years 2011, 2012, 2014, and 2017 for the timespan of controlled pollination ±2 weeks

7.2.2 Harvest

For each treatment, four containers with four plants each were harvested 2 days after pollination (2 DAP) and at physiological maturity. For the mature plants, straw and grain dry weight were determined as well as kernel number per cob and single kernel weight. At the intermediate harvests, shoots were cut at the base and plant height was measured. The cob was separated from the stalk, and kernels were cut off from the rachis with a knife and immediately shock-frozen in liquid N_2. Fresh weight of shoots and kernels was recorded. For laboratory analyses the frozen kernels were ground under liquid N_2 using mortar and pestle and stored at −80 °C until the following analyses were performed:

Sugar analysis: A subsample of frozen, ground kernels was either lyophilized or dried at 80 °C for 48 h. Subsequently, sucrose, glucose, and fructose concentrations were determined in water extracts.

Enzyme extraction and measurements of acid invertase activity: Frozen and ground samples were extracted with HEPES buffer (pH 7.2). After centrifugation, the supernatant was shock-frozen in liquid N_2 and stored at −80 °C. All extracts were desalted with Econo-Pac® 10 DG columns from BIO-RAD. For determination of acid invertase activities (EC 3.2.1.26), desalted extracts were mixed with Na acetate (pH 4.8) and sucrose. Incubation was carried out at 30 °C for 30 min. Generated glucose was determined and activity rates were calculated (μmol glucose g^{-1} fresh mass min^{-1}).

Isolation of plasma membrane vesicles and measurements of H^+-ATPase activity: Detailed description of isolation of plasma membrane vesicles from maize kernels and measurements of hydrolytic and pumping activity of plasma membrane H^+-ATPase (EC 7.1.2.1) in these vesicles can be obtained from Jung et al. (2017).

7.2.3 Statistical Analysis

Means ± standard errors (SE) were calculated from four replicates per treatment and year. After Student's two-sided *t*-test (Microsoft Office Excel 2010), significant differences between control and NaCl treatment are given for each year ($*p \leq 5\%$, $**p \leq 1\%$, $***p \leq 0.1\%$). Additionally, a two-way ANOVA was conducted to analyze the data for variance of stress treatment and year effects, using *RStudio*. Multiple comparisons were done with the post hoc *Tukey* test ($p \leq 5\%$). Significant differences are indicated by different small letters.

7.3 Grain Yield at Maturity and Its Determinants

At maturity, the cobs of the NaCl-treated plants were shorter and showed distinct kernel abortion mainly in the apical part (Fig. 7.1d). During the 4 years of container experiments, the grain yield was always significantly lower under salt stress with reductions between 30 and 51% (Fig. 7.3a). In the NaCl treatment, no difference in

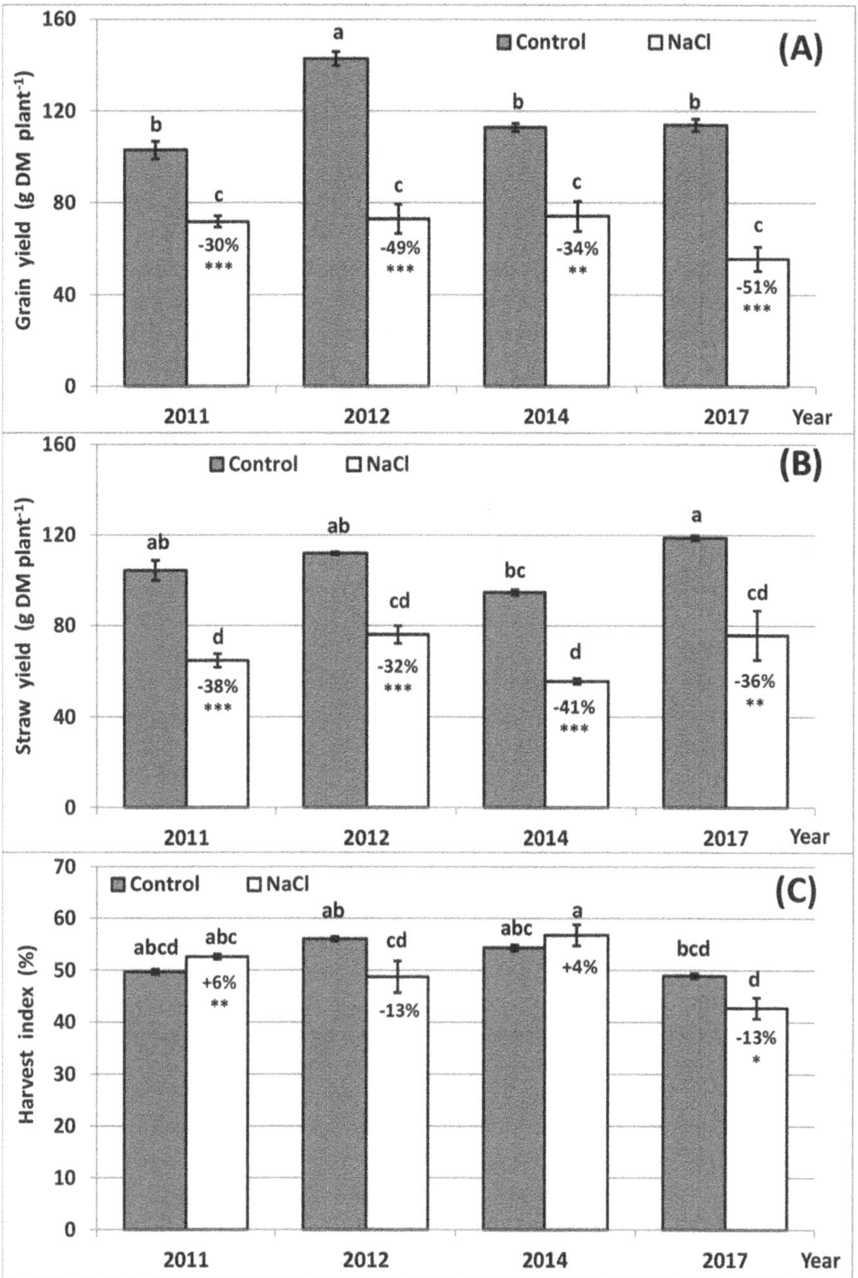

Fig. 7.3 Grain dry matter yield (**a**), straw dry matter yield (**b**), and harvest index (**c**) of maize cultivar Pioneer 3906 grown in 4 years in container technique under control and saline conditions, harvested at maturity; data show means of 4 replicates ± *SE*; for each year differences in percentage between control and stress treatment are given and, if significant after Student's *t*-test, indicated by $*p \leq 5\%$, $**p \leq 1\%$, and $***\ p \leq 0.1\%$; for Tukey test after two-way ANOVA: significant differences are indicated by different small letters ($p \leq 5\%$)

grain yields among experimental years was observed; under control conditions the grain yield in 2012 was significantly higher than in the other years (Fig. 7.3a). In addition, the straw yield was significantly reduced under salt stress (32–41%; Fig. 7.3b). In 2017, the control plants had the highest straw yield, which lays significantly above that in 2014.

The harvest index, which is the ratio of grain yield to total aboveground biomass at physiological maturity, was significantly affected by NaCl treatment in 3 years: in 2011 the harvest index was increased by 6% under salinity, whereas in 2012 and 2017, it was decreased by 13% (Fig. 7.3c). On a physiological basis, the harvest index is an indicator of the relative investment of plant resources in reproductive plant parts and can be used as a measure of reproductive efficiency (Unkovich et al. 2010). With an increase of the harvest index, an improvement of nutrient and water-use efficiency can be expected. More assimilates and nutrients are allocated to grains as the harvested product and less biomass is contained in the often unused shoot residues. Thus, in 2011 the reproductive efficiency of grain maize was improved under salt stress, whereas in 2012 and 2017, it decreased. In the control plants, no significant year effects were observed for the harvest index, whereas in the NaCl treatment, the highest value was achieved in 2014 and the lowest in 2017 (Fig. 7.3c). In 2014 the particularly small straw yield is responsible for the high harvest index, whereas in 2017 the comparably small grain yield caused the low value (Fig. 7.3a–c; NaCl treatment).

For maize, the grain yield is determined by the kernel number per plant and by the single kernel weight. During the 4 experimental years, under salt stress the kernel number was always significantly reduced between 30 and 50% (Fig. 7.4a). There were no significant differences in kernel number among years, neither in the control nor in the salt treatment. In 2011, 2012, and 2014, the kernel weight was slightly but not significantly enhanced under salt stress, whereas in 2017 it was significantly reduced by 30% (Fig. 7.4b). In 2014 the kernel weight of both treatments was significantly smaller than in 2012. The smallest kernel weight was achieved under salt stress in 2017 with significant differences to the weights of the respective treatment in 2011 and 2012 (Fig. 7.4b).

It can be summarized that in 3 years (2011, 2012, 2014), the reduction in grain yield under salt stress was exclusively caused by a reduced kernel number. Decisive for kernel setting is the time around pollination when kernel abortion occurs under stress conditions. Therefore, 2 days after controlled pollination (2 DAP), intermediate harvests were conducted in order to elucidate physiological processes relevant for kernel setting and yield performance of maize (Sect. 7.5). In a later developmental stage, under salt stress Pioneer 3906 was able to maintain grain filling of the remaining kernels to a similar extent as in the control treatment. Probable reasons for the reduced kernel weight in the NaCl treatment in 2017 will be discussed later (Sect. 7.6).

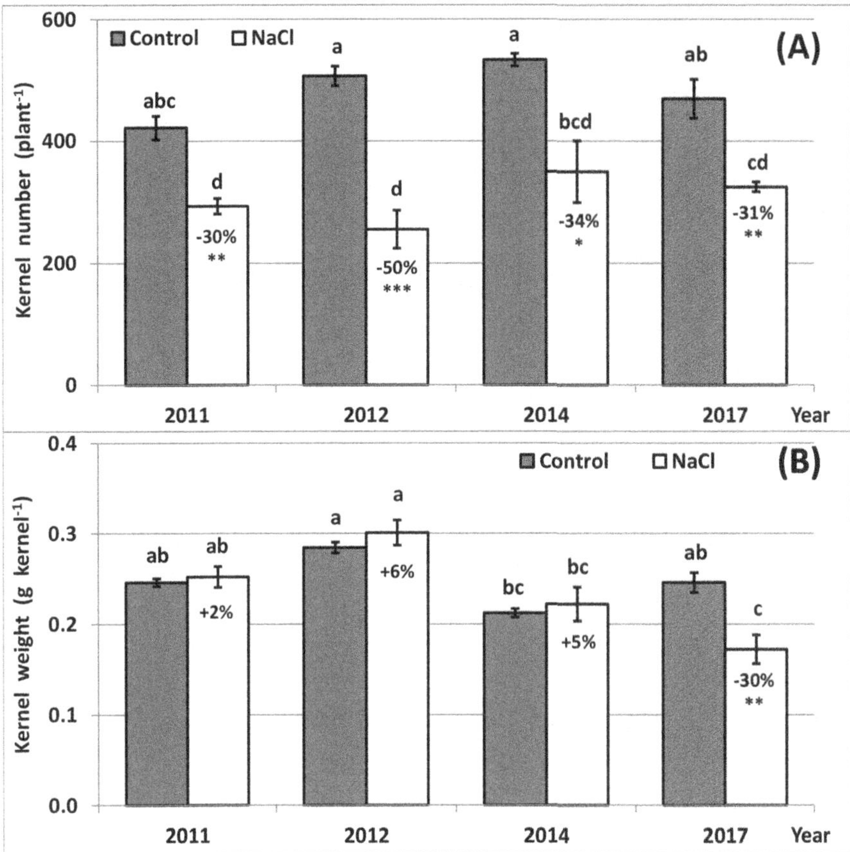

Fig. 7.4 Number of kernels (**a**) and single kernel weight (**b**) of maize cultivar Pioneer 3906 grown in 4 years in container technique under control and saline conditions, harvested at maturity; data show means of 4 replicates $\pm$ *SE*; for each year differences in percentage between control and stress treatment are given and, if significant after Student's *t*-test, indicated by $*p \leq 5\%$, $**p \leq 1\%$, and $*** \ p \leq 0.1\%$; for Tukey test after two-way ANOVA: significant differences are indicated by different small letters ($p \leq 5\%$)

7.4 Water Consumption and Water-Use Efficiency

The water consumption of salt-treated maize plants was significantly reduced by 35–53% in comparison to the control plants (Fig. 7.5a). This is in accordance with the smaller vegetative growth (Fig. 7.1b, c) and the reduced leaf area, which resulted in a significantly reduced transpiration rate under salt stress (Hütsch et al. 2015). In 2011 and 2012, the plants needed significantly more water in both treatments in comparison to 2014 and 2017.

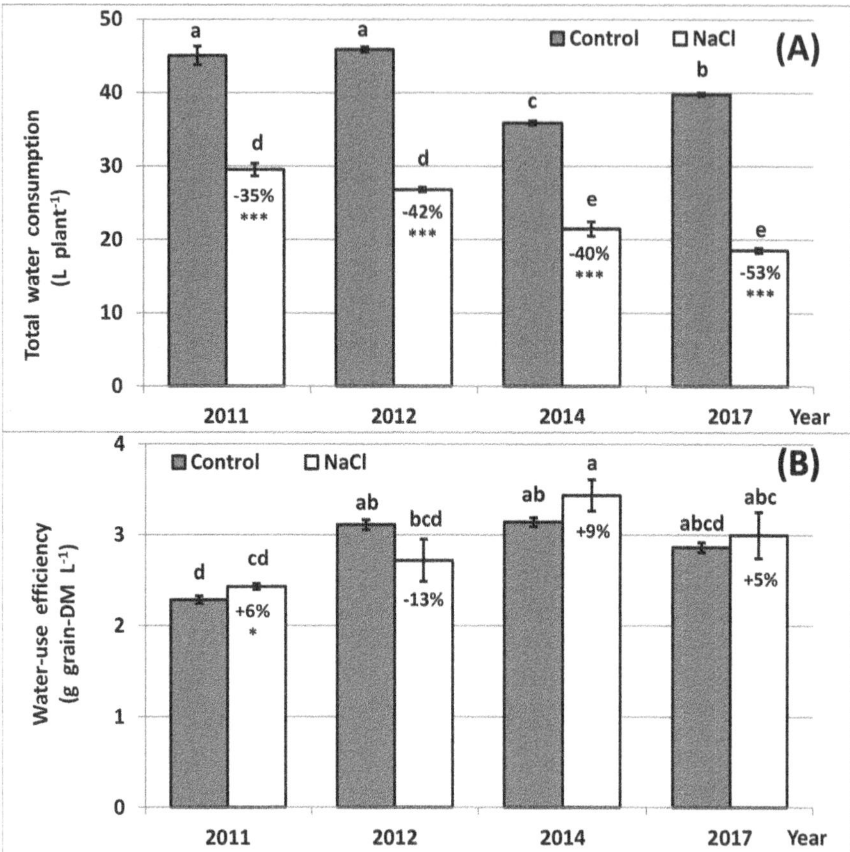

Fig. 7.5 Total water consumption (**a**) and water-use efficiency (**b**) of maize cultivar Pioneer 3906 grown in 4 years in container technique under control and saline conditions, determined at maturity; data show means of 4 replicates ± *SE*; for each year differences in percentage between control and stress treatment are given and, if significant after Student's *t*-test, indicated by $*p \leq 5\%$, and $***p \leq 0.1\%$; for Tukey test after two-way ANOVA: significant differences are indicated by different small letters ($p \leq 5\%$)

Water-use efficiency, which is defined as WUE_{grain} = grain dry matter/total water consumption, showed a significant increase under salt stress in 2011, whereas no treatment effect was observed in the other years (Fig. 7.5b). Thus, in the NaCl treatment, the maize plants used water at least as efficiently as the control plants. The WUE_{grain} of the control plants was significantly smaller in 2011 than in 2012 and 2014; under salt stress significantly smaller values were obtained in 2011 and 2012 in comparison to 2014 (Fig. 7.5b).

Climate change models predict decreased precipitation in many of the world's cropping regions; thus irrigation agriculture will increasingly take place under water scarcity. If the irrigation water is contaminated with salts, this will aggravate salinization of agricultural land. Apart from decreased precipitation, with higher atmospheric CO_2 concentrations, an increase in ambient temperature takes place, and thus the water vapor pressure deficit of the air increases. This results in considerable increases in transpiration, strengthening the need for irrigation in order to obtain reasonably high yields.

Apart from an efficient water use, the distribution of available water throughout the entire growth period is decisive for yield development. Excessive vegetative growth can aggravate water limitations by using too much water before flowering (Passioura and Angus 2010), as sufficient water availability during the reproductive period is particularly important for kernel set and yield performance (Yang and Grassini 2014). This points to the particular importance of rainfall distribution for yield performance, as precipitation close to the critical period of kernel setting would be more beneficial than in earlier or later parts of the season (Andrade et al. 1999; Hay and Gilbert 2001; Otegui et al. 1995; Passioura and Angus 2010; Zhang et al. 2014).

7.5 Physiologically Relevant Parameters During Kernel Setting (2 DAP)

For maize, carbon assimilation during kernel set and grain filling is a prerequisite for good kernel development, as remobilization of reserves is unimportant (Cliquet et al. 1990; Schussler and Westgate 1994). Thus, in maize approximately 50% of the total dry matter is accumulated before flowering, with the remaining 50% being fixed during the grain-filling period (Lee and Tollenaar 2007). A reduction in kernel number is frequently observed when maize is subjected to abiotic stress. Kernel setting is often terminated in a timeframe of just a few days after pollination. Especially kernels at the apical end of cobs are susceptible to kernel abortion, and abiotic stress can increase the number of aborted kernels (Oury et al. 2016b). Therefore, the physiological causes for kernel abortion under salt stress were investigated at 2 DAP.

7.5.1 Shoot Growth and Kernel Development

At the intermediate harvests (2 DAP), plant heights were significantly reduced under salt stress in comparison to the control (between 21 and 30%; Fig. 7.6a). Similarly, shoot fresh mass was significantly smaller under salt stress with even stronger reductions (between 30 and 45%; Fig. 7.6b). Significant year effects occurred with highest values of plant height as well as shoot fresh mass in 2012 and lowest plant height in 2014 and of shoot fresh mass in 2011. It is well known that under salt stress the vegetative shoot biomass is strongly reduced in comparison to control conditions (Hütsch et al. 2014, 2015; Jung et al. 2017; Munns 1993, 2011; Schubert 2011).

For kernel fresh mass, the salt stress effect varied depending on the year when the experiment was conducted. In 2012 and 2014, kernel fresh mass was significantly

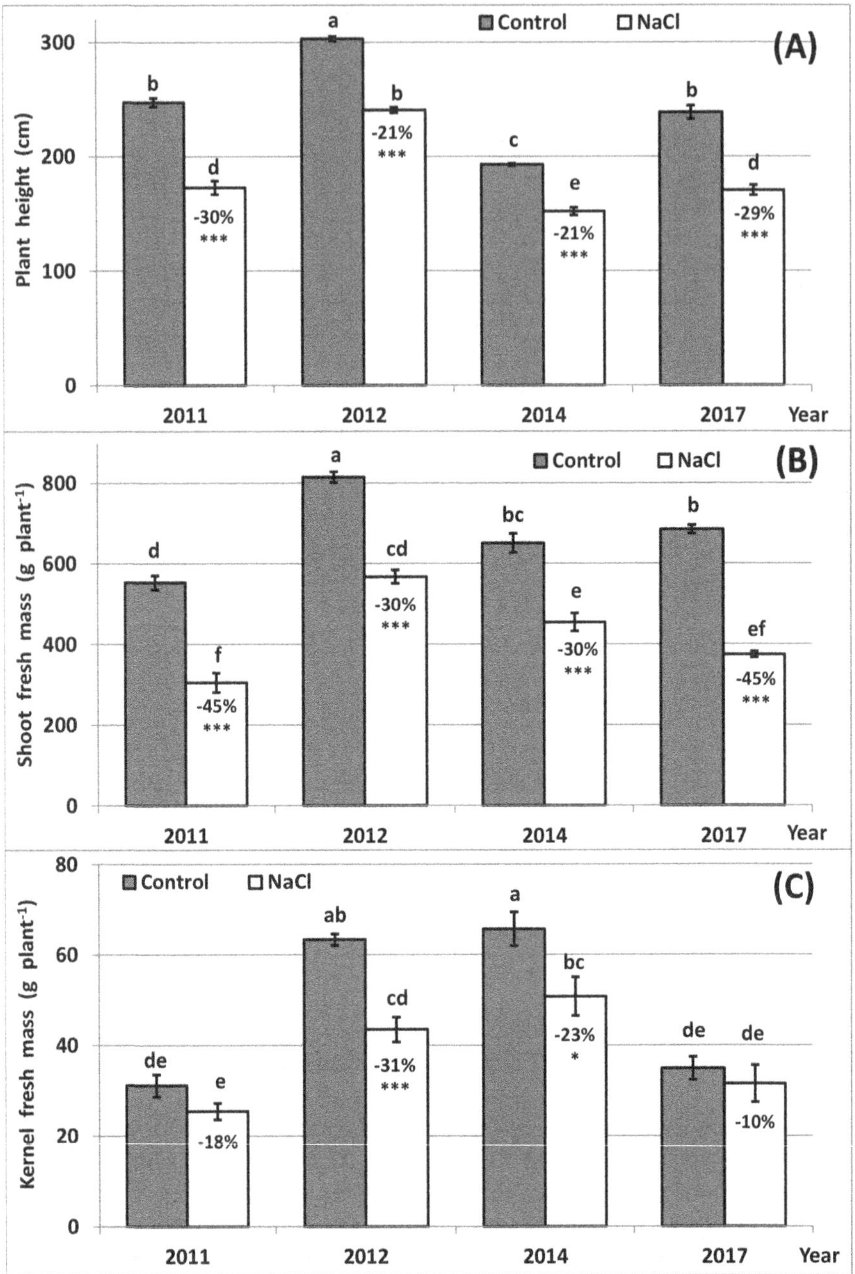

Fig. 7.6 Plant height (**a**), shoot fresh mass (**b**), and kernel fresh mass (**c**) of maize cultivar Pioneer 3906 grown in 4 years in container technique under control and saline conditions, harvested 2 days after pollination; data show means of 4 replicates ± *SE*; for each year differences in percentage between control and stress treatment are given and, if significant after Student's *t*-test, indicated by $*p \leq 5\%$, and $***p \leq 0.1\%$; for Tukey test after two-way ANOVA: significant differences are indicated by different small letters ($p \leq 5\%$)

decreased in the saline treatment by 31% and 23%, respectively (Fig. 7.6c). However, in 2011 and 2017, no salt stress effect on kernel fresh mass has occurred yet 2 days after pollination. The kernel fresh mass of both treatments was significantly smaller in 2011 and 2017 than in 2012 and 2014 (Fig. 7.6c). Overall kernel fresh mass 2 DAP was less affected by salt stress than shoot fresh mass, except for the year 2012 (Fig. 7.6b, c).

7.5.2 Assimilate Availability in Developing Kernels

At the intermediate harvest, the sucrose concentrations in the developing kernels were always significantly higher under salt stress than in the control (Fig. 7.7a). Thus, under stress conditions sufficient sucrose as main transport metabolite in the phloem had reached the kernels, and availability of carbon assimilates did not limit kernel development during the 4 experimental years. Also the concentrations of the hexoses glucose and fructose were significantly increased in the kernels of the NaCl treatment (Fig. 7.7b, c; exception: no change in glucose concentration in 2011). The availability of hexoses for feeding the developing kernels was also not restricted. Differences in sugar concentrations among the 4 years occurred (Fig. 7.7a–c).

Although almost all of the dry matter allocated to the grain has to be fixed during the timespan between pollination and harvest at maturity, there is evidence that the phase of kernel setting and early grain filling is not limited by restricted assimilate availability (Henry et al. 2015; Hütsch et al. 2015; Oury et al. 2016a). This is particularly true for the newer stay-green maize hybrids with delayed leaf senescence (Rajcan and Tollenaar 1999a, b). The increased leaf longevity was associated with an improvement of the ratio of assimilate supply (source) and demand (sink) during grain filling (Rajcan and Tollenaar 1999b). Additionally, it should be considered that maize as a C_4 plant is very efficient in CO_2 assimilation in comparison to C_3 species such as wheat, barley, or rice. Even under salt and drought stress, kernel setting was not source-limited, as significant accumulation of sucrose as the major transport metabolite in phloem was found in the developing kernels shortly after pollination (Fig 7.7a; Hütsch et al. 2014, 2015; Jung et al. 2017). Henry et al. (2015) also found an accumulation of sucrose and hexoses in kernel tissue at pollination and 3 DAP under salt stress. Accordingly, Below et al. (1981) found that the rate of supply of photosynthates to the cob was probably not a yield-limiting factor for any of the five field-grown maize hybrids investigated. Sampling at intervals during the grain-filling period showed that the capacity of the leaves to produce photosynthates through the first half of the grain-filling period exceeded the needs of the cob and/or the transport system (Swank et al. 1982).

While overall photosynthesis is often reduced under salt stress, this reduction seldom impairs photoassimilate availability in sink tissues (Ruan et al. 2012; Henry et al. 2015; Hütsch et al. 2015, 2016; Oury et al. 2016a). Since there was no limitation in photoassimilate supply, source limitation can be ruled out as a reason for kernel abortion.

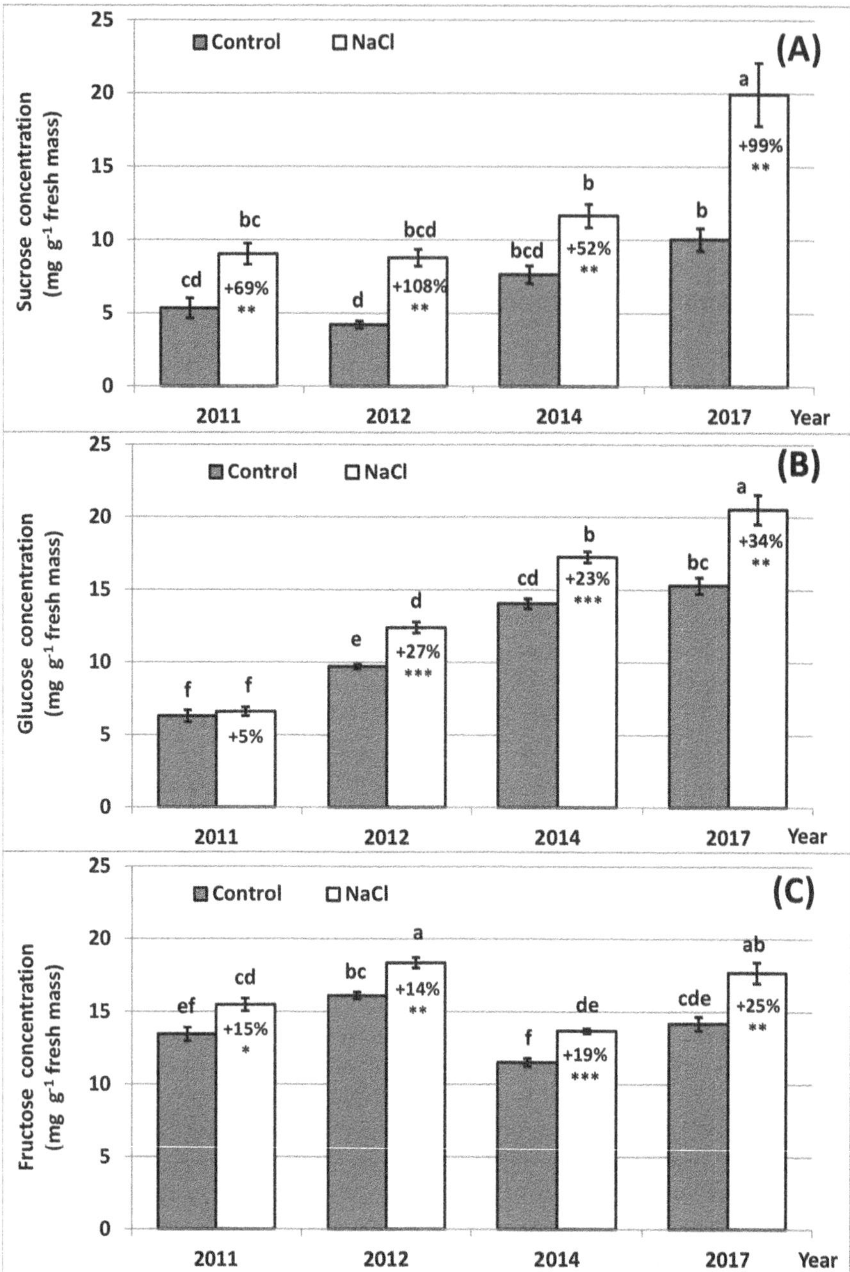

Fig. 7.7 Sucrose (**a**), glucose (**b**), and fructose (**c**) concentrations in kernel fresh mass of maize cultivar Pioneer 3906 grown in 4 years in container technique under control and saline conditions, harvested 2 days after pollination; data show means of 4 replicates ± *SE*; for each year differences in percentage between control and stress treatment are given and, if significant after Student's *t*-test, indicated by $*p \leq 5\%$, $**p \leq 1\%$, and $***p \leq 0.1\%$; for Tukey test after two-way ANOVA: significant differences are indicated by different small letters ($p \leq 5\%$)

7.5.3 Activity of Key Enzymes in Developing Kernels

A possible reason for reduced kernel setting is inhibited activity of the two enzymes, acid invertase and plasma membrane H^+-ATPase. As described in the previous chapter, source limitation is most unlikely; thus the subsequent paragraph concentrates on investigations of these two enzymes which both are important determinants for sink activity of the developing maize kernels.

During the days around pollination, apoplastic loading of hexoses into ovaries is essential to maintain kernels (Andersen et al. 2002; Mäkelä et al. 2005; McLaughlin and Boyer 2004; Schussler and Westgate 1994, 1995; Zinselmeier et al. 1995, 1999). The hexoses in the developing kernel are needed for biosynthetic purposes and for energy supply. There are no symplastic connections between the parental and the daughter cells; thus the apoplastic pathway is compulsory (Tang and Boyer 2013). Sucrose imported via phloem has to be hydrolyzed by acid invertase, which not only supplies hexoses for carrier-driven import into ovaries but also prevents retrieval of sucrose by phloem (Hütsch and Schubert 2017). For both of these reasons, acid invertase activity may be regarded as a key enzyme to establish sink strength in maize kernels shortly after pollination (Cheng et al. 1996; Chourey et al. 2006; Miller and Chourey 1992).

In our studies, acid invertase in developing maize kernels under salt stress showed either a significant reduction by 52% in comparison to the control in 2011 or no significant change in the years 2012 and 2014 (Hütsch et al. 2015; Jung et al. 2017). Inhibition of acid invertase activity in maize kernels under water limitation has been observed by several other authors (Zinselmeier et al. 1995, 1999; Andersen et al. 2002; Mäkelä et al. 2005). However, the lower acid invertase activity did not cause a decrease in the availability of hexoses in the kernels (Fig. 7.7b, c). The delivery of hexoses by acid invertase activity did obviously not limit the kernel development and finally yield performance. Other factors must have contributed to the observed differences. Possibly, the plasma membrane H^+-ATPase activity, which is inhibited in maize leaves under salt stress (Zörb et al. 2005; Pitann et al. 2009; Wakeel et al. 2010), was also reduced in the kernels, causing a pH increase in the apoplast. This pH increase may not only reduce in vivo acid invertase activity by shifting the pH toward less favorable conditions but also restricts the uptake of hexoses as they are taken up from the developing kernels via H^+ cotransport. This transport is driven by the pH gradient which is established by the plasma membrane H^+-ATPase (Bihmidine et al. 2013; Sondergaard et al. 2004; Zhao et al. 2000).

Further research focused on the role of plasma membrane H^+-ATPase activity for maize kernel development (Jung et al. 2017). Salt stress effects on kernel development of Pioneer 3906 were investigated at 2 DAP. It was shown for the first time that maize kernel plasma membrane H^+-ATPase activity in vitro was negatively affected under salt stress, pointing to a strong involvement in kernel abortion under these environmental constraints. The inhibited enzyme activity probably reduced the pH gradient at the plasma membrane, limiting the energization of hexose carriers, which resulted in the measured accumulation of hexoses in the apoplast of the kernel tissue. The hexoses cannot enter the cytoplasm of the developing kernels resulting in diminished metabolic activity and a poor energy status, eventually leading to kernel abortion.

7.6 Temperature Conditions During Vegetation

In contrast to the controlled water supply according to the demand of the maize plants, the temperature in the vegetation hall varied like the ambient conditions. Of particular importance for fertilization of the maize ovaries is the time around pollination. In Fig. 7.2 the average daily temperature in the 4 experimental years is shown for the timespan of 2 weeks before and 2 weeks after controlled pollination. Silking occurred always earlier in control treatments than under salt stress; thus pollination of control plants was also performed earlier. In 2011, the timespan between pollination of both treatments was 7 days, and the temperature at pollination was approximately 18 °C and thus comparable low (Fig. 7.2, blue line). In 2012, pollination of the salt-treated plants was only 2 days delayed, and the temperature was around 25 °C (Fig. 7.2, black line). In 2014, the difference in pollination time was 3 days, and the temperature was about 22 °C (Fig. 7.2, green line). In 2017, silking of control plants occurred rather early (63 DAS in comparison to 72/73 DAS in the other years), and the timespan until salt-stressed plants were ready for pollination was 13 days (Fig. 7.2, yellow line).

In 2017, the temperature dropped by 10 °C after pollination of the control plants and increased again until pollination of the salt-stressed plants. At maturity, all 16 control plants had developed a cob with good kernel setting, whereas 56% of the salt-stressed plants produced no cob or several small, barren cobs. This was not a matter of available pollen at late silking stage, as all plants were hand-pollinated with pollen from control donor plants. Instead, in the NaCl treatment, some maize plants did not even reach the silking stage. The temperature drop during silking of the salt-stressed plants could have caused the delayed or missing silk appearance. Divergent results of kernel analyses in 2017 from the 3 other years can partly be explained by this temperature decrease and the resulting large timespan between pollination of control and stressed plants.

Silking may also be inhibited under salt stress by a lack of hexose uptake. Apart from kernels the silks are also sink tissues on the cob, which have no symplastic connections to the mother plant. The silks also belong to the daughter cell tissue of the kernel, like embryo and endosperm (Bedinger and Fowler 2009). Every single kernel has an elongated stigma (the silks) that originates near the embryo and grows from there out of the husk leaves till it gets pollinated or reaches its final size. Since it is growing quite fast (approximately 3.8 cm d^{-1}), nearly exclusively by cell-extension growth, the silk cells also have a high demand for hexoses for their metabolism. The silk fresh weight was reduced in the salt stress treatment on the day of pollination, indicating that the growth process was inhibited (Jung et al. 2017). Oury et al. (2016b) studied the reduction of silk elongation under drought stress and found a reduced upregulation on the transcript level of enzymes for cell wall-modifying enzymes that are involved in cell-extension growth. Since the kernel development in the osmotic phase of salt stress is similar to the development during drought stress (Hütsch et al. 2015), it could be assumed that a similar modification also occurred in the present experiments. Future studies should focus on the reasons for reduced silk extension growth under salt stress. It is not unreasonable to speculate that the plasma membrane H^+-ATPase also plays a major role in growth reduction of silks by means of reduced apoplastic acidification, reducing cell-extension growth in this plant organ (Jung et al. 2017).

7.7 Final Evaluation of 4 Years of Container Experiments

In order to finally evaluate the effects of salt stress and experimental year on the parameters presented in this paper, the results of the two-way ANOVA are summarized in Table 7.1. At maturity and at the intermediate harvests, both factors had highly significant impacts on most of the measured parameters. Only harvest index, single kernel weight, and water-use efficiency were not affected by salt stress, and the impact of year on kernel number was only weak. At maturity, interactions between salt stress and year were particularly pronounced for grain yield and total water consumption; no interactions were obtained for straw dry weight, kernel number, and water-use efficiency. Concerning the grain yield determinants, kernel number was mainly affected by salt stress, whereas kernel weight was exclusively affected by year of experiment. This again strengthens the observation that salt stress has an impact primarily on kernel setting, whereas year effects such as temperature

Table 7.1 Analysis of the measured parameters at maturity and at the intermediate harvests for variance of stress treatment and year effects and their interactions according to two-way ANOVA

Parameter	Salt stress	Year of experiments	Interactions
	Harvest at maturity		
Grain yield	***	***	***
Straw dry weight	***	***	ns
Harvest index	ns	***	**
Kernel number	***	*	ns
Single kernel weight	ns	***	**
Total water consumption	***	***	***
Water use efficiency	ns	***	ns
	Intermediate harvests		
Plant height	***	***	**
Shoot fresh weight	***	***	*
Kernel fresh weight	***	***	*
Sucrose concentration	***	***	*
Glucose concentration	***	***	***
Fructose concentration	***	***	ns
Acid invertase	**	***	**

Significances are indicated by $*p \leq 5\%$, $**p \leq 1\%$, and $***p \leq 0.1\%$ or *ns* not significant

variations during grain filling are responsible for single kernel weights. At the intermediate harvests, both factors were equally important for development of the measured parameters (Table 7.1).

7.8 Conclusions

In 4 years of container experiments, it was clearly demonstrated that grain yield reductions of maize under salt stress are determined by kernel number and not by single kernel weight. Thus, decisive for yield performance is the time around pollination, when kernel setting takes place. Under salt stress the availability of assimilates, namely, sucrose, was always enhanced in the developing kernels. Thus, source limitation as reason for yield depressions can be ruled out. Although acid invertase activity, as a determinant for sink activity, was reduced or unchanged (in vitro) under saline conditions, it did not limit kernel development as the hexose concentrations were increased. There is strong evidence that the transport of hexoses from the apoplast into the cytosol of the developing kernel is restricted, as H^+-ATPase activity in the plasma membrane of the sink cells was significantly reduced under salt stress (in vitro). Plasma membrane H^+-ATPase establishes the pH gradient which is necessary for H^+-cotransport of the hexoses from the apoplast into the cells. Thus, kernel development and hence grain yield performance under salt stress seem to be limited by a transport problem, caused by inhibition of plasma membrane H^+-ATPase.

Acknowledgments Many thanks are due to our coworkers at the Institute of Plant Nutrition in Giessen for cultivation of the maize plants and for laboratory analyses as well as for many helpful discussions. Financial support by the German Research Foundation (DFG) to BWH is gratefully acknowledged.

References

Abbate PE, Andrade FH, Culot JP (1995) The effect of radiation and nitrogen on number of grains in wheat. J Agric Sci 124:351–360

Andersen MN, Asch F, Wu Y, Jensen CR, Næsted H, Mogensen VO, Koch KE (2002) Soluble invertase expression is an early target of drought stress during the critical, abortion-sensitive phase of young ovary development in maize. Plant Physiol 130:591–604

Andrade FH, Vega C, Uhart S, Dirilo A, Cantarero M, Valentinuz O (1999) Kernel number determination in maize. Crop Sci 39:453–459

Bedinger PA, Fowler JE (2009) The corn male gametophyte. In: Bennetzen JL, Hake SC (eds) Handbook corn its biology. Springer, New York, pp 57–77

Below FE, Christensen LE, Reed AJ, Hageman RH (1981) Availability of reduced N and carbohydrates for ear development of maize. Plant Physiol 68:1186–1190

Bihmidine S, Hunter CTI, Johns CE, Koch KE, Braun DM (2013) Regulation of assimilate import into sink organs: update on molecular drivers of sink strength. Front Plant Sci 4:177. https://doi.org/10.3389/fpls.2013.00177

Cárcova J, Uribelarrea M, Borrás L, Otegui ME, Westgate ME (2000) Synchronous pollination within and between ears improved kernel set of maize. Crop Sci 40:1056–1061

Cheng WH, Taliercio EW, Chourey PS (1996) The Miniature 1 seed locus of maize encodes a cell wall invertase required for normal development of endosperm and maternal cells in the pedicel. Plant Cell 8:971–983
Chourey PS, Jain M, Li Q-B, Carlson SJ (2006) Genetic control of cell wall invertases in developing endosperm of maize. Planta 223:159–167
Cliquet J-B, Deléens E, Mariotti A (1990) C and N mobilization from stalk and leaves during kernel filling by ^{13}C and ^{15}N tracing in *Zea mays* L. Plant Physiol 94:1547–1553
Hay RKM, Gilbert RA (2001) Variation in the harvest index of tropical maize: evaluation of recent evidence from Mexico and Malawi. Ann Appl Biol 138:103–109
Henry C, Bledsoe SW, Griffiths CA, Kollman A, Paul MJ, Sakr S, Lagrimini LM (2015) Differential role for trehalose metabolism in salt-stressed corn. Plant Physiol 169:1072–1089
Ho LC (1988) Metabolism and compartmentation of imported sugars in sink organs in relation to sink strength. Annu Rev Plant Physiol Plant Mol Biol 39:355–378
Ho LC, Gifford RM (1984) Accumulation and conversion of sugars by developing wheat grains V The endosperm apoplast and apoplastic transport. J Exp Bot 35:58–73
Hütsch BW, Schubert S (2017) Harvest index of maize (*Zea mays* L.): are there possibilities for improvement? In DL Sparks, ed. Adv Agron 146:37–82
Hütsch BW, Saqib M, Osthushenrich T, Schubert S (2014) Invertase activity limits grain yield of maize under salt stress. J Plant Nutr Soil Sci 177:278–286
Hütsch BW, Jung S, Schubert S (2015) Comparison of salt and drought-stress effects on maize growth and yield formation with regard to acid invertase activity in the kernels. J Agron Crop Sci 201:353–367
Hütsch BW, Osthushenrich T, Faust F, Kumar A, Schubert S (2016) Reduced sink activity in growing shoot tissues of maize under salt stress of the first phase may be compensated by increased PEP-carboxylase activity. J Agron Crop Sci 202:384–393
Jung S, Hütsch BW, Schubert S (2017) Salt stress reduces kernel number of corn by inhibiting plasma membrane H^+-ATPase activity. Plant Physiol Biochem 113:198–207
Lee EA, Tollenaar M (2007) Physiological basis of successful breeding strategies for maize grain yield. Crop Sci 47(S3):S202–S215
Mäkelä P, McLaughlin JE, Boyer JS (2005) Imaging and quantifying carbohydrate transport in the developing ovaries of maize. Ann Bot 96:939–949
McLaughlin JE, Boyer JS (2004) Glucose localization in maize ovaries when kernel number decreases at low water potential and sucrose is fed to the stems. Ann Bot 94:75–86
Miller ME, Chourey PS (1992) The maize invertase-deficient miniature-1 seed mutation is associated with aberrant pedicel and endosperm development. Plant Cell 4:297–305
Mohanty SK, Swain MR (2019) Bioethanol production from corn and wheat: food, fuel, and future. In: Ray RC, Ramachandran S (eds) Bioethanol production from food crops. Elsevier Inc., Amsterdam, pp 45–59
Munns R (1993) Physiological processes limiting plant growth in saline soils: some dogmas and hypotheses. Plant Cell Environ 16:15–24
Munns R (2011) Plant adaptations to salt and water stress: differences and commonalities. Adv Bot Res 57:1–32
Oladosu G, Kline K, Uria-Martinez R, Eaton L (2011) Sources of corn for ethanol production in the United States: a decomposition analysis of the empirical data. Biofuels Bioprod Biorefin 5:640–653
Otegui ME, Andrade FH, Suero EE (1995) Growth, water use, and kernel abortion of maize subjected to drought at silking. Field Crop Res 40:87–94
Oury V, Caldeira CF, Prodhomme D, Pichon J-P, Gibon Y, Tardieu F, Turc O (2016a) Is change in ovary carbon status a cause or a consequence of corn ovary abortion in water deficit during flowering? Plant Physiol 171:997–1008
Oury V, Tardieu F, Turc O (2016b) Ovary apical abortion under water deficit is caused by changes in sequential development of ovaries and in silk growth rate in corn. Plant Physiol 171:986–996
Passioura JB, Angus JF (2010) Improving productivity of crops in water-limited environments. Adv Agron 106:37–75

Pitann B, Schubert S, Mühling KH (2009) Decline in leaf growth under salt stress is due to an inhibition of H^+-pumping activity and increase in apoplastic pH of maize leaves. J Plant Nutr Soil Sci 172:535–543
Qadir M, Noble AD, Schubert S, Thomas RJ, Arslan A (2006) Sodicity induced land degradation and its sustainable management: problems and prospects. Land Degrad Dev 17:661–676
Rajcan I, Tollenaar M (1999a) Source: sink ratio and leaf senescence in maize: I. Dry matter accumulation and partitioning during grain filling. Field Crop Res 60:245–253
Rajcan I, Tollenaar M (1999b) Source: sink ratio and leaf senescence in maize: II. Nitrogen metabolism during grain filling. Field Crop Res 60:255–265
Ruan Y-L, Patrick JW, Bouzayen M, Osorio S, Fernie AR (2012) Molecular regulation of seed and fruit set. Trends Plant Sci 17:656–665
Schubert S (2011) Salt resistance of crop plants: physiological characterization of a multigenic trait. In: Hawkesford MJ, Barraclough P (eds) The molecular and physiological basis of nutrient use efficiency in crops. John Wiley & Sons, Inc., Chichester, pp 443–455
Schubert S, Neubert A, Schierholt A, Sümer A, Zörb C (2009) Development of salt-resistant maize hybrids: the combination of physiological strategies using conventional breeding methods. Plant Sci 177:196–202
Schussler JR, Westgate ME (1994) Increasing assimilate reserves does not prevent kernel abortion at low water potential in maize. Crop Sci 34:1569–1576
Schussler JR, Westgate ME (1995) Assimilate flux determines kernel set at low water potential in maize. Crop Sci 35:1074–1080
Sondergaard TE, Schulz A, Palmgren MG (2004) Energization of transport processes in plants. Roles of the plasma membrane H^+-ATPase. Plant Physiol 136:2475–2482
Sümer A, Zörb C, Yan F, Schubert S (2004) Evidence of sodium toxicity for the vegetative growth of maize (*Zea mays* L.) during the first phase of salt stress. J Appl Bot 78:135–139
Swank JC, Below FE, Lambert RJ, Hageman RH (1982) Interaction of carbon and nitrogen metabolism in the productivity of maize. Plant Physiol 70:1185–1190
Tang A-C, Boyer JS (2013) Differences in membrane selectivity drive phloem transport to the apoplast from which maize florets develop. Ann Bot 111:551–562
Thorne JH (1985) Phloem unloading of C and N assimilates in developing seeds. Annu Rev Plant Physiol 36:317–343
Unkovich M, Baldock J, Forbes M (2010) Variability in harvest index of grain crops and potential significance for carbon accounting: examples form Australian agriculture. Adv Agron 105:173–219
Wakeel A, Hanstein S, Pitann B, Schubert S (2010) Hydrolytic and pumping activity of H^+-ATPase from leaves of sugar beet (*Beta vulgaris* L.) as affected by salt stress. J Plant Physiol 167:725–731
Wang N, Fisher DB (1995) Sucrose release into the endosperm cavity of wheat grains apparently occurs by facilitated diffusion across the nucellar cell membranes. Plant Physiol 109:579–585
Yang H, Grassini P (2014) Quantifying and managing corn water use efficiencies under irrigated and rainfed conditions in Nebraska using the hybrid-maize simulation model. Adv Agric Syst Model 5:113–138
Zhang S, Sadras V, Chen X, Zhang F (2014) Water use efficiency of dryland maize in the loess plateau of China in response to crop management. Field Crop Res 163:55–63
Zhao R, Dielen V, Kinet J-M, Boutry M (2000) Cosuppression of a plasma membrane H^+-ATPase isoform impairs sucrose translocation, stomatal opening, plant growth, and male fertility. Plant Cell 12:535–546
Zinselmeier C, Westgate ME, Schussler JR, Jones RJ (1995) Low water potential disrupts carbohydrate metabolism in maize (*Zea mays* L.) ovaries. Plant Physiol 107:385–391
Zinselmeier C, Jeong B-R, Boyer JS (1999) Starch and the control of kernel number in maize at low water potentials. Plant Physiol 121:25–35

Zörb C, Stracke B, Tramnitz B, Denter D, Sümer A, Mühling KH, Yan F, Schubert S (2005) Does H^+ pumping by plasmalemma ATPase limit leaf growth of maize (Zea mays L.) during the first phase of salt stress? J Plant Nutr Soil Sci 168:550–557

Birgit W. Hütsch scientist and lecturer at the Institute of Plant Nutrition at Justus Liebig University Giessen, Germany. Her main research areas are ecological and yield aspects of plant physiology, in particular salt and drought stress effects on maize and improvement of maize harvest index, as well as heat stress effects on wheat. She is author of 35 refereed publications and 43 further publications.

Stephan Jung is a scientist working at the research center for horticulture of the chamber of agriculture North Rhine-Westphalia, Germany. He did his PhD on the physiological effects of salt stress and Mg deficiency on maize at the Institute of Plant Nutrition at Justus Liebig University Giessen, Germany. His main research areas are plant nutrition and salt stress research.

Marleen Steinbach is Quality Manager in the Food Industry. Prior to her current position, she worked as a research assistant at the Institute of Plant Nutrition at Justus Liebig University Giessen, Germany. Her topics of research were the influence of salt stress, phytohormones, and climate effects on maize. Mrs. Steinbach is a nutritional scientist and published her graduation project as part of an international cooperation.

Sven Schubert is Chair of the Institute of Plant Nutrition at Justus Liebig University Giessen, Germany. His main research areas are nutrient acquisition of plants, salt and acidity stress, and membrane biochemistry. Besides 120 refereed publications and 100 further publications, he is author of the two German textbooks *Pflanzenernährung* (Plant Nutrition) and *Biochemie* (Biochemistry).

Part II

Microalgae and Engineered Crops for Production of Biofuels and High-Value Products

Bioproduction from Microalgal Resources

8

Osamu Iwata and Keiichi Mochida

Abstract

Microalgae have been attracting increasing attention as a renewable energy source and feedstock because of their potential for use in the production of bio-based fuels and materials. In this chapter, we provide an overview of bioproduction based on microalgae species. Specifically, we describe the taxonomic distribution of major industrially exploited microalgae species and highlight their utilities and recent advances. We also introduce recent advances in breeding and engineering techniques to improve the productivity of microalgae to enhance their biomass use.

Keywords

Microalgae · Bioproduction · Biomass

8.1 Introduction

Microalgae have been attracting increasing attention as a renewable energy source and feedstock because of their potential for use in the production of bio-based fuels and materials. Microalgae species often show higher productivity per unit area and time compared to terrestrial plant species in terms of biomass use (Chisti 2008). Because of their high CO_2 assimilation ability, scalable advantage in mass culture, and metabolic diversity, the industrial use of mass-cultured microalgae may contribute to carbon capture and utilization (CCU), which is a key technology for climate

O. Iwata · K. Mochida (✉)
RIKEN Center for Sustainable Resource Science, Bioproductivity Informatics Research Team
RIKEN Cluster for Science, Technology, and Innovation Hub, RIKEN Baton Zone Program, Microalgae Production Control Technology Laboratory, Yokohama, Japan
e-mail: iwata@euglena.jp; keiichi.mochida@riken.jp

A. Kumar et al. (eds.), *Climate Change, Photosynthesis and Advanced Biofuels*,
https://doi.org/10.1007/978-981-15-5228-1_8

change mitigation (Wijffels et al. 2010), as well as for ensuring a sustainable shift to a bio-economy (Williams and Laurens 2010). Moreover, metabolic diversity has resulted from evolutionary diversity in the broad taxonomic distribution of photosynthetic organisms, providing opportunities for exploiting various types of metabolism and compounds for fuel and feedstock production (Georgianna and Mayfield 2012; Jagadevan et al. 2018).

The establishment of a mass culture system for microalgae species is an essential step toward microalgal domestication for biomass use. Early studies of microalgae mass culture, aimed at developing a new food source, began in the 1950s mainly using *Chlorella*, a freshwater microalga (Burlew 1953). Methods for microalgae mass culture are mainly classified as open systems and closed reactors (Narala et al. 2016). Raceway pond systems and circular pond systems with paddle wheels have been widely used in commercial microalgae mass culture because of their low cost of construction and easy operation and maintenance compared to closed reactors. However, open pond-based microalgae mass cultures exhibit considerable limitations, such as contamination risks, uneven light intensity, evaporation loss, low diffusion of atmospheric CO_2, and dependency on weather conditions (Ugwu et al. 2008). To overcome the limitations of open pond systems, various types of closed photobioreactor systems have been widely used for the mass culture of microalgae. These systems are composed of transparent materials such as glass, plastic tubes, and plates, as well as polyethylene and polyvinyl chloride bags, with internal or external illumination, and controlled gas exchange and growth media circulation. In contrast to the open pond systems that require land space, closed photobioreactor systems can be flexibly designed (Chisti 2008). Thus, closed photobioreactor systems overcome the major drawbacks of open pond systems but are typically costly to set up and maintain (Barry et al. 2016). Recently, ocean culture systems have attracted increased attention for potential use in the commercial-scale cultivation of marine microalgae because of their advantages, such as the use of ocean wave-based mixing, utilizing nutrients in the seawater, and area availability (Greene et al. 2016).

Cascading use is an essential principle in the industrial utilization of biomass resources to ensure efficient use of raw materials and economic profitability. "Biomass 5Fs" is a keyword that explains the principle of cascading use of biomass resources, which represents the priority of higher-value uses in specialty areas that allow the recycling of products, after which materials are streamed to lower-value commodities, such as biofuels (Fig. 8.1). To promote the cascading use of microalgae biomass, a comprehensive understanding of the metabolic systems of target microalgae species is essential (Hu et al. 2008).

To improve the productivity of microalgae species industrially exploited for biomass use, it is important to develop a breeding framework. To improve the various traits of microalgae, such as productivity of specific metabolites and biomass, conventional chemical and radiation mutagenesis and genetic engineering have been widely used (Mystikou et al. 2016). Moreover, recent remarkable advances in biotechnologies, such as genome editing and synthetic biology, have provided opportunities to engineer metabolic systems and install artificially designed

Fig. 8.1 Biomass of 5Fs

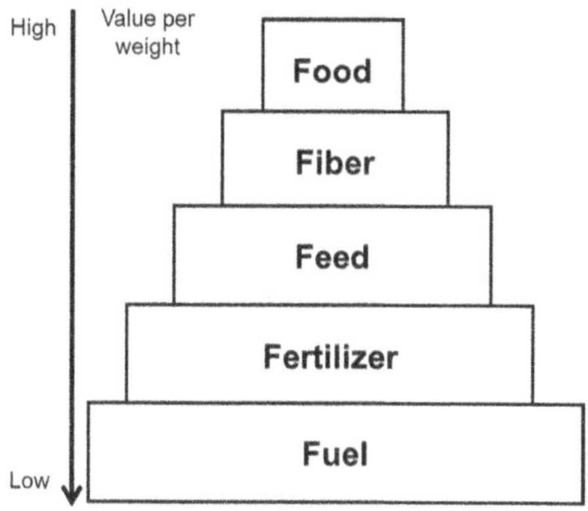

Table 8.1 Typical industrial microalgae and their applications

	Genus or common name	Application or functional ingredient
Prokaryotes (cyanobacteria)	*Arthrospira (Spirulina)*	Pigment (phycobillin), dietary supplement
	Nostoc	Food
	Aphanothece (Suizenji nori)	Food, fiber, cosmetics
Eukaryotes	*Auranthiochytrium*	DHA
	Botryococcus	Fuel
	Chaetoceros	Feed
	Chlorella	Dietary supplement
	Dunaliella	Cosmetics, dietary supplement
	Euglena	Dietary supplement, cosmetics, fuel
	Haematococcus	Pigment (astaxanthin)
	Nannochloropsis	Feed

metabolic processes to extend the utilities of microalgal species in industrial applications.

In this chapter, we provide an overview of bioproduction based on microalgae species. First, we describe the taxonomic distribution of industrially exploited microalgae species. Next, we highlight the utilities and recent advances of major industrially exploited microalgae species (Table 8.1). Moreover, we describe the recent advances in breeding and engineering techniques aimed at improving the productivity of microalgae to enhance their biomass use.

8.2 Taxonomic Distribution of Microalgae

Microalgae are a diverse group of photosynthetic unicellular organisms and include photosynthetic unicellular eukaryotes and cyanobacteria. It is well-known that chloroplasts originated from endosymbiosis between an ancestral organism of cyanobacteria and a non-photosynthetic host (cyanobacterial primary endosymbiosis), which occurred in the common ancestor of the supergroup Archaeplastida (Primoplantae) comprised of Viridiplantae (green algae and plants), Rhodophyta (red algae), and Glaucophyta (glaucophyte algae) (Rockwell et al. 2014; Howe et al. 2008). After Archaeplastida emerged, photosynthesis spread widely among other eukaryotic groups via secondary (with green algae or red algae) and tertiary endosymbiotic plastid acquisition. The secondary plant groups that originated via endosymbiosis with green algae include Chlorarachniophyta and Euglenophyta, while those with red algae include Cryptophyta, Heterokontophyta, Haptophyta, Dinophyta, and Chromerida (Petersen et al. 2006).

8.3 Industrially Exploited Microalgae

8.3.1 Industrially Exploited Cyanobacteria

Cyanobacteria (blue-green algae) is a bacteria phylum including widely distributed prokaryotes that perform oxygenic photosynthesis, including some industrially exploited species. Due to the long history of *Cyanobacteria* as a model organism and its industrial potential, numerous studies have evaluated their roles in basic research and industrial applications (Singh et al. 2016; Sharma et al. 2011). Free-floating filamentous cyanobacteria of the genus *Arthrospira*, such as *A. platensis* and *A. maxima*, have a long history as dietary supplements, referred to as spirulina (Furmaniak et al. 2017), because of their high nutritional value and digestibility (Muys et al. 2019, Wild et al. 2018). Moreover, their extracts have been used as a source of natural blue color in foods (Mysliwa-Kurdziel and Solymosi 2017). Furmaniak et al. reviewed the genetics and cultivation methods, as well as the application to human health of *Arthrospira* (Furmaniak et al. 2017). *Aphanothece sacrum* is a cyanobacterium found in Japan that is consumed as a luxury ingredient in the local cuisine of the Kyushu area. Okajima et al. found that Suizenji nori produces a megamolecular polysaccharide, referred to as sacran (Okajima et al. 2009), which is potentially useful as a moisturizing agent. Another cyanobacterium, *Nostoc commune* is a terrestrial cyanobacterium that shows tolerance to abiotic stresses such as desiccation, UV irradiation, and oxidation because of its 3D extracellular matrix (Wright et al. 2005). Ishikurage is also known as a potential supplement in nitrogenous fertilizer due to its ability to fix atmospheric *nitrogen.* Panjiar et al. reviewed these, along with other cyanobacteria, for their utilities as functional foods (Panjiar et al. 2017).

Cyanobacteria are also potential cell factories for biofuel and value-added material production because of their high productivity, tractability in genetic engineering, and associated genome resources (Nozzi et al. 2013). Through genetic engineering of the cyanobacterium *Synechococcus* sp. strain PCC 7942 with pyruvate decarboxylase (pdc) and alcohol dehydrogenase II (adh) from *Zymomonas mobilis*, Deng and Coleman demonstrated in 1999 that the engineered cyanobacterium synthesized ethanol (Deng and Coleman 1999). Dexter et al. reviewed the advances in metabolic engineering-based ethanol production in cyanobacteria over the 15 years since its initial report (Dexter et al. 2015). Moreover, the potential applications of cyanobacterial hydrogen gas production have been discussed for a long time, and these species have recently attracted increasing attention as sustainable energy carriers (McKinlay and Harwood 2010, Krishnan et al. 2018). For example, to increase H_2 production in cyanobacteria, Baebprasert performed genetic engineering of a nitrate assimilation pathway in the cyanobacterium *Synechocystis* sp. strain PCC 6803 (Baebprasert et al. 2011). Moreover, cyanobacterial production of bioplastics has advanced through metabolic engineering approaches and optimization of growth conditions.

8.3.2 Industrially Exploited Green Microalgae

Chlorophyta contains several microalgal species that have been industrially exploited for a long time. Chlorella (*Chlorella vulgaris*), which has been used as a model organism for studying photosynthesis, is among the microalgae with a long history of commercial-scale mass culture for industrial use. Recently, it has been applied in the production of biofuels such as biodiesel (Jain et al. 2019) and biogas (Sakarika and Kornaros 2019). Dunaliella (*Dunaliella salina*) has been widely used in β-carotene production (García-González et al. 2005; Lamers et al. 2008). Haematococcus (*Haematococcus pluvialis*) is a rich source of natural astaxanthin, an antioxidant useful as a food supplement and in the production of cosmetics (Shah et al. 2016). Botryococcus (*Botryococcus braunii*), a freshwater colonial microalga, has attracted attention as a potential source of biofuel because of its high productivity of hydrocarbons (Metzger and Largeau 2005). This species has also been proposed as an alternative source of carotenoids (Ambati et al. 2018).

8.3.3 Industrially Exploited Heterokontophyta

Diatoms are one of the major unicellular eukaryotic algal groups that play crucial roles in the global carbon cycle. They are estimated to be responsible for 20% of CO_2 fixation through photosynthesis. Levitan et al. reviewed the potential of diatoms for use in biofuel production and discussed their genetic engineering for producing biofuels (Levitan et al. 2014). *Nannochloropsis* spp. are unicellular marine microalgae capable of accumulating large amounts of lipids such as eicosapentaenoic acid ($C_{20}H_{30}O_2$) and docosahexaenoic acid ($C_{22}H_{32}O_2$), for commercial applications (Martins et al. 2013), both of which are ω-3 polyunsaturated fatty acids important in the aquaculture of fish and food supplements.

8.3.4 Industrially Exploited *Euglena*

Euglena gracilis, a unicellular photosynthesizing flagellate in the supergroup Excavata, is rich in nutrients and accumulates crystalized β-1,3-glucan and paramylon (Inui et al. 1982). Moreover, *E. gracilis* converts paramylon to wax esters that mainly consist of myristic acid (C14:0) and myristyl alcohol (C14:0), both of which have been suggested as sources for biojet fuel, rather than the fatty acids produced by most of other microalgae due to their lower freezing point (Furuhashi et al. 2014).

8.4 Microalgae Breeding to Improve Their Productivity in Biorefinery

8.4.1 Conventional Breeding Through Mutagenesis

To improve traits related to bioproduction in microalgae, such as the productivity of specific chemicals and adaptation ability to different environments, conventional chemical and radiation mutagenesis has been widely employed. As examples of chemical mutagenesis in microalgae breeding, Tanadul et al. used ethyl methanesulfonate to induce random mutation in *Chlorella* sp. to enhance their lipid productivity (Tanadul et al. 2018), and Huang et al. used *N*-methyl-*N*′-nitro-*N*-nitrosoguanidine to increase the production of zeaxanthin, lutein, and β-carotene in *Chlorella zofingiensis* (Huang et al. 2018). As examples of irradiation mutagenesis, Ahmed and Scenk demonstrated that UV-C radiation increased the sterol productivity of *Pavlova lutheri* (Ahmed and Schenk 2017), and Liu et al. increased the lipid productivity of *Chlorella* sp. by fast neutron irradiation (Liu et al. 2016). Moreover, Yamada et al. demonstrated that heavy-ion-beam-based breeding was useful for increasing lipid productivity in *Euglena* (Yamada et al. 2016), and Fang et al. demonstrated atmospheric and room temperature plasma is a useful method for preparing a mutant library of *Spirulina platensis* (Fang et al. 2013).

8.4.2 Synthetic Biology of Microalgae for Biorefinery

In the genetic engineering of microalgae, their competence for transformation is a decisive factor in engineering target functions (Guihéneuf et al. 2016). Additionally, associated genome information, such as whole genome sequences, gene annotation, and transcriptome datasets, are crucial for designing transgenic DNA constructs and RNAs for genome editing. Moreover, genome-scale metabolic models provide a framework for model-driven design of metabolic traits to improve the productivity of target products (Georgianna and Mayfield 2012). Banerjee et al. recently compared the different approaches for increasing lipid biogenesis in microalgae (Banerjee et al. 2016). Moreover, combinatorial use of metabolic engineering techniques and high-throughput screening technologies, such as intelligent cell sorting technologies, will accelerate the design of microalgae with improved traits (Nitta et al. 2018), enabling the development of sustainable CO_2 bioconversion technologies.

References

Ahmed F, Schenk PM (2017) UV-C radiation increases sterol production in the microalga *Pavlova lutheri*. Phytochemistry 139:25–32. https://doi.org/10.1016/j.phytochem.2017.04.002

Ambati RR, Gogisetty D, Aswathnarayana Gokare R, Ravi S, Bikkina PN, Su Y et al (2018) Botryococcus as an alternative source of carotenoids and its possible applications - an overview. Crit Rev Biotechnol 38:541–558. https://doi.org/10.1080/07388551.2017.1378997

Baebprasert W, Jantaro S, Khetkorn W, Lindblad P, Incharoensakdi A (2011) Increased H2 production in the cyanobacterium Synechocystis sp. strain PCC 6803 by redirecting the electron supply via genetic engineering of the nitrate assimilation pathway. Metab Eng 13:610–616. https://doi.org/10.1016/j.ymben.2011.07.004

Banerjee C, Dubey KK, Shukla P (2016) Metabolic engineering of microalgal based biofuel production: prospects and challenges. Front Microbiol 7:432. https://doi.org/10.3389/fmicb.2016.00432

Barry A, Wolfe A, English C, Ruddick C, Lambert D (2016) National algal biofuels technology review. https://doi.org/10.2172/1259407

Burlew JS (1953) Algal culture from laboratory to pilot plant. AIBS Bull 3:11–11. https://doi.org/10.1093/aibsbulletin/3.5.11

Chisti Y (2008) Biodiesel from microalgae beats bioethanol. Trends Biotechnol 26:126–131. https://doi.org/10.1016/j.tibtech.2007.12.002

Deng MD, Coleman JR (1999) Ethanol synthesis by genetic engineering in cyanobacteria. Appl Environ Microbiol 65:523–528. http://www.ncbi.nlm.nih.gov/pubmed/9925577

Dexter J, Armshaw P, Sheahan C, Pembroke JT (2015) The state of autotrophic ethanol production in cyanobacteria. J Appl Microbiol 119:11–24. https://doi.org/10.1111/jam.12821

Fang M, Jin L, Zhang C, Tan Y, Jiang P, Ge N et al (2013) Rapid mutation of *Spirulina platensis* by a new mutagenesis system of atmospheric and room temperature plasmas (ARTP) and generation of a mutant library with diverse phenotypes. PLoS One 8:1–12. https://doi.org/10.1371/journal.pone.0077046

Furmaniak MA, Misztak AE, Franczuk MD, Wilmotte A (2017) Edible cyanobacterial genus *Arthrospira*: actual state of the art in cultivation methods, genetics, and application in medicine. Front Microbiol 8:1–21. https://doi.org/10.3389/fmicb.2017.02541

Furuhashi T, Ogawa T, Nakai R, Nakazawa M, Okazawa A, Padermschoke A et al (2014) Wax ester and lipophilic compound profiling of *Euglena gracilis* by gas chromatography-mass spectrometry: toward understanding of wax ester fermentation under hypoxia. Metabolomics 11:175–183. https://doi.org/10.1007/s11306-014-0687-1

García-González M, Moreno J, Manzano JC, Florencio FJ, Guerrero MG (2005) Production of *Dunaliella salina* biomass rich in 9-cis-beta-carotene and lutein in a closed tubular photobioreactor. J Biotechnol 115:81–90. https://doi.org/10.1016/j.jbiotec.2004.07.010

Georgianna DR, Mayfield SP (2012) Exploiting diversity and synthetic biology for the production of algal biofuels. Nature 488:329–335. https://doi.org/10.1038/nature11479

Greene C, Huntley M, Archibald I, Gerber L, Sills D, Granados J et al (2016) Marine microalgae: climate, energy, and food security from the sea. Oceanography 29:10–15. https://doi.org/10.5670/oceanog.2016.91

Guihéneuf F, Khan A, Tran LP (2016) Genetic engineering: a promising tool to engender physiological, biochemical, and molecular stress resilience in green microalgae. Front Plant Sci 7:1–8. https://doi.org/10.3389/fpls.2016.00400

Howe CJ, Barbrook AC, Nisbet RER, Lockhart PJ, Larkum AWD (2008) The origin of plastids. Philos Trans R Soc Lond Ser B Biol Sci 363:2675–2685. https://doi.org/10.1098/rstb.2008.0050

Hu Q, Sommerfeld M, Jarvis E, Ghirardi M, Posewitz M, Seibert M et al (2008) Microalgal triacylglycerols as feedstocks for biofuel production: perspectives and advances. Plant J 54:621–639. https://doi.org/10.1111/j.1365-313X.2008.03492.x

Huang W, Lin Y, He M, Gong Y, Huang J (2018) Induced high-yield production of zeaxanthin, lutein, and β-carotene by a mutant of *Chlorella zofingiensis*. J Agric Food Chem 66:891–897. https://doi.org/10.1021/acs.jafc.7b05400

Inui H, Miyatake K, Nakano Y, Kitaoka S (1982) Wax ester fermentation in *Euglena gracilis*. FEBS Lett 150:89–93. https://doi.org/10.1016/0014-5793(82)81310-0

Jagadevan S, Banerjee A, Banerjee C, Guria C, Tiwari R, Baweja M (2018) Biotechnology for biofuels recent developments in synthetic biology and metabolic engineering in microalgae towards biofuel production. Biotechnol Biofuels 11:1–21. https://doi.org/10.1186/s13068-018-1181-1. BioMed Central

Jain D, Ghonse SS, Trivedi T, Fernandes GL, Menezes LD, Damare SR et al (2019) CO2 fixation and production of biodiesel by *Chlorella vulgaris* NIOCCV under mixotrophic cultivation. Bioresour Technol 273:672–676. https://doi.org/10.1016/j.biortech.2018.09.148

Krishnan A, Qian X, Ananyev G, Lun DS, Dismukes GC (2018) Rewiring of cyanobacterial metabolism for hydrogen production: synthetic biology approaches and challenges. Adv Exp Med Biol 1080:171–213. https://doi.org/10.1007/978-981-13-0854-3_8

Lamers PP, Janssen M, De Vos RCH, Bino RJ, Wijffels RH (2008) Exploring and exploiting carotenoid accumulation in *Dunaliella salina* for cell-factory applications. Trends Biotechnol 26:631–638. https://doi.org/10.1016/j.tibtech.2008.07.002

Levitan O, Dinamarca J, Hochman G, Falkowski PG (2014) Diatoms: a fossil fuel of the future. Trends Biotechnol 32:117–124. https://doi.org/10.1016/j.tibtech.2014.01.004

Liu S, Xu J, Chen W, Fu H, Ma LY, Xu H et al (2016) Enhancement of lipid productivity in green microalgae *Chlorella* sp. via fast neutron irradiation. Biomass Bioenergy 91:196–203. https://doi.org/10.1016/j.biombioe.2016.05.013

Martins DA, Pereira H, Ben-hamadou R, Abu-salah KM, Arabia S (2013) Alternative sources of n-3 long-chain polyunsaturated fatty acids in marine microalgae dulce. Mar Drugs 11:2259–2281. https://doi.org/10.3390/md11072259

McKinlay JB, Harwood CS (2010) Photobiological production of hydrogen gas as a biofuel. Curr Opin Biotechnol 21:244–251. https://doi.org/10.1016/j.copbio.2010.02.012

Metzger P, Largeau C (2005) *Botryococcus braunii*: a rich source for hydrocarbons and related ether lipids. Appl Microbiol Biotechnol 66:486–496. https://doi.org/10.1007/s00253-004-1779-z

Muys M, Sui Y, Schwaiger B, Lesueur C, Vandenheuvel D, Vermeir P et al (2019) High variability in nutritional value and safety of commercially available Chlorella and Spirulina biomass indicates the need for smart production strategies. Bioresour Technol 275:247–257. https://doi.org/10.1016/j.biortech.2018.12.059

Mysliwa-Kurdziel B, Solymosi K (2017) Phycobilins and phycobiliproteins used in food industry and medicine. Mini Rev Med Chem 17:1173–1193. https://doi.org/10.2174/1389557516666160912180155

Mystikou A, Alzahmi A, Salehi-ashtiani K (2016) Algal cell factories: approaches, applications, and potentials. Mar Drugs 14:1–19. https://doi.org/10.3390/md14120225

Narala RR, Garg S, Sharma KK, Thomas-hall SR (2016) Comparison of microalgae cultivation in photobioreactor, open raceway pond, and a two-stage hybrid system. Front Energy Res 4:1–10. https://doi.org/10.3389/fenrg.2016.00029

Nitta N, Sugimura T, Isozaki A, Mikami H, Hiraki K, Sakuma S et al (2018) Intelligent image-activated cell sorting. Cell 175:266–276.e13. https://doi.org/10.1016/j.cell.2018.08.028

Nozzi NE, Oliver JWK, Atsumi S (2013) Cyanobacteria as a platform for biofuel production. Front Bioeng Biotechnol 1:1–6. https://doi.org/10.3389/fbioe.2013.00007

Okajima MK, Miyazato S, Kaneko T (2009) Cyanobacterial megamolecule sacran efficiently forms LC gels with very heavy metal ions. Langmuir 25:8526–8531. http://www.ncbi.nlm.nih.gov/pubmed/20050044

Panjiar N, Mishra S, Yadav AN, Verma P (2017) Functional foods from cyanobacteria. In: Gupta VK, Treichel H, Shapaval V, Antonio de Oliveira L, Tuohy MG (eds) Microbial functional foods and nutraceuticals. https://doi.org/10.1002/9781119048961.ch2

Petersen J, Teich R, Brinkmann H, Cerff R (2006) A "green" phosphoribulokinase in complex algae with red plastids: evidence for a single secondary endosymbiosis leading to haptophytes, cryptophytes, heterokonts, and dinoflagellates. J Mol Evol 62(2):143–157

Rockwell NC, Lagarias JC, Bhattacharya D, Durnford DG (2014) Primary endosymbiosis and the evolution of light and oxygen sensing in photosynthetic eukaryotes. Front Ecol Evol 2:1–13. https://doi.org/10.3389/fevo.2014.00066

Sakarika M, Kornaros M (2019) *Chlorella vulgaris* as a green biofuel factory: comparison between biodiesel, biogas, and combustible biomass production. Bioresour Technol 273:237–243. https://doi.org/10.1016/j.biortech.2018.11.017

Shah MR, Liang Y, Cheng JJ, Daroch M (2016) Astaxanthin-producing green from single cell to high value commercial products. Front Plant Sci 7:531. https://doi.org/10.3389/fpls.2016.00531

Sharma NK, Tiwari SP, Tripathi K, Rai AK (2011) Sustainability and cyanobacteria (blue-green algae): facts and challenges. J Appl Phycol 23:1059–1081. https://doi.org/10.1007/s10811-010-9626-3

Singh JS, Kumar A, Rai AN, Singh DP (2016) Cyanobacteria: a precious bio-resource in agriculture, ecosystem, and environmental sustainability. Front Microbiol 7:1–19. https://doi.org/10.3389/fmicb.2016.00529

Tanadul O-U-M, Noochanong W, Jirakranwong P, Chanprame S (2018) EMS-induced mutation followed by quizalofop-screening increased lipid productivity in *Chlorella* sp. Bioprocess Biosyst Eng 41:613–619. https://doi.org/10.1007/s00449-018-1896-1

Ugwu CU, Aoyagi H, Uchiyama H (2008) Photobioreactors for mass cultivation of algae. Bioresour Technol 99:4021–4028. https://doi.org/10.1016/j.biortech.2007.01.046

Wijffels RH, Barbosa MJ, Oswald WJ, Golueke CG, Usui N, Ikenouchi M et al (2010) An outlook on microalgal biofuels. Science 329:796–799. https://doi.org/10.1126/science.1189003

Wild KJ, Steingaß H, Rodehutscord M (2018) Variability in nutrient composition and in vitro crude protein digestibility of 16 microalgae products. J Anim Physiol Anim Nutr (Berl) 102:1306–1319. https://doi.org/10.1111/jpn.12953

Williams PJLB, Laurens LML (2010) Microalgae as biodiesel & biomass feedstocks: review & analysis of the biochemistry, energetics & economics. Energy Environ Sci 3:554. https://doi.org/10.1039/b924978h

Wright DJ, Smith SC, Joardar V, Scherer S, Jervis J, Warren A et al (2005) UV irradiation and desiccation modulate the three-dimensional extracellular matrix of *Nostoc commune* (cyanobacteria). J Biol Chem 280:40271–40281. https://doi.org/10.1074/jbc.M505961200

Yamada K, Suzuki H, Takeuchi T, Kazama Y, Mitra S, Abe T et al (2016) Efficient selective breeding of live oil-rich *Euglena gracilis* with fluorescence-activated cell sorting. Sci Rep 6:26327. https://doi.org/10.1038/srep26327

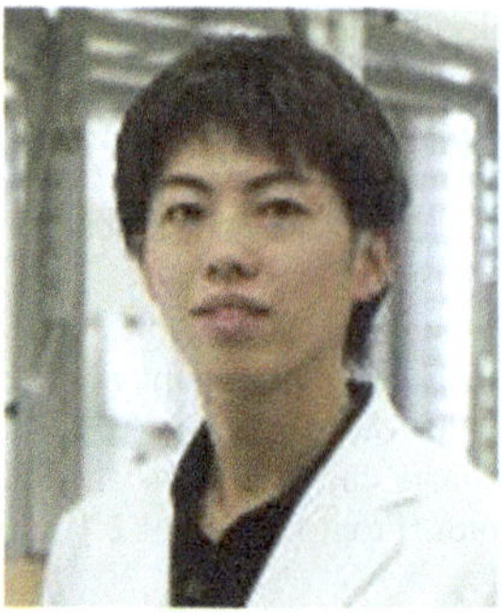

Osamu Iwata is a scientist working at the Euglena Co., Ltd. He aims to develop technologies for breeding and improvement of productivity of useful microalgae species, such as *Euglena* for foods, feeds, bio-based materials, and biofuels.

Keiichi Mochida is the leader of the Bioproductivity Informatics Research Team at RIKEN Center for Sustainable Resource Science, Japan. He aims to identify useful genes to improve productivity in useful species for biomass resources, such as grasses and micro algae.

Hydrogen Photoproduction in Green Algae: Novel Insights and Future Perspectives

9

Martina Jokel, Sergey Kosourov, and Yagut Allahverdiyeva

Abstract

Molecular hydrogen (H_2) is a promising energy carrier for a future sustainable economy. There are a number of different approaches for the industrial production of H_2 fuel. These include steam reforming, water electrolysis, and coal gasification. Nevertheless, the renewable production of H_2 remains a challenge. Some photosynthetic green algae possess hydrogenase enzyme(s) and naturally photoproduce H_2 gas. Due to the high sensitivity of hydrogenases to O_2 and also to other cellular metabolic hindrances, H_2 photoproduction is not yet efficient enough for industrial applications. This chapter summarizes different protocols that have been developed thus far for the production of H_2 in algal cultures, including two novel and promising approaches, and discusses the advantages and disadvantages of these methods.

Keywords

Hydrogen production · Nutrient deprivation · Chlamydomonas · Pulse-illumination protocol · Hydrogenase

9.1 Introduction

The rapid increase of global CO_2 emissions (NASA 2019) and scarcity of natural resources necessitates a transition from the fossil-based economy of the past to a sustainable bioeconomy of the future. This transition will rely on the continued development of renewable low- or non-carbon technologies. Molecular hydrogen

M. Jokel · S. Kosourov · Y. Allahverdiyeva (✉)
Molecular Plant Biology Unit, Department of Biochemistry, University of Turku, Turku, Finland
e-mail: martjok@utu.fi; sergey.kosourov@utu.fi; allahve@utu.fi

A. Kumar et al. (eds.), *Climate Change, Photosynthesis and Advanced Biofuels*,
https://doi.org/10.1007/978-981-15-5228-1_9

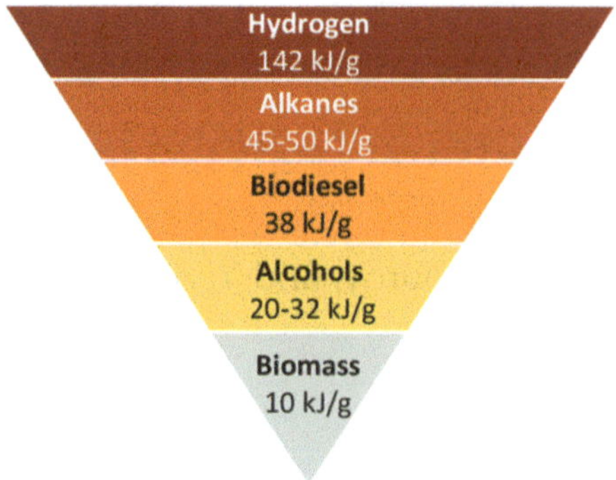

Fig. 9.1 Energy density pyramid of different fuels

(H_2) can be cleanly combusted to yield the highest energy density of currently employed fuels, and its production can also be carbon-neutral (Fig. 9.1).

These qualities make H_2 an ideal fuel for meeting future energy demands. Along with other renewable approaches, biological H_2 photoproduction by photosynthetic microorganisms has been intensively studied over recent years. The first report of biological H_2 photoproduction was based on anaerobically adapted *Scenedesmus obliquus* cultures (Gaffron and Rubin 1942). However, the unicellular green alga *Chlamydomonas reinhardtii* soon became the most studied model phototroph. *C. reinhardtii* has been well characterized as a green alga capable of producing H_2 by allocating photosynthetic electrons from reduced ferredoxin (Fd) to a [Fe-Fe]-hydrogenase under specific conditions (Fig. 9.2).

The theoretical light energy-to-H_2 energy conversion efficiency (LHCE) is around 10–13% in green algae (Bolton 1996; Ghirardi et al. 2006), which is high enough for industrial applications. Despite great potential, there are many metabolic hindrances and technological barriers to the application of algae for H_2 photoproduction at industrial levels. The major metabolic bottlenecks include (1) the high sensitivity of the [Fe-Fe]-hydrogenase to O_2 (Torzillo et al. 2015), (2) the non-dissipated proton gradient across the thylakoid membrane (Ghirardi and Mohanty 2010), and (3) alternative electron transport pathways competing for photosynthetic reducing power (Godaux et al. 2015). Currently, several protocols are available for the induction of algal H_2 photoproduction: the dark-to-light transition protocol (Gaffron and Rubin 1942), the nutrient deprivation protocol (Melis et al. 2000; Volgusheva et al. 2015; reviewed in Gonzalez-Ballester et al. 2015), the substrate (CO_2 and acetate) limitation protocol (Nagy et al. 2018a), and the pulse-illumination protocol (Kosourov et al. 2018). In this chapter, we present a critical examination of the benefits and limitations of these different H_2 photoproduction approaches.

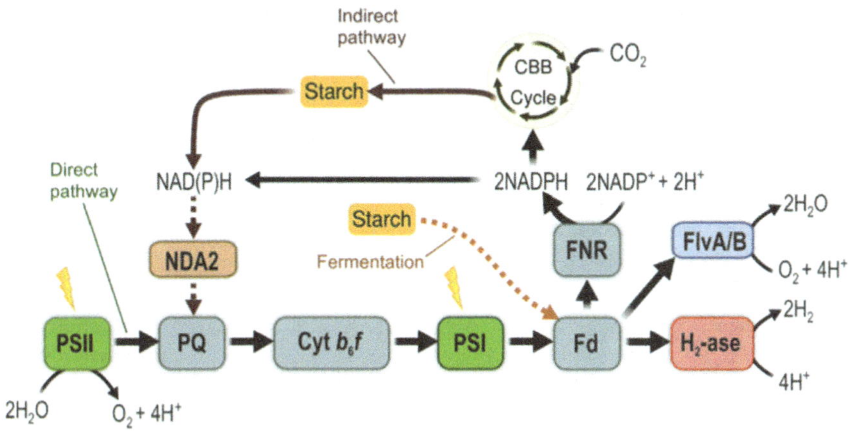

Fig. 9.2 Electron transfer pathways involved in H_2 production in green algae. The direct water photolysis (or PSII-dependent) pathway involves photosynthetic electron transfer (PET) from the water-oxidizing photosystem (PS)II via different photosynthetic redox cofactors to ferredoxin (Fd) and finally to the [Fe-Fe]-hydrogenase (H_2ase) which reduces protons to molecular H_2 (under anoxic conditions). Photosynthetically reduced Fd acts as branching point transferring electrons (1) to ferredoxin/$NADP^+$ oxidoreductase (FNR) to reduce $NADP^+$ to NADPH which in turn fuels CO_2 fixation and cell metabolism, (2) to flavodiiron proteins (FLVA/B) to reduce O_2 to water, (3) to PGR5/PGRL1 pathway, and (4) to the H_2ase. The indirect pathway injects electrons stemming from the breakdown of accumulated carbohydrates into the PET chain at the plastoquinone (PQ) pool level via the NAD(P)H dehydrogenase (NDA2) and is, therefore, bypassing PSII. It is important to note that when cultures are grown in the presence of acetate, glycerate 3-phosphate may also feed the Calvin-Benson-Bassham (CBB) cycle. A third pathway (fermentation) donates electrons originating from the starch breakdown directly to Fd and the [Fe-Fe]-hydrogenase (H_2ase); therefore it does not involve light-dependent reactions

9.2 Transient H_2 Photoproduction During a Dark-to-Light Transition in Anaerobic Cultures

Dark-adapted anaerobic algal cultures are capable of catalyzing direct water biophotolysis during subsequent exposure to light: the [Fe-Fe]-hydrogenase enzyme catalyzes H_2 photoproduction with electrons derived from reduced Fd, originally stemming from photosynthetic water splitting (Gaffron and Rubin 1942; Boichenko and Hoffmann 1994; also see the reproduced protocol in Kosourov et al. 2018). Direct water biophotolysis occurs at a high rate (up to 300 μmol H_2 mg Chl^{-1} h^{-1}) but lasts only for a very short period due to the inhibition of [Fe-Fe]-hydrogenase by photosynthetically accumulated O_2 (a scheme of this protocol is depicted in Fig. 9.3).

This process can be extended in the presence of 3-(3′,4′-dichlorophenyl)-1,1-dimethylurea (DCMU), a specific inhibitor of electron transport at photosystem (PS) II (Florin et al. 2001), or shifting the culture to low light (below the compensation point) and/or high cell density conditions (Aparicio et al. 1985; Degrenne et al. 2011). Nevertheless, the long-term acclimation of algae to hypoxia may result in

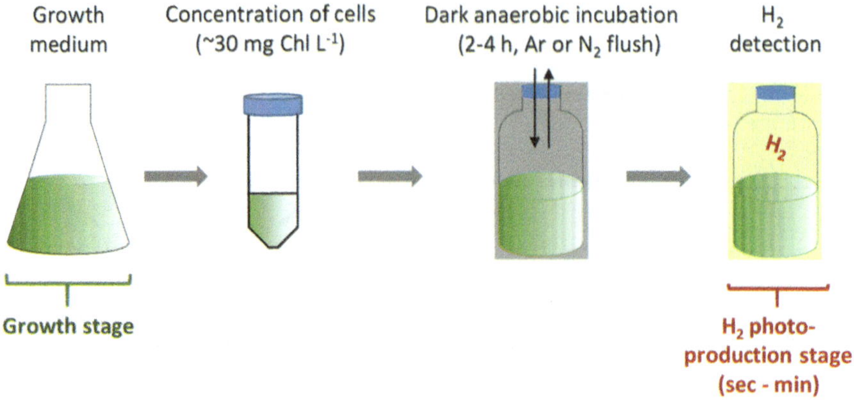

Fig. 9.3 Dark anaerobic adaptation protocol for H_2 photoproduction. The photosynthetically active algal culture is concentrated (to ~30 mg Chl L^{-1}), flushed with Ar or N_2 to remove remaining O_2, and simultaneously dark-adapted for 2–4 h. The H_2 photoproduction during the illumination lasts from several seconds to minutes before the photosynthetic O_2 level reaches to the level of inhibition of the hydrogenase activity

sustained H_2 production via the direct mechanism (Scoma et al. 2014). Under these conditions, however, H_2 photoproduction in algae occurs exclusively (in the presence of DCMU) or predominantly (in the case of low light) via the indirect pathway and, thus, cannot be sustained for an extended period, due to limitations on metabolic reserves of starch and protein (Gfeller and Gibbs 1984; Degrenne et al. 2011; Scoma et al. 2014). Even though the maximum efficiency for direct water biophotolysis is high enough (10–13%, Ghirardi et al. 2006, Bolton 1996) to satisfy industrial applications, the efficient production only lasts for a very short time and, therefore, cannot be considered a cost-effective process.

9.3 Different Nutrient Deprivation Protocols for H_2 Photoproduction

A two-stage sulfur (S)-deprivation method was introduced by Melis et al. (2000) which soon became the most commonly applied protocol to induce long-term H_2 photoproduction. Later on, other deprivation protocols omitting different key nutrients from algae cultures were introduced (Ghirardi et al. 2000; Gonzalez-Ballester et al. 2015). The main strategy of two-stage protocols is the temporary separation of H_2 production, derived by the O_2-sensitive hydrogenase, from water splitting at PSII. The first "biomass" stage of the process occurs in full medium suitable for intensive biomass accumulation. Here, the algae culture is photosynthetically active, oxidizing water to O_2 and storing part of the produced carbohydrates like starch.

During the second "production" stage, the algal cells are deprived of certain nutrients, which induces specific metabolic changes shifting the cultures from O_2 to H_2 photoproduction via the downscaling of photosynthetic O_2 production and establishment of anoxia (Melis et al. 2000; He et al. 2012; Philipps et al. 2012; Batyrova et al. 2012, 2015; Papazi et al. 2014; Volgusheva et al. 2015, 2017). The interplay of the metabolic acclimation processes to nutrient deprivation causes the culture to pass through five distinct stages: (1) active photosynthesis, (2) O_2 consumption, (3) anaerobiosis, (4) H_2 production, and (5) termination (Kosourov et al. 2002; Tsygankov et al. 2006; Volgusheva et al. 2013). While these stages occur in all nutrient deprivation protocols, they vary considerably in duration, especially for the aerobic active photosynthesis phase (1).

The S-deprivation protocol (Melis et al. 2000) leads to the fast inactivation of the O_2-evolving PSII complex (Wykoff et al. 1998; Volgusheva et al. 2007; Nagy et al. 2018b). Additionally, S-deprivation impairs de novo protein synthesis and significantly lowers the abundance of the PSII reaction center protein D1, which incorporates the S-containing amino-acid methionine (Wykoff et al. 1998; Melis et al. 2000; Zhang et al. 2002). Other changes in the photosynthetic machinery include state-2 transition and increased cyclic electron transport, which further contribute to the reduction of PSII activity and diminished O_2 evolution (Wykoff et al. 1998; Melis 2007). Mitochondrial respiration is not affected by S-deprivation, perhaps even slightly increased, during the early phases of the process, ensuring the rapid removal of O_2 from sealed cultures (Melis et al. 2000; Zhang and Melis 2002). A recent study has shown that A-type flavodiiron (Flv) proteins function as a powerful electron sink by redirecting photosynthetic electrons to O_2 (Jokel et al. 2018 and Fig. 9.2), assisting in diminishing the O_2 level inside the chloroplast at the onset of anaerobiosis (Jokel et al. 2015). Simultaneous reduction of RuBisCo protein levels and a decline in CO_2 fixation and other metabolic processes during nutrient deprivation lead to the situation where electrons stemming from residual PSII activity become available for direct H_2 photoproduction (Zhang et al. 2002; Winkler et al. 2010).

Studies have estimated that depending on the phase of S-deprivation,

- Algae produce from 60 to 90% of H_2 in a PSII-dependent manner (Kosourov et al. 2003; Fouchard et al. 2005; Hemschemeier et al. 2008; Antal et al. 2009, 2011; Volgusheva et al. 2013). Another share of photosynthetic H_2 is produced via a PSII-independent pathway, which relies largely on the degradation of starch and the injection of electrons via NAD(P)H dehydrogenase 2 (NDA2) into the photosynthetic electron transport chain at the plastoquinone pool level (Melis et al. 2000; Chochois et al. 2009; Mignolet et al. 2012). Starch accumulation occurs during the early phase of S-deprivation and is more pronounced with the presence of acetate in the medium, which also significantly improves H_2 production yield (Fouchard et al. 2005; Tsygankov et al. 2006; Kosourov et al. 2007). Just a small fraction of H_2 is produced by the fermentative pathway, indicated by the accumulation of formate and ethanol during S-deprivation (Kosourov et al. 2003; Timmins et al. 2009; Noth et al. 2013). Successful S-deprivation could be achieved using a double-chemostat photobioreactor system, which separates the photosynthetic and H_2 production stages in two continuous-flow photobioreactors, thus engaging a continuous H_2 production process (Fedorov et al. 2005; Kosourov et al. 2010). H_2 production can be also sustained by green algae in acetate-rich wastewater (Hwang et al. 2018).

Several efforts have been made to advance H_2 production by testing other nutrient deprivations, including nitrogen (N)-, phosphorus (P)-, potassium (Na)-, and magnesium (Mg)-deprivation (He et al. 2012; Philipps et al. 2012; Batyrova et al. 2012, 2015; Papazi et al. 2014; Volgusheva et al. 2015, 2017). Like S-deprivation, the depletion of these nutrients induces reduced photosynthetic O_2 production activity, the accumulation of starch, establishment of anoxia, and, subsequently, H_2 production.

Upon N-deprivation, algal cells demonstrate an increased accumulation of starch and lipids and a slower decline in PSII activity, which results in a substantially longer aerobic phase, as compared to S-deprivation (Philipps et al. 2012). However, N-deprivation also results in the degradation of several photosynthetic proteins including subunits of the cytochrome (Cyt) *b6f* complex. This severely limits electron transport and possibly reduces the availability of electrons for H_2 production (Bulté and Wollman 1992). Indeed, Philipps et al. (2012) reported a lower H_2 production rate as compared to S-deprivation.

P-deprivation also yields less H_2 than S-deprivation (Batyrova et al. 2012, 2015), and like N-deprivation, P-deprivation induces a slower decline in PSII activity and lengthens the aerobic phase (Batyrova et al. 2012). More detailed studies are needed to better understand the acclimation processes and to clarify why P-deprivation is a less effective stimulus for H_2 production than S-deprivation. Nevertheless, this approach has value in the induction of H_2 photoproduction in marine algae where S-deprivation is not applicable (Batyrova et al. 2015).

Importantly, Mg-deprivation results in similar H_2 production rates to S-deprivation, but H_2 production can be sustained for a substantially longer time period (Volgusheva et al. 2015). As Mg is a central element of chlorophyll molecules, its deprivation results in smaller light-harvesting complex (LHC) antenna which increases the light conversion efficiency of cells in dense culture. Similar to other nutrient deprivations, Mg-deprivation reduces PSII activity less than S-deprivation, and these additional PSII-derived electrons are the probable reason for the longer H_2 production phase (Volgusheva et al. 2015, 2017).

The combination of simultaneous N/S-, N/P-, or even N/S/P-deprivation protocols has been tested in *Chlorella* and reported to produce more H_2 than S-deprivation alone, showing that further medium optimization could enhance H_2 production (He et al. 2012; Pongpadung et al. 2018).

Despite their demonstrated potential, nutrient deprivation protocols bear several drawbacks which hinder their industrial feasibility in terms of efficiency (Dubini and Ghirardi 2015; Oey et al. 2016). A major issue is that the protocols require the transfer of cells into nutrient-depleted media, which involves several laborious and expensive washing and centrifugation steps (a scheme for nutrient deprivation protocols is depicted in Fig. 9.4). To simplify the culture harvesting and transferring processes required for nutrient deprivation, cellular immobilization has been proposed (Laurinavichene et al. 2006; Kosourov and Seibert 2009; Canbay et al. 2018). Immobilization of algae in thin Ca^{2+}- alginate films has the additional benefit of enhancing light conversion efficiency by improving light utilization under low light (Kosourov et al. 2017) and allowing H_2 photoproduction under aerobic conditions (Kosourov and Seibert 2009).

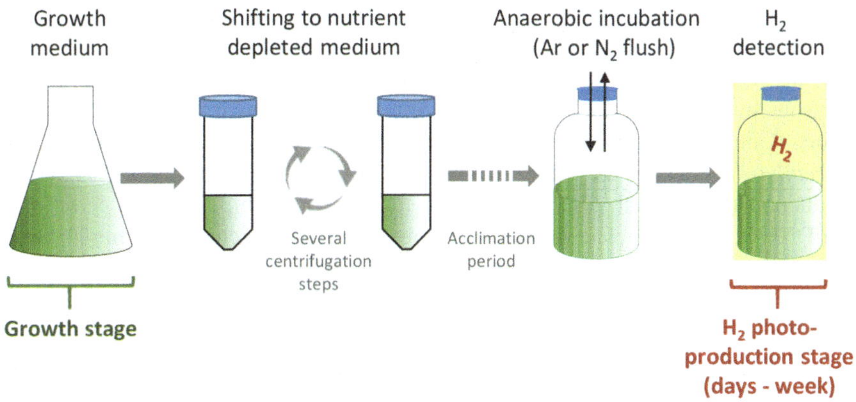

Fig. 9.4 Nutrient deprivation protocol for H_2 photoproduction. The photosynthetically active algal culture is shifted from full to nutrient-depleted medium, which involves several centrifugation steps. Depending on which nutrient is depleted, the culture needs an aerobic acclimation period of 24 h to 2 weeks. Then the culture is flushed with Ar or N_2 to remove remaining O_2. The H_2 photoproduction during the following illumination lasts several days up to a week (for Mg-deprivation) until the culture is terminated

While in the case of S-deprivation, a relatively short acclimation phase of 1–3 days is sufficient to induce H_2 production, other nutrient deprivations may require weeks (*e.g.*, Mg requires 1 week and P a minimum of 2 weeks) before fully anoxic cultures and subsequent H_2 production can be reached (Batyrova et al. 2012; Volgusheva et al. 2015). Reaching full P-deprivation is particularly challenging, as intracellular P reserves delay true P-deficiency for days (Siderius et al. 1996; Komine et al. 2000). Reaching full deprivations without nutrient contamination depends on very clean equipment and the careful handling of cultures, which may be problematic in large-scale H_2 production. Another drawback is the dependency of efficient H_2 production on the presence of acetate, which decreases the direct light-to-H_2 conversion efficiency (Kosourov et al. 2002; Fouchard et al. 2005; see also Tolstygina et al. 2009 for efficient H_2 production at photoautotrophic S-deprivation conditions). Furthermore, the degradation of many proteins of the photosynthetic apparatus and diminished metabolic function leaves nutrient-deprived cultures ultimately unviable for an efficient industrial production process. These drawbacks deem nutrient deprivation an unlikely option for a feasible large-scale industrial H_2 production and make the pursuit of other approaches necessary.

9.4 Substrate Limitation of the Calvin-Benson-Bassham (CBB) Cycle

As mentioned in Sect. 9.3 of this chapter, nutrient deprivation protocols developed so far negatively affect both fitness and the biocatalytic lifetime of the cells. Recently, more efficient approaches to photoautotrophic H_2 production by green algae have been developed (Nagy et al. 2018a current section; Kosourov et al. 2018 described in Sect. 9.5). It was decades ago that Gaffron and Rubin (1942) suggested that the H_2 production pathway is challenged by competition for photosynthetic electrons by the CO_2 fixation pathways. Several studies have since supported this theory (Cinco et al. 1993; Hemschemeier et al. 2008). Indeed, a mutant strain of *C. reinhardtii* lacking the large subunit of RuBisCO has exhibited enhanced H_2 production (Pinto et al. 2013). Likewise, the genetic manipulation of ferredoxin 1 (FDX1) to enhance electron transport toward the hydrogenase enzyme in green alga confirmed the existence of electron competition between hydrogenase and other metabolic pathways (Rumpel et al. 2014; Eilenberg et al. 2016).

The H_2 production protocol proposed by Nagy et al. (2018a) relies on the elimination of the competing CBB cycle by limiting the availability of carbon substrates (CO_2 and acetate) and by the employment of a highly concentrated culture to ensure strong respiration. The protocol involves a dark anaerobic adaptation of several hours, flushing with N_2 at regular intervals (removing both CO_2 and O_2 from the culture), and induction of H_2 production during the following light period (a scheme for the "substrate limitation" protocol is depicted in Fig. 9.5).

In this approach, the amounts of photosynthetic proteins (PsbA, PetB, PsaA) remain steady, with RuBisCO levels the only photosynthetic protein showing a moderate decrease. Moreover, PSII was shown to remain relatively active (F_V/F_M slowly decreased during the 96 h of H_2 production but remained relatively high at 0.4, even though the over-reduction of the photosynthetic intersystem chain resulted in a decreased O_2 evolution activity). Importantly, the accumulation of O_2 over time, as a function of residual PSII water-oxidizing activity, induced the inhibition of hydrogenase. Therefore, an iron-salt-based O_2 absorbent was applied to maintain an

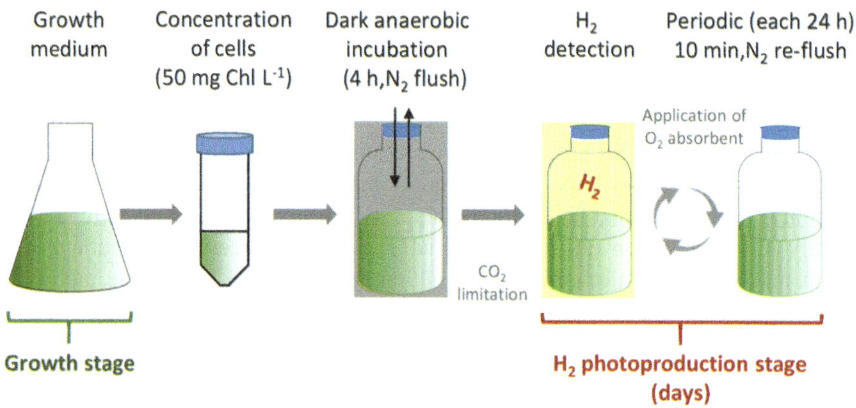

Fig. 9.5 Substrate limitation of the CBB cycle for H_2 photoproduction. The photosynthetically active algae culture grown in minimal medium is concentrated to 50 mg Chl L^{-1} and then dark-adapted for 4 h. Simultaneously, the culture is flushed with N_2 to remove the remaining CO_2 and O_2. The H_2 photoproduction during the following illumination lasts several days if the culture is regularly (every 24 h) reflushed with N2. To increase H_2 production, photosynthetically produced O_2 is efficiently removed by addition of an O_2 absorbent in the headspace of the production vial

anaerobic culture throughout the production process. The use of O_2 scavengers, as distinct from the "substrate limitation" protocol, but toward the induction of long-term and enhanced H_2 photoproduction has been reported previously (Healey 1970; Ma et al. 2011; Wei et al. 2017). Photoautotrophic H_2 production yields obtained under substrate limitation conditions with additional O_2 absorbent were comparable and even higher than those achieved by the S-deprivation of mixotrophic algae (Nagy et al. 2018a).

Nevertheless, it is important to emphasize that most studies with S-deprived algae have been performed in sealed photobioreactors with very limited headspace, where the H_2 content exceeds 80% (Melis et al. 2000). Thus, in contrast to the protocol of Nagy et al. (2018a), the process was driven against a high H_2 partial pressure, which negatively affects H_2 photoproduction yields (Greenbaum et al. 2001; Kosourov et al. 2012).

The H_2 production protocol based on substrate limitation has several advantages over other nutrient deprivation protocols, as it does not require costly media exchanges, nor long adaptation periods, and is not dependent on acetate or other carbon sources. These factors enhance the light-to-H_2 conversion efficiency (reaching 1.6% during the first 30 min of illumination in the presence of the O_2 absorbent) and reduce the risk of bacterial contamination of the culture. Furthermore, the cells remain viable throughout the H_2 production process and resume growth when recovered, and thus they can be reused for another cycle of H_2 production (Nagy et al. 2018a). However, this approach requires reflushing with N_2 at regular intervals to ensure continuous H_2 production and the application of an O_2 absorbent, which could affect the cost-effectiveness of the process at large scale. This H_2 production protocol (EU patent application PCT/EP2018/053115) holds promise for future commercial H_2 production by microalgae, but the protocol needs to be adapted to larger-scale production conditions.

9.5 The Pulse-Illumination Protocol for H_2 Photoproduction

In pursuit of a method for avoiding significant stress to algal cells, researchers from the University of Turku recently suggested an alternative approach for sustaining H_2 photoproduction activity (Kosourov et al. 2018). This method exploits the ability of anaerobic dark-adapted cultures to produce H_2 on a shift from dark to light conditions. However, in contrast to the traditional dark adaptation protocol (mentioned in Sect. 9.2), the new method allows the H_2 production process to be sustained for up to several days by applying a train of strong light pulses superimposed on darkness or low background illumination.

It has long been known that the CBB cycle in dark-adapted algae is not immediately active upon the shift to light (Pedersen et al. 1966; Ghysels et al. 2013; Godaux et al. 2015) and that light activation requires a few seconds, which affects photosynthetic productivity (Graham et al. 2017). This physiological condition creates a unique possibility to control the distribution of photosynthetic electrons to the [Fe-Fe]-hydrogenase enzyme, which acts as an alternative sink for reductants upon

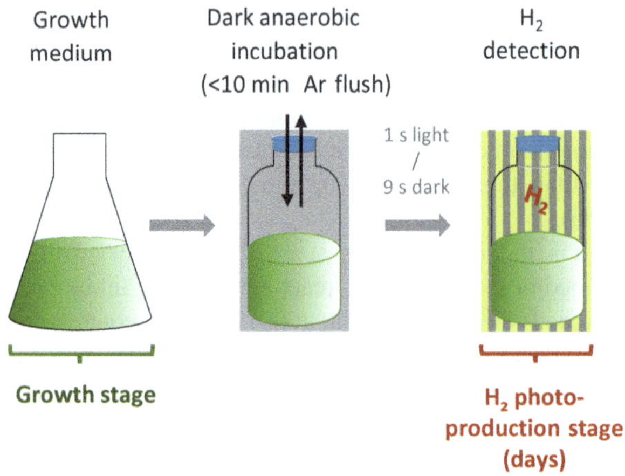

Fig. 9.6 Pulse-illumination protocol for H_2 photoproduction. The photosynthetically active algal culture is dark-adapted for less than 10 min under Ar atmosphere. Application of periodic light pulses on dark or dim-light background (3 µmol photons m^{-2} s^{-1}) induces H_2 photoproduction, which lasts for several days. Application of recurring growth recovery phases (24-h incubation at aerobic conditions) in between H_2 production phases improves cell fitness and extends H_2 photoproduction for about 3 weeks (unpublished data)

illumination and prevents the accumulation of excess electrons in the photosynthetic electron transport chain (Appel and Schulz 1998). Kosourov et al. (2018) proposed that the application of short light pulses to algae, which are short enough for preventing activation of the CBB cycle, will funnel electrons toward hydrogenase. It was shown that a train of short (1–5 s) light pulses interrupted by longer (3–9 s) dark phases indeed induced efficient and long-term H_2 photoproduction in cultures of the green alga *C. reinhardtii*. This was observed under both mixotrophic and photoautotrophic conditions (Kosourov et al. 2018). Yet, in both cases, only a very short preadaptation of algae under dark anaerobic conditions (about 5 min) is required (a scheme for the pulse-illumination protocol is depicted in Fig. 9.6). The same protocol also worked when light pulses were superimposed over low background illumination (3 µmol photons m^{-2} s^{-1}) in place of darkness. As expected, the pulse-illuminated algae were unable to fix CO_2 and produce biomass, suggesting the successful diversion of photosynthetic reductants from the photosynthetic light reactions to H_2 photoproduction. The algal cultures were also unable to accumulate O_2 due to efficient respiration, especially during longer dark periods. As a consequence, such pulse-illuminated algae could spontaneously establish anaerobiosis in a low O_2 environment and subsequently photoproduce H_2 gas.

Despite the apparent simplicity of this approach, the pulse-illumination protocol demonstrated three remarkable phenomena: (1) the presence of non-active [Fe-Fe]-hydrogenase in aerobically grown algae, thus partially confirming data obtained by Liran et al. 2016; (2) the activation of this enzyme within a few seconds after establishment of anaerobiosis; and (3) the presence of a strong H_2 uptake activity in cultures on the switch to darkness, thus proving the reversible nature of [Fe-Fe]-

hydrogenase functioning in intact algal cells (Kosourov et al. 2012). These three observations support the important role of the [Fe-Fe]-hydrogenase enzyme in algal energy metabolism under anaerobic conditions, indicating the value of continued screening for H_2-producing strains and the potential of designing whole-cell biocatalysts for the efficient production of H_2 biofuel. As shown in Fig. 9.6, the protocol is very simple and noninvasive to algae and can be applied almost immediately in any research laboratory. Since maximum H_2 photoproduction activities in algal cultures are observed during the first 10 h, the protocol can be easily adapted to the normal photoperiod.

9.6 Comparison of Different H_2 Photoproduction Approaches in Terms of Light-to-Hydrogen Conversion Efficiency

The maximum theoretical solar energy conversion efficiency of C3 photosynthesis, before photorespiration and respiration, is ca. 12.6% for the total incident solar radiation at ground level (Zhu et al. 2008). Assuming full elimination of photorespiration under anaerobic conditions and 2–2.5% energy loss to respiration, green algae may produce H_2 with a potential efficiency of around 10%. However, considering an increase in the maintenance cost of anaerobiosis and the light condition limiting the rate of photosynthesis, a value of around 7–8% may be more realistic. Indeed, under ambient sunlight conditions, this value would be much lower due to a significant dissipation of energy in photosynthetic antennae complexes (Melis 2012).

Working with dark-adapted algae, Greenbaum (1988) reported PAR efficiencies around 24% (or ~10% of the total solar spectrum) obtained in short-term experiments, thus reaching the theoretical limit. Values close to the theoretical limit or above were also demonstrated by Boichenko and coauthors (for review see Boichenko et al. 2004). It is important to note, however, that all these experiments were performed under extremely low light irradiation (<1 W m^{-2}) and, thus, might be affected by the domination of the photofermentative component of the H_2 photoproduction mechanism. In such cases, efficiency will be overestimated by the additional energy stored in organic matter. Indeed, when Greenbaum (1988) increased the incident light intensity to 1000 W m^{-2}, the PAR efficiencies were well below 1% for the same experimental setup.

Photoheterotrophic S-deprived *C. reinhardtii* cells in suspension culture require intermediate light intensities of around 30–40 μmol photons m^{-2} s^{-1} (~6–9 W m^{-2} PAR) to reach maximum H_2 photoproduction yields (Laurinavichene et al. 2004). Laurinavichene et al. (2004) noticed that a decrease of light intensity results in a sharp decline of H_2 production activity. This decline also occurs with increased light intensity, but more gradually than in low light. Nevertheless, most reports on efficiency were performed at light intensities exceeding 100 μmol photons m^{-2} s^{-1}. Therefore, it is not surprising that energy conversion efficiencies calculated based on incident light have not been very high. At 200 μmol photons m^{-2} s^{-1} PAR, Ghirardi (2006) reported 0.24% (~0.1% of sunlight). With improved mixing in the photobioreactor, wild-type S-deprived *C. reinhardtii* is able to produce H_2 with a

PAR efficiency of around 0.8% (1.6% for the maximum rate) at 140 μmol photons m^{-2} s^{-1} (Giannelli et al. 2009). Similar efficiencies (0.9% for the whole process and 1.5% for the maximum rate) were demonstrated for *C. reinhardtii* cells entrapped in thin alginate films, but at light intensities of around 60 μmol photons m^{-2} s^{-1} PAR (Kosourov and Seibert 2009). There are several differences to be considered when comparing suspension to immobilized cultures. Firstly, immobilized cells do not require mixing. Secondly, the calculations made for suspension cultures only consider the period of H_2 photoproduction, without taking into account the photosynthetic O_2 consumption and anaerobic stages totaling around 22–50 h (dependent on the condition, Kosourov et al. 2002). Thirdly, algae entrapped in alginate films did not decrease H_2 photoproduction yields with decreased light intensity until at least 30 μmol photons m^{-2} s^{-1} PAR (Ghirardi 2015). This means that *C. reinhardtii* cultures entrapped in thin alginate films can utilize light more efficiently under low light conditions. The situation, however, is much worse under autotrophic conditions, where the same immobilized cultures produce H_2 with a PAR efficiency not exceeding 0.14% for a wide range of light intensities (Kosourov et al. 2017).

In the substrate limitation protocol which redirected energy from the CBB cycle, extremely high cell densities (50 mg Chl L^{-1}) and high light intensities (320 μmol photons m^{-2} s^{-1} PAR) were applied to autotrophic *C. reinhardtii* cultures (Nagy et al. 2018a). The experiment showed a maximum PAR efficiency of 1.43% during the first 15 min, which decreased to 0.4% within the first hour. The quoted values were, however, based on calculations using the upper H_2 gas combustion energy of 285.8 kJ mol^{-1}. Replacement of this value with the standard Gibb's energy for the water-oxidizing reaction of 237.2 kJ mol^{-1} will give 1.19% and 0.33%, respectively. Application of O_2 absorbents further improved the efficiency of H_2 production in *C. reinhardtii* cultures, which in the best case was around 4% (3.3% for G^o) during the first hour and decreased to 0.51% during 24 h of H_2 photoproduction.

In contrast to the substrate limitation and nutrient deprivation protocols, the pulse-illumination approach requires normal physiological conditions (Kosourov et al. 2018). As described in Sect. 9.5, the growing cultures are simply transferred to anaerobic conditions under pulse illumination (180 μmol photons m^{-2} s^{-1} PAR). Here, the cells produced H_2 with the maximum efficiency of 1.7% measured during the first 4 h and an overall efficiency of 0.5% measured throughout the whole experiment (54 h). Since the cells do not accumulate O_2 in the medium, the protocol does not require periodic reflushing of the culture with Ar or the application of extra O_2 absorbents, like in the substrate limitation protocol.

9.7 Conclusion

Although the direct comparison of photosynthetic efficiencies of H_2 photoproduction obtained in different setups is often obstructed by differences in experimental conditions, it becomes clear from the above comparison that both the duration of the process and its efficiency improve over time. Strategies to further improve H_2 photoproduction require more effort in the heterologous expression of

O_2-resistant hydrogenases; bypassing of competing pathways by increasing the affinity of hydrogenase; modifying photosynthetic electron transport, including a "photosynthetic control" to funnel more photosynthetic electrons toward hydrogenase; and the design of special photobioreactors to stimulate efficient H_2 production. The combination of immobilization and "pulse-illumination" and/or "substrate limitation" approaches may also result in a protocol less dependent on extra energy input in the form of centrifugation, mixing, and culture reflushing.

References

Antal TK, Volgusheva AA, Kukarskih GP et al (2009) Relationships between H_2 photoproduction and different electron transport pathways in sulfur-deprived *Chlamydomonas reinhardtii*. Int J Hydrog Energy 34:9087–9094

Antal TK, Krendeleva TE, Rubin AB (2011) Acclimation of green algae to sulfur deficiency: underlying mechanisms and application for hydrogen production. Appl Microbiol Biotechnol 89:3–15

Aparicio PJ, Azuara MP, Ballesteros A, Fernández VM (1985) Effects of light intensity and oxidized nitrogen sources on hydrogen production by *Chlamydomonas reinhardtii*. Plant Physiol 78:803–806

Appel J, Schulz R (1998) Hydrogen metabolism in organisms with oxygenic photosynthesis: hydrogenases as important regulatory devices for a proper redox poising? J Photochem Photobiol B Biol 47:1–11

Batyrova KA, Tsygankov AA, Kosourov SN (2012) Sustained hydrogen photoproduction by phosphorus-deprived *Chlamydomonas reinhardtii* cultures. Int J Hydrog Energy 37:8834–8839

Batyrova K, Gavrisheva A, Ivanova E et al (2015) Sustainable hydrogen photoproduction by phosphorus-deprived marine green microalgae *Chlorella sp*. Int J Mol Sci 16:2705–2716

Boichenko VA, Hoffmann P (1994) Photosynthetic hydrogen production in prokaryotes and eukaryotes: occurrence, mechanism, and functions. Photosynthetica 30:527–552

Boichenko VA, Greenbaum E, Seibert M (2004) Hydrogen production by photosynthetic microorganisms. In: Barber J, Archer MD (eds) Molecular to global photosynthesis: photoconversion of solar energy. Imperial College Press, London, pp 397–451

Bolton JR (1996) Solar photoproduction of hydrogen: a review. Sol Energy 57:37–50

Bulté L, Wollman FA (1992) Evidence for a selective destabilization of an integral membrane protein, the cytochrome *b6/f* complex, during gametogenesis in *Chlamydomonas reinhardtii*. Eur J Biochem 204:327–336

Canbay E, Kose A, Oncel SS (2018) Photobiological hydrogen production via immobilization: understanding the nature of the immobilization and investigation on various conventional photobioreactors 3. Biotech 8:244

Chochois V, Dauvillée D, Beyly A et al (2009) Hydrogen production in *Chlamydomonas*: photosystem II-dependent and -independent pathways differ in their requirement for starch metabolism. Plant Physiol 151:631–640

Cinco RM, MacInnis JM, Greenbaum E (1993) The role of carbon dioxide in light-activated hydrogen production by *Chlamydomonas reinhardtii*. Photosynth Res 38:27–33

Degrenne B, Pruvost J, Legrand J (2011) Effect of prolonged hypoxia in auto-trophic conditions in the hydrogen production by the green microalga *Chlamydomonas reinhardtii* in photobioreactor. Bioresour Technol 102:1035–1043

Dubini A, Ghirardi ML (2015) Engineering photosynthetic organisms for the production of biohydrogen. Photosynth Res 123:241–253

Eilenberg H, Weiner I, Ben-Zvi O et al (2016) The dual effect of a ferredoxin-hydrogenase fusion protein in vivo: successful divergence of the photosynthetic electron flux towards hydrogen production and elevated oxygen tolerance. Biotechnol Biofuels 9:182

Fedorov AS, Kosourov S, Ghirardi ML, Seibert M (2005) Continuous hydrogen photoproduction by *Chlamydomonas reinhardtii*. In: Twenty-Sixth symposium on biotechnology for fuels and chemicals. Humana Press, Totowa, pp 403–412

Florin L, Tsokoglou A, Happe T (2001) A novel type of iron hydrogenase in the green alga *Scenedesmus obliquus* is linked to the photosynthetic electron transport chain. J Biol Chem 276:6125–6132

Fouchard S, Hemschemeier A, Caruana A et al (2005) Autotrophic and mixotrophic hydrogen photoproduction in sulfur-deprived *Chlamydomonas* cells. Appl Environ Microbiol 71:6199–6205

Gaffron H, Rubin J (1942) Fermentative and photochemical production of hydrogen in algae. J Gen Physiol 26:219–240

Gfeller RP, Gibbs M (1984) Fermentative metabolism of *Chlamydomonas reinhardtii*. Plant Phys 75:212–218

Ghirardi ML (2006) Hydrogen production by photosynthetic green algae. Indian J Biochem Biophys 43:201–210

Ghirardi ML (2015) Algal systems for hydrogen photoproduction. Golden, CO (United States). https://doi.org/10.2172/1222789

Ghirardi M, Mohanty P (2010) Oxygenic hydrogen photoproduction–current status of the technology. Curr Sci 98:499–507

Ghirardi ML, Kosourov S, Tsygankov A, Seibert M (2000) Two-phase photobiological algal H_2-production system. In: Proceedings of the 2000 U.S. DOE hydrogen program review. San Ramon, California, pp 282–294

Ghirardi ML, Cohen J, King P, et al (2006) [FeFe]-hydrogenases and photobiological hydrogen production. In: Vayssieres L (ed) International Society for Optics and Photonics, p 63400X

Ghysels B, Godaux D, Matagne RF et al (2013) Function of the chloroplast hydrogenase in the microalga *Chlamydomonas*: the role of hydrogenase and state transitions during photosynthetic activation in anaerobiosis. PLoS One 8:e64161

Giannelli L, Scoma A, Torzillo G (2009) Interplay between light intensity, chlorophyll concentration and culture mixing on the hydrogen production in sulfur-deprived *Chlamydomonas reinhardtii* cultures grown in laboratory photobioreactors. Biotechnol Bioeng 104:76–90

Godaux D, Bailleul B, Berne N, Cardol P (2015) Induction of photosynthetic carbon fixation in anoxia relies on hydrogenase activity and proton-gradient regulation-like1-mediated cyclic electron flow in *Chlamydomonas reinhardtii*. Plant Physiol 168:648–658

Gonzalez-Ballester D, Jurado-Oller JL, Fernandez E (2015) Relevance of nutrient media composition for hydrogen production in *Chlamydomonas*. Photosynth Res 125:395–406

Graham PJ, Nguyen B, Burdyny T, Sinton D (2017) A penalty on photosynthetic growth in fluctuating light. Sci Rep 7:12513

Greenbaum E (1988) Energetic efficiency of hydrogen photoevolution by algal water splitting. Biophys J 54:365–368

Greenbaum E, Blankinship SL, Lee JW, Ford RM (2001) Solar photobiochemistry: simultaneous photoproduction of hydrogen and oxygen in a confined bioreactor. J Phys Chem B 105:3605–3609

He M, Li L, Zhang L, Liu J (2012) The enhancement of hydrogen photoproduction in *Chlorella protothecoides* exposed to nitrogen limitation and sulfur deprivation. Int J Hydrog Energy 37:16903–16915

Healey FP (1970) The mechanism of hydrogen evolution by *Chlamydomonas moewusii*. Plant Physiol 45:153–159

Hemschemeier A, Jacobs J, Happe T (2008) Biochemical and physiological characterization of the pyruvate formate-lyase Pfl1 of *Chlamydomonas reinhardtii*, a typically bacterial enzyme in a eukaryotic alga. Eukaryot Cell 7:518–526

Hwang J-H, Church J, Lim J, Lee WH (2018) Photosynthetic biohydrogen production in a wastewater environment and its potential as renewable energy. Energy 149:222–229

Jokel M, Kosourov S, Battchikova N et al (2015) *Chlamydomonas* flavodiiron proteins facilitate acclimation to anoxia during sulfur deprivation. Plant Cell Physiol 56:1598–1607

Jokel M, Johnson X, Peltier G et al (2018) Hunting the main player enabling *Chlamydomonas reinhardtii* growth under fluctuating light. Plant J 94:822–835

Komine Y, Eggink LL, Park H, Hoober JK (2000) Vacuolar granules in *Chlamydomonas reinhardtii*: polyphosphate and a 70-kDa polypeptide as major components. Planta 210:897–905

Kosourov S, Seibert M (2009) Hydrogen photoproduction by nutrient-deprived *Chlamydomonas reinhardtii* cells immobilized within thin alginate films under aerobic and anaerobic conditions. Biotechnol Bioeng 102:50–58

Kosourov S, Tsygankov A, Seibert M, Ghirardi ML (2002) Sustained hydrogen photoproduction by *Chlamydomonas reinhardtii*: effects of culture parameters. Biotechnol Bioeng 78:731–740

Kosourov S, Seibert M, Ghirardi ML (2003) Effects of extracellular pH on the metabolic pathways in sulfur-deprived, H_2-producing *Chlamydomonas reinhardtii* cultures. Plant Cell Physiol 44:146–155

Kosourov S, Patrusheva E, Ghirardi ML et al (2007) A comparison of hydrogen photoproduction by sulfur-deprived *Chlamydomonas reinhardtii* under different growth conditions. J Biotechnol 128:776–787

Kosourov S, Ghirardi ML, Seibert M (2010) Multistage microbial system for continuous hydrogen production. US Patent #7732174

Kosourov S, Batyrova KA, Petushkova EP, Tsygankov AA, Ghirardi ML, Seibert M (2012) Maximizing the hydrogen photoproduction yields in *Chlamydomonas reinhardtii* cultures: the effect of the H_2 partial pressure. Int J Hydrog Energy 37:8850–8858

Kosourov S, Murukesan G, Seibert M, Allahverdiyeva Y (2017) Evaluation of light energy to H_2 energy conversion efficiency in thin films of cyanobacteria and green alga under photoautotrophic conditions. Algal Res 28:253–263

Kosourov S, Jokel M, Aro E-M, Allahverdiyeva Y (2018) A new approach for sustained and efficient H_2 photoproduction by *Chlamydomonas reinhardtii*. Energy Environ Sci 11:1431–1436

Laurinavichene T, Tolstygina I, Tsygankov A (2004) The effect of light intensity on hydrogen production by sulfur-deprived *Chlamydomonas reinhardtii*. J Biotechnol 114:143–151

Laurinavichene TV, Fedorov AS, Ghirardi ML et al (2006) Demonstration of sustained hydrogen photoproduction by immobilized, sulfur-deprived *Chlamydomonas reinhardtii* cells. Int J Hydrog Energy 31:659–667

Liran O, Semyatich R, Milrad Y, Eilenberg H, Weiner I, Yacoby I (2016) Microoxic niches within the thylakoid stroma of air-grown Chlamydomonas reinhardtii protect [FeFe]-hydrogenase and support hydrogen production under fully aerobic environment. Plant Physiol 172:264–271

Ma W, Chen M, Wang L, Wei L, Wang Q (2011) Treatment with $NaHSO_3$ greatly enhances photobiological H_2 production in the green alga *Chlamydomonas reinhardtii*. Bioresour Technol 102:8635–8638

Melis A (2007) Photosynthetic H_2 metabolism in *Chlamydomonas reinhardtii* (unicellular green algae). Planta 226:1075–1086

Melis A (2012) Photosynthesis-to-fuels: from sunlight to hydrogen, isoprene, and botryococcene production. Energy Environ Sci 5:5531–5539

Melis A, Zhang L, Forestier M et al (2000) Sustained photobiological hydrogen gas production upon reversible inactivation of oxygen evolution in the green alga *Chlamydomonas reinhardtii*. Plant Physiol 122:127–136

Mignolet E, Lecler R, Ghysels B et al (2012) Function of the chloroplastic NAD(P)H dehydrogenase Nda2 for H_2 photoproduction in sulphur-deprived *Chlamydomonas reinhardtii*. J Biotechnol 162:81–88

Nagy V, Podmaniczki A, Vidal-Meireles A et al (2018a) Water-splitting-based, sustainable and efficient H_2 production in green algae as achieved by substrate limitation of the Calvin–Benson–Bassham cycle. Biotechnol Biofuels 11:69

Nagy V, Vidal-Meireles A, Podmaniczki A et al (2018b) The mechanism of photosystem-II inactivation during sulphur deprivation-induced H_2 production in *Chlamydomonas reinhardtii*. Plant J 94:548–561

NASA (2019). https://climate.nasa.gov/vital-signs/carbon-dioxide. Accessed 3 May 2019

Noth J, Krawietz D, Hemschemeier A, Happe T (2013) Pyruvate:ferredoxin oxidoreductase is coupled to light-independent hydrogen production in *Chlamydomonas reinhardtii*. J Biol Chem 288:4368–4377

Oey M, Sawyer AL, Ross IL, Hankamer B (2016) Challenges and opportunities for hydrogen production from microalgae. Plant Biotechnol J 14:1487–1499

Papazi A, Gjindali A-I, Kastanaki E et al (2014) Potassium deficiency, a "smart" cellular switch for sustained high yield hydrogen production by the green alga *Scenedesmus obliquus*. Int J Hydrog Energy 39:19452–19464

Pedersen TA, Kirk M, Bassham JA (1966) Light-dark transients in levels of intermediate compounds during photosynthesis in air-adapted *Chlorella*. Physiol Plant 19:219–231

Philipps G, Happe T, Hemschemeier A (2012) Nitrogen deprivation results in photosynthetic hydrogen production in *Chlamydomonas reinhardtii*. Planta 235:729–745

Pinto TS, Malcata FX, Arrabaça JD et al (2013) Rubisco mutants of *Chlamydomonas reinhardtii* enhance photosynthetic hydrogen production. Appl Microbiol Biotechnol 97:5635–5643

Pongpadung P, Zhang L, Sathasivam R et al (2018) Stimulation of hydrogen photoproduction in *Chlorella sorokiniana* subjected to simultaneous nitrogen limitation and sulfur- and/or phosphorus-deprivation. J Pure Appl Microbiol 12:1719–1727

Rumpel S, Siebel JF, Farès C et al (2014) Enhancing hydrogen production of microalgae by redirecting electrons from photosystem I to hydrogenase. Energy Environ Sci 7:3296–3301

Scoma A, Durante L, Bertin L, Fava F (2014) Acclimation to hypoxia in *Chlamydomonas reinhardtii*: can biophotolysis be the major trigger for long-term H_2 production? New Phytol 204:890–900

Siderius M, Musgrave A, Ende H et al (1996) *Chlamydomonas eugametos* (Chlorophyta) stores phosphate in polyphosphate bodies together with calcium. J Phycol 32:402–409

Timmins M, Thomas-Hall SR, Darling A et al (2009) Phylogenetic and molecular analysis of hydrogen-producing green algae. J Exp Bot 60:1691–1702

Tolstygina IV, Antal TK, Kosourov SN et al (2009) Hydrogen production by photoautotrophic sulfur-deprived *Chlamydomonas reinhardtii* pre-grown and incubated under high light. Biotechnol Bioeng 102:1055–1061

Torzillo G, Scoma A, Faraloni C, Giannelli L (2015) Advances in the biotechnology of hydrogen production with the microalga *Chlamydomonas reinhardtii*. Crit Rev Biotechnol 35:485–496

Tsygankov AA, Kosourov SN, Tolstygina IV et al (2006) Hydrogen production by sulfur-deprived *Chlamydomonas reinhardtii* under photoautotrophic conditions. Int J Hydrog Energy 31:1574–1584

Volgusheva A, Zagidullin VE, Antal TK et al (2007) Examination of chlorophyll fluorescence decay kinetics in sulfur deprived algae *Chlamydomonas reinhardtii*. Biochim Biophys Acta 1767:559–564

Volgusheva A, Styring S, Mamedov F (2013) Increased photosystem II stability promotes H_2 production in sulfur-deprived *Chlamydomonas reinhardtii*. Proc Natl Acad Sci U S A 110:7223–7228

Volgusheva A, Kukarskikh G, Krendeleva T et al (2015) Hydrogen photoproduction in green algae *Chlamydomonas reinhardtii* under magnesium deprivation. RSC Adv 5:5633–5637

Volgusheva A, Jokel M, Allahverdiyeva Y et al (2017) Comparative analyses of H_2 photoproduction in magnesium and sulfur starved *Chlamydomonas reinhardtii* cultures. Physiol Plant 161:124–137

Wei L, Yi J, Wang L, Huang T, Gao F, Wang Q, Ma W (2017) Light intensity is important for hydrogen production in NaHSO3-treated *Chlamydomonas reinhardtii*. Plant Cell Physiol 58:451–457

Winkler M, Hemschemeier A, Jacobs J et al (2010) Multiple ferredoxin isoforms in *Chlamydomonas reinhardtii*—their role under stress conditions and biotechnological implications. Eur J Cell Biol 89:998–1004

Wykoff DD, Davies JP, Melis A, Grossman AR (1998) The regulation of photosynthetic electron transport during nutrient deprivation in *Chlamydomonas reinhardtii*. Plant Physiol 117:129–139

Zhang L, Melis A (2002) Probing green algal hydrogen production. Philos Trans R Soc B Biol Sci 357:1499–1507. https://doi.org/10.1098/rstb.2002.1152

Zhang L, Happe T, Melis A (2002) Biochemical and morphological characterization of sulfur-deprived and H_2-producing *Chlamydomonas reinhardtii* (green alga). Planta 214:552–561

Zhu XG, Long SP, Ort DR (2008) What is the maximum efficiency with which photosynthesis can convert solar energy into biomass? Curr Opin Biotechnol 19:153–159

Martina Jokel graduated the PhD at the Molecular Plant Biology group at the University of Turku in Finland studying bioenergy and alternative electron transport via flavodiiron proteins in *Chlamydomonas*. Currently she works as a postdoctoral researcher at the University of Turku focusing on high-value products from microalgae.

Sergey Kosourov is a senior researcher in the Molecular Plant Biology unit at the University of Turku, PhD in plant physiology, and an expert in photobiological hydrogen production. Current research is focusing on thin-layer immobilization of photosynthetic algae for improved light utilization and enhanced hydrogen production yields.

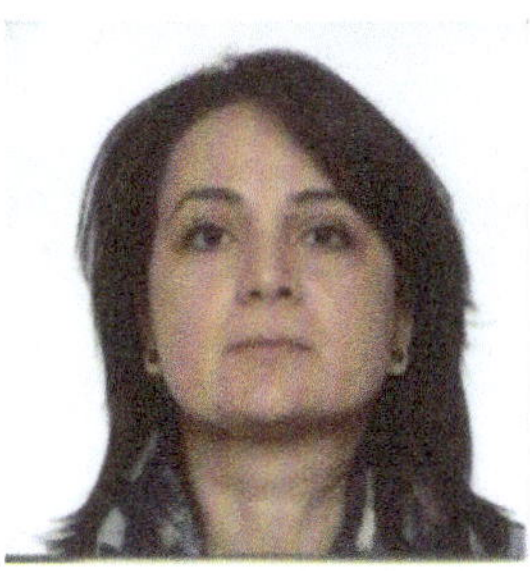

Yagut Allahverdiyeva is an associate professor at the University of Turku. Her research focuses on fundamental (bioenergetics /biochemistry) and applied photosynthesis (sustainable bioproduction, waste-water treatment) in cyanobacteria and green algae.

Synthetic Biofuels and Greenhouse Gas Mitigation

10

Jyoti Porwal, Suheel K. Porwal, Raman Singh, and Kuldeep Singh

Abstract

Increasing awareness among the masses, environment, and the depleting natural oil reservoirs, a replacement for the fossil fuels is urgently required. Presently, biofuels are having increased scientific and societal attention, due to factors such as the need for high energy security, foreign exchange savings, oil price fluctuation, and concern over greenhouse gas (GHG) emissions arising from fossil fuels. Conventional fuel generated severe environmental impacts across the globe. Thus, increasing drastic greenhouse gas emission levels and decreasing crude oil depletion need to arise to study toward an alternative for fuel. Biodiesel and bioethanol are primary biofuels, yet they have limitations toward the feedstock and production process. Synthetic biofuel can be produced from any type of biomass. So they have the diversity of feedstocks and pathways. It may be sustainable, renewable alternative fuel over fossil fuels. It may be boon in GHG mitigation, especially world level carbon dioxide (CO_2) reduction problem.

Keywords

Synthetic biofuels · Greenhouse gas mitigation · Biodiesel · Bioenergy

J. Porwal
CSIR-Indian Institute of Petroleum, Dehradun, Uttarakhand, India
e-mail: jporwal@iip.res.in

S. K. Porwal
DIT University, Dehradun, Uttarakhand, India
e-mail: suheelporwal@dituniversity.edu.in

R. Singh · K. Singh (✉)
Maharishi Markandeshwar (Deemed to be University), Mullana, Haryana, India
e-mail: raman@orgsyn.in; singh@orgsyn.in

A. Kumar et al. (eds.), *Climate Change, Photosynthesis and Advanced Biofuels*,
https://doi.org/10.1007/978-981-15-5228-1_10

10.1 Introduction

The world is a spectator of melting ice peaks, temperature rise, and increasing natural disasters. It is accepted that one of the main reasons for the huge change in climate is the consumption of fossil fuels excessively and the release of greenhouse gases (GHGs) into the atmosphere.

As per the US Department of Energy, a biofuel such as ethanol produces less carbon dioxide (up to 48%) than conventional gasoline. The biodiesel releases only one-fourth the amount of carbon dioxide than the conventional diesel, thus making it a much more eco-friendly as compared to fossil fuels (Hassan and Kalam 2013; Science 2011).

The current biofuel production trends are promising and increasing steadily. The biofuel markets are now expanding in the European Union, USA, India, Brazil, China, and Argentina and contributing to their economies. However, the big question arises: will the current biofuel production rate be able to meet the accelerating transportation fuel demands (Ferry et al. 2012; Popp et al. 2014)?

The total estimated production of biomass in the world is 150 billion tons annually. An increase in the production of biofuels in the recent years and the usage of edible commodities like maize, sugarcane, and vegetable oil has led to the worldwide apprehension toward the future of biofuels and to the "food vs. fuel" debate (Prasad and Ingle 2019; Tenenbaum 2008). The second-generation biofuels, however, are produced from renewable, cheap, and sustainable feedstocks, for example, nonedible oilseeds (Kaul et al. 2011), agriculture waste, rice husk, citrus peel, corn stover, sawdust, bagasse, straw, and rice peel, and are attracting ever-increasing attention (Dahman et al. 2019).

Regarding resource conservation and CO_2 emission reduction targets, the use of biomass as the only renewable carbon source is to be explored and developed for its contribution in this field (Dahmen et al. 2012). The bioliq project (funded by FNR, Agency of Renewable Raw Materials) aims at the large-scale production of synthetic biofuels from any kind of dry biomass including purpose-grown crops, for example, from short rotation plantations as well as residues from agriculture, forestry, and certain types of organic waste (Henrich et al. 2009). Biomass such as straw, hay, residual wood, and so forth usually exhibit low volumetric energetic densities, thus limiting collection area and transportation distances. In the bioliq process, biomass is pretreated for energy densification in regionally distributed fast pyrolysis plants (Dahmen et al. 2012).

The present chapter summarizes about synthetic biofuel and its effect on greenhouse gas mitigation. This study includes synthetic biofuel, source, types, production pathways, greenhouse effect, its prospects, and challenges in greenhouse gas mitigation.

10.2 Synthetic Biofuels

The biofuel or bio-renewable fuel is referred to as solid, liquid, or gaseous fuels that are predominantly produced from biomass (Chhetri and Islam 2008). Synthetic fuels from biomass are called synthetic biofuels (also referred to as BtL, biomass to liquids). It may arise as a future motor fuel consumption to a considerable extent. Liquid biofuels fall into the following categories: (a) bioalcohols (Hacisaligoglu 2009; Hansen et al. 2005), (b) vegetable oils (Chhetri and Islam 2008; Kaul et al. 2011) and biodiesels (Kaul et al. 2010; Porwal et al. 2012), and (c) biocrude and synthetic oils (Demirbas 2008). Biofuels have an important role because they substitute petroleum fuels. Further, the demand for biofuels will rise in the time ahead. Biofuels are alternative fuel sources to petroleum. Biofuels are generally offering many prime concerns, including sustainability, reduction of greenhouse gas emissions, regional development, agriculture, social structure, and security of supply (Demirbas and Balat 2006).

The biggest difference between biofuels and petroleum feedstocks is oxygen content. Biofuels are nonpolluting, accessible, sustainable, locally available, and reliable fuel obtained from renewable sources. Sustainability of renewable energy systems must favor both human and ecosystem health in the long term; aim on tolerable emissions should look well into the future. In the nearest future, electricity generation from biofuels is a promising method. The future of biomass electricity lies in biomass-integrated gasification/gas turbine technology, which offers high energy conversion efficiencies (Quadrelli and Peterson 2007).

Liquid biofuels as a transportation fuel have recently attracted vast attention all over the world because of its sustainability, renewability, local availability, regional development, creating rural jobs, biodegradability, and reduction of greenhouse gas emissions. Biofuels have a significant role in energy security. In the developing countries, policy drivers for biofuels have attracted in rural development and economic opportunities. The European Union and France are in the third and second rank for the production of biofuels behind Brazil and the USA. In Europe, Germany is the largest one, and France is the second-largest producer (Balat 2007).

Synthetic fuels from biomass will replace part of our fossil energy sources and will contribute to an efficient mix of renewable energies. A wide range of different fuels from second-generation feedstock such as gasoline, diesel, and kerosene, BtL (biomass-to-liquid) fuels offer various advantages. Almost any type of vegetable biomass material whose origin and needs do not clash with those of plants grown for the food and feed industry can be used for biofuel production. Dry, lignocellulosic residual biomass (residual wood, straw) from agricultural, forestry, and landscaping is particularly suited. The synthetic fuels are fully compatible with conventional fuels and can be used as a drop-in fuel. Besides being used in the blend form, they can be used as stand-alone products. The quality of high-performance fuels or fuel components should improve the combustion properties and reduce emissions significantly. Biofuels are capable of reducing CO_2, as requested by the European Commission. The 60% goal can easily be achieved by BtL fuel production because heat and electrical power are important by-products of the BtL biomass conversion process, which can provide the energy needs for the production process (Budarin et al. 2013; Bulkowska et al. 2016).

10.3 Sources of Biofuels

Alternative fuels can be made from natural, renewable sources such as biomass, vegetable oil, fats, algal lipids, waste materials, etc. The most commonly used oils are soybean, sunflower, rapeseed, palm, canola, cottonseed, and *Jatropha* (Shalaby 2013; Porwal et al. 2015). Since the prices of edible vegetable oils are higher than that of diesel fuel and in the developing countries, edible oils are not good source due to the food versus fuel debate, therefore waste vegetable oils and nonedible crude vegetable oils are preferred as potential biofuel feedstocks (Singh and Singh 2010). The use of edible oil to produce biofuel in India is also not feasible due to the big gap between demand and supply of these oils. Under Indian conditions, only such plants can be considered for biofuel, which produces nonedible oil in appreciable quantity and can be grown on a large scale on nonagriculture and wastelands. Animal fats, although mentioned frequently, have not been studied to the same extent, like vegetable oils because of natural property differences. Animal fats contain a higher level of saturated fatty acids; therefore, they are solid at room temperature (Ma and Hanna 1999).

Any vegetable crop or agriculture waste may be a feedstock for biofuel production. Biofuels sources in different countries include *ethanol* (often made from *corn* in the USA and *sugarcane* in Brazil), biodiesel (*vegetable oils* and liquid animal fats), *green diesel* (derived from *algae* and other plant sources), and *biogas* (*methane* derived from animal manure and other digested organic materials) (Welker et al. 2015).

10.4 Types of Synthetic Biofuels

Types of renewable hydrocarbon biofuels (Ma and Hanna 1999) include:

Bio-gasoline—It is also known as renewable gasoline or "green" gasoline; bio-gasoline is a transportation fuel derived from biomass and suitable for use in spark-ignition engines. It meets the EN 228 in Europe and ASTM D4814 specification in the USA.

Green diesel—Also called "renewable" diesel, green diesel is a biomass-derived transportation fuel. It is suitable for diesel engines. It meets the EN 590 in Europe and ASTM D975 specification in the USA.

Green diesel is distinct from biodiesel in chemical structure. Biodiesel is a mono-alkyl ester, while green diesel is chemically similar to petrodiesel. Thus biodiesel has different physical properties and hence different fuel specifications (EN 14214 and ASTM D6751). The processes of production of both fuels are very different. Biodiesel is produced via transesterification reaction, while renewable diesel is produced through various processes such as gasification, pyrolysis, hydrotreating (isomerization), and other thermochemical and biochemical routes. Green diesel and biodiesel are different in feedstock sources. Biodiesel is produced mainly from lipid materials (such as animal fats, vegetable oils, algae, and grease), whereas green diesel is produced from lipids and cellulosic biomass (such as woody biomass, crop residues, and dedicated energy crops).

Bio-jet fuel—It is also referred to as "renewable jet fuel" or aviation biofuel. Bio-jet fuel is a biomass-derived fuel and alternative for petroleum-based aviation fuel. It can be blended up to 50% with conventional commercial and military jet (or aviation turbine) fuel. It follows the specification ASTM D7566.

A blend in the range between 10 and 50% of synthetic paraffin kerosene (SPK) fuels can be used depending on the fuel type with conventional and aviation turbine (or military jet) fuel. Other synthetic kerosene with aromatics (SKA) fuels can be used interchangeably with fossil fuels. Blending is necessitating with SPK fuels because of their deficit sufficient aromatic hydrocarbons, which are present in conventional jet fuel. Aromatics should be limited in jet fuel to prevent smoke formation during combustion, but a minimum aromatic content is required to increase fuel density and elastomer swelling in aircraft fuel systems.

The ASTM D7566 standard approved the following fuel categories:

- Hydrogenated esters and fatty acids (HEFA) fuels derived from animal fats, algae, used cooking oil, and vegetable oils (e.g., camelina) (HEFA-SPK).
- Fischer-Tropsch (FT) fuels derived from solid biomass resources (e.g., wood residues) (FT-SKA).
- Fischer-Tropsch (FT) fuels with aromatics derived from solid biomass resources (e.g., wood residues) (FT-SKA).
- Synthetic iso-paraffin (SIP) derived from fermented hydro-processed sugar, formerly known as direct-sugar-to-hydrocarbon fuels. Up to 10% blends are permitted for this fuel (SIP-SPK).
- Alcohol-to-jet (ATJ) fuels derived from isobutanol and a maximum 30% level of blending (ATJ-SPK).

10.5 Bioethanol

Cellulosic ethanol offers promise because cellulose fibers, a major and universal component in plant cell walls, can be used to produce ethanol. As per the International Energy Agency (IEA), cellulosic ethanol could permit ethanol fuels to play a much big role in the future than previously thought.

The basic steps for large-scale production of ethanol are microbial (yeast) fermentation of sugar, distillation, dehydration, and denaturing (optional). Before the fermentation process, some crops require saccharification or hydrolysis of carbohydrates such as cellulose and starch into sugars. The saccharification of cellulose is called cellulosis. Enzymes are used to convert starch into sugar (Jain 2019).

10.6 Biomass-to-Liquid [BtL] Fuel

Biomass-to-liquid fuel is for high-quality fuels or fuel components produced from renewable, sustainable biomass. Biomass is distributed widely. Large quantities of biomass are required for large-scale fuel production plants by economies of scale. In first step, pretreatment of biomass produces a high energy density intermediate energy carrier (bioliqSynCrude). bioliqSynCrude can be transported to the central unit for the production of synthetic fuel over long distances. The fuel can be used either as drop-in fuels or as stand-alone products. These fuels are fully compatible with existing gasoline or diesel-type fuels. Almost any type of dry biomass can be utilized for this process; a focus is set on by-products and residues of agriculture, forestry, or landscaping.

The bioliquid pilot plant covers the complete process chain required for producing customized fuels from residual biomass. For energy densification of the biomass, fast pyrolysis is applied. The liquid pyrolysis oil and solid char obtained can be processed into intermediate fuels of high energy density. Fuel and chemical production from syngas requires high pressures. Therefore, syngas production is already performed at pressures up to 80 bar by entrained flow gasification. Gas cleaning and conditioning are conducted at the same pressure at high temperatures allowing for optimal heat recovery and thus improved energy efficiency. In the bioliquid pilot plant, the purified syngas is firstly converted into dimethyl ether and then further to gasoline (Ceccarelli 2018).

10.7 Production Pathways

Any biomass can be converted into fuels through a variety of thermal, chemical, and biological processes. These products are similar to petroleum-based fuels like gasoline, diesel, or jet fuel in chemical makeup and are therefore considered infrastructure-compatible fuels. Renewable hydrocarbon biofuels can be synthesized from any biomass source. These sources include lipid materials (such as vegetable oils, animal fats, greases, and algae) and cellulosic material (such as woody biomass, crop residues, and dedicated energy crops). A variety of methods are exploring to produce synthetic hydrocarbon biofuels. Production plants may be either stand-alone or centralized at petroleum refineries (Hughes et al. 2013). Synthetic fuels have several advantages because they can be used without modification in existing engines and fuel supply. In addition, synthetic biofuels are necessarily cleaner than traditional fuels owing to the removal of all contaminants to avoid poisoning the catalysts used in the processing. There are several thermal and chemical processes, which can be used to produce synthetic hydrocarbons (Reisch and Th_gersen 2015).

The kind of pathways has been explored for the renewable hydrocarbon biofuel production (Porwal et al. 2015); they include:

Traditional hydrotreating of lipid—Hydrotreating involves reacting the feedstock (lipids) with hydrogen under elevated temperatures and pressures in the presence of a catalyst.

Biological conversion of sugar—This pathway carried out by a biochemical reconstruct process with the addition of some organisms that convert the sugar molecules to hydrocarbons.

Catalytic conversion of sugars—This pathway includes a series of catalytic reactions process to transform a carbohydrate-rich stream into H-C fuels.

Gasification process—In this process, biomass is thermally converted to syngas (H_2 + CO) and catalytically produced hydrocarbon-based fuels.

Pyrolysis process—In this process, the chemical decomposition of organic materials in the absence of oxygen occurs at elevated temperatures. This process makes liquid pyrolysis oil that can be upgraded to produce hydrocarbon fuels.

Hydrothermal processing—This process initiates chemical decomposition of biomass or wet waste materials at high pressure and moderate temperature to produce a bio-oil that may be further converted to hydrocarbon fuels by catalysis.

Presently, commercial-scale production of renewable hydrocarbon biofuels in the USA is limited. There were four commercial facilities at the end of 2016, with a combined capacity of 280 million gallons (Diamond Green Diesel, Cetane Energy, and Renewable Energy Group in Louisiana producing diesel and AltAir Fuels in California producing jet fuel).

10.8 Greenhouse Gas [GHG] Effect

Fossil fuels pollute the environment badly. Unfortunately, burning fossil fuels emits greenhouse gases and harmful particles into the atmosphere, which results in adverse effects for humans as well as earth as a whole. Burning coal releases toxic particles like sulfur dioxide and heavy metals into the atmosphere.

Greenhouse gases, such as carbon dioxide (CO_2), nitrous oxide (N_2O), and methane (CH_4), trap the heat in the atmosphere. With higher than natural concentrations, they lead to unnatural warming. Greenhouse gas concentrations have increased fastly and for which human activities are the primary reason. The result is worldwide, unnatural warming that's driving other changes in our environment (Sher 1998). Major greenhouse gases and their characteristics were shown in Table 10.1 (US EPA). Carbon dioxide (CO_2) is one of the primary greenhouse gases emitted through human activities.

In the USA, carbon dioxide emissions increased by about 2.9% between 1990 and 2017. Since the largest source of greenhouse gas emissions is the combustion of fossil fuel in the USA. Changes in CO_2 emissions due to the combustion of fossil fuel are influenced by many short-term and long-term factors. The factors include

Table 10.1 Major long-lived greenhouse gases (GHGs) and their characteristics (Quadrelli and Peterson 2007)

Name of greenhouse gases	Emission	An average lifetime in the atmosphere	100-year global warming potential
Carbon dioxide (CO_2)	It is emitted primarily through the burning of fossil fuels (oil, coal, and natural gas), trees, woods, and solid waste. Changes due to land use also play a role. Soil degradation and deforestation add CO_2 to the atmosphere, while forest regrowth takes it out of the atmosphere	It cannot be represented with a single value because the gas is not destroyed over time	Carbon dioxide
Methane (CH_4)	It is emitted in the production and transportation of oil, natural gas, and coal. Methane emissions also result from agricultural practices and livestock and from the anaerobic decay of organic waste in municipal solid waste landfills	12.4 years[a]	Methane
Nitrous oxide (N_2O)	It is produced during agricultural and industrial activities, along with combustion of solid waste and fossil fuels	121 years[a]	265–298
Fluorinated gases (CFCs)	These are a group of gases that contain fluorine such as perfluorocarbons, hydrofluorocarbons, and sulfur hexafluoride, among other chemicals. These are emitted from various industrial and commercial processes, household uses. They do not occur naturally. Sometimes these gases are used as substitutes for ozone-depleting substances such as chlorofluorocarbons (CFCs)	Few weeks to thousands of years	Varies (the highest is sulfur hexafluoride at 23,500)

This table shows 100-year global warming potentials (GWPs), which describe the effects that occur over 100 years after a particular mass of a gas is emitted

[a]Perturbation lifetimes

economic growth, population growth, changing energy prices, new technologies, changing behavior, and seasonal temperatures. Between 1990 and 2017, the increase in CO_2 emissions corresponded with increased energy use by an expanding economy and population.

10.9 Energy Consumption and Greenhouse Gas Emission

The EU's renewable energy targets are one important part of the combined efforts to decarbonize the energy system. Although the RES targets are expressed in relative terms (i.e., as a share related to the future levels of energy consumption), progressing toward them can effectively displace fossil fuels and complement the GHG reduction policies for the energy system. As energy efficiency improvements—another key dimension of the EU's decarbonization efforts—gradually reduce our total energy needs, the growing share of renewables results in a progressively larger displacement of nonrenewable alternatives.

The following sections estimate the gross effect of renewable energy on fossil fuel consumption and its associated GHG emissions and then—statistically—on primary energy consumption. The estimates were made by comparing actual growth in renewable energy since 2005 with a counterfactual scenario in which this growth would be delivered from nonrenewable energy sources. This assumes that renewable energy growth since 2005 has substituted for an equivalent amount of energy supplied by other sources. The method is reported in detail in the EEA report Renewable Energy in Europe—approximated recent growth and knock-on effects (Reisch and Th_gersen 2015; European Environmental Agency 2016).

10.10 Gross Avoided GHG Emissions

According to the European Environmental Agency (EEA), the growth in the consumption of renewable energy (including biofuels) after 2005 resulted in an estimated 362 Mt of gross avoided CO_2 emissions at the EU level in 2013 and 380 Mt in 2014—a 5% increase compared with 2013. This yearly amount is comparable to the GHG emissions of Poland. The contribution from renewable electricity (283 Mt CO_2 in 2014, or 75% of all gross avoided emissions) is considerably larger than that of renewable heating and cooling (57 Mt CO_2 in 2014, or 15% of all gross avoided emissions) and biofuels in transport (39 Mt CO_2 or around 10% of total gross avoided emissions). This is because the increase in renewable electricity has reduced the need for uses of solid fuels—the most carbon-intensive fossil fuels—in the power sector. In 2014, the gross avoided CO_2 emissions corresponded to a 9% reduction in the total GHG emissions.

The growth in the consumption of renewable energy after 2005 helped the EU achieve an estimated gross reduction of CO_2 emissions of 362 Mt in 2013 and 380 Mt in 2014—an amount that is equivalent to the yearly GHG emissions of Poland. Three-quarters of these effects have taken place in energy-intensive industrial sectors under the EU Emissions Trading Scheme (ETS), where the increase in renewable electricity decreased the need for the most carbon-intensive fossil fuels. Overall, the gross avoided CO_2 emissions corresponded to a 7% reduction in total EU GHG emissions in 2013, increasing to an estimated 9% in 2014 (European Environmental Agency 2016).

10.11 Role of Biofuels in the Mitigation of Greenhouse Gases (GHGs)

The amount of atmospheric carbon is currently increasing at a rate of 4.3 ± 0.1 gigatons (C) per year, mainly as a result of human activity (Smith and Klosek 2001). Multiple lines of scientific evidence show that this increased amount of carbon in the atmosphere is warming the global climate system (Hannula 2016). To limit warming under 2 °C, the European Council in 2011 reconfirmed the EU objective of reducing greenhouse gas (GHG) emissions by 80–95% by 2050 compared to 1990. The European Council also endorsed a binding EU target of at least 40% domestic reduction in GHG emissions by 2030 compared to 1990. The target will be achieved collectively by the EU in the most cost-effective manner possible, in which reductions in the ETS and non-ETS sectors amount to 43% by 2030 compared to 30% in 2005, respectively. According to an IEA roadmap study (Liu et al. 2011), biofuels (i.e., fuels produced from renewable plant matter) could provide 27% of total transport fuel consumption by 2050 while avoiding around 2.1 gigatons of CO_2 emissions per year if sustainably produced. However, meeting this demand would require 65 exajoules of biofuel feedstock, occupying around 100 million hectares of land, which was considered challenging by the study given the growing competition for land for food and fiber.

10.12 Environmental Impact of Biofuels

In nature, through the evolutionary process, plants and other organisms developed a quite ingenious capacity to use sunlight to convert atmospheric CO_2 into biomass. This is achieved through a phenomenon known as photosynthesis, which uses chlorophylls and other pigments in the process of conversion of CO_2 into biomass in a sufficiently stable form. Biomass is basically chemical energy. This occurs through a series of physical processes and chemical reactions, resulting in biomass (Cortez et al. 2012).

Synthetic biofuel is a type of fuel, in which energy is derived from biological carbon fixation. These fuels are derived from biomass conversion (such as solid biomass, liquid fuels, and various biogases). Although fossil fuels also have their origin in ancient carbon fixation, they are not considered as biofuels by the generally accepted definition because they have carbon that has been considered "out" of the C cycle for a very long time. Presently, biofuels are gaining increased public and scientific attention, due to the factors such as the need for increased energy security, oil price spikes, foreign exchange savings, concern over GHG emissions from fossil fuels, and support from government subsidies.

Subsidies and incentives are provided independently from the environmental impact that ethanol may have during its entire life cycle, therefore supporting biofuel production in the USA. In 2001, the European Commission launched a policy to promote the use of biofuels for transport in order to reduce GHG emissions and the environmental impact of transport, as well as to increase the security of supply, technological innovation, and agricultural diversification (Zarrilli 2006).

Renewable energies are important sources for this transition because they mitigate GHG emissions. It lowers environmental constraints associated with conventional energy production and reduces the reliance on fossil fuels. Other benefits associated with the growth of RES include the reduction of fossil fuel imports, the diversification of energy supply, and the creation of jobs, skills, and innovation in local markets and progressive sectors with significant growth potential. From a life-cycle perspective, the environmental pressures arising from renewable energy technologies are 3–10 times lower than from fossil fuel-based systems (Owusu and Asumadu-Sarkodie 2016). However, as with all industrial activities, renewable energy projects too may increase health and environmental pressures, especially when project designs and technologies do not take into account local considerations (Parry et al. 2017). Developing a strong renewable energy base in Europe has implications for Europe's competitiveness and export potential.

10.13 Need for Biofuel Over Conventional Fuels in the Future

Energy is a critical factor in developing countries for economic growth as well as for social development and human welfare and has a vital role in all developmental activities. The economic development of various countries is hindered due to a paucity of energy. Over two billion people in the world are still deprived of electrical energy. The conventional sources of energy are not enough to provide energy to the developing world, as energy usage has doubled owing to rising populations, expanding economies, energy-intensive industries, urbanization, a quest for modernization, and improved quality of life. At the same time, the world energy scenario depicts a grim picture. The adverse effects on the environment caused by the production and consumption of energy also have generated severe environmental impacts across the globe. The by-products of conventional energy sources, such as SO_2, NO_x, CO_2, and other air pollutants, cause acid rain and health problems to humans. The greenhouse gases (GHGs) have exacerbated global warming. According to data collected by Frances Moore of the Earth Policy Institute, emissions of GHG grew by 3.1% from 2000 to 2006. The five largest emitters of energy-related CO_2 are China, the USA, the European Union, Russia, and India. They together account for almost two-thirds of global CO_2 emissions. Without clean energy solutions to reduce the world's carbon footprint, CO_2 emissions could increase twofold between 2000 and 2030. At this rate, it would be impossible to avoid an increase in temperature of 3 °C above the preindustrial era. A less than 2 °C increase in temperature would cause a dangerous change in climate (Smith and Klosek 2001).

World energy-related CO_2 emissions have increased from 30.2 billion metric tons in 2008 to 35.2 billion metric tons in 2020, and it is assumed to be 43.2 billion metric tons in 2035 an increase of 43% over the projection period. With the increasing economic growth and continued strong reliance on fossil fuels expected for most non-organizations for Economic Co-operation and Development (non-OECD) economies under current policies, much of the projected increase in CO_2 emissions occurs among the developing non-OECD nations. In 2008, non-OECD emissions exceeded OECD emissions by 24%, while they are projected to exceed OECD emissions by more than 100% by 2035. Furthermore, with respect to total GHG emissions in absolute terms, a study estimates that India's emissions in 2031 will be between 4.0 billion metric tons of CO_2-eq and 7.3 billion metric tons of CO_2-eq (Chhetri and Islam 2008).

Apart from these environmental and social problems, conventional energy sources are not sustainable and nonrenewable in nature. These limited reserves are likely to become exhausted in the future. The fluctuating prices of petroleum products are also a matter of real concern. Therefore, alternative energy sources will be the need of the hour. Renewable energy sources are the least costly and most feasible solution, as they are unlimited. The limited supply, exhaustible nature, and environmental concerns over fossil fuel resources have resulted in a search for eco-friendly and inexhaustible renewable energy sources all around the world due to their various benefits as inexhaustible, environmentally friendly, and sustainable energy resources (Smith and Klosek 2001).

10.14 Challenges for the Promotion of Biofuels Over Fossil Fuels

A challenging issue regarding the promotion of biofuels over fossil fuels is the achievement of sustainability by considering the three interrelated pillars of sustainability, namely, economic, environmental, and social (Fig. 10.1).

The term "sustainable development" was defined in 1987 by the Brundtland Commission as "development that meets the needs of the present generation without compromising the ability of future generations to meet their own needs." This definition is accepted as a standard and a starting point for most who aim to define the concept of sustainability (Buytaert et al. 2011). In order to achieve sustainability, the environmental impacts of each phase of the biofuel supply chain (i.e., production or collection of biomass feedstock, feedstock processing, conversion to biofuel, and

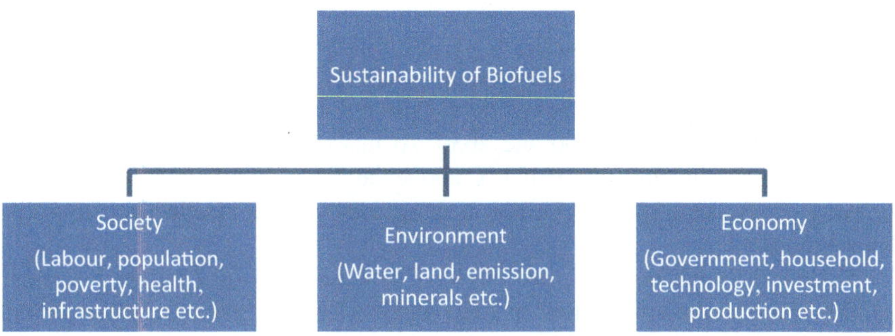

Fig. 10.1 Three pillars of sustainability of biofuels

end-product distribution) must be evaluated using well-defined criteria. The evaluation is always difficult due to the great number of factors that are weighted differently by the involved stakeholders (i.e., farmers, manufacturers, policymakers, economic development agencies, local communities, etc.).

The key question is how to measure biofuel sustainability in such a complex system with a diversity of feedstocks, a large number of biofuel pathways, and variations on specific interests of the stakeholders. The answer lies within the establishment of environmental and other indicators, which enable the assessing of the sustainability of different types of bioenergy systems. The indicators should, however, apply to both large installations and local sites and also should be useful to diverse stakeholders (Fokaides and Christoforou 2016). According to Silva Lora et al. (2011), a sustainable biofuel shall meet at least the following requirements (Silva Lora et al. 2011):

- To be carbon neutral
- Not to affect the quality, quantity, and rational use of available natural resources
- Not to have undesirable social consequences
- To contribute to society economically

10.15 Conclusion

Biofuel can yield net GHG benefits and socioeconomic benefits. Multiple scientific evidence show that an increased amount of carbon in the atmosphere is causing global warming and disturbs the climate system. The carbon footprint describes the total greenhouse gas emission. To limit warming under 2 °C, the European Council in 2011 reconfirmed the EU objective of reducing greenhouse gas (GHG) emissions by 80–95% by 2050 compared to 1990. The adverse effects on the environment caused by the production and consumption of energy have generated drastic environmental impacts across the world. Apart from these environmental and social problems, conventional energy sources are nonrenewable in nature. These limited reserves are likely to become exhausted in the future. Environmental concern, health hazards, and depleting sources are looking at an alternative such as biofuel. Synthetic biofuel may be a promising alternative for fossil fuel because it can be produced from any type of biomass. So they have the diversity of feedstocks and pathways for the production and can be used as stand-alone or in blend form with conventional fuel. It may be sustainable, renewable alternative fuel over fossil fuels. It can fulfill the need of the day by reducing carbon dioxide (CO_2) level in all over the world and may be helpful as a big step toward greenhouse gas mitigation.

References

Balat M (2007) An overview of biofuels and policies in the European Union. Energy Sources Part B 2(2):167–181

Budarin V, Shuttleworth PS, Lanigan B, Clark JH (2013) Nanocatalysts for biofuels. In: Polshettiwar V, Asefa T (eds) Nanocatalysis synth. appl. Wiley-VCH, Weinheim, pp 595–614
Bulkowska K, Gusiatin ZM, Klimiuk E, Pawlowski A, Pokoj T (2016) Biomass for biofuels. CRC Press, Boca Raton
Buytaert V, Muys B, Devriendt N, Pelkmans L, Kretzschmar JG, Samson R (2011) Towards integrated sustainability assessment for energetic use of biomass: a state of the art evaluation of assessment tools. Renew Sustain Energy Rev 15:3918–3933. Elsevier Ltd
Ceccarelli C (2018) (IKFT). Bioliq—the bioliq® process. Ceccarelli, Christina (IKFT)
Chhetri AB, Islam MR (2008) Towards producing a truly green biodiesel. Energy Sources Part A 30 (8):754–764
Cortez L, Leite RC de C (2012) Relation Between Biofuels versus Fossil Fuels. In: Petroleum Engineering Downstream. Encyclopedia of Life Support Systems. http://www.eolss.net/sample-chapters/c08/E6-185-21.pdf
Dahman Y, Dignan C, Fiayaz A, Chaudhry A (2019) An introduction to biofuels, foods, livestock, and the environment. In: Biomass, biopolymer-based materials, and bioenergy. Elsevier, Amsterdam, pp 241–276
Dahmen N, Dinjus E, Kolb T, Arnold U, Leibold H, Stahl R (2012) State of the art of the bioliq® process for synthetic biofuels production. Environ Prog Sustain Energy 31(2):176–181
Demirbas A (2008) Conversion of corn stover to chemicals and fuels. Energy Sources Part A 30 (9):788–796
Demirbas MF, Balat M (2006) Recent advances on the production and utilization trends of bio-fuels: a global perspective. Energy Convers Manag 47(15–16):2371–2381
European Environmental Agency (2016) Renewable energy in Europe— recent growth and knock-on effects
Ferry MS, Hasty J, Cookson NA (2012) Synthetic biology approaches to biofuel production. Biofuels 3(1):9–12
Fokaides PA, Christoforou E (2016) Life cycle sustainability assessment of biofuels. In: Handbook of biofuels production. Elsevier, Amsterdam, pp 41–60
Hacisaligoglu S (2009) Ethanol–gasoline and ethanol–diesel fuel blends. Energ Edu Sci Technol 22:31–46
Hannula I (2016) Hydrogen enhancement potential of synthetic biofuels manufacture in the European context: a techno-economic assessment. Energy 104:199–212
Hansen AC, Zhang Q, Lyne PWL (2005 Feb) Ethanol-diesel fuel blends—a review. Bioresour Technol 96(3):277–285
Hassan MH, Kalam MA (2013) An overview of biofuel as a renewable energy source: development and challenges. Proc Eng 56:39–53
Henrich E, Dahmen N, Dinjus E (2009) Cost estimate for biosynfuel production via biosyncrude gasification. Biofuels Bioprod Biorefin 3(1):28–41
Hughes SR, Gibbons WR, Moser BR, Rich JO (2013) Sustainable multipurpose biorefineries for third-generation biofuels and value-added co-products. In: Biofuels—Economy, Environment and Sustainability. InTech, London
Jain S (2019) The current and future perspectives of biofuels. In: Biomass, biopolymer-based materials, and bioenergy. Elsevier, Amsterdam, pp 495–517
Kaul S, Porwal J, Garg MO (2010) Parametric study of Jatropha seeds for biodiesel production by reactive extraction. JAOCS J Am Oil Chem Soc 87(8):903–908
Kaul S, Sharma G, Porwal J, Bisht N (2011) Effect of low frequency ultrasonic assisted extraction on the quality of seed oils of Indian origin. Fuel Process Technol 92(10):1813–1820
Liu G, Larson ED, Williams RH, Kreutz TG, Guo X (2011) Making Fischer–Tropsch fuels and electricity from coal and biomass: performance and cost analysis. Energy Fuel 25(1):415–437
Ma F, Hanna MA (1999) Biodiesel production: a review1Journal Series #12109, Agricultural Research Division, Institute of Agriculture and Natural Resources, University of Nebraska–Lincoln.1. Bioresour Technol 70(1):1–15

Owusu PA, Asumadu-Sarkodie S (2016) A review of renewable energy sources, sustainability issues and climate change mitigation. Dubey S, editor. Cogent Eng 3(1):1167990

Parry I, Pittel K, Vollebergh H (2017) Energy tax and regulatory policy in Europe: reform priorities. MIT Press, Cambridge

Popp J, Lakner Z, Harangi-Rákos M, Fári M (2014) The effect of bioenergy expansion: food, energy, and environment. Renew Sust Energ Rev 32:559–578

Porwal J, Bangwal D, Garg M, Kaul S, Harvey A, Lee J et al (2012) Reactive-extraction of pongamia seeds for biodiesel production. J Sci Ind Res (India) 71:822–828

Porwal J, Behra B, Ponnekanti N, Bangwal D, Kaul S (2015) An integrated analytical approach for the compositional evaluation of different stages of fully ripened *Jatropha curcas* seed oil. Eur J Lipid Sci Technol 117(3):398–405

Prasad S, Ingle AP (2019) Impacts of sustainable biofuels production from biomass. In: Sustainable bioenergy. Elsevier, Amsterdam, pp 327–346

Quadrelli R, Peterson S (2007) The energy–climate challenge: recent trends in CO_2 emissions from fuel combustion. Energy Policy 35(11):5938–5952

Reisch LA, Th_gersen J (2015) Handbook of research on sustainable consumption. Edward Elgar Publishing, Cheltenham

Science E (2011) Biofuels: alternative Feedstocks and conversion processes. Elsevier Science, Amsterdam

Shalaby AE (2013) Biofuel: sources, extraction and determination. In: Liquid, gaseous solid biofuels-convers. tech. InTech, London

Sher E (1998) Environmental aspects of air pollution. In: Handbook of air pollution from internal combustion engines. Elsevier, Amsterdam, pp 27–41

Silva Lora EE, Escobar Palacio JC, Rocha MH, Grillo Renó ML, Venturini OJ, Almazán del Olmo O (2011) Issues to consider, existing tools and constraints in biofuels sustainability assessments. Energy 36(4):2097–2110. Elsevier Ltd

Singh SP, Singh D (2010) Biodiesel production through the use of different sources and characterization of oils and their esters as the substitute of diesel: a review. Renew Sustain Energy Rev 14:200–216

Smith AR, Klosek J (2001) A review of air separation technologies and their integration with energy conversion processes. Fuel Process Technol 70(2):115–134. Elsevier

Tenenbaum DJ (2008) Food vs. fuel: diversion of crops could cause more hunger. Environ Health Perspect 116(6):A254–A257

Welker C, Balasubramanian V, Petti C, Rai K, DeBolt S, Mendu V (2015) Engineering plant biomass lignin content and composition for biofuels and bioproducts. Energies 8(8):7654–7676

Zarrilli S (2006) The emerging biofuels market: regulatory, trade and development implications.

Jyoti Porwal joined CSIR-IIP, Dehradun, in 2008. She has completed her doctoral degree in chemistry from UTU, Dehradun, in 2019. Her area of research interest includes biofuels, biodiesel, biolubricants, bioadditives, and applied chemistry. She also visited University of Newcastle, Newcastle upon Tyne, United Kingdom (UK) twice in 2009 and 2011, respectively, under the prestigious UKIERI collaborative Project, Standard Award SA07-0094 (2008–2011) on reactive extraction for biofuel production. Previously, she has also a research experience of 3 years at CSIR-CDRI and CSIR-CIMAP, Lucknow, respectively.

Suheel K. Porwal joined as an assistant professor in the Department of Chemistry, DIT University, Dehradun, in January 2015. After completed his doctorate in chemistry from IIT Madras in 2011, he has received a Postdoctoral Fellowship at CWRU Cleveland, Ohio, USA (2011–2014). He has also been honored with best thesis award in PhD Chemistry [Prof. C. N. Pillai Prize for Organic & Biochemistry] by IIT Madras in 2012. During his doctoral and postdoctoral study, he has published many research articles in peer-reviewed journals. His area of research interest includes green additives for fuels and lubricants, synthetic methodologies, and natural product chemistry.

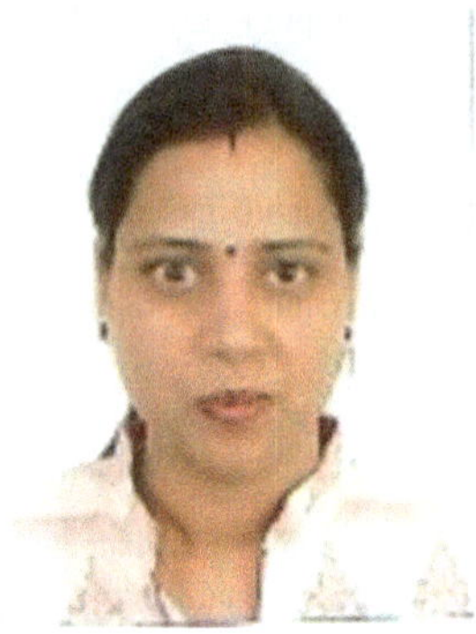

Raman Singh joined MMDU, Mullana, as an assistant professor in 2014. Before joining MMDU, she worked at National Dope Testing Laboratory, New Delhi. She has completed her doctoral research in organic chemistry from Jiwaji University, Gwalior, in 2007. She has published many research articles in peer-reviewed journals. Her area of research interest includes green synthesis, biodiesel, and bio lubricants. She is a member of Indian Science Congress Association and Graduate Women in Science, USA.

Kuldeep Singh is an associate professor at Maharishi Markandeshwar (Deemed to be University), Mullana, India. He obtained his PhD in 2007 at Indian Institute of Technology Bombay, Mumbai (India). Postdoctoral scientist with Stéphane Meunier, Laboratoire de Synthèse Bioorganique, Illkirch (France). His research activities focus on (a) the development of new synthetic methodologies using metathesis, coupling reactions, and C–H activation, (b) the design and synthesis of new molecules against antibacterial drug resistance, and (c) natural products, nanotechnology, biofuels. He is a member of the Chemical Research Society of India, the Royal Society of Chemistry, and the American Chemical Society.

Synthetic Biology and Future Production of Biofuels and High-Value Products

11

Ashwani Kumar

Abstract

Synthetic biology aims to build increasingly complex biological systems from standard interchangeable parts. The ideal microorganism for biofuel production may produce a single fermentation product and might possess high substrate utilization and processing capacities. Such microorganisms may also possess fast and deregulated pathways for sugar transport, good tolerance to inhibitors and product, and high metabolic fluxes. The choice to produce such an organism lies between engineering natural function and importing biosynthetic capacity which is affected by current progress in metabolic engineering and synthetic biology. Synthetic biology is bringing together engineers and biologists to design and build novel biomolecular components, networks, and pathways and to use these constructs to rewire and reprogram organisms. Recent findings that plant metabolic pathways can be reconstituted in heterologous hosts and metabolism in crop plants can be engineered to improve the production of biofuels have given new hope for molecular biological approaches in improving food and biofuel production. The de novo engineering of genetic circuits, biological modules, and synthetic pathways is beginning to address these crucial problems and is being used in related practical applications.

Keywords

Electron transport · Photosynthetic process · PSII · Transmembrane complex

A. Kumar (✉)
Department of Botany, University of Rajasthan, Jaipur, India

A. Kumar et al. (eds.), *Climate Change, Photosynthesis and Advanced Biofuels*,
https://doi.org/10.1007/978-981-15-5228-1_11

11.1 Introduction

Public concerns over environmental pollution, greenhouse gas emissions, and the shortage of raw oils are increasing, and considerable attention is turning toward alternative, renewable sources of chemical products to reduce both dependency on oil reserves and carbon dioxide emissions into the environment (US Energy Information Administration 2012; Arslan et al. 2012; Kawaguchi et al. 2016; Scheffers et al. 2016; Kumar et al. 2018; Kumar 2020). Analysis by Rogelj et al. (2011) confirms that if the mechanisms needed to enable an early peak in global emissions followed by steep reductions are not put in place, there is a significant risk that the 2 °C target will not be achieved. Long et al. (2015) reported the global food demand of the future by engineering crop photosynthesis and yield potential. Recent reviews on synthetic biology have provided excellent information about the development of synthetic biology (Barber 2009; Khalil and Collins 2010; Erb and Zarzycki 2016; Bhansali and Kumar 2018; Kumar et al. 2019).

The production of numerous sustainable chemicals using engineered microbes has a potential environmental impact with a significant reduction in greenhouse gas emissions (GGEs) while offering the potential of advanced products with improved properties (Wu et al. 2015; Lynch 2016).

Environmental applications of synthetic biology include microbes that sense, report, and degrade toxic chemicals (Hillson et al. 2007; Chen et al. 2014). Besides, it has the capability to produce a variety of chemical products ranging from therapeutics to plastics and biofuels (Fortman et al. 2008; Lee et al. 2012; Beller et al. 2015; Sitepu et al. 2014; Yu et al. 2015; Bhansali and Kumar 2018; French 2019).

Biofuels are environmentally friendly and sustainable sources. Their production including bioethanol, biobutanol, and biodiesel has gained considerable interest (Jiang et al. 2019). Bioethanol was regarded as one of the most promising biofuels, particularly as a carbon-neutral liquid transportation fuel (Jiang et al. 2019). Artificial microbial consortia are specifically constructed to broaden the scope of feedstocks, enhance the productivity of target bio-products, etc. (Jiang et al. 2019). Next-generation biofuels and green chemicals will be produced from lignocellulosic materials, such as agricultural residues, woody energy crops, and municipal solid waste, which are abundant and inexpensive (Carroll and Somerville 2009; Green 2011; Kumar 2020).

The natural fermentation produces alcohols such as ethanol and propanol, lacking the energy density of petroleum fuels (Mackenzie 2013). According to Connor and Atsumi (2010), some of the next-generation biofuels depend on highly precise modification and can produce energy-dense hydrocarbon by introduction of "foreign genes and pathways into central metabolism" of well-studied model organisms such as yeasts and bacteria (Mackenzie 2013).

Engineering of biological systems has emerged as one of the most exciting recent technologies (Nielsen and Keasling 2011; Kumar 2014; Farr et al. 2014; Guo et al. 2016; Gall et al. 2017). The complex oleochemicals that cannot be obtained from natural sources because of low abundance can be produced by introducing novel synthetic biochemical pathways into platform chassis (Marella et al. 2018).

Jang et al. (2012) reviewed systems metabolic engineering which allows systematic changes of metabolic pathways toward desired goals including enhancement of product concentration, yield, and productivity. Guo et al. (2016) reviewed the development of metabolic engineering and synthetic biology and microbial production of fatty alcohols from renewable feedstock in both *Escherichia coli* and *Saccharomyces cerevisiae*. The boundaries and overlap between metabolic engineering and synthetic biology are often blurry as practitioners often work in both fields, which also share common tools (Couto et al. 2018).

The integration of protein engineering, systems biology, and synthetic biology into metabolic engineering has extended strain engineering from local modification to system-wide optimization. Powerful omics technologies, such as genomics, transcriptomics, proteomics, and fluxomics, have been combined for in-depth understanding of glycerol metabolism and regulation of microorganism at the system level (Wang et al. 2003; Liao et al. 2011; Beckers et al. 2016; Salazar et al. 2009; Kumar 2015; Kumar et al. 2018, 2019).

11.2 Sugar Is the Next Oil

Plant metabolic pathways can be reconstituted in heterologous hosts, and metabolism in crop plants can be engineered to improve the production of biofuels. According to Sanford et al. (2016), the theme of "sugar is the next oil" connects chemical, biological, and thermochemical conversions of renewable feedstocks to products which are drop-in replacements for petroleum-derived chemicals, bio-polymers (Wang et al. 2015; Dai and Nielsen 2015), or are new to market chemicals/materials.

11.3 Bugs to Synthetic Biofuels

Lee et al. (2008) proposed the term bugs to synthetic biofuels. Gaida et al. (2016) reported for the first time the production of n-butanol directly from crystalline cellulose using a single engineered organism—*Clostridium cellulolyticum*, a bacterium. According to Becker and Wittmann (2016), *E. coli* has also entered the precious market of high-value molecules and is becoming a flexible, efficient production platform for various therapeutics, pre-biotics, nutraceuticals, and pigments. This is enabled by systems metabolic engineering concepts that integrate systems biology and synthetic biology into the design and engineering of powerful *E. coli* cell factories.

An artificial *Escherichia coli* binary culture was constructed for the direct conversion of hemicellulose into ethanol. Short-chain alcohols can also be produced in *E. coli* from 2-keto acids, common intermediates in amino acid biosynthetic pathways. By expressing genes in *E. coli*, six short-chain alcohols including

1-propanol, 1-butanol, isobutanol, 2-methyl-1-butanol, 3-methyl-1-butanol, and 2-phenylethanol were produced by non-fermentative pathways (Atsumi et al. 2008a, b; Liao et al. 2016).

11.3.1 Xylose Utilization

Efficient xylose utilization is one of the most important prerequisites for developing an economic microbial conversion process of terrestrial lignocellulosic biomass into biofuels and biochemical (Kwak and Jin 2017). Kwak and Jin (2017) reported a robust ethanol-producing yeast *Saccharomyces cerevisiae* has been engineered with heterologous xylose assimilation pathways. A two-step oxidoreductase pathway consisting of NAD(P)H-linked xylose reductase and NAD^+-linked xylitol dehydrogenase and a one-step isomerase pathway using xylose isomerase have been employed to enable xylose assimilation in engineered *S. cerevisiae* (Alper and Stephanopoulos 2009) (Fig. 11.1).

11.3.2 Xylose Fermenting

Native *Saccharomyces cerevisiae* (Scer) does not consume xylose but can be engineered for xylose consumption with a minimal set of assimilation enzymes, including xylose reductase (Xyl1) and xylitol dehydrogenase (Xyl2) from the xylose-fermenting *Pichia stipitis* (Psti) (Jeffries 2006; Van Vleet and Jeffries 2009). However, xylose fermentation remains slow and inefficient in Scer, especially under anaerobic conditions when NADH cannot be recycled for NAD^+-dependent Xyl2 (Jeffries 2006). Therefore, improving xylose utilization in industrially relevant yeasts is essential for producing economically viable biofuels from cellulosic material (Wohlbach et al. 2011). Yeasts engineered to ferment xylose do so slowly and cannot utilize xylose until glucose is completely consumed (Fig. 11.1). Ha et al. (2011) engineered yeasts to coferment mixtures of xylose and cellobiose (see also Diao et al. 2013).

The development of xylose-utilizing strains of *Saccharomyces cerevisiae* has improved the prospects of lignocellulosic biorefinery, enabling the creation of full-scale second-generation bioethanol production plants worldwide (Diao et al. 2013; Jansen et al. 2017). Tran et al. (2018) successfully developed a high-performance xylose-fermenting strain of *S. cerevisiae*, XUSE, through CRISPR–Cas9-mediated rational engineering and evolutionary engineering. According to Tran et al. (2018), for further engineering, XUSE could serve as a promising platform strain for lignocellulosic biorefinery (see also Estrela and Cate 2016).

11.4 Biosynthetic Pathways of Biofuels

Different pathways of carbon feedstocks are shown by Liao et al. (2016) (Fig. 11.2).

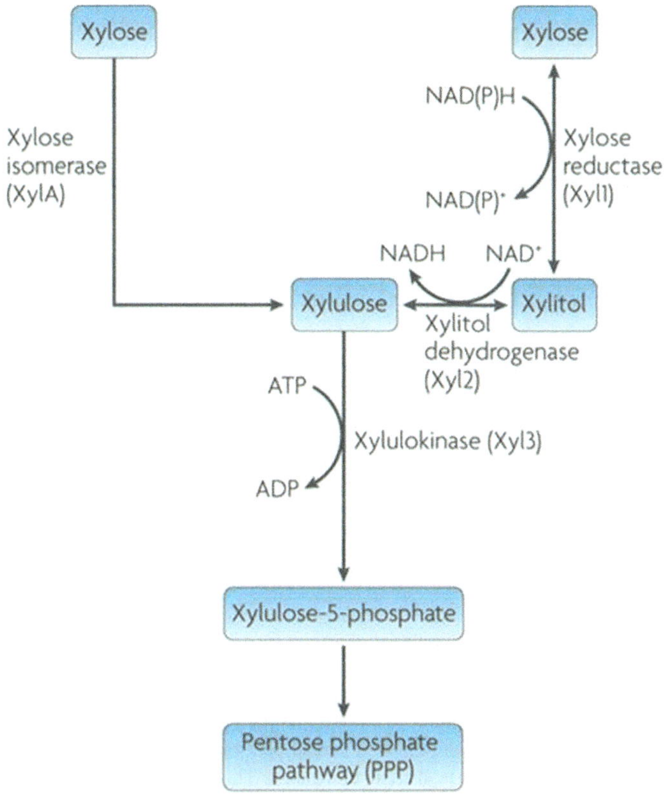

Fig. 11.1 Two routes to xylose assimilation. When xylose enters *Saccharomyces cerevisiae*, it can be incorporated into the pentose phosphate pathway through either the three-enzyme pathway containing a xylitol intermediate or a two-step process that uses a fungal or bacterial xylose isomerase gene. The two-step process bypasses the need for the reducing power that is incorporated in NAD- and NADP-reducing partners and has been shown to improve ethanol production. Xylulose 5-phosphate is formed by both pathways and can enter into central carbon metabolism through the transketolase and transaldolase reactions. (Source: Alper, H. & Stephanopoulos, G. (2009). Engineering for biofuels: exploiting innate microbial capacity or importing biosynthetic potential. *Nature Reviews. Microbiology 7*: 715–723. Retrieved from https://doi.org/10.1038/nrmicro2186. Reproduced with license number 46456408400514)

Different pathways can be assembled to produce molecules not currently used as fuels, but with likely suitable properties, including fatty alcohols (Steen et al. 2010; Feng et al. 2014), methyl ketones (Goh et al. 2012, 2014), γ-hydroxy and dicarboxylic acids (Clomburg et al. 2015), and other fatty acid-derived products.

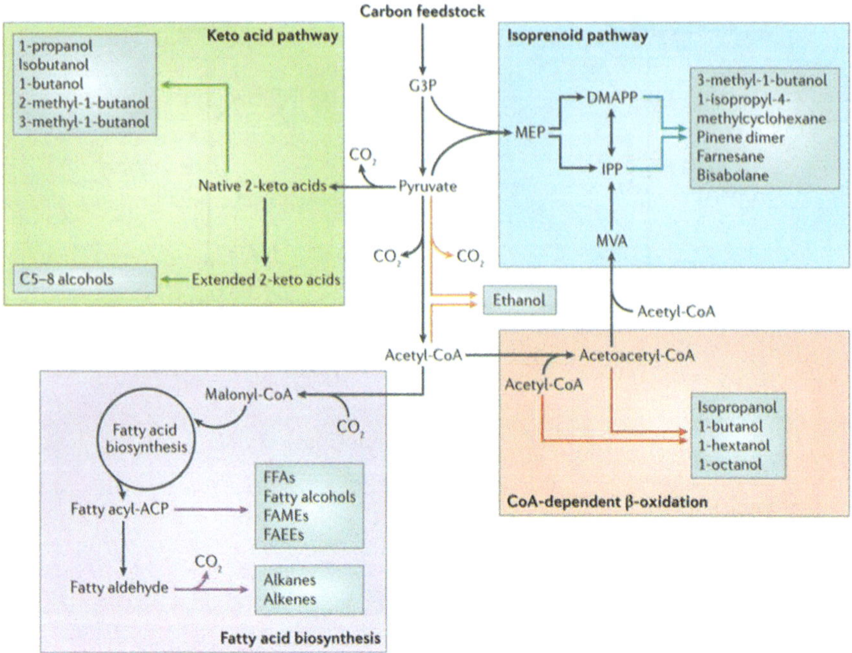

Fig. 11.2 Biosynthetic pathways of biofuels. Ethanol is produced from either pyruvate or acetyl-CoA (orange arrows), with acetaldehyde as a common intermediate. The keto acid pathway (green arrows) can be used to produce both branched and straight-chain alcohols. It uses parts of amino acid biosynthesis pathways for keto acid chain elongation. This is followed by decarboxylation and reduction of the keto acid, analogous to the conversion of pyruvate to ethanol. Fatty acid synthesis (purple arrows) extends acyl-acyl carrier proteins (ACPs) in a cyclical manner, using malonyl-CoA as a precursor. Fatty acyl-ACPs may be converted into free fatty acids (FFAs) with acyl-ACP thioesterase. FFAs can be esterified to esters, such as fatty acid methyl esters (FAMEs) or fatty acid ethyl esters (FAEEs), reduced to fatty alcohols, or reduced to fatty aldehydes followed by decarbonylation to alkanes and alkenes. The CoA-dependent pathway (red arrows) uses reverse β-oxidation chemistry for the production of higher alcohols or decarboxylation of the precursor acetoacetyl-CoA for the production of isopropanol. Isopentenyl pyrophosphate (IPP) and dimethylallyl pyrophosphate (DMAPP), the universal precursors of isoprenoid biofuel biosynthesis (blue arrows), may be produced either through the mevalonate (MVA) or methylerythritol 4-phosphate (MEP) pathway. *G3P* glyceraldehyde-3-phosphate. Metabolic engineering for the production of biofuels has been reviewed by Kumar (2010), Kumar (2013), and Kumar (2015). (Source: Liao, J. C., Mi, L., Pontrelli, S., & Luo, S. (2016). Fuelling the future: microbial engineering for the production of sustainable biofuels. *Nature Publishing Group*, Nature Review. Microbiology *14*(5): 288–304. https://doi.org/10.1038/nrmicro.2016.32. Reproduced under license number 4645730007098)

11.5 Metabolic Engineering

Martien and Amador-Noguez (2017) suggested the major goals of metabolic engineering for microbial biofuel production are (1) to direct metabolic flux toward maximum biofuel generation, (2) to enable the use of economical feedstock such as lignocellulose, and (3) to improve stress tolerance to inhibitors produced during pre-processing or biofuel production (Fig. 11.3). Metabolic engineering is a process of optimizing native metabolic pathways and regulatory networks or assembling heterologous metabolic pathways for the production of targeted molecules using molecular, genetic, and combinatorial approaches (Zhu and Jackson 2015). A common strategy of metabolic engineering is to increase the endogenous supply of

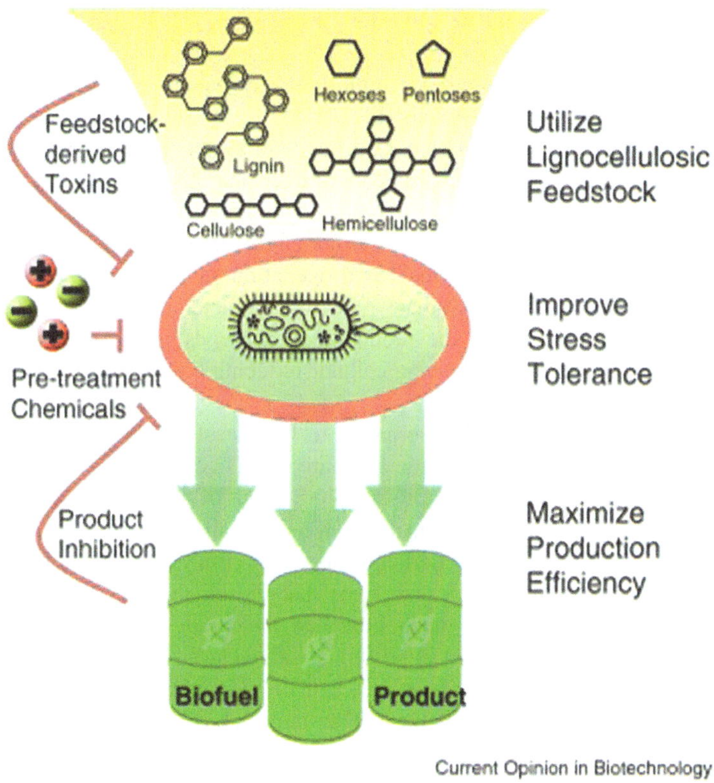

Fig. 11.3 The major goals of metabolic engineering for microbial biofuel production are (1) to direct metabolic flux toward maximum biofuel generation, (2) to enable the use of economical feedstock such as lignocellulose, and (3) to improve stress tolerance to inhibitors produced during pre-processing or biofuel production. The studies featured in this review apply knowledge gained from metabolomics-based methods to achieve these goals. (Source: Martien J.I., and Amador-Noguez D. (2017). Recent applications of metabolomics to advance microbial biofuel production. *Current Opinion in Biotechnology* 43: 118–126. https://doi.org/10.1016/j.copbio.2016.11.006. Reproduced with permission Licence number 4666750205840)

precursor metabolites to improve pathway productivity (Leonard et al. 2010). Several excellent reviews on systems metabolic engineering and synthetic biology have highlighted the motivation and need for pathway balancing (Lee et al. 2008; Völler and Budisa 2017).

Maximizing microbial biofuel production from plant biomass (i.e., lignocellulosic biomass or plant dry matter) requires reprogramming metabolism to ensure a seamless supply of carbon, energy (e.g., ATP), and reducing power (e.g., NAD(P)H) toward engineered biofuel pathways (Martien and Amador-Noguez 2017). Nature exploits a very limited set of just 20 canonical alpha-L-amino acids (cAAs) for the ribosomal translation of peptides and proteins. Reprogramming this process enables the incorporation of additional ncAAs capable of delivering a variety of novel chemical and biophysical properties into target proteins or protein-based complex structures (Agostini et al. 2017). Significant progress has been achieved in understanding and engineering the de novo lipid biosynthesis in *Y. lipolytica* (Zhu and Jackson 2015).

Jones et al. (2015) reviewed metabolic pathway balancing and its role in the production of biofuels and chemicals (Fig. 11.4).

Chae et al. (2017) reviewed recent advances in systems metabolic engineering which analyzes various omics data together, rather than just a single type of omics. The multiomics approach can be used to elucidate various phenomena in a metabolically engineered strain and to identify further engineering targets.

They further resorted to chemical hydrogenation of bisabolene into the final product bisabolane with the ultimate goal of complete microbial production of bisabolane. This will require the reduction of terpenes in vivo using designer reductases and, potentially, balancing cellular reducing equivalents (Peralta-Yahya et al. 2011).

Bisabolane as a biosynthetic alternative to D2 diesel fuel. Peralta-Yahya et al. (2011) identified a novel biosynthetic alternative to D2 diesel fuel, bisabolane, and engineered microbial platforms for the production of its immediate precursor, bisabolene (Fig. 11.5). Peralta-Yahya and Keasling 2010 hypothesized that for a fully reduced monocyclic sesquiterpene, bisabolane may serve as a biosynthetic alternative to diesel (Figs. 11.5 and 11.6). D2 diesel, the fuel for compression ignition engines, is a mixture of linear, branched, and cyclic alkanes with an average carbon length of 16 (Fortman et al. 2008). Bisabolane has a carbon length (C15) close to the average carbon length of diesel (C16). To our knowledge, there are no reports of bisabolane as a biosynthetic alternative to D2 diesel. Source: Peralta-Yahya, P. P., Ouellet, M., Chan, R., Mukhopadhyay, A., Keasling, J. D., & Lee, T. S. (2011). Identification and microbial production of a terpene-based advanced biofuel. *Nature Communications* 2: 483–488. https://doi.org/10.1038/ncomms1494.

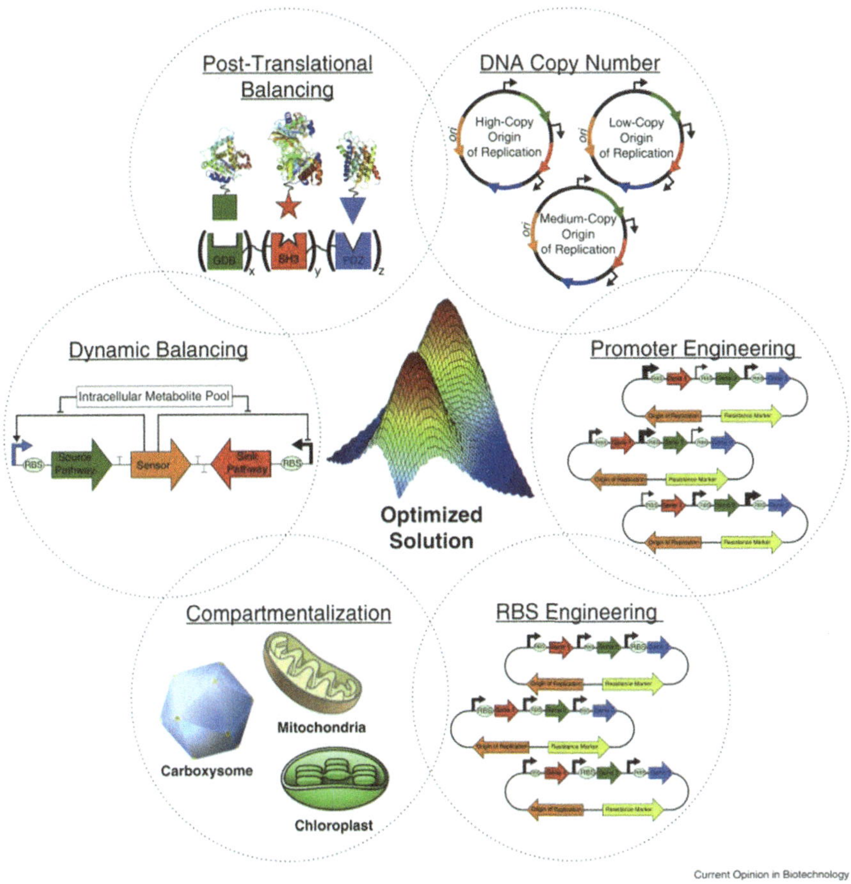

Fig. 11.4 Six major approaches to optimize metabolic pathways in common laboratory organisms such as *E. coli* and *S. cerevisiae*. The left- and right-hand sides of the figure represent modern and classical approaches, respectively. Modern techniques can be summarized as dynamic metabolite monitoring and balancing through critical intermediate chemicals, spatial organization of enzymes by using synthetic scaffolds or fusion proteins, and organelle-level compartmentalization of both metabolites and pathway enzymes to take advantage of elevated concentrations of substrates and enzymes. On the other hand, classical techniques include utilizing plasmid copy number or chromosomal integration modularity by combinational approach; gene expression level control through promoter engineering, including synthetic hybrid promoters (e.g., regulation through toxic chemicals or specific precursors); and lastly, ribosome binding site engineering for each different pathway gene to optimize and normalize their translational efficiencies. (Source: Jones, J.A., Ö. Duhan Toparlak and Mattheos AG Koffas (2015). Metabolic pathway balancing and its role in the production of biofuels and chemicals. *Current Opinion in Biotechnology*, *33*, 52–59. https://doi.org/10.1016/j.copbio.2014.11.013. Reproduced with permission no 4671031226483)

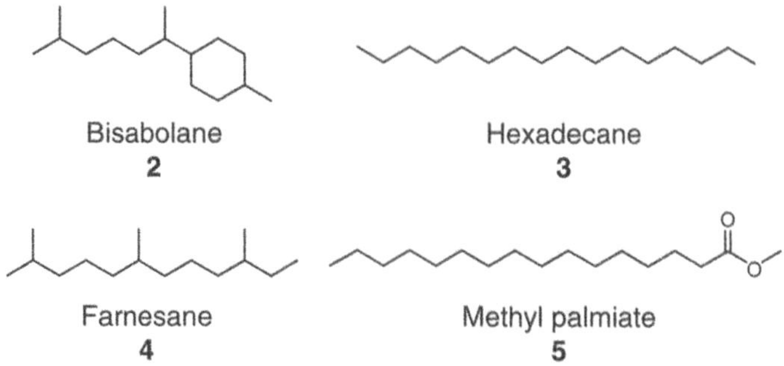

Fig. 11.5 Chemical structures of fuels. Bisabolane (**2**); hexadecane (**3**), a representative molecule for diesel; farnesane (**4**); and methyl palmitate (**5**), a representative molecule for fatty acid methyl esters. (Source: Peralta-Yahya, P. P., Ouellet, M., Chan, R., Mukhopadhyay, A., Keasling, J. D., & Lee, T. S. (2011). Identification and microbial production of a terpene-based advanced biofuel. *Nature Communications* 2: 483–488. https://doi.org/10.1038/ncomms1494. This is an open-access article distributed under the terms of the Creative Commons CC-BY license, which permits unrestricted use, distribution, and reproduction in any medium, provided the original work is properly cited)

11.5.1 Lycopene

Ma et al. (2019) established a heterologous lycopene pathway in strain YZL141 (Fig. 11.2) by genomic integration of genes encoding GGPP synthase (CrtE), phytoene synthase (CrtB), and phytoene desaturase (CrtI) from different sources. Ma et al.'s (2019) findings are the first, describing lipid-metabolic engineer to promote lycopene overproduction in a non-oleaginous organism (Figs. 11.7 and 11.8).

Using systematic traditional engineering methods, Ma et al. (2019) established high-yield heterologous lycopene biosynthesis in *S. cerevisiae*. Their results confirmed the successful development of an oleaginous biorefinery platform in *S. cerevisiae* that enabled the efficient overproduction of the intracellular lipophilic natural product lycopene.

Efforts to increase terpenoid production in *E. coli* previously focused on (1) overexpression of pathway enzymes and (2) optimizing the expression of enzymes by codon bias (Leonard et al. 2010; Lindberg et al. 2009; Dueber et al. 2009; Tyo et al. 2009). Thus, in addition to metabolic engineering, the molecular reprogramming of key metabolic nodes such as prenyltransferase (GGPPS) and terpenoid synthase (LPS) through protein engineering is required to achieve substantial overproduction of a desired terpenoid product (Keeling and Bohlmann 2006; Tholl 2006; Keeling and Bohlmann 2006; Christianson 2008; Leonard et al. 2010; Peralta-Yahya and Keasling 2010; Kumar 2013).

There are two main precursors which are isopentenyl pyrophosphate (IPP) and dimethylallyl pyrophosphate (DMAPP). There are two pathways to generate isoprenoids: the mevalonic acid pathway (MVA, for some bacteria, plants, and higher eukaryotes) and the 2-C-methyl-d-erythritol 4-phosphate/1-deoxy-d-

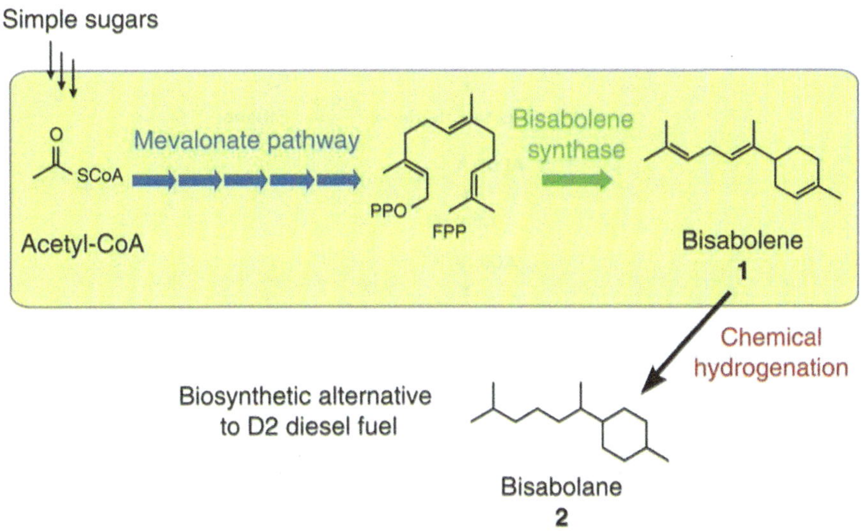

Fig. 11.6 Bisabolane from chemical hydrogenation of microbially produced bisabolene. The engineered microbe (yellow box) converts simple sugars into acetyl-CoA via primary metabolism. A combination of metabolic engineering of the heterologous mevalonate pathway to convert acetyl-CoA into FPP and enzyme screening to identify a terpene synthase to convert FPP into bisabolene (1) is used to produce bisabolene at high titers. Chemical hydrogenation of biosynthetic bisabolene leads to bisabolane (2), a biosynthetic alternative to D2 diesel. (Source: Peralta-Yahya, P.P., Ouellet, M., Chan, R., Mukhopadhyay, A., Keasling, J.D. and Lee, T.S. (2011). Identification and microbial production of a terpene-based advanced biofuel. *Nature Communications 2:* 483–488. https://doi.org/10.1038/ncomms1494. This is an open-access article distributed under the terms of the Creative Commons CC-BY license, which permits unrestricted use, distribution, and reproduction in any medium, provided the original work is properly cited)

xylulose5-phosphate pathway (DXP, for plants and most of the bacterial strains). The end products of both pathways are the precursors of all terpenoids, some with pharmaceutical relevance such as taxol, artemisinin, and lycopene (Figs. 11.7 and 11.8).

The fully reduced form of the linear terpene farnesene is being pursued as an alternative biosynthetic diesel in the market (Renninger and McPhee 2008).

Generally, butanol was synthesized through traditional acetone–butanol–ethanol (ABE) fermentation process by solventogenic *Clostridium* sp. (Jin et al. 2011; Campos-Fernández et al. 2012; Zheng et al. 2015; Trindade and Santos 2017; Sun et al. 2018; Shanmugam et al. 2018). However, according to Jiang et al. (2018), most *Clostridia* could not directly utilize polysaccharides, such as lignocellulose due to the inexpression of polysaccharide-degrading enzymes. Even though metabolic engineering has provided different alternatives such as improved solvent tolerance and non-acetone-forming strains, systems biology-guided strain engineering and synthetic biology can lead to sustained industrial viability (Birgen et al. 2019).

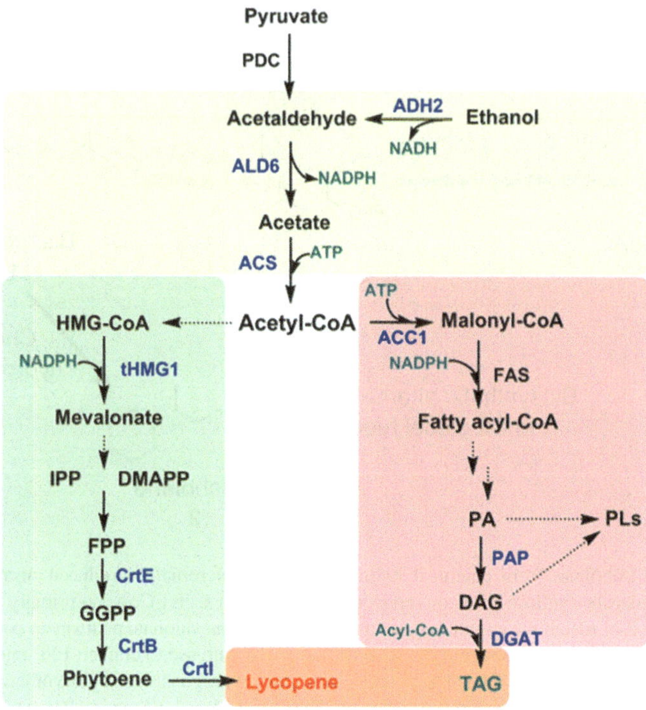

Fig. 11.7 Simplified schematic representation of key fluxes in lycopene biosynthesis coupled with TAG metabolism in *S. cerevisiae*. The acetyl-CoA-producing pathway is highlighted in a yellow rectangle. Reactions associated with TAG synthesis are highlighted in a red rectangle. Lycopene-biosynthetic flux is highlighted in a green rectangle. *PDC* pyruvate decarboxylase, *ADH2* alcohol dehydrogenase, *ALD6* acetaldehyde dehydrogenase, *ACS* acetyl-CoA synthetase, *tHMG1* truncated 3-hydroxy-3-methylglutaryl-CoA reductase, *CrtE* geranylgeranyl diphosphate synthase, *CrtB* phytoene synthase, *CrtI* phytoene desaturase, *ACC1* acetyl-CoA carboxylase, *FAS* fatty acyl-CoA synthetases, *PAP* phosphatidate phosphatase, *DGAT* acyl-CoA: diacylglycerol acyltransferase, *HMG-CoA* 3-hydroxy-3-methyl-glutaryl-CoA, *IPP* isopentenyl diphosphate, *DMAPP* dimethylallyl diphosphate, *FPP* farnesyl diphosphate, *GGPP* geranylgeranyl diphosphate, *PA* phosphatidic acid, *PLs* phospholipids, *DAG* diacylglycerol, *TAG* triacylglycerol. (Source: Ma, T., Shi, B., Ye, Z., Li, X., Liu, M., Chen, Y. & Nielsen, J. (2019). Lipid engineering combined with systematic metabolic engineering of *Saccharomyces cerevisiae* for high-yield production of lycopene. *Metabolic Engineering 52*: 134–142. https://doi.org/10.1016/j.ymben.2018.11.009. Reproduced under license number 4651230668162)

Isobutanol which is a promising second-generation biofuel candidate is already formed as a by-product in fermentations with the yeast *Saccharomyces cerevisiae*, although only in very small amounts (Hammer and Avalos 2017; Wess et al. 2019). Wess et al. (2019) reported that overexpressing a cytosolic isobutanol synthesis pathway and by blocking non-essential isobutanol competing pathways, they could achieve the highest yield ever obtained with *S. cerevisiae* in shake flask cultures.

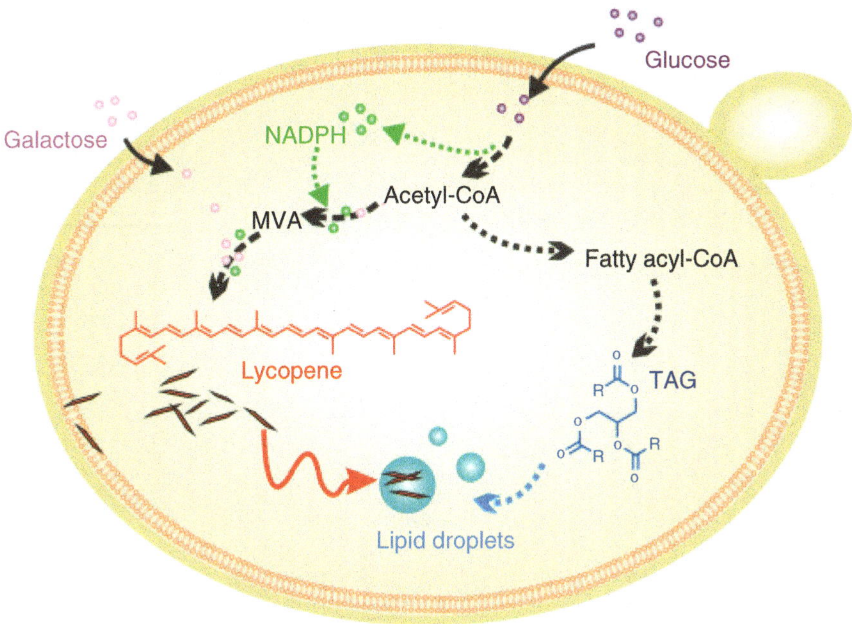

Fig. 11.8 Lycopene biosynthesis in *S. cerevisiae. S. cerevisiae* takes up glucose from the extracellular environment, and glucose metabolism results in acetyl-CoA accumulation and the release of NADPH. For lycopene production, acetyl-CoA is used in the endogenous MVA pathway and heterologous carotenoid pathway. Lycopene is distributed in lipid structures (e.g., phospholipid membranes and LDs). For TAG production, acetyl-CoA is used for endogenous fatty acid biosynthesis. TAGs are incorporated into LDs to store energy and dissolve lycopene crystals. Purple spheres represent glucose particles, pink spheres represent galactose, green spheres represent NADPH, and blue spheres represent LDs. Dotted lines represent multiple reactions. (Source: Ma, T., Shi, B., Ye, Z., Li, X., Liu, M., Chen, Y., and Nielsen, J. (2019). Lipid engineering combined with systematic metabolic engineering of *Saccharomyces cerevisiae* for high-yield production of lycopene. *Metabolic Engineering 52*: 134–142. https://doi.org/10.1016/j.ymben.2018.11.009. Reproduced under license number 4651230668162 from RightsLink)

11.5.2 Production of Fatty Acid- and Polyketide-Derived Biofuels

Recently, with the development of metabolic engineering and synthetic biology, microbial production of fatty alcohols from renewable feedstock has been achieved successfully in *E. coli*. Metabolic pathways used for the production of fatty acid- and polyketide-derived biofuels have been presented by Peralta-Yahya et al. (2012) (Fig. 11.9).

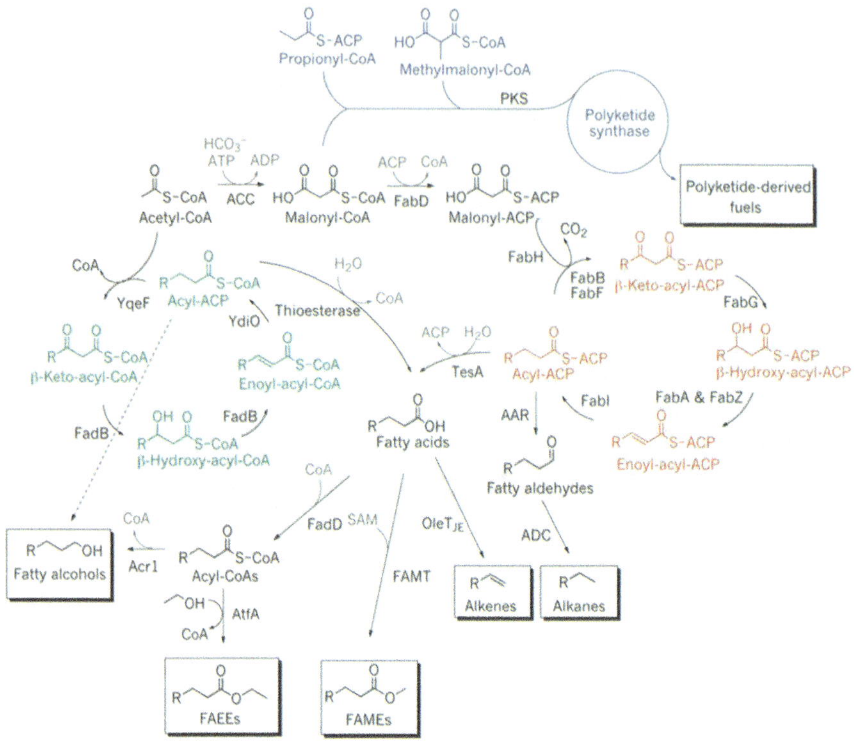

Fig. 11.9 Metabolic pathways used for the production of fatty acid- and polyketide-derived biofuels. The fatty acid biosynthetic cycle is in red, the reversal of the β-oxidation cycle is in green, and polyketide synthase is in blue. *AAR* acyl-ACP reductase, *ACC* acetyl-CoA carboxylase, *Acrl* acyl-CoA reductase, *ADC* aldehyde decarbonylase, *AtfA* wax ester synthase, *FabB* β-keto-acyl-ACP synthase I, *FabD* malonyl-CoA:ACP transacylase, *FabF* β-keto-acyl-ACP synthase II, *FabG* β-keto-acyl-ACP reductase, *FabH* β-keto-acyl-ACP synthase III, *FabA and FabZ* β-hydroxyacyl-ACP dehydratase, *FabI* enoyl-acyl-ACP reductase, *FadB* enoyl-CoA hydratase/3-hydroxyacyl-CoA dehydrogenase, *FadD* acyl-CoA synthase, *FAMT* fatty acid methyltransferase, *$OleT_{JE}$* *Jeotgalicoccus* sp. terminal olefin-forming fatty acid decarboxylase, *PKS* polyketide synthase, *TesA* acyl-ACP thioesterase, *YdiO* enoyl-CoA reductase, *YqeF* thiolase. (Source: Peralta-Yahya P.P. et al.(2012). Microbial engineering for the production of advanced biofuels. https://doi.org/10.1038/nature488320-328. Reproduced with license no. 4643340791481)

11.5.3 Synthetic Enzymatic Pathways for the Production of High-Yield Hydrogen

Natural and genetically modified microorganisms cannot produce hydrogen with a yield of more than 4 H_2 per glucose, that is, the Thauer limit (Thauer et al. 2008; Zhang 2011, 2015) (Fig. 11.10), although a theoretical yield is 12 H_2 per glucose. Nature cannot evolve such high-yield hydrogen generation pathways due to two reasons. First, the theoretical yield of hydrogen production is an endothermic reaction so that it cannot co-generate ATP. Second, if a small fraction of reduced

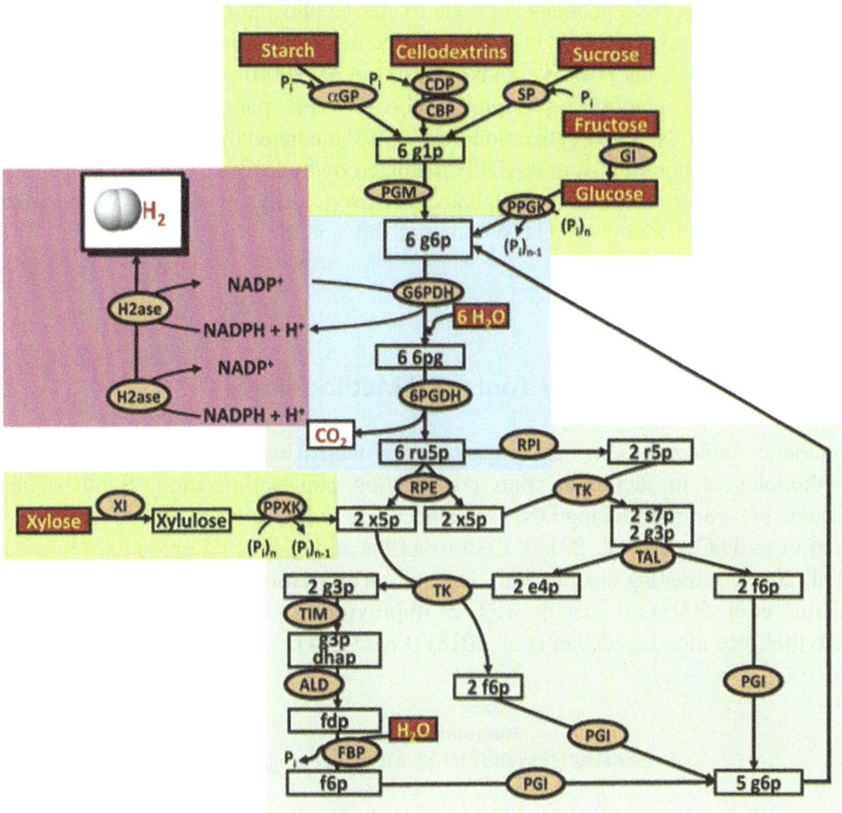

Fig. 11.10 Scheme of in vitro synthetic enzymatic pathways for the production of high-yield hydrogen from a variety of carbohydrates—starch, cellodextrin, sucrose, glucose, fructose, and xylose as well as water. The pathways are compiled and modified from References: Martín del Campo et al. 2013; Myung et al. 2014; Rollin et al. 2016; Ye et al. 2009; Zhang et al. 2007). The enzymes are αGP, alpha-glucan phosphorylase; CDP, cellodextrin phosphorylase; CBP, cellobiose phosphorylase; SP, sucrose phosphorylase; GI, glucose isomerase; XI, xylose isomerase; PPGK, polyphosphate glucokinase; PPXK, polyphosphate xylulokinase; PGM, phosphoglucomutase; G6PDH, glucose-6-phosphate dehydrogenase; 6PGDH, 6-phosphogluconate dehydrogenase; RPI, ribose 5-phosphate isomerase; RPE, ribulose-5-phosphate 3-epimerase; TK, transketolase; TAL, transaldolase; TIM, triose phosphate isomerase; ALD, (fructose-bisphosphate) aldolase; FBP, fructose bisphosphatase; PGI, phosphoglucose isomerase; and H_2ase. P_i and $(P_i)_n$ are inorganic phosphate and polyphosphate with a degree of polymerization of n. The metabolites are g1p, glucose-1-phosphate; g6p, glucose-6-phosphate; ru5p, ribulose 5-phosphate; x5p, xylulose 5-phosphate; r5p, ribose 5-phosphate; s7p, sedoheptulose 7-phosphate; g3p, glyceraldehyde 3-phosphate; e4p, erythrose 4-phosphate; dhap, dihydroxyacetone phosphate; fdp, fructose-1,6-diphosphate; and f6p, fructose 6-phosphate. (Source: Zhang, Y. P. (2015). Production of biofuels and biochemicals by in vitro synthetic biosystems: Opportunities and challenges. *Biotechnology Advances 33*(7): 1467–1483. https://doi.org/10.1016/j.biotechadv.2014.10.009. Reproduced with license number 4652950482642)

NAD(P)H was used to generate ATP via oxidative phosphorylation (Swartz 2013), the presence of oxygen would inhibit oxygen-sensitive hydrogenase activity greatly.

Woodward and his coworkers (Woodward et al. 2000) produced nearly 12 H_2 from the costly glucose 6-phosphate (G-6-P). This pathway comprised three modules: (1) two NADPH generation from G-6-P mediated by two dehydrogenases, (2) hydrogen generation from NADPH mediated by hydrogenase, and (3) regeneration of G-6-P from ribulose 5-phosphate. However, costly substrate G-6-P prevents its potential application so that Woodward did not file a patent for this in vitro synthetic pathway.

11.5.4 Synthetic Biology Tools and Methodologies

Synthetic biology today encompasses an increasing number of tools and methodologies to facilitate strain construction and optimization. Synthesizing, sequencing, and introducing DNA sequences into living cells are cheaper and easier than ever (DiCarlo et al. 2013). Codon optimization, directed evolution (Korman et al. 2013), screening enzyme libraries, and incorporating non-natural amino acids (Cirino et al. 2003) all provide ways of improving or generating novel enzymatic activities (see also Jagadevan et al. 2018) (Fig. 11.11).

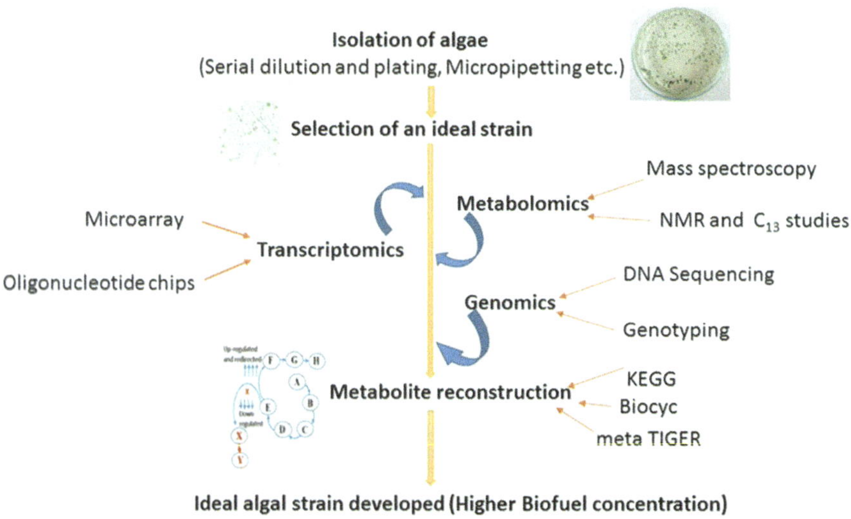

Fig. 11.11 Pictorial representation of the overall process toward biofuel production in microalgae using synthetic biology approach (i.e., isolation, selection of an ideal strain, redirecting the metabolism to maximize synthesis of the targeted biofuel). (Source: Jagadevan, S., Banerjee, A., Banerjee, C., Guria, C., Tiwari, R., & Baweja, M. (2018). Biotechnology for Biofuels Recent developments in synthetic biology and metabolic engineering in microalgae towards biofuel production. *Biotechnology for Biofuels 11*: 1–21. https://doi.org/10.1186/s13068-018-1181-1. Used under creative commons license)

The major challenge of the modern era is the transition to a bio-based economy. Biofuels are a key part of this landscape, but challenges to efficiently and cost-effectively produce biofuels still remain (Tyner 2012; Taheripour et al. 2012).

The standard of skill and expertise in synthetic biology and metabolic engineering has made significant strides over the past 25 years, and now the production of numerous chemical products with a range of market applications is available (Lynch 2016). Tatsis and O'Connor (2016) demonstrated with examples how the metabolic pathways of plants can be successfully harnessed using several metabolic engineering approaches. According to O'Connor (2015), one approach to harness plant metabolic pathways is to reconstitute the biosynthetic genes into a heterologous organism.

Hybrid processes: Hybrid processes combine the biochemical and chemical processes to enhance competitiveness of bio-based products (Beerthuis et al. 2015) such as polymers, and bioplastics will grow their market share by synergizing and collaborating with the chemical process industry (Babu et al. 2013). Creating the necessary process flow sheets, assessing cost sensitivities, and identifying bottlenecks upfront by the use of modeling, simulation, and techno-economic analysis will aid in a successful scale-up (Earhart et al. 2012; Claypool and Ramon 2013; Claypool et al. 2014; Harrison et al. 2015).

Reducing cell wall digestibility: Lignin concentration also increases with the maturation of plants and is associated with reduced cell wall digestibility (Jung and Deetz 1993). Cell wall lignification creates an access barrier to potentially digestible wall material by microorganisms if cells have not been physically ruptured. Traditional breeding has focused on increasing total dry matter digestibility rather than cell wall digestibility, which has resulted in minimal reductions in cell wall lignification (see Kumar et al. 2018). While major reductions in lignin concentration have been associated with poor plant fitness, smaller reductions in lignin provided measurable improvements in digestibility without significantly impacting agronomic fitness (Jung et al. 2012; see also Kumar et al. 2018).

The engineering of proteins along with pathways is the key strategy in achieving microbial biosynthesis and overproduction of pharmaceutical, chemical products, and biofuels.

11.5.5 Exploiting Diversity and Synthetic Biology for the Production of Algal Biofuels

Engineering of algal metabolism has an important role in the improvement of growth and biomass accumulation (Angermayr et al. 2009; US DOE 2010; Georgianna and Stephen 2012; Case and Atsumi 2016; Meyer et al. 2016; Shih et al. 2014). Manipulating the primary carbon-fixing enzyme Rubisco could also increase efficiency. The cultivation of algae in industrial photobioreactors or agricultural ponds aims to harvest as much solar energy as possible (Figs. 11.13 and 11.14) Efforts to improve photosynthetic efficiency have not been specific to algae; as a strategy, it has been proposed for increasing the yield of land plants to keep pace with increasing

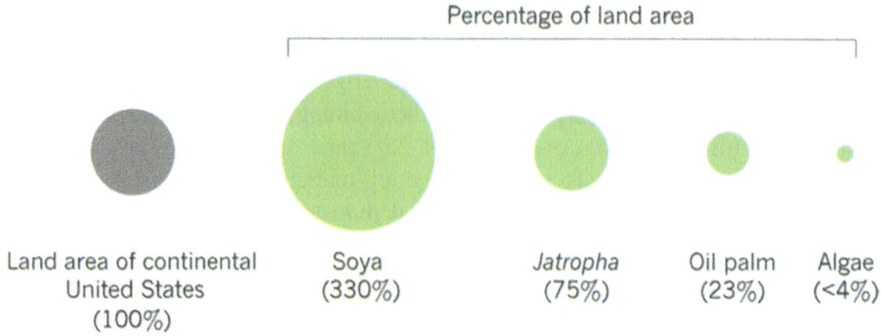

Fig. 11.12 Comparison of oleaginous crops. The United States consumes 25% of the world's petroleum. The land area needed to replace all domestic and imported petroleum used in the United States is shown as a percentage relative to the land area of the United States. The area required for algae is estimated to be significantly less than for any other biomass source (Dismukes et al. 2008). (Source: Georgianna, D. R. & Stephen, P. (2012). Exploiting diversity and synthetic biology for the production of algal biofuels, Nature *488*: 330–335. https://doi.org/10.1038/nature11479. Reproduced under license number 4646381493445 from RightsLink)

food demand where usable cropland is limited (US DOE 2010). Jagadevan et al. (2018) reviewed the upcoming field of microalgae employed as a model system for synthetic biology applications and highlighted the importance of genome-scale reconstruction models and kinetic models, to maximize the metabolic output by understanding the intricacies of algal growth (see also Georgianna and Stephen 2012) (Figs. 11.12, 11.13, and 11.14).

11.5.6 Biofuel from Protein Sources

According to Huo et al. (2011), biofuels are currently produced from carbohydrates and lipids in the feedstock. They suggested the use of proteins to synthesize fuels. Huo et al. (2011) applied metabolic engineering to generate *Escherichia coli* that can deaminate protein hydrolysates, enabling the cells to convert proteins to C4 and C5 alcohols at 56% of the theoretical yield (Huo et al. 2011) (Fig. 11.15).

Liu et al. (2017) reviewed the production of organic acids, especially carboxylic acids, as renewable sources of chemical products to substitute fossil fuels. They have been applied in a wide range of industries, including pharmaceutical, food, cosmetic, polymer, detergent, and textile (Becker and Wittmann 2016; Huo et al. 2011). The more economical and sustainable production of organic acids can be expected with the combination of these modern engineering techniques (Liu et al. 2017; Giessen and Silver 2017).

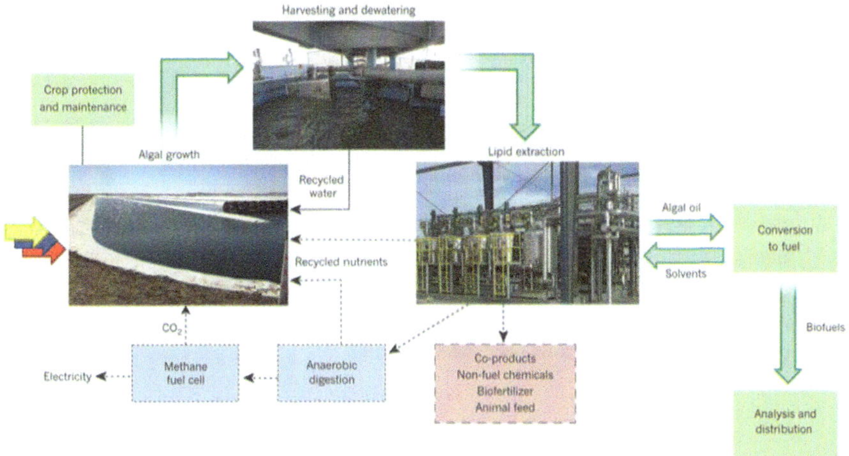

Fig. 11.13 Algal biofuel production: Light, water, and nutrients (yellow, blue, and red arrows) are required for algal growth in ponds. Some of the processes involved in algal biofuel production are common to most systems (green arrows). After fuel molecule extraction, there are alternative uses for algal biomass (dashed arrows); many of these can produce co-products that are beneficial for economic and life cycle analysis considerations. (Images courtesy of Sapphire Energy, San Diego, California). (Source: Georgianna, D. R., & Stephen, P. (2012). Exploiting diversity and synthetic biology for the production of algal biofuels, Nature *488*: 330–335. https://doi.org/10.1038/nature11479. Reproduced under license number 4646381493445 from RightsLink)

Fig. 11.14 Algae cultivation methods: (**a**) Algal ponds of 0.5 ha and 1 ha are part of the first commercial-scale algal biofuel facility in the United States at Sapphire Energy's Integrated Algal BioRefinery. They cover an area of 400 m wide by 1600 m long at a location near Columbus, New Mexico. (**b**) A single one-million-liter paddle-wheel-driven pond from the Columbus facility. (**c**) A pilot-scale flat panel photobioreactor developed at the Laboratory for Algae Research and Biotechnology at Arizona State University in Mesa (image courtesy of Q. Hu). (**d**) A commercial-scale tubular photobioreactor designed and constructed by IGV and operated by Salata in Germany (image courtesy of C. Grewe). (**e**) An industrial-scale fermentation tank for heterotrophic cultivation of microalgae at Martek Biosciences, part of DSM in Heerlen, the Netherlands (image courtesy of D. Dong). (Source: Georgianna, D. R. and Stephen, P. (2012). Exploiting diversity and synthetic biology for the production of algal biofuels. *Nature 488*: 330–335. https://doi.org/10.1038/nature11479. Reproduced under license number 4646381493445 from RightsLink)

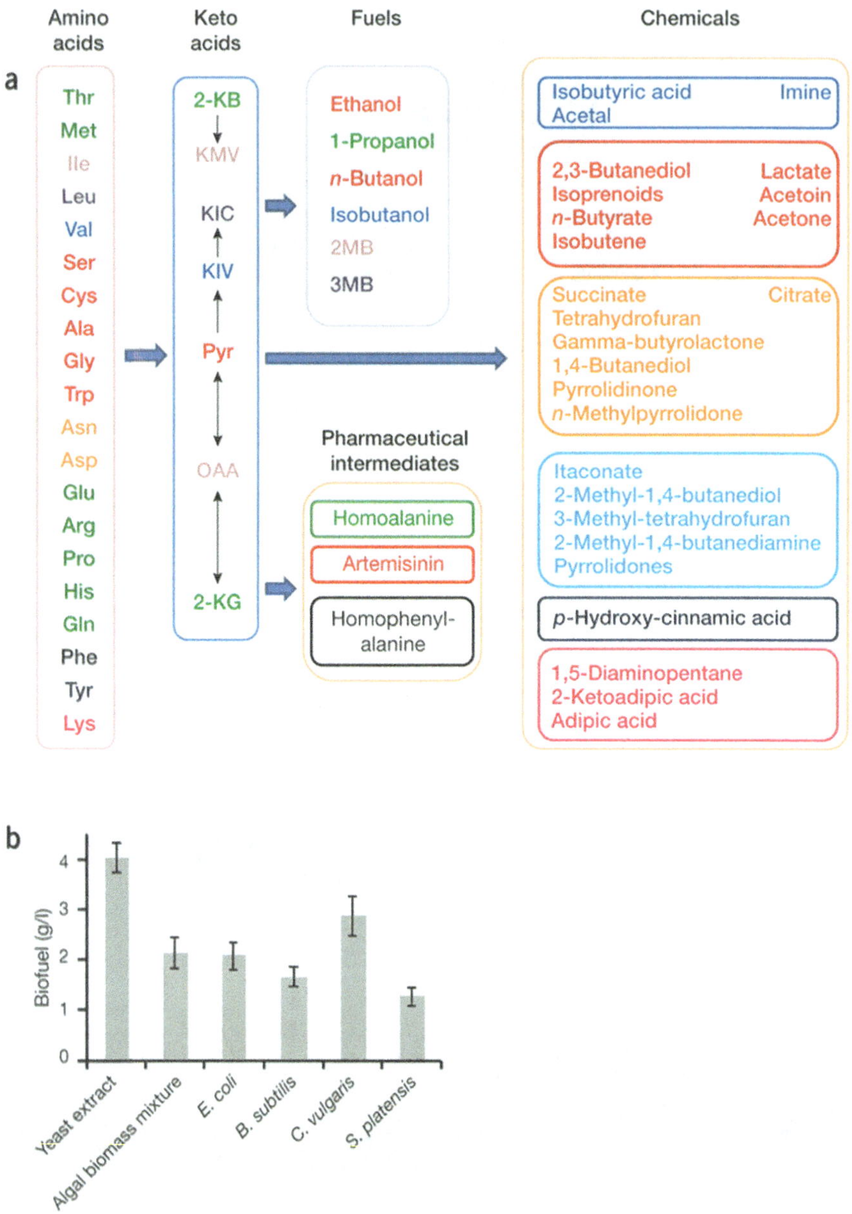

Fig. 11.15 Biofuel production and biorefining scheme from algal or bacterial protein sources: (**a**) The proposed protein-based biorefinery scheme. Amino acids are deaminated to various keto acids, which are then used to produce fuels, chemicals, and pharmaceutical intermediates. The colors link products and intermediates to the amino acids from which they are derived. (**b**) Biofuel (EtOH, iBOH, 2 MB, 3 MB) produced from the engineered *E. coli* strain YH83 grown in flasks using algal or bacterial cell hydrolysates. Small laboratory-scale reactors (1 L or 30 L) were used to grow bacterial and algal cells individually. The algal biomass mixture includes *C. vulgaris*, *P. purpureum*, *S. platensis*, and *S. elongatus*. All protein sources were adjusted to contain 21.6 g/L peptides and amino acids. Error bars indicate s.d. ($n = 3$). *OAA* oxaloacetate, *2-KB*

11.5.7 Metabolic Engineering in Methanotrophic Bacteria

Methane is 38-fold more effective at promoting global warming than carbon dioxide on a molar basis over a span of 20 years (Howarth 2015). Thus, harnessing methane is one of the most important near-term goals for biochemical engineering (Lee and Kim 2015). Methane as natural gas or biogas is the least expensive source of carbon for (bio)chemical synthesis (Kalyuzhnaya et al. 2015).

Methanotrophs are bacteria that grow on methane as their sole carbon and energy source. Methanotrophic bacteria and microbes converting methane into value-added products are both promising approaches for taking advantage of methane as a future bio-feedstock. There is resurgent interest in mitigating methane in the atmosphere as a greenhouse gas (Shindell et al. 2012) and in part its abundance, its low cost, and its potential to create liquid value-added products (Conrado and Gonzalez 2014). The activation of methane by a single species, *Methanosarcina acetivorans*, creates possibilities for metabolic engineering for anaerobic methane conversion to other products (Santos et al. 2011; Fei et al. 2014; see review Kalyuzhnaya et al. 2015; Soo et al. 2016; Mcanulty et al. 2017). It might also be possible to engineer strains that grow directly on cellulosic biomass, or other abundant and inexpensive substrates, such as methane or CO_2 (Espaux et al. 2015).

Despite these optimistic signs, a significant number of gaps in the fundamental knowledge of methanotrophy need to be filled to allow the potential of these systems to be fully reached (Kalyuzhnaya et al. 2015).

11.5.8 Engineered Microbial Biofuel Production and Recovery Under Supercritical Carbon Dioxide

Supercritical carbon dioxide ($scCO_2$) has been used for the depolymerization of lignocellulosic biomass to release fermentable sugars (Luterbacher et al. 2010). Brock et al. (2019) proposed a high-pressure fermentation strategy, coupled with in situ extraction using the abundant and renewable solvent supercritical carbon dioxide ($scCO_2$), which is also known for its broad microbial lethality to avoid end-product toxicity, culture contamination, and energy-efficient product recovery. They reported the domestication and engineering of a $scCO_2$-tolerant strain of *Bacillus megaterium*, to produce branched alcohols that have potential use as biofuels (Brock et al. 2019).

Fig. 11.15 (continued) 2-ketobutyrate. (Source: Huo, Y.-X., Cho, K. M., Rivera, J. G. L., Monte, E., Shen, C. R., Yan, Y. & Liao, J. C. (2011). Conversion of proteins into biofuels by engineering nitrogen flux. *Nature Biotechnology* 29(4): 346–351. https://doi.org/10.1038/nbt.1789. Reproduced with permission under license number 4646190098001 from RightsLink)

11.5.9 Solar-to-Chemical and Solar-to-Fuel Technology

Recent researches in solar-to-chemical and solar-to-fuel technology describe the use of solar energy to convert CO_2 to desired chemicals and fuels. The direct conversion of carbon dioxide to chemicals and fuels presents a sustainable solution for reducing greenhouse gas emissions and sustaining our supply of energy (Liao et al. 2016). According to Woo (2017), ultimately solar energy must be used for CO_2 reduction and conversions to provide a sustainable system, and this system is now available in the forms of solar-to-chemical (S2C) and solar-to-fuel (S2F) technologies. The S2C and S2F technology must be developed to capture and convert the essential feedstocks using only three inputs (CO_2, H_2O, and solar energy) to produce the desired value-added chemicals and fuels. Woo (2017) reviewed carbon capture utilization (CCU) for the reduction of greenhouse gas emissions.

Photoautotrophic cyanobacterial platforms have been extensively developed on this principle, producing a diverse range of alcohols, organic acids, and isoprenoids directly from CO_2 (Savakis and Hellingwerf 2015). Recent breakthroughs in the metabolic engineering of cyanobacteria, adoption of the light-harvesting mechanisms from nature, photovoltaics-derived water-splitting technologies have been integrated with microbial biotechnology to produce desired chemicals (Woo 2017).

Photosynthetic organisms (including cyanobacteria) have been engineered to produce value-added chemicals, providing a number of promising S2C and S2F platforms. Thus, hybrid systems comprising an electrochemical in situ hydrogen-evolution reaction at the electrode and the biological CO_2 fixation using autotrophic bacteria have been suggested as an alternative S2C and S2F platform.

11.5.10 Implementing CRISPR–Cas Technologies for Obtaining High-Value Products

Several approaches of rebalancing or rewiring of the metabolic network and the use of dynamic metabolic control strategies to conditionally reduce essential competitive fluxes have yielded better results. Liu et al. (2013) reviewed recent advances that allow more precise regulation of gene expression in plants, including synthetic promoters, transcriptional activators, and repressors.

The use of newer gene silencing technologies, including CRISPR interference, makes transcriptional tuning an attractive platform for any desired microbe (Lynch 2016). Success in using CRISPR–Cas9 for gene targeting in laboratory *S. cerevisiae* strains was first demonstrated in 2013 (DiCarlo et al. 2013) Estrela and Cate (2016) reviewed the use of CRISPR–Cas9 technology for energy biotechnology in *S. cerevisiae.* They further reported that recently, other bacteria have been successfully edited, such as *Streptomyces* (Cobb et al. 2015; Huang et al. 2015; Tong et al. 2015), *Lactobacillus reuteri* (Oh and van Pijkeren 2014), *Taumatella citrea* (Jiang et al. 2015), *Streptococcus pneumoniae*, and *E. coli* (Jiang et al. 2015).

In metabolic engineering, of photosynthetic, cyanobacteria can use CO_2 as a building block to synthesize carbon-based chemicals. In recent years, clustered regularly interspaced short palindromic repeats (CRISPR)-dependent approaches have rapidly gained popularity for engineering cyanobacteria. Behler et al. (2018) reviewed CRISPR-based tools for the metabolic engineering of cyanobacteria. Rather than utilizing CRISPR-based genome editing, CRISPR interference (CRISPRi) offers an alternative, viable approach for cyanobacterial engineering which relies on an enzymatically inactive dead Cas9 (dCas9) (Yao et al. 2016). Increased understanding of various CRISPR mechanisms and systems will undoubtedly inspire more advanced approaches for the engineering of biological hosts such as cyanobacteria (Behler et al. 2018).

Yeasts are widely used host organisms in biotechnology to produce fine chemicals, industrial biocatalysts, biopharmaceuticals, food additives, and renewable biofuels (Kim et al. 2015). Within 5 years, the CRISPR–Cas system has emerged as the dominating tool for genome engineering while also changing the speed and efficiency of metabolic engineering in conventional (*Schizosaccharomyces pombe* and *Saccharomyces cerevisiae*) and non-conventional (*Candida albicans*, *Yarrowia lipolytica*, *Pichia pastoris* syn. *Komagataella phaffii*, *Kluyveromyces lactis*, and *C. glabrata*) yeasts (Raschmanová et al. 2018).

11.6 Discussion

Metabolic pathway optimization is generally a very challenging endeavor because of the complex regulation that cells have evolved to maintain homeostasis and robustness (Nielsen and Keasling 2016: Wang et al. 2017). In vitro synthetic biosystems provide several other biomanufacturing advantages, such as easy product separation, open process control, fast reaction rate, broad reaction condition, tolerance to toxic substrates, etc. According to Lynch (2016), many challenges still remain; these recent efforts further support the potential of this discipline in making a significant impact in the production of high-volume industrial products, with the potential to displace petroleum with more sustainable alternatives. According to Woo (2017), synthetic biology-inspired metabolic engineering of next-generation microbes will be established to accommodate more efficient S2C and S2F platforms.

Hence, rather than trying to understand how synthetic biology is shaped by commercial forces, it might be better to understand sciences like synthetic biology as co-emerging with new market regimes and forms. Energy-rich parts of the world look to the Global South. As many observers have pointed out, biofuel crops compete with food crops and through deforestation reduce biodiversity more generally (Chakravorty et al. 2009; Shaik and Kumar 2014; Kumar et al. 2018).

According to Mackenzie (2013), in synthetic biology, this conflict between food and fuel is mentioned as something that must be avoided in the development of advanced biofuels by using microbes to produce fuel without relying too heavily on feedstocks or other inputs that compete with agriculture.

11.7 Conclusion

Sustainable large-scale production of biofuels will require the integration of knowledge across many disciplines. In the short term, the major research opportunities for plant biologists seem to be in identifying promising species, knowing paths of biofuel production, and altering genes to produce more or insert missing links or synthesize required protein into organisms. Large parts of next-generation biofuels exist in partial realizations: metabolic models, research projects, pilot plants, and various other technologies in testing. As the industrial reality of synthetic biology, next-generation biofuels can also prompt us to consider synthetic biology less from the perspective of epistemic value and more from the perspective of the mode of existence of technical objects.

Acknowledgments The author acknowledges support from Alexander von Humboldt Foundation (Germany) to work in the laboratory of Late Professor Dr. K-H. Neumann and later on Professor Dr. Sven Schubert at the Institut für Pflanzenernährung, Justus Liebig University, Gießen, Germany, and the Japanese Society for Promotion of Science (JSPS) fellowship to work with Professor Shinjiro Ogita during the course of investigations spreading over 20 years. Financial support from the Department of Biotechnology, Govt. of India, New Delhi, and infrastructural support for providing 50 ha land area for experimentation by the University of Rajasthan Syndicate are gratefully acknowledged. I also acknowledge my research scholars and technical staff who have worked in the lab and field with me for investigations presented here. I also thank my wife Professor Vijay R. Kumar for going through the manuscript and making suggestions.

I also acknowledge with thanks the different figures with legends reproduced with permissions from the following publications: Figure 11.1. Alper, H. & Stephanopoulos, G. (2009). Engineering for biofuels: exploiting innate microbial capacity or importing biosynthetic potential, Nature Reviews. Microbiology *7*: 715–723. Retrieved from https://doi.org/10.1038/nrmicro2186. Reproduced with license number 46456408400514. Figure 11.2. Liao, J. C., Mi, L., Pontrelli, S. & Luo, S. (2016). Fuelling the future: microbial engineering for the production of sustainable biofuels. *Nature Publishing Group*, Nature Review. Microbiology *14*(5): 288–304. https://doi.org/10.1038/nrmicro.2016.32. Reproduced under license number 4645730007098. Figure 11.3. Martien J.I. and Amador-Noguez D. (2017). Recent applications of metabolomics to advance microbial biofuel production. *Current Opinion in Biotechnology*, 43, 118–126. https://doi.org/10.1016/j.copbio.2016.11.006. Reproduced with permission license number 4666750205840. Figure 11.4. Jones, J.A., Toparlak, Duhan, Ö., and Koffas, Mattheos AG (2015). Metabolic pathway balancing and its role in the production of biofuels and chemicals. *Current Opinion in Biotechnology 33*, 52–59. https://doi.org/10.1016/j.copbio.2014.11.013. Reproduced with permission no 4671031226483. Figures 11.5 and 11.6. Peralta-Yahya, P. P., Ouellet, M., Chan, R., Mukhopadhyay, A., Keasling, J. D., & Lee, T. S. (2011). Identification and microbial production of a terpene-based advanced biofuel. *Nature Communications 2*(May): 483–488. https://doi.org/10.1038/ncomms1494. This is an open-access article distributed under the terms of the Creative Commons CC-BY license, which permits unrestricted use, distribution, and reproduction in any medium, provided the original work is properly cited. Figures 11.7 and 11.8. Ma, T., Shi, B., Ye, Z., Li, X., Liu, M., Chen, Y. & Nielsen, J. (2019). Lipid engineering combined with systematic metabolic engineering of *Saccharomyces cerevisiae* for high-yield production of lycopene. *Metabolic Engineering 52*(November 2018): 134–142. https://doi.org/10.1016/j.ymben.2018.11.009. Reproduced under license number 4651230668162. Figure 11.9. Peralta-Yahya P. P. et al. (2012). Microbial engineering for the production of advanced biofuels. https://doi.org/10.1038/nature488320–328. Reproduced with license no. 4643340791481. Figure 11.10. Zhang, Y. P. (2015). Production of biofuels and biochemicals by in vitro synthetic biosystems: Opportunities and challenges. *Biotechnology Advances 33*(7): 1467–1483. https://doi.org/10.1016/j.biotechadv.

2014.10.009. Reproduced with license number 4652950482642. Figure 11.11. Jagadevan, S., Banerjee, A., Banerjee, C., Guria, C., Tiwari, R., & Baweja, M. (2018). Biotechnology for Biofuels Recent developments in synthetic biology and metabolic engineering in microalgae towards biofuel production. *Biotechnology for Biofuels* 11: 1–21. https://doi.org/10.1186/s13068-018-1181-1. Used under creative commons license. Figures 11.12, 11.13, and 11.14. Georgianna, D. R. & Stephen, P. (2012). Exploiting diversity and synthetic biology for the production of algal biofuels. Nature *488*: 330-335. https://doi.org/10.1038/nature11479. Reproduced under license number 4646381493445. Figure 11.15. Huo, Y.-X., Cho, K. M., Rivera, J. G. L., Monte, E., Shen, C. R., Yan, Y. & Liao, J. C. (2011). Conversion of proteins into biofuels by engineering nitrogen flux. *Nature Biotechnology 29*(4): 346–351. https://doi.org/10.1038/nbt.1789. Reproduced with permission under license number 4646190098001 from RightsLink.

References

Agostini F, Voller J-S, Koksch B, Acevedo-Rocha CG, Kubyshkin V, Budisa N (2017) Xenobiology meets enzymology: exploring the potential of unnatural building blocks in biocatalysis. Angew Chem Int Ed Engl. https://doi.org/10.1002/anie.201610129. [Epub ahead of print]

Alper H, Stephanopoulos G (2009) Engineering for biofuels: exploiting innate microbial capacity or importing biosynthetic potential. Nat Rev Microbiol 7:715–723

Angermayr SA, Hellingwerf KJ, Lindblad P, de Mattos MJ (2009) Energy biotechnology with cyanobacteria. Curr Opin Biotechnol 20:257–263

Arslan D, Steinbusch KJJ, Diels L, De Wever H, Buisman CJN, Hamelers HVM (2012) Effect of hydrogen and carbon dioxide on carboxylic acids patterns in mixed culture fermentation. Bioresour Technol 118:227–234

Atsumi S, Hanai T, Liao JC (2008a) Non-fermentative pathways for synthesis of branched-chain higher alcohols as biofuels. Nature 451:86–89

Atsumi S, Canna AF, Connora MR, Shena CR, Smitha KM, Brynildsena MP, Choua KJY, Hanai T, Liao JC (2008b) Metabolic engineering of *Escherichia coli* for 1-butanol production. Metab Eng 10(6):305–311. https://doi.org/10.1016/j.ymben.2007.08.003

Babu RP, O'Connor KO, Seeram R (2013) Current progress on bio-based polymers and their future trends. Prog Biomater 2:1–16

Barber J (2009) Photosynthetic energy conversion: natural and artificial. Chem Soc Rev 38:185–196

Becker J, Wittmann C (2016) Systems metabolic engineering of *Escherichia coli* for the heterologous production of high value molecules a veteran at new shores. Curr Opin Biotechnol 42:178–188. https://doi.org/10.1016/j.copbio.2016.05.004

Beckers V, Castro IP, Tomasch J, Wittmann C (2016) Integrated analysis of gene expression and metabolic fluxes in PHA—producing *Pseudomonas putida* grown on glycerol. Microb Cell Factories 15:73

Beerthuis R, Rothenber G, Shiju NR (2015) Catalytic routes towards acrylic acid, adipic acid, and ε-caprolactam starting from biorenewables. Green Chem 17:1341–1361

Behler J, Vijay D, Hess WR, Akhtar MK (2018) CRISPR-based technologies for metabolic engineering in cyanobacteria. Trends Biotechnol 36(10):996–1010. https://doi.org/10.1016/j.tibtech.2018.05.011

Beller HR, Lee TS, Katz L (2015) Natural products as biofuels and bio-based chemicals: fatty acids and isoprenoids. Nat Prod Rep 32:1508–1526

Bhansali S, Kumar A (2018) Synthetic and semisynthetic metabolic pathways for biofuel production. In: Kumar A, Ogita S, Yau Y-Y (eds) Biofuels: greenhouse gas mitigation and global warming, Next generation biofuels and role of biotechnology. Springer, Heidelberg, pp 421–432

Birgen C, Dürre P, Preisig HA, Wentzel A (2019) Butanol production from lignocellulosic biomass: revisiting fermentation performance indicators with exploratory data analysis. Biotechnol Biofuels 12:167. https://doi.org/10.1186/s13068-019-1508-6

Brock JT, Freedman AJE, Tompsett GA, Muse SK, Allen AJ, Jackson LA, Thompson JR (2019) Engineered microbial biofuel production and recovery under supercritical carbon dioxide. Nat Commun 10(1):587. https://doi.org/10.1038/s41467-019-08486-6

Campos-Fernández J, Arnal JM, Gómez J, Dorado MP (2012) A comparison of performance of higher alcohols/diesel fuel blends in a diesel engine. Appl Energy 95:267–275

Carroll A, Somerville C (2009) Cellulosic biofuels. Annu Rev Plant Biol 60:165–182

Case AE, Atsumi S (2016) Cyanobacterial chemical production. J Biotechnol 231:106–114

Chae TU, Choi SY, Kim JW, Ko Y (2017) Science direct recent advances in systems metabolic engineering tools and strategies. Curr Opin Biotechnol 47:67–82. https://doi.org/10.1016/j.copbio.2017.06.007

Chakravorty U, Hubert M, Nostbakken L (2009) Fuel versus food. Annu Rev Res Econ 1:645–663. https://doi.org/10.1146/annurev.resource.050708.144200

Chen J, Sun S, Li C-Z, Zhu Y-G, Rosen BP (2014) Biosensor for organoarsenical herbicides and growth promoters. Environ Sci Technol 48:1141–1147

Christianson DW (2008) Unearthing the roots of the terpenome. Curr Opin Chem Biol 12 (2):141–150

Cirino PC, Tang Y, Takahashi K, Tirrell DA, Arnold FH (2003) Global incorporation of norleucine in place of methionine in cytochrome P450 BM-3 heme domain increases peroxygenase activity. Biotechnol Bioeng 83:729–734

Claypool JT, Ramon DR (2013) Development and validation of a technoeconomic analysis tool for early-stage evaluation of bio-based chemical production processes. Bioresour Technol 150:486–495

Claypool JT, Ramon DR, Jarboe LR, Nielsen DR (2014) Technoeconomic evaluation of bio-based styrene production by engineered *Escherichia coli*. J Ind Microbiol Biotechnol 2014 (41):1211–1216

Clomburg JM, Blankschien MD, Vick JE, Chou A, Kim S, Gonzalez R (2015) Integrated engineering of b-oxidation reversal and v-oxidation pathways for the synthesis of medium chain v-functionalized carboxylic acids. Metab Eng 28:202–212

Cobb RE, Wang Y, Zhao H (2015) High-efficiency multiplex genome editing of Streptomyces species using an engineered CRISPR/Cas system. ACS Synth Biol 4:723–728

Connor MR, Atsumi S (2010) Synthetic biology guides biofuel production. J Biomed Biotechnol 2010:1–9

Conrado RJ, Gonzalez R (2014) Chemistry. Envisioning the bioconversion of methane to liquid fuels. Science 343:621–623

Couto JM, Mcgarrity A, Russell J, Sloan WT (2018) The effect of metabolic stress on genome stability of a synthetic biology chassis *Escherichia coli* K12 strain. Microb Cell Factories 17:1–10. https://doi.org/10.1186/s12934-018-0858-2

Dai Z, Nielsen J (2015) Advancing metabolic engineering through systems biology of industrial microorganisms. Curr Opin Biotechnol 36:8–15

Diao L, Liu Y, Qian F, Yang J, Jiang Y, Yang S (2013) Construction of fast xylose-fermenting yeast based on industrial ethanol-producing diploid *Saccharomyces cerevisiae* by rational design and adaptive evolution. BMC Biotechnol 13:110

DiCarlo JE, Norville JE, Mali P, Rios X, Aach J, Church GM (2013) Genome engineering in *Saccharomyces cerevisiae* using CRISPR-Cas systems. Nucleic Acids Res 41:4336–4343

Dismukes GC, Carrieri D, Bennette N, Ananyev GM, Posewitz MC (2008) Aquatic phototrophs: efficient alternatives to land-based crops for biofuels. Curr Opin Biotechnol 19:235–240

Dueber JE et al (2009) Synthetic protein scaffolds provide modular control over metabolic flux. Nat Biotechnol 27(8):753–759

Earhart AJJF, Forijj APC, Patel MK (2012) Replacing fossil based PET with biobased PEF: process analysis, energy and GHG balance. Energy Environ Sci 5:6407–6422

Erb TJ, Zarzycki J (2016) Biochemical and synthetic biology approaches to improve photosynthetic CO_2-fixation. Curr Opin Chem Biol 34:72–79

Espaux L, Mendez-perez D, Li R, Keasling JD (2015) Synthetic biology for microbial production of lipid-based biofuels. Curr Opin Chem Biol 29:58–65. https://doi.org/10.1016/j.cbpa.2015.09.009

Estrela R, Cate JHD (2016) Energy biotechnology in the CRISPR-Cas9 era. Curr Opin Biotechnol 38:79–84. https://doi.org/10.1016/j.copbio.2016.01.005

Farr G, Blancquaert D, Capell T, Van Der Straeten D, Christou P, Zhu C (2014) Engineering complex metabolic pathways in plants. Annu Rev Plant Biol 65:187–223. https://doi.org/10.1146/annurev-arplant-050213-035825

Fei Q, Guarnieri MT, Tao L, Laurens LM, Dowe N, Pienkos PT (2014) Bioconversion of natural gas to liquid fuel: opportunities and challenges. Biotechnol Adv 32:596–614

Feng X, Lian J, Zhao H (2014) Metabolic engineering of *Saccharomyces cerevisiae* to improve 1-hexadecanol production. Metab Eng 27:10–19

Fortman JL et al (2008) Biofuel alternatives to ethanol: pumping the microbial well. Trends Biotechnol 26:375–381

French KE (2019) Harnessing synthetic biology for sustainable development. Nat Sustain 2:250–252

Gaida SM, Liedtke A, Heinz A, Jentges W, Engels B, Jennewein S (2016) Metabolic engineering of *Clostridium cellulolyticum* for the production of n-butanol from crystalline cellulose. Microb Cell Fact 15:6. https://doi.org/10.1186/s12934-015-0406-2

Gall DL, Ralph J, Donohue TJ, Noguera DR (2017) Biochemical transformation of lignin for deriving valued commodities from lignocellulose. Curr Opin Biotechnol 45:120–126. https://doi.org/10.1016/j.copbio.2017.02.015

Georgianna DR, Stephen P (2012) Exploiting diversity and synthetic biology for the production of algal biofuels. Nature 488:330–335. https://doi.org/10.1038/nature11479

Giessen TW, Silver PA (2017) Engineering carbon fixation with artificial protein organelles. Curr Opin Biotechnol 46:42–50. https://doi.org/10.1016/j.copbio.2017.01.004

Goh E-B, Baidoo EEK, Keasling JD, Beller HR (2012) Engineering of bacterial methyl ketone synthesis for biofuels. Appl Environ Microbiol 78:70–80

Goh E-B, Baidoo EEK, Burd H, Lee TS, Keasling JD, Beller HR (2014) Substantial improvements in methyl ketone production in *E. coli* and insights on the pathway from in vitro studies. Metab Eng 26:67–76

Green EM (2011) Fermentative production of butanol—the industrial perspective. Curr Opin Biotechnol 22:337–343

Guo W, Sheng J, Zhao H, Feng X (2016) Metabolic engineering of *Saccharomyces cerevisiae* to produce 1-hexadecanol from xylose. Microb Cell Fact 15:1–11. https://doi.org/10.1186/s12934-016-0423-9

Ha S, Galazka JM, Rin S, Choi J, Yang X, Seo J (2011) Engineered *Saccharomyces cerevisiae* capable of simultaneous cellobiose and xylose fermentation. PNAS 108:504–509. https://doi.org/10.1073/pnas.1010456108/-/DCSupplemental.www.pnas.org/cgi/doi/10.1073/pnas.1010456108

Hammer SK, Avalos JL (2017) Uncovering the role of branched-chain amino acid transaminases in *Saccharomyces cerevisiae* isobutanol biosynthesis. Metab Eng 44:302–312. https://doi.org/10.1016/j.ymben.2017.10.001

Harrison R, Todd P, Rudge S, Petrides D (2015) Bioprocess design and economics chapter in bioseparations science and engineering. Oxford Press, Oxford. isbn:978-0-19-539181-7

Hillson NJ, Hu P, Andersen GL, Shapiro L (2007) *Caulobacter crescentus* as a whole-cell uranium biosensor. Appl Environ Microbiol 73:7615–7621

Howarth RW (2015) Methane emissions and climatic warming risk from hydraulic fracturing and shale gas development: implications for policy. Energy Emission Control Technol 3:45–45

Huang H, Zheng G, Jiang W, Hu H, Lu Y (2015) One-step high-efficiency CRISPR/Cas9-mediated genome editing in Streptomyces. Acta Biochim Biophys Sin 47:231–243

Huo Y-X, Cho KM, Rivera JGL, Monte E, Shen CR, Yan Y, Liao JC (2011) Conversion of proteins into biofuels by engineering nitrogen flux. Nat Biotechnol 29(4):346–351. https://doi.org/10.1038/nbt.1789

Jagadevan S, Banerjee A, Banerjee C, Guria C, Tiwari R, Baweja M (2018) Recent developments in synthetic biology and metabolic engineering in microalgae towards biofuel production. Biotechnol Biofuels 11:1–21. https://doi.org/10.1186/s13068-018-1181-1

Jang Y-S, Park JM, Choi S, Choi YJ, Seung DY, Cho JH, Lee SY (2012) Engineering of microorganisms for the production of biofuels and perspectives based on systems metabolic engineering approaches. Biotechnol Adv 30(5):989–1000. https://doi.org/10.1016/j.biotechadv.2011.08.0

Jansen MLA, Bracher JM, Papapetridis I, Verhoeven MD, de Bruijn H, de Waal PP, van Maris AJA, Klaassen P, Pronk JT (2017) *Saccharomyces cerevisiae* strains for second-generation ethanol production: from academic exploration to industrial implementation. FEMS Yeast Res 17:fox044

Jeffries TW (2006) Engineering yeasts for xylose metabolism. Curr Opin Biotechnol 17:320–326

Jiang Y, Chen B, Duan C, Sun B, Yang J, Yang S (2015) Multigene editing in the *Escherichia coli* genome via the CRISPR-Cas9 system. Appl Environ Microbiol 81:2506–2514

Jiang YJ, Chen TP, Dong WL, Zhang M, Zhang WM, Wu H, Ma JF, Jiang M, Xin FX (2018) The draft genome sequence of *Clostridium beijerinckii* NJP7, a unique bacterium capable of producing isopropanol-butanol from hemicellulose through consolidated bioprocessing. Curr Microbiol 75(3):305–308

Jiang Y, Wu R, Zhou J, He A, Xu J, Xin F, Zhang W (2019) Recent advances of biofuels and biochemicals production from sustainable resources using co-cultivation systems. Biotechnol Biofuels 12:155. https://doi.org/10.1186/s13068-019-1495-7

Jin C, Yao M, Liu H, Lee CFF, Ji J (2011) Progress in the production and application of n-butanol as a biofuel. Renew Sust Energ Rev 15(8):4080–4106

Jones JA, Toparlak ÖD, Koffas MAG (2015) Metabolic pathway balancing and its role in the production of biofuels and chemicals. Curr Opin Biotechnol 33:52–59. https://doi.org/10.1016/j.copbio.2014.11.013

Jung H.G and D.A. Deetz (1993). Cell wall lignification and degradability, In: H.G. Jung, D.R. Buxton, R.D. Hatfield, et al. (eds.) Forage cell wall structure and digestibility, ASA-CSSA-SSSA, Madison, pp. 315–346

Jung H-JG, Samac DA, Sarath G (2012) Modifying crops to increase cell wall digestibility. Plant Sci 185–186:65–77. https://doi.org/10.1016/j.plantsci.2011.10.014

Kalyuzhnaya MG, Puri AW, Lidstrom ME (2015) Metabolic engineering in methanotrophic bacteria. Metab Eng 29:142–152. https://doi.org/10.1016/j.ymben.2015.03.010

Kawaguchi H, Hasunuma T, Ogino C, Kondo A (2016) Bioprocessing of bio-based chemicals produced from lignocellulosic feedstocks. Curr Opin Biotechnol 42:30–39. https://doi.org/10.1016/j.copbio.2016.02.031

Keeling CI, Bohlmann J (2006) Genes, enzymes, and chemicals of terpenoid diversity in the constitutive and induced defence of conifers against insects and pathogens. New Phytol 170 (4):657–675

Khalil AS, Collins JJ (2010) Synthetic biology: applications come of age. Nat Publ Group 11 (5):367–379. https://doi.org/10.1038/nrg2775

Kim H, Yoo SJ, Kang HA (2015) Yeast synthetic biology for the production of recombinant therapeutic proteins. FEMS Yeast Res 15:1–16. https://doi.org/10.1111/1567-1364.12195

Korman TP, Sahachartsiri B, Charbonneau DM, Huang GL, Beauregard M, Bowie JU (2013) Dieselzymes: development of a stable and methanol tolerant lipase for biodiesel production by directed evolution. Biotechnol Biofuels 6:70

Kumar A (2010) Plant genetic transformation and molecular markers. Jaipur, Pointer Publishers, 288 p

Kumar A (2013) Biofuels utilisation: an attempt to reduce GHG's and mitigate climate change. In: Nautiyal S, Rao K, Kaechele H, Raju K, Schaldach R (eds) Knowledge Systems of Societies for adaptation and mitigation of impacts of climate change, Environmental science and engineering. Springer, Berlin, pp 199–224

Kumar A (2014) Biotechnology for biofuels: lignocellulosic ethanol production. J Pharm Sci Innov 3(6):495–498. https://doi.org/10.7897/2277-4572.036203

Kumar A (2015) Metabolic engineering of plants. In: Bahadur B, VenkatRajam M, Sahijram L, Krishnamurthy KV (eds) Plant biology and biotechnology. Springer, Heidelberg, pp 517–526

Kumar A (2020) Synthetic and semi-synthetic metabolic pathways for biofuel production. In: Biofuels: greenhouse gas mitigation and global warming. Springer, New Delhi, pp 421–432

Kumar A, Ogita S, Yau YY (eds) (2018) Biofuels: greenhouse gas mitigation and global warming. Next generation biofuels and role of biotechnology. Springer, Heidelberg, 432 p. isbn: 978-81-322-3761-72

Kumar A, Bhansali S, Gupta N, Sharma M (2019) Bioenergy and climate change: greenhouse gas mitigation. In: Rastegari AA, Yadav AN, Gupta A (eds) Prospects of renewable bioprocessing in future energy systems, Biofuel and biorefinery technologies, vol 10. Springer, Cham, pp 269–290

Kwak S, Jin YS (2017) Production of fuels and chemicals from xylose by engineered *Saccharomyces cerevisiae*: a review and perspective. Microb Cell Fact 16:1–15. https://doi.org/10.1186/s12934-017-0694-9

Lee SY, Kim HU (2015) Systems strategies for developing industrial microbial strains. Nat Biotechnol 33:1061–1072

Lee SK, Chou H, Ham TS, Lee TS, Keasling JD (2008) Metabolic engineering of microorganisms for biofuels production: from bugs to synthetic biology to fuels. Curr Opin Biotechnol 19:556–563. https://doi.org/10.1016/j.copbio.2008.10.014

Lee JW et al (2012) Systems metabolic engineering of microorganisms for natural and non-natural chemicals. Nat Chem Biol 8:536–546

Leonard E, Kumaran P, Thayer K, Xiao W, Mo JD et al (2010) Combining metabolic and protein engineering of a terpenoid biosynthetic pathway for overproduction and selectivity control. Proc Natl Acad Sci U S A 107:13654–13659. https://doi.org/10.1073/pnas.1006138107/-/DCSupplemental.www.pnas.org/cgi/doi/10.1073/pnas.1006138107

Liao Y-C, Huang T-W, Chen F-C, Charusanti P, Hong JSJ, Chang H-Y, Tsai S-F, Palsson BO, Hsiung CA (2011) An experimentally validated genome-scale metabolic reconstruction of *Klebsiella pneumoniae* MGH 78578, iYL1228. J Bacteriol 193:1710–1717

Liao JC, Mi L, Pontrelli S, Luo S (2016) Fuelling the future: microbial engineering for the production of sustainable biofuels. Nat Rev Microbiol 14:288–304

Lindberg P, Park S, Melis A (2009) Engineering a platform for photosynthetic isoprene production in cyanobacteria, using *Synechocystis* as the model organism. Metab Eng 12(1):70–79

Liu W, Yuan JS, Stewart CN Jr (2013) Advanced genetic tools for plant biotechnology. Nat Publ Group 14(11):781–793. https://doi.org/10.1038/nrg3583

Liu J, Li J, Shin H, Liu L, Du G, Chen J (2017) Bioresource technology. Bioresour Technol 239:412–421. https://doi.org/10.1016/j.biortech.2017.04.052

Long SP, Marshall-Colon A, Zhu XG (2015) Meeting the global food demand of the future by engineering crop photosynthesis and yield potential. Cell 161:56–66

Luterbacher JS, Tester JW, Walker LP (2010) High-solids biphasic CO_2-H_2O pretreatment of lignocellulosic biomass. Biotechnol Bioeng 107:451–460

Lynch MD (2016) Into new territory: improved microbial synthesis through engineering of the essential metabolic network. Curr Opin Biotechnol 38:106–111. https://doi.org/10.1016/j.copbio.2016.01.009

Ma T, Shi B, Ye Z, Li X, Liu M, Chen Y, Nielsen J (2019) Lipid engineering combined with systematic metabolic engineering of *Saccharomyces cerevisiae* for high-yield production of lycopene. Metab Eng 52:134–142. https://doi.org/10.1016/j.ymben.2018.11.009

Mackenzie A (2013) Synthetic biology and the technicity of biofuels. Stud Hist Philos Biol Biomed Sci 44(2):190–198. https://doi.org/10.1016/j.shpsc.2013.03.014

Marella ER, Holkenbrink C, Siewers V, Borodina I (2018) Engineering microbial fatty acid metabolism for biofuels and biochemicals. Curr Opin Biotechnol 50:39–46. https://doi.org/10.1016/j.copbio.2017.10.002

Martien JI, Amador-Noguez D (2017) Recent applications of metabolomics to advance microbial biofuel production. Curr Opin Biotechnol 43:118–126. https://doi.org/10.1016/j.copbio.2016.11.006

Martín del Campo JS, Rollin J, Myung S, Chun Y, Chandrayan S, Patiño R et al (2013) High-yield production of dihydrogen from xylose by using a synthetic enzyme cascade in a cell-free system. Angew Chem Int Ed 52(17):4587–4590

Mcanulty MJ, Poosarla VG, Li J, Soo VWC, Zhu F, Wood TK (2017) Metabolic engineering of *Methanosarcina acetivorans* for lactate production from methane. Biotechnol Bioeng 114:852–861. https://doi.org/10.1002/bit.26208

Meyer MT, McCormick AJ, Griffiths H (2016) Will an algal CO_2-concentrating mechanism work in higher plants? Curr Opin Plant Biol 31:181–188

Myung S, Rollin J, You C, Sun F, Chandrayan S, Adams MWW et al (2014) In vitro metabolic engineering of hydrogen production at theoretical yield from sucrose. Metab Eng 24(1):70–77

Nielsen J, Keasling JD (2011) Synergies between synthetic biology and metabolic engineering. Nat Biotechnol 29:693–695

Nielsen J, Keasling JD (2016) Engineering cellular metabolism. Cell 164(6):1185–1197. https://doi.org/10.1016/j.cell.2016.02.004

O'Connor SE (2015) Engineering of secondary metabolism. Annu Rev Genet 49:71–94

Oh JH, van Pijkeren JP (2014) CRISPR-Cas9-assisted recombineering in *Lactobacillus reuteri*. Nucleic Acids Res 42:e131

Peralta-Yahya PP, Keasling JD (2010) Advanced biofuel production in microbes. Biotechnol J 5:147–162

Peralta-Yahya PP, Ouellet M, Chan R, Mukhopadhyay A, Keasling JD, Lee TS (2011) Identification and microbial production of a terpene-based advanced biofuel. Nat Commun 2:483–488. https://doi.org/10.1038/ncomms1494

Peralta-Yahya PP et al (2012) Microbial engineering for the production of advanced biofuels. Nature 488:320–328. https://doi.org/10.1038/nature488320-328

Raschmanová H, Weninger A, Glieder A, Kovar K, Vogl T (2018) Implementing CRISPR-Cas technologies in conventional and non- conventional yeasts: current state and future prospects. Biotechnol Adv 36(3):641–665. https://doi.org/10.1016/j.biotechadv.2018.01.006

Renniger N, McPhee D (2008) Fuel compositions comprising farnesane and farnesane derivatives and method of making and using same. US Patent No. 7399323

Rogelj J, Hare W, Lowe J, van Vuuren DP, Riahi K, Matthews B, Meinshausen M et al (2011) Emission pathways consistent with a 2°C global temperature limit. Nat Clim Chang 1 (8):413–418. https://doi.org/10.1038/nclimate1258

Rollin JA, Ye XH, Del Campo JM, Adams MWW, Zhang Y-HP (2016) Novel hydrogen detection apparatus along with bioreactor systems. Adv Biochem Eng Biotechnol 152:35–51. https://doi.org/10.1007/10_2014_274

Salazar M, Vongsangnak W, Panagiotou G, Andersen MR, Nielsen J (2009) Uncovering transcriptional regulation of glycerol metabolism in Aspergilli through genome-wide gene expression data analysis. Mol Gen Genomics 282:571–586

Sanford K, Chotani G, Danielson N, Zahn JA (2016) Scaling up of renewable chemicals. Curr Opin Biotechnol 38:112–122. https://doi.org/10.1016/j.copbio.2016.01.008

Santos F, Boele J, Teusink B (2011) A practical guide to genome-scale metabolic models and their analysis. Methods Enzymol 500:509–532

Savakis P, Hellingwerf KJ (2015) Engineering cyanobacteria for direct biofuel production from CO_2. Curr Opin Biotechnol 33:8–14

Scheffers BR, De Meester L, Bridge TC, Hoffmann AA, Pandolfi JM, Corlett RT, Butchart SH, Pearce-Kelly P, Kovacs KM, Dudgeon D et al (2016) The broad footprint of climate change from genes to biomes to people. Science 354(6313):719

Shaik N, Kumar A (2014) Energy crops for biofuel and food security. J Pharm Sci Innov 3:507–515

Shanmugam S, Sun C, Zeng X, Wu YR (2018) High-efficient production of biobutanol by a novel *Clostridium* sp. strain WST with uncontrolled pH strategy. Bioresour Technol 256:543–547

Shih PM, Zarzycki J, Niyogi KK, Kerfeld CA (2014) Introduction of a synthetic CO_2-fixing photorespiratory bypass into a cyanobacterium. J Biol Chem 289:9493–9500

Shindell D, Kuylenstierna JCI, Vignati E, Dingenen R, Amann M, Klimont Z, Anenberg S, Muller N, Janssens-Maenhout G, Raes F, Schwartz J, Faluvegi G, Pozzoli L, Kupiainen K, Höglund-Isaksson L, Emberson L, Streets D, Ramanathan V, Hicks K, Oanh NT, Milly G, Williams M, Demkine V, Fowler D (2012) Simultaneously mitigating near-term climate change and improving human health and food security. Science 335:183–189

Sitepu IR, Garay LA, Sestric R, Levin D, Block DE, Bruce GJ, Boundy-Mills KL (2014) Oleaginous yeasts for biodiesel: current and future trends in biology and production. Biotechnol Adv 32(7):1336–1360. https://doi.org/10.1016/j.biotechadv.2014.08.003

Soo VW, McAnulty MJ, Tripathi A, Zhu F, Zhang L, Hatzakis E, Smith PB, Agrawal S, Nazem-Bokaee H, Gopalakrishnan S, Salis HM, Ferry JG, Maranas CD, Patterson AD, Wood TK (2016) Reversing methanogenesis to capture methane for liquid biofuel precursors. Microb Cell Fact 15:11

Steen EJ, Kang Y, Bokinsky G, Hu Z, Schirmer A, McClure A, Del Cardayre SB, Keasling JD (2010) Microbial production of fatty-acid- derived fuels and chemicals from plant biomass. Nature 463:559–562

Sun C, Zhang S, Xin FX, Shanmugam S, Wu YR (2018) Genomic comparison of Clostridium species with the potential of utilizing red algal biomass for biobutanol production. Biotechnol Biofuels 11:42

Swartz JR (2013) Cell-free bioprocessing. Chem Eng Prog 11:40–45

Taheripour TMF, Zhuang Q, Tyner WE, Lu X (2012) Biofuels, cropland expansion, and the extensive margin. Energy Sustain Soc 2:25

Tatsis EC, O'Connor SE (2016) New developments in engineering plant metabolic pathways. Curr Opin Biotechnol 42:126–132. https://doi.org/10.1016/j.copbio.2016.04.012

Thauer RK, Kaster AK, Seedorf H, Buckel W, Hedderich R (2008) *Methanogenic archaea*: ecologically relevant differences in energy conservation. Nat Rev Microbiol 6:579–591

Tholl D (2006) Terpene synthases and the regulation, diversity and biological roles of terpene metabolism. Curr Opin Plant Biol 9(3):297–304

Tong Y, Charusanti P, Zhang L, Weber T, Lee SY (2015) CRISPR-Cas9 based engineering of actinomycetal genomes. ACS Synth Biol 4:1020–1029

Tran P, Hoang N, Ko JK, Gong G, Um Y, Lee SM (2018) Biotechnology for biofuels genomic and phenotypic characterization of a refactored xylose-utilizing *Saccharomyces cerevisiae* strain for lignocellulosic biofuel production. Biotechnol Biofuels 11:1–13. https://doi.org/10.1186/s13068-018-1269-7

Trindade WRDS, Santos RGD (2017) Review on the characteristics of butanol, its production and use as fuel in internal combustion engines. Renew Sust Energ Rev 69:642–651

Tyner WE (2012) Biofuels and agriculture: a past perspective and uncertain future. Int J Sustain Dev World Ecol 19:389–394

Tyo KE, Ajikumar PK, Stephanopoulos G (2009) Stabilized gene duplication enables long- term selection-free heterologous pathway expression. Nat Biotechnol 27(8):760–765

US DOE (2010) National algal biofuels technology roadmap. United States Department of Energy

US Energy Information Administration (2012). http://www.eia.gov. (US Energy Information Administration)

Van Vleet JH, Jeffries TW (2009) Yeast metabolic engineering for hemicellulosic ethanol production. Curr Opin Biotechnol 20:300–306

Völler J-S, Budisa N (2017) Coupling genetic code expansion and metabolic engineering for synthetic cells. Curr Opin Biotechnol 48:1–7. https://doi.org/10.1016/j.copbio.2017.02.002

Wang W, Sun J, Hartlep M, Deckwer W-D, Zeng A-P (2003) Combined use of proteomic analysis and enzyme activity assays for metabolic pathway analysis of glycerol fermentation by *Klebsiella pneumoniae*. Biotechnol Bioeng 83:525–536

Wang J, Guleria S, Koffas MA, Yan Y (2015) Microbial production of value-added nutraceuticals. Curr Opin Biotechnol 37:97–104

Wang C, Pfleger BF, Kim S (2017) Reassessing *Escherichia coli* as a cell factory for biofuel production. Curr Opin Biotechnol 45:92–103. https://doi.org/10.1016/j.copbio.2017.02.010

Wess J, Brinek M, Boles E (2019) Improving isobutanol production with the yeast *Saccharomyces cerevisiae* by successively blocking competing metabolic pathways as well as ethanol and glycerol formation. Biotechnol Biofuels 12:173. https://doi.org/10.1186/s13068-019-1486-8

Wohlbach DJ, Kuo A, Sato TK, Potts KM, Salamov AA, Labutti KM, Sun H (2011) Comparative genomics of xylose-fermenting fungi for enhanced biofuel production. PNAS 108:13213. https://doi.org/10.1073/pnas.1103039108/-/DCSupplemental.www.pnas.org/cgi/doi/10.1073/pnas.1103039108

Woo HM (2017) Solar-to-chemical and solar-to-fuel production from CO_2 by metabolically engineered microorganisms. Curr Opin Biotechnol 45:1–7. https://doi.org/10.1016/j.copbio.2016.11.017

Woodward J, Orr M, Cordray K, Greenbaum E (2000) Enzymatic production of biohydrogen. Nature 405:1014–1015

Wu J et al (2015) Enhancing flavonoid production by systematically tuning the central metabolic pathways based on a CRISPR interference system in *Escherichia coli*. Sci Rep 5:13477

Yao L et al (2016) Multiple gene repression in cyanobacteria using CRISPRi. ACS Synth Biol 5:207–212

Ye X, Wang Y, Hopkins RC, Adams MWW, Evans BR, Mielenz JR et al (2009) Spontaneous high-yield production of hydrogen from cellulosic materials and water catalyzed by enzyme cocktails. ChemSusChem 2(2):149–152

Yu L, Xu M, Tang I-C, Yang S-T (2015) Metabolic engineering of\n Clostridium tyrobutyricum\n for n-butanol production through co-utilization of glucose and xylose. Biotechnol Bioeng 112 (10):2134–2141. https://doi.org/10.1002/bit.25613

Zhang Y-HP (2011) Simpler is better: high-yield and potential low-cost biofuels production through cell-free synthetic pathway biotransformation (SyPaB). ACS Catal 1:998–1009

Zhang Y-HP (2015) Production of biofuels and biochemicals by *in vitro* synthetic biosystems: opportunities and challenges. Biotechnol Adv 33(7):1467–1483. https://doi.org/10.1016/j.biotechadv.2014.10.009

Zhang Y-HP, Evans BR, Mielenz JR, Hopkins RC, Adams MWW (2007) High-yield hydrogen production from starch and water by a synthetic enzymatic pathway. PLoS One 2(5):e456

Zheng T, Olson DG, Tian L, Bomble YJ, Himmel ME, Lo J, Hon S, Shaw AJ, van Dijken JP, Lynd LR (2015) Cofactor specificity of the bifunctional alcohol and aldehyde dehydrogenase (AdhE) in wild-type and mutant *Clostridium thermocellum* and *Thermoanaerobacterium saccharolyticum*. J Bacteriol 197(15):2610–2619

Zhu Q, Jackson EN (2015) Metabolic engineering of *Yarrowia lipolytica* for industrial applications. Curr Opin Biotechnol 36:65–72. https://doi.org/10.1016/j.copbio.2015.08.010

Ashwani Kumar Professor *Emeritus*, Department of Botany, University of Rajasthan, Jaipur, India, Studied photosynthetic apparatus in vitro and in vivo initially working with Professor Dr. K H Neumann at Institute of Plant Nutrition at Justus Liebig University Giessen, Germany, and subsequently with Professor Dr. Sven Schubert on physiology role of enzymes in salinity stress resistance with support from Alexander von Humboldt Fellowship. JSPS visiting Professor Japan. Guided 39 students for PhD, published 220 research papers and 23 books. Presently, he is the president of the Indian Botanical Society.

Part III

Genetic Resources and Engineering Methods to Improve Crop Plants

Kinetics Genetics and Heterosis

12

James A. Birchler

Abstract

Heterosis is the phenomenon that hybrids exhibit greater biomass and fertility than either of the parents. A consensus has not coalesced about the genetic and molecular basis despite over a century of investigation. Some recent studies of the so-called single-gene heterosis and large-scale genomic/phenomic studies have suggested that genes typically involved with quantitative traits seem to be the major determinants of heterosis rather than complementation of random mutations in a variety of functions. As usual, once a new insight emerges, one can recognize prescient experimental results in the classical studies that foreshadow such realizations. Here, we trace the history of these prescient results and relate them to new ideas about how quantitative traits are evaluated with a special emphasis on heterosis. Many questions remain in terms of the details of the mechanism. The prospects for genetic engineering and gene editing to foster heterosis are challenging but some possibilities are emerging.

Keywords

Hybrid vigor · Progressive heterosis · Tetraploidy

12.1 Introduction

It has been known for centuries that the offspring of different parents of plants will produce exuberant growth, enhanced fertility, greater yield, and other characteristics that exceed both parents (Darwin 1868). This phenomenon is called heterosis or hybrid vigor and is the basis of production of many agricultural crops including

J. A. Birchler (✉)
Division of Biological Science, University of Missouri-Columbia, Columbia, MO, USA
e-mail: BirchlerJ@Missouri.edu

A. Kumar et al. (eds.), *Climate Change, Photosynthesis and Advanced Biofuels*,
https://doi.org/10.1007/978-981-15-5228-1_12

maize and rice among the top three contributors (the third being wheat) to human calorie consumption worldwide. Heterosis has been studied from many different angles including genetically, physiologically, genomically, and metabolically, among other approaches (Chen 2013). From the myriad of studies, there is a remarkable lack of synthesis that has emerged. Some investigators argue that different underlying processes are at work and thus there is no need to even attempt to produce a consensus or synthesis. However, taking an evolutionary perspective would suggest that there is likely to be a common theme to any aspect of plant biology that fosters superior reproductive success as does heterosis. If a process has a genetic component, as every indication there is one for heterosis, and genetic variation exists for the process, superior reproductive success will result in selection. Yet, a common theme does not necessarily imply that the same genes, the same molecules, or the same metabolites are responsible for each instance of heterosis within or between species but rather that there may be parallels in process. In this article, we will focus on the genetic aspects of heterosis and argue that new data intertwined with cryptic evidence from the distant corners of heterosis research over the years suggest that hybrid vigor shares with generalized quantitative traits a dosage component to the behavior of alleles and genomes in hybrids.

A popular genetic explanation of heterosis was suggested almost exactly a century ago (Jones 1917). It posits that there are recessive homozygous alleles in each parent and that these are different between the two parents. Then, in the hybrid, the recessive alleles from the different parents will be complemented and additive across loci. It is quite clear that heterosis for different characteristics *can* be independent of others (Flint-Garcia et al. 2009; Yao et al. 2013) and thus this explanation seems to be straightforward and simple. So, why has there not been a consensus that surrounds this idea? Because there have been numerous lines of evidence that it does not seem to explain, in particular, the apparent purging of detrimental alleles from inbred lines does not necessarily ameliorate heterosis to any degree (East 1936; Duvick 1999). Here we will discuss these issues.

In mutagenesis studies, completely recessive mutations are the predominant form of lesions in genes (Stadler 1928a, b). Indeed, if we imagine, for the sake of illustration that two dwarf mutations in different genes in different lines were homozygous and then crossed together, the F1 hybrid would clearly exceed both parents. A similar scenario would be found for other types of mutations for other characteristics as well. Thus, complementation of completely recessive alleles will occur in hybrids.

However, many quantitative trait loci behave somewhat differently. Quantitative genetics characterizes allelic effects as additive (i.e., incompletely dominant between the parents), partially dominant (intermediate between incompletely dominant and completely dominant), dominant (complete complementation), or overdominant (exceeding both parents) (Fisher 1918; Tanksley 1993). The default assumption is that alleles will be additive or, in other words, they are incompletely dominant. In yet other words, the effective dosage of the alleles has an impact on the phenotype. On the complementation hypothesis, it would seem, therefore, that the generalized action of genes affecting quantitative traits would not be responsible for heterosis because the hybrids would be intermediate between the parents rather than exceeding them. However, below we will point out evidence that there is indeed a dosage component to heterosis.

12.2 The Case for an Interrelationship of QTL and Aneuploidy Dosage Effects

In the above narrative, we have described a dichotomy of qualitative characteristics and quantitative characteristics. This dichotomy assumes that qualitative traits are controlled by alleles that do not exhibit a dosage effect, whereas quantitative traits are controlled by alleles with dosage effects, which are obviously an oversimplification. While we recognize that there will be overlap, we make the argument below for a generalized distinction based on several lines of evidence.

First of all, from classical studies of adding extra chromosomes to the genotype of plants (aneuploidy) compared to changing the whole genome in copy number (ploidy), there is a greater dosage impact for segments of the genome than varying the whole genome in concert (Blakeslee et al. 1920; Blakeslee 1934; Birchler and Veitia 2012; Henry et al. 2015). Multiple aneuploidies can affect the same traits in much the same fashion as the genetics of quantitative traits (Guo and Birchler 1994). Maize is particularly amenable to studies of aneuploidy with both segmental additions and subtractions from the genome being possible. Both are detrimental (Birchler and Veitia 2012; Lee et al. 1996; Sheridan and Auger 2008; Brunelle and Sheridan 2014). A recent collection of vegetatively propagated heterozygous deletions in popular also illustrates the multi-aneuploid impact on plant phenotype (Henry et al. 2015). Indeed, as quantitative trait loci were cloned and studied, it was clear that many exhibit a dosage effect on the phenotype (Frary et al. 2000; Liu et al. 2002; Cong et al. 2002, 2008).

As genes were molecularly identified that were responsible for aneuploidy effects and quantitative traits, a generalized pattern emerged that they are typically transcription factors or components of signal transduction pathways (Tanksley 1993; Birchler et al. 2001; Weber et al. 2007). Because of this dosage component to both, it is a reasonable conclusion that there is a common basis for the effect of multiple aneuploidies and multiple quantitative trait loci on individual plant characteristics (Birchler and Veitia 2012; Guo and Birchler 1994; Birchler et al. 2001).

Why are quantitative characteristics so often impacted by dosage-sensitive genes? Studies of the action of components of multi-subunit complexes (Bray and Lay 1997) illustrate how this could be the case (Veitia 2002). Gene regulation in eukaryotes is mediated by several different oligomeric complexes, and their assembly could explain this phenomenon (Birchler et al. 2005; Birchler and Veitia 2007; Veitia et al. 2008; Birchler and Veitia 2010; Veitia et al. 2013) (Fig. 12.1).

Studies in baker's yeast illustrate that changing individually the dosage of genes involved in macromolecular complexes will affect the fitness of those strains (Papp et al. 2003). Also, data from humans indicate that transcription factors and signal transduction components have dosage effects with regard to numerous clinical

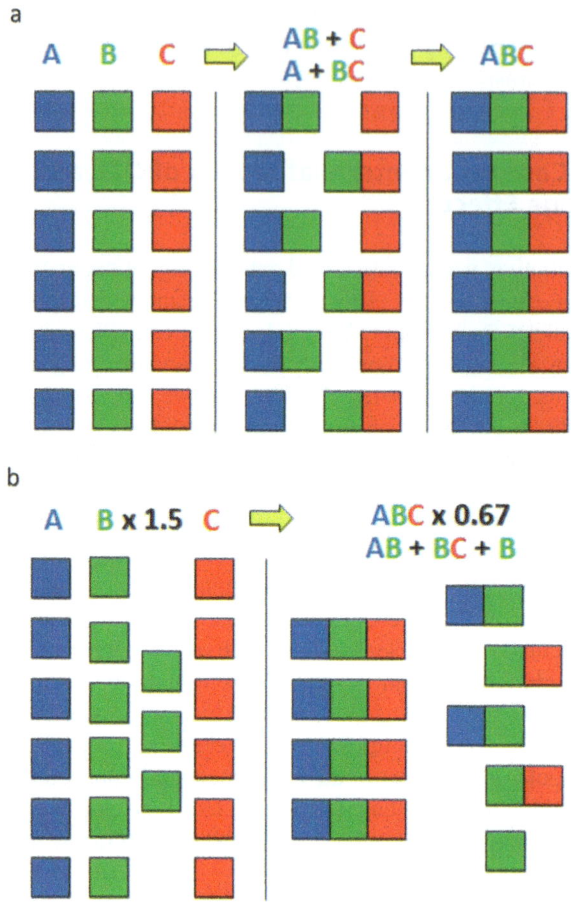

Fig. 12.1 Kinetics genetics: how changing the concentration of a bridge subunit of an oligomeric regulatory complex affect the completion of the whole. (**a**) One can imagine a regulatory complex composed for simplicity of three subunits to produce ABC that controls numerous target genes. To form the complex, A can join with B followed by joining with C, and B can join with C to form BC followed by joining with A. When the concentration of the individual subunits, A, B, and C, are similar, the reaction can produce a certain concentration of ABC. (**b**) However, if the bridge subunit, B, is increased by a factor of 1.5, which might occur in a trisomic or CNV, the completion of the whole complex ABC is inversely affected such that only two third of the whole is formed together with unproductive partial complexes. Target genes of the ABC regulatory complex would be reduced in expression. (Image by Adam Johnson)

conditions indicating a concentration-dependent impact on the phenotype (Veitia 2002; Kondrashov and Koonin 2004; Schuster-Bockler et al. 2010).

Further evidence for the dosage sensitivity of these classes of genes comes from evolutionary genomics. The genome sequences of many species indicate that whole-genome duplication (WGD) has been a frequent occurrence in most eukaryotic lineages including yeast, protozoa, vertebrates, and especially plants (Bowers et al.

2003; Maere et al. 2005; Blanc and Wolfe 2004; Freeling and Thomas 2006; Thomas et al. 2006; Freeling et al. 2008; Freeling 2009; Chang et al. 2010; Li et al. 2016a). Following such events, there is fractionation of the genome back to a near diploid state by gene deletion. There are, however, retentions of certain classes of genes preferentially compared to others. The classes of genes more likely retained are typically those involved with macromolecular complexes including transcription factors and signal transduction components. In contrast, transcription factors and signal transduction components are preferentially underrepresented in copy number variants (CNV) in populations (Schuster-Bockler et al. 2010; Maere et al. 2005; Freeling et al. 2008). Thus, the genomic balance illustrated by the contrasting effects of aneuploidy and ploidy experimentally is played out over evolutionary time and in populations. The evolutionary genomic results illustrate the functional significance of the dosage-dependent regulatory machinery in eukaryotes.

Why are dosage effects characteristic of regulatory processes? One potential answer is that the absolute amount of gene products is critical to the expression in the phenotype. However, the combination of the relative effects of changing the genome in aneuploidy/ploidy comparisons and the parallel results from evolutionary genomics suggests that the stoichiometry of the components is often effective. Modeling of the assembly of macromolecular complexes illustrates that changing the concentration of bridge molecules relative to its various interactors will change the kinetics for the completion of the full complex because partial unproductive complexes will accumulate without a path toward full assembly (Bray and Lay 1997; Veitia 2002; Veitia et al. 2008). This concept was modeled to explain haploinsufficiency of alleles (i.e., dosage sensitivity) for human clinical conditions (Veitia 2002). The involvement of the relative stoichiometry for gene regulatory processes is consistent with the genetic balance issues from aneuploidy/ploidy and how genomes react to WGD and CNV.

The above considerations of the dosage sensitivity of regulatory processes are further supported by the finding in several organisms that gene expression is modulated more by aneuploidy than by ploidy change. Studies in maize indicate that aneuploidy can modulate many genes for each aneuploidy and each aneuploidy can have overlapping effects on the same genes similar to the way that multiple QTL affect a single characteristic (Guo and Birchler 1994; Birchler and Newton 1981). The global effects in aneuploidy are not cumulative in whole ploidy changes in parallel with phenotypic and evolutionary genomic results.

These considerations of dosage and stoichiometry from several disciplines have led to the concept of "kinetics genetics" in which a tweaking of the historical classifications of quantitative traits might be useful (Birchler et al. 2016). Kinetics genetics notes that changing the stoichiometry of interacting regulatory complexes will change the outcome of how alleles will produce a phenotypic effect. This changing stoichiometry could result from natural variation in expression of the various subunits of macromolecular components, from CNV or from larger aneuploidy. If parents differ in their quantitative expression of regulatory gene alleles, the hybrid level of expression would typically be intermediate; however, the manner in which the whole complex is completed could be intermediate or not, and the trait could even exceed both parents when the expression of the subunit component in question is indeed intermediate (Birchler et al. 2016). Thus, the whole range of additive, partial dominance, dominance, and overdominance of "allelic" action in quantitative genetics could, in fact, reflect how the allelic expression interplays in trans with other interacting gene products as much as the specific allelic expression per se (Fig. 12.2).

12.3 What Is the Accelerating Evidence for a Dosage Component to Heterosis?

With the above narrative as background, we now turn our attention to the evidence that there is a strong dosage component to heterosis. Several studies on heterosis in tomato point to a dosage component. Segmental introgression lines of portions of wild tomato genetic material into domesticated tomato show that heterotic effects were often overdominant, whereas other quantitative characteristics were not (Semel et al. 2006). Single-gene heterosis was found for a tomato with the *Single Flower Truss* (*Sft*) gene in that the apparent amount of gene product in a mutant/normal heterozygote fosters greater yield than either parent (Krieger et al. 2010). Further dissection of this example illustrated that the background genotype of other dosage-sensitive genes has an impact on the heterotic action of *Sft* (Jiang et al. 2013; Park et al. 2014). In another example, heterozygotes of the *Shell* gene in oil palm suggested by its mode of action via interaction with other regulators is such that it is dosage-sensitive in producing single-gene heterosis (Singh et al. 2013). In an exhaustive study in rice, sequencing and phenotyping of hundreds of lines of parents and heterotic hybrids and their progeny provided numerous candidate genes involved with heterosis (Huang et al. 2016; Li et al. 2016b). None of them showed complete dominance but rather exhibited partial dominance or overdominance. Different heterotic hybrids did not show the same set of candidate responsible genes. However, what was in common across heterotic combinations was that the responsible genes were various types of regulatory genes typically involved with macromolecular complexes and those that typically exhibit dosage effects. The heterozygotes of the candidate genes produced superior performance relative to the parents, but homozygotes of the different alleles did not produce a linear effect on the phenotype although the classification of a heterotic effect in some cases was operational rather than an absolute effect on the phenotype. In all of these cases, there was no example that showed the complementation of a completely recessive effect by a dominant allele. While examples of null or leaky alleles of candidate genes were detrimental, the heterozygotes were not equivalent to the homozygotes of the alternative allele as occurs with conventional complementation. The effect on the phenotype is not linear. Kinetics genetics provides one possible explanation for such nonlinearity (Fig. 12.2).

From a different approach of a whole-genome dosage study, diploid and triploid inbreds and hybrids were produced for maize and examined for heterotic behavior (Yao et al. 2013). In diploids, reciprocal crosses between different inbreds will

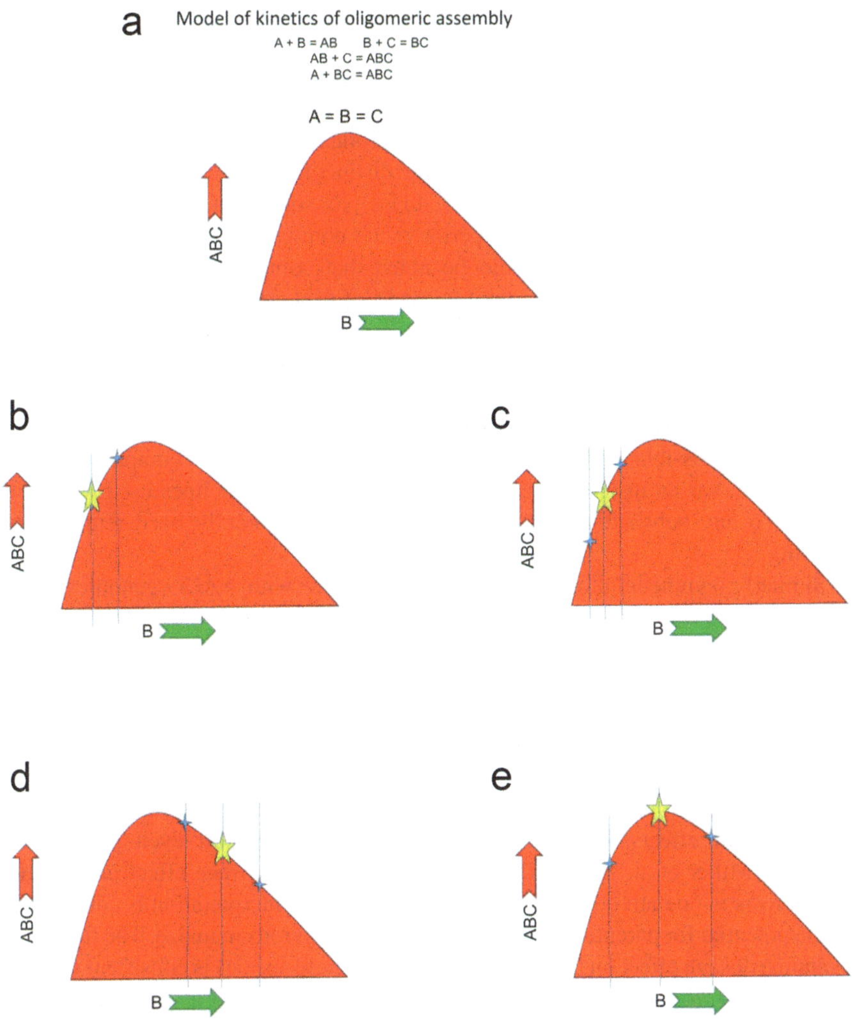

Fig. 12.2 Illustration of the potential nonlinearity of allelic action due to kinetics genetics. (**a**) A curve is shown that illustrates the production of complex ABC as described in Fig. 12.1, depending on the concentration of subunit B. At low concentrations of B, the completion of ABC increases positively with the concentration of B, but once the apex of equal expression of A = B = C is passed, the completion of ABC is reduced. (**b**) In circumstances in which the context of the concentration of B is not effective at limiting the assembly of ABC, the level of expression of two different alleles (illustrated by the two blue vertical lines and blue stars) would be such that a complete dominance would occur and the hybrid (illustrated by the yellow star) would show the equivalent expression/phenotype as one of the parents. (**c**) If the two parents have different expression and the hybrid expression is additive, the hybrid expression/phenotype would fall on the curve at an intermediate position to determine an additive or partial dominant response. (**d**) Also, at higher concentrations, two alleles could have additive expression, and the expression/phenotype would be additive or partially dominant. (**e**) If the two parents have expression on either side of the peak, the hybrid could have additive expression at the gene product level but the assembly of ABC would be greater than both, producing an overdominant effect. Other possibilities include situations in which allelic expression is not additive, which could produce other cases of nonlinear production of the full ABC complex

produce the same genotype with just the two alleles originating from different sexes of the parents. Imperceptible differences in heterosis are typically seen in diploid reciprocal crosses as was the case in this study. Triploids, on the other hand, can have two types of hybrids (actually more are possible) such as AAB and BBA. When these types of hybrids were examined for a heterotic response, the degree of heterosis in the two types of triploid hybrids was quite distinct for most characteristics (which were largely independent as noted above). In fact, for some characteristics, one of the two types of hybrids did not show superior values compared to the parents, while the other type of hybrid did. It should be noted that complementation of complete recessives would be the same in the two types of triploid hybrids. If such complementation were the sole basis of heterosis, the prediction would be that heterosis would be equal in these triploid hybrids. Further, the data were analyzed to test the idea that the differences of the two types of hybrids were on a common foundation in the two that might result from complementation of complete recessives (Fig. 12.3).

It was not possible to detect such a foundation suggesting that complementation, in this case, was an insignificant contributor to heterosis. The lines used were elite inbreds so any substantial detrimental alleles have apparently been removed by breeding.

Similarly, synthetic allotetraploid *Brassica napus* (with AACC genomes) composed of chromosome sets from diploids *B. napus* (AA) and *B. oleracea* (CC) were used to produce triploids via backcrossing to diploid progenitors with different doses of contributing genomes (Tan et al. 2016). Again, the degree of heterosis, depending on the characteristic, was distinctly different in the two types of triploid hybrids. In some cases, one triploid hybrid showed strong heterosis, while the other showed little, similarly to maize.

In *Arabidopsis*, reciprocal diploid hybrids show some differences, presumably from parental effects, but reciprocal triploid hybrids exhibit even greater size disparity (Miller et al. 2012). Thus, it is possible that a dosage effect of genomes on heterosis is operative in this species as well but the interpretation is confounded by the potential for parental effects, which were not obvious in maize. The finding of a genome dosage effect on heterosis in triploids in these species is consistent with the single-gene dosage effects on heterosis seen in tomato, oil palm, and rice as noted above.

12.4 Digging Up the Treasures of Old

When new results emerge in science, it is often the case that one can look back in the history of the discipline and find prescient supporting results that were ignored or misunderstood. Eighty years ago, East (1936) laid out the case that the simple complementation of detrimental recessives was inadequate to explain heterosis. First, he noted that the improvement of inbred lines did not diminish a heterotic response. Further, it was noted that as a general rule, hybrids between different species or genera showed increasing heterosis with increasing phylogenetic distance (East 1936; Gravatt 1914; Karpechenko 1927; Li et al. 2008). It is difficult to explain

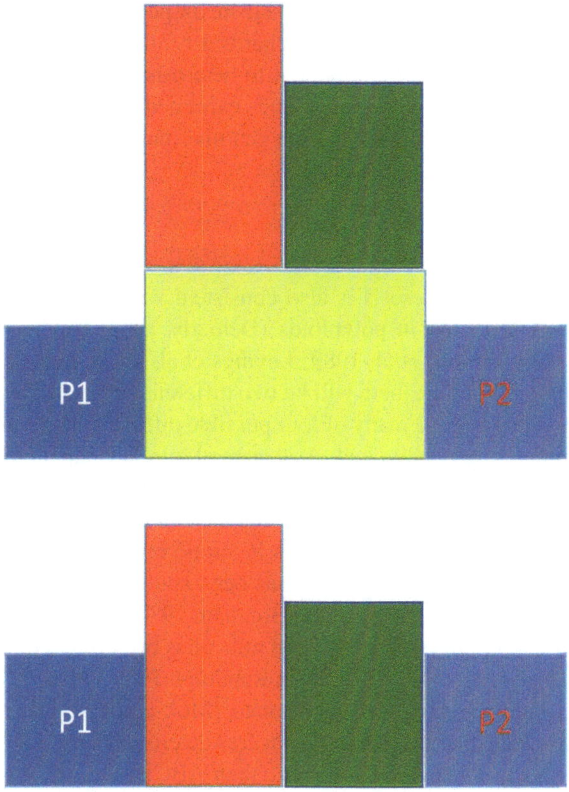

Fig. 12.3 A dosage component to heterosis. In the study by Yao et al. (2013), diploid and triploid inbreds and hybrids were assessed for heterotic phenotypes. The two diploid reciprocal hybrids were very similar to each other, but the two types of triploid hybrids were distinct. A test was performed to examine the question of whether there was a foundation of complementation of complete recessive slightly deleterious alleles from opposite parents on which a dosage effect rests. In the top, this scenario is depicted. The two parents are shown in blue and the different triploid hybrids in red and green. The hypothetical complementation foundation is shown in yellow. This situation could be consistent with a difference in the triploid hybrids and constitute complementation and dosage effects together. At the bottom is depicted the scenario in which the difference between the two types of triploid hybrids is primarily affected by genomic dosage effects with no detectable complementation of complete recessive alleles. An analysis of the data indicated no detectable complementation contribution as shown at the bottom

this result by postulating that with increasing phylogenetic distance, there are an increasing number of *homozygous* detrimental recessive mutations in the different lineages. Such a postulate would require the proliferation of homozygotes in populations also containing heterozygotes that would by definition have greater reproductive success. Heterotic hybrids would outcompete the homozygous genotype for numbers of progeny entering the next generation. Indeed, this postulate flies in the face of the fact that plants have evolved innumerable means to foster cross-pollination such as many forms of self-incompatibility, dioecy, protandry, protogyny, pollinator attractions, etc. (Richards 1997). But more specifically, East (1936) foreshadowed a dosage component to heterosis in describing how different crosses in tobacco between different ploidies exhibited different levels of heterosis depending on the degree of diversity of the genomes present and their copy number.

12.5 Evidence from Polyploids

A dosage component to heterosis is also consistent with the phenomenon of progressive heterosis observed in polyploids (Demarly 1963; Dunbier and Bingham 1975; Gallais 1984; Groose et al. 1989; Levings et al. 1967; Sockness and Dudley 1989a, b). In diploid hybrids, there will be two different genomes present. However, in tetraploids, for example, as many of four possible different alleles at any one locus can be placed into a single individual. As a general rule, with an increasing diversity of alleles in tetraploids, the greater is the heterotic effect. In other words, AABB and CCDD are heterotic as in diploids, but crossing the two hybrids together to produce a potential ABCD will cause an even greater heterotic response. This phenomenon is referred to as "progressive heterosis." It has been studied most rigorously in tetraploid alfalfa (Demarly 1963; Dunbier and Bingham 1975; Gallais 1984; Groose et al. 1989; Bingham 1980), but the same type of effect has been found in maize (Levings et al. 1967; Sockness and Dudley 1989a, b; Randolph 1935; Chase 1980; Riddle and Birchler 2008; Riddle et al. 2010) and potato (Mok and Peloquin 1975). Further distinctions have been made in that ABBB is less heterotic than AABB, which is less heterotic than ABCC which is less heterotic than ABCD (Groose et al. 1989; Chase 1980). Again, the dosage of genomes is impacting heterosis and is not predicted by the complementation model for complete recessives, particularly when comparing the ABBB and AABB genotypes.

Polyploidy is generally thought to be associated with robust plants. However, careful analysis of increasing ploidy with maintenance of homozygosity indicates that in many, if not, most plant species examined, the stature of plants declines with increasing ploidy despite the fact that cell size correlates with ploidy (Randolph 1935, 1942; Rhoades and Dempsey 1966; Yao et al. 2011; Abel and Becker 2007; Riddle et al. 2006; Stupar et al. 2007). However, with the maximization of diverse genomes in polyploids, the vigor and stature do in fact increase with ploidy (Riddle and Birchler 2008). One can illustrate this with the wheat species. Diploid, allotetraploid, and allohexaploid wheat increase in robustness with increasing diversity of genomes and ploidy. However, even hexaploid wheat can exhibit heterosis with a further increase in diversity of genomes (Briggle 1963). And when common hexaploid wheat is crossed with its allodecaploid relative, wheatgrass, enormous hybrid plants are produced (Li et al. 2008). These results parallel those of progressive heterosis observed under experimental conditions.

The flip side to progressive heterosis from the classical literature is that inbreeding rates in matched diploid and tetraploid accessions are very similar. The first such studies were in autotetraploid alfalfa (Williams 1931; Tysdal et al. 1942;

Busbice and Wilsie 1966), and related results have been realized in autotetraploid maize (Sockness and Dudley 1989a; Alexander and Sonnemaker 1961; Rice and Dudley 1974) as well as autotetraploid and autohexaploid crested wheatgrass (Dewey 1966, 1969). When diploid and tetraploid inbreeding were compared, the tetraploid curve declined as rapidly with each generation of inbreeding as did the diploid (Busbice and Wilsie 1966). Based on the frequency of gametes produced from diploids and tetraploids of comparable genotype (AB versus AABB), for any one gene, heterozygous diploids will produce half of the progeny that are homozygous, but, in tetraploids, for any one gene, a heterozygote will produce only 1/18 of the progeny that are homozygous. (The frequency actually depends to some degree on the gene in question and its distance from the centromere. Recombination between the gene and the centromere can change the specific distribution of alleles to gametes.) In both ploidies, one half of the selfed progeny is again heterozygous A/a versus AA/aa. However, Aaaa and AAAa are also produced in the tetraploid (Fig. 12.4). If homozygosis of detrimental fully recessive alleles were the underlying basis of heterosis, then the progression would be predicted to be much slower in tetraploids than in diploids. Interestingly, Busbice and Wilsie (1966) in their study of this phenomenon postulated that the dosage of alleles could come close to explaining this phenomenon. If Aaaa and AAAa genotypes are less heterotic than AAaa as indicated (Groose et al. 1989; Chase 1980), then a more similar rate of inbreeding depression would be predicted. Thus, the patterns of progressive heterosis and inbreeding depression in tetraploids produce a consistent picture.

12.6 Kinetics Genetics and Heterosis

The genes in tomato and rice implicated in heterosis do not exhibit linear effects, suggesting a manifestation of kinetics genetics. In other words, homozygotes of each allele and the heterozygotes neither show complete dominance/recessive behavior nor a linear dosage effect. There are many levels at which these nonlinear effects could be manifested. Clearly, this could occur at the level of gene expression between alleles as has been extensively documented. However, what has been less well appreciated is that the interaction of the protein products across genes will not necessarily produce a linear effect compared to the encoding messenger RNA. This circumstance can occur by differential degradation of RNAs and proteins as one contributing factor. However, it can also be manifested by the assembly of multi-subunit complexes particularly those involved with gene regulatory and developmental processes as illustrated by the dosage sensitivity of many developmental regulators (Birchler and Veitia 2012; Birchler et al. 2001). The change in stoichiometry of individual subunits will alter the kinetics of assembly of the whole, particularly with regard to those subunits that serve as bridges between other subunits (Bray and Lay 1997). The impact on the stoichiometry need not reflect the steady-state level of protein subunits but might in fact be affected by the rate of their synthesis. These regulatory and developmental biological considerations potentially intersect with the observations that heterotic effects of single genes are very context-

	AA'	Aa	Aa'	A'a	A'a'	aa'
AA'	AA AA	AA Aa	AA Aa	AA Aa	AA Aa	AA aa
Aa	AA Aa	Aa Aa	Aa Aa	Aa Aa	Aa Aa	Aa aa
Aa'	AA Aa	Aa Aa	Aa Aa	Aa Aa	Aa Aa	Aa aa
A'a	AA Aa	Aa Aa	Aa Aa	Aa Aa	Aa Aa	Aa aa
A'a'	AA Aa	Aa Aa	Aa Aa	Aa Aa	Aa Aa	Aa aa
aa'	AA aa	aa Aa	aa Aa	aa Aa	aa Aa	aa aa

Segregation in an Autotetraploid (AA' aa')

Fig. 12.4 The distribution of gametes and resulting zygotes from self-pollination of an AAaa heterozygous autotetraploid. Gametes from an autotetraploid are diploid containing two alleles. One sixth of the gametes are homozygous AA for a gene near the centromere, and another sixth are homozygous for the recessive aa. Four sixths of the gametes from either parent are heterozygous. With random joining of gametes, homozygous recessive tetraploids would be present in 1/36 of the progeny. In contrast to selfing of an A/a diploid with 1/4 being recessive, the homozygosis of recessive alleles is quite different between diploids and tetraploids with the prediction that inbreeding depression would be quite different between the two ploidies. However, in the tetraploid 8/36 have a shift to AAAa and 8/36 have a shift to Aaaa. If this change in allelic dosage has an impact on heterosis as indicated by results from progressive heterosis studies (Chase 1980), the similarity of inbreeding depression curves between diploids and autotetraploids can more easily be explained. After Birchler (2012)

dependent (Jiang et al. 2013; Park et al. 2014). The single-gene heterosis examples operate in particular backgrounds as do the genes implicated in rice hybrids. In the latter case, genes implicated in affecting heterosis in one hybrid circumstance do not behave in the same type of action in other types of hybrids. Postulated roles of circadian rhythm (Ni et al. 2009), chromatin modifications (Kawanabe et al. 2016), and small RNA involvement (Groszman et al. 2011) in heterosis are not mutually exclusive to the concept of a dosage component to heterosis and its relationship in general to the control of quantitative traits.

Wang et al. (2015, 2017) took *Arabidopsis* hybrids that exhibited strong biomass heterosis and selfed them for several generations with selection for the greatest biomass. They were able to achieve homozygous lines that showed biomass comparable to hybrids, which they dubbed "hybrid mimics." While the different hybrid mimics had some differences in performance, they had a related constellation of combinations of genomic regions from the parental lines. Comparisons of gene expression in the F1 and hybrid mimics revealed commonalities and suggested interactions among genomic regions selected in the hybrid mimics. Auxin response genes were upregulated in both hybrids and hybrid mimics, suggesting an increase in auxin levels (Wang et al. 2017).

12.7 Will It Be Possible to Eventually Genetically Engineer Heterotic Mimics?

If hybrids have a change or disruption of regulatory macromolecular complexes from trans dominant interactions of differing subunits (Singh et al. 2013; Veitia and Vaiman 2011), why is the result usually more robust growth as opposed to equal manifestation of under- and overdominance? In some cases, this might be merely what is classified as heterotic such as flowering time, which in some species is earlier in hybrids but in others later. Yet, biomass shows a fairly consistent increase in hybrids, so are genes involved with control of cell proliferation (Guo et al. 2010) particularly prone to show disruption in hybrids? Many studies on global gene expression have been conducted in hybrids and their parents in many plant species during the past decade (Schnable and Springer 2013). A wide range of results have been found, but a common central theme has not emerged that might illuminate a shared impact on regulatory complexes. However, these studies have assayed diverse tissues and used different normalization assumptions that could have obscured any patterns. This is an open question in the heterosis field that needs a creative experimental attack.

As more information becomes available about the relative rheostat among regulatory factors, it might become feasible to engineer crops for heterotic expression using genetic engineering and gene editing techniques (Kremling et al. 2018). The lessons noted above would suggest that particular relative expression of critical subunits of regulatory factors might elicit a desired response in the phenotype. The available evidence suggests that the total heterotic effect is a conglomerate so such directed experimentation would have to be performed for various characteristics. The context would be important in each case and must be carefully constructed. Nevertheless, the engineering of improved crops using information learned from heterosis experiments is an admirable goal notwithstanding its challenges.

Acknowledgments Related research to this topic in the author's laboratory is funded by the National Science Foundation grant IOS-1545780.

References

Abel S, Becker HC (2007) The effect of autopolyploidy on biomass production in homozygous lines of *Brassica rapa* and *Brassica oleracea*. Plant Breed 126:642–643

Alexander DE, Sonnemaker EH (1961) Inbreeding depression in autotetraploid maize. Maize Genetics Cooperation Newsletter 35:45

Bingham ET (1980) Maximizing heterozygosity in autotetraploids. In: Lewis WH (ed) Polyploidy: biological relevance. Plenum, New York, pp 471–489
Birchler JA (2012) Genetic consequences of polyploidy in plants. In: Soltis PS, Soltis DE (eds) Polyploidy and genome evolution. Springer, Berlin
Birchler JA, Newton KJ (1981) Modulation of protein levels in chromosomal dosage series of maize: the biochemical basis of aneuploid syndromes. Genetics 99:247–266
Birchler JA, Veitia RA (2007) The gene balance hypothesis: from classical genetics to modern genomics. Plant Cell 19:395–402
Birchler JA, Veitia RA (2010) The gene balance hypothesis: implications for gene regulation, quantitative traits and evolution. New Phytol 186:54–62
Birchler JA, Veitia RA (2012) Gene balance hypothesis: connecting issues of dosage sensitivity across biological disciplines. Proc Natl Acad Sci U S A 109:14746–14753
Birchler JA, Bhadra U, Bhadra MP, Auger DL (2001) Dosage-dependent gene regulation in multicellular eukaryotes: implications for dosage compensation, aneuploid syndromes, and quantitative traits. Dev Biol 234:275–288
Birchler JA, Riddle NC, Auger DL, Veitia RA (2005) Dosage balance in gene regulation: biological implications. Trends Genet 21:219–226
Birchler JA, Johnson AF, Veitia RA (2016) Kinetics genetics: incorporating the concept of genomic balance into an understanding of quantitative traits. Plant Sci 245:128–134
Blakeslee AF (1934) New Jimson weeds from old chromosomes. J Hered 25:81–108
Blakeslee AF, Belling J, Farnham ME (1920) Chromosomal duplication and Mendelian phenomena in Datura mutants. Science 52:388–390
Blanc G, Wolfe KH (2004) Functional divergence of duplicated genes formed by polyploidy during Arabidopsis evolution. Plant Cell 16:1679–1691
Bowers JE, Chapman BA, Rong J, Paterson AH (2003) Unravelling angiosperm genome evolution by phylogenetic analysis of chromosomal duplication events. Nature 422:433–438
Bray D, Lay S (1997) Computer-based analysis of the binding steps in protein complex formation. Proc Natl Acad Sci U S A 94:13493–13498
Briggle LW (1963) Heterosis in wheat—a review. Crop Sci 3:407–412
Brunelle DC, Sheridan WF (2014) The effects of varying chromosome arm dosage on maize plant morphogenesis. Genetics 198:171–180
Busbice TH, Wilsie CP (1966) Inbreeding depression and heterosis in autotetraploids with application to *Medicago sativa* L. Euphytica 15:52–67
Chang PL, Dilkes BP, McMahon M, Comai L, Nuzhdin SV (2010) Homoeolog-specific retention and use in allotetraploid *Arabidopsis suecica* depends on parent of origin and network partners. Genome Biol 11:R125
Chase SS (1980) Studies of monoploids, diploids and tetraploids of maize in relation to heterosis and inbreeding depression. In: Proceedings of the Argentine Society of Genetics
Chen ZJ (2013) Genomic and epigenetic insights into the molecular bases of heterosis. Nat Rev Genet 14:471–482
Cong B, Liu J, Tanksley SD (2002) Natural alleles at a tomato fruit size quantitative trait locus differ by heterochronic regulatory mutations. Proc Natl Acad Sci U S A 99:13606–13611
Cong B, Barrero LS, Tanksley SD (2008) Regulatory change in YABBY-like transcription factor led to evolution of extreme fruit size during tomato domestication. Nat Genet 40:800–804
Darwin C (1868) The variation of animals and plants under domestication. John Murray, London
Demarly Y (1963) Genetique des tetraploids et amelioration des plants. Ann Amelior Plant 13:307–400
Dewey DR (1966) Inbreeding depression in diploid, tetraploid and hexaploid crested wheatgrass. Crop Sci 6:144–147
Dewey DR (1969) Inbreeding depression in diploid and induced-autotetraploid crested wheatgrass. Crop Sci 9:592–595
Dunbier MW, Bingham ET (1975) Maximum heterozygosity in alfalfa: results using haploid-derived autotetraploids. Crop Sci 15:527–531

Duvick DN (1999) Heterosis: feeding people and protecting natural resources. In: Coors JG, Pandey S (eds) Genetics and exploitation of heterosis in crops. American Society of Agronomy, Inc. and Crops Science Society of America, Inc, Madison, pp 19–29

East EM (1936) Heterosis. Genetics 21:375–397

Fisher RA (1918) The correlation between relatives on the supposition of Mendelian inheritance. Trans Royal Soc Edinburgh 52:399–433

Flint-Garcia SA, Buckler ES, Tiffin P, Ersoz E, Springer NM (2009) Heterosis is prevalent for multiple traits in diverse maize germplasm. PLoS One 4:e7433

Frary A, Nesbitt TC, Grandillo S, Knaap E, Cong B, Liu J, Meller J, Elber R, Alpert KB, Tanksley SD (2000) fw2.2: a quantitative trait locus key to the evolution of tomato fruit size. Science 289:85–88

Freeling M (2009) Bias in plant gene content following different sorts of duplication: tandem, whole-genome, segmental, or by transposition. Annu Rev Plant Biol 60:433–453

Freeling M, Thomas BC (2006) Gene-balanced duplications, like tetraploidy, provide predictable drive to increase morphological complexity. Genome Res 16:805–814

Freeling M, Lyons E, Pedersen B, Alam M, Ming R, Lisch D (2008) Many or most genes in Arabidopsis transposed after the origin of the order Brassicales. Genome Res 18:1924–1937

Gallais A (1984) An analysis of heterosis vs. inbreeding effects with an autotetraploid cross-fertilized plant: *Medicago sativa* L. Genetics 106:123–137

Gravatt FA (1914) Radish-cabbage hybrid. J Hered 5:269–272

Groose RW, Talbert LE, Kojis WP, Bingham ET (1989) Progressive heterosis in autotetraploid alfalfa: studies using two types of inbreds. Crop Sci 29:1173–1177

Groszman M, Greaves IK, Albertyn ZI, Scofield GN, Peacock WJ, Dennis ES (2011) Changes in 24-nt siRNA levels in Arabidopsis hybrids suggest an epigenetic contribution to hybrid vigor. Proc Natl Acad Sci U S A 108:2617–2622

Guo M, Birchler JA (1994) Trans-acting dosage effects on the expression of model gene systems in maize aneuploids. Science 266:1999–2002

Guo M, Rupe MA, Dieter JA, Zou J, Spielbauer D, Duncan KE, Howard RJ, Hou Z, Simmons CR (2010) *Cell Number Regulator1* affects plant and organ size in maize: implications for crop yield enhancement and heterosis. Plant Cell 22:1057–1073

Henry IM, Zinkgraf MS, Groover AT, Comai L (2015) A system for dosage-based functional genomics in poplar. Plant Cell 27:2370–2383

Huang X, Yang S, Gong J, Zhao Q, Fang Q, Zhan Q, Zhao Y, Li W, Cheng B, Xia J, Chen N, Huang T, Zhang L, Fan D, Chen J, Zhou C, Lu Y, Weng Q, Han B (2016) Genomic architecture of heterosis for yield traits in rice. Nature 537:629–633

Jiang K, Liberatore KL, Park SJ, Alvarez JP, Lippman ZB (2013) Tomato yield heterosis is triggered by a dosage sensitivity of the florigen pathway that fine-tunes shoot architecture. PLoS Genet 9:e1004043

Jones DF (1917) Dominance of linked factors as a means of accounting for heterosis. Genetics 2:466–479

Karpechenko GD (1927) Polyploid hybrids of *Raphanus sativus* L. x *Brassica oleracea* L. Bull Appl Bot 17:305–410

Kawanabe T, Ishikura S, Miyaji N, Sasaki T, Wu LM, Itabashi E, Takada S, Shimizu M, Takasaki-Yasuda T, Osabe K, Peacock WJ, Dennis ES, Fujimoto R (2016) Role of DNA methylation in hybrid vigor in *Arabidopsis thaliana*. Proc Natl Acad Sci U S A 113:E6704–E6711

Kondrashov FA, Koonin EV (2004) A common framework for understanding the origin of genetic dominance and evolutionary fates of gene duplications. Trends Genet 20:287–290

Kremling KAG, Chen SY, Su MH, Lepak NK, Romay MC, Swarts KL, Lu F, Lorant A, Bradbury PJ, Buckler ES (2018) Dysregulation of expression correlates with rare-allele burden and fitness loss in maize. Nature 555:520–523

Krieger U, Lippman ZB, Zamir D (2010) The flowering gene SINGLE FLOWER TRUSS drives heterosis for yield in tomato. Nat Genet 42:459–463

Lee EA, Darrah LL, Coe EH (1996) Dosage effects on morphological and quantitative traits in maize aneuploids. Genome 39:898–908

Levings CS, Dudley JW, Alexander DE (1967) Inbreeding and crossing in autotetraploid maize. Crop Sci 7:72–73
Li Z, Li B, Tong Y (2008) The contribution of distant hybridization with decaploid *Agropyron elongatum* to wheat improvement in China. J Genet Genomics 35:451–456
Li Z, Defoor J, Tasdighian S, Maere S, Van de Peer Y, De Smet R (2016a) Gene duplicability of core genes is highly consistent across all angiosperms. Plant Cell 28:326–344
Li D, Huang Z, Song S, Xin Y, Mao D, Lv Q, Zhou M, Tian D, Tang M, Wu Q, Liu X, Chen T, Song X, Fu X, Zhao B, Liang C, Li A, Liu G, Li S, Hu S, Cao X, Yu J, Yuan L, Chen C, Zhu L (2016b) Integrated analysis of phenome, genome, and transcriptome of hybrid rice uncovered multiple heterosis-related loci for yield increase. Proc Natl Acad Sci U S A 113:E6026–E6035
Liu J, Van Eck J, Cong B, Tanksley SD (2002) A new class of regulatory genes underlying the cause of pear-shaped tomato fruit. Proc Natl Acad Sci U S A 99:13302–13306
Maere S, De Bodt S, Raes J, Casneuf T, Van Montagu M, Kuiper M, Van de Peer Y (2005) Modeling gene and genome duplications in eukaryotes. Proc Natl Acad Sci U S A 102:5454–5459
Miller M, Zhang C, Chen ZJ (2012) Ploidy and hybridity effects on growth vigor and gene expression in *Arabidopsis thaliana* hybrids and their parents. G3 2:505–513
Mok DWS, Peloquin SJ (1975) Breeding value of 2n pollen (diploandroids) in tetraploid x diploid crosses in potato. Theor Appl Genet 46:307–314
Ni Z, Kim ED, Ha M, Lackey E, Liu J, Zhang Y, Sun Q, Chen ZJ (2009) Altered circadian rhythms regulate growth vigor in hybrids and allopolyploids. Nature 457:327–331
Papp B, Pal C, Hurst LD (2003) Dosage sensitivity and the evolution of gene families in yeast. Nature 424:194–197
Park SJ, Jiang K, Tal L, Yichie Y, Gar O, Zamir D, Eshed Y, Lippman ZB (2014) Optimization of crop productivity in tomato using induced mutations in the florigen pathway. Nat Genet 46:1337–1342
Randolph LF (1935) Cytogenetics of tetraploid maize. J Agric Res 50:591–606
Randolph LF (1942) The influence of heterozygosis on fertility and vigor in autotetraploid maize. Genetics 27:163
Rhoades MM, Dempsey E (1966) Induction of chromosome doubling at meiosis by the elongate gene in maize. Genetics 54:505–522
Rice JS, Dudley JW (1974) Gene effects responsible for inbreeding depression in autotetraploid maize. Crop Sci 14:390–393
Richards AJ (1997) Plant breeding systems, 2nd edn. Chapman and Hall, London
Riddle NC, Birchler JA (2008) Comparative analysis of inbred and hybrid maize at the diploid and tetraploid levels. Theor Appl Genet 116:563–576
Riddle NC, Kato A, Birchler JA (2006) Genetic variation for the response to ploidy change in *Zea mays* L. Theor Appl Genet 114:101–111
Riddle NC, Jiang H, An L, Doerge RW, Birchler JA (2010) Gene expression analysis at the intersection of ploidy and hybridity in maize. Theor Appl Genet 120:341–353
Schnable PS, Springer NM (2013) Progress toward understanding heterosis in crop plants. Annu Rev Plant Biol 64:71–88
Schuster-Bockler B, Conrad D, Bateman A (2010) Dosage sensitivity shapes the evolution of copy-number varied regions. PLoS One 5:e9474
Semel Y, Nissenbaum J, Menda N, Zinder M, Krieger U, Issman N, Pleban T, Lippman Z, Gur A, Zamir D (2006) Overdominant quantitative trait loci for yield and fitness in tomato. Proc Natl Acad Sci U S A 103:12981–12986
Sheridan WF, Auger DL (2008) Chromosome segmental dosage analysis of maize morphogenesis using B-A-A translocations. Genetics 180:755–769
Singh R, Low ET, Ooi LC, Ong-Abdullah M, Ting NC, Nagappan J, Nookiah R, Amiruddin MD, Rosli R, Manaf MA, Chan KL, Halim MA, Azizi N, Lakey N, Smith SW, Budiman MA, Hogan M, Bacher B, Van Brunt A, Wang C, Ordway JM, Sambanthamurthi R, Martienssen RA (2013) The oil palm SHELL gene controls oil yield and encodes a homologue of SEEDSTICK. Nature 500:340–344

Sockness BA, Dudley JW (1989a) Performance of single and double cross autotetraploid maize hybrids with different levels of inbreeding. Crop Sci 29:875–879
Sockness BA, Dudley JW (1989b) Morphology and yield of isogenic diploid and tetraploid maize inbreds and hybrids. Crop Sci 29:1029–1032
Stadler LJ (1928a) Mutations in barley induced by x-rays and radium. Science 68:186–187
Stadler LJ (1928b) Genetic effects of X-rays in maize. Proc Natl Acad Sci U S A 14:69–75
Stupar RM, Bhaskar PB, Yandell BS, Rensink WA, Hart AL, Ouyang S, Veilleux RE, Busse JS, Erhardt RJ, Buell CR, Jiang J (2007) Phenotypic and transcriptomic changes associated with potato autopolyploidization. Genetics 176:2055–2067
Tan C, Pan Q, Cui C, Xiang Y, Ge X, Li Z (2016) Genome-wide gene/genome dosage imbalance regulates gene expressions in synthetic *Brassica napus* and derivatives (AC, AAC, CCA, CCAA). Front Plant Sci 7:1432
Tanksley SD (1993) Mapping polygenes. Annu Rev Genet 27:205–233
Thomas BC, Pedersen B, Freeling M (2006) Following tetraploidy in an Arabidopsis ancestor, genes were removed preferentially from one homeolog leaving clusters enriched in dose-sensitive genes. Genome Res 16:934–946
Tysdal HM, Kiesselbach TA, Westover HL (1942) Alfalfa breeding (Research Bulletin: Bulletin of the Agricultural Experiment Station of Nebraska No. 124)
Veitia RA (2002) Exploring the etiology of haploinsufficiency. Bioessays 24:175–184
Veitia RA, Vaiman D (2011) Exploring the mechanistic bases of heterosis from the perspective of macromolecular complexes. FASEB J 25:476–482
Veitia RA, Bottani S, Birchler JA (2008) Cellular reactions to gene dosage imbalance: genomic, transcriptomic and proteomic effects. Trends Genet 24:390–397
Veitia RA, Bottani S, Birchler JA (2013) Gene dosage effects: nonlinearities, genetic interactions, and dosage compensation. Trends Genet 29:385–393
Wang L, Greaves IK, Groszmann M, Wu LM, Dennis ES, Peacock WJ (2015) Hybrid mimics and hybrid vigor in Arabidopsis. Proc Natl Acad Sci U S A 112:E4959–E4967
Wang L, Wu LM, Greaves IK, Zhu A, Dennis ES, Peacock WJ (2017) PIF4-controlled auxin pathway contributes to hybrid vigor in Arabidopsis thaliana. Proc Natl Acad Sci U S A 114: E3555–E3562
Weber A, Clark RM, Vaughn L, Sanchez-Gonzalez Jde J, Yu J, Yandell BS, Bradbury P, Doebley J (2007) Major regulatory genes in maize contribute to standing variation in teosinte (*Zea mays* ssp. *parviglumis*). Genetics 177:2349–2359
Williams RD (1931) Self- and cross-fertility and flowering habits of certain herbage grasses and legumes. Welsh Plant Breeding Station Bulletin Series H. No. 12, pp 217–220
Yao H, Kato A, Mooney B, Birchler JA (2011) Phenotypic and gene expression analysis of a ploidy series of maize inbred Oh43. Plant Mol Biol 75:237–251
Yao H, Dogra Gray A, Auger DL, Birchler JA (2013) Genomic dosage effects on heterosis in triploid maize. Proc Natl Acad Sci U S A 110:2665–2669

James A. Birchler is Curators' Distinguished Professor of Biological Sciences at the University of Missouri, Columbia. He is an elected fellow of the National Academy of Sciences, American Association for the Advancement of Science, a fellow of the National Academy of Inventors, and was named as an Einstein Professor of the Chinese Academy of Sciences. His laboratory works on questions of genome balance, development of plant biotechnology tools, and heterosis.

13 Genome Information Resources to Improve Plant Biomass Productivity

Keiichi Mochida and Lam-Son Phan Tran

Abstract

Recent advances in analytical platforms and information techniques have enabled us to develop genomic resources and tools in various plant species useful for biomass feedstocks. In this chapter, we provide an overview of recent advances in genome resource development in biomass plants. Specifically, we focus on grass species such as maize, sugarcane, sorghum, switchgrass, and *Miscanthus* spp. as well as oil crops such as soybean, sunflower, *Jatropha*, and oil palm, highlighting genome-based efforts and information resources to improve their biomass productivity.

Keywords

Grass · Oil crop · Biomass · Genome

13.1 Introduction

Technologies for biorefinery are expected to facilitate building the bio-based economy to ensure our sustainable development, thus reducing our dependence on fossil resources (Ragauskas et al. 2006; OECD 2009). Bioconversion processes in biorefineries generate bio-based products, such as biofuel, bioenergy, and bio-based chemicals from biomass input as feedstock (Clark et al. 2012). In 1941, Henry Ford demonstrated a prototype car with its body and fenders built from bioplastic derived from soybean (*Glycine max*), wheat (*Triticum aestivum*), and corn (*Zea mays*) (Kay and Bud 2006). As plant varieties with useful traits for biorefining may provide a promising source for biomass feedstock from terrestrial

K. Mochida (✉) · L.-S. P. Tran
RIKEN Center for Sustainable Resource Science, Yokohama, Japan
e-mail: keiichi.mochida@riken.jp; son.tran@riken.jp

A. Kumar et al. (eds.), *Climate Change, Photosynthesis and Advanced Biofuels*,
https://doi.org/10.1007/978-981-15-5228-1_13

areas (De Bhowmick et al. 2018), the improvement of traits related to biomass utilities and productivity is crucial to ensure the downstream bioconversion process. Therefore, genomic and information resources integrating genome-scale outcomes from plant species for biomass use have increasingly become important to assist in their molecular breeding.

Recent advances in analytical platforms and informatics have enabled us to rapidly develop genome-scale resources and tools not only in model species but also in crops (Mochida and Shinozaki 2013). Remarkable innovations in the area of high-throughput sequencing have enabled us to decipher entire genetic codes from any organism, population, or biological community, and applications for transcriptomics and epigenomics are expanding dramatically (Mochida and Shinozaki 2010, 2011, 2013). For efficient utilization of genome-scale datasets to improve the biomass productivity of plants (Clifton-Brown et al. 2019), useful bioinformatics tools have become increasingly important, which improve our analytical throughput in molecular breeding of plants and the heuristic power to identify genes associated with plant biomass productivity.

Terrestrial biomass usually includes grass species, woody species, and oil crops. The grass family includes promising species for lignocellulosic feedstocks with useful biological properties, such as high yield, rapid growth, high water, and nutrition use efficiency, and the ability to adapt to broader environments (Tubeileh et al. 2016). Oil crops supply biomass for the production of biodiesel fuel (BDF), which can be used not only in pure form but also blended with petroleum diesel at any concentration. Advances in high-throughput sequencing technologies have accelerated reference and draft genome sequencing projects in various plant species, including a number of grass and oil crop species like soybean (Biswas et al. 2008) and *Jatropha* (Meher et al. 2013). Moreover, high-throughput sequencing-based applications, such as whole-genome resequencing and exome sequencing to identify genome-scale polymorphisms, have assisted us in exploring genetic diversities associated with traits related to biomass productivity through quantitative genetics approaches like genome-wide association study (GWAS) and quantitative trait locus (QTL) analyses and in promoting genomic-assisted breeding to estimate breeding values and genetic gains for ideal traits through genomic selection.

In this chapter, we provide an overview of recent advances in genome resource development in biomass plants. We describe recent progress in genomic resources and their applications in grass plants and oil crops. Throughout the chapter, we highlight recent genome-based efforts and available information resources to improve plant biomass productivity.

13.2 Information Resources in Grasses

Panicoideae, a subfamily in the Poaceae family, contains several promising grass species for biomass use. Maize and sugarcane have long histories as feedstocks for bioethanol production (Waclawovsky et al. 2010; van der Weijde et al. 2013) due to their high yield and ease of conversion into alcohol. Sorghum biomass is also a promising source for biomass use, owing to its high yield, wide environmental adaptability, and high nitrogen and water use efficiency (Ordonio et al. 2016). In addition to these crops, several perennial grass species have been proposed as potential sources for biomass use. Here, we describe genomic and information resources developed for these biomass grass plants (Table 13.1).

13.3 Maize

Maize (*Zea mays*) is a staple crop that provides a large proportion of human caloric intake and animal feed globally (Schnable et al. 2009). Unlike the year 2000, in which less than 5% of US corn was used in bioethanol production, 40% went toward producing ethanol in 2014 (https://www.extension.iastate.edu/agdm/crops/outlook/cornbalancesheet.pdf). Therefore, corn-based ethanol production has emerged as a competitor for grain use and land used for the production of other staple food crops. In addition to its features as an important crop species, maize has a long history serving as a model genetic system. There are numerous genomic resources and omics spectrums for maize, such as MaizeGDB, a web-accessible database for maize genetics and genomics that serves as a community portal to integrate broad genomic resources. In 2009, the reference assembly and annotation of the B73 maize genome, which has been revised three times, were released using the bacterial artificial chromosome (BAC)-by-BAC sequencing strategy (Schnable et al. 2009). The latest assembly of the B73 maize genome was constructed based on PacBio single-molecule real-time sequencing and optical mapping (Jiao et al. 2017), in which 39,591 protein-encoding genes and 6812 non-coding genes have been annotated. After the maize reference genome was mapped, Gore et al. (2009) reported a first-generation haplotype map (maize HapMap) based on genotype datasets of 27 diverse maize inbred lines (Gore et al. 2009), whereas Chia et al. (2012) reported a second-generation maize map (HapMap2) based on 55 million SNPs detected in 103 lines across pre-domesticated and domesticated varieties (Chia et al. 2012). These maps have provided invaluable resources for gene discovery and molecular breeding to improve productivity in maize. More recently, to demonstrate the feasibility of high-resolution genetic mapping for constructing the maize "pan-genome" covering the entire gene set of all individuals of a single species, Lu et al. mapped 26 million tags generated by reduced representation sequencing of 14,129 maize inbred lines (Lu et al. 2015). A nested association mapping population composed of 25 maize recombinant inbred lines and their genome-scale polymorphism datasets have provided the opportunity to identify genes related to complex traits (Buckler et al. 2009; McMullen et al. 2009). Another multi-parental genetic population, called the multi-parent advanced generation intercross (MAGIC) population, has facilitated the genetic analysis of complex traits, owing to its advantage that eliminates population structure (Holland 2015). Combining these useful genetic resources and high-throughput genotyping applications, diverse traits have been dissected through large-scale quantitative genetics approaches, as recently reviewed by Xiao et al. (2017).

Table 13.1 Information resources for potential biomass grasses

		Zea mays	*Saccharum* spp.	*Sorghum bicolor*	*Panicum* spp.	*Miscanthus* spp.
Genome sequence and annotation						
Phytozome	https://phytozome.jgi.doe.gov/pz/portal.html	*Zea mays Ensembl-18*		*Sorghum bicolor* v3.1.1		*Miscanthus sinensis* v7.1
Ensembl Plants	http://plants.ensembl.org/index.html	B73_RefGen_v4		Sorghum_bicolor_NCBIv3	PhalliiHAL_v2.1 PHallii_v3.1	
Functional classification						
KEGG	https://www.genome.jp/kegg/kegg_ja.html	T01088		T01086		
PlantCyc	https://www.plantcyc.org/	CORNCYC 9.0		SORGHUMBICOLORCYC 6.0	SWITCHGRASSCYC 6.0 PHALLIICYC 1.0	MSINENSISCYC 1.0
InterMine	https://intermine.org/	MaizeMine				
Integrated database		MaizeGDB	The Sugarcane Genome Hub	Sorghum Functional Genomics Database	Switchgrass Functional Genomics Server	

KEGG Kyoto Encyclopedia of Genes and Genomes

13.4 Sugarcane

Modern cultivated sugarcane (*Saccharum* spp.) is one of the most valuable crops for the global sugar economy and for bioethanol (de Souza Dias et al. 2015). Because modern sugarcane is an allopolyploid species generated through interspecific hybridization between *Saccharum officinarum* and *S. spontaneum*, the nuclear genome of its typical cultivars has been estimated to be 7.88 Gb (2n = 8x = 80), which is composed of sub-genomes from these two species (Thirugnanasambandam et al. 2018). Exploiting conserved synteny between *Saccharum* species and sorghum, Garsmeur et al. constructed an assembly of a 382 Mb single tiling path of 4660 sugarcane BAC, identified 25,316 protein-encoding gene models (Garsmeur et al. 2018), and provided a web-accessible database, The Sugarcane Genome Hub (http://sugarcane-genome.cirad.fr/), to integrate and allow access to the associated dataset. Although its complex and large genome size has been a long-standing challenge for decoding its whole-genome sequence, an international consortium comprising more than 100 scientists from 16 institutions eventually decoded the whole-genome sequence of a haploid *S. spontaneum* using multiple methods, including Illumina short-read sequencing, BAC library sequencing, PacBio sequencing, and Hi-C library sequencing, and alleles of 35,525 genes were annotated (Zhang et al. 2018). In addition to the reference assembly, they resequenced 64 *S. spontaneum* genomes and identified balancing selection of sequences in the rearranged chromosomal regions (Zhang et al. 2018). These lines of information on sugarcane genomes and annotated genes have provided crucial data for trait improvements in sugarcane breeding.

13.5 Sorghum

Sorghum (*Sorghum bicolor*) is another species in Saccharinae plants, which has been widely used as food, feed, dietary fiber, and feedstock for biofuel and is widely grown owing to its ability to adapt to hot and dry conditions (Ordonio et al. 2016). In 2009, the sequencing team of *S. bicolor* deciphered its 730 Mb genome using Sanger sequencing of plasmids, fosmids, and two BAC libraries (Paterson et al. 2009). The most recent update release (v3.0) includes 34,118 protein-encoding genes and 1449 non-coding genes in the genome annotation. To develop genome-scale polymorphism data in sorghum, Bekele et al. resequenced five genetically diverse *S. bicolor* genotypes, including three sweet sorghums and two grain sorghums, and identified over one million high-quality SNPs (Bekele et al. 2013). Furthermore, by analyzing the genetic variation of diverse sorghum germplasm comprising 971 accessions from worldwide collections using genotyping-by-sequencing, Morris et al. revealed its population structure and performed GWAS analyses on plant height components and inflorescence architecture (Morris et al. 2013). More recently, Deschamps et al. demonstrated that the combined approach of long-read nanopore sequencing and optical mapping was useful to construct a chromosome-scale de novo assembly of the repeat-rich *Sorghum* Tx430 genome (Deschamps et al. 2018). Moreover, aiming to establish reference phenome datasets for sorghum research, the Transportation Energy Resources from Renewable Agriculture Phenotyping Reference Platform (TERRA-REF) has provided genotype data from more than 350 sorghum accessions, as well as their phenotype datasets acquired by using a high-resolution field scanner system (https://terraref.org/). These genome and phenome resources in sorghum have provided a research platform beneficial for gene discovery and molecular breeding to improve its traits for biomass use.

13.6 Switchgrass

Switchgrass (*Panicum virgatum*), a native perennial tetraploid grass, is a potential biomass crop for biofuel production (Bouton 2007). Regarding genome resources and tools for switchgrass before 2013, Nageswara-Rao et al. provided a comprehensive review that considered the biotechnology tools available for genetic improvement of switchgrass as well as its genomic resources (Nageswara-Rao et al. 2013). In the latest version of Phytozome (v. 12.1), a release assembly annotation of the switchgrass AP13 genome is available, which comprises approximately 1.23 Gb arranged in 319,670 contigs generated from the assembly of Roch454 reads and contains 98,007 protein-encoding genes (https://phytozome.jgi.doe.gov/pz/portal.html#!info?alias=Org_Pvirgatum). On the other hand, to analyze polymorphisms in genic regions of switchgrass based on datasets of its Sanger sequence and pyrosequencing-derived transcript sequence, Evans et al. designed exome capture probes in the SeqCap platform (the Roche-NimbleGen probe set, 120911_Switchgrass_GLBRC_R_EZ_HX1) and applied the exome capture probes to assess nucleotide polymorphisms and copy number variation in switchgrass (Evans et al. 2014). The same authors then applied the exome capture probe set to analyze a switchgrass panel composed of 547 individuals from 45 upland and 21 lowland populations and developed the switchgrass HapMap v.1 set, which includes 1,377,841 SNPs (Evans et al. 2015). The exome capture set was used to investigate genetic polymorphisms of various switchgrass populations (Ramstein et al. 2016, 2018; Taylor et al. 2018). Using this probe set, Grabowski et al. examined genetic association based on polymorphisms in an association panel comprising 509 genotypes with different flowering times in switchgrass (Grabowski et al. 2017). Regarding whole-genome assemblies from the species closest to switchgrass, a chromosome-scale assembly of broomcorn millet (*P. miliaceum*), which contains 55,930 protein-encoding genes and 339 microRNA genes, has been recently published using a combination of short-read sequencing, single-molecule real-time sequencing, Hi-C, and a high-density genetic map (Shi et al. 2019). As for other *Panicum* species, whole-genome sequences of two perennial grasses, the *P. hallii* accessions, were sequenced and compared, which were further used to investigate the population structure of *P. hallii* and eQTL analysis to dissect the genetic basis of its drought responses.

13.7 *Miscanthus* spp.

Miscanthus, a genus of the tribe Andropogoneae in Panicoideae, contains C4 grasses that are promising for biomass use because of their high biomass yields and adaptive ability in cold environments (Lee and Kuan 2015). Phytozome (ver. 12), the first chromosome-scale assembly of *M. sinensis* doubled haploid DH1 (IGR-2011-001) that is now accessible, is composed of 2 Gb sequence data arranged in 19 chromosomes and some unmapped scaffolds with 67,789 protein-encoding genes (https://phytozome.jgi.doe.gov/pz/portal.html#!info?alias=Org_Msinensis_er). This assembly is expected to provide reference genome information for its relatives, including the interspecific triploid *M. x giganteus* (Mxg) between *M. sacchariflorus* and *M. sinensis*. Swaminathan et al. performed survey sequencing of the Mxg genome and its small RNA (Swaminathan et al. 2010). Clark et al. investigated a total of 1513 genotypes of *Miscanthus* species from Japan using restriction site-associated DNA sequencing (RAD-seq) of 20,704 SNPs and 10 plastid microsatellites, revealing their population structure (Clark et al. 2015). Clark et al. also reported the population structure of *Miscanthus* species from eastern Russia using RAD-seq and simple sequence markers (Clark et al. 2016). In addition, multiple reports on transcriptome analyses in *Miscanthus* species that examine the transcriptome diversity in populations are available for readers (Xu et al. 2016; Yan et al. 2017; Zhu et al. 2017; Xing et al. 2018).

13.8 Information Resources in Oil Crops

Several existing oil crops have been considered as potential biomass sources of feedstock for biofuel and bioplastic production. Here, we briefly introduce the current situation in the development of genomic resources in soybean, sunflower, *Jatropha*, and oil palm (Table 13.2), which have been used not only as human food and animal feed but also as biomass feedstock.

13.9 Soybean

Soybean (*Glycine max*) is one of the most important leguminous crops for human food, animal feed, and biofuel and bioplastic production. Its entire 1.1 Gb genome has been deciphered, in which 46,430 protein-encoding genes were annotated in 2010 (Schmutz et al. 2010). The latest genomic information available in Phytozome (https://phytozome.jgi.doe.gov/pz/portal.html#!info?alias=Org_Gmax) represents an assembly of approximately 978 Mb, which is mostly assigned to its 20 chromosomes and 56,044 protein-encoding genes. Lam et al. reported resequencing data of 17 wild and 14 cultivated soybeans and identified linkage disequilibrium and 205,614 tag SNPs (Lam et al. 2010). Subsequently, Zhou et al. reported a resequencing of 302 wild and cultivated accessions, representing the population stratification associated with geographical regions (Zhou et al. 2015).

Table 13.2 Information resources for potential oil crops

		Soybean	Sunflower	Jatropha	Oil palm
Genome sequence and annotation					
Phytozome	https://phytozome.jgi.doe.gov/pz/portal.html	*Glycine max* Wm82.a2.v1	*Helianthus annuus* r1.2		
Ensembl Plants	https://plants.ensembl.org/index.html	Glycine_max_v2.1	HanXRQr1.0		
Functional classification					
KEGG	https://www.genome.jp/kegg/kegg_ja.html	T01710	T05101	T03922	T03921
PlantCyc	https://www.plantcyc.org/	SOYCYC 9.0	SUNFLOWERCYC 1.0		
InterMine	https://intermine.org/	SoyMine			
Integrated database		SoyBase	Sunflower Genome Database INRA Sunflower Bioinformatics Resources	JCDB	DRDB PalmXplore MyOPGP

KEGG Kyoto Encyclopedia of Genes and Genomes

Li et al. published a whole-genome de novo assembly of an annual wild soybean, *G. soja*, that is closest to the cultivated soybean (Li et al. 2014). More recently, Xie et al. published a reference-grade assembly and annotation of a wild soybean accession, W05 (Xie et al. 2019). For soybean research, SoyBase serves as an information portal resource by providing a one-stop shop for various genome tools and resources in soybean (https://soybase.org/).

13.10 Sunflower

Sunflower (*Helianthus annuus*) is an oil crop with the potential for biofuel production (Badouin et al. 2017). The international sunflower genome sequencing consortium published its reference genome (3.9 Gb) containing 52,243 protein-encoding genes (https://phytozome.jgi.doe.gov/pz/portal.html#!info?alias=Org_Hannuus_er). A member of the sequencing consortium, Institut National de la Recherche Agronomique (INRA), has provided associated genome resources on a website called "INRA Sunflower Bioinformatics Resources" (https://www.heliagene.org/). Additionally, the sunflower genome database (https://sunflowergenome.org/) provides various genome and transcriptome datasets.

13.11 *Jatropha*

Jatropha (physic nut [*Jatropha curcas*]) is a member of Euphorbiaceae and has attracted wide attention as a feedstock for biodiesel production owing to its high seed oil content, rapid growth, and high adaptive ability in adverse environments (Kumar and Sharma 2008; Abdulla et al. 2011; Montes and Melchinger 2016). Its first whole-genome assembly was reported in 2011, which represents approximately 285.9 Mb and 21,225 annotated genes, and has been accessible from the Jatropha Genome Database (http://www.kazusa.or.jp/jatropha/statistics.html). In 2015, Wu et al. published a whole-genome assembly of *Jatropha* comprising 320.5 Mb in total scaffold length with 27,172 protein-encoding genes, in which 81.7% of the assembled sequences were anchored to the 11 linkage group based on 1208 markers (Wu et al. 2015). Recently, Xia et al. reported a high-density linkage map composed of 3422 SNPs and InDel markers in *Jatropha*, and they applied the map to identify QTLs related to its agronomic traits (Xia et al. 2018). More recently, using genotyping-by-sequencing-based polymorphism discovery, Vandepitte et al. sequenced 175 *Jatropha* genotypes and identified 25,715 SNPs that were used to examine genetic differences between toxic and nontoxic genotypes (Vandepitte et al. 2019).

13.12 Oil Palm

Oil palm (*Elaeis guineensis*, African oil palm) is a perennial monocot of the family Arecaceae, which has a long cultivation history in tropical and subtropical areas as a source for edible and non-edible applications (Corley and Tinker 2015). Singh et al. reported the *E. guineensis* genome in 2013, which contained approximately 1.5 Gb of assembled sequences and 34,802 annotated genes (Singh et al. 2013). Jin et al. published a draft genome assembly of an elite Dura palm (a major oil palm variety) and reported the resequencing results of an additional 17 oil palm trees (Jin et al. 2016). Aiming to facilitate oil palm breeding based on genomic resources, the Malaysian Oil Palm Genome Programme (MyOPGP) has provided genome sequences from *E. guineensis* and *E. oleifera* and their associated database (http://genomsawit.mpob.gov.my/index.php?track=30&nu=1).

13.13 Conclusions and Future Perspectives

The available genomic resources for potential biomass sources and oil plants have dramatically advanced in recent years. Rapidly increasing whole reference genome information and genome-scale and population-scale variation datasets of the biomass and oil plant species will facilitate gene discovery research programs and breeding processes to improve their biomass productivity and utilities. High-throughput plant phenotyping combined with the available genome-wide polymorphism data will enable us to accelerate the genetic factors associated with agronomically important traits related to biomass and oil contents in biomass and oil crops through quantitative genetics approaches (Mochida et al. 2018). Moreover, integration of such genomic information with other omics data, such as transcriptomes and metabolomes, will contribute to our understanding of cellular systems underlying genotype/phenotype relationships associated with the productivity of biomass and oil crops. Therefore, information resources that integrate data from the various large-scale analytical platforms and analytical tools will help us to achieve improvements in the productivity of biomass and oil crops for the development of a sustainable society with a biomaterial-based economy.

References

Abdulla R, Chan ES, Ravindra P (2011) Biodiesel production from *Jatropha curcas*: a critical review. Crit Rev Biotechnol 31(1):53–64. https://doi.org/10.1016/j.enconman.2013.12.058

Badouin H, Gouzy J, Grassa CJ, Murat F, Staton SE, Cottret L, Lelandais-Brière C, Owens GL, Carrère S, Mayjonade B, Legrand L, Gill N, Kane NC, Bowers JE, Hubner S, Bellec A, Bérard A, Bergès H, Blanchet N, Boniface M-C, Brunel D, Catrice O, Chaidir N, Claudel C, Donnadieu C, Faraut T, Fievet G, Helmstetter N, King M, Knapp SJ, Lai Z, Le Paslier M-C, Lippi Y, Lorenzon L, Mandel JR, Marage G, Marchand G, Marquand E, Bret-Mestries E, Morien E, Nambeesan S, Nguyen T, Pegot-Espagnet P, Pouilly N, Raftis F, Sallet E, Schiex T, Thomas J, Vandecasteele C, Varès D, Vear F, Vautrin S, Crespi M, Mangin B, Burke JM,

Salse J, Muños S, Vincourt P, Rieseberg LH, Langlade NB (2017) The sunflower genome provides insights into oil metabolism, flowering and Asterid evolution. Nature 546 (7656):148–152. https://doi.org/10.1038/nature22380

Bekele WA, Wieckhorst S, Friedt W, Snowdon RJ (2013) High-throughput genomics in sorghum: from whole-genome resequencing to a SNP screening array. Plant Biotechnol J 11 (9):1112–1125. https://doi.org/10.1111/pbi.12106

Biswas A, Sharma BK, Willett JL, Erhan SZ, Cheng HN (2008) Soybean oil as a renewable feedstock for nitrogen-containing derivatives. Energy Environ Sci 1(6):639–644. https://doi.org/10.1039/b809215j

Bouton JH (2007) Molecular breeding of switchgrass for use as a biofuel crop. Curr Opin Genet Dev 17(6):553–558. https://doi.org/10.1016/j.gde.2007.08.012

Buckler ES, Holland JB, Bradbury PJ, Acharya CB, Brown PJ, Browne C, Ersoz E, Flint-Garcia S, Garcia A, Glaubitz JC, Goodman MM, Harjes C, Guill K, Kroon DE, Larsson S, Lepak NK, Li H, Mitchell SE, Pressoir G, Peiffer JA, Rosas MO, Rocheford TR, Romay MC, Romero S, Salvo S, Villeda HS, Da Silva HS, Sun Q, Tian F, Upadyayula N, Ware D, Yates H, Yu J, Zhang Z, Kresovich S, McMullen MD (2009) The genetic architecture of maize flowering time. Science 325(5941):714–718. https://doi.org/10.1126/science.1174276

Chia JM, Song C, Bradbury PJ, Costich D, De Leon N, Doebley J, Elshire RJ, Gaut B, Geller L, Glaubitz JC, Gore M, Guill KE, Holland J, Hufford MB, Lai J, Li M, Liu X, Lu Y, McCombie R, Nelson R, Poland J, Prasanna BM, Pyhäjärvi T, Rong T, Sekhon RS, Sun Q, Tenaillon MI, Tian F, Wang J, Xu X, Zhang Z, Kaeppler SM, Ross-Ibarra J, McMullen MD, Buckler ES, Zhang G, Xu Y, Ware D (2012) Maize HapMap2 identifies extant variation from a genome in flux. Nat Genet 44(7):803–807. https://doi.org/10.1038/ng.2313

Clark JH, Luque R, Matharu AS (2012) Green chemistry, biofuels, and biorefinery. Annu Rev Chem Biomol Eng 3:183–207. https://doi.org/10.1146/annurev-chembioeng-062011-081014

Clark LV, Ryan Stewart J, Nishiwaki A, Toma Y, Kjeldsen JB, Jørgensen U, Zhao H, Peng J, Yoo JH, Heo K, Yu CY, Yamada T, Sacks EJ (2015) Genetic structure of *Miscanthus sinensis* and *Miscanthus sacchariflorus* in Japan indicates a gradient of bidirectional but asymmetric introgression. J Exp Bot 66(14):4213–4225. https://doi.org/10.1093/jxb/eru511

Clark LV, Dzyubenko E, Dzyubenko N, Bagmet L, Sabitov A, Chebukin P, Johnson DA, Kjeldsen JB, Petersen KK, Jørgensen U, Yoo JH, Heo K, Yu CY, Zhao H, Jin X, Peng J, Yamada T, Sacks EJ (2016) Ecological characteristics and in situ genetic associations for yield-component traits of wild Miscanthus from eastern Russia. Ann Bot 118(5):941–955. https://doi.org/10.1093/aob/mcw137

Clifton-Brown J, Harfouche A, Casler MD, Dylan Jones H, Macalpine WJ, Murphy-Bokern D, Smart LB, Adler A, Ashman C, Awty-Carroll D, Bastien C, Bopper S, Botnari V, Brancourt-Hulmel M, Chen Z, Clark LV, Cosentino S, Dalton S, Davey C, Dolstra O, Donnison I, Flavell R, Greef J, Hanley S, Hastings A, Hertzberg M, Hsu TW, Huang LS, Iurato A, Jensen E, Jin X, Jørgensen U, Kiesel A, Kim DS, Liu J, McCalmont JP, McMahon BG, Mos M, Robson P, Sacks EJ, Sandu A, Scalici G, Schwarz K, Scordia D, Shafiei R, Shield I, Slavov G, Stanton BJ, Swaminathan K, Taylor G, Torres AF, Trindade LM, Tschaplinski T, Tuskan GA, Yamada T, Yeon Yu C, Zalesny RS, Zong J, Lewandowski I (2019) Breeding progress and preparedness for mass-scale deployment of perennial lignocellulosic biomass crops switchgrass, miscanthus, willow and poplar. GCB Bioenergy 11(1):118–151. https://doi.org/10.1111/gcbb.12566

Corley RHV, Tinker PB (2015) The oil palm, 5th edn. Wiley-Blackwell, Hoboken. isbn:978-1-118-95329-7

De Bhowmick G, Sarmah AK, Sen R (2018) Lignocellulosic biorefinery as a model for sustainable development of biofuels and value added products. Bioresour Technol 247:1144–1154. https://doi.org/10.1016/j.biortech.2017.09.163

de Souza Dias MO, Maciel Filho R, Mantelatto PE, Cavalett O, Rossell CEV, Bonomi A, Leal MRLV (2015) Sugarcane processing for ethanol and sugar in Brazil. Environ Dev 15:35–51. https://doi.org/10.1016/j.envdev.2015.03.004

Deschamps S, Zhang Y, Llaca V, Ye L, Sanyal A, King M, May G, Lin H (2018) A chromosome-scale assembly of the sorghum genome using nanopore sequencing and optical mapping. Nat Commun 9(1):4844. https://doi.org/10.1038/s41467-018-07271-1

Evans J, Kim J, Childs KL, Vaillancourt B, Crisovan E, Nandety A, Gerhardt DJ, Richmond TA, Jeddeloh JA, Kaeppler SM, Casler MD, Buell CR (2014) Nucleotide polymorphism and copy number variant detection using exome capture and next-generation sequencing in the polyploid grass *Panicum virgatum*. Plant J 79(6):993–1008. https://doi.org/10.1111/tpj.12601

Evans J, Crisovan E, Barry K, Daum C, Jenkins J, Kunde-Ramamoorthy G, Nandety A, Ngan CY, Vaillancourt B, Wei CL, Schmutz J, Kaeppler SM, Casler MD, Buell CR (2015) Diversity and population structure of northern switchgrass as revealed through exome capture sequencing. Plant J 84(4):800–815. https://doi.org/10.1111/tpj.13041

Garsmeur O, Droc G, Antonise R, Grimwood J, Potier B, Aitken K, Jenkins J, Martin G, Charron C, Hervouet C, Costet L, Yahiaoui N, Healey A, Sims D, Cherukuri Y, Sreedasyam A, Kilian A, Chan A, Van Sluys MA, Swaminathan K, Town C, Bergès H, Simmons B, Glaszmann JC, Van Der Vossen E, Henry R, Schmutz J, D'Hont A (2018) A mosaic monoploid reference sequence for the highly complex genome of sugarcane. Nat Commun 9(1):2638. https://doi.org/10.1038/s41467-018-05051-5

Gore MA, Chia JM, Elshire RJ, Sun Q, Ersoz ES, Hurwitz BL, Peiffer JA, McMullen MD, Grills GS, Ross-Ibarra J, Ware DH, Buckler ES (2009) A first-generation haplotype map of maize. Science 326(5956):1115–1117. https://doi.org/10.1126/science.1177837

Grabowski PP, Evans J, Daum C, Deshpande S, Barry KW, Kennedy M, Ramstein G, Kaeppler SM, Buell CR, Jiang Y, Casler MD (2017) Genome-wide associations with flowering time in switchgrass using exome-capture sequencing data. New Phytol 213(1):154–169. https://doi.org/10.1111/nph.14101

Holland JB (2015) MAGIC maize: a new resource for plant genetics. Genome Biol 16(1):163. https://doi.org/10.1186/s13059-015-0713-2

Jiao Y, Peluso P, Shi J, Liang T, Stitzer MC, Wang B, Campbell MS, Stein JC, Wei X, Chin CS, Guill K, Regulski M, Kumari S, Olson A, Gent J, Schneider KL, Wolfgruber TK, May MR, Springer NM, Antoniou E, McCombie WR, Presting GG, McMullen M, Ross-Ibarra J, Dawe RK, Hastie A, Rank DR, Ware D (2017) Improved maize reference genome with single-molecule technologies. Nature 546(7659):524–527. https://doi.org/10.1038/nature22971

Jin J, Lee M, Bai B, Sun Y, Qu J, Rahmadsyah AY, Lim CH, Suwanto A, Sugiharti M, Wong L, Ye J, Chua NH, Yue GH (2016) Draft genome sequence of an elite Dura palm and whole-genome patterns of DNA variation in oil palm. DNA Res 23(6):527–533. https://doi.org/10.1093/dnares/dsw036

Kay LE, Bud R (2006) The uses of life: a history of biotechnology. Technol Cult. https://doi.org/10.2307/3106295

Kumar A, Sharma S (2008) An evaluation of multipurpose oil seed crop for industrial uses (*Jatropha curcas* L.): a review. Ind Crops Prod 28:1–10. https://doi.org/10.1016/j.indcrop.2008.01.001

Lam HM, Xu X, Liu X, Chen W, Yang G, Wong FL, Li MW, He W, Qin N, Wang B, Li J, Jian M, Wang J, Shao G, Wang J, Sun SS, Zhang G (2010) Resequencing of 31 wild and cultivated soybean genomes identifies patterns of genetic diversity and selection. Nat Genet 42(12):1053–1059. https://doi.org/10.1038/ng.715

Lee WC, Kuan WC (2015) Miscanthus as cellulosic biomass for bioethanol production. Biotechnol J 10(6):840–854. https://doi.org/10.1002/biot.201400704

Li YH, Zhou G, Ma J, Jiang W, Jin LG, Zhang Z, Guo Y, Zhang J, Sui Y, Zheng L, Zhang SS, Zuo Q, Shi XH, Li YF, Zhang WK, Hu Y, Kong G, Hong HL, Tan B, Song J, Liu ZX, Wang Y, Ruan H, Yeung CKL, Liu J, Wang H, Zhang LJ, Guan RX, Wang KJ, Wen-bin L, Chen SY, Chang RZ, Jiang Z, Jackson SA, Li R, Qiu LJ (2014) De novo assembly of soybean wild relatives for pan-genome analysis of diversity and agronomic traits. Nat Biotechnol 32(10):1045–1052. https://doi.org/10.1038/nbt.2979

Lu F, Romay MC, Glaubitz JC, Bradbury PJ, Elshire RJ, Wang T, Li Y, Li Y, Semagn K, Zhang X, Hernandez AG, Mikel MA, Soifer I, Barad O, Buckler ES (2015) High-resolution genetic mapping of maize pan-genome sequence anchors. Nat Commun 6:6914. https://doi.org/10.1038/ncomms7914

McMullen MD, Kresovich S, Villeda HS, Bradbury P, Li H, Sun Q, Flint-Garcia S, Thornsberry J, Acharya C, Bottoms C, Brown P, Browne C, Eller M, Guill K, Harjes C, Kroon D, Lepak N, Mitchell SE, Peterson B, Pressoir G, Romero S, Rosas MO, Salvo S, Yates H, Hanson M, Jones E, Smith S, Glaubitz JC, Goodman M, Ware D, Holland JB, Buckler ES (2009) Genetic properties of the maize nested association mapping population. Science 325(5941):737–740. https://doi.org/10.1126/science.1174320

Meher LC, Churamani CP, Arif M, Ahmed Z, Naik SN (2013) *Jatropha curcas* as a renewable source for bio-fuels—a review. Renew Sust Energ Rev 26:397–407. https://doi.org/10.1016/j.rser.2013.05.065

Mochida K, Shinozaki K (2010) Genomics and bioinformatics resources for crop improvement. Plant Cell Physiol 51(4):497–523. https://doi.org/10.1093/pcp/pcq027

Mochida K, Shinozaki K (2011) Advances in omics and bioinformatics tools for systems analyses of plant functions. Plant Cell Physiol 52(12):2017–2038. https://doi.org/10.1093/pcp/pcr153

Mochida K, Shinozaki K (2013) Unlocking Triticeae genomics to sustainably feed the future. Plant Cell Physiol 54(12):1931–1950

Mochida K, Koda S, Inoue K, Hirayama T, Tanaka S, Nishii R, Melgani F (2018) Computer vision-based phenotyping for improvement of plant productivity: a machine learning perspective. Gigascience 8(1)

Montes JM, Melchinger AE (2016) Domestication and breeding of Jatropha curcas L. Trends Plant Sci 21(12):1045–1057. https://doi.org/10.1016/j.tplants.2016.08.008

Morris GP, Ramu P, Deshpande SP, Hash CT, Shah T, Upadhyaya HD, Riera-Lizarazu O, Brown PJ, Acharya CB, Mitchell SE, Harriman J, Glaubitz JC, Buckler ES, Kresovich S (2013) Population genomic and genome-wide association studies of agroclimatic traits in sorghum. Proc Natl Acad Sci U S A 110(2):453–458. https://doi.org/10.1073/pnas.1215985110

Nageswara-Rao M, Soneji JR, Kwit C, Stewart CN (2013) Advances in biotechnology and genomics of switchgrass. Biotechnol Biofuels 6(1):77. https://doi.org/10.1186/1754-6834-6-77

OECD (2009) The bioeconomy to 2030: designing a policy agenda. OECD Publishing, Paris. https://doi.org/10.1787/9789264056886-en

Ordonio R, Ito Y, Morinaka Y, Sazuka T, Matsuoka M (2016) Molecular breeding of *Sorghum bicolor*, a novel energy crop. In: International review of cell and molecular biology. https://doi.org/10.1016/bs.ircmb.2015.09.001

Paterson AH, Bowers JE, Bruggmann R, Dubchak I, Grimwood J, Gundlach H, Haberer G, Hellsten U, Mitros T, Poliakov A, Schmutz J, Spannagl M, Tang H, Wang X, Wicker T, Bharti AK, Chapman J, Feltus FA, Gowik U, Grigoriev IV, Lyons E, Maher CA, Martis M, Narechania A, Otillar RP, Penning BW, Salamov AA, Wang Y, Zhang L, Carpita NC, Freeling M, Gingle AR, Hash CT, Keller B, Klein P, Kresovich S, McCann MC, Ming R, Peterson DG, Mehboob-Ur-Rahman, Ware D, Westhoff P, Mayer KFX, Messing J, Rokhsar DS (2009) The *Sorghum bicolor* genome and the diversification of grasses. Nature 457(7229):551–556. https://doi.org/10.1038/nature07723

Ragauskas AJ, Williams CK, Davison BH, Britovsek G, Cairney J, Eckert CA, Frederick WJ, Hallett JP, Leak DJ, Liotta CL, Mielenz JR, Murphy R, Templer R, Tschaplinski T (2006) The path forward for biofuels and biomaterials. Science 311(5760):484–489. https://doi.org/10.1126/science.1114736

Ramstein GP, Evans J, Kaeppler SM, Mitchell RB, Vogel KP, Buell CR, Casler MD (2016) Accuracy of genomic prediction in Switchgrass (*Panicum virgatum* L.) improved by accounting for linkage disequilibrium. G3 (Bethesda) 6(4):1049–1062. https://doi.org/10.1534/g3.115.024950

Ramstein GP, Evans J, Nandety A, Saha MC, Brummer EC, Kaeppler SM, Buell CR, Casler MD (2018) Candidate variants for additive and interactive effects on bioenergy traits in Switchgrass

(L.) identified by genome-wide association analyses. Plant Genome 11(3). https://doi.org/10.3835/plantgenome2018.01.0002

Schmutz J, Cannon SB, Schlueter J, Ma J, Mitros T, Nelson W, Hyten DL, Song Q, Thelen JJ, Cheng J, Xu D, Hellsten U, May GD, Yu Y, Sakurai T, Umezawa T, Bhattacharyya MK, Sandhu D, Valliyodan B, Lindquist E, Peto M, Grant D, Shu S, Goodstein D, Barry K, Futrell-Griggs M, Abernathy B, Du J, Tian Z, Zhu L, Gill N, Joshi T, Libault M, Sethuraman A, Zhang XC, Shinozaki K, Nguyen HT, Wing RA, Cregan P, Specht J, Grimwood J, Rokhsar D, Stacey G, Shoemaker RC, Jackson SA (2010) Genome sequence of the palaeopolyploid soybean. Nature 463(7278):178–183. https://doi.org/10.1038/nature08670

Schnable PS, Ware D, Fulton RS, Stein JC, Wei F, Pasternak S, Liang C, Zhang J, Fulton L, Graves TA, Minx P, Reily AD, Courtney L, Kruchowski SS, Tomlinson C, Strong C, Delehaunty K, Fronick C, Courtney B, Rock SM, Belter E, Du F, Kim K, Abbott RM, Cotton M, Levy A, Marchetto P, Ochoa K, Jackson SM, Gillam B, Chen W, Yan L, Higginbotham J, Cardenas M, Waligorski J, Applebaum E, Phelps L, Falcone J, Kanchi K, Thane T, Scimone A, Thane N, Henke J, Wang T, Ruppert J, Shah N, Rotter K, Hodges J, Ingenthron E, Cordes M, Kohlberg S, Sgro J, Delgado B, Mead K, Chinwalla A, Leonard S, Crouse K, Collura K, Kudrna D, Currie J, He R, Angelova A, Rajasekar S, Mueller T, Lomeli R, Scara G, Ko A, Delaney K, Wissotski M, Lopez G, Campos D, Braidotti M, Ashley E, Golser W, Kim H, Lee S, Lin J, Dujmic Z, Kim W, Talag J, Zuccolo A, Fan C, Sebastian A, Kramer M, Spiegel L, Nascimento L, Zutavern T, Miller B, Ambroise C, Muller S, Spooner W, Narechania A, Ren L, Wei S, Kumari S, Faga B, Levy MJ, McMahan L, Van Buren P, Vaughn MW, Ying K, Yeh CT, Emrich SJ, Jia Y, Kalyanaraman A, Hsia AP, Barbazuk WB, Baucom RS, Brutnell TP, Carpita NC, Chaparro C, Chia JM, Deragon JM, Estill JC, Fu Y, Jeddeloh JA, Han Y, Lee H, Li P, Lisch DR, Liu S, Liu Z, Nagel DH, McCann MC, Sanmiguel P, Myers AM, Nettleton D, Nguyen J, Penning BW, Ponnala L, Schneider KL, Schwartz DC, Sharma A, Soderlund C, Springer NM, Sun Q, Wang H, Waterman M, Westerman R, Wolfgruber TK, Yang L, Yu Y, Zhang L, Zhou S, Zhu Q, Bennetzen JL, Dawe RK, Jiang J, Jiang N, Presting GG, Wessler SR, Aluru S, Martienssen RA, Clifton SW, McCombie WR, Wing RA, Wilson RK (2009) The B73 maize genome: complexity, diversity, and dynamics. Science 326(5956):1112–1115. https://doi.org/10.1126/science.1178534

Shi J, Ma X, Zhang J, Zhou Y, Liu M, Huang L, Sun S, Zhang X, Gao X, Zhan W, Li P, Wang L, Lu P, Zhao H, Song W, Lai J (2019) Chromosome conformation capture resolved near complete genome assembly of broomcorn millet. Nat Commun 10(1):464. https://doi.org/10.1038/s41467-018-07876-6

Singh R, Ong-Abdullah M, Low ETL, Manaf MAA, Rosli R, Nookiah R, Ooi LCL, Ooi SE, Chan KL, Halim MA, Azizi N, Nagappan J, Bacher B, Lakey N, Smith SW, He D, Hogan M, Budiman MA, Lee EK, Desalle R, Kudrna D, Goicoechea JL, Wing RA, Wilson RK, Fulton RS, Ordway JM, Martienssen RA, Sambanthamurthi R (2013) Oil palm genome sequence reveals divergence of interfertile species in old and new worlds. Nature 500(7462):335–339. https://doi.org/10.1038/nature12309

Swaminathan K, Alabady MS, Varala K, De Paoli E, Ho I, Rokhsar DS, Arumuganathan AK, Ming R, Green PJ, Meyers BC, Moose SP, Hudson ME (2010) Genomic and small RNA sequencing of *Miscanthus* × *giganteus* shows the utility of sorghum as a reference genome sequence for Andropogoneae grasses. Genome Bio 11(2):R12l. https://doi.org/10.1186/gb-2010-11-2-r12

Taylor M, Tornqvist C-E, Zhao X, Grabowski P, Doerge R, Ma J, Volenec J, Evans J, Ramstein GP, Sanciangco MD, Buell CR, Casler MD, Jiang Y (2018) Genome-wide association study in pseudo-F2 populations of Switchgrass identifies genetic loci affecting heading and Anthesis dates. Front Plant Sci 9:1250. https://doi.org/10.3389/fpls.2018.01250

Thirugnanasambandam PP, Hoang NV, Henry RJ (2018) The challenge of analyzing the sugarcane genome. Front Plant Sci 9:616. https://doi.org/10.3389/fpls.2018.00616

Tubeileh A, Rennie TJ, Goss MJ (2016) A review on biomass production from C4 grasses: yield and quality for end-use. Curr Opin Plant Biol 31:172–180. https://doi.org/10.1016/j.pbi.2016.05.001

van der Weijde T, Alvim Kamei CL, Torres AF, Vermerris W, Dolstra O, Visser RGF, Trindade LM (2013) The potential of C4 grasses for cellulosic biofuel production. Front Plant Sci 4:107. https://doi.org/10.3389/fpls.2013.00107

Vandepitte K, Valdés-Rodríquez OA, Sánchez-Sánchez O, De Kort H, Martinez-Herrera J, García-Pérez E, De Meyer T, Pérez-Vázquez A, Muys B, Honnay O (2019) High SNP diversity in the non-toxic indigenous *Jatropha curcas* germplasm widens the potential of this upcoming major biofuel crop species. Ann Bot 124(4):645–652. https://doi.org/10.1093/aob/mcz008

Waclawovsky AJ, Sato PM, Lembke CG, Moore PH, Souza GM (2010) Sugarcane for bioenergy production: an assessment of yield and regulation of sucrose content. Plant Biotechnol J 8 (3):263–276. https://doi.org/10.1111/j.1467-7652.2009.00491.x

Wu P, Zhou C, Cheng S, Wu Z, Lu W, Han J, Chen Y, Ni P, Wang Y, Xu X, Huang Y, Song C, Wang Z, Shi N, Zhang X, Fang X, Yang Q, Jiang H, Chen Y, Li M, Chen F, Wang J, Wu G (2015) Integrated genome sequence and linkage map of physic nut (*Jatropha curcas* L.), a biodiesel plant. Plant J 81(5):810–821. https://doi.org/10.1111/tpj.12761

Xia Z, Zhang S, Wen M, Lu C, Sun Y, Zou M, Wang W (2018) Construction of an ultrahigh-density genetic linkage map for *Jatropha curcas* L. and identification of QTL for fruit yield. Biotechnol Biofuels 11:3. https://doi.org/10.1186/s13068-017-1004-9

Xiao Y, Liu H, Wu L, Warburton M, Yan J (2017) Genome-wide association studies in maize: praise and stargaze. Mol Plant 10(3):359–374. https://doi.org/10.1016/j.molp.2016.12.008

Xie M, Chung CY-L, Li M-W, Wong F-L, Wang X, Liu A, Wang Z, Leung AK-Y, Wong T-H, Tong S-W, Xiao Z, Fan K, Ng M-S, Qi X, Yang L, Deng T, He L, Chen L, Fu A, Ding Q, He J, Chung G, Isobe S, Tanabata T, Valliyodan B, Nguyen HT, Cannon SB, Foyer CH, Chan T-F, Lam H-M (2019) A reference-grade wild soybean genome. Nat Commun 10(1):1216. https://doi.org/10.1038/s41467-019-09142-9

Xing S, Tao C, Song Z, Liu W, Yan J, Kang L, Lin C, Sang T (2018) Coexpression network revealing the plasticity and robustness of population transcriptome during the initial stage of domesticating energy crop *Miscanthus lutarioriparius*. Plant Mol Biol 97(6):489–506. https://doi.org/10.1007/s11103-018-0754-5

Xu Q, Zhu C, Fan Y, Song Z, Xing S, Liu W, Yan J, Sang T (2016) Population transcriptomics uncovers the regulation of gene expression variation in adaptation to changing environment. Sci Rep 6:25536. https://doi.org/10.1038/srep25536

Yan J, Song Z, Xu Q, Kang L, Zhu C, Xing S, Liu W, Greimler J, Züst T, Li J, Sang T (2017) Population transcriptomic characterization of the genetic and expression variation of a candidate progenitor of Miscanthus energy crops. Mol Ecol 26(21):5911–5922. https://doi.org/10.1111/mec.14338

Zhang J, Zhang X, Tang H, Zhang Q, Hua X, Ma X, Zhu F, Jones T, Zhu X, Bowers J, Wai CM, Zheng C, Shi Y, Chen S, Xu X, Yue J, Nelson DR, Huang L, Li Z, Xu H, Zhou D, Wang Y, Hu W, Lin J, Deng Y, Pandey N, Mancini M, Zerpa D, Nguyen JK, Wang L, Yu L, Xin Y, Ge L, Arro J, Han JO, Chakrabarty S, Pushko M, Zhang W, Ma Y, Ma P, Lv M, Chen F, Zheng G, Xu J, Yang Z, Deng F, Chen X, Liao Z, Zhang X, Lin Z, Lin H, Yan H, Kuang Z, Zhong W, Liang P, Wang G, Yuan Y, Shi J, Hou J, Lin J, Jin J, Cao P, Shen Q, Jiang Q, Zhou P, Ma Y, Zhang X, Xu R, Liu J, Zhou Y, Jia H, Ma Q, Qi R, Zhang Z, Fang J, Fang H, Song J, Wang M, Dong G, Wang G, Chen Z, Ma T, Liu H, Dhungana SR, Huss SE, Yang X, Sharma A, Trujillo JH, Martinez MC, Hudson M, Riascos JJ, Schuler M, Chen LQ, Braun DM, Li L, Yu Q, Wang J, Wang K, Schatz MC, Heckerman D, Van Sluys MA, Souza GM, Moore PH, Sankoff D, VanBuren R, Paterson AH, Nagai C, Ming R (2018) Allele-defined genome of the autopolyploid sugarcane *Saccharum spontaneum* L. Nat Genet 50(11):1565–1573. https://doi.org/10.1038/s41588-018-0237-2

Zhou Z, Jiang Y, Wang Z, Gou Z, Lyu J, Li W, Yu Y, Shu L, Zhao Y, Ma Y, Fang C, Shen Y, Liu T, Li C, Li Q, Wu M, Wang M, Wu Y, Dong Y, Wan W, Wang X, Ding Z, Gao Y, Xiang H, Zhu B,

Lee SH, Wang W, Tian Z (2015) Resequencing 302 wild and cultivated accessions identifies genes related to domestication and improvement in soybean. Nat Biotechnol 33(4):408–414. https://doi.org/10.1038/nbt.3096

Zhu C, Liu W, Kang L-F, Xu Q, Xing S-L, Fan Y-Y, Song Z-H, Yan J, Li J-Q, Sang T (2017) Haplotypes phased from population transcriptomes detecting selection in the initial adaptation of Miscanthus lutarioriparius to stressful environments. Plant Genome 10(2). https://doi.org/10.3835/plantgenome2016.11.0119

Keiichi Mochida is the leader of the Bioproductivity Informatics Research Team at RIKEN Center for Sustainable Resource Science, Japan. He aims to identify useful genes to improve productivity in useful species for biomass resources, such as grasses and micro algae.

Lam-Son Phan Tran is the leader of the Stress Adaptation Research Unit at RIKEN Center for Sustainable Resource Science, Japan. He aims to elucidate the roles of plant growth regulators and their interactions in plant stress responses, as well as to enhance crop productivity under environmental stress conditions.

RNA Interference: For Improving Traits and Disease Management in Plants

14

Sanjeev Kumar, Avishek Dey, Yuan-Yeu Yau, Mona Easterling, and Lingaraj Sahoo

Abstract

Plant disease and reduced yield are both significant threats to modern agriculture. Cultivation managers continue to find control of disease and yield elusive. Plant biologists have adopted various methods of engineering for generating plants resistant to insect pests, viruses, nematodes, fungi, salinity, and drought. Most utilize resistance based on RNA silencing, a powerful genetic engineering tool with a robust history of enhancing plant growth, crop yield, and disease resistance for the past two decades. Genetically engineered plants expressing small RNAs are increasingly vital and likely to provide future effective strategies. Rapid application and other advantages of RNAi make it a novel gene therapy against drought, salinity, fungus, virus, and bacteria. RNAi also has the potential to improve plant metabolic traits through chromatin remodeling, gene expression via mRNA degradation, and inhibition of translation. A processed product dsRNA known as small interfering RNAs and microRNAs guides the silencing mechanism. Use of tissue-specific or inducible gene silencing, with appropriate promoters, to silence multiple genes simultaneously should allow genetic engineers to protect crops against disease and improve traits. The focus of this chapter is a general discussion of RNAi development, including its role in trait improvement and disease management in plants.

S. Kumar · A. Dey · L. Sahoo (✉)
Department of Bioscience and Bioengineering, Indian Institute of Technology Guwahati, Guwahati, Assam, India
e-mail: ls@iitg.ac.in

Y.-Y. Yau (✉) · M. Easterling
Department of Natural Sciences, Northeastern State University, Broken Arrow, OK, USA
e-mail: mona.easterling@tulsacc.edu

A. Kumar et al. (eds.), *Climate Change, Photosynthesis and Advanced Biofuels*,
https://doi.org/10.1007/978-981-15-5228-1_14

Keywords

RNAi · Dicer · PTGS · Disease management · Virus resistance · Metabolic engineering

14.1 Introduction

RNA is ribonucleic acid. RNAs can be divided into two major groups: *coding* RNAs and *noncoding* RNAs (ncRNAs) (Storz 2002). Coding RNAs (mRNA) are translated into proteins. Noncoding RNAs are not. ncRNAs have other important cell functions. These include (1) ribosomal RNAs (rRNAs) which function in protein synthesis with the ribosomal subunits, (2) transfer RNAs (tRNAs) which function in protein synthesis as transport molecules for amino acids, (3) small nucleolar RNAs (snoRNAs) which function in both pre-rRNA processing and rRNA modification, and (4) regulatory RNAs which regulate vital gene expression in organism growth and development. Some examples of regulatory RNAs are (a) small interfering RNAs (siRNAs), ~21–22 nucleotides (nt) in length; (b) microRNAs (miRNAs), ~19–25 nt; (c) transfer RNA-derived small RNAs (tsRNAs), ~18–40 nt (Li et al. 2018); (d) Piwi-interacting RNA (piRNA); 26–31 nt; (e) small nuclear RNAs or snRNAs (~150 nt; function in mRNA splicing); and (f) long noncoding RNA or LncRNAs (>200 nt). Small regulatory RNAs containing 70–300 nucleotides were discovered long ago, but due to their small nt size, "small regulatory RNAs" (such as siRNAs and miRNAs) are recent discoveries (Grosshans and Filipowicz 2008). It is very likely other types of regulatory RNAs remain undiscovered. This chapter will focus on small RNA-induced RNA silencing.

RNA silencing is one of the most important mechanisms regulating gene expression. The most common and well-studied example is *RNA interference* (or RNAi). It is often induced by smaller nt RNA molecules. RNAi is a blanket term, referring to RNA silencing caused by siRNAs, the first to be discovered, or miRNA.

Both siRNA and miRNA induce RNAi, and research indicate some similarities. For example, there are known shared molecular mechanisms. One known similarity is use of "Dicer" enzyme in both pathways. The fruit fly has two Dicer proteins encoded by *Dcr*-1 and *Dcr*-2 genes. *Dcr*-1 is vital for mRNA-triggered silencing, and *Dcr*-2 is the major miRNA-producing enzyme (Tijsterman and Plasterk 2004). Both small molecules regulate gene expression in different ways. Gene regulation by siRNA is specific to a particular gene and induces degradation of the complementary messenger RNA, while miRNA silences by degrading mRNA or blocking translation. In addition, siRNA is double-stranded, and miRNA is a single-stranded molecule. Discovery of siRNA involved induction through exogenous materials (vectors like viruses or exogenous transgenes on a DNA construct), while miRNAs are derived endogenously (Mack 2007). Evolutionary evidence indicates miRNA is younger than siRNA. Lower organisms, like plants and fungi, use siRNAs to help generate viral immunity, but this is not the role of siRNA in higher organisms like humans and mice. In mammals, interferon response provides viral immunity to hosts. miRNA has a widespread role in growth and development. In this chapter, the review will focus on siRNA-induced RNAi and siRNA-based RNAi technology.

RNAi phenomenon was observed in the early 1990s. Plants were the organism of discovery in 1990, followed by fungi in 1992. *Pathogen-derived resistance* (PDR) producing virus resistance in plants was described in 1993. However, RNAi was not defined as the source of this phenomenon until 1998, within nematode *Caenorhabditis elegans*. Biological researchers attach tremendous significance to observations of viral RNA degradation by RNAi and the mechanism and the enzymes in RNAi pathway. This presents the potential for designing artificial siRNAs to generate virus-resistant crops or using siRNAs to study gene function of desired crop genes via the knockdown process. Because of its effectiveness and relative ease of use, RNAi technique has developed into a powerful genetic and reverse engineering tool for both basic and applied research. In agriculture currently, RNAi is used to improve plant yield, to protect crops from viral, bacterial, fungal pathogens, insect pests, nematodes, and parasitic weeds.

This chapter provides an overview of RNAi with a focus on plants. Topics involving RNAi include the historical relevance of discovery, the molecular mechanism, the uses as a molecular tool, requirements of vector design, limitations of technology, and possibilities for future applications. Case studies using RNAi technology for enhancement of crop yield, to improve selected characteristics, and for resistance to disease and pests will be discussed. The most recent developments using RNAi technology in agriculture will also be reviewed. Applications using a combination of RNAi technology and nanotechnology and improvement of plant biofuel feedstock for better biofuel production using RNAi technology will also be discussed.

14.2 What Is RNAi?

RNA interference (RNAi) is a natural cellular process. RNAi is a double-stranded (ds) RNA-induced, sequence-specific RNA degradation mechanism. RNAi is also known by different terms, "co-suppression" in plants and "quelling" in fungi (see "Discovery of RNAi" paragraph). However, these terms are now collectively referred to as *RNA silencing* (Waterhouse et al. 2001; Hannon 2002; Plasterk 2002). A common feature of RNA silencing is production of double-stranded, small RNAs (sRNAs) 21–26 nucleotides in length. These particular specificity determinant sRNAs act to downregulate individual gene activity, by reducing or switching off expression of particular genes (Tijsterman et al. 2002) (Fig. 14.1). RNAi also operates as a natural antiviral system in most eukaryotes, including mammals. Some important features of RNAi include the following: (a) the interfering agent is double-stranded RNA rather than single-stranded antisense RNA, (b) there is a high degree of specificity in gene silencing with less effort, (c) it is highly potent and effective (a few dsRNA molecules per cell are sufficient for effective interference), (d) it avoids problems with abnormalities caused by a knocked-out gene in early stages, (e) target gene silencing can be introduced at

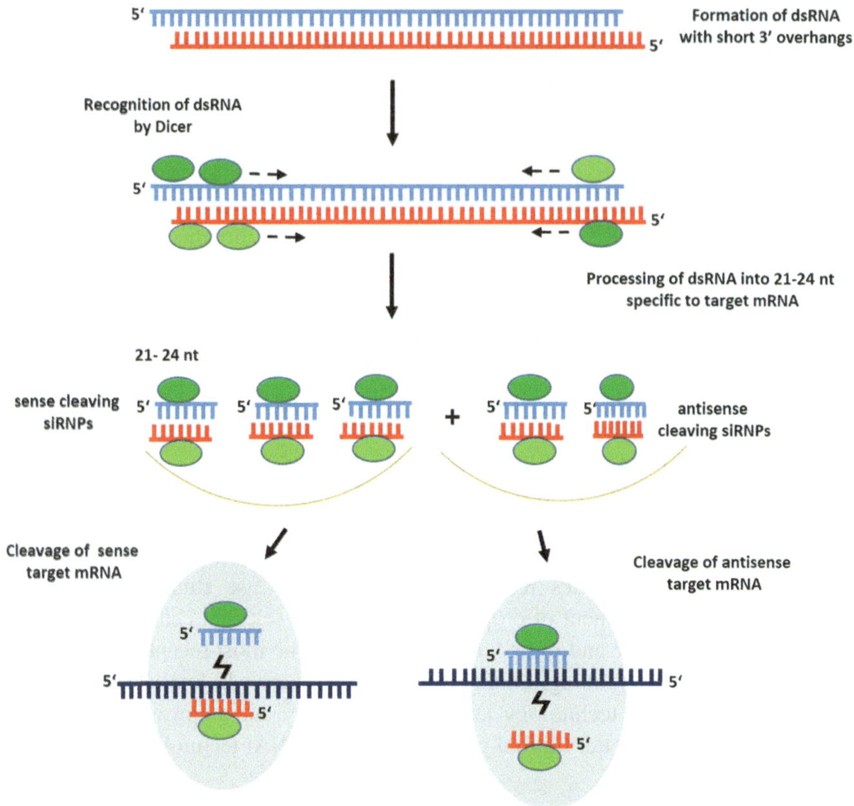

Fig. 14.1 Model for RNAi-mediated target RNA degradation pathway: The dsRNA processing proteins containing an RNA-binding domain and a dsRNA-specific endonuclease domain are indicated as light and dark green ovals. The dark green color marked protein domain binds in the 5′–3′ direction and the light green color marked protein domain in the 3′–5′ direction. The RNA-induced silencing complex (RISC) is shown as large gray oval. A conformational change is proposed to occur in the RISC before target RNA cleavage because the cleavage site of the target mRNA is displaced by 10–12 nt relative to the dsRNA processing site. The cleaved target RNA is directed into the processing pathway where it will be sequentially degraded

different developmental stages in systemic silencing, and (f) the effect of silencing is passed through generations.

14.3 Discovery of RNAi

The phenomenon now referred to as RNA interference (RNAi) was first observed and described in the 1990s during genetic transformation experiments (Table 14.1). The first RNAi event was observed in plants (Napoli et al. 1990; van der Krol et al. 1990). Napoli et al. (1990) were first to describe the phenomenon in plant *petunia* in

Table 14.1 Events related to RNAi technology

1971
RNA-dependent RNA polymerase (RdRP) isolated from plants (Chinese cabbage)
1983
First transgenic plants created (with *Agrobacterium*)
1985
Concept of pathogen-derived resistance (PDR) was proposed to produce virus-resistant crops. Coat protein genes were used for plant genetic transformation to produce viru-resistant crops (Sanford and Johnston 1985)
1987
Gene gun created for genetic transformation
1990
Co-suppression phenomenon reported in plants
1992
Quelling phenomenon reported in fungus
1992
PDR experiments unexpectedly showed untranslatable constructs conferring resistance in transgenic plants (Lindbo and Dougherty 1992a, b)
1993-1994
Co-suppression/quelling phenomenon via PDR described in rapid degradation of both transgene mRNA and viral RNA (Lindbo et al. 1993)
1995
First proof-of-concept VIGS experiment reported in tobacco *PHYTOENE DESATURASE (PDS)* gene (Kumagai et al. 1995). VIGS technique employs an engineered virus to reduce specific, endogenous gene activity based on PTGS for the study of gene functions (for review, see Ruiz et al. 1998; Becker and Lange 2010)
1996
Prins and Goldbach (1996) summarized PDR successful generation of TMV-, PVX-, CMV-resistant plants
1997
"A similarity between viral defense and gene silencing in plants" published in *Science* (v. 276, p1558–1560). This article describes the discovery of a "new form of plant defense" capable of targeting viruses and specific degradation of their RNA
1998 (January 2)
Founding AGO protein, AGO1, described in model plant *Arabidopsis*
1998 (February 19)
RNA interference (RNAi) phenomenon observed in nematode (*C. elegans*). siRNA discovered to be determinant. The paper was published in *Nature* (Fire et al. 1998).
1998 (December 23)
RNAi phenomenon reported in *Drosophila* and published in *Cell* (Kennerdell and Carthew 1998)
1999
Post-transcriptional gene silencing (PTGS) reported in plants (Hamilton and Baulcombe 1999). Complementary RNA found in experiment

(continued)

Table 14.1 (continued)

2000
Two papers published in *Cell*, showing RdRP enzyme required for RNAi in plants. Authors describe transgenically expressed ssRNA was first copied into short dsRNA, in a RdRP enzyme-dependent process
2001
Generic vectors (pHANNIBAL, pKANNIBAL, and pHELLSGATE) constructed to generate a highly effective ihpRNA silencing construct (Wesley et al. 2001)
2004
Two DICER enzymes found in *Drosophila* with two different roles
2006
Nobel Prize in Physiology or Medicine awarded jointly to Andrew Z. Fire and Craig C. Mello for discovery of RNA interference
2013
RNA interference found to function in mammal antiviral immunity (Li et al. 2013). Antiviral RNA interference described in mammalian cell lines (*Science* 342, pp. 235-238)

petal pigmentation genes. Similar phenomenon was observed in fungus *Neurospora crassa* in genetic transformation experiments (Pandit and Russo 1992). In both cases, the common feature was gene expression in both transgene and homologous endogenous genes was silenced in some transgenic plant lines and fungus strains. Coordinated silencing of both ectopic transgenes and endogenous homologous genes prompted the descriptive terms *co-suppression* in plants (Napoli et al. 1990) and *quelling* in fungi (Pandit and Russo 1992).

Discovery of quelling in *Neurospora crassa* by Pandit and Russo (1992) yielded some surprising results. (1) Methylation was observed in earlier studies involving strains with multiple transgene integration. So researchers first checked the DNA level of the gene, and they found no methylation. (2) The pre-mRNA levels in quelled strains and wild type were similar. (3) Mature mRNA was diminished in quelled strains versus wild-type strains. (4) Transgenic fragments containing the whole gene sequence or part of the gene sequence invoked the quelling effect, although the latter induced a milder effect. (5) Use of partial genes for transformation required no specific part of a gene to induce the quelling effect. Different parts of the same gene were shown to induce the quelling effect. (6) Antisense mRNA did not induce a quelling effect. Antisense mRNA is produced when a transgene is inserted in an antisense orientation near an endogenous promoter. (7) The "effector" causing quelling could diffuse out of cells. These observations led researchers to propose quelling effector was some type of RNA molecule.

While both co-suppression in transgenic plants and quelling in transgenic fungus were being observed, researchers conducting viral plant experiments using *pathogen-derived resistance* (PDR) noticed a similar phenomenon. Scientists Sanford and Johnston first proposed PDR in 1985, where use of genes from target pathogens conferred host resistance. It was considered one way to produce virus-resistant plants (Wilson 1993; Baulcombe 1994). Using PDR, virus essential protein genes were overexpressed in plant hosts through genetic transformation. The overexpression conferred virus-resistant immunity, as viral coat protein genes were successfully used to produce virus-resistant plants (Fitch et al. 1992). In PDR, similar to previously described, co-suppression and quelling phenomenon were also observed in the host plants.

Further studies revealed RNAi to be an ancient plant mechanism for viral defense capable of degrading dsRNA derived from viruses in a sequentially oriented manner within infected cells (Matzke et al. 2001a, b; Zamore 2001; Vance and Vaucheret 2001; Hannon 2002). RNA silencing limits viral replication and impedes viral ability to infect distant tissues (Anandalakshmi et al. 1998). The evolution of RNAi as a whole plant defense system begins as a unidirectional, mobile-signal-directing RNA degradation in a sequence-specific manner at distant tissues (Voinnet and Baulcombe 1997). Antiviral RNAi is developed by first converting the viral genome to dsRNA that Dicer enzyme processes into small dsRNA, approximately 20s nucleotides (nt). These small molecules are known as small interfering RNA or siRNA. This siRNA guides another nuclease complex, RISC, to cleave homologous single-stranded viral RNAs. Suppression of RNAi is essential for efficient viral infection. Plant viruses were found to express a complex of RNAi suppressor proteins to protect their genome against the antiviral effect of RNAi degradation (Li and Ding 2001). In plants, it was hypothesized siRNA-directed sequence-specific nuclear DNA methylation, RNA-directed RNA degradation, and intercellular signaling to trigger silencing in distant tissues (Matzke et al. 2001a, b). This process requires proteins from the Argonaute family (Hammond et al. 2001). After 2001, antiviral RNAi was described in fungus, nematode, insect cells, and mammals. Antiviral RNAi in mammals was not confirmed until 2018 (Li et al. 2002; Cullen 2002; Li et al. 2013; Ding et al. 2018; Berkhout 2018). Despite these findings, the precise molecular mechanism of RNAi suppression remains unclear.

RNAi phenomenon was observed in plants, fungi, and PDR, before the effector inducing the process was described in 1998. Drs. Andrew Z. Fire and Craig C. Mello discovered this particular RNA molecule while studying nematode *Caenorhabditis elegans*. Their research provided evidence that small dsRNA (siRNA) induced a RNAi reaction, when injecting dsRNA into *C. elegans* (Fire et al. 1998). These dsRNA injections led to an efficient sequence-specific silencing, and this group called the phenomenon *RNA interference* (or RNAi). Their magnificent work won the Nobel Prize in Physiology or Medicine in 2006. Their seminal work was titled "Discovery of RNA interference—gene silencing by double-stranded RNA" per the Nobel Prize website (https://www.nobelprize.org/nobel_prizes/medicine/laureates/2006/). After the discovery of siRNA in *C. elegans*, it became feasible to explore the RNAi mechanism in various biological systems. Since that time, RNA interference has been observed in almost all eukaryotes. The few exceptions include *Drosophila*, a model organism and insect that does not display a robust, systemic RNAi response.

Today, co-suppressing in plants, quelling in fungi, reduction of viral RNA in PDR, and RNAi in animal *C. elegans* are known to share a common fundamental mechanism, a conserved regulatory mechanism for gene expression. Collectively, they are called *RNA silencing* or *RNA interference*. RNA silencing belongs to the category of gene silencing known as post-transcriptional gene silencing (PTGS) (Vaucheret et al. 2001; Waterhouse et al. 2001; Matzke et al. 2001a, b; Plasterk

2002; Hannon 2002). This should not be confused with the other category of gene silencing that represses transcription, known as transcriptional gene silencing (TGS).

The interesting RNAi phenomenon discovered in *C. elegans* was limited to lower organisms. Its delivery of longer dsRNA for RNAi acted as a nonspecific inhibitor in mammalian cells. Within a few years, engineered Dicer-like synthetic RNAs demonstrated the ability to induce sequence-specific gene silencing in human cells. Since these synthetic molecules did not initiate nonspecific gene silencing pathways, this siRNA became a suitable, novel tool to knock down specific target genes in mammalian cells (Elbashir et al. 2001). Naturally expressed small hairpin RNAs, known as miRNAs, were also found to be processed by Dicer-like enzymes and ultimately function in a manner similar to RNAi (Grishok et al. 2001; Ketting et al. 2001).

14.4 Molecular Mechanism of siRNA-Based RNAi

RNA silencing is a conserved mechanism in most eukaryotes that functions through sequence-specific inhibition of gene expression. Two kinds of small RNAs in the RNAi pathway have been described, siRNAs (short interfering RNA) and miRNAs (microRNA). siRNAs are generated from longer dsRNA, while miRNAs are derived from stem-loop precursors. Both classes must be processed by a RNase III-like enzyme called Dicer (DCL) (Jaskiewicz and Filipowicz 2008; Dunoyer et al. 2010; Praveen et al. 2012). Emily Bernstein (2001), a graduate student in Greg Hannon's lab at Cold Spring Harbor Laboratory, is credited with discovery of the enzyme and naming it Dicer (Bernstein et al. 2001; https://en.wikipedia.org/wiki/Dicer). Long dsRNAs are processed into small dsRNAs, approximately 21–24 base pairs in length (Zamore et al. 2000). siRNA fragments carry a two-base overhang on the 3′ end of each strand, sometimes referred to as a passenger strand, target strand, or guide strand (Fig. 14.1). Double-stranded siRNAs are subsequently incorporated into a *RNA-induced silencing complex* (RISC) containing an Argonaute (AGO) protein. Each AGO contains a sRNA-binding domain with endonucleolytic activity capable of cleaving target RNAs (Williams and Rubin 2002). Research has shown the protein complex initially recognizes both strands of double-stranded siRNA. Following recognition, the passenger strand of dsRNA is cleaved. This occurs during RISC assembly following the rules for siRNA-guided cleavage of a target RNA (Leuschner et al. 2006). With the passenger strand gone, the remaining guide strand siRNA guides RISC to homologous mRNAs to cleave them endonucleolytically (Meister and Tuschl 2004).

The number of Dicer proteins in species varies. For example, humans and *C. elegans* each encode one Dicer protein; *Drosophila* has two Dicer proteins, while plants utilize at least four Dicer-like proteins (Tijsterman and Plasterk 2004; Fukudome and Fukuhara 2017). In *Arabidopsis*, four Dicer-like proteins have been described. Poplar has reported five, and rice contains six Dicer-like proteins (Margis et al. 2006). Extensive genetic studies in plants reveal each Dicer-like protein participates in a specific gene silencing pathway with some redundancy (Fukudome and Fukuhara 2017). AGO proteins in RISC provide the catalytic component, containing an RNase H-like domain responsible for target mRNA degradation (Filipowicz 2005). Primitive plants encode only a few AGO proteins. The higher plant, *A. thaliana*, encodes ten AGO proteins, designated AGO1 to AGO10. These ten *A. thaliana* AGO proteins belong to three phylogenetic clades (Vaucheret 2008; Zhang et al. 2015).

RNA silencing pathways are triggered by self-complementary fold-back structures or by production of double-stranded RNA (dsRNA). Dicer digests longer dsRNA into small dsRNAs (siRNAs), which can initiate or amplify RNAi. Sijen et al. (2001) reported siRNA serving as a template for RNA-dependent RNA polymerase transformation of target ssRNA into dsRNA. How is long dsRNA initially formed? In plants, these long dsRNA molecules are synthesized by the enzyme *RNA-dependent RNA polymerase* (RdRP; EC 2.7.7.48) (Bartel 2004; Baulcombe 2004; Carthew and Sontheimer 2009) [in Willmann et al. 2011]. RdRP enzyme synthesizes siRNA-producing dsRNA molecules by using a single-stranded RNA (ssRNA) molecule as the template (Willmann et al. 2011). RdRP enzyme was reported in some plants as early the 1970s (Astier-Manifacier and Cornuet 1971). However, their link to RNA silencing in plants was discovered in a 1993 study of virus resistance in plants (Lindbo et al. 1993). Lindbo et al. (1993) proposed that RdRP copies a sense transgene into complementary RNA and induces a RNAi response. RdRP catalytic activity in tomato was later studied in vitro. This study showed transcription of short, single-stranded RNA molecules into precise full-length, complementary RNA templates. In the model plant *Arabidopsis thaliana*, researchers identified several genetic loci required for post-transcriptional gene silencing (Dalmay et al. 2000). One of the required loci, *SDE1*, is a plant homolog of *QDE-1* in *Neurospora crassa* encoding RNA-dependent RNA polymerase (Dalmay et al. 2000). At the same time, another group reported similar results in plant species *Arabidopsis* (Mourrain et al. 2000). Both Dalmay and Mourrain's papers were published in the same issue of *Cell*. More recently, RdRP tobacco gene, *NtRDRP1*, was isolated. Researchers found *NtRDRP1* could be induced either by viral infection or by treatment with SA (including its biologically active analogs) (Xie et al. 2001).

RdRP proteins have been found in a number of plants, including tobacco, tomato, cowpea, and cucumber (Astier-Manifacier and Cornuet 1978; Dorssers et al. 1982; Duda et al. 1973; Duda 1979; Takanami and Fraenkel-Conrat 1982). Plant species differ in their number of RdRP enzymes. For example, six RdRPs were identified in *Arabidopsis* (Wassenegger and Krczal 2006). A large body of evidence shows RdRP plant enzymes have additional important biological functions aside from RNAi. Three of the best characterized are in *Arabidopsis* (RdRP1, 2, and 6) have been shown to have other biological functions (Willmann et al. 2011).

RdRP proteins found in RNA viruses, plants, fungi, protists, and some lower animals are absent in *Drosophila*, mice, and humans (Willmann et al. 2011). Mammals have no RdRP (Watanabe et al. 2008). Researchers found they are derived from retrotransposons or some genomic regions producing transcripts capable of forming dsRNA structures. Inverted repeat structures, bidirectional transcription, antisense transcripts, and pseudogenes from various loci are sources of the dsRNAs (Watanabe et al. 2008).

The process is closely linked to post-transcriptional gene regulation by miRNAs, as the last step is inhibition of translation initiation. The two processes share many of the same components. The presence of RdRP interacts with the RISC complex to generate new dsRNA, using hybridized siRNA strands as primers. Newly synthesized dsRNA interacts with Dicer enzyme generating new siRNAs, also known as secondary siRNAs (Fig. 14.1). Once they are introduced into the cell, dsRNA effects can persist throughout development. In fact, dsRNA can be exported to neighboring cells, spreading the knockout effect (Daniel and John 2008).

14.5 Popular Vectors Used for RNAi and Components of RNAi Vector

Double-stranded RNA is an effective trigger of gene silencing in vertebrate, invertebrate, and plant systems. It follows that RNAi can be artificially triggered by exogenous introduction of either dsRNA- or shRNA- expressing constructs. For example, RNAi has been achieved by introducing DNA constructs expressing self-complementary (hairpin) RNA via plant transformation. The process requires expressed RNA to contain sequences homologous to genes targeted for silencing (Waterhouse and Helliwell 2003; Helliwell and Waterhouse 2003). In plants, an effective vector design for gene silencing would allow for easy cloning and increase the frequency of silenced plants. The early generation of RNAi vector pHANNIBAL is one example. This generic vector, pHANNIBAL, was generated to facilitate production of intron-spliced hairpin RNA (ihpRNA). The construct encoded a hairpin RNA (hpRNA) that consisted of an inverted repeat fragment from a target gene sequence separated by an intron. The ihpRNA-mediated gene silencing is highly efficient in a number of plants, with up to 100% transformation efficiency reported. The silenced plants produced a particular ihpRNA construct with differing degrees of silencing, with target gene silencing close to 100% reported.

14.5.1 The pHANNIBAL and pKANNIBAL Vectors

pHANNIBAL is a vector containing bacterial ampicillin resistance, GenBank accession number AJ311872, measuring ~5.8 kb. Its sister vector pKANNIBAL measures ~6.0 kb, contains kanamycin resistance, and has the GenBank accession number AJ311873. Both are popular RNAi intermediate constructs (Wesley et al. 2001). Figure 14.2 shows schematic map of the hairpin RNAi construct within pKANNIBAL. Both vectors were designed to accept PCR fragment insertion in either sense or antisense orientation. These are traditional digestion- and ligation-based vectors. Building the ihpRNA-expressing vector begins by amplifying PCR fragments from a gene of interest. These will be cloned, in both the sense and antisense orientations, using conventional restriction digestion and inserted using

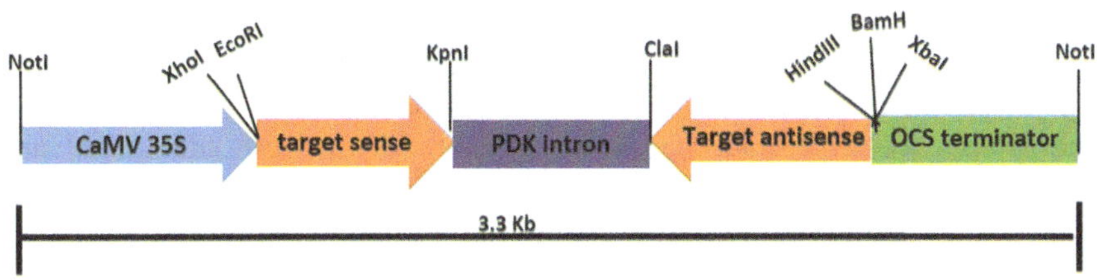

Fig. 14.2 Schematic map of hairpin RNAi construct of pKANNIBAL

DNA ligation techniques (Wesley et al. 2001). pHANNIBAL was first used to silence the pigment biosynthesis gene, chalcone synthase (*CHS*), in *Arabidopsis*. A 741 nt piece of the CHS coding region was amplified from *A. thaliana* (Landsberg erecta L.). The primers used added both *Xho*I and *Kpn*I sites to the ends of one product and both *Xba*I and *Bam*HI sites to the ends of the other product. The two amplified PCR products were ligated into a pHANNIBAL vector backbone (Wesley et al. 2001). Authors reported 21 of the 23 CHS-HANNIBAL transformed plants with pronounced silencing (Wesley et al. 2001). The pKANNIBAL vector allows for PCR-amplified product from a target gene to be directly cloned without prior restriction digestion. This can be accomplished with commercially available 3′-T-overhang vectors, such as pGEM-T Easy (Promega) or pTZ57R/T (Fermentas). The resulting product can be sub-cloned into pKANNIBAL vector. In plants, *Not* I fragment [(35S promoter)-(target gene in sense orientation)-(PDK intron)-(target gene in antisense orientation)-(OCS terminator)] from pHANNIBAL or pKANNIBAL containing hpRNA cassettes can be sub-cloned into a convenient binary vector, such as pART27. RNAi vectors pHANNIBAL and pKANNIBAL have been reported successful with a number of transforming events, involving multiple gene targets. Researchers report both to be efficient and effective for silencing one or multiple genes. Hairpin construction usually takes approximately 2 weeks.

14.5.2 The pHELLSGATE Vectors

Inverted-repeat transgene constructs encoding hairpin RNA (hpRNA) have been reported effective by a number of research groups. Production of hpRNA requires a vector, such as pHANNIBAL, capable of directional insertion of a gene fragment into an inverted repeat conformation separated by an intron. The intron-spaced inverted repeat cassette can be assembled either by "pull-through" PCR or by ligating the four fragments: vector backbone, target gene sense fragment, target gene antisense fragment, and intron spacer. Constructing with a pHANNIBAL- or pKANNIBAL-based vector is relatively slow and inefficient, so more efficient cloning systems are used. Site-specific recombination (SSR) systems can be used to facilitate cloning efficiency. pHELLSGATE vector is one such RNAi cloning vector (Wesley et al. 2001; Helliwell et al. 2002). pHELLSGATE is a binary

Gateway® donor vector, GenBank accession number AJ311874, designed for plant gene silencing research with intron-containing hairpin RNA. The pHELLSGATE vector system is based on a PCR product flanked by *attB1* and *attB2* sites recombined with a plasmid carrying a cassette with two *attP* sites (*attP1* and *attP2*) separated by an intron sequence. The system requires incubation with BP Clonase and produces inverted-repeat constructs with high efficiency.

14.6 RNAi Experiment Example

(a) **Assign Target Region for Gene Silencing**
Assigning a target region for effective gene silencing is necessary. Any host gene can be chosen for downregulation per the following requirements. These criteria should always be kept in mind for assigning a target region:
1. siRNA targeted sequence is usually 21 nt in length.
2. Avoid regions within 50–100 bp of start codon and termination codon.
3. Avoid intron regions.
4. Avoid stretches of four or more identical bases (e.g., AAAA, CCCC).
5. Avoid regions with GC content <30% or >60%.
6. Avoid repeats and sequences displaying low complexity.
7. Avoid SNP sites.
8. Perform a BLAST homology search to avoid off-target effects on other genes or sequences.
9. Design negative controls as a scrambled sequence of target.
10. A/U at 5′ end of antisense strand.
11. G/C at 5′ end of sense strand.
12. Minimum of five A/U residues in 5′-terminal third of antisense strand.
13. Absence of any GC stretch more than 9 nt in length.

(b) **Prepare Hairpin RNAi Constructs**
Prepare hairpin RNAi construct using a plant intermediate vector pHANNIBAL or pKANNIBAL. Target gene DNA can be cloned in sense strand (5′–3′) using *Xho*I/*Eco*RI/*Kpn*I restriction sites and in antisense strand (3′–5′) using *Cla*I/*Hind*III/*Bam*HI/*Xba*I linked by a PDK intron. PCR amplify the target gene DNA from source using target gene-specific primers. Cassette of target-gene-hairpin loop, target (sense)-intron-target (antisense), is inserted between CaMV35S promoter and OCS terminator. In the case of silencing PDS gene, construction of target-gene-hairpin-loop plant binary vector, pART27, requires cloning whole cassette [CaMV35S-PDS (sense)-intron-PDS (antisense)-OCS terminator] from source plasmid (pHANNIBAL or pKANNIBAL). *Not*I enzyme is used to mobilize cassette into pART27, plant vector. The pART27-PDS-hairpin-loop vector was subsequently used for *Agrobacterium*-mediated genetic transformation for targeted gene silencing in host plants.

(c) **Agroinfiltration Using Plant RNAi Constructs**
Agrobacterium carrying RNAi constructs can be used for either production of stably transformed transgenic lines (Vu et al. 2013; Singh et al. 2015; Kumar

Fig. 14.3 Representation of *Agrobacterium* infiltration of hairpin RNAi or agroinfectious dimeric constructs for checking the siRNA expression for various targets and inducing viral symptoms into the plants

et al. 2017) or transient assays. One example is agroinfiltration, which the hairpin RNAi constructs are first transformed into *Agrobacterium* using electroporation. The transformed *Agrobacterium* are cultured in YEP medium containing appropriate antibiotics and grown overnight at 28 °C to mid-log growth phase ($OD_{600} = 0.6$). *Agrobacterium* cultures are then centrifuged for cell harvest. Pellets are resuspended in an equal volume solution [10 mM 2-(*N*-morpholino)ethanesulfonic acid (MES) buffer and 10 mM $MgCl_2$, pH 5.6, and 200 μM acetosyringone]. Acetosyringone is a phenolic compound known to increase *Agrobacterium* transformation efficiency (Baker et al. 2005). Resuspended cells can be kept in a shaking incubator at 90 rpm at 28 °C for 1 h and subsequently used to infiltrate abaxial surfaces of young trifoliate leaves of control host plants 3–4 weeks in age (Fig. 14.3). Inoculated plants should be maintained under appropriate greenhouse conditions for subsequent analysis. RNAi constructs can also be delivered into plant cells by other methods, such as particle bombardment (Schweizer et al. 2000).

(d) **Molecular Validation of siRNA Accumulation in Plants**

Molecular validation of siRNA accumulation can be analyzed by RT-PCR or Northern blot hybridization. These methods should be performed on inoculated plants and controls. For this approach, total RNA should be isolated from leaves using a standard RNA extraction protocol. For Northern hybridization-based siRNA detection, PAGE gels with fractionated RNA are electroblotted to a Hybond-N^+ membrane. The membrane is then subjected to hybridization with a probe specific to the target gene. Membranes can be processed for signal detection using radiolabeled (such as ^{32}P) or enzyme-labeled substrates per standard procedure. The siRNA probes can be obtained from cloned target gene. These clones are available from hpRNA construct preparation. A suitable vector will have both a sense (T7) and an antisense (SP6) direction-specific promoter. Cloned segments can undergo in vitro transcription with T7 RNA polymerase for sense strand and SP6 RNA polymerase for antisense strand, per kit manufacturer's instructions. The enzyme-/radiolabeled transcript of target gene specific [mixture of sense (T7) and antisense (SP6)] can be hydrolyzed and denatured at 100 °C for 5 min. Following this step, the labeled transcript is added to a fresh aliquot of DIG Easy Hyb buffer for membrane hybridization lasting 18 h at 42 °C. Hybridized blots then undergo a post-hybridization wash at 42 °C and protocol for chemiluminescence or radiolabeled signal detection per manufacturer's instructions.

(e) **Analyzing Correlation Between siRNA Accumulation and Level of Resistance**
Northern blot hybridization can determine levels of processed target gene-specific hpRNA in transgenic plants, which shows variable levels of siRNA accumulation correlated with gene silencing trait improvement. For example, viral resistant cowpea plants were generated overexpressing both viral Rep (*AC1*/*AC4*) and transcriptional activator (*AC2*) proteins. Transgenic cowpea plants with higher siRNA accumulation showed maximum viral resistance following viral challenge. This was evidenced by absence of viral genomic components in assayed plants and absence of viral symptoms. Multiple transcript forms, specific to *AC2* and *AC4* in Northern blot, may be due to formation of intermediate mRNA products. Dicer-like enzyme action acts during post-transcriptional gene silencing leading to various levels of resistance (Kumar et al. 2017).

14.7 Applications of RNAi Technology in Crop Breeding

14.7.1 RNAi as a Tool for Plant Gene Function Analysis

RNAi knockdown technology applied effectively to silence plant genes makes it applicable to genome-wide analysis of gene function in a wide range of species and gene varieties. RNAi knockdown has successfully identified various gene functions involved in plant development (Qiao et al. 2007), abiotic stress (Senthil-Kumar et al. 2008, 2010), several biotic stresses (Nora et al. 2009), secondary metabolism (Wagner and Kroumova 2008), and symbiosis (Limpens et al. 2004). Desired target genes can be cloned as an inverted repeat. In an hpRNA-producing vector, it is commonly spaced with an unrelated sequence and driven by a strong constitutive promoter, often *35S CaMV* for dicots or maize 1 promoter for monocots. Intron spacers demonstrate almost 100% gene silencing in transgenic plants (Wesley et al. 2001). However, the underlying mechanism by which intron spacers increase silencing efficiency is unknown (Helliwell et al. 2002). Whether dsDNA is introduced directly or transient hpRNA produced through vector plasmids by particle bombardment, both have been shown to induce RNAi in plants (Klahre et al. 2002). Both methods are useful for gene function analysis. In some cases, stable transformation requires a quite difficult transgenic approach.

14.7.2 Engineering for Novel Traits in Plants

RNA silencing is a major area of investigation leading to exciting new discoveries in several areas. It has been successfully used for controlling gene expression by silencing target genes or their promoters, engineering novel traits, analyzing gene function, and engineering host defense. Plants are an important natural resource providing food, fiber, wood, oils, dyes, pharmaceuticals, and secondary metabolites. RNAi has been a tool for genetic engineering of starches, storage proteins, oils, and secondary metabolites. A variety of genes have been targeted for downregulation by small RNAs, including transcription factors controlling key metabolic pathways in many developmental processes. Many valuable secondary metabolites are scarcely produced, making sufficient quantities difficult to obtain. Metabolic engineering with RNAi provides a promising approach for overcoming some of these limitations. Additional plant examples of RNAi metabolic engineering are reduced caffeine production targeting *CaMxMt 1* gene in coffee bean (Ogita 2004), increasing stearic acid and oleic acid seed oil targeting *ghSAD-1* and *ghFAD2-1* genes in cotton to remove the need for hydrogenation (Liu et al. 2002). *Polyphenol oxidase* gene associated with enzymatic browning in potato to increase shelf life (Wesley et al. 2001). *BP1* gene reducing petals in oilseed rape to improve photosynthesis, and *DET1* gene for increased carotenoid and flavonoid content in tomato to benefit consumer health (Davuluri et al. 2005). Tobacco's *CHI* gene responsible for flower color was targeted (Nishihara 2005); amylase content increased by targeting a starch-branching enzyme in maize (Chai et al. 2005). *1-Aminocyclopropane-1-carboxylate oxidase* was targeted to reduce ethylene sensitivity and slow ripening in tomato to increase shelf life (Xiong 2004). *Lol p1* and *Lol p2* were targeted in ryegrass (*Lolium* spp.) to reduce potential allergic response (Petrovska 2005). *ACR2* gene encodes arsenic reductase and was targeted for phytoremediation of soils in *Arabidopsis* (Dhankher 2006).

14.7.3 Metabolic Engineering

Small RNA are key mediators in post-transcriptional downregulation of protein-coding genes. Higher plants produce a wide variety of secondary metabolites. Plant metabolites include more than 25,000 terpenoids, 8000 phenolic compounds, and 12,000 alkaloids (Croteau et al. 2000). These provide important pharmaceuticals, dyes, and fragrances. Several strategies have been proposed to enhance production of secondary metabolites in plants, but the main obstacle is identifying desired genes for molecular engineering. Synthesis of isoquinoline alkaloid is a well-characterized pathway in secondary plant metabolism, and many important key biosynthetic plant genes have been isolated (Hashimoto and Yamada 2003; Shitan et al. 2003). Biosynthesis of isoquinoline alkaloid provides a suitable model for metabolic engineering. (S)-Reticuline is the branch point intermediate involved in biosynthesis of many types of morphine, codeine, papaverine, and berberine in isoquinoline alkaloid biosynthesis. Furthermore, cDNA encoding berberine bridge enzyme (BBE) has been isolated from California poppy cells (Hauschild et al. 1998). This provides a convenient means to close the reticuline accumulation pathway. However, a previous attempt at closing the pathway with antisense RNAi-mediated suppression of BBE in California poppy was unsuccessful, although the end product, sanguinarine, was considerably reduced (Park et al. 2003). RNA interference (RNAi) technology was to downregulate the target enzyme and allow the intermediate to accumulate. BBE RNAi transgenic California poppy cells showed remarkable reduction of BBE expression and reticuline accumulation. This is the first successful report of targeted gene silencing for production of an important secondary metabolic precursor (Sato 2005).

14.7.4 Improving Grain Yield, Biomass, and Biofuel Production

RNAi technology has emerged as an attractive tool for improving grain yield and biomass production in plants by altering the lignin composition (Hisano et al. 2009). Resultant products can be used to improve digestibility characteristics in forage for better livestock production (Marino 2008). Downregulation of specific cytochrome P450 enzymes involved in lignin synthesis has improved forage quality in alfalfa, by increasing digestibility of fodder (Reddy et al. 2005). Biomass with reduced ligning also has an important application in the paper pulp industry. Genetically engineering agronomic traits requires a specific degree of modification by suppression of target genes. RNAi is capable of this specific degree of gene suppression. As a result, RNAi has effectively improved a number of agronomic plant traits through targeted gene downregulation (Miki et al. 2005; Dixon et al. 2007). A backcross breeding program allows for the improved RNAi trait to be introgressed with high-yielding target genotypes. Due to its ease of plant breeding program integration, the RNAi approach has enormous potential to facilitate new genetic studies in hybrids.

Energy derived from cellulosic biomass largely resides in plant cell walls. Cellulosic biomass is difficult to break into simple sugars, because of lignin and complexity of cell wall structure. Recalcitrance of plant material is a major obstacle for converting lignocellulosic biomass to ethanol. GE-reduced lignin content has effectively overcome cell wall recalcitrance when bioconverting to ethanol. Transgenic downregulation of major lignin synthesis genes has led to reduced lignin content, increased dry matter degradability, and improved cellulose accessibility for cellulose degradation (Chen and Dixon et al. 2007). Downregulation of plant lignin genes, such as shikimate hydroxycinnamoyl transferase (HCT), cinnamate-4-hydroxylase (C4H), and 4-coumarate-coA ligase (4CL), drastically reduced total lignin content. This increased degradability of dry matter and improved cellulose accessibility for cellulose degradation (Hisano et al. 2009). Lignin modification eliminates the need for acid pretreatment (Chen and Dixon 2007). The cost of pretreatment has been reduced or eliminated for biomass from low-lignin producing transgenic plants. This greatly reduces production cost of biofuel (Hisano et al. 2011). Lignin monomer biosynthetic pathways are conserved across several species. Detailed applicable knowledge is available for modifying lignin composition in major biofuel crops, like switchgrass and *Miscanthus*. While manipulating lignin content, proper care must be taken to prevent severely limiting crop productivity. Earlier studies have shown altering lignin composition does not compromise plant fitness. On the contrary, some cases resulted in resistance to certain phytopathogens (Peltier et al. 2009). Recently, hairpin RNAi technology (HGS-hpRNAi) successfully reduced lignin content in switchgrass (Hisano et al. 2009). Commercial products from these studies are expected in the near future.

14.7.5 Improving Crop Nutritional Value

Plants are the principal source of human food and livestock feed. Many efforts have been employed to improve plant nutritional content by breeding and genetic engineering. Classical plant breeding is based on selection of natural or induced gene variations. Genetic engineering has advantages over classical breeding, providing an increased scope of genes and mutation types for manipulation. It also provides means to control spatial or temporal expression patterns for particular genes of interest. RNAi technology has improved nutritional quality in several plants, by increasing amino acid synthesis, by modifying oil's fatty acid composition, by reducing coffee plant caffeine content, and by generating a dominant high-lysine maize variant. Hairpin RNA was used for RNAi of cotton to nutritionally improve an increase of high-oleic and high-stearic cottonseed oils. This genetically improved cottonseed oil contains heart-healthy, essential fatty acids produced by downregulation of fatty acid desaturase genes encoding *stearoyl-acyl carrier* protein Δ9-desaturase and *oleoyl-phosphatidylcholine ω6-desaturase* (Liu et al. 2002). The amount of caffeine in coffee plants has been markedly reduced by RNAi-mediated suppression of caffeine synthase gene (Ogita et al. 2003). RNAi has successfully generated a dominant high-lysine maize variant by knockdown of the 22-kD maize zein storage protein, which is poor in lysine content (Segal 2003). Use of RNAi did not alter general function of O_2, a maize basic leucine zipper transcriptional factor controlling expression of a storage protein subset. Downregulation of 22-kD maize lysine-poor zein gene generated higher-quality, normal-size maize seeds rich in proteins (Guiliang and Gad 2004). Cereals high in amylose content (AC) and resistant starch (RS) offer potential health benefits, and RNAi generated high-amylose wheat (Regina et al. 2006). In the study, suppression of *SBEIIb* expression alone had no effect on amylose content; however, suppression of both *SBEIIa* and *SBEIIb* expressions resulted in starch containing >70% amylose. When mice were fed this >70% amylose wheat, their large bowel function improved (Regina et al. 2006).

14.7.6 Improving Crop Drought Tolerance

Drought is a major abiotic constraint creating a variety of physiological, biochemical, and metabolic changes in plants that limits growth and productivity in rain-fed agriculture (Chaves and Oliveira 2004). Drought resistance has been improved by both traditional breeding and biotechnology-based approaches (Valliyodan and Nguyen 2006). Vain (2007) made it possible to genetically engineer drought-tolerant plants by establishing stable genetic transformation protocols (Vain 2007). Functional relevance of stress-responsive genes can be elucidated by either overexpression or downregulation of specific genes. Widely utilized downregulation approaches include chemical, transposon, T-DNA insertional mutagenesis, irradiation, and PTGS-based methods for conferring stress tolerance (Watson et al. 2005). Both chemical and radiation mutagenesis conveniently generate mutants but are difficult and time-consuming procedures. T-DNA insertion mutagenesis can cause loss of function via ectopic activation of neighboring genes and disruption of coding sequence or UTR. This procedure is often time-consuming, and defining the number of insertions can be difficult. Transposon tagging shares many of the same limitations. To overcome some of these difficulties, RNAi or antisense approach techniques have been used to develop a stable knockout line to assess relevance of abiotic stress-responsive genes (Abbasi et al. 2007). Antisense expressing or RNAi construct transgenic plants are generated to analyze function of stress-responsive genes based upon phenotypes resulting from loss of function. Moreover, transgenic overexpressors are very useful for functional analyses of stress-inducible genes. They also demonstrate improved stress tolerance in plants generated by gene transfer.

14.7.7 RNAi for Crop Pest Management

RNA interference has a proven role in functional genomic insect research and shows considerable potential for insect pest control. Previously, transgenic plants producing dsRNAs directed against insect genes showed an economically important advantage of enhanced resistance to both cotton bollworm (*Helicoverpa armigera*; Lepidoptera) and western corn rootworm (WCR; *Diabrotica virgifera virgifera* LeConte; Coleoptera). This was made possible by two factors: (1) identification of a suitable insect target and (2) dsRNA delivery and expression in plants in amounts sufficient to produce intact dsRNA for insect uptake. For practical applications, it will be vital to minimize insecticidal effects on non-target insects, so specificity of RNAi-mediated insecticidal impacts is an important consideration. In this example, dsRNAs directed against three target genes (*β-tubulin*, *V-ATPase A*, and *V-ATPase E*) demonstrated effective RNAi response in WCR resulting in high larval mortality. Strongest effects were reported against gene encoding V-type *ATPase A*. Rapid knockdown of endogenous mRNA occurred within 24 h of ingestion with specific RNAi response via low concentrations of dsRNA. Baum (2007) reported these results on insect control through dsRNA-feeding experiments. This study provides evidence for potential insect pest control use of RNAi in crop protection. It demonstrates the possibility of gene silencing in insects via consumption of plant material that expresses hairpin dsRNA constructs against chosen target insect genes. Reduction of corn root damage was reported in transgenic maize plants producing vacuolar H^+ ATPase dsRNA following infestation with western corn rootworm (*D. virgifera virgifera*) (Hanneke and Guy 2010). In another report, model plants *Nicotiana tabacum* and *Arabidopsis thaliana* were modified with cytochrome P450 gene of *H. armigera* (Mao et al. 2007). Insect cotton bollworm larvae fed on transgenic leaves displayed reduced cytochrome P450 mRNA levels and impeded larval growth. The first bioassays using transgenic plants crops show delayed development of insects and reduced damage to plants (Baum 2007; Mao et al. 2007). However, efficiency must improve in RNAi insect control models before they are considered a tenable pesticide alternative. DvSnf7 is an essential western corn rootworm (WCR, *Diabrotica virgifera virgifera* LeConte) gene. It encodes an essential intracellular trafficking protein. DvSnf7 dsRNA effectively kills WCR larvae through oral delivery (Bolognesi et al. 2012). DvSnf7 dsRNA can be expressed in corn plants. When WCR larvae consume the corn-expressed DvSnf7 dsRNA, it enters larvae's digestion system. By disruption of critical rootworm gene expression, the dsRNA eventually kills the larvae. In 2017, the US EPA approved the first "RNAi insecticide" using DvSnf7 dsRNA (https://www.epa.gov/newsreleases/epa-registers-innovative-tool-control-corn-rootworm). SmartStax PRO maize, a line of GM corn developed collaboratively by agricultural giants, Monsanto and Dow, is expected to feature DvSnf7 dsRNA. SmartStax PRO expresses Cry3Bb1, Cry34Ab1/Cry35Ab1, and DvSnf7 genes (Head et al. 2017). Head et al. (2017) report the addition of DvSnf7 to SmartStax PRO reduces root damage under high WCR densities. It also prolongs Cry3Bb1 and Cry34Ab1/Cry35Ab1 durability.

14.7.8 RNAi to Produce Virus Resistance in Plants

Plant virus is a major productivity constraint for a wide range of economically important crops worldwide. There are two standard approaches for developing genetically engineered virus resistance that depend on the source of genes used. Genes can be derived from the pathogenic virus specifically (PDR) or from other sources. Studies of pathogen-derived resistance (PDR) in plants are responsible for this approach. For PDR, a partial or complete pathogen (virus) gene is introduced into the plant. This interferes with one or more essential steps in the viral life cycle. For example, Namba et al. (1991) demonstrated expression of a coat protein gene derived from *cucumber mosaic virus* (CMV) provided transgenic tobacco plant protection from several CMV strains (Namba et al. 1991). Namba et al. (1992) also observed transgenic plants expressing a coat protein gene of watermelon mosaic virus II or zucchini yellow mosaic virus protected against six potyviruses (Namba et al. 1992). Expression of other specific viral genes (e.g., coat proteins, replicases, movement proteins) in host plants can also provide protection against associated viruses (Beachy 1997). Non-pathogen-derived resistance is based on utilization of host resistance genes and other genes responsible for adaptive host processes. This type of resistance is activated after a pathogen attack, to obtain viral resistance (Dasgupta et al. 2003). In the last few years, there have been considerable advances in understanding post-transcriptional gene silencing (PTGS). This type of gene silencing is believed to be a natural, antiviral defense system in plants activated as a double-stranded RNA (dsRNA) response. In plants, dsRNA or self-complementary hairpin RNA transcribed from engineered inverted repeats potently induced gene silencing response, when directed against transgenes. Furthermore, plants transformed with hairpin RNAi constructs producing RNAs capable of duplex formation were found to induce host immunity against targeted viruses with almost 100% efficiency (Smith et al. 2000). Transient expression of hairpin RNA effectively blocks viral replication, multiplication, and spread in non-transgenic plants. Hairpin-dependent interference by transiently expressed constructs is a potential tool for studying PTGS onset in the context of a viral infection (Tenllado et al. 2004). Topical plant application of pathogen-specific double-stranded RNA (dsRNA) for viral resistance is an attractive alternative (Mitter et al. 2017). The major challenge for this method is the instability of naked dsRNA spread on plant leaves, which limits the application. For this reason, Mitter et al. (2017) loaded naked dsRNA into designer, nontoxic, degradable, layered double hydroxide (LDH) clay nanosheets before spreading on plant leaves (Mitter et al. 2017). Their report showed LDH successfully attached to leaves and degraded and released dsRNA. This allowed for dsRNA uptake into plant cells, successfully silencing homologous RNA.

14.7.9 Using RNAi for Management Against Parasitic Weeds

Parasitic weeds, from families Orobanchaceae (*Aeginetia*, *Orobanche*, broomrape) and Scrophulariaceae (*Alectra*, *Striga*, witchweed) are considered the most serious agricultural pests of economic importance in many parts of the world. Some parasitic weeds are specific to sorghum (*Sorghum bicolor* (L.) Moench), maize (*Zea mays* L.), pearl millet (*Pennisetum americanum* L.), rice (*Oryza sativa* L.), sugarcane (*Saccharum officinarum* L.), and others (http://www.fao.org/docrep/006/y5031e/y5031e0a.htm). Parasitic weeds are responsible for both enormous crop yield loss and subsequent economic loss. Parasitic weed management is extremely difficult. Conventional control methods are not efficient enough to suffice. No genetic loci conferring resistance to crop plants have been identified, preventing traditional genetic engineering options. Recently, RNAi has been explored as an alternative. RNAi has the potential to be a genetic tool for engineering parasitic weed resistance. The general mechanism allows for transformation of a host plant via plasmid carrying double-stranded hairpin RNA (hpRNA). This hpRNA targets one or more parasitic genes for silencing. Due to the specific design involvement against parasitic gene sequences, no phenotypic effect is expected on host cells.

Haustorium is a specialized, invasive organ found on some parasites capable of penetrating host tissue to obtain nutrients (Yoshida et al. 2016). Haustorium studies report a dramatic effect on parasites taking up parasite-specific RNAi from the host (Yoder et al. 2009). Transgenic RNAi tomato plants expressing mannose-6-phosphate reductase gene (*M6PR*) hpRNA construct have been effective against *Orobanche aegyptiaca* (Egyptian broomrape) by causing death to tubercles attached to tomato. *Orobanche aegyptiaca* is an obligate holoparasite plant from the family Orobanchaceae. This parasitic plant infects many economically important crops (Eizenberg et al. 2002). Other studies show parasites can take up specific RNAi from host via haustorium (John et al. 2009). These studies clearly demonstrate future potential for plant parasite control using RNAi technology.

14.7.10 RNAi as a Tool for the Genetic Improvement of Crop Plants

Genetic improvement of crop plants is dependent on trait stability from generation to generation. Suppression of phenotypes by PTGS provides genetic enhancements with stable inheritance patterns (Mitsuhara et al. 2002). hpRNA transgenes in *Arabidopsis* show stable inheritance to T_5 generation (Stoutjesdijk et al. 2002). The first commercially useful RNAi cultivar was produced in rice mutant line *LGC-1* with low glutelin content1 and increased prolamin (Kusaba et al. 2003). *Lgc1* mutant was produced by a tail-to-tail inverted repeat of functional *GluB* gene and truncated *GluB* gene. This combination is thought to produce a functional *GluB* hairpin RNA with a dsRNA region responsible for silencing low-protein trait in LGC-1, useful for patients on dietary protein restrictions. Interestingly, the mutant trait demonstrated stability for more than 20 generations, suggesting stable inheritance of hpRNA-induced RNAi for suppression of gene expression (Kusaba et al. 2003).

14.8 Practical Limitations of RNAi Technology

Apart from the many advantages of RNAi technology are off-target effects, the major disadvantage of siRNA. Although it has been reported, several chemical modifications can reduce siRNA off-target effect (Xu et al. 2006). This effect has potential for negative impact on transgenic plants generated through RNAi technology. RNAi for insect control can also have unintended impacts on beneficial organisms like pollinators. This ultimately warrants long-term research to elucidate the factors contributing to off-target silencing for prevention of potential environmental problems. Additionally, several reports suggest RNAi off-target effects have potential to silence plant genes. Due to constant production, dsRNA has a high threshold level in plant cell. This dsRNA is not utilized when insect pests or viral hosts are absent, which increases access of off-target endogenous plant mRNA to RISC complex. In a future hypothetical scenario, a community of RNAi transgenic plant, insect, weed, and nematode will share the same environment. This community interaction will magnify off-target effect and alter the balance of plant-pest interactions. This could potentially threaten future ecological balance and impact natural selection. Off-target silencing is influenced by sequence identity between RNAi trigger and target mRNA, trigger gene region, trigger size, and transitive nature of RNAi. Prevention of these off-target effects can be achieved by use of a specific RNAi-silencing trigger sequence in the vector, or by use of tissue-specific and inducible methods of silencing. siRNA Scan is publicly available, web-based computational tool (http://bioinfo2.noble.org/RNAiScan.htm) provided by researchers for identification of plant genes with potential for off-target effect during PTGS (Xu et al. 2006). This software can be effectively adapted to RNAi construct design for prevention of off-target silencing.

This summary lists major disadvantages to consider when designing effective siRNA gene silencing:

1. Off-target effect
2. Appropriate delivery system
   (a) *Agrobacterium* method is widely adapted for RNAi plant transformation:
      - It is time-consuming.
      - Gene integration is uncertain.
      - Expression of small RNA is uncertain.
3. Induction of unintended plant immune response
4. Variable or incomplete knockdowns and nonspecificity of reagents (Krausz and Korn 2008)
5. Unmodified siRNA easily degraded by RNase:
   (a) Poor chemical stability and low half-life in circulation
6. Transcripts with high turnover are sometimes difficult to achieve RNA-mediated silencing

14.9 Future Perspectives of RNAi Technology

RNAi technology has recently become an incredibly functional tool. This technology will retain its potential as a genomic tool in the coming decades for large-scale, knockdown lines being developed in major crop plants. There is a wide array of applications for RNA silencing in single-celled organisms. *Chlamydomonas reinhardtii*, photosynthetic alga, enables high-throughput plant gene function identification. This would allow for analysis of various abiotic stress tolerance genes. A stress-specific cDNA library could be cloned into a gene silencing RNAi vector (e.g., sense strand overexpression to induce RNAi) providing a potential source for development of knockdown in various species. Potential of RNAi for engineering crop plants for tolerance to abiotic stresses has not yet been exploited except for drought. Despite a few limitations, RNAi crop improvement is expected to be significant in the near future. This is especially relevant for control of various insect pests and diseases. Moreover, stable RNAi transgenic plants can be easily integrated into plant breeding programs. Plant breeders can use this technology to incorporate insect or disease resistance into high-yielding commercial hybrids or varieties.

Additionally, an agronomically superior cultivar could be engineered for additional plant fitness, perhaps stress tolerance, using RNAi technology.

References

Abbasi AR, Hajirezaei M, Hofius D, Sonnewald U, Voll LM (2007) Specific roles of a- and g-tocopherol in abiotic stress responses of transgenic tobacco. Plant Physiol 143:1720–1738

Anandalakshmi R, Pruss GJ, Ge X, Marathe R, Smith TH, Vance VB (1998) A viral suppressor of gene silencing in plants. Proc Natl Acad Sci U S A 95:13079–13084

Astier-Manifacier S, Cornuet P (1971) RNA-dependent RNA polymerase in Chinese cabbage. Biochim Biophys Acta 232:484–493

Astier-Manifacier S, Cornuet P (1978) Purification and molecular weight of an RNA-dependant RNA polymerase from *Brassicae oleracea* var. Botrytis. C R Acad Sci Hebd Seances Acad Sci D 287:1043–1046

Baker C, Jacyn NM, Mock Bruce D, Whitaker DP, Roberts CP, Rice KL, Deahl Aver'yanov AA (2005) Involvement of acetosyringone in plant-pathogen recognition. Biochem Biophys Res Commun 328:130–136

Bartel DP (2004) MicroRNAs: genomics, biogenesis, mechanism, and function. Cell 116 (2):281–297

Baulcombe DC (1994) Novel strategies for engineering virus resistance in plants. Curr Opin Biotechnol 5:117–124

Baulcombe DC (2004) RNA silencing in plants. Nature 431:356

Baum JA (2007) Control of coleopteran insect pests through RNA interference. Nat Biotechnol 25:1322–1326

Beachy RN (1997) Mechanisms and applications of pathogen-derived resistance in transgenic plants. Curr Opin Biotechnol 8:215–220

Becker A, Lange M (2010) VIGS—genomics goes functional. Trends Plant 15(1):1–4

Berkhout B (2018) RNAi-mediated antiviral immunity in mammals. Curr Opin Virol 32:9–14

Bernstein E, Caudy AA, Hammond SM, Hannon GJ (2001) Role for a bidentate ribonuclease in the initiation step of RNA interference. Nature 409:363–366

Bolognesi R, Ramaseshadri P, Anderson J, Bachman P, Clinton W, Flannagan R, Ilagan O, Lawrence C, Levine S, Moar W, Mueller G, Tan J, Uffman J, Wiggins E, Heck G, Segers G (2012) Characterizing the mechanism of action of double-stranded RNA activity against western corn rootworm (*Diabrotica virgifera virgifera* LeConte). PLoS One 7(10):e47534

Carthew RW, Sontheimer EJ (2009) Origins and mechanisms of miRNAs and siRNAs. Cell 136 (4):642–655

Chai XJ, Wang PW, Guan SY, Xu YW (2005) Reducing the maize amylopectin content through RNA interference manipulation. Zhi Wu Sheng Li Yu Fen Zi Sheng Wu Xue Xue Bao 31:625–630

Chaves MM, Oliveira MM (2004) Mechanisms underlying plant resilience to water deficits: prospects for water-saving agriculture. J Exp Bot 55:2365–2384

Chen F, Dixon RA (2007) Lignin modification improves fermentable sugar yields for biofuel production. Nat Biotechnol 25:759–761

Croteau R, Kutchan TM, Lewis NG (2000) Natural products (secondary metabolites). In: Buchanan B, Gruissem W, Jones R (eds) Biochemistry and molecular biology of plants. American Society of Plant Physiologists, Rockville, pp 1250–1319

Cullen BR (2002) RNA interference: antiviral defense and genetic tool. Nat Immunol 3:597–599

Dalmay T, Hamilton A, Rudd S, Angell S, Baulcombe DC (2000) An RNA-dependent RNA polymerase gene in *Arabidopsis* is required for posttranscriptional gene silencing mediated by a transgene but not by a virus. Cell 101:543–553

Daniel RGP, John AG (2008) RNAi-mediated crop protection against insects. Trends Biotechnol 26:393–400
Dasgupta I, Malathi VG, Mukherjee SK (2003) Genetic engineering for virus resistance. Curr Sci 84:341–354
Davuluri GR, Van Tuinen A, Fraser PD, Manfredonia A, Newman R, Burgess D, Bramley PM (2005) Fruit-specific RNAi-mediated suppression of DET1 enhances carotenoid and flavonoid content in tomatoes. Nat Biotechnol 23:890–895
Dhankher OP (2006) Hyperaccumulation of arsenic in the shoots of *Arabidopsis* silenced for arsenate reductase (ACR2). Proc Natl Acad Sci U S A 103:5413–5418
Ding SW, Han Q, Wang J, Li WX (2018) Antiviral RNA interference in mammals. Curr Opin Immunol 54:109–114
Dixon RA, Bouton JH, Narasimhamoorthy B, Saha M, Wang ZY, May GD (2007) Beyond structural genomics for plant science. Adv Agron 95:77–161
Dorssers L, Zabel P, van der Meer J, van Kammen A (1982) Purification of a host-encoded RNA-dependent RNA polymerase from cowpea mosaic virus-infected leaves. Virology 116:236–249
Duda CT (1979) Synthesis of double-stranded RNA. II. Partial purification and characterization of an RNA-dependent RNA polymerase in healthy tobacco leaves. Virology 92:180–189
Duda CT, Zaitlin M, Siegel A (1973) In vitro synthesis of double-stranded RNA by an enzyme system isolated from tobacco leaves. Biochim Biophys Acta 319:62–71
Dunoyer P, Christopher AB, Gregory S, Wang Y, Florence J, Abdelmalek A, Christophe H, Olivier V (2010) An endogenous, systemic RNAi pathway in plants. EMBO J 29:1699–1712
Eizenberg H, Golan S, Joel DM (2002) First report of the parasitic plant *Orobanche aegyptiaca* infecting olive. Plant Dis 86:814
Elbashir SM, Harborth J, Lendeckel W, Yalcin A, Weber K, Tuschl T (2001) Duplexes of 21-nucleotide RNAs mediate RNA interference in cultured mammalian cells. Nature 411:494–498
Filipowicz W (2005) RNAi: the nuts and bolts of the RISC machine. Cell 122:17–20
Fire A, Xu S, Montgomery MK, Kostas SA, Driver SE, Mello CC (1998) Potent and specific genetic interference by double-stranded RNA in *Caenorhabditis elegans*. Nature 391:806–811
Fitch MMM, Manshardt RM, Gonsalves D, Slightom JL, Sanford JC (1992) Virus resistant papaya plants derived from tissues bombarded with the coat protein gene of papaya ringspot virus. Nat Biotechnol 10:1466–1472
Fukudome A, Fukuhara T (2017) Plant dicer-like proteins: double-stranded RNA-cleaving enzymes for small RNA biogenesis. J Plant Res 130:33–44
Grishok A, Pasquinelli AE, Conte D, Li N, Parrish S, Ha I, Baillie DL, Fire A, Ruvkun G, Mello CC (2001) Genes and mechanisms related to RNA interference regulate expression of the small temporal RNAs that control *C. elegans* developmental timing. Cell 106:23–34
Grosshans H, Filipowicz W (2008) Molecular biology: the expanding world of small RNAs. Nature 451:414–416
Guiliang T, Gad G (2004) Using RNAi to improve plant nutritional value: from mechanism to application. Trends Biotechnol 22:463–469
Hamilton AJ, Baulcombe DC (1999) A species of small antisense RNA in posttranscriptional gene silencing in plants. Science 286:950–952
Hammond SM, Boettcher S, Caudy AA, Kobayashi R, Hannon GJ (2001) Argonaute2, a link between genetic and biochemical analyses of RNAi. Science 293:1146–1150
Hanneke H, Guy S (2010) Mechanisms of dsRNA uptake in insects and potential of RNAi for pest control: a review. J Insect Physiol 56:227–235
Hannon GJ (2002) RNA interference. Nature 418:244–251
Hashimoto T, Yamada Y (2003) New genes in alkaloid metabolism and transport. Curr Opin Biotechnol 14:163–168

Hauschild K, Pauli HH, Kutchan TM (1998) Isolation and analysis of a gene bbe1 encoding the berberine bridge enzyme from the California poppy *Eschscholzia californica*. Plant Mol Biol 36:473–478

Head GP, Carroll MW, Evans SP, Rule DM, Willse AR, Clark TL, Storer NP, Flannagan RD, Samuel LW, Meinke LJ (2017) Evaluation of SmartStax and SmartStax PRO maize against western corn rootworm and northern corn rootworm: efficacy and resistance management. Pest Manag Sci 73:1883–1899

Helliwell C, Waterhouse P (2003) Constructs and methods of high throughput gene silencing in plants. Methods 30:289–295

Helliwell CA, Wesley SV, Wielopolska AJ, Waterhouse PM (2002) High-throughput vectors for efficient gene silencing in plants. Funct Plant Biol 29:1217–1225

Hisano H, Nandakumar R, Wang ZY (2009) Genetic modification of lignin biosynthesis for improved biofuel production. In Vitro Cell Develop Biol Plant 45:306–313

Hisano H, Nandakumar R, Wang ZY (2011) Genetic modification of lignin biosynthesis for improved biofuel production. In: Biofuels. Springer, New York, pp 223–235

Jaskiewicz L, Filipowicz W (2008) Role of dicer in posttranscriptional RNA silencing. Curr Top Microbiol Immunol 320:77–97

John IY, Pradeepa G, Biao W, Natalya T, Alexey AT (2009) Engineering host resistance against parasitic weeds with RNA interference. Pest Manag Sci 65:460–466

Kennerdell JR, Carthew RW (1998) Use of dsRNA-mediated genetic interference to demonstrate that frizzled and frizzled 2 act in the wingless pathway. Cell 95:1017–1026

Ketting RF, Fischer SE, Bernstein E, Sijen T, Hannon GJ, Plasterk RH (2001) Dicer functions in RNA interference and in synthesis of small RNA involved in developmental timing in *C. elegans*. Genes Dev 15:2654–2659

Klahre U, Crété P, Leuenberger SA, Iglesias VA, Meins FJ (2002) High molecular weight RNAs and small interfering RNAs induce systemic posttranscriptional gene silencing in plants. Proc Natl Acad Sci U S A 18:11981–11986

Krausz E, Korn K (2008) High-content siRNA screening for target identification and validation. Expert Opin Drug Discov 3:551–564

Kumagai MH, Donson J, della-Cioppa G, Harvey D, Hanley K, Grill LK (1995) Cytoplasmic inhibition of carotenoid biosynthesis with virus-derived RNA. Proc Natl Acad Sci U S A 92 (5):1679–1683

Kumar S, Tanti B, Patil BL, Mukherjee SK, Sahoo L (2017) RNAi-derived transgenic resistance to Mungbean yellow mosaic India virus in cowpea. PLoS One 12(10):e0186786

Kusaba M, Miyahara K, Iida S, Fukuoka H, Takano T, Sassa H, Nishio T (2003) *Low glutelin content1*: a dominant mutation that suppresses the *glutelin* multigene family via RNA silencing in rice. Plant Cell 15:1455–1467

Leuschner PJ, Ameres SL, Kueng S, Martinez J (2006) Cleavage of the siRNA passenger strand during RISC assembly in human cells. EMBO Rep 7:314–320

Li WX, Ding SW (2001) Viral suppressors of RNA silencing. Curr Opin Biotechnol 12:150–154

Li H, Li WX, Ding SW (2002) Induction and suppression of RNA silencing by an animal virus. Science 296:1319–1321

Li Y, Lu J, Han Y, Fan X, Ding SW (2013) RNA interference functions as an antiviral immunity mechanism in mammals. Science 342:231–234

Li S, Xu Z, Sheng J (2018) tRNA-derived small RNA: a novel regulatory small non-coding RNA. Gene (Basel) 9:246

Limpens E, Ramos J, Franken C, Raz V, Compaan B, Franssen H, Bisseling T, Geurts R (2004) RNA interference in *Agrobacterium rhizogenes*-transformed roots of *Arabidopsis* and *Medicago truncatula*. J Exp Bot 55:983–992

Lindbo JA, Dougherty WG (1992a) Untranslatable transcripts of the tobacco etch virus coat protein gene sequence can interfere with tobacco etch virus replication in transgenic plants and protoplasts. Virology 189:725–733

Lindbo JA, Dougherty WG (1992b) Pathogen-derived resistance to a potyvirus: immune and resistant phenotypes in transgenic tobacco expressing altered forms of a potyvirus coat protein nucleotide sequence. Mol Plant Microbe Interact 5(2):144–153

Lindbo JA, Silva-Rosales L, Proebsting WM, Dougherty WG (1993) Induction of a highly specific antiviral state in transgenic plants: implications for regulation of gene expression and virus resistance. Plant Cell 5:1749–1759

Liu Q, Singh SP, Green AG (2002) High-stearic and high-oleic cottonseed oils produced by hairpin RNA-mediated post-transcriptional gene silencing. Plant Physiol 129:1732–1743

Mack GS (2007) MicroRNA gets down to business. Nat Biotechnol 25:631–638

Mao YB, Cai WJ, Wang JW, Hong GJ, Tao XY, Wang LJ, Huang YP, Chen XY (2007) Silencing a cotton bollworm P450 monooxygenase gene by plant-mediated RNAi impairs larval tolerance of gossypol. Nat Biotechnol 25:1307–1313

Margis R, Fusaro AF, Smith NA, Curtin SJ, Watson JM, Finnegan EJ, Waterhouse PM (2006) The evolution and diversification of dicers in plants. FEBS Lett 580(10):2442–2450

Marino M (2008) Profile of Richard Dixon. Proc Natl Acad Sci U S A 105:2263–2265

Matzke MA, Matzke AJ, Kooter JM (2001a) RNA: guiding gene silencing. Science 293:1080–1083

Matzke MA, Matzke AJ, Pruss GJ, Vance VB (2001b) RNA-based silencing strategies in plants. Curr Opin Genet Dev 11:221–227

Meister G, Tuschl T (2004) Mechanisms of gene silencing by double-stranded RNA. Nature 431:343–349

Miki D, Itoh R, Shimamoto K (2005) RNA silencing of single and multiple members in a gene family of rice. Plant Physiol 138:1903–1913

Mitsuhara I, Shirasawa SN, Iwai T, Nakamura S, Honkura R, Ohashi Y (2002) Release from post-transcriptional gene silencing by cell proliferation in transgenic tobacco plants: possible mechanism for noninheritance of the silencing. Genetics 160:343–352

Mitter N, Worrall EA, Robinson KE, Li P, Jain RG, Taochy C, Fletcher SJ, Carroll BJ, Lu GQ, Xu ZP (2017) Clay nanosheets for topical delivery of RNAi for sustained protection against plant viruses. Nat Plants 3:16207

Mourrain P, Béclin C, Elmayan T, Feuerbach F, Godon C, Morel JB, Jouette D, Lacombe AM, Nikic S, Picault N, Rémoué K, Sanial M, Vo TA, Vaucheret H (2000) Arabidopsis SGS2 and SGS3 genes are required for posttranscriptional gene silencing and natural virus resistance. Cell 101:533–542

Namba S, Ling K, Gonsalves C, Gonsalves D, Slightom JL (1991) Expression of the gene encoding the coat protein of cucumber mosaic virus (CMV) strain WL appears to provide protection to tobacco plants against infection by several different CMV strains. Gene 107:181–188

Namba S, Ling K, Gonsalves C, Slightom JL, Gonsalves D (1992) Protection of transgenic plants expressing the coat protein gene of watermelon mosaic virus ii or zucchini yellow mosaic virus against six potyviruses. Phytopathology 82:940–946

Napoli C, Lemieux C, Jorgensen R (1990) Introduction of a chimeric Chalcone synthase gene into petunia results in reversible co-suppression of homologous genes in trans. Plant Cell 2:279–289

Nishihara M (2005) Flavonoid components and flower color change in transgenic tobacco plants by suppression of chalcone isomerase gene. FEBS Lett 579:6074–6078

Nora S, MichÈLe Z, Asuka I, Biao D, Ming-Bo W, Gabi K, Michael W (2009) RNAi-mediated resistance to *Potato spindle tuber viroid* in transgenic tomato expressing a viroid hairpin RNA construct. Mol Plant Pathol 10:459–469

Ogita S (2004) Application of RNAi to confirm theobromine as the major intermediate for efficient biosynthesis in coffee plants with potential for construction of decaffeinated varieties. Plant Mol Biol 54:931–941

Ogita S, Uefuji H, Yamaguchi Y, Koizumi N, Sano H (2003) RNA interference: producing decaffeinated coffee plants. Nature 423:823–823

Pandit NN, Russo VE (1992) Reversible inactivation of a foreign gene, hph, during the asexual cycle in *Neurospora crassa* transformants. Mol Gen Genet 234:412–422

Park SU, Yu M, Facchini PJ (2003) Modulation of berberine bridge enzyme levels in transgenic root cultures of California poppy alters the accumulation of benzophenanthridine alkaloids. Plant Mol Biol 51:153–164

Peltier AJ, Hatfield RD, Grau CR (2009) Soybean stem lignin concentration relates to resistance to *Sclerotinia sclerotiorum*. Plant Dis 93:149–154

Petrovska N (2005) Transgenic ryegrasses (*Lolium* spp.) with down-regulation of main pollen allergens. Mol Breed 14:489–501

Plasterk RH (2002) RNA silencing: the genome's immune system. Science 296:1263–1265

Praveen G, Deepmala G, Monika M, Kumar V, Jyoti B, Sudesh KY (2012) MicroRNAs and their role in plants during abiotic stresses. In: Ahmad P, Prasad MNV (eds) Environmental adaptations and stress tolerance of plants in the era of climate change, vol 12. Springer, New York, pp 265–278

Prins M, Goldbach R (1996) RNA-mediated virus resistance in transgenic plants. Arch Virol 141:2259–2276

Qiao F, Yang Q, Wang CL, Fan YL, Wu XF, Zhao KJ (2007) Modification of plant height via RNAi suppression of OsGA20ox2 gene in rice. Euphytica 158:35–45

Reddy MSS, Chen F, Shadle G, Jackson L, Aljoe H, Dixon RA (2005) Targeted down-regulation of cytochrome P450 enzymes for forage quality improvement in alfalfa (*Medicago sativa* L.). Proc Natl Acad Sci U S A 102:16573–16578

Regina A, Bird A, Topping D, Bowden S, Freeman J, Barsby T, Kosar-Hashemi B, Li Z, Rahman S, Morell M (2006) High-amylose wheat generated by RNA interference improves indices of large-bowel health in rats. Proc Natl Acad Sci U S A 103:3546–3551

Ruiz MT, Voinnet O, Baulcombe DC (1998) Initiation and maintenance of virus-induced gene silencing. Plant Cell 10:937–946

Sanford JC, Johnston SA (1985) The concept of parasite-derived resistance—deriving resistance genes from the parasite's own genome. J Theor Biol 113:395–405

Sato F (2005) RNAi and functional genomics. Plant Biotechnol 22:431–442

Schweizer P, Pokorny J, Schulze-Lefert P, Dudler R (2000) Double-stranded RNA interferes with gene function at the single-cell level in cereals. Plant J 24:895–903

Segal G (2003) A new opaque variant of maize by a single dominant RNA-interference-inducing transgene. Genetics 165:387–397

Senthil-Kumar M, Rame Gowda HV, Hema R, Mysore KS, Udayakumar M (2008) Virus-induced gene silencing and its application in characterizing genes involved in water-deficit-stress tolerance. J Plant Physiol 165:1404–1421

Senthil-Kumar M, Hema R, Suryachandra TR, Ramegowda HV, Gopalakrishna R, Rama N, Udayakumar M, Mysore KS (2010) Functional characterization of three water deficit stress-induced genes in tobacco and *Arabidopsis*: an approach based on gene down regulation. Plant Physiol Biochem 48:35–44

Shitan N, Bazin I, Dan K, Obata K, Kigawa K, Ueda K, Sato F, Forestier C, Yazaki K (2003) Involvement of CjMDR1, a plant multidrug-resistance-type ATP-binding cassette protein, in alkaloid transport in *Coptis japonica*. Proc Natl Acad Sci U S A 100:751–756

Sijen T, Fleenor J, Simmer F, Thijssen KL, Parrish S, Timmons L, Plasterk RH, Fire A (2001) On the role of RNA amplification in dsRNA-triggered gene silencing. Cell 107:465–476

Singh A, Taneja J, Dasgupta I, Mukherjee SK (2015) Development of plants resistant to tomato geminiviruses using artificial trans-acting small interfering RNA. Mol Plant Pathol 16:724–734

Smith NA, Singh SP, Wang MB, Stoutjesdijk PA, Green AG, Waterhouse PM (2000) Total silencing by intron-spliced hairpin RNAs. Nature 407:319–320

Storz G (2002) An expanding universe of noncoding RNAs. Science 296:1260–1263

Stoutjesdijk PA, Singh SP, Liu Q, Hurlstone CJ, Waterhouse PA, Green AG (2002) hpRNA-mediated targeting of the Arabidopsis FAD2 gene gives highly efficient and stable silencing. Plant Physiol 129:1723–1731

Takanami Y, Fraenkel-Conrat H (1982) Comparative studies on ribonucleic acid dependent RNA polymerases in cucumber mosaic virus infected cucumber and tobacco and uninfected tobacco plants. Biochemistry 21:3161–3167

Tenllado F, Llave C, Díaz-Ruíz JR (2004) RNA interference as a new biotechnological tool for the control of virus diseases in plants. Virus Res 102:85–96

Tijsterman M, Plasterk RH (2004) Dicers at RISC: the mechanism of RNAi. Cell 117:1–3

Tijsterman M, Ketting RF, Plasterk RH (2002) The genetics of RNA silencing. Annu Rev Genet 36:489–519

Vain P (2007) Thirty years of plant transformation technology development. Plant Biotechnol J 5:221–229

Valliyodan B, Nguyen HT (2006) Understanding regulatory networks and engineering for enhanced drought tolerance in plants. Curr Opin Plant Biol 9:189–195

van der Krol AR, Mur LA, Beld M, Mol JNM, Stuitje AR (1990) Flavonoid genes in petunia: addition of a limited number of gene copies may lead to a suppression of gene expression. Plant Cell 2:291–299

Vance V, Vaucheret H (2001) RNA silencing in plants: defense and counter-defense. Science 292:2277–2280

Vaucheret H (2008) Plant ARGONAUTES. Trends Plant Sci 13:350–358

Vaucheret H, Beclin C, Fagard M (2001) Post-transcriptional gene silencing in plants. J Cell Sci 114:3083–3091

Voinnet O, Baulcombe DC (1997) Systemic signaling in gene silencing. Nature 389:553

Vu TV, Choudhury NR, Mukherjee SK (2013) Transgenic tomato plants expressing artificial microRNAs for silencing the pre-coat and coat proteins of a begomovirus, *Tomato leaf curl New Delhi* virus, show tolerance to virus infection. Virus Res 172:35–45

Wagner GJ, Kroumova AB (2008) The use of RNAi to elucidate and manipulate secondary metabolite synthesis in plants. In: Ying SY (ed) Current perspectives in microRNAs (miRNA). Springer, Dordrecht, pp 431–459

Wassenegger M, Krczal G (2006) Nomenclature and functions of RNA-directed RNA polymerases. Trends Plant Sci 11:142–151

Watanabe T, Totoki Y, Toyoda A, Kaneda M, Kuramochi-Miyagawa S, Obata Y, Chiba H, Kohara Y, Kono T, Nakano T, Surani MA, Sakaki Y, Sasaki H (2008) Endogenous siRNAs from naturally formed dsRNAs regulate transcripts in mouse oocytes. Nature 453:539–543

Waterhouse PM, Helliwell CA (2003) Exploring plant genomes by RNA-induced gene silencing. Nat Rev Genet 4:29–38

Waterhouse PM, Wang MB, Lough T (2001) Gene silencing as an adaptive defense against viruses. Nature 411:834–842

Watson JM, Fusaro AF, Wang M, Waterhouse PM (2005) RNA silencing platforms in plants. FEBS Lett 579:5982–5987

Wesley SV, Helliwell CA, Smith NA, Wang M, Rouse DT, Liu Q, Gooding PS, Singh SP, Abbott D, Stoutjesdijk PA, Robinson SP (2001) Construct design for efficient, effective and high-throughput gene silencing in plants. Plant J 27:581–590

Williams RW, Rubin GM (2002) *ARGONAUTE1* is required for efficient RNA interference in *Drosophila* embryos. Proc Natl Acad Sci U S A 99:6889–6894

Willmann MR, Endres MW, Cook RT, Gregory BD (2011) The functions of RNA-dependent RNA polymerases in *Arabidopsis*. Arabidopsis Book 9:e0146. (American Society of Plant Biologists)

Wilson TMA (1993) Strategies to protect crop plants against viruses: pathogen-derived resistance blossoms. Proc Natl Acad Sci U S A 90:3134–3141

Xie Z, Fan B, Chen C, Chen Z (2001) An important role of an inducible RNA-dependent RNA polymerase in plant antiviral defense. Proc Natl Acad Sci U S A 98:6516–6521

Xiong AS (2004) Different effects on ACC oxidase gene silencing triggered by RNA interference in transgenic tomato. Plant Cell Rep 23:639–646

Xu P, Zhang Y, Kang L, Roossinck MJ, Mysore KS (2006) Computational estimation and experimental verification of off-target silencing during posttranscriptional gene silencing in plants. Plant Physiol 142:429–440
Yoder JI, Gunathilake P, Wu B, Tomilova N, Tomilov AA (2009) Engineering host resistance against parasitic weeds with RNA interference. Pest Manag Sci 65:460–466
Yoshida S, Yoshida S, Cui S, Ichihashi Y, Shirasu K (2016) The Haustorium, a specialized invasive organ in parasitic plants. Annu Rev Plant Biol 67:643–667
Zamore PD (2001) RNA interference: listening to the sound of silence. Nat Struct Biol 8:746–750
Zamore PD, Tuschl T, Phillip AS, David PB (2000) RNAi: double-stranded RNA directs the ATP-dependent cleavage of mRNA at 21 to 23 nucleotide intervals. Cell 101:25–33
Zhang H, Xia R, Meyers BC, Walbot V (2015) Evolution, functions, and mysteries of plant ARGONAUTE proteins. Curr Opin Plant Biol 27:84–90

Sanjeev Kumar is working as a DST-SERB National Postdoctoral Fellow in the Department of Biosciences and Bioengineering, IIT Guwahati, India. His research project aims to elucidate the fundamental mechanism of drought tolerance induced through microRNA regulations in mung bean.

Avishek Dey is working as a DST INSPIRE Faculty in the Department of Biosciences and Bioengineering, IIT Guwahati, India. His research project aims to generate aphid (*L. erysimi* Kalt.)-resistant marker-free transgenic mustard through RNAi-mediated gene silencing approach.

Yuan-Yeu Yau obtained a B.Sc. in Botany at National Taiwan University (NTU) and received Master and Ph.D. from University of Wisconsin-Madison, USA. He then worked at University of California, Berkeley, and Plant Gene Expression Center (USDA-ARS) at Albany (California, USA), the Chinese Academy of Sciences (Professor), and Northeastern State University (USA). He works in areas of plant biotechnology to develop clean-gene technology using microbial site-specific recombination system. He also worked on cotton transformation and production of stroke drug using molecular farming. He is also an active member of *ResearchGate*, a professional network for scientists and researchers.

Mona Easterling is currently an Assistant Professor of Biology at Tulsa Community College. Her Master's research involved use of Bxb1 recombination systems for the removal of selectable marker genes.

Lingaraj Sahoo is working as a Professor at Department of Biosciences and Bioengineering, IIT Guwahati, India, and Adjunct Professor at Gifu University, Japan. His research area are "biotic stress tolerance in grain legumes, climate-resilient crops (grain legumes and oil seeds) for sustainable agriculture, functional food and biofortification of grain legumes for healthcare, bioresource and biofuel, and academia-industry linkage for translational bioresource and food technology."

Current Transformation Methods for Genome-Editing Applications in Energy Crop Sugarcane

15

Chakravarthi Mohan, J. Ashwin Narayan, Mona Esterling, and Yuan-Yeu Yau

Abstract

Sugarcane is an important worldwide cash crop used for both sugar and ethanol production. Improvement of sugarcane through conventional breeding practices has been limited by its complex polyploid genome. Considerable improvements have been made with sugarcane through transgenic technology. Biolistic and *Agrobacterium* transformation methods have been regularly employed to develop transgenic sugarcane containing key agronomic traits. Transgenic sugarcane has been approved for commercial cultivation in both Indonesia and Brazil. Hightransformation efficiencies are on the horizon due to improved methods. Recent advances using CRISPR (clustered regularly interspaced short palindrome repeats)/Cas9 (CRISPR-associated) genome-editing systems require a robust transformation method. Efficient sugarcane transformation protocols will be vital in harnessing the potential of this cash crop. This chapter describes recent advances in sugarcane transformation and highlights novel improvement strategies to enhance target gene expression.

C. Mohan (✉)
Department of Genetics and Evolution, Federal University of São Carlos, São Carlos, Brazil

J. Ashwin Narayan
Division of Crop Improvement, ICAR—Sugarcane Breeding Institute, Coimbatore, India

M. Esterling
Department of Natural Sciences, Northeastern State University, Broken Arrow, OK, USA

Tulsa Community College, Northeast Campus, Tulsa, OK, USA

Y.-Y. Yau (✉)
Department of Natural Sciences, Northeastern State University, Broken Arrow, OK, USA

A. Kumar et al. (eds.), *Climate Change, Photosynthesis and Advanced Biofuels*,
https://doi.org/10.1007/978-981-15-5228-1_15

Keywords

Agrobacterium · Biolistic · Genetic transformation · Genome editing · Sugarcane · Transgenic

15.1 Introduction

Sugarcane (*Saccharum* spp. hybrids) is a tall, perennial C4 grass. It has high biomass with an ability to store large amounts of sucrose making it an important cash crop worldwide. Increased demand for sugarcane and its by-products worldwide contribute to its importance. In food industries and as a biofuel energy crop, sugarcane makes a critical economic contribution with many benefits to be gained. Higher crop yield, higher sucrose recovery, disease resistance, and abiotic stress tolerance are targeted improvements for sugarcane. Although conventional breeding techniques have achieved significant progress, they are hindered by large genome size, high polyploidy, low fertility, complex environmental interactions, and difficulty in selecting desired traits. A growing human population and the limitations of cultivable land necessitate more reliable and efficient methods for increasing the quantity and quality of crop production. In order to achieve this goal, researchers utilize genetic engineering to produce transgenic plants containing desirable and improved traits. Genetic modification can boost crop yield, reduce the cost of production, increase biotic and abiotic stress tolerance for plants, and reduce the use of herbicides.

The advance of genome-editing technologies, CRISPR-Cas9 in particular, has revolutionized plant biology. New techniques increase precision, accuracy, ease, and versatility of genome editing. It is now easier to knock genes into genomes, knock genes out of genomes, or replace a target gene in a host plant genome—as evidenced from a variety of reports in diverse crop plants. One critical constraint of genome-editing techniques is in vector delivery, which plant transformation protocols depend on for maximized efficiency. This chapter discusses the merits and demerits of various methods employed in sugarcane transformation and highlights novel strategies utilized to achieve higher transformation efficiency and enhanced target gene expression.

15.2 Diversity of Sugarcane

The genus *Saccharum* is part of the Poaceae family and includes six species, namely, *S. spontaneum*, *S. robustum*, *S. officinarum*, *S. barberi*, *S. sinense*, and *S. edule*. *S. spontaneum* is the most primitive of *Saccharum* species with a chromosome number of $2N = 40$. India is the geographic origin and center of diversity for *S. spontaneum*. *Saccharum robustum* is a wild species derived from an introgression of *S. spontaneum*. In Southeast Asia and New Guinea, *S. officinarum* (noble canes) with chromosome number $2N = 80$ is cultivated for its high sucrose and low fiber content. For a very long time, the production of sugar in India relied on *S. barberi* ($2N = 81$–154), while China used *S. sinense* ($2N = 116$–120). These species include clones that are well adapted to the abiotic stresses specific to their geographic area of origin. *S. edule* with $2N = 60$, 70, and 80 is a separate species of sugarcane, cultivated as a vegetable (Alix et al. 1998). In addition to these six species, several native species have been identified worldwide, 25 species from Asia, 6 from North America, 4 from Central America, 2 from Africa, and 1 from Australia. A total of seven species native to Brazil were initially described in 1982 (Smith et al. 1982). Three of those species were later identified in a 2001 study as *S. asperum*, *S. angustifolium*, and *S. villosum* (Filgueiras and Lerina 2001), and these scientific names are now accepted and listed in The Plant List (2010) and (Kumar et al. 2016).

15.3 Genome Complexity of Sugarcane

The complex aneuploidy nature of sugarcane, with approximately 80–120 chromosomes (D'Hont et al. 1998), generates challenges for genetic manipulation. Ploidy level for sugarcane is more than 10. This stands in disparity with other monocot plant species of rice and maize. Sugarcane has a large and complex genome size of approximately 10 Gb (Narayan et al. 2017). In comparison, the genome size of rice comprises only 4.3% of the sugarcane genome (i.e., 430 Mb). The genome of sugarcane is roughly four times larger than that of its close relative, maize (2300 Mb). Due to its high polyploid and aneuploid nature, a complete set of homologous genes is predicted to range from 10 to 12 alleles (Souza et al. 2011). Introgression of the highly polymorphic *S. spontaneum* genome is considered a major source of genetic variability among modern sugarcane cultivars (D'Hont et al. 1996). Advances in genomic tools, bioinformatic approaches, and next-generation sequencing technologies have facilitated the extrication of these insights into the complex genome of sugarcane (Riaño-Pachón and Mattiello 2017; Garsmeur et al. 2018). Such studies will be crucial to sugarcane genome assembly.

15.4 Genetic Transformation of Sugarcane

The development of a novel sugarcane variety requires stringent performance and quality tests. Conventional breeding techniques require approximately 15 years to complete this rigorous process (Gazaffi et al. 2010). Another barrier for sugarcane crop development is its inconsistent production of fertile seed, which hampers conventional breeding methods (Singh et al. 2013; Masoabi et al. 2018). In order to meet growing consumer demand for sugarcane and associated by-products, genetic manipulation is a promising alternative.

Sugarcane was first transformed genetically by polyethylene glycol (PEG) induced direct uptake of DNA using protoplasts (Chen et al. 1987). Earlier methods of sugarcane transformation include electroporation, particle gun, and other PEG-mediated transformation. In the early years, *Agrobacterium*-mediated transformation was not used for monocots, as they are not a natural target for this method. However, the development of Acetosyringone-adding protocols enabled *Agrobacterium*-mediated transformation in monocot plants. In 1998, sugarcane transformation with *Agrobacterium* was first demonstrated. Since that time, several researchers have transformed sugarcane using this approach.

Transgenic sugarcane has been developed over subsequent years for various input and output traits that include herbicide resistance (Manickavasagam et al. 2004), drought tolerance, salinity stress tolerance (Augustine et al. 2015a, b, c; Kumar et al. 2014), and borer resistance (Kalunke et al. 2009; Christy et al. 2009; Cristofoletti et al. 2018). Major reports of transgenic sugarcane are listed in Table 15.1. Sugarcane is a potential candidate for molecular farming applications, particularly one where the desired gene can be targeted to vacuoles, allowing proteins to be produced and purified from sugarcane juice (Palaniswamy et al. 2016). Commercial cultivation of genetically modified sugarcane has been approved in Indonesia (drought resistance) and Brazil (borer resistance). It is also in the pipeline for approval in several other countries.

15.5 Explant Types and Preparation

Since the first transgenic sugarcane was produced in 1987 through PEG-mediated transformation, there has been a constant improvement in modes of genetic transformation, as well as in tissue culture mediums. Before genetic transformation begins, explants must be sterilized using surface-sterilizing agents, namely, sodium hypochlorite (NaClO), mercuric chloride ($HgCl_2$) and ethanol (EtOH). Many explant types have been used for transformation, including apical meristem, young leaf sheath, seed, and embryogenic callus. Sterilization protocols vary based on the explant used and the surface-sterilizing agent used. Generally, plants are surface sterilized using 70% ethanol for 1 min followed by a rinse with sterilized distilled water and ending with an exposure to either 0.1% $HgCl_2$ for 3–5 min or NaClO for approximately 3 min with either followed by a sterilized, distilled water rinse (Arvinth et al. 2010).

After surface sterilization, explants are subjected to a chosen transformation protocol. Following transformation, the explants must be maintained in an appropriate plant tissue culture medium following a well-defined procedure. Generally, Murashige and Skoog (MS) medium containing appropriate growth hormones and antibiotics is used for the different stages of transformation. For biolistic transformation, embryogenic calli are used as explants. To develop embryogenic calli, MS medium containing 2,4-dichlorophenoxyacetic acid (3 mg/L), 10% coconut milk and 100 mg/L myoinositol along with 2% sucrose is the preferred medium (Arvinth et al. 2010). For *Agrobacterium*-mediated transformation, explants are placed in MS medium containing 200 μM phenolic compound acetosyringone to induce *Agrobacterium* infection. After co-cultivation, explants are subjected to several rounds of selection and regeneration to obtain transgenic sugarcane. In addition to the information found in the next section, more detailed protocols for sugarcane

Table 15.1 List of transgenic sugarcane developed for specific traits

	Method of transformation	Explant	Gene for transfer	Trait	Promoter	References
1	PEG-mediated	Protoplasts	APH(3′)II	Kanamycin resistance	CaMV gene VI	Chen et al. (1987)
2	Electroporation	Protoplasts	Neomycin phosphotransferase	Kanamycin resistance	CaMV35S	Rathus and Birch (1992)
3	Particle gun	Embryogenic calli	Neomycin phosphotransferase	Kanamycin resistance	Emu	Bower and Birch (1992)
4	Electroporation	Embryogenic calli	β-glucuronidase	GUS gene expression	CaMV35S	Arencibia et al. (1995)
5	Particle gun	Embryogenic calli	Luciferase, β-glucuronidase (GUS)	Reporter gene expression	CaMV35S	Bower et al. (1996)
6	*Agrobacterium*	Meristematic tissues and calli	β-glucuronidase (GUS)	GUS gene expression	CaMV35S	Arencibia et al. (1998)
7	Particle gun	Embryogenic calli	Green fluorescent protein (GFP)	GFP gene expression	Maize ubiquitin	Elliott et al. (1999)
8	Electroporation	Embryogenic calli	β-glucuronidase (GUS)	GUS gene expression	CaMV35S	Seema et al. (2001)
9	*Agrobacterium*	Axillary shoot	Phosphinothricin acetyltransferase (*bar*)	Herbicide resistance	CaMV35S	Manickavasagam et al. (2004)
10	Particle gun	Embryogenic calli	β-glucuronidase (GUS)	GUS gene expression	Maize ubiquitin	Kaur et al. (2007)
11	Particle gun	Embryogenic calli	Resveratrol synthase	Synthesize resveratrol	Maize ubiquitin	Xu et al. (2008)
12	*Agrobacterium*	Meristematic leaf whorls	Cry1Aa3	Borer resistance	CaMV35S	Kalunke et al. (2009)
13	Particle gun	Embryogenic calli	Aprotinin	Top borer resistance	Maize ubiquitin	Christy et al. (2009)

(continued)

Table 15.1 (continued)

	Method of transformation	Explant	Gene for transfer	Trait	Promoter	References
14	Particle gun	Embryogenic calli	Cry1Ab	Shoot borer resistance	Maize ubiquitin	Arvinth et al. (2010)
	Agrobacterium	Meristematic leaf whorls				
15	Particle gun	Basal part of leaf roll	Green fluorescent protein (GFP)	GFP expression	Maize ubiquitin, Emu	Van der Vyver (2010)
16	*Agrobacterium*	Embryogenic calli	Luciferase	Luciferase gene expression	Osa promoter	Jackson et al. (2013)
	Particle gun					
17	Particle gun	Meristematic leaf whorls	Yellow fluorescent protein (YFP), β-glucuronidase	Reporter gene expression	Maize ubiquitin	Gao et al. (2013)
18	Particle gun	Embryogenic calli	Glyphosate-tolerant gene	Herbicide tolerance	CaMV35S	Nasir et al. (2014)
19	*Agrobacterium*	Embryogenic calli	Cyanofluorescent protein (CFP), green fluorescent protein (GFP)	Reporter gene expression	NM	Dong et al. (2014)
20	*Agrobacterium*	Meristematic leaf whorls	*Arabidopsis* Vacuolar pyrophosphatase	Drought and salinity stress tolerance	CaMV35S	Kumar et al. (2014)
21	*Agrobacterium*	Sugarcane setts	β-glucuronidase (GUS)	GUS expression	CaMV35S	Mayavan et al. (2015)
22	*Agrobacterium*	Meristematic leaf whorls	Heat-shock protein 70 (HSP70)	Drought and salinity Stress tolerance	Port Ubi 2.3	Augustine et al. (2015a)
23	*Agrobacterium*	Meristematic leaf whorls	Pea DNA helicase 45 (PDH45)	Drought and salinity Stress tolerance	Port Ubi 2.3	Augustine et al. (2015b)
24	*Agrobacterium*	Meristematic leaf whorls	DREB2	Drought and salinity Stress tolerance	Port Ubi 2.3	Augustine et al. (2015c)
	Particle gun	Embryogenic calli	Pyramiding of DREB2 and PDH45		Port Ubi 2.3	
25	*Agrobacterium*	Meristematic leaf whorls	Aprotinin	Vacuolar targeting	Port Ubi882	Palaniswamy et al. (2016)

26	Particle gun	Embryogenic calli	Cry1Ac	Insect resistance	CaMV35S	Gao et al. (2016)
27	*Agrobacterium*	Embryogenic calli	Cry1A(b)	Insect resistance	CaMV35S	Islam et al. (2016)
28	*Agrobacterium*	Embryogenic calli	Cry1Ab and EPSPS	Insect and herbicide resistance	Maize ubiquitin	Wang et al. (2017)
29	Particle gun	Embryogenic calli	Chitinase II	Red rot resistance	Maize ubiquitin	Tariq et al. (2018)
30	*Agrobacterium*	Embryogenic calli	Cry1Ab and Cry2Ab	Borer resistance	CaMV35S	Cristofoletti et al. (2018)

NM not mentioned in the paper

transformation methods can be obtained through published papers (Taparia et al. 2012a; Dong et al. 2014; Basso et al. 2017).

15.6 Methods of Sugarcane Transformation

When cultivation began, conventional breeding was the only technique available to develop improved crop varieties, and certain limitations were reached. With the introduction of plant tissue culture and genetic transformation, scientists began utilizing biotechnology as a tool to quickly develop transgenic plants containing desirable trait modifications. Several genes have been successfully transferred to sugarcane using an array of methods, including electroporation, PEG treatment, particle bombardment (biolistic), and *Agrobacterium*-mediated transformation. The latter two methods are more popular at this time. Recently, Mayavan et al. (2015) demonstrated a technique *in planta* for sugarcane transformation using an *Agrobacterium*-mediated method (see Sect. 15.6.4). Developing a reliable, reproducible transformation technique with high transformation efficiency has been a major challenge, and current methods are relatively less efficient (Rakoczy-Trojanowska 2002). At this time, a demand for alternative, highly efficient strategies exists. The following chapter section describes four different methods of sugarcane transformation and briefly highlights the respective advantages and limitations.

15.6.1 Electroporation

Electroporation-mediated transformation utilizes protoplasts, which are plant cells with the cell wall removed. This method subjects the protoplast to a high-voltage electric impulse. The impulse generates reverse cell permeability, allowing a plasmid or other DNA to penetrate protoplast cells. Successful stable transformation of sugarcane using electroporation was first reported in 1992 (Chowdhury and Vasil 1992; Rathus and Birch 1992; Arencibia et al. 1992). This technique, using electrical impulse, generates cell hyperactivity and increases transformation efficiency. As a result, it yields more positive transformants. Critical factors affecting the efficiency of this method include the concentration of plasmid or other DNA molecules used, the duration of electric impulse, the intensity of the electric field, and the composition of electroporation medium.

15.6.2 PEG-Mediated Transformation

PEG-mediated gene delivery system uses polyethylene glycol and is a method widely used for direct gene transfer. PEG-mediated gene transformation was the first methodology reported to yield stable sugarcane transformants (Chen et al. 1987). As with electroporation, plant protoplasts are used for PEG-mediated transformation. Protoplasts are treated with polyethylene glycol in the presence of divalent calcium cations (Ca^{++}) to destabilize the cell membrane and allow uptake of naked DNA. The use of this method is limited, due to a history of producing results with very low transformation efficiency and poor reproducibility. The transformation frequency for PEG-mediated transformation is approximately 8 per 10^7 cells.

15.6.3 Biolistic Transformation

Since its 1988 development by J. C. Sanford (Sanford 1988), the biolistic transformation has been routinely used for gene transfer in diverse plant species. This method is referenced in a variety of ways, including particle bombardment, gene gun, and direct DNA delivery. Biolistic transformation involves the delivery of coated microcarriers, gold or tungsten particles that range from 1 to 4 μm in diameter. The coated particles are propelled at high speed into target explants, releasing the DNA within target cells, where it will be integrated into the genome of the plant. This technique enables researchers to perform both stable and transient transformation. Transgenic sugarcane plants were first developed using bombardment techniques in 1992 with embryogenic calli as explants (Bower and Birch 1992). As a commonly used method, it offers many advantages over other methods. Biolistic transformation allows direct gene transfer to specific or nonspecific tissues with high efficiency and limitations on host choice. It does not require any intrinsic vector (or vector backbone), meaning linear transgene DNA of any size can be directly bombarded into host cells (Christou 1992). Biolistic transformation requires expensive equipment and skilled personnel. Disadvantages of this delivery method include multiple transgene insertions and cell damage.

15.6.4 *Agrobacterium*-Mediated Transformation

Agrobacterium tumefaciens is a rod-shaped, gram-negative soil bacterium responsible for the crown-gall disease. It infects plant tissue by transferring a T-DNA region present in its Ti plasmid (Chilton 1977). Researchers disarmed *Agrobacterium* strains, for their use in gene transfer technology in plant transformation (Fraley 1985). *Agrobacterium*-mediated transformation was first used exclusively in dicotyledonous plants since monocotyledons were outside the natural host range of this bacterium (De Cleene and Deley 1976). Later, an improved protocol adding acetosyringone was developed, allowing monocot plants to be effectively transformed. Since the first transgenic sugarcane plant line was developed using this technique (Arencibia et al. 1998), many viable protocols from different researchers across the world have been standardized (Singh et al. 2013). A simple, reliable protocol for sugarcane transformation using *Agrobacterium* was described by Arvinth et al. (2010). This efficient method has a history of major advantages that include high transformation numbers, stable expression, reduced transgene silencing, integration of low-copy numbers of the target gene, and low cost. Figure 15.1 depicts the stages involved with developing transgenic sugarcane through

Fig. 15.1 Different stages of *Agrobacterium*-mediated sugarcane (Co86032) genetic transformation. (**a**) Sterilization of explants; (**b**) co-cultivation of sugarcane *leaf whorls* infected with *Agrobacterium*; (**c**) selection of transformants on selection medium; (**d**) regeneration of putative transgenic shoots from callus; (**e**) transgenic sugarcane in rooting medium; and (**f**) hardening of transgenics in soil

Agrobacterium-mediated transformation. Recently, Mayavan et al. (2015) reported an *Agrobacterium*-based transformation protocol *in planta* using sugarcane setts (stem cuttings or sections of the stalks). This method yielded 29.6% transformation efficiency with a reduced time frame. The authors validated the protocol using five different sugarcane varieties and were able to demonstrate both reproducibility and genotype-independence with this protocol.

15.7 Factors Affecting Transformation Efficiency

Despite considerable progress in sugarcane transformation strategies, several factors remain, which impact the efficiency of transformation. Irrespective of the method used, the choice between targeting tissues and explants will greatly influence plant transformation outcomes. The particle bombardment method is highly influenced by the DNA concentration and quality, gold or tungsten particle size, pressure level, and distance between the microcarriers in the bombardment equipment and target tissue

(Wang et al. 1988; Southgate et al. 1995). The use of minimal-expression cassettes and reduced DNA concentration are beneficial in achieving low-copy-number transgenic plants (Taparia et al. 2012b). In contrast, *Agrobacterium*-mediated transformation efficacy depends on the *Agrobacterium* strain used, co-cultivation time, acetosyringone concentration, and chosen selection markers (Cheng et al. 2004; Opabode 2006). Several published studies have optimized these parameters and have established efficient transformation protocols for sugarcane (Joyce et al. 2010; Anderson and Birch 2012). Transformation efficiency between the different methods will depend on a variety of factors that have been extensively reviewed (Anunanthini et al. 2017).

15.8 Promoters and Transgene Expression

Promoters or upstream regulatory sequences precede gene coding regions and are key determinants of the strength and specificities of transgene expression (Potenza et al. 2004; Chakravarthi et al. 2015). For sugarcane transformation, the maize ubiquitin promoter (M-*Ubi*1) has been the workhorse promoter for more than two decades. M-*Ubi*1 is routinely used for driving target genes, as it confers a uniform and constitutive expression (Christensen et al. 1992). The 2 kb size of M-*Ubi*1 promoter extends from −899 bp 5′ of the transcription start site (+1) to about 1093 bp 3′ of the transcription start site and includes a TATA box sequence located at −30, two overlapping sequences similar to the consensus heat shock element found in heat-inducible genes located at −214 and −204 position from the transcription start site, an 83 bp untranslated exon sequence 3′ adjacent to the transcription start site, and a 1 kb intron from 84 to 1093 position. However, Wang et al. (2005) reported that sustained transgene expression throughout the sugarcane growth cycle was not conferred by the M-*Ubi*1 promoter. Promoters from viruses, namely, *sugarcane bacilliform virus* promoter (Braithwaite et al. 2004), *banana streak virus* promoter (Schenk et al. 1999), and CaMV35S:Zmubi1 tandem promoter (Groenewald and Botha 2008) drove higher expression in mature sugarcane. Recently, Kinkema et al. (2014a, b) employed the enhanced maize ubiquitin promoter and maize carboxylase promoter which conferred fivefold and fourfold higher expression than the M-*Ubi*1 promoter, respectively. The use of truncated promoters and promoters with intron sequences have also facilitated enhanced transgene expression in sugarcane (Philip et al. 2013; Chakravarthi et al. 2015). Apart from constitutive promoters, several stem-specific and inducible promoters conferring high expression have been isolated and characterized from sugarcane and its wild genera (Krishnan and Mohan 2017). Meticulous deployment of promoters capable of high expression, which can be regulated, will improve transgene expression, as well as avoid transgene-silencing issues in this polyploid crop.

15.9 Selection of Transgenic Plants

Another decisive stage in plant transformation is the selection of transformed tissues. The regeneration of non-transformed cells/tissues should be avoided. Typically, antibiotics or herbicide-resistant genes are used as selection markers, to facilitate the easy screening of positive transformants in vitro culture medium. In sugarcane transformation, the *npt*II (neomycin phosphotransferase) and *hpt* (hygromycin phosphotransferase) genes are widely used as antibiotic selectable markers (Enriquez-Obregon et al. 1998; Zhangsun et al. 2007; Chakravarthi et al. 2015). Among the herbicide genes, *bar* (bialaphos resistance) gene has been predominantly used to select transformed sugarcane tissues (Gallo-Meagher and Irvine 1996; Leibbrandt and Snyman 2003; Basso et al. 2017). Bialaphos contains phosphinothricin (PPT), which is an inhibitor of glutamine synthetase. The *bar* gene was originally cloned from *Streptomyces hygroscopicus* and encodes a modifying enzyme phosphinothricin acetyltransferase. Manickavasagam et al. (2004) transformed sugarcane axillary bud explants and compared the efficacy of different selection agents: kanamycin, geneticin, and PPT. PPT was determined to be more effective. Another positive selection system utilizes an *E. coli* phosphomannose isomerase gene in sugarcane transformation (Jain et al. 2007). However, this system has the disadvantage of a high escape rate (~44%) and can lead to false-positive events. Recently, a mutated acetolactate synthase gene from tobacco was successfully tested in producing transgenic sugarcane (Van der Vyver et al. 2013). However, with a 20% escape recorded, it is a less-efficient selection system. The use of antibiotics and herbicide genes as selection markers to generate genetically modified crops has raised serious public concern and prompted scientists to develop alternate selection systems (Yau and Stewart 2013). Incorporating other systems, such as a site-specific recombination system, into a transformation system can eliminate selectable marker genes and generate marker-free transgenic plants (Wang et al. 2011). Recently, Zhao et al. (2019) have incorporated the FLPe-*FRT* site-specific recombination system to successfully remove the *npt*II selectable marker gene from the transgenic sugarcane genome. The removal of the selectable marker gene will also facilitate re-transformation (Yau and Stewart 2013).

15.10 Novel Strategies for Improving Transgene Expression

Sugarcane transformation methods have greatly improved over the past decades. Researchers have enhanced transformation efficiency and reduced the duration of existing protocols. This section of the chapter reviews various advances in recent years with sugarcane genetic engineering. These notable achievements, if utilized, will impact sugarcane improvement through biotechnology applications. One of the first reports optimized parameters for *Agrobacterium*-mediated transformation, including *Agrobacterium* strains, co-cultivation time, and selection system (Joyce et al. 2010). The authors reported a simple and reproducible protocol using the *npt*II gene as a plant selectable marker. In this study, a 4-day co-cultivation period produced the highest number of sugarcane transgenic events. In a later study, the authors compared the field performance of *Agrobacterium*-mediated transformation transgenic sugarcane to that obtained by biolistic methods over 3 years (Joyce et al. 2014). While combined evaluation showed growth reduction, an individual event-based evaluation showed transgenic events produced similar yield compared to control (non-transformed) plants. Their study also revealed that transgene expression was not influenced by either method and low-copy-number events were generated by both the *Agrobacterium*-mediated and bombardment methods.

Jackson et al. (2013) evaluated both *Agrobacterium*-mediated transformation and biolistic methods for transformation efficiency and strength of transgene expression. While both methods could generate high-expressing single-copy-number transgenic lines, a higher number of transformants with low-copy-number transgene insertions and reporter gene expression were achieved with *Agrobacterium*-mediated transformation. The authors recommended these methods could be utilized best when combined with a rapid early screening of transformants for low-copy-number and high transcript expression.

Biolistic transformation in sugarcane using minimal-expression cassettes (MC) proved to be efficient in yielding stable transgene expression with low-copy-number integration (Taparia et al. 2012a, b). Using MCs eliminates transgene silencing, which is a major barrier in sugarcane, and avoids the usage of prokaryotic backbone sequences. Minimal-expression cassettes provide a simpler gene integration strategy. Taparia et al. also established a rapid transformation protocol capable of generating stable transgenic sugarcane lines within a period of 3 months.

Although considerable success has been achieved by *Agrobacterium*-mediated sugarcane transformation, the method is limited by low transformation efficiency, genotype dependency, and variability between experiments. Minimal handling of callus before and during co-cultivation; the use of super binary vector with extra *vir*B, *vir*C, and *vir*G gene copies; and the use of AGL1 *Agrobacterium* strain are critical parameters that contribute to an efficient sugarcane transformation with high transformation efficiency (Anderson and Birch 2012). A single-step, direct-regeneration protocol that avoided the callus phase by using agro-infected sugarcane spindle leaf-roll segments was recently reported (Sandhu et al. 2016). Basso et al. (2017) reported an improved *Agrobacterium*-mediated sugarcane transformation protocol with a transformation efficiency of 2.2% that can be adopted for other sugarcane genotypes.

Dong et al. (2014) used desiccation during co-cultivation to yield higher transformation efficiency. This robust, reproducible protocol could be applied on an industrial scale and has been tried with several sugarcane varieties in laboratories worldwide. Another notable achievement was made by Lowe et al. (2016). Overexpression of the maize morphogenic regulators *Baby boom* (*Bbm*) and *Wuschel2* (*Wus2*) in maize, sorghum, rice, and sugarcane could obtain higher transformation frequencies. Apart from these transformation methods, the usage of high-efficiency promoters and codon-optimized, sugarcane-specific target genes also enhances transgene expression (Kinkema et al. 2014a). A few effective genetic insulators have been characterized in plants (Hily et al. 2009; Singer et al. 2010, 2011, 2012;

Yang et al. 2011; Zhang et al. 2012). These genetic insulators include EXOB from bacteriophage λ and TBS (the transformation booster sequence) from Petunia hybrid. These insulators can block enhancer-promoter interaction activity in plants, increased the transgene expression level, and reduced the line-to-line variation in transgene expression (Singer et al. 2011, 2012). Zhao et al. (2019) used insulators EXOB and TBS in their transformation construct to transform sugarcane. They have found the line-to-line variation decreased, and the average transgene expression level of insulator-containing lines was nearly twice higher compared to the lines without insulators (Zhao et al. 2019). Besides, the transformation construct also brought a landing pad (a mutated *lox76* site) into the genome inserting at a specific location for future gene stacking through Cre-*lox* site-specific recombination system (Zhao et al. 2019). Gene integrating at a predetermined locus can ensure predictable gene expression (Hou et al. 2014). Genes stacking at the same locus would also greatly simplify trait introgression into cultivars (Hou et al. 2014).

15.11 Perspectives and Final Remarks

Over the years, sugarcane genetic engineering has advanced rapidly. Transgenic sugarcane, containing key traits such as herbicide resistance, disease resistance, and abiotic stress tolerance, has been developed and tested in both laboratory and field conditions. The recent approval of commercialization of transgenic sugarcane in Indonesia and Brazil has provided a boost to other countries where transgenic sugarcane lines are under field trials (Parisi et al. 2016; Kennedy et al. 2018). Advances in next-generation sequencing have paved the way for sequencing of the sugarcane genome, which will aid researchers in functional genomics studies in sugarcane. Efficient and reliable transformation strategies could yield high transgene expression, low- or single-copy number events, and stable expression over subsequent generations. Each advance will greatly benefit sugarcane crop improvement. While CRISPR-Cas9-based, genome-editing techniques have been widely applied in monocot crops, such as rice and wheat (Jiang et al. 2013; Upadhyay et al. 2013), the technology is still in its infancy in sugarcane. As genome editing relies on an efficient delivery system, a robust transformation strategy is highly critical for more genome-editing approaches in this complex polyploid.

Acknowledgments CM is grateful to the São Paulo Research Foundation for the postdoctoral and visiting research fellowships (FAPESP, Proc. 2015/10855-9 and Proc. 2018/11544-5). ANJ acknowledges CSIR, New Delhi, India, for the award of Senior Research Fellowship.

References

Alix K, Baurens FC, Paulet F, Glaszmann JC, D'hont A (1998) Isolation and characterization of a satellite DNA family in the *Saccharum* complex. Genome 41:854–864

Anderson DJ, Birch RG (2012) Minimal handling and super-binary vectors facilitate efficient, *Agrobacterium*-mediated, transformation of sugarcane (*Saccharum* spp. hybrid). Trop Plant Biol 5:183–192

Anunanthini P, Kumar SR, Sathishkumar R (2017) Factors affecting genetic transformation efficiency in sugarcane. In: Sugarcane biotechnology: challenges and prospects. Springer, Cham, pp 61–73

Arencibia A, Molina P, Gutierrez C, Fuentes A, Greenidge V, Menéndez E, De la Riva G, Selman G (1992) Regeneration of transgenic sugarcane (*Saccharum officinarum* L.) plants from intact meristematic tissues transformed by electroporation. Biotecnología Aplicada 9:156–165

Arencibia A, Molina PR, de la Riva G, Selman-Housein G (1995) Production of transgenic sugarcane (*Saccharum officinarum* L.) plants by intact cell electroporation. Plant Cell Rep 14:305–309

Arencibia AD, Carmona ER, Tellez P, Chan MT, Yu SM, Trujillo LE, Oramas P (1998) An efficient protocol for sugarcane (*Saccharum* spp. L.) transformation mediated by *Agrobacterium tumefaciens*. Transgenic Res 7:213–222

Arvinth S, Arun S, Selvakesavan RK, Srikanth J, Mukunthan N, Kumar PA, Premachandran MN, Subramonian N (2010) Genetic transformation and pyramiding of aprotinin-expressing sugarcane with cry1Ab for shoot borer (*Chilo infuscatellus*) resistance. Plant Cell Rep 29:383–395

Basso MF, da Cunha BADB, Ribeiro AP, Martins PK, de Souza WR, de Oliveira NG et al (2017) Improved genetic transformation of sugarcane (*Saccharum* spp.) embryogenic callus mediated by *Agrobacterium tumefaciens*. Curr Prot Plant Biol 2:221–239

Bower R, Birch RG (1992) Transgenic sugarcane plants via microprojectile bombardment. Plant J 2:409–416

Bower R, Elliott AR, Potier BA, Birch RG (1996) High-efficiency, microprojectile-mediated cotransformation of sugarcane, using visible or selectable markers. Mol Breeding 2:239–249

Braithwaite KS, Geijskes RJ, Smith GR (2004) A variable region of the sugarcane bacilliform virus (SCBV) genome can be used to generate promoters for transgene expression in sugarcane. Plant Cell Rep 23:319–326

Chakravarthi M, Philip A, Subramonian N (2015) Truncated ubiquitin 5′ regulatory region from *Erianthus arundinaceus* drives enhanced transgene expression in heterologous systems. Mol Biotechnol 57:820–835

Chen WH, Gartland KMA, Davey MR, Sotak R, Gartland JS, Mulligan BJ, Power JB, Cocking EC (1987) Transformation of sugarcane protoplasts by direct uptake of a selectable chimeric gene. Plant Cell Rep 6:297–301

Cheng M, Lowe BA, Spencer TM, Ye X, Armstrong CL, Cheng M (2004) Factors influencing *Agrobacterium*-mediated transformation of monocotyledonous species. In Vitro Cell Dev Biol Plant 40:31–45

Chilton MD (1977) Successful integration of T-DNA in plants. Cell 11:263–271

Chowdhury MKU, Vasil IK (1992) Stably transformed herbicide resistance callus of sugarcane via microprojectile bombardment of cell suspension cultures and electroporation of protoplasts. Plant Cell Rep 11:494–498

Christensen AH, Sharrock RA, Quail PH (1992) Maize polyubiquitin genes: structure, thermal perturbation of expression and transcript splicing, and promoter activity following transfer to protoplasts by electroporation. Plant Mol Biol 18:675–689

Christou P (1992) Genetic-transformation of crop plants using microprojectile bombardment. Plant J 2:275–281

Christy LA, Arvinth S, Saravanakumar M, Kanchana M, Mukunthan N, Srikanth J, Thomas G, Subramonian N (2009) Engineering sugarcane cultivars with bovine pancreatic trypsin inhibitor (aprotinin) gene for protection against top borer (*Scirpophaga excerptalis* Walker). Plant Cell Rep 28:175–184

Cristofoletti PT, Kemper EL, Capella AN, Carmago SR, Cazoto JL, Ferrari F, Galvan TL, Gauer L, Monge GA, Nishikawa MA, Santos NZ (2018) Development of transgenic sugarcane resistant to sugarcane borer. Trop Plant Biol 11(1–2):17–30

D'Hont A, Grivet L, Feldmann P, Rao PS, Berding N, Glaszmann JC (1996) Characterization of the double genomic structure of modern sugarcane cultivars, *Saccharum* spp, by molecular cytogenetics. Mol Gen Genet 250:405–413

D'Hont A, Ison D, Alix K, Roux C, Glaszmann JC (1998) Determination of basic chromosome numbers in the genus *Saccharum* by physical mapping of ribosomal RNA genes. Genome Res 41:221–225

De Cleene M, Deley J (1976) The host range of crown gall. Bot Rev 42:389–466

Dong S, Delucca P, Geijskes RJ, Ke J, Mayo K, Mai P, Sainz M, Caffall K, Moser T, Yarnall M, Setliff K (2014) Advances in *Agrobacterium*-mediated sugarcane transformation and stable transgene expression. Sugar Tech 16:366–371

Elliott AR, Campbell JA, Dugdale B, Brettell RIS, Grof CPL (1999) Green-fluorescent protein facilitates rapid *in vivo* detection of genetically transformed plant cells. Plant Cell Rep 18:707–714

Enriquez-Obregon GA, Vazquez PRI, Prieto SDL, Riva-Gustavo ADI, Selman HG (1998) Herbicide resistant sugarcane (*Saccharum officinarum* L.) plants by *Agrobacterium*-mediated transformation. Planta 206:20–27

Filgueiras TS, Lerina R (2001) Saccharum. In: Longhi-Wagner HM (ed) Poacaeae, Flora Fanerogâmica do Estado de São Paulo. Fapesp-Hucitec, São Paulo

Fraley RT (1985) Development of disarmed Ti plasmid vector system for plant transformation. Biotechnology 3:629–635

Gallo-Meagher M, Irvine JE (1996) Herbicide resistant transgenic sugarcane plants containing the bar gene. Crop Sci 36:1367–1374

Gao SJ, Damaj MB, Park JW, Beyene G, Buenrostro-Nava MT, Molina J, Wang X, Ciomperlik JJ, Manabayeva SA, Alvarado VY, Rathore KS (2013) Enhanced transgene expression in sugarcane by co-expression of virus-encoded RNA silencing suppressors. PLoS One 8:e66046

Gao S, Yang Y, Wang C, Guo J, Zhou D, Wu Q, Su Y, Xu L, Que Y (2016) Transgenic sugarcane with a *cry1Ac* gene exhibited better phenotypic traits and enhanced resistance against sugarcane borer. PLoS One 11:e0153929

Garsmeur O, Droc G, Antonise R, Grimwood J et al (2018) A mosaic monoploid reference sequence for the highly complex genome of sugarcane. Nat Commun 9:2638

Gazaffi R, Oliveira KM, Souza AP, Garcia AAF (2010) The importance of the germplasm in developing agro-energetic profile sugarcane cultivars. In: Cortez LAB (ed) Sugarcane bioethanol: R&D for productivity and sustainability. Blucher, São Paulo, p 333–343. isbn 9788521205302

Groenewald JH, Botha FC (2008) Down-regulation of pyrophosphate: fructose 6-phosphate 1-phosphotransferase (PFP) activity in sugarcane enhances sucrose accumulation in immature internodes. Transgenic Res 17:85–92

Hily JM, Singer SD, Yang Y, Liu Z (2009) A transformation booster sequence (TBS) from *Petunia hybrid* functions as an enhancer blocking insulator in *Arabidopsis thaliana*. Plant Cell Rep 28:1095–1104

Hou L, Yau YY, Wei J, Han Z, Dong Z, Ow DW (2014) An open-source system for *in planta* gene stacking by Bxb1 and Cre recombinases. Mol Plant 7:1756–1765

Islam N, Laksana C, Chanprame S (2016) *Agrobacterium*-mediated transformation and expression of Bt gene in transgenic sugarcane. J ISSAAS 22:84–95

Jackson MA, Anderson DJ, Birch RG (2013) Comparison of *Agrobacterium* and particle bombardment using whole plasmid or minimal cassette for production of high-expressing, low-copy transgenic plants. Transgenic Res 22:143–151

Jain M, Chengalrayan K, Abouzid A, Gallo M (2007) Prospecting the utility of a PMI/mannose selection system for the recovery of transgenic sugarcane (*Saccharum* spp. hybrid) plants. Plant Cell Rep 28:581–590

Jiang W, Zhou H, Bi H, Fromm M, Yang B, Weeks DP (2013) Demonstration of CRISPR/Cas9/sgRNA-mediated targeted gene modification in *Arabidopsis*, tobacco, sorghum and rice. Nucleic Acid Res 41:188

Joyce P, Kuwahata M, Turner N, Lakshmanan P (2010) Selection system and co-cultivation medium are important determinants of *Agrobacterium*-mediated transformation of sugarcane. Plant Cell Rep 29:173–183

Joyce P, Hermann S, O'Connell A, Dinh Q, Shumbe L, Lakshmanan P (2014) Field performance of transgenic sugarcane produced using *Agrobacterium* and biolistics methods. Plant Biotechnol J 12:411–424

Kalunke RM, Kolge AM, Babu KH, Prasad DT (2009) *Agrobacterium* mediated transformation of sugarcane for borer resistance using Cry 1Aa3 gene and one-step regeneration of transgenic plants. Sugar Tech 11:355–359

Kaur A, Gill MS, Gill R, Gosal SS (2007) Standardization of different parameters for 'particle gun' mediated genetic transformation of sugarcane (*Saccharum officinarum* L.). Indian J Biotechnol 6:31–34

Kennedy RD, Cheavegatti-Gianotto A, de Oliveira WS, Lirette RP, Hjelle JJ (2018) A general safety assessment for purified food ingredients derived from biotechnology crops: case study of brazilian sugar and beverages produced from insect-protected sugarcane. Front Bioeng Biotechnol 6:45

Kinkema M, Geijskes J, Palupe A, Shand K, Coleman HD, Brinin A, Williams B, Sainz M, Dale JL (2014a) Improved molecular tools for sugar cane biotechnology. Plant Mol Biol 84:497–508

Kinkema M, Geijskes RJ, Shand K, Coleman HD, De Lucca PC, Palupe A et al (2014b) An improved chemically inducible gene switch that functions in the monocotyledonous plant sugar cane. Plant Mol Biol 84:443–454

Krishnan SR, Mohan C (2017) Methods of sugarcane transformation. In: Sugarcane biotechnology: challenges and prospects. Springer, Cham, pp 51–60

Kumar T, Khan MR, Abbas Z, Ali GM (2014) Genetic improvement of sugarcane for drought and salinity stress tolerance using *Arabidopsis* vacuolar pyrophosphatase (AVP1) gene. Mol Biotechnol 56:199–209

Kumar U, Priyanka, Kumar S (2016) Genetic improvement of sugarcane through conventional and molecular approaches. In: Molecular breeding for sustainable crop improvement. Springer, Cham, pp 325–342

Leibbrandt NB, Snyman SJ (2003) Stability of gene expression and agronomic performance of a transgenic herbicide-resistant sugarcane line in South Africa. Crop Sci 43:671–678

Lowe K, Wu E, Wang N, Hoerster G, Hastings C, Cho MJ et al (2016) Morphogenic regulators baby boom and Wuschel improve monocot transformation. Plant Cell 28:1998–2015

Manickavasagam M, Ganapathi A, Anbazhagan VR, Sudhakar B, Selvaraj N, Vasudevan A, Kasthurirengan S (2004) *Agrobacterium*-mediated genetic transformation and development of herbicide-resistant sugarcane (*Saccharum* species hybrids) using axillary buds. Plant Cell Rep 23:134–143

Masoabi M, Lloyd J, Kossmann J, van der Vyver C (2018) Ethyl methanesulfonate mutagenesis and *in vitro* polyethylene glycol selection for drought tolerance in sugarcane (*Saccharum* spp.). Sugar Tech 20:50–59

Mayavan S, Subramanyam K, Jaganath B, Sathish D, Manickavasagam M, Ganapathi A (2015) *Agrobacterium*-mediated *in planta* genetic transformation of sugarcane setts. Plant Cell Rep 34:1835–1848

Narayan JA, Manoj VM, Kaur L, Appunu C (2017) Unravelling the sugarcane genome: progress made so far and challenges ahead. In: Sugarcane biotechnology: challenges and prospects. Springer, Cham, pp 33–49

Nasir IA, Tabassum B, Qamar Z, Javed MA, Tariq M, Farooq AM, Butt SJ, Qayyum A, Husnain T (2014) Herbicide-tolerant sugarcane (*Saccharum officinarum* L.) plants: an unconventional method of weed removal. Turk J Biol 38:439–449

Opabode JT (2006) Agrobacterium-mediated transformation of plants: emerging factors that influence efficiency. Biotechnol Mol Biol Rev 1:12–20

Palaniswamy H, Syamaladevi DP, Mohan C, Philip A, Petchiyappan A, Narayanan S (2016) Vacuolar targeting of r-proteins in sugarcane leads to higher levels of purifiable commercially equivalent recombinant proteins in cane juice. Plant Biotechnol J 14:791–807

Parisi C, Tillie P, Rodriguez-Cerezo E (2016) The global pipeline of GM crops out to 2020. Nat Biotechnol 34:31–36

Philip A, Syamaladevi DP, Chakravarthi M, Gopinath K, Subramonian N (2013) 5′ regulatory region of ubiquitin 2 gene from *Porteresia coarctata* makes efficient promoters for transgene expression in monocots and dicots. Plant Cell Rep 32:1199–1210

Potenza C, Aleman L, Sengupta-Gopalan C (2004) Targeting transgene expression in research, agricultural, and environmental applications: promoters used in plant transformation. In Vitro Cell Dev Biol Plant 40:1–22

Rakoczy-Trojanowska M (2002) Alternative methods of plant transformation—a short review. Cell Mol Biol Lett 7:849–858

Rathus C, Birch RG (1992) Stable transformation of callus from electroporated sugarcane protoplasts. Plant Sci 82:81–89

Riaño-Pachón DM, Mattiello L (2017) Draft genome sequencing of the sugarcane hybrid SP80-3280. F1000Res 6:861

Sandhu JS, Kaur M, Kaur A, Kalia A (2016) Single step direct transgenic plant regeneration from adventive embryos of agro-infected sugarcane (*Saccharum* spp.) spindle leaf roll segments with assured genetic fidelity. Plant Cell Tissue Org Cult 125:149–162

Sanford JC (1988) The biolistic process. Trends Biotechnol 6:299–302

Schenk PM, Sagi L, Remans T, Dietzgen RG, Bernard MJ, Graham MW, Manners JM (1999) A promoter from sugarcane bacilliform badnavirus drives transgene expression in banana and other monocot and dicot plants. Plant Mol Biol 39:1221–1230

Seema G, Pande HP, Lal J, Madan VK (2001) Plantlet regeneration of sugarcane varieties and transient GUS expression in calli by electroporation. Sugar Tech 3:27–33

Singer SD, Hily JM, Liu Z (2010) A 1-kb bacteriophage lambda fragment functions as an insulator to effectively block enhancer promoter interactions in *Arabidopsis thaliana*. Plant Mol Biol Rep 28:69–76

Singer SD, Cox KD, Liu Z (2011) Enhancer-promoter interference and its prevention in transgenic plants. Plant Cell Rep 30:723–731

Singer SD, Liu Z, Cox KD (2012) Minimizing the unpredictability of transgene expression in plants: the role of genetic insulators. Plant Cell Rep 31:13–25

Singh RK, Kumar P, Tiwari NN, Rastogi J, Singh SP (2013) Current status of sugarcane transgenic: an overview. Adv Genet Eng 2:112

Smith LB, Wasshausen DC, Klein RM (1982) Gramíneas. Gêneros: 85. Paspalum até Zea. Flora Ilustrada Catarinense. Herbário Barbosa Rodrigues, Itajaí 504

Southgate EM, Davey MR, Power JB, Marchant R (1995) Factors affecting the genetic engineering of plants by microprojectile bombardment. Biotechnol Adv 13:631–651

Souza GM, Berges H, Bocs S, Casu R, D'Hont A, Ferreira JE, Henry R, Ming R, Potier B, Van Sluys MA, Vincentz M (2011) The sugarcane genome challenge: strategies for sequencing a highly complex genome. Trop Plant Biol 4:145–156

Taparia Y, Gallo M, Altpeter F (2012a) Comparison of direct and indirect embryogenesis protocols, biolistic gene transfer and selection parameters for efficient genetic transformation of sugarcane. Plant Cell Tissue Organ Cult 111:131–141

Taparia Y, Fouad WM, Gallo M, Altpeter F (2012b) Rapid production of transgenic sugarcane with the introduction of simple loci following biolistic transfer of a minimal expression cassette and direct embryogenesis. In Vitro Cell Dev Biol Plant 48:15–22

Tariq M, Khan A, Tabassum B, Toufiq N, Bhatti M, Riaz S, Nasir I, Husnain T (2018) Antifungal activity of chitinase II against *Colletotrichum falcatum* Went. causing red rot disease in transgenic sugarcane. Turk J Biol 42:45–53

The Plant List (2010) Version 1. http://www.theplantlist.org/

Upadhyay SK, Kumar J, Alok A, Tuli R (2013) RNA-guided genome editing for target gene mutations in wheat. G3 Genes 3:2233–2238
Van der Vyver C (2010) Genetic transformation of the euploid *Saccharum officinarum* via direct and indirect embryogenesis. Sugar Tech 12:21–25
Van der Vyver C, Conradie T, Kossmann J, Lloyd J (2013) *In vitro* selection of transgenic sugarcane callus utilizing a plant gene encoding a mutant form of acetolactate synthase. In Vitro Cell Dev Biol Plant 49:198–206
Wang Y, Klein TM, Fromm M, Cao J, Sanford JC, Wu R (1988) Transient expression of foreign genes in rice, wheat and soybean cells following particle bombardment. Plant Mol Biol 11:433–439
Wang ML, Goldstein C, Su W, Moore PH, Albert HH (2005) Production of biologically active GM-CSF in sugarcane: a secure biofactory. Transgenic Res 14:167–178
Wang Y, Yau YY, Perkins-Balding D, Thomson JG (2011) Recombinase technology: applications and possibilities. Plant Cell Rep 30:267–285
Wang WZ, Yang BP, Feng XY, Cao ZY, Feng CL, Wang JG, Xiong GR, Shen LB, Zeng J, Zhao TT, Zhang SZ (2017) Development and characterization of transgenic sugarcane with insect resistance and herbicide tolerance. Front Plant Sci 8:1535
Xu LP, Que YX, Xu JS, Fang SR, Zhang MQ, Chen YQ, Chen RK (2008) Establishment of genetic transformation system and obtaining transgenic sugarcane (var. badila) transformed with RS gene. Sugar Tech 10:128–132
Yang Y, Singer SD, Liu Z (2011) Evaluation and comparison of the insulation efficiency of three enhancer-blocking insulators in plants. Plant Cell Tiss Org 105:405–414
Yau YY, Stewart CN (2013) Less is more: strategies to remove marker genes from transgenic plants. BMC Biotechnol 13:36
Zhang Y, Zheng Y, Xiao N, Wang L, Zhang Z, Fang R, Chen X (2012) Functional analysis of the HS185 regulatory element in the rice HSP70 promotor. Mol Biol Rep 39:1649–1657
Zhangsun D, Luo S, Chen R, Tang K (2007) Improved *Agrobacterium*-mediated genetic transformation of GNA transgenic sugarcane. Biologia 62:386–393
Zhao Y, Kim JY, Karan R, Jung JH, Pathak B, Williamson B, Kannan B, Wang D, Fan C, Yu W, Dong S, Srivastava V, Altpeter F (2019) Generation of a selectable marker free, highly expressed single copy locus as landing pad for transgene stacking in sugarcane. Plant Mol Biol 100:247–263

Chakravarthi Mohan is pursuing his postdoctoral research at the department of Genetics and Evolution, Federal University of São Carlos, Brazil, funded by São Paulo Research Federation (FAPESP), where he aims to develop transgenic sugarcane for weevil resistance through RNAi approach.

J Ashwin Narayan is currently a Research Scholar in Indian Council of Agricultural Research (ICAR)-Sugarcane Breeding Institute (SBI), Coimbatore, funded by Council for Scientific and Industrial Research. His doctoral research is based on developing water-deficit stress-tolerant sugarcane through genetic engineering.

Mona Esterling is currently an assistant professor of Biology at Tulsa Community College. Her master's research involved use of Bxb1 Recombination Systems for the removal of Selectable-Marker Genes.

Yuan-Yeu Yau obtained a BSc in Botany at National Taiwan University (NTU), and received Master and PhD from University of Wisconsin-Madison, USA. He then worked at University of California-Berkeley and Plant Gene Expression Center (USDA-ARS) at Albany (California, USA), Chinese Academy of Sciences (Professor), and Northeastern State University (USA). He works in areas of plant biotechnology to develop clean-gene technology using microbial site-specific recombination system. He also worked on cotton transformation and production of stroke drug using molecular farming. He is also an active member of *ResearchGate*, a professional network for scientists and researchers.

Development of Transgenic Sugarcane for Insect Resistance

16

J. Ashwin Narayan, Chakravarthi Mohan, Mona Esterling, and Yuan-Yeu Yau

Abstract

Sugarcane is an economically important crop across the world. Demand for sugar and sugarcane by-products increases in both domestic and industrial sectors daily. In order to meet this demand, sustainable sugarcane yield improvement is inevitable. Yield loss in sugarcane due to insect pests ranges from 10 to 30%. Despite application of insecticides, pesticides, other chemicals, and different integrated pest management (IPM) techniques, scope for improvement remains. With genetic engineering, considerable success has been achieved in sugarcane trait improvement. To this end, this chapter focuses on development of transgenic sugarcane for disease and pest resistance.

Keywords

Bioassay · Insect · Pest · Sugarcane · Transgenic

J. Ashwin Narayan
Division of Crop Improvement, ICAR—Sugarcane Breeding Institute, Coimbatore, India

C. Mohan (✉)
Department of Genetics and Evolution, Federal University of São Carlos, São Carlos, Brazil

M. Esterling
Tulsa Community College, Northeast Campus, Tulsa, OK, USA
e-mail: mona.easterling@tulsacc.edu

Y.-Y. Yau (✉)
Department of Natural Sciences, Northeastern State University, Broken Arrow, OK, USA

A. Kumar et al. (eds.), *Climate Change, Photosynthesis and Advanced Biofuels*,
https://doi.org/10.1007/978-981-15-5228-1_16

16.1 Introduction

Sugarcane (*Saccharum sp.* Hybrid) is an important food and industrial crop used in production of sugar and biofuel worldwide. It belongs to the Poaceae family and is grown in both tropical and subtropical regions. More than 70% of world sugar is produced from sugarcane (Mustafa et al. 2018). Along with sugar production, sugarcane is a major source for biofuel production. Currently, biofuel is the best alternative to fossil fuel. These facts make sugarcane an important crop both domestically and industrially. With increasing demand, it is necessary to increase both production and yield of the sugarcane and, subsequently, sugar. One major factor affecting sugarcane plants is stress, both biotic and abiotic stress. Researchers around the world are tirelessly working to develop an improved variety of sugarcane with better biotic and abiotic stress resistance. Reaching this goal will improve sugar recovery and improve yield.

Sugarcane has a large and complex genome. Due to crop complexity, conventional breeding takes many growing seasons and poses challenges in trait selection. With a consistent increase in sugarcane demand, there is an urgent need to produce improved sugarcane varieties at a pace not possible by conventional breeding. The best alternative to conventional breeding is genetic transformation. Improved sugarcane varieties can be produced in less time with required traits successfully incorporated in the crop.

The major biotic stress factor affecting growth, yield, and productivity of sugarcane is the insect pest. Sugarcane is a long-growing crop, increasing the likelihood of an insect pest attack. Integrated pest management (IPM) includes cultural practices, physical control, mechanical control, and chemical control which have been used to control pests in sugarcane. It has been suggested that a combination of IPM strategy and a genetic transformation technique could reduce damage caused to sugarcane by insect pests (Allsopp and Manners 1997; Legaspi and Mirkov 2000; Falco and Silva-Filho 2003). This chapter will discuss the development of genetic transformation techniques and progress in production of insect-tolerant sugarcane.

16.2 Insect Pathogens

Insects are considered the most diverse group of living things and the most adaptable living organism on the planet. Their population rate per square meter is higher than most animals. Some insects are parasitoids, some are pollinators, and some produce valuable products like honey and silk. Insect pests comprise only 0.5% of the insect species, but this small percentage damages humans, animals, and crops (Dhaliwal et al. 2015). Insect pests cause major yield and production loss in the sugarcane. Pests are broadly classified into three categories: borers, sucking pests, and subterranean pests (Srikanth et al. 2011). Borers include early shoot borer, pink borer, top-shoot borer, root borer, internode borer, and stalk borer. White woolly aphid, black bug, whitefly, pyrilla (*Pyrilla perpusilla*, commonly known as Sugarcane plant hopper), mealybug, and mite are categorized as sucking pests. Termites and white grubs form the subterranean pest group. Loss of sugarcane crop productivity is high in both developed and developing countries. However, it is difficult to assess the real loss number due to insect pests, and there are few studies reporting the pest loss. In 1989, the estimated worldwide yield loss due to insect pests in sugarcane exceeded 10% (Ricaud and Ryan 1989). Mexican rice borer *E. loftini* causes sugar production losses of $10–$20 million annually in the Lower Rio Grande Valley of Texas (Samad and Leyva 1998). Woolly aphid *C. lanigera* is blamed for 18.3% yield loss when it attacks in the sixth month of the growth cycle (Mukunthan et al. 2008). Around 25–30% yield loss in sugarcane is attributed to sugarcane borers (Kalunke et al. 2009). From recent reports, it is estimated a 1% rise sugarcane borer infestations results in 0.2–1.1% yield loss (Cristofoletti et al. 2018).

16.3 Integrated Pest Management (IPM)

Insect management around the world is rapidly changing. Farmers do not depend exclusively on cultural methods and biological control, as insecticides and chemicals are used to suppress pests (Casida and Quistad 1998). IPM began following evidence that increased use of insecticides and chemical pesticides harmed both the environment and other living organisms (Hellmich et al. 2008). IPM was first introduced as a term in 1967 by R.F. Smith and R. van den Bosch. During 1970 to 1980, IPM was accepted and adopted by several governments around the world. The USA was an early IPM adopter in 1972. IPM was declared as official ministerial policy by India in 1985. According to FAO, IPM can be defined as the careful consideration of all available pest control techniques and subsequent integration of appropriate measures that discourage the development of pest populations and keep pesticides and other interventions at levels that are economically justified and reduce or minimize risks to human health and the environment. IPM emphasizes growth of a healthy crop with the least possible disruption to agro-ecosystems and encourages natural pest control mechanisms (http://www.fao.org/agriculture/crops/core-themes/theme/pests/ipm/en/). IPM provides several different strategies to manage and control pest in sugarcane such as cultural practices, mechanical practices, genetic practices, regulatory practices, biological practices, and chemical practices. Because of adoption of pest management practices, the average annual number of insecticide uses for control of the sugarcane borer has decreased dramatically (LSU AgCenter). For example, in 1960, the average number of annual insecticide applications was 12; now it is slightly less than 1 (LSU AgCenter; https://www.lsuagcenter.com).

16.4 Transgenic Sugarcane

The first sugarcane genetic transformation was carried out with protoplasts by polyethylene glycol (PEG)-induced direct uptake of DNA (Chen et al. 1987). Since that time, transgenic sugarcane has been produced using a variety of techniques and with differing trait targets. Constant improvement in procedures has led to success in production of transgenic sugarcane with enhanced traits and increased yield compared to conventional sugarcane. For example, transgenic sugarcane with improved sucrose accumulation has been produced and reported (Botha et al. 2001). Herbicide resistance was conferred on sugarcane by the transfer of genes such as phosphinothricin acetyltransferase and glyphosate-tolerant gene (Manickavasagam et al. 2004; Nasir et al. 2014). Abiotic stress resistance was introduced by PDH45, DREB2, and HSP70 gene insertion using *Agrobacterium*-mediated transformation and particle bombardment techniques. Resulting transgenic plants showed resistance to drought and salinity stress (Augustine et al. 2015a, b, c). More recently, red rot-resistant sugarcane lines were developed by transfer of the chitinase II gene (Tariq et al. 2018). In this study, chitinase II gene, driven by a polyubiquitin promoter, showed resistance towards red rot causative agent, *C. falcatum*. Two of the four transgenic sugarcane lines showed high endochitinase activity of 0.72 and 0.58 U/mL and strong resistance against *C. falcatum* compared to control plants. The wild-type control plants died 3 weeks after infection (Tariq et al. 2018). *Agrobacterium*-mediated transformation and particle bombardment method provided the majority of the transformations in this study.

16.5 Transgenic Sugarcane for Insect Resistance

There are a variety of genetic transformation techniques involving different genes that are used to protect sugarcane from insect pests. These include proteinase inhibitors (PI), plant lectins, ribosome-inactivating proteins, plant secondary metabolites, delta endotoxins, and vegetative insecticidal protein from *Bacillus thuringiensis* (Bt) and related species. Each category can be used alone or in combination with Bt genes (Bates et al. 2005). Some major reports of transgenic sugarcane developed for insect pest resistance are listed in Table 16.1. Genetic engineered crops have been produced via introduction of Cry toxin engineering and are often referred to as Bt-crops (e.g., Bt-cotton, Bt-corn, or Bt-sugarcane). Each Bt-crop produces Cry-toxin protein, which kills feeding insects. This can protect crops and subsequently lead to increased yield without the use of chemical insecticides. The mechanism of Cry toxin is activated once leaves of the Bt-crop are consumed by the insect. The Cry toxin protein is solubilized and activated by insect gut proteases. Activated toxin binds to receptors, allowing gut membrane insertion and pore formation. The gut wall is ultimately broken down, allowing spores and gut bacteria to enter the body cavity of the insect. Both the spores and gut bacteria proliferate in the body of the insect (Pardo-Lopez et al. 2013). Affected insects stop feeding immediately, leading to death of larva. Many crop yields have benefited from Cry gene incorporation, but the most famous are Bt-cotton and Bt-brinjal (eggplant). There are many reports of Cry gene introduction into sugarcane crops and resulting experiments.

Insect-pest tolerance was reported in sugarcane beginning in 1997. Initial studies used a truncated version of BTK HD-1 Cry1A(b) gene driven by CaMV 35S promoter. Transformed sugarcane obtained resistance against *Diatraea saccharalis*,

Table 16.1 List of transgenic sugarcane developed for insect pest resistance (modified from Srikanth et al. 2011)

	Method of transformation	Explant	Gene for transfer	Target pest	Trait	Promoter	References
1	Electroporation	Embryogenic calli	*cry1A(b)*	*Diatraea saccharalis*	Stem borer resistance	CaMV 35S	Arencibia et al. (1997)
2	Electroporation	Embryogenic calli	*cry1A(b)*	*Diatraea saccharalis*	Stem borer resistance	CaMV 35S	Arencibia et al. (1999)
3	Particle bombardment	Embryogenic calli	*cry1A(b)*	*Diatraea saccharalis*	Stem borer resistance	Maize PEPC Maize pith	Braga et al. (2001, 2003)
4	Particle bombardment	Embryogenic calli	Soybean Kunitz trypsin inhibitor (SKTI) and soybean Bowman-Birk inhibitor (*SBBI*)	*Diatraea saccharalis*	Borer resistance	Maize Ubi-1	Falco and Silva-Filho (2003)
5	Particle bombardment	Embryogenic calli	Synthetic *cry1Ac* (s-*cry1Ac*)	*Proceras venosatus*	Stem borer resistance	Maize Ubi-1	Weng et al. (2006)
6	*Agrobacterium* mediated	Embryogenic calli	Snowdrop lectin	*Ceratovacuna lanigera*	Sucking pests resistance	RSs-1 / Maize Ybi-1	Zhangsun et al. (2007)
7	*Agrobacterium* mediated	Embryogenic calli	Fusion *Amaranthus viridis* agglutinin and SKTI genes	*Diatraea saccharalis*	Insect resistance	Maize Ubi-1	Deng et al. (2008)
8	*Agrobacterium* mediated	Meristematic leaf whorls	*cry1Aa3*	*Scirpophaga excerptalis*	Borer resistance	CamV 35S	Kalunke et al. (2009)
9	Particle bombardment	Embryogenic calli	Aprotinin	*Scirpophaga excerptalis*	Top borer resistance	Maize Ubi-1	Christy et al. (2009)
10	Particle bombardment	Embryogenic calli	*cry1Ab*	*Chilo infuscatellus*	Stem borer resistance	Maize Ubi-1	Arvinth et al. (2010)
	Agrobacterium mediated	Meristematic leaf whorls					

(continued)

Table 16.1 (continued)

	Method of transformation	Explant	Gene for transfer	Target pest	Trait	Promoter	References
11	Particle bombardment	Embryogenic calli	Modified *cry1Ac* (m-*cry1Ac*)	*Proceras venosatus*	Stem borer	Maize Ubi-1	Weng et al. (2011)
12	Particle bombardment	Embryogenic calli	*cry1Ac*	*Diatraea saccharalis*	Insect resistance	CamV 35S	Gao et al. (2016)
13	Particle bombardment	Embryogenic calli	CaneCPI-1	*Sphenophorus levis*	Insect resistance	Maize Ubi-1	Schneider et al. (2017)
13	*Agrobacterium* Mediated	Embryogenic calli	*cry1Ab* and *EPSPS*	*Helicoverpa armigera*	Insect and herbicide resistance	Maize ubiquitin	Wang et al. (2017)
14	Particle bombardment	Embryogenic calli	Chitinase II	*Colletotrichum falcatum*	Red rot resistance	Maize ubiquitin	Tariq et al. (2018)
15	Particle bombardment	Embryogenic calli	*cry1Ac*	*Diatraea saccharalis*	Stem resistance	CaMV 35S	Zhou et al. (2018)
16	*Agrobacterium* mediated	Embryogenic calli	*cry1Ab* and *cry2Ab*	*Diatraea saccharalis*	Borer resistance	CaMV 35S	Cristofoletti et al. (2018)

a sugarcane stem borer (Arencibia et al. 1997). This was the first report of sugarcane stem borer (SCSB)-resistant sugarcane. It was also suggested that transforming gene *cry1A(b)* to obtain SCSB-resistant sugarcane provided advantages to conventional breeding, since the sugarcane population lacks a corresponding gene in its gene pool (Arencibia et al. 1997). In 1999, Arencibia et al. evaluated 42 transgenic sugarcane clones expressing Cry1A(b) under field conditions by artificially infecting them with stem borer. Results demonstrated 5 transgenic clones out of 42 showing high tolerance to the stem borer. Further analysis confirmed the genomic changes and found a small number of qualitative traits to differ (Arencibia et al. 1999).

Researchers had changed and increased GC content of Cry1Ac gene (*cry1Ac*) to 47.5% from 37.4%, by following the sugarcane codon usage pattern, to generate "synthetic" *cry1Ac* gene (s-*cry1Ac*). This increase in GC content enhanced the expression of insecticidal protein. Maize ubi-1 promoter was used to drive the s-*cry1Ac* gene. Resistance of transgenic sugarcane was checked against *Proceras venosatus*. Expression of s-Cry1Ac toxin protein in transgenic sugarcane lines was sevenfold higher than expression of Cry1A(b) toxin protein reported earlier by Arencibia et al. in 1997 (Weng et al. 2006). When GC content of the *cry1Ac* gene was further increased to 54.8%, forming "modified" *cry1Ac* gene (m-*cry1Ac*), the expression level of m-Cry1Ac toxin protein increased fivefold over the previous study using s-*cry1Ac* gene (Weng et al. 2011). Expression of Cry1A(b) protein was 0.59–1.35 ng/mg as s-Cry1Ac was 1–10 ng/mg total soluble protein in leaves. The expression of m-Cry1Ac protein was even higher at 2.2–50 ng/mg of soluble proteins. These studies confirm *cry* gene with high GC content enhancing the expression by following the sugarcane codon usage pattern.

A research group from Brazil showed *cry1A(b)* gene from *Bacillus thuringiensis* performs better against *Diatraea saccharalis*, stem borer, both in laboratory and field condition in sugarcane. All resulting transgenic sugarcane plants showed high expression levels of *cry1A(b)* gene in their tissue, eliminating or reducing borer damage compared to conventional varieties (Braga et al. 2001, 2003).

The toxicity of Cry1Aa, Cry1Ab, and Cry1Ac toxin proteins was evaluated against neonate larvae of sugarcane shoot borer *C. infuscatellus* using in vitro bioassays with a diet surface contamination method. Of the three proteins, Cry1Ab demonstrated the most toxic effect. Codon modification protocol specific to sugarcane enhanced GC content and removed polyadenylation signals. The resulting codon-optimized *cry1Ab* was used in subsequent transformation, and transgenic lines were evaluated against shoot borer in greenhouse conditions. *cry1Ab* gene was expressed alone in a predominant cultivar using *Agrobacterium*-mediated transformation. In addition, a transgenic line was also generated using gene pyramiding to stack *cry1Ab* with transgenic-expressing *bovine-pancreatic-trypsin inhibitor*, known as aprotinin (Arvinth et al. 2010). Maize Ubi-1 promoter was used for both *cry1Ab* and aprotinin. Expression of Cry1Ab toxin protein was relatively high, comprising 0.007–1.73% of total soluble leaf proteins. Aprotinin expression varied between 0.15 and 0.24% of total soluble leaf proteins (Arvinth et al. 2010). Higher levels of transgene expression could be attributed to the deployment of monocot-specific ubiquitin promoter, high GC content (48.5%), and low copy number of transgenes. In vivo screening of transgenic sugarcane lines with neonate larvae of *C. infuscatellus* showed a negative correlation between Cry1Ab content and sugarcane dead heart damage to infested plants. "Dead heart" is a condition in that when sugarcane is attacked by stem borer, injury is often first noticed when the youngest partially unfurled leaf of the plant begins to wither and die (LSU AgCenter). Higher correlation was reported between pyramid-stacked Cry1Ab content and dead-heart damage levels than in Cry1Ab alone which suggested a possible synergistic effect between aprotinin and Cry toxin (Arvinth et al. 2010).

It was also suggested transgenic lines be used as parents in a conventional breeding program for imparting pest resistance. Crosses between a Cry1Ab transgenic line expressing 0.13% toxin and an untransformed cultivar can produce progeny with varying Cry toxin content between 0.15 and 1.19%. This indicates a possible tenfold increase in Cry content in progeny (Arvinth et al. 2009). In the past 3 years, many research groups have been transforming Cry gene into sugarcane for insect-pest resistance. In 2016, medium-copy number of *cry1Ac* gene for transgenic sugarcane was reported to perform better than high- or low-copy number. Medium-copy number lines express Cry1Ac protein at an optimum level to increase borer resistance and simultaneously maintain a sucrose level less than or equal to control plants (Gao et al. 2016). Transgenic lines with high-expression Cry1Ac protein consumes more plant energy and has a negative effect on agronomic traits (Gao et al. 2016). During sugarcane field trials, transgenic lines showed greater resistance and less visible borer damage than untransformed sugarcane control lines (Fig. 16.1). Later studies suggested low-copy number Cry1Ac lines are not effective or sufficient to improve insect resistance in sugarcane. High-copy number Cry1Ac lines display deterioration of yield and quality, due to *cry* gene's high level of energy consumption (Zhou et al. 2018).

Apart from *cry* gene, both plant and animal proteinase inhibitors (PIs) have also been used to produce transgenic sugarcane resistant to borers. Transgenic sugarcane lines expressing soybean Kunitz trypsin inhibitor (SKTI) and soybean Bowman-Birk inhibitor (SBBI), driven by the Maize Ubi-1 promoter, were first produced in 2003 (Falco and Silva-Filho 2003). In insect feeding trails, lines were evaluated against *D. saccharalis*. Also, in vivo screening was used on greenhouse sugarcane plants infested with neonate larvae. Results suggest SKTI and SBBI transgenic sugarcane hinders insect growth and metabolism, leading to reduction of larva weight (Falco and Silva-Filho 2003). In another study, aprotinin driven by Maize Ubi-1 promoter was transformed into two sugarcane cultivars (CoC 92061 and Co 86032) using particle bombardment. Aprotinin more effectively inhibited gut proteinases in *S. excerptalis* over *Chilo* genus. There was maximum weight reduction of 99.8% in *S. excerptalis* larvae but low mortality (Christy et al. 2009).

Sugarcane *cysteine peptidase inhibitor 1* (CaneCPI-1) gene was transformed into sugarcane for overexpression, by a group of Brazilian researchers. Embryogenic calli from sugarcane cultivar SP80-185 were utilized for biolistic transformation. Transgenic sugarcane resistant to *Sphenophorus levis* larvae was produced. The relative expression study demonstrated an eightfold increase in CaneCPI-1 gene expression in one transgenic line. Immunoblot assays and feeding bioassays were

Fig. 16.1 Degree of damage between transgenic lines and untransformed sugarcane control plants when infected by stem borer (Source from Gao et al. 2016)

performed to check the efficiency of this transgenic sugarcane against *Sphenophorus levis*. It was concluded that transgenic sugarcane expressing CaneCPI-1 exhibits resistance to *Sphenophorus levis*. The weight of *S. levis* larvae fed on transgenic plants was reduced approximately 50% compared to larvae fed on the untransformed plants. Ultimately, transgenic plants exhibited less damage than untransformed plants, since the CaneCPI-1 gene inhibits cysteine peptidase (Schneider et al. 2017).

In one study, transgenic sugarcanes were produced with both borer resistance and herbicide tolerance (Wang et al. 2017). This was achieved using Cry1Ab for borer resistance and glyphosate-tolerant EPSPS for herbicide tolerance, respectively. The selection marker used was PMI. Results were promising, and transgenic lines showed significant resistance/tolerance compared to untransformed controls. Field trials of selected transgenic lines showed that none of the stalks from the transgenic lines showed cane borer damage compared to 31% of control-line stalks, showing

Fig. 16.2 Field trial evaluations of transgenic lines against borers and herbicides. (**a**) Stem of the transgenic lines without being affected by cane borer. (**b**) Stalk of transgenic lines showing no symptom of growth of cane borer. (**c**) Stem of the untransformed sugarcane destroyed badly by cane borer. (**d**) Stalk of the untransformed sugarcane control affected by cane borer. (**e**) Transgenic lines growing healthily after spraying of 0.2% roundup for 10 days. (**f**) Untransformed sugarcane control died after spraying of 0.2% roundup for 10 days (source from Wang et al. 2017)

cane borer damage (Fig. 16.2a–d). During the same time period, herbicide tolerance was also tested using 0.2% roundup spray for 10 days. This resulted in the elimination of untransformed control sugarcane. The transgenic sugarcane survived and grew healthily (Fig. 16.2e, f) (Wang et al. 2017).

16.6 RNA Interference (RNAi)

RNA interference (RNAi) is a process, first discovered in *Caenorhabditis elegans*. Exogenously applied, double-stranded RNA (dsRNA) can silence or degrade endogenously expressed complementary mRNA transcripts within a cell. This results in sequence-specific gene suppression (Fire et al. 1998). Since its discovery, RNAi technique has been used as a tool to protect plants from pests (Huvenne and Smagghe 2010). It is more widely used to develop virus-resistant plants to control diseases (Stevens et al. 2012). *CiHR3* dsRNA can be expressed in bacteria or synthesized in vitro. When fed to *Chilo infuscatellus*, *CiHR3* dsRNA triggered silencing of molt-regulating transcription factor *CiHR3*. This causes abnormal growth and development of insects, resulting in weight loss within 7 days of treatment. During the third instar stage, larvae of *Chilo infuscatellus* experience significant reduction of body weight feeding on CiHR3 dsRNA compared to larvae feeding on negative control. Twenty-five percent of CiHR3 dsRNA-fed insects died prior to entering the fourth instar stage. Those entering the fifth instar larval stage were significantly reduced in weight, and 20% of those insects failed to enter the pupal stage. This result suggests feeding-based RNAi is a better method for control of *Chilo infuscatellus* (Zhang et al. 2012).

Small RNAs produced by transgenic sugarcane plants can induce RNAi upon sugarcane mosaic virus (SCMV) and suppress the infection caused by this virus (Aslam et al. 2018). The coat protein (CP) gene of SCMV was targeted by expression of short hairpin RNAs (shRNAs) to provide SCMV resistance. In the study, two stable shRNAs (shRNA2 and shRNA4) were used for transformation into two sugarcane cultivars (SPF-234 and NSG-311) with particle bombardment. The degree of resistance was found variable among the two sugarcane cultivars, but it revealed that transgenic plants expressing shRNA4 were almost immune to SCMV infection (Aslam et al. 2018). The results suggest a RNAi strategy with a right shRNA can be an effective approach to development of SCMV-resistant, transgenic sugarcane plants (Aslam et al. 2018).

16.7 Bioassays of Transgenic Sugarcane Plants for Insect Resistance

Bioassays are experiments, or a series of experiments, used to estimate potencies of drugs, poisons, or other therapeutic preparations. These preparations may be derived from animals or plants or any substance that has been introduced into a biological system (Finney 1979). In transformed sugarcane, bioassays are performed to determine the level of gene expression. Many studies involve Cry gene transformed into sugarcane. Some examples follow. Thirty-day transgenic sugarcane plants and control plants were assessed using an antifungal bioassay. In this assay, PDA agar block (0.5 cm^2) of *C. falcatum* was placed in a container with both transgenic and control sugarcane. The inoculated plants were placed at 25 ± 2 °C under a photoperiod of 16 h light and 8 h dark. Plants were observed for any morphological symptoms particular to *C. falcatum* and survival for up to 15 days. This assay showed no inhibition of fungal growth in control plants and a maximum of 56% inhibition of *C. falcatum* in transgenic plants (Tariq et al. 2018). Back in 1997, a similar kind of assay was used. Positive transgenic sugarcane was challenged with sugarcane stem borer (*Diatraea saccharalis*) larvae. In this assay, plants were placed in a glass tube containing distilled water under sterile condition. Damage caused by larvae was studied and scored accordingly (Arencibia et al. 1997). Another example, analysis of Cry toxin protein against neonate shoot borer larvae (*C. infuscatellus*) in vitro bioassay through diet-surface contamination method, showed Cry1Ab to be more toxic than Cry1Aa and Cry1Ac (Arvinth et al. 2010). Gut assays demonstrated aprotinin is more effective towards *S. excerptalis* (Christy et al. 2009).

Snowdrop lectin (*Galanthus nivalis* agglutinin; GNA) transgenic sugarcane plants resist woolly aphids. These were tested by *fecundity-and-survival* bioassay. Middle leaves of transgenic and non-transgenic control sugarcane plants were inoculated with second-instar larvae. Progeny quantity of each mother aphid was recorded. The larvae were removed every day until mother aphids died to calculate fecundity. Aphid population was investigated using six second-instar larvae inoculated onto leaves of transgenic plants and the non-transgenic control plants separately. This assay result suggested a decrease in overall fecundity (production of offspring) and survival (population density) in the transgenic plant lines (Zhangsun et al. 2007).

Feeding bioassays are also used to assay the resistance against insects or pests. *Sphenophorus levis* eggs were placed on an artificial diet and cultivated to the 10-day larvae stage. Manual perforation was made to inoculate the stem base of each transgenic and non-transgenic plant with larvae. The holes were sealed with adhesive tape to prevent entry of other insects. Ten days after inoculation of larva, plants were analyzed for damage caused by larval feeding. Larvae lengths and weights were measured. Larvae feeding on untransformed control sugarcane averaged 74 mg, but those feeding on transgenic sugarcane clones 1 and 17 had an average weight of 38.2 mg and 35.6 mg, respectively (Schneider et al. 2017). A feeding bioassay was also used by Zhang et al. to test dsRNA-mediated gene silencing (RNAi). *C. infuscatellus* larvae were fed with two vectors for the *CiHR3* gene, one bacteria-expressed dsRNA and one in vitro synthesized dsRNA. This bioassay was carried out for seven consecutive days. The weight of larvae and death rate were recorded. There was a significant reduction of weight in larvae fed with dsRNA of the *CiHR3* gene compared to the negative control (water-treated larvae) (Fig. 16.3). No difference between bacteria-expressed dsRNA and in vitro synthesized dsRNA was observed (Zhang et al. 2012).

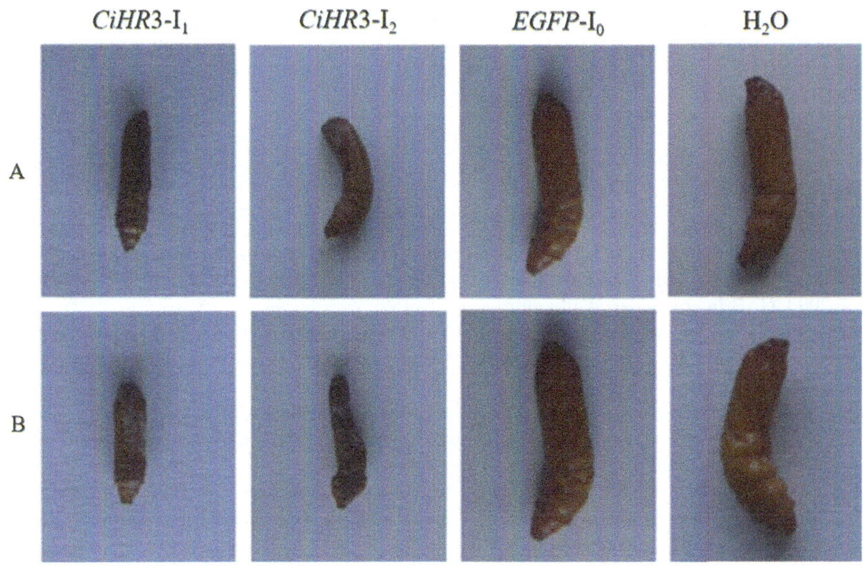

Fig. 16.3 Feeding *CiHR3* dsRNA to the fifth instar larvae of *C. infuscatellus* caused larval body weight to significantly decrease. (**a**) Bacterial expressed dsRNA. (**b**) In vitro synthesized dsRNA (Source from Zhang et al. 2012)

16.8 Conclusion

Sugarcane is an important, worldwide food and industrial crop. Continuous increases in sugar demand, ethanol production, and paper industries necessitate maintaining and even improving yield and production for sugarcane crops. Loss of sugarcane yield to differing biotic and abiotic stressors must be minimized and controlled. Loss due to pests is high around the world, and only moderate success has been achieved using integrated pest management (IPM) techniques. Conventional breeding of sugarcane for pest resistance or other beneficial traits is constrained. Genome complexity contributes to the problem. It is time-consuming (~14 years) and difficult to select for a particular trait. The continuous increase in demand for sugar and sugarcane necessitates a quick and reliable method for development of sugarcane varieties with improved resistance to differing stresses, particularly pests. The best alternative to conventional breeding is genetic transformation of sugarcane. With improved methodology and techniques, great progress has been made in the field of transgenic sugarcane. With novel techniques, such as the CRISPR-Cas system, the potential for improvement in this field is high, and improved varieties of sugarcane can be obtained.

Acknowledgments CM thanks the São Paulo Research Foundation (FAPESP, Proc. 2015/10855-9) for the postdoctoral research fellowship. ANJ acknowledges Council of Scientific and Industrial Research (CSIR), New Delhi, India, for the award of Senior Research Fellowship (SRF) (09/706/0003/2018-EMR-I).

References

Allsopp PG, Manners JM (1997) Novel approaches for managing pests and diseases in sugarcane. In: Keating BA, Wilson JR (eds) Intensive sugarcane production: meeting the challenge beyond 2000. CAB International, Wallingford, pp 173–188

Arencibia A, Vázquez RI, Prieto D, Téllez P, Carmona ER, CoegoA HL, De la Riva GA, Selman-Housein G (1997) Transgenic sugarcane plants resistant to stem borer attack. Mol Breed 3:247–255

Arencibia AD, Carmona ER, Cornide MT, Castiglione S, O'Relly J, Chinea A, Oramas P, Sala F (1999) Somaclonal variation in insect-resistant transgenic sugarcane (*Saccharum* hybrid) plants produced by cell electroporation. Transgenic Res 8:349–360

Arvinth S, Selvakesavan RK, Subramonian N, Premachandran MN (2009) Transmission and expression of transgenes in progeny of sugarcane clones with *Cry1*Ab and aprotinin genes. Sugar Tech 11:290–295

Arvinth S, Arun S, Selvakesavan RK, Srikanth J, Mukunthan N, Kumar PA, Premachandran MN, Subramonian N (2010) Genetic transformation and pyramiding of aprotinin-expressing sugarcane with *cry1Ab* for shoot borer (*Chilo infuscatellus*) resistance. Plant Cell Rep 29:383–395

Aslam U, Tabassum B, Nasir IA, Khan A, Husnain T (2018) A virus-derived short hairpin RNA confers resistance against sugarcane mosaic virus in transgenic sugarcane. Transgenic Res 27:203–210

Augustine SM, Narayan JA, Syamaladevi DP, Appunu C, Chakravarthi M, Ravichandran V, Subramonian N (2015a) Erianthus arundinaceus HSP70 (EaHSP70) overexpression increases drought and salinity tolerance in sugarcane (*Saccharum* spp. hybrid). Plant Sci 232:23–34

Augustine SM, Narayan JA, Syamaladevi DP, Appunu C, Chakravarthi M, Ravichandran V, Tuteja N, Subramonian N (2015b) Overexpression of EaDREB2 and pyramiding of EaDREB2 with the pea DNA helicase gene (PDH45) enhance drought and salinity tolerance in sugarcane (*Saccharum* spp. hybrid). Plant Cell Rep 34:247–263

Augustine SM, Narayan JA, Syamaladevi DP, Appunu C, Chakravarthi M, Ravichandran V, Tuteja N, Subramonian N (2015c) Introduction of pea DNA helicase 45 into sugarcane (*Saccharum* spp. hybrid) enhances cell membrane thermostability and upregulation of stress-responsive genes leads to abiotic stress tolerance. Mol Biotechnol 57:475–488

Bates SL, Zhao JZ, Roush RT, Shelton AM (2005) Insect resistance management in GM crops: past, present and future. Nat Biotechnol 23:57–62

Botha FC, Sawyer BJB, Birch RG (2001) Sucrose metabolism in the culm of transgenic sugarcane with reduced soluble acid invertase activity. Proc Int Soc Sugar Cane Technol 24:588–591

Braga DPV, Arrigoni EDB, Burnquist WL, Silva Filho MC, Ulian EC, Hogarth DM (2001) A new approach for control of *Diatraea saccharalis* (Lepidoptera: Crambidae) through the expression of an insecticidal *Cryla(b)* protein in transgenic sugarcane. Proc Int Soc Sug Cane Technol Congr 24:331–336

Braga DPV, Arrigoni EDB, Silva Filho MC, Ulian EC (2003) Expression of the *Cry1*Ab protein in genetically modified sugarcane for the control of *Diatraea saccharalis* (Lepidoptera: Crambidae). J New Seeds 5:209–221

Casida JE, Quistad GB (1998) Golden age of insecticide research: past, present, or future? Annu Rev Entomol 43:1–16

Chen WH, Gartland KMA, Davey MR, Sotak R, Gartland JS, Mulligan BJ, Power JB, Cocking EC (1987) Transformation of sugarcane protoplasts by direct uptake of a selectable chimeric gene. Plant Cell Rep 6:297–301

Christy LA, Arvinth S, Saravanakumar M, Kanchana M, Mukunthan N, Srikanth J, Thomas G, Subramonian N (2009) Engineering sugarcane cultivars with bovine pancreatic trypsin inhibitor (aprotinin) gene for protection against top borer (*Scirpophaga excerptalis* Walker). Plant Cell Rep 28:175–184

Cristofoletti PT, Kemper EL, Capella AN, Carmago SR, Cazoto JL, Ferrari F, Galvan TL, Gauer L, Monge GA, Nishikawa MA, Santos NZ (2018) Development of transgenic sugarcane resistant to sugarcane borer. Trop Plant Biol 11:17–30

Deng ZN, Wei YW, Lu WL, Li YR, Suprasanna P (2008) Fusion insect resistant gene mediated by matrix attachment region (MAR) sequence in transgenic sugarcane. Sugar Tech 10:87–90

Dhaliwal GS, Jindal V, Mohindru B (2015) Crop losses due to insect pests: global and Indian scenario. Indian J Entomol 77:165–168

Falco MC, Silva-Filho MC (2003) Expression of soybean proteinase inhibitors in transgenic sugarcane plants: effects on natural defense against *Diatraea saccharalis*. Plant Physiol Biochem 41:761–766

Finney DJ (1979) Bioassay and the practice of statistical inference. In: International Statistical Review/Revue Internationale de Statistique, pp 1–12

Fire A, Xu S, Montgomery MK, Kostas SA, Driver SE, Mello CC (1998) Potent and specific genetic interference by double-stranded RNA in *Caenorhabditis elegans*. Nature 391:806–811

Gao S, Yang Y, Wang C, Guo J, Zhou D, Wu Q, Su Y, Xu L, Que Y (2016) Transgenic sugarcane with a *Cry1Ac* gene exhibited better phenotypic traits and enhanced resistance against sugarcane borer. PLoS One 11:e0153929

Hellmich RL, Albajes R, Bergvinson D, Prasifka JR, Wang ZY, Weiss MJ (2008) The present and future role of insect-resistant genetically modified maize in IPM. In: Integration of insect-resistant genetically modified crops within IPM programs. Springer, Dordrecht, pp 119–158

Huvenne H, Smagghe G (2010) Mechanisms of dsRNA uptake in insects and potential of RNAi for pest control: a review. J Insect Physiol 56:227–235

Kalunke RM, Kolge AM, Babu KH, Prasad DT (2009) *Agrobacterium* mediated transformation of sugarcane for borer resistance using *Cry1Aa3* gene and one-step regeneration of transgenic plants. Sugar Tech 11:355–359

Legaspi JC, Mirkov TE (2000) Evaluation of transgenic sugarcane against stalk borers. In: Allsopp PG, Suasaard W (eds) Proceedings of the International Society Sugar Cane Technology, Sugarcane Entomol Workshop, Khon Kaen, Thailand, vol 4, pp 68–71

Manickavasagam M, Ganapathi A, Anbazhagan VR, Sudhakar B, Selvaraj N, Vasudevan A, Kasthurirengan S (2004) *Agrobacterium*-mediated genetic transformation and development of herbicide-resistant sugarcane (*Saccharum* species hybrids) using axillary buds. Plant Cell Rep 23:134–143

Mukunthan N, Srikanth J, Singaravelu B, Asokan S, Kurup NK, Goud YS (2008) Assessment of woolly aphid impact on growth, yield and quality parameters of sugarcane. Sugar Tech 10:143–149

Mustafa G, Joyia FA, Anwar S, Parvaiz A, Khan MS (2018) Biotechnological interventions for the improvement of sugarcane crop and sugar production. In: Sugarcane-Technology and Research, Alexandre Bosco De Oliveira, IntechOpen

Nasir IA, Tabassum B, Qamar Z, Javed MA, Tariq M, Farooq AM, Butt SJ, Qayyum A, Husnain T (2014) Herbicide-tolerant sugarcane (*Saccharum officinarum* L.) plants: an unconventional method of weed removal. Turk J Biol 38:439–449

Pardo-Lopez L, Soberon M, Bravo A (2013) *Bacillus thuringiensis* insecticidal three-domain Cry toxins: mode of action, insect resistance and consequences for crop protection. FEMS Microbiol Rev 37:3–22

Ricaud C, Ryan CC (1989) Leaf scald. In: Ricaud C, Ryan CC, Egan BT, Gillaspie AG Jr, Hughes CG (eds) Diseases of sugarcane. Elsevier, Amsterdam, pp 39–53

Samad MA, Leyva A (1998) Integrated pest management and transgenic plants for insect control in Texas sugarcane. In: Proceedings of the Inter-American Suarcane Seminar, Crop Prod Mechanization, 9–11 September, 1998, Miami, Florida, USA. pp 127–131, pp 397–401

Schneider VK, Soares-Costa A, Chakravarthi M, Ribeiro C, Chabregas SM, Falco MC, Henrique-Silva F (2017) Transgenic sugarcane overexpressing CaneCPI-1 negatively affects the growth and development of the sugarcane weevil *Sphenophorus levis*. Plant Cell Rep 36:193–201

Srikanth J, Subramonian N, Premachandran MN (2011) Advances in transgenic research for insect resistance in sugarcane. Trop Plant Biol 4:52

Stevens J, Dunse K, Fox J, Evans S, Anderson M (2012) Biotechnological approaches for the control of insect pests in crop plants. In: Soundararajan RP (ed) Pesticides-advances in chemical and botanical pesticides. IntechOpen, pp 269–308

Tariq M, Khan A, Tabassum B, Toufiq N, Bhatti M, Riaz S, Nasir I, Husnain T (2018) Antifungal activity of chitinase II against Colletotrichum falcatum Went. causing red rot disease in transgenic sugarcane. Turk J Biol 42:45–53

Wang WZ, Yang BP, Feng XY, Cao ZY, Feng CL, Wang JG, Xiong GR, Shen LB, Zeng J, Zhao TT, Zhang SZ (2017) Development and characterization of transgenic sugarcane with insect resistance and herbicide tolerance. Front Plant Sci 8:1535

Weng LX, Deng HH, Xu JL, Li Q, Wang LH, Jiang ZD, Zhang HB, Li QW, Zhang LH (2006) Regeneration of sugarcane elite breeding lines and engineering of strong stem borer resistance. Pest Manag Sci 62:178–187

Weng LX, Deng HH, Xu JL, Li Q, Zhang YQ, Jiang ZD, Li QW, Chen JW, Zhang LH (2011) Transgenic sugarcane plants expressing high levels of modified *Cry1*Ac provide effective control against stem borers in field trials. Transgenic Res 20:759–772

Zhang YL, Zhang SZ, Kulye M, Wu SR, Yu NT, Wang JH, Zeng HM, Liu ZX (2012) Silencing of molt-regulating transcription factor gene, *CiHR3*, affects growth and development of sugarcane stem borer, *Chilo infuscatellus*. J Insect Sci 12:91

Zhangsun DT, Luo SL, Chen RK, Tang KX (2007) Improved *Agrobacterium*-mediated genetic transformation of GNA transgenic sugarcane. Biologia (Bratislava) 62:386–393

Zhou D, Liu X, Gao S, Guo J, Su Y, Ling H, Wang C, Li Z, Xu L, Que Y (2018) Foreign *Cry1Ac* gene integration and endogenous borer stress-related genes synergistically improve insect resistance in sugarcane. BMC Plant Biol 18:342

J. Ashwin Narayan is currently a Research Scholar in Indian Council of Agricultural Research (ICAR)-Sugarcane Breeding Institute (SBI), Coimbatore, funded by Council for Scientific and Industrial Research. His doctoral research is based on developing water-deficit stress-tolerant sugarcane through genetic engineering.

Chakravarthi Mohan is pursuing his postdoctoral research at the department of Genetics and Evolution, Federal University of São Carlos, Brazil, funded by São Paulo Research Federation (FAPESP), where he aims to develop transgenic sugarcane for weevil resistance through RNAi approach.

Mona Esterling is currently an assistant professor of Biology at Tulsa Community College. Her master's research involved use of Bxb1 Recombination Systems for the removal of Selectable-Marker Genes.

Yuan-Yeu Yau obtained a BSc in Botany at National Taiwan University (NTU), and received Master and PhD from University of Wisconsin-Madison, USA. He then worked at University of California-Berkeley and Plant Gene Expression Center (USDA-ARS) at Albany (California, USA), Chinese Academy of Sciences (Professor), and Northeastern State University (USA). He works in areas of plant biotechnology to develop clean-gene technology using microbial site-specific recombination system. He also worked on cotton transformation and production of stroke drug using molecular farming. He is also an active member of *ResearchGate*, a professional network for scientists and researchers.

17

Rapid *Agrobacterium*-Mediated Transformation of Tobacco Cotyledons Using Toothpicks

Yuan-Yeu Yau, Mona Easterling, and Lindsey Brennan

Abstract

Tobacco is a model plant for genetic transformation, with leaf disk transformation being the most commonly used method for its transformation. One disadvantage of leaf disk transformation is obtaining an adequately sized leaf. Cotyledons from young seedlings are considered too small and fragile to use. In an attempt to overcome this drawback, a protocol was developed using toothpicks as a tool to inoculate cotyledons ~2 mm in diameter. *Agrobacterium tumefaciens* LBA4404 hosting two different plasmids (pC35.BNK.2 or pRB140-Bxb1-op) was used for transformation. Fifty-six putative transgenic shoots (T_0) were obtained from pC35.BNK.2 transformation. Among them, 38 (68%) grew roots in kanamycin-containing medium. Approximately 35% of transgenic lines contained a single-copy transgenic locus based on Mendelian inheritance analysis and chi-square (χ^2) test of T_1 seedlings from 17 lines. To simplify the protocol, water-prepared *Agrobacterium* inoculum was used in pRB140-Bxb1-op (containing *gus* gene) transformation. This resulted in ~35% putative T_0 transgenic lines stained strong blue with GUS histochemical staining assay. Both sets of results demonstrate toothpick inoculation to be an effective approach for *Agrobacterium*-mediated tobacco cotyledon transformation. This reduces wait time required in existing leaf disk transformation method using mature leaves. In the removal of step requiring submersion of explants in *Agrobacterium* liquid culture, the protocol also has advantages by minimizing *Agrobacterium* overgrowth and maintaining explant fitness for later tissue culturing.

Y.-Y. Yau (✉) · M. Easterling · L. Brennan
Department of Natural Sciences, Northeastern State University, Broken Arrow, OK, USA
e-mail: mona.easterling@tulsacc.edu

A. Kumar et al. (eds.), *Climate Change, Photosynthesis and Advanced Biofuels*,
https://doi.org/10.1007/978-981-15-5228-1_17

Keywords

Agrobacterium · Cotyledon transformation · Leaf disk transformation · Tobacco · Toothpick

17.1 Introduction

Tobacco is a model plant for *Agrobacterium*-mediated genetic transformation due to the simplicity of its transformation procedures. The traditional technique does not require expensive machinery or complicated procedures. Some significant advantages of using tobacco for genetic transformation are the following:

1. Tobacco plants can be easily regenerated from tobacco leaf pieces through organogenesis (Constantin et al. 1977).
2. Short acclimatization time and a high trainspotting survival rate: up to 100% of the in vitro-raised plants transferred from lab to greenhouse condition were successfully established ex vitro. The acclimatization is brief, taking only a matter of days (Chandra et al. 2010).
3. Ability to maintain a hemizygous state with T-DNA cassette transfer: the inserted T-DNA cassette can be maintained at hemizygous status in a sterile environment through simple vegetative propagation. This can be achieved through the cutting of tips or stems and reproduction in solid MS medium (Murashige and Skoog 1962) without phytohormone. The simplicity of the hemizygous T-DNA cassette is necessitated in some research projects. For example, hemizygous T-DNA structure has been used for site-specific deletion or integration experiments using site-specific recombinases (Hou et al. 2014).
4. Easy crossing: due to large flower size, hand-pollination is easily accomplished.
5. Longevity: by removing the flowering buds or tips, plants continue growing in greenhouse conditions for extended periods of time, which provides supplemental experimental material, particularly for the W38 tobacco species.
6. Prolific seed production for sustaining lines and testing results.
7. Increased biomass with the potential for molecular farming to produce recombinant proteins, due to tobacco's high biomass yield (Twyman et al. 2003).
8. A model plant for agroinfiltration transient assays (Ma et al. 2012).

Due to the variety of advantages mentioned above, most plant research scientists view tobacco as a prime choice for genetic transformation in proof-of-concept experiments with the added benefit of multiple, practical uses.

Currently, leaf disk transformation is the most frequently used method for tobacco genetic transformation (Horsch et al. 1985). However, by using this method, adequate size of *true leaf* (as opposed to cotyledons) tissue is required for cutting into leaf disks (usually ~1 cm in diameter). It is not practical to use very young tobacco cotyledons, for example, 10-day-old cotyledons (~2 mm in diameter), as experimental material. The tiny size and fragility of cotyledons limit the use of forceps for manipulation, and it is difficult to handle such tiny tissue disks in transformation solutions as well. Therefore, waiting for the true leaves to reach a bigger size before transformation use is necessary. This process can take a couple of months from seeds to use. In addition, in a retransformation project (performing the second transformation on a transgenic line), the seeds of the first-transformed lines needed to be selected on an antibiotic-containing medium beforehand, to ensure the presence of the first transgenic cassette. During the process of selection, the growth of surviving (antibiotic-resistant) seedlings is generally slow and stunted. This is especially true when the selection agent is hygromycin, as hygromycin is generally more toxic than kanamycin. To prevent the loss of transgenic plants, seedlings are usually transferred to fresh medium which does not contain antibiotics to stabilize their growth following selection. Using true leaves, the wait can usually be measured in weeks for plant material to be sufficient in size for use in a second transformation. Using cotyledons of the surviving seedlings following selection for second-run transformation would save weeks of time and funds. Therefore, it is desirable to develop a method for transforming cotyledons (instead of true leaves) from the earliest stage of tobacco seedlings.

In this study, we explore the possibility of using cotyledons from young tobacco seedlings for transformation by using sterile toothpicks to deliver the *Agrobacterium* for infection. Two experiments were conducted: (1) retransformed cotyledons of transgenic tobacco lines (previously transformed with binary vector, pN6.Bxb1, which contains a **hygromycin**-resistant gene) with another binary vector (pC35. BNK.2) containing **kanamycin**-resistant gene and (2) repeat procedures mentioned in (1) again with another plasmid pRB140-Bxb1-op (with a *gus* gene), using *Agrobacterium* prepared in water (Fig. 17.1a). The final transgenic lines should contain both T-DNA cassettes and both hygromycin- and kanamycin-resistant genes.

17.2 Materials and Methods

17.2.1 Plant Materials and Tissue Culture Conditions

Seeds of transgenic tobacco (*Nicotiana tabacum* L. cultivar "Petit Havana" SR1) pN6.Bxb1 were germinated. These T_1 seeds were from a previous project (Thomson et al. 2012). The project was a functional study of site-specific recombination system Bxb1-*att* in plants. Seeds were sterilized with 70% ethanol for 2 min and bleach (sodium hypochlorite) [30% (v/v) and drops of Triton X-100] for 20 min and washed thoroughly with autoclaved distilled water. Sterilized seeds were germinated on MS medium, which contains MS mineral salts (Murashige and Skoog 1962; Cat. No. M524, *Phyto*Technology Lab), 3% (w/v) sucrose, 1× Gamborg's vitamin solution (Cat. No. G1019, Sigma-Aldrich), 0.8% agar, and 45 μg/mL hygromycin (Cat. # H3274, Sigma, USA). Plates were sealed with medical air-permeable tape (Micropore™ Surgical Tape; 3M Health Care, USA) (Clarke et al. 1992) and placed in a 25 °C growth chamber with 16 h/8 h (light/dark) photoperiod. Seedlings that

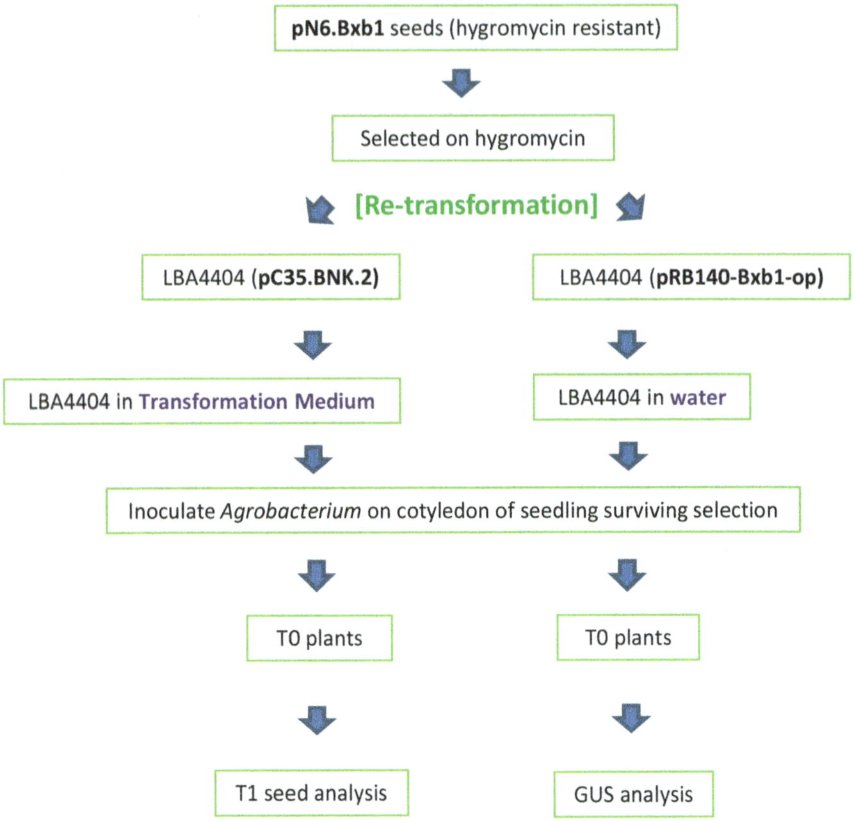

Fig. 17.1 Outline of genetic transformation of this study

displayed stunted growth, a pale green to yellowish cotyledons, and inhibition of hypocotyl extension were considered susceptible to hygromycin. Seedlings with healthy green cotyledons and roots are considered hygromycin-resistant. Cotyledons of survived seedlings were used for genetic transformation.

17.2.2 *Agrobacterium* Strain and Binary Vectors

Detailed procedures for constructing pC35.BNK.2 were described earlier (Yau et al. 2011). The construct of binary vector pRB140-Bxb1-op (Fig. 17.1b) was also described previously (Yau et al. 2012). pC35.BNK.2 uses pCambia2300 as a backbone, which carries the *nptII* gene (confers kanamycin resistance). *Agrobacterium tumefaciens* LBA4404 was used to host either pC35.BNK.2 or pRB140-Bxb1-op for genetic transformation. The vectors were electroporated into ElectroMax™ *Agrobacterium tumefaciens* LBA4404 competent cells (Cat. No. 18313-015, Invitrogen, USA) separately using an electroporator (Multiporator®,

Eppendorf). Forty microliters LBA4404 competent cells and 3 μL plasmid were mixed and then transferred into a 1-mm-gap electroporation cuvette (Cat. No. 94000100-5, Eppendorf, USA). The electroporation was carried out using a manufacturer preloaded program designated for bacterial electroporation (2000V, time constant: 5.0 ms). One milliliter of LB liquid medium was then added into the electroporation cuvette and mixed with the electroporated competent cells. The mixture was then transferred to a Falcon® 14 mL polypropylene round-bottom tube (Becton Dickinson Labware, USA) and incubated at 30 °C for 3 h with a 225 rpm shaking. After 3 h, the bacterial culture of 20 μL, 50 μL, or 100 μL was spread onto LB + streptomycin (100 μg/mL) + kanamycin (50 μg/mL) plates. The streptomycin was used to select *A. tumefaciens* LBA4404 cells' disarmed Ti pAL4404, and kanamycin was used to select transformed bacteria. Plates were placed in a 30 °C incubator for 2–3 days to produce colonies.

17.2.3 Prepare *Agrobacterium* for Transformation

For the first experiment, single colonies (derived from pC35.BNK.2 electroporation) from plates were picked with a 15 cm sterile Cotton Tipped Applicator (Puritan Medical Products Company, Guilford, Maine, USA) and streaked on LB plates containing antibiotics streptomycin and kanamycin and allowed to grow at 30 °C for 1 day. For tobacco genetic transformation, *Agrobacterium* grown overnight was scraped from the plates with a sterile inoculation loop and suspended in 100 μL Transformation Medium [MS mineral salts, 3% (w/v) sucrose, 1× Gamborg's vitamin solution, 3 μg/mL 6-benzylaminopurine hydrochloride (Cat. No. B5920, Sigma-Aldrich), and 100 μM acetosyringone (AS) (Cat. No. D134406, Sigma-Aldrich)]. Transformation Medium was adjusted to pH 5.8 with 0.1 N KOH or HCl and autoclaved at 121 °C and 120 Kpa (1 PSI = 6.89 Kpa) for 20 min. AS was dissolved in 70% ethanol and added to the cooled autoclaved medium (Jones et al. 2005). AS is a phenolic compound that stimulates the induction of *Agrobacterium* virulence genes and improves the transformation efficiency (Nadolska-Orczyk and Orczyk 2000). *Agrobacterium* colonies were suspended in Transformation Medium and were then diluted ten times with the same medium for genetic transformation. For the second experiment, colonies (derived from pRB140-Bxb1-op electroporation) were directly picked and dissolved in autoclaved water (not Transformation Medium) for toothpick inoculation.

17.2.4 *Agrobacterium*-Mediated Transformation

The process of using sterile toothpicks to inoculate *Agrobacterium* to tobacco cotyledons was summarized in Fig. 17.2. Two-week-old seedlings surviving hygromycin selection were pulled out from plates and placed in another sterile plate to cut off the roots in an ESCO Horizontal Airstream® Laminar Flow hood (ESCO, USA) (Fig. 17.2a, b). Cotyledons, ~2.5 mm in diameter, were gently bruised

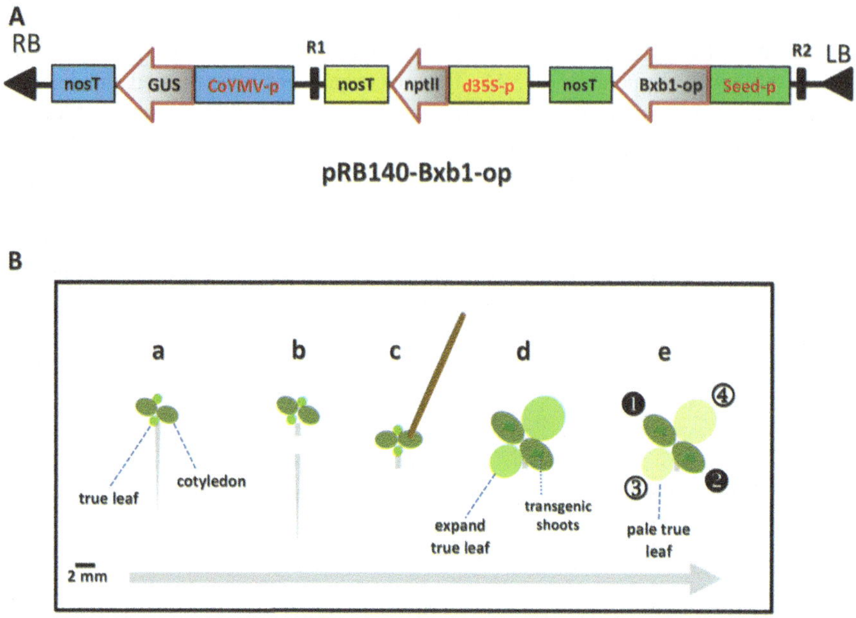

Fig. 17.2 (**a**) Linear (partial) schematic cassette of binary vector pRB140-Bxb1-op T-DNA. *LB/RB* left/right border of *Agrobacterium*, *nosT* nopaline synthase (NOS) terminator, *GUS* β-glucuronidase gene, *CoYMV-p* Commelina yellow mottle virus promoter, *nptII* neomycin phosphotransferase II gene, *d35S-p* double cauliflower mosaic virus 35S promoter, *Bxb1-op* codon-optimized site-specific recombinase gene *bxb1*, and *seed-p* seed-specific promoter. *R1/R2* Bxb1-*att* site-specific recombination system recognition sites. (**b**) Schematic representation of *Agrobacterium* inoculation of tobacco cotyledons with sterile toothpicks for genetic transformation. (a) Two-week-old seedlings with two large cotyledons (dark green oval) and two small true leaves (light green oval). (b) Cut root, a small portion of the root still remaining with the two cotyledons and two true leaves. (c) Inoculation of *Agrobacterium* with a sterile toothpick. (d) The size of both cotyledons and true leaves expands quickly. Putative transgenic shoots appear from the two *Agrobacterium*-infected cotyledons. (e) Putative transgenic shoots grow and true leaves (circles 3 and 4) become yellowish after antibiotic selection

near the center with a sterilized (autoclaved) point-ended toothpick. The toothpick had been dipped into the *Agrobacterium* (containing pC35.BNK.2) suspension described above (Fig. 17.2c). After inoculation, the cotyledons (on the leftover stem) were placed abaxial on co-cultivation medium [Transformation Medium solidified with 0.8% agar (Cat. No. A7921, Sigma)] for 3 days in the dark and then transferred to selection medium [Transformation Medium + cefotaxime/carbenicillin (500 μg/mL) + kanamycin (100 μg/mL), and solidified with 0.8% agar], with the leftover stem sticking into the medium (Fig. 17.2d). Mixture of 50% (w/w) cefotaxime (Cat. No. C380, *Phyto*Technology Lab., USA) and 50% (w/w) carbenicillin (Cat. No. C346, *Phyto*Technology Lab., USA) were used together to remove *Agrobacterium*. Plates were sealed with an air-exchangeable 3M Micropore™ tape and placed in a growth chamber with a 16 h light/8 h dark photoperiod. Subculturing was carried out every 2 weeks. For pRB140-Bxb1-op transformation, the procedures were similar to that of pC35.BNK.2 transformation, but the *Agrobacterium* solution was prepared by simply suspending bacterial colonies in autoclaved water (not Transformation Medium). pRB140-Bxb1-op contains a GUS gene. GUS expression in putative transgenic plants was evaluated and documented.

17.2.5 Kanamycin Selection of Putative Transformants from Secondary Transformation

Two weeks into kanamycin selection, the transformed cotyledons were separated from the stem with a sterilized scalpel and placed on freshly prepared selection (kanamycin) medium. Putative transgenic shoots from the bruised region of the cotyledons were allowed to grow further. Shoots 1 cm in length were cut and transferred into rooting medium [MS mineral salts, 3% (w/v) sucrose, 1× Gamborg's vitamin solution, 0.8% agar] supplemented with 100 μg/mL kanamycin and 400 μg/mL of a mixture of cefotaxime and carbenicillin. One to two putative transgenic shoots were excised from every single cotyledon. Rooted plants were allowed to grow to 5 cm in a Magenta® Plant Tissue boxes and then transferred to soil.

17.2.6 Genomic DNA Extraction

A portion (a 1/4 size cap of a 1.5 mL microcentrifuge tube) of each leaf excised from putative transgenic plants or controls in the Magenta® boxes was harvested into 1.5 mL microcentrifuge tubes. 400 μL grinding buffer (200 mM Tris-HCl, pH 5.7, 250 mM NaCl, 25 mM EDTA, and 0.5% SDS) was added to each tube and ground with a Kontes pellet pestle® (VWR, Batavia, IL, USA) driven by an overhead stirrer (Cat. No. 2572101, IKA Works Inc., USA). The ground samples were centrifuged for 5 min at maximum speed (16,800 ×*g*) with an Eppendorf benchtop centrifuge (centrifuge model 5418). 300 μL of supernatant was transferred to a new microcentrifuge tube, and 300 μg/mL of isopropanol was added to precipitate genomic DNA. After inverting several times, the mixture was centrifuged for an additional 15 min. Once the supernatant was discarded, 70% ethanol was added to wash the DNA pellet. After the ethanol was discarded, the microcentrifuge tubes containing DNA samples were allowed to air-dry 20 min before being resuspended in 50 μL of sterilized water for PCR. Concentrations of DNA samples were measured using a NanoDrop™ 2000 Spectrophotometer (Thermo Scientific, USA).

17.2.7 PCR Analysis

Extracted genomic DNA from leaf tissues of putative transgenic lines and controls were used in PCR amplification for GUS gene. GUS gene (*gusA*)-specific primers **GUS-2**: 5′-CGTTTCGATGCGGTCACTCATTACG-3′ (forward primer) and **GUS-3**: 5′-TCTCCTGCCAGGCCAGAAGTTCTT-3′ (reverse primer) were designed and purchased from Invitrogen (USA). Promega Go*Taq*® Flexi DNA polymerase kit was used for amplification. Each PCR reaction contained 3 μL (approximately 300 ng) of genomic DNA, 2 μL 2.5 mM dNTPs, 2 μL 25 mM $MgCl_2$, 5 μL 5× PCR buffer, 1 μL of each primer (10 μM), 0.12 μL polymerase, and water for a total volume of 25 μL. The thermocycler program used an initial denaturation at 94 °C for 4 min, followed by 35 cycles of 94 °C (30 s), 65 °C (30 s), and 72 °C (1 min 20 s) and a final extension step at 72 °C for 2 min. All PCR were performed on an Eppendorf's Mastercycler Gradient® PCR machine (Eppendorf, USA). The PCR products were separated on a 1% TAE agarose gel (Cat. No. 820723, MP Biomedicals, USA) stained with ethidium bromide (Cat. No. E3050, TechNova, USA). The gel was photographed with a GelDoc-It™ Imaging System (Ultra-Violet Products LLC., USA).

17.2.8 GUS Histochemical Assay

Putative transgenic lines and controls were tested for β-glucuronidase (GUS) expression according to Jefferson et al. (1987). GUS was assayed by placing leaf tissues in the wells of a 96-well plate containing GUS staining solution [1 mM 5-bromo-4-chloro-3-indoxyl-β-D-glucuronide (X-gluc)] (Gold Biotechnology, Inc., St. Louis, MO, USA), 100 mM sodium phosphate buffer pH 7.0, 0.5 mM potassium ferricyanide, 0.5 mM potassium ferrocyanide, and 0.1% Triton X-100. After vacuum filtration for 10 min, the plate was incubated at 37 °C overnight. To check GUS staining, the chlorophyll of the leaf tissue was removed by repeated washing in 70%. Chlorophyll interferes with the observation of stained blue color. Stained leaf tissues were examined under a dissecting microscope and scored for blue coloration.

17.2.9 Mendelian Inheritance Analysis of T_1 Seedlings

T_1 seeds derived from kanamycin-resistant T_0 putative transgenic lines were sterilized with ethanol and bleach and then placed on the germination medium [MS mineral salts, 3% (w/v) sucrose, 1× Gamborg's vitamin solution (Gamborg et al. 1968)] supplemented with 100 μg/mL kanamycin and 200 μg/mL of mixture of cefotaxime and carbenicillin. Plates were placed in a growth chamber with a 16 h light/8 h dark photoperiod. Antibiotic-resistant or susceptible plant seedlings were counted and documented for Mendelian inheritance analysis.

17.2.10 Statistical Analysis

The test of the "goodness of fit" of Mendelian ratio 3:1 (the ratio of resistant to susceptible seedlings) was carried out with the chi-square (χ^2) test to estimate the number of single-locus transgenic lines at $p = 0.05$ level.

17.3 Results and Discussion

To check and produce material for transformation, seeds of four previously pN6. Bxb1-transformed lines were selected either on kanamycin- or hygromycin-containing medium. The transgenic ones should survive hygromycin selection and die from kanamycin selection (Table 17.1). After 10-day selection, all seedlings died on *kanamycin*-containing plates (Fig. 17.3a), and resistant seedlings were observed on *hygromycin*-containing plates (Fig. 17.3b). The resistant plant seedlings grew normally with green cotyledons (and two tiny true leaves), having main roots growing and extending into the selection medium. The mature cotyledons are around 2 mm in diameter. In contrast, the growth of hygromycin-sensitive seedlings was stunted and has pale green (or yellowish) cotyledons (Fig. 17.3b). The cotyledons were curvy and the roots could not grow and extend into the selection medium. Seedlings at the 2-week-old stage were used for calculating the numbers and ratio of the resistant/susceptible seedlings to estimate transgene copy number (Table 17.1). Resistant seedlings were also used for secondary genetic transformation.

Seedlings of 2-week-old surviving hygromycin selection were pulled out with a pair of forceps and placed in a sterile Petri dish to excise the roots ~1.5 mm below the cotyledons (Figs. 17.2b and 17.3c, d). Removal of the most part of root eased transformation manipulation. The cotyledons were then used for *A. tumefaciens* strain LBA4404 (harboring pC35.BNK.2 or pRB140-Bxb1-op) transformation. Both pC35.BNK.2 and pRB140-Bxb1-op contain the *nptII* gene which confers resistance to antibiotic kanamycin for transgenic plants. Toothpicks were dipped in the *Agrobacterium* suspension and used to gently bruise the central area of the two cotyledons of the seedlings (Figs. 17.2d and 17.3e). After inoculation, the root-cut seedlings were placed on the selection medium abaxially, with the left root stuck into the medium (Fig. 17.3f). After 10 days of selection, embryoids were observed from the embryogenic calli around the wound area (Figs. 17.2d and 17.3g), and embryoid

Table 17.1 T_1 seeds of pN6-Bxb1 transgenic lines were selected on MS medium containing antibiotic hygromycin. The survived plants were starting materials for secondary genetic transformation

pN6-Bxb1 transformants (T_0 lines)	Total T_1 seedlings	Resistant	Susceptible	Ratio
pN6-Bxb1-3	162	123	39	~3:1
pN6-Bxb1-6	131	98	33	~3:1
pN6-Bxb1-10	116	103	13	≠~3:1
pN6-Bxb1-11	121	89	32	~3:1

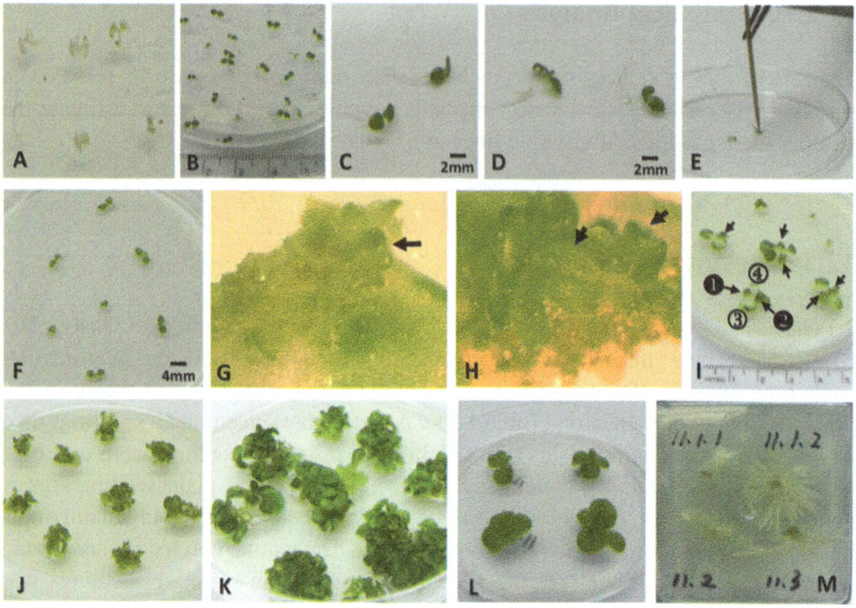

Fig. 17.3 Toothpick inoculation of tobacco cotyledon. (**a**) Seedlings on kanamycin-containing MS plates. (**b**) Seedlings survived hygromycin selection. (**c**, **d**) Part of the main root was excised. (**e**) *Agrobacterium* inoculation with a sterile toothpick. (**f**) Two days of co-cultivation on MS medium and then on selection medium. (**g**) Somatic embryoids observed 10 days after selection. (**h**) Germination of embryoids. (**i–k**) Putative transgenic shoots grew on the *Agrobacterium*-infected regions of explants (on selection medium). (**l**) Individual shoots were cut at the base and transferred to rooting medium with selection agent. (**m**) Roots (plant on the right side) were observed in the selection medium

germination was seen at 2 weeks (Fig. 17.3h). A dissecting microscope is needed to observe the embryoids and their germination. The pair of true leaves (Fig. 17.3i ③ and ④) grew and expanded rapidly. The sizes became bigger than the cotyledons over days (Fig. 17.3i ① and ②). However, under kanamycin selection, the expanded true leaves turned yellowish in color, while the cotyledons with putative transgenic shoots remained green (Figs. 17.2e and 17.3i). The putative transgenic shoots may provide kanamycin-detoxified proteins to cross protect the rest of the cotyledons and contribute to the green color of the cotyledons bearing these shoots. Cotyledons with putative transgenic shoots (Fig. 17.3i ① and ②) were excised from the leftover stem and placed in fresh selection medium to allow those putative transgenic shoots to grow further (Fig. 17.3j–k). One to two putative transgenic shoots (several of them on each infected cotyledon) were then chosen and cut at the base after they reached 1 cm and placed in the rooting medium (with 100 μg/mL kanamycin) (Fig. 17.3l). To ensure the two putative transgenic shoots excised from each infected cotyledon were two independent transformation events, two shoots separated by a greater distance were carefully chosen. Roots could be seen within a week. Rooted plants around

5 cm in height were then transferred to soil for flowering and seed setting (Fig. 17.3m).

For pC35.BNK.2 transformation, a total of 56 putative transgenic shoots were excised from the infected cotyledons and allowed to grow in rooting medium with selection. Among the 56 putative transgenic shoots, 38 of them (68%) grew roots in 100 μg/mL kanamycin-containing medium. Thirty-two plants were randomly chosen and transplanted to soil. Among them, 28 of the plants survived transplanting and grew into adult plants. Except for 1 plant with stunted growth, the other 27 plants showed wild-type architecture, were fertile, and set T_1 seeds. T_1 seeds derived from 21 T_0 transgenic plants were randomly chosen and plated out on kanamycin-containing MS medium. Among them, 17 lines showed resistant/susceptible segregation for kanamycin selection (Fig. 17.4b, c), while wild-type seedlings were dead (Fig. 17.4a). Three lines [pN6.Bxb1 #3 (15.2), pN6.Bxb1 #11 (7.1), and pN6.Bxb1 #11 (10)] had no surviving seedlings from kanamycin selection, indicating that their parental T_0 lines were either selection escapees or gene-silenced transgenic plants (Table 17.2). T_1 seeds of line pN6.Bxb1#11 (18) did not germinate. Using the data of kanamycin selection on the T_1 seed from 21 putative transgenic plants, the chi-square (χ^2) "goodness of fit test" for 3:1 ratio (resistant/susceptible) were performed at p-value = 0.05 level. The results indicated that six individual lines

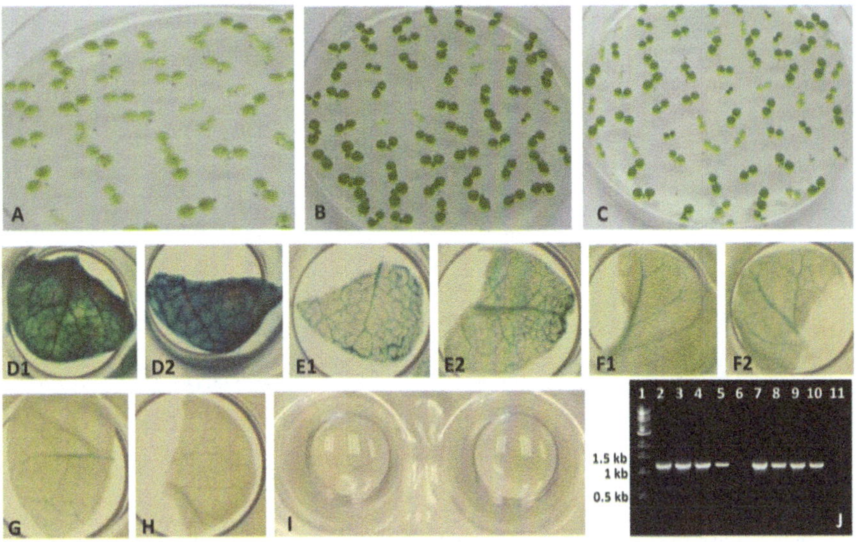

Fig. 17.4 Mendelian inheritance analysis of T_1 seeds from putative T_0 transgenic lines (pC35. BNK.2 transformants) on kanamycin medium (**a–c**) and GUS staining patterns of T_0 transgenic lines from pRB140-Bxb1-op transformation (**d–j**). (**a**) Kanamycin selection: wild-type seedlings. (**b**, **c**) T_1 seedlings of putatively pC35.BNK.2-transformed T_0 lines. (**d–h**) GUS staining of leaf tissues from putatively pRB140-Bxb1-op-transformed transgenic lines. (**i**) GUS staining of wild-type plants. (**j**) PCR results of GUS gene from pRB140-Bxb1-op-transformed putative transgenic lines. Lane 1, DNA size markers; lanes 2–9, individual transgenic plants; lane 10, positive control; lane 11, water (negative control)

Table 17.2 Effects of kanamycin (100 μg/mL) on root growth of putative T_0 transgenic lines (in rooting medium) and T_1 seeds (in germination medium) from pC35.BNK.2 transformation. One or two T_0 putative transgenic lines were randomly chosen from *Agrobacterium*-mediated cotyledon transformation for analysis. Segregation (resistant vs. susceptible to kanamycin) ratio of T_1 generation on kanamycin-containing media was also analyzed. Goodness of fit for 3:1 ratio was determined using the chi-square (χ^2) test, at *p-value* = 0.05 level

Parental lines	T_0 putative transgenic lines	Rooting in kan medium (T_0)	Kanamycin selection (T_1 seeding R:S ratio)
pN6.Bxb1 #3			
	5	+[a] (seeds)[b]	N/A[c]
	8.2	+ (seeds)	***83:30 (3:1)***
	9	+ (seeds)	68:11
	11	+ (seeds)	130:9
	12.1	+ [loss of plant]	N/A
	12.2	+ (seeds)	***72:25 (3:1)***
	15.2	+ (seeds)	0:114
	20	+ (seeds)	12:0
	21	+ (seeds)	100:3
pN6.Bxb1 #11			
	1.2	+ (seeds)	68:4
	3	+ (seeds)	118:14
	5	+ (seeds)	***35:13 (3:1)***
	7.1	+ (seeds)	0:66
	7.2	+ (seeds)	N/A
	8.1	+ (seeds)	103:21
	8.2	+ (seeds)	***91:30 (3:1)***
	9	+ (seeds)	107:18
	10	+ (seeds)	0:111
	11	+ (seeds)	N/A
	13	+ (seeds)	52:6
	14	+ (seeds)	36:7
	15	+ (seeds)	N/A
	16	+ (seeds)	***110:34 (3:1)***
	17	+ (seeds)	***99:29 (3:1)***
	18	+ (seeds)	No germination
	19	+ (seeds)	69:15
	20	+ (seeds)	N/A
	21	+ (seeds)	N/A

[a]"+" indicates that roots grew into kanamycin medium
[b]Indicates that T_0 plants set T_1 seeds
[c]N/A: not applicable

should have single-locus transgene integration (indicated "3:1" in Table 17.2), while the remaining lines showed multiple-locus transgene integration (Table 17.2).

In summary, in this transformation study, stable transgenic lines can be obtained by using sterile toothpicks as a tool to deliver *Agrobacterium*. 81% (17 out of 21) T_0

Table 17.3 Results from GUS staining of leaf tissues from putatively pRB140-Bxb1-op-transformed individual transgenic lines. Water-prepared *Agrobacterium* culture used for transformation

Transgenic lines	GUS staining	Tissue with staining[a]
1	**Dark blue**	**All tissue** [Fig. 17.4d][b]
2	**Strong blue**	**Veins** [Fig. 17.4e]
3	**Strong blue**	**Veins** [Fig. 17.4e]
4	No blue staining appear	All tissue [Fig. 17.4h]
5	Very faint light blue	Veins [Fig. 17.4h]
6	Light blue	Veins [Fig. 17.4f]
7	Light blue	Veins [Fig. 17.4f]
8	Light blue	Veins [Fig. 17.4f]
9	**Dark blue**	**All tissue** [Fig. 17.4d]
10	**Dark blue**	**All tissue** [Fig. 17.4d]
11	Very faint light blue	Veins [Fig. 17.4h]
12	Light blue	Veins [Fig. 17.4f]
13	Light blue	Veins [Fig. 17.4f]
14	**Strong blue**	**Veins** [Fig. 17.4e]
15	Very light blue	Veins [Fig. 17.4g]
16	N/A[c]	N/A
17	Very light blue	Veins [Fig. 17.4g]
18	N/A[c]	N/A
19	Light blue	Veins [Fig. 17.4f]
Positive control[d]	Blue	Transgenic seeds

[a]Types of tissues stained blue
[b]Refer to photograph showed in Fig. 17.4
[c]No putative transgenic plants produced from cotyledons of these seedlings. Callus formed
[d]Previous GUS gene (*gus*) transformed T_1 seeds (Yau et al. 2011) were used for positive control

transgenic lines, which were randomly chosen for analysis, showed stable insertion of *nptII* transgene and conferred kanamycin resistance in the pC35.Bxb1.2 transformation experiment. 35% (6 out of the 17) of the stable transformants demonstrated single-locus transgene integration deduced from the chi-square (χ^2) test. Transgenic plants with single-copy transgene insertions are preferred over those having multiple transgene copies because the latter is prone to gene silencing (Tang et al. 2007). It was reported that the frequency of single-copy transgene insertion in *Arabidopsis* for *Agrobacterium*-mediated transformation was 15% (De Paepe et al. 2009). Although demonstrated in a different species, this method has generated a higher percentage (35%) of single-locus transgene insertion transformants.

For pRB140-Bxb1-op transformation, using water-prepared *Agrobacterium* inoculum, we have also obtained putative transgenic lines (Table 17.3), at least six independent lines with dark blue GUS staining on their leaf tissues (representatives of two plants presented in Fig. 17.4d1, d2). Other GUS staining patterns were also observed (Fig. 17.4e1, e2, f1, f2, g, h). Leaf tissue of wild-type plants did not stain blue (Fig. 17.4i). PCR of GUS gene amplification of those plant tissues showed the expected size (Fig. 17.4j). The data indicated that transgenic plants can be produced using *Agrobacterium* suspension prepared with only water (*Agrobacterium* colonies resuspend in water). We did not track these plants into adulthood. The purpose of this experiment was just to see whether transgene (*gus*) could be stably transformed and expressed using toothpick inoculation. Water-prepared *Agrobacterium* inoculum also successfully generated stable transgenic plants. This can save the labor of medium preparation and funds for the cost of the medium. However, transformation efficiency between using medium-prepared or water-prepared *Agrobacterium* suspension was not compared in this study. This would require further experiments.

No *Agrobacterium* overgrowth was observed in these experiments. Overgrowth is one of the major problems of plant genetic transformation, and *Agrobacterium* can be seen to grow out of control on explants and eventually destroy the explants (Liu et al. 2016). How to control *Agrobacterium* overgrowth is a frequently asked question in *ResearchGate*, the largest professional network for scientists (https://www.researchgate.net/). In most cases, once overgrowth occurs, it is impossible to reverse. The best-known solution is to begin again with another transformation experiment. Instead of submerging the whole leaf disk in *Agrobacterium* culture, this protocol uses *Agrobacterium* inoculation only on a small area of the cotyledons. This practice could potentially minimize the *Agrobacterium* overgrowth problem. There is no requirement for washing the infected explant (cotyledons) in antibiotic solution. Submersion of leaf explants in liquids can also have a negative impact on the fitness of tissues for later use in culture. Successful genetic transformation using explant "cut-edge" for *Agrobacterium* inoculation has also been reported in cotton (Sunikumar and Rathore 2001; Yau 2017). Although toothpick inoculation had been used for screening *Agrobacterium* clones on 4- to 8-month-old tobacco true leaves for VIGS study (https://www.plantsci.cam.ac.uk/research/davidbaulcombe/methods/vigs), to our best knowledge, this is the first report of using toothpick inoculation for tobacco cotyledon transformation.

In a species as well reported as tobacco, we did not perform Southern assay for these studies. Instead, stable gene expression (ex. from *nptII* and *gus*) results were the focus in this report.

17.4 Conclusions

Toothpick inoculation method has demonstrated positive results with using both medium-based and water-based *Agrobacterium* suspension for the purpose of infecting of very young tobacco seedlings. Transgenic plants resistant to kanamycin were obtained by using sterile toothpicks to inoculate *Agrobacterium* on cotyledons. T_1 generation of stable transgenic lines segregated for kanamycin resistance and susceptibility was observed from pC35.BNK.2 transformation. GUS-positive transgenic lines obtained from pRB140-Bxb1-op transformation were also observed. The previous leaf disk transformation method involving the use of the true leaves of tobacco plants requires waiting on seedlings to develop leaves of adequate size for traditional leaf disk transformation. This study demonstrates waiting for true leaves is not required for performing genetic transformation of tobacco. This can help researchers save a minimum of several weeks' wait time, by removing the need to wait for true leaves to be obtained from seeds. Also of note, when hygromycin resistance is a characteristic of transformed seeds, researchers must wait an extended period of time for true leaves to appear, since seedlings positive for selection are weakened in selection media. Growth of seedlings positive for selection will be slow, increasing the time necessary for true leaves to appear, even after transfer to media free of antibiotics. This innovative method involving sterile toothpicks allows researchers to perform a genetic transformation on the cotyledons of tobacco seedlings. A vital benefit to this technique is the reduction of wait time in the production of genetically transformed tobacco plants. By avoiding the step of submerging the explants in *Agrobacterium* suspension can also minimize the occurrence of *Agrobacterium* overgrowth. The explants can also be better maintaining fitness for later tissue culturing.

Acknowledgments This work was partially supported by Northeastern State University Faculty Research Committee Grant P120000. The authors are grateful to Dr. Kevin Wang for his assistance in this work. The authors also wish to thank Dr. C. Neal Stewart, Jr. for useful discussions and suggestions.

Authors' Contributions YYY designed the experiment, constructed the plasmids, collected data, and interpreted the research results. YYY supervised ME and LB and prepared and submitted the manuscript. ME and LB provided technical assistance with plant tissue culture, medium preparation, sample collection, PCR, and GUS analysis. ME also participated with manuscript preparation and editing.

Conflict of Interest The authors declare no conflict of interest.

References

Chandra S, Bandopadhyay R, Kumar V, Chandra R (2010) Acclimatization of tissue cultured plantlets: from laboratory to land. Biotechnol Lett 32:1199–1205

Clarke MC, Wei W, Lindsey K (1992) High-frequency transformation of *Arabidopsis thaliana by Agrobacterium tumefaciens*. Plant Mol Biol Rep 2:178–189

Constantin MJ, Henke RR, Mansur MA (1977) Effect of activated charcoal on callus growth and shoot organogenesis in tobacco. In Vitro 13:293–296

De Paepe A, De Buck S, Hoorelbeke K, Nolf J, Peck I, Depicker A (2009) High frequency of single-copy T-DNA transformants by floral dip in CRE-expressing Arabidopsis plants. Plant J 59:517–527

Gamborg OL, Miller RA, Ojima K (1968) Nutrient requirement of suspension cultures of soybean root cells. Exp Cell Res 50:151–158

Horsch RB, Fry JE, Hoffmann NL, Eichholtz D, Rogers SG, Fraley RT (1985) A simple and general method for transferring genes into plants. Science 227:1229–1231

Hou L, Yau YY, Wei J, Han Z, Dong Z, Ow DW (2014) An open-source system for in planta gene stacking by Bxb1 and Cre recombinases. Mol Plant 7:1756–1765

Jefferson RA, Kavanagh TA, Bevan MW (1987) GUS fusions: beta-glucuronidase as a sensitive and versatile gene fusion marker in higher plants. EMBO J 6:3901–3907

Jones HD, Doherty A, Wu H (2005) Review of methodologies and a protocol for the *Agrobacterium*-mediated transformation of wheat. Plant Methods 1:5

Liu Y, Miao J, Traore S, Kong D, Liu Y, Zhang X, Nimchuk ZL, Liu Z, Zhao B (2016) SacB-SacR gene cassette as the negative selection marker to suppress *Agrobacterium* overgrowth in *Agrobacterium*-mediated plant transformation. Front Mol Biosci 3:70

Ma L, Lukasik E, Gawehns F, Takken FL (2012) The use of agroinfiltration for transient expression of plant resistance and fungal effector proteins in *Nicotiana benthamiana* leaves. Methods Mol Biol 835:61–74

Murashige T, Skoog F (1962) A revised medium for rapid growth and bioassay with tobacco tissue culture. Physiol Plant 15:473–497

Nadolska-Orczyk A, Orczyk W (2000) Study of the factors influencing *Agrobacterium*-mediated transformation of pea (*Pisum sativum* L.). Mol Breed 6:185–194

Sunikumar G, Rathore KS (2001) Transgenic cotton: factors influencing Agrobacterium-mediated transformation and regeneration. Mol Breed 8:37–52

Tang W, Newton RJ, Weidner DA (2007) Genetic transformation and gene silencing mediated by multiple copies of a transgene in eastern white pine. J Exp Bot 58:545–554

Thomson JG, Chan R, Smith J, Thilmony R, Yau YY, Wang Y, Ow DW (2012) The Bxb1 recombination system demonstrates heritable transmission of site-specific excision is *Arabidopsis*. BMC Biol 18:237–248

Twyman RM, Stoger E, Schillberg S, Christou P, Fischer R (2003) Molecular farming in plants: host systems and expression technology. Trends Biotechnol 21:570–578

Yau YY (2017) A pictorial guide to cotton genetic transformation. https://doi.org/10.13140/RG.2.2.23673.36961/7

Yau YY, Wang Y, Thomson JG, Ow DW (2011) Method for Bxb1-mediated site-specific integration in planta. In: Birchler JA (ed) Methods in molecular biology, vol 701. Humana Press, Totowa, pp 147–166

Yau YY, Alonzo E, Lindsey H, Wang K (2012) Bxb1-mediated site specific recombination for DNA deletion in tobacco using a seed promoter. In: Abstract, annual meeting of American Society Plant Biologist, Austin, TX, USA

Yuan-Yeu Yau obtained a BSc in Botany at National Taiwan University (NTU), and received Master and PhD from University of Wisconsin-Madison, USA. He then worked at University of California-Berkeley and Plant Gene Expression Center (USDA-ARS) at Albany (California, USA), Chinese Academy of Sciences (Professor), and Northeastern State University (USA). He works in areas of plant biotechnology to develop clean-gene technology using microbial site-specific recombination system. He also worked on cotton transformation and production of stroke drug using molecular farming. He is also an active member of *ResearchGate*, a professional network for scientists and researchers.

Mona Easterling is currently an assistant professor of Biology at Tulsa Community College. Her master's research involved use of Bxb1 Recombination Systems for the removal of Selectable-Marker Genes.

Lindsey Brennan is currently an andrologist at a fertility center in Tulsa, Oklahoma, USA. She was a student research assistance participating in a plant genetic transformation program at Northeastern State University.

Genetic Improvement of *Jatropha curcas* L. Through Conventional and Biotechnological Tools

18

Sujatha Mulpuri and Srinivasan Nithiyanantham

Abstract

Jatropha curcas (Jatropha), an oilseed plant with a multitude of uses, is considered as a potential biofuel crop. The limited information of this species, low and inconsistent yields, lack of high genetic variability, and susceptibility to biotic and abiotic stresses hamper selective breeding. *J. curcas* is a shrub/tree and genetic improvement and domestication are time-consuming when compared to annual food crops. Most of the programs are dependent upon the germplasm available in undomesticated condition, and the wish list for genetic improvement of the crop is exhaustive. The crop has a history of 500 years and is a new entrant for domestication. Research progress witnessed during the past few years indicate the possibility for widening the genetic base of *J. curcas* through conventional breeding methods complemented with mutation breeding, interspecific hybridization, and biotechnological tools. Genetic diversity analysis using molecular markers unarguably confirmed the Central American and Mexican regions as the treasure troves of *J. curcas* genetic diversity which need to be exploited in varietal development and hybrid breeding programs. Mutation breeding coupled with functional genomics and gene editing techniques will accelerate the development of novel germplasm with desirable traits. In interspecific hybridization, crosses involving *J. integerrima* were extensively studied necessitating the need for exploitation of other economically important species for trait incorporation. With a modest estimate of 6–8 years of concerted efforts, improved germplasm with desired attributes could be made available, and such improved germplasm can be used to replace the already established plantations with unproductive yields in a phased manner.

S. Mulpuri (✉)
ICAR-Indian Institute of Oilseeds Research, Hyderabad, India

S. Nithiyanantham
Tierra Seed Science Private Limited, Hyderabad, India

A. Kumar et al. (eds.), *Climate Change, Photosynthesis and Advanced Biofuels*,
https://doi.org/10.1007/978-981-15-5228-1_18

Keywords

Biotechnological tools · Genetic improvement · Genomic resources · Interspecific hybridization · Jatropha · Mutation breeding · Selective hybridization

18.1 Introduction

Among the oil-bearing tree species, *Jatropha curcas* L. is considered to be the most prospective candidate for biofuel due to its drought endurance ability, life span of 40–50 years, low cost of seeds, high oil content (35%), its ideal size for small farm agroforestry systems, shorter gestation period with a possibility of regular seed harvests within 3–4 years of establishment, its natural spread and distribution, non-allelopathic effects on arable crops, relatively long storability of dried seeds, and above all the plant size that makes the seed collection more convenient (Jones and Miller 1991; Francis et al. 2005; Daniel and Hegde 2007). In the controversy of food vs. fuel, *J. curcas* has emerged as a potential feedstock for biodiesel production as it is predominantly nonedible (no competition for food crops), grows on marginal lands (no competition for arable land), and desirable energetic properties of the oil. Seed oil characteristics of *Jatropha* are superior to others for biodiesel production; has less than 2% of free fatty acids, higher oil to biodiesel conversion ratio, higher flash point, and can be used directly in diesel engines without modification. Perennial crops are preferred over annual crops on marginal lands due to their deep root system which helps in storing more carbon, besides effective utilization of water and nutrients and maintaining soil quality (King et al. 2009).

Despite the advantages the crop offers, *J. curcas* cultivation has not made much headway due to the use of wild and undomesticated material with unpredictable and varying yield patterns, lack of access to germplasm, nonavailability of quality planting materials, low oil content, the presence of toxic and carcinogenic compounds, high male to female flower ratio, asynchronous flowering associated with nonsynchronous fruit maturation, the plant height which is not ideal for mechanization and susceptibility to abiotic (drought, salinity, frost) and biotic (leaf spots, collar rot, root grub, stem and capsule borer, webber) stresses. The large variations in fruit and seed yield of the candidate plus trees are due to the cross-pollinated nature of the crop and the initial variability has been found insignificant when the plants are grown on common site thus indicating low genetic variability. As a live fence, it served as a protection function, and as a biodiesel crop it has a production function where seed yield in a sustainable manner assumes importance. Hence, systematic breeding approaches are required for broadening the genetic base of the cultivated germplasm and development of breeding materials and cultivars with high seed productivity.

18.2 Breeding Objectives

The major concerns with regard to the material being evaluated are the poor productivity, low genetic diversity for key traits, continuous flowering necessitating multiple harvests which are labor-intensive, susceptibility to biotic and abiotic stresses, seed quality to comply with the biodiesel standard, reduced seed toxicity, and enhanced oil content. Thus, the major breeding objectives to meet the demands of the industry, growers, and various stakeholders are as follows.

- Development of high-yielding varieties coupled with yield stability across diverse environments
- Development of dual-purpose varieties (edible accessions with high protein content and high oil content)
- Increase in seed oil content and oil productivity for use as a biofuel feedstock
- Improvement of seed quality with regard to the fatty acid composition for improvisation of fuel properties and elimination of toxic compounds and anti-nutritional factors from seeds to enhance the use as food and feed
- Development of early maturing genotypes and plants with reduced height and altered architecture amenable for mechanical harvest and to fit in different cropping systems as a sole or intercrop.
- Development of pistillate lines with the good combining ability for exploitation in hybrid breeding programs
- Improve the tolerance to diseases and pests
- Improve the tolerance to abiotic stresses as the crop is targeted to semi-arid and marginal growing conditions

To achieve these, various strategies individually and in combination need to be used for widening the genetic base and accelerating the breeding process (Fig. 18.1).

18.3 Germplasm Collection and Evaluation

Realizing that the exploitable genetic diversity in the cultivar germplasm is the base for the development of improved cultivars, the genetic resources of *J. curcas* were adequately characterized and conserved. Knowledge about the extent of genetic diversity among the naturally occurring populations within and outside of the accepted "Center of Origin" in the world not only provides clues about the domestication route but also enables identification of desirable accessions for utilization in the breeding programs. Genetic diversity existing in the crop was determined using morphological characteristics and molecular markers. Although grown extensively in the tropics and subtropics (Fig. 18.2), all the studies unequivocally established the existence of rich genetic diversity in the populations growing wild in Mexico and the Central American region.

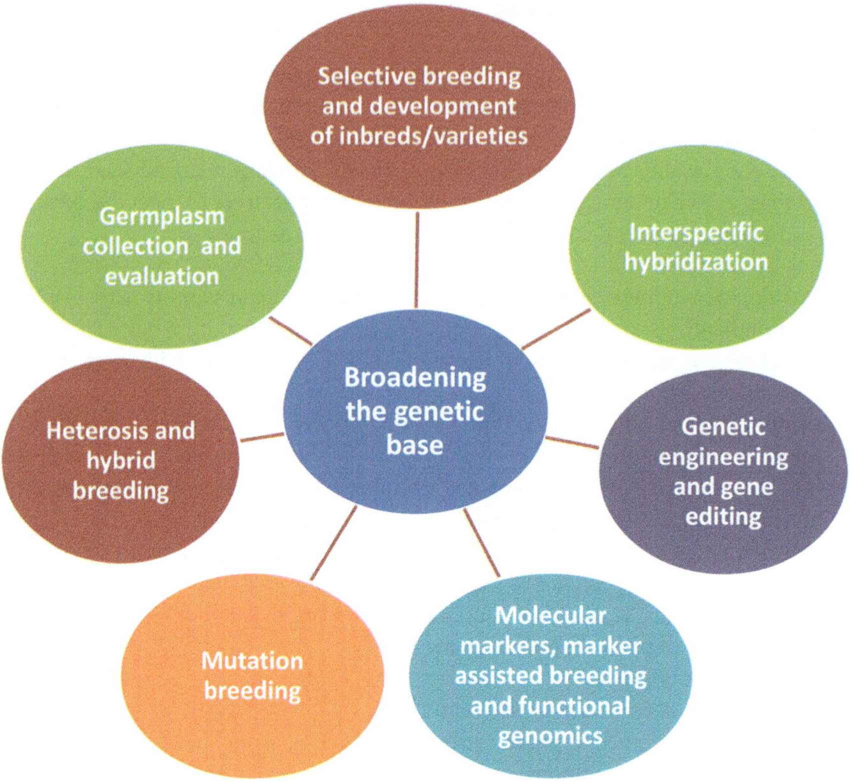

Fig. 18.1 Strategies for genetic improvement of *Jatropha curcas*

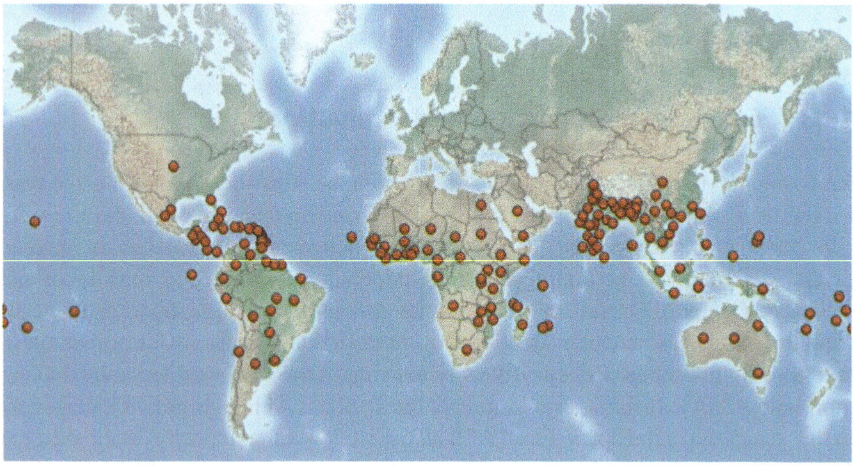

Fig. 18.2 Distribution map of *Jatropha curcas*. (Source: https://www.cabi.org/isc/datasheet/28393)

18.3.1 Germplasm and Gene Banks

Plant genetic resources comprise a diversity of genetic material contained in traditional varieties, modern cultivars, crop wild relatives, and other wild species. A good and systematic collection of germplasm will be the one that includes material from wide and diverse locations including those in the wild and natural habitats. With the increased emphasis on the utilization of germplasm from the center of origin and other regions, there is a threat of genetic drift and genetic erosion. There is a need to strengthen programs for the conservation of *J. curcas* both in situ and ex situ. In situ conservation of genetic resources should be in the natural habitats in the Central American and Mexican regions either as wild and uncultivated plant communities or in backyards. Ex situ conservation of seed in seed banks and field banks should also be done. Immediate attention needs to be given for the assessment of intra- and inter-accessional variability in the germplasm available in the center of origin and then subject the material for purification and maintenance of the germplasm. Large-sized fruits, heavy bearing, altered canopy structure, pistillate types, high seed mass (70–75 mg), oil content (>35%), low toxic (curcin and phorbol esters) types should be the main criterion for selection and multiplication.

According to Heller (1996), germplasm conservation centers were in three locations which include CATIE, Costa Rica (3 provenances), CNSF, Burkina Faso under medium-term storage (12 provenances) and INIDA, Cape Verde (1 provenance), and accessions from the center of origin were represented with only five provenances from two countries. Subsequently, no information is available about these provenances. Attempts were made in several countries to collect germplasm and conserve it ex situ. In Mexico, CEPROBI-IPN, CIBA-IPN, UNAM, Chapingo, have started collecting seeds anticipating threat to the large genetic diversity through the introduction of toxic genotypes from other countries. Otherwise, the genetic base for *J. curcas* could be dramatically reduced; landraces in the farmer's field could be degrading and disappearing. Some of the wild plants in Mexico were used by at least three generations. CEPROBI-IPN has started ex situ conservation for the last 4 years and obtained plants from Yautepac, Morelos, Huitzilan, Pueblillo Puebla, Veracruz with yields above 3 kg per bush (dry seed), multi-bud break bushes which bear branches from the base of the plant and nontoxic genotypes. Some of these promising accessions are being evaluated since 2006 in Chiapa de Corzo, Chiapas (5 ha), and also in several regions in Morelos and Veracruz states (Martinez-Herrera, personal communication). In 2007, INIFAP, Mexico, began the collection, conservation, and characterization of genetic resources of *J. curcas* in various states of the Republic of Mexico. INIFAP, Mexico, maintains more than four hectares of plantations at its Rosario Izapa Experimental Station that make up the National Germplasm Bank of *J. curcas* with 422 accessions from Chiapas, Oaxaca, Yucatan, Tamaulipas, Veracruz, Puebla, Guerrero, Morelos, San Luis Potosí, Michoacán, and Jalisco. An exclusive Germplasm Bank with 25 accessions of nontoxic Jatropha collected in the states of Puebla, Oaxaca, and Veracruz is also maintained. Characterization of the 422 accessions of diverse geographic origins based on morphological, biochemical, and genetic variability within the germplasm through molecular biology confirmed a large genetic base (Pecina-Quintero et al. 2011). Pecina-Quintero et al. (2011) showed that the diversity index of Jatropha's germplasm in Chiapas was as high as 60%, and further analysis of genetic diversity based on amplified fragments length polymorphism (AFLP) has suggested that Chiapas could be the center of origin of *J. curcas* and that domestication was carried out in the states that border the Gulf of Mexico (Pecina-Quintero et al. 2014).

In India, *Jatropha* germplasm including few wild species has been introduced from Brazil, Mali, Australia, Ghana, Nigeria, Mexico, Nicaragua, and Nepal, but these include only one or two accessions. Attempts were made in India under different operational networks involving the National Bureau for Plant Genetic Resources (NBPGR) for collection and assemblage of germplasm (2399 accessions) existing in different agro-ecological regions of the country, and promising genotypes (518 CPTs) were cryopreserved at NBPGR (Anonymous 2008). Ex situ conservation is done by storing the seeds with about 5% (w/w) moisture content and at −18 °C for achieving longevity. The DBT JatrophaNET (www.dbtjatropha.gov.in) has developed a directory for germplasm and descriptors for passport data, collection, management, multiplication/regeneration, environment and site evaluations, plant characters, and reaction to biotic and abiotic stresses. The distribution and morphological diversity of 100 accessions from 100 sites in 5 districts of the Southeast coastal zone of India were studied using geographic information systems (Sunil et al. 2009). The grid maps (DIVA-GIS) generated based on distribution pattern, plant height, number of primary branches, collar length, number of fruits per cluster, and oil content showed diversity with regard to flowering period and fruits per cluster and good variability and richness for oil content. The system also facilitated the identification of gaps in the collection and spotting the diversity richness. Such mapping studies ought to be extended to the assessment of diversity in the germplasm from the center of origin.

In China, a 250-ha *J. curcas* germplasm was established in Panxi area, Sichuan Province, with 70 accessions from Philippines, Thailand, Indonesia, Mali, and from the naturally wild populations in Guangxi, Yunnan, Sichuan, Guizhou, and Hainan provinces (Chen 2007; Li et al. 2007). Germplasm from different countries and areas were collected and assessed for desirable traits in China (Chen 2007). The seed oil content ranged from 15 to 40%. The collections were used to select superior varieties for direct release or were transferred into local, adapted cultivars for increasing the oil content and stress-enduring ability (Chen 2007). Depending on seasonal variation, climate, and land topography, large differences in plant growth, canopy structure, flowering and fruiting ability, number of flowering flushes, fruit and seed set and seed characters were observed (Li et al. 2007).

The germplasm bank at EMBRAPA, Brazil, is one of the largest in the world in terms of an absolute number of accessions (~200) and plants (>2000) and is thought to represent most of the genetic variability of the species in Brazil (Rosado et al. 2010). The PRI through the Global Jatropha Evaluation Programme (GJEP) has assembled more than 200 accessions from 39 countries (Montes Osorio et al. 2008b). Indonesia has built a germplasm repository from 15 provenances (Hasnam 2007).

The minimal descriptors have been developed to assist in germplasm collection and evaluation. Some of the descriptive traits include growth form, number of primary branches, height at which branching is initiated, internodal length, petiole color, petiole length, leaf color of young and fully expanded leaves, leaf size, leaf shape and reticulation, number of leaf lobes, plant branching pattern (convergent, divergent), inflorescence type (compact, lax), male to female flower ratio, presence/absence of co-florescence, unisexual/bisexual flowers, flower color, fruit size and shape, fruit coat color, pedicel length, number of fruits per inflorescence, seed color, seed shape, and number of fruit locules (Sunil et al. 2013).

Morphological diversity: The key to the success of any breeding program lies in an adequate genetic variability and availability of accessions with desired traits and maximum diversity. Knowledge, access, and exploitation of available genetic diversity in domesticated and wild relatives are essential for broadening the genetic base of cultivars to increase crop stability and performance and effective management and use of genetic resources. Attempts are being made to assess the extent of variability in *J. curcas* germplasm using morphological (qualitative and quantitative) traits. These include plant height, canopy diameter, stem girth, collar diameter, oil content on seed basis, oil content on kernel basis, seed to kernel ratio, seed mass (single seed/100 seeds), seed yield per plant, ratio of female to male flowers, composition of seeds in terms of fatty acid profile, ash, and protein. Forty clonal lines investigated for intraspecific variability in Thailand revealed no morphological differences (Sakaguchi and Somabhi 1987). Likewise, morphological differences were not significant in 58 samples characterized in China (Sun et al. 2008). Phenotypic variations were not distinct, but seed characters and chemical composition were found to be highly variable (Ginwal et al. 2005; Kaushik et al. 2007). Ginwal et al. (2005) reported variability in seed characters for germplasm from Central India. Analysis of 1000 samples representing 12 states of India showed significant variation in oil content (25–44%) and kernel and seed coat ratio (0.36–2.12) with accessions from Uttaranchal recording maximum frequency of accessions (73%) with high oil content (Kaushik et al. 2006). Significant variability has been reported in seed size, 100 seed weight (49.2–69.2 g) and oil content (28–38.8%) of 24 accessions of *J. curcas* collected from different agroclimatic zones of Haryana state, India (Kaushik et al. 2007). The phenotypic coefficient of variation was higher than the genotypic coefficient of variation indicating the strong influence of the environment. Genotype–environment interaction was significant for vegetative and generative development (Heller 1996). High estimates of broad-sense heritability were recorded for seed dimensions and seed weight indicating the heritable nature of the variability present, and genetic gain was recorded for oil content revealing the additive gene action (Rao et al. 2008). Martinez-Herrera et al. (2006) described differences in morphological characteristics from four different agroclimatic regions of Mexico (Castillo de Teayo, Pueblilio, Coatzacoalcos, and Yautepec). These constitute regions with average rainfall ranging from 900 to 2500 mm and from semi-hot to hot-humid. The accessions showed differences in crude protein (31–34.5%), lipid (55–58%), fiber (3.9–4.5%), and gross energy (31.1–31.6 MJ/kg dry mass). Proximate composition of 25 accessions from different agroclimatic zones showed a wide variation in crude protein and crude lipid, and maximum kernel oil content (64%) was recorded in a nontoxic accession from Huitzilan, Puebla state (Martinez-Herrera et al. 2007). Pant et al. (2006) observed variation in vegetative and generative characters due to elevation. Altitude had a significant and positive effect on various oil yield components, including the number of branches per tree, number of fruits per branch, number of fruits and seeds per tree, but a significant reduction was observed in kernel oil content with 43.1% at a lower altitude as against 30.6% at higher elevations. Seed yield and seed oil content of *J. curcas* sampled from 167 sites in south and southwest China were significantly variable and ranged from 660 to 4500 kg/ha and from 18.8 to 47.95%, respectively (Sun 2008). Rao et al. (2008) observed significant variation among 29 accessions from different regions in Andhra Pradesh state, India, in terms of seed dimensions, plant growth (plant height, number of branches), flowering attributes (number of flowers, ratio of female to male flowers, days taken from flowering to fruiting and fruiting to maturity), yield characteristics, 100 seed weight (57.0–79.1 g), and oil content (29.9–37.1%). The highest variance was observed for 100 seed weight and oil content. Oil content on seed basis varied between 36.1 and 53.1% in Indonesian accessions (Hasnam 2007). Variation was recorded in seed mass (560–745 mg) and kernel oil content (42–57%) in 25 accessions from 23 field sites across Madagascar (Graham 2006). At PRI, Netherlands, fruit weight varied between 9.7 and 16.99 g and maximum frequency ranged from 12.6 to 15.5 g (Montes Osorio et al. 2008b). Fruit thickness varied between 26.9 and 31.95 mm, and the maximum frequency was between 26.9 and 28.9 mm. Phenotypic variation for a number of fruits was very high and varied from a low number to a very high number of fruits in the global germplasm. Accessions from Andhra Pradesh and Chhattisgarh possessed a higher amount of oil and disclosed a higher amount of molecular polymorphism (Tatikonda et al. 2009). *J. curcas* exhibits environmental elasticity, and few of which may have a genetic basis too. Gross phenotypic variations like plant architecture were found in material from Latin America.

The seed oil of *J. curcas* is rich in unsaturated fatty acids (oleic acid, linoleic acid) with a predominance of oleic acid (Heller 1996; Akintayo 2004; Martinez-Herrera et al. 2006). The fatty acid composition of seed oil is reported to be under the environmental influence (Martinez-Herrera et al. 2006; King et al. 2009). Analysis of fatty acid composition in four provenances from different agroclimatic regions showed that samples from Veracruz were rich in oleic acid while those from Morelos were rich in linoleic acid (Martinez-Herrera et al. 2006). Significant variation was observed in 23 accessions from Madagascar for relative amounts of oleic and linoleic acid, and oleic acid-rich samples were obtained from sites close to sea level with mean annual temperatures higher than the mountainous regions (King et al. 2009). A wide variation was observed in the proximate composition of seeds of 72 accessions from 13 countries (Basha et al. 2009). Levels of crude protein (18.8–34.5%), kernel oil content (45.4–64.5%), and ash content (3.2–6.7%) varied significantly but were not associated with the geographical structure.

Genotypic variation has been observed for phorbol ester content as well. The content of phorbol esters in the nontoxic accessions varied from provenance to provenance (Makkar et al. 1998b). The seeds of Quintana Roo were of better quality with higher levels of protein, lipid, and ash and lower levels of phorbol esters, trypsin, lectin activities, and saponins in the raw meal than those in meals from nontoxic provenance from Veracruz state (Makkar et al. 1998b).

Most of the germplasm evaluation studies are being done with material collected from plus trees of different agroecological regions, different aged plants (3–20 years), and propagated through seeds or vegetative cuttings. Comparison of yield-contributing traits based on such material results in erroneous conclusions about the superiority of the identified clone as it is strongly influenced by the method of propagation (vegetative propagation or direct seeding), soil type (fertile or marginal; irrigated or rainfed; fertilized or not), climatic conditions, age of the plant and plant density (unstipulated spacing between plants), collection period, drying and storage, maturity of the trees, etc. Hence, evaluation should be done with germplasm subjected to common agronomic practices under similar edapho-climatic conditions, and the superiority of the identified accessions should be confirmed through multilocation trials.

Molecular diversity: Molecular markers are generally employed in the management of genetic resources to estimate genetic diversity, identify duplicates in the collection, to devise appropriate conservation strategies for successful utilization of genetic resources, understand the population structure, and resolve taxonomic relationships. Conventionally, diversity is assessed by measuring phenotypic variation but morphological characterization and expression of quantitative traits are subjected to strong environmental influence. Following the advances in molecular biology during the last two decades, a host of molecular marker systems have been developed for the assessment of genetic diversity which differs with respect to technical requirements, level of polymorphism detected, reproducibility, and cost. Molecular markers are reliable indicators of genetic diversity as they are less influenced by the environment and scan the differences at the whole genome level. DNA markers play an important role in the establishment of molecular fingerprints for distinct and most divergent accessions as well.

Initial studies on determining the molecular diversity were with random amplified polymorphic DNA—RAPD (Basha and Sujatha 2007; Ganesh Ram et al. 2008; Gupta et al. 2008; Padmidiamarri et al. 2009; Ikbal et al. 2010; Murty et al. 2013; Saptadi et al. 2017) and inter-simple sequence repeat—ISSR (Basha and Sujatha 2007; Kumar et al. 2008; Basha et al. 2009; Maghuly et al. 2015; Pazeto et al. 2015; Santos et al. 2016) markers which are economical, rapid, and does not require prior information about the genome. Subsequently, amplified fragment length polymorphism—AFLP (Tatikonda et al. 2009; Sun et al. 2008; Maghuly et al. 2015), simple sequence repeat—SSR (Sun et al. 2008; Montes et al. 2014; Yue et al. 2014; Santos et al. 2016; Saptadi et al. 2017), SPAR (Ranade et al. 2008), start codon targeted polymorphism—SCoT (Sujatha et al. 2013a), and single nucleotide polymorphism—SNP (Montes et al. 2014; Maghuly et al. 2015; Anggraeni et al. 2018) markers were employed to determine the genetic diversity.

Analysis of genetic diversity in 42 germplasm lines collected from different regions in India using molecular markers revealed low inter-accessional variability

(Basha and Sujatha 2007). Following this work, molecular markers were used for the assessment of genetic variation in local populations from different countries (Reviewed by Wei et al. 2012). Studies of Ranade et al. (2008) indicated low genetic diversity in Indian accessions and distinctness of germplasm from the North-Eastern part of India. AFLP profile of *J. curcas* collected from different agro-ecological regions of India revealed a narrow genetic base (DBTIndia 2007). Sun et al. (2008) characterized 58 accessions (56 from China and 2 from Malaysia) using SSR and AFLP markers and opines that the source of *J. curcas* in China may be the same as that in India, and the germplasm in Southeast Asian countries could probably have a common ancestry. Grativol et al. (2011) and Cai et al. (2010) reported higher genetic diversity in China and Brazil populations based on the ISSR markers. Comparison of 17 accessions from three countries, viz., Thailand (14), India (2), and Nigeria (1) showed clustering of all accessions despite the geographical divergence (Kohli et al. 2008) confirming the similar ancestry for accessions from Asian and African region. At TLL, 192 samples from Asia, South America, and Africa were genotyped using SSR and AFLP markers. Preliminary studies with SSR markers revealed low diversity with mostly monomorphic homozygous bands (Hong 2008).

Genetic variation reported in the molecular studies was mainly due to the inclusion of wild species (Ganesh Ram et al. 2008; Ranade et al. 2008) or geographically isolated germplasm (Sujatha et al. 2005; Basha and Sujatha 2007; Padmidiamarri et al. 2009). Diversity analysis with local germplasm thus indicates the need for widening the genetic base of *J. curcas* through the introduction of accessions with a broader geographical background (Basha and Sujatha 2007; Ranade et al. 2008). Cross-pollinating species have a significantly higher genetic diversity compared to self-pollinating species. Although a predominantly out-breeding species, *J. curcas* exhibited lower genetic variation in local populations which could probably be due to its propagation through vegetative cuttings and or apomixis. Further, some of the studies were confined to a few accessions (<10) and a limited number (<10) of primers. It is important to have a cautious approach while carrying out the genotyping assays and need to consider the minimum population size, the number of data points, the polymorphism information content of the markers being employed, besides understanding the population structure in terms of its geographic isolation and mode of reproduction to draw meaningful conclusions. Regardless of the number of accessions used, the robustness of the primer and number of marker data points, the local accessions from the respective countries clustered together confirming the existence of low genetic variation in the *J. curcas* ecotypes being genotyped.

Based on the observations of the existence of low genetic variation in local populations in different countries, the need for the assessment of genetic diversity in global germplasm has been realized. Sujatha et al. (2005) demonstrated genetic distinctness between toxic Indian and nontoxic Mexican accessions using molecular markers. Studies of Basha et al. (2009), with a representative set of 72 accessions from 13 countries, showed clear separation of accessions from the Mexican region with those from the rest of the world. The study indicated the possible spread of 1 or 2 toxic accessions from the Mexican region to all the countries and the existence of rich diversity in the Mexican germplasm. Under the GJEP program at PRI, Netherlands, 60 accessions were studied for genetic variation using AFLP and NBS (motif binding) markers (Montes Osorio et al. 2008b; van Loo et al. 2008). High diversity was detected in accessions from the Central American region and low diversity in accessions from Africa and Asia (Luis et al. 2014; Montes Osorio et al. 2014; Trebbi et al. 2015). Genotyping studies of Guatemalan accessions separated the accessions into regions with high, average, and low yields which indicate the existence of large untapped genetic resources in *J. curcas* (van Loo et al. 2008). It is envisaged to subject the global germplasm comprising at least 400 samples from different countries for genotyping under the GJEP program (www.jatropha.wur.nl) which should provide more information on the allelic diversity in the center of origin, genetic relatedness of the accessions in the Central and Meso-American regions, and serve as a valuable resource for trait-based gene transfer. Results using AFLP and SSR markers indicate very little genetic variation between accessions from India, Ghana, Tanzania, and Madagascar but significant variation with Mexican accessions (Graham 2006), and the same observation was confirmed with SCoT polymorphism (Sujatha et al. 2013a). Cluster analysis based on 20 mixed traits grouped 57 accessions from nine countries into five clusters with all the nontoxic accessions forming a separate cluster, while the toxic accessions were grouped into four clusters (Francis et al. 2018). Analysis of global diversity thus confirms the observation of Heller (1996) who showed the distribution and spread of *J. curcas* in the tropical belt via the Cape Verde islands. All the studies unequivocally establish the fact that the Central American and Meso-American regions harbor accessions with useful and novel genes that provide a good basis for widening the genetic base of *J. curcas*.

Although molecular markers disclose variation, molecular measures of genetic diversity have a very limited ability to predict quantitative genetic variability (Sun et al. 2008). Hence, morphological characterization and estimates of molecular diversity need to be combined to identify divergent material for breeding. Jatropha accessions showed higher variability in phenotypic and yield trait parameters and conversely these accessions exhibited low genetic diversity (Yi et al. 2010; Saadaoui et al. 2015; Anggraeni et al. 2018). On the other hand, the Central American and Mexican populations showed higher genetic diversity with useful traits and novel genes. These accessions should be used for breeding to improve the genetic base of Jatropha. Alves et al. (2013) conducted a joint analysis based on both phenotypic and molecular diversities in 117 Brazilian accessions which revealed that the genetic diversity was 156% and 64% higher than the diversity estimated from phenotypic and molecular data, respectively.

Thus the assessment of *J. curcas* germplasm using morphological and molecular markers indicate the following:

- Low phenotypic and genotypic diversity in local populations in Asian regions and close clustering of accessions from Africa and Asia indicating a common genetic base.

- Phenotypic diversity was not associated with genotypic diversity indicating a strong influence of the environment.
- Preliminary studies at global germplasm analysis using molecular markers confirmed the availability of rich allelic diversity in South American, Mexican, and Meso-American regions.
- Variations are reported for low and high number of fruits, tree architecture, toxicity (in terms of phorbol ester levels), seed mass, and seed oil content.

18.4 Breeding Methods and Varietal Development

J. curcas is a cross-pollinated crop, and genetic improvement has to be based on populations (Heller 1996). The following breeding methods could be adopted for the genetic improvement of the crop.

- Mass selection would be the simplest breeding method for selecting superior plants and making composites. It is an ideal method for improving seed yield and oil content for each generation.
- If the populations are large, they can be stepwise improved to have genetic gain through additive genetic variation.
- Recurrent selection with concurrent cycles of selection with or without progeny tests can be adopted for the development of superior material for stabilizing productivity in various production systems.
- Exploitation of heterosis and development of superior hybrids that are suitable for different agroclimatic conditions.

18.4.1 Selection Criteria and Character Association

Seed-derived seedlings are reported to segregate for vegetative growth, branching pattern, flowering time, fruit and seed yields, inflorescence size, the ratio of male to female flowers, and oil content. Hence, there is a need to initiate breeding programs to improve Jatropha as a predictable crop. Quantitative characters such as yield and its determinants are strongly influenced by the environment, and it is imminent to estimate the genotypic (GCV) and phenotypic (PCV) variance. Components that contribute to oil yield per hectare in Jatropha are the number of female flowers per inflorescence, number of capsules per shrub, 100 seed weight, and oil content of seeds. Highly significant positive correlations were recorded for 100/1000 seed weight, crude fat, and crude fiber content which indicate interesting possibilities for selecting and combining the traits with high yield (Heller 1996; Rao et al. 2008). A high kernel to seed coat ratio was found desirable for better oil yield (Kaushik et al. 2006). A strong correlation existed between plant height, branch length, number of branches and collar diameter (Kumar et al. 2008) and male to female flower ratio, and yield (Rao et al. 2008). The dry matter distribution ratio between fruit coat and seed weight is a good selection criteria for seed yield and hence genotypes that partition more dry matter to fruits than the stems and leaves have to be selected (Jongschaap et al. 2007). Increasing the number of branches and, as a consequence, an increased number of inflorescences resulting in increased seed yield could be a possibility with heavy branching types. Such self-propagating types with more branches are reported from China and Mexico. Selections from early plantations of Jatropha can be made based on seed yield, female to male flower ratio, number of branches, plant height, days taken from fruiting to maturity, 100 seed weight, and oil content as the broad-sense heritability was high for these traits (Rao et al. 2008). Female flowers on an inflorescence may not be a useful criterion for the selection of CPTs as adverse conditions of soil and temperature lead to growth retardation, flower drop, and fruit fall. Oil content is strongly influenced by environment, and differences in oil content of the same plant harvested at different times are highly significant.

18.4.2 Varieties

Most of the varieties reported are selections from natural populations. The Cape Verde variety is the one that is spread all over the world (Heller 1996; Henning 2006). A Jatropha variety in Nicaragua has fewer, but larger fruits but the yield per ha seem to be the same as that of other varieties (Henning 2006). Most of the accessions designated as varieties are those that were collected from plants growing wild. The toxic and nontoxic varieties evaluated by Makkar et al. (1998a) were wild varieties collected from Nicaragua, Cape Verde, Nigeria, and Mexico. Subsequently, breeding principles were applied, and elite varieties with improved productivity are being developed, and the success of selective breeding relies on the usable genetic variability. Several years of precise selection, concerted efforts, and high capital investments are required until a new cultivar is commercialized.

In Indonesia, mass selection was practiced with two cycles of simple recurrent selection (Hasnam 2007). Superior plants based on vegetative and reproductive characters coupled with heavy bearing were composited to constitute new improved populations. Seed yield increased from 0.36 to 0.97 t/ha in cycle-1 and 2.2 t/ha in cycle-2 on Lampung provenance. Likewise, seed yield has been improved from 0.43 to 1.0 t/ha in cycle-1 and 1.9 t/ha on cycle-2 on West Nusa Tenggara provenance. Thus, simple recurrent selection increased seed yield by 169% and 146% than the original population in Lampung and West Nusa Tenggara provinces, respectively. The improved population IP-1 was released in 2006 with a yield of around 4–6 t/ha and IP-2 in 2007 with an estimated yield of 7–8 t/ha under high input conditions.

Edible (nontoxic) varieties are available in Mexico and found in Papantla, Castillo de Teayo and Pueblillo regions from Veracruz state, Yucatan Peninsula, Totonacapan, Yautepac in Morelos state, and Quintana Roo state (Schmook and Seralta-Peraza 1997; Makkar et al. 1998a, b; Martinez-Herrera et al. 2006). The level of phorbol esters in the toxic (3–6 mg/g) and nontoxic varieties (0.01–0.02 mg/g) differs and is 20-fold higher in toxic varieties. However, the levels of other anti-nutrients like trypsin inhibitor (14.6–28.7 mg trypsin inhibited/g), lectin

(25.6–52.2 units), and phytate (8.4–10%) did not differ much between the toxic and nontoxic genotypes (Makkar et al. 1998a). These compounds do not contribute to acute toxicity but might aggravate adverse effects. The amino acid composition of meals from both the toxic and nontoxic genotypes was similar and was not affected by agroclimatic conditions (Martinez-Herrera et al. 2006). These two types showed variations in fatty acid profiles, and toxic accessions are rich in oleic acid while nontoxic accessions have high linoleic acid. The seeds of the nontoxic varieties are consumed by humans after roasting. The consumption of raw seeds produces cramps and uneasy feeling in the stomach. This nontoxic variety of Jatropha devoid of phorbol esters could be a potential source of oil for human consumption, and the seed cake can be a good protein source for humans as well as for livestock (Becker and Makkar 1998). Toxic and low-toxic genotypes look morphologically similar except for slow growth of the nontoxic genotypes during the first year of planting. During the second year, growth differences were insignificant. Toxic plants are distinguished from nontoxic plants based on the round shape with slightly larger seeds as compared to the elongated seeds of the nontoxic types. In regions where nontoxic seeds are predominant, traditional use as food is common. In regions where both nontoxic and toxic plants exist, toxic plants are differentiated only by eating their seed if they produce toxic diarrhea and vomiting. All varieties were susceptible to pest attack, and there were no differences in reaction between toxic and nontoxic accessions (Montes Osorio et al. 2008b). The level of molecular polymorphism between toxic and nontoxic accessions was high (Sujatha et al. 2005; Basha et al. 2009; Padmidiamarri et al. 2009) which clearly indicates that the accessions could differ in other traits as well. In Mexico, some states have opted to plant toxic genotypes as they are more resistant to attack by pests and in the states where nontoxic accessions are available, preference is given to plant them as it is traditionally used as food and also as poultry feed (Martinez-Herrera, personal communication).

Alfredo and Quintero (2017) released three clonal varieties (Gran Victoria, Dona Aurelia-100% pistillates, and Don Rafael—a male line) in Mexico based on seed yield, oil content, growth habit, and the higher number of female flowers to meet the demand of the industry and the growers.

Yi et al. (2014) reported the development of the variety JOS2 which was early flowering (143 days after sowing), better self branching, good uniformity among the plants with a high number of inflorescences, and a low male to female flower ratio (14.6). It is an open-pollinated seed population harvested from plus trees (5–10% of the population) with high productivity and subsequently subjected to systematic mass selection for two cycles. The yields obtained were 900 g/plant and 1720 g/plant during the first and second years of planting, respectively.

In India, the Sardar Krishinagar Agricultural University has evolved four genotypes, viz., Chatrapati, Urlikanchan, Liansray, and Sardarkrishinagar Big based on selections from a large germplasm collection and evaluation in replicated trials (Gour 2006). Of these, the first-ever *Jatropha* variety SDAUJ I (Chatrapati) with 49.2% kernel oil and an average yield of 1000–1100 kg per ha under rainfed conditions has been identified for commercial cultivation in the semi-arid and arid regions of Gujarat and Rajasthan in India (www.icar.org).

In China, 250 accessions collected from different regions and assembled in Sichaun province were screened for genotypes with high oil yield and good quality oil. Of these, two improved varieties, viz., CSC high oil content 63# (Chuan R-Sc-Jc-002-2005) and CSC high toxin #1 (Chuan R-Sc-Jc-001-2005), were approved by the Sichuan Province Forest Improved Variety Certification Commission for planting in a large area with a validity of 10 years (Chen 2007; Chen et al. 2008). The fruits of CSC high oil content 63# are small and almost round with thin pericarps and seed coats. Suitable regions for cultivation of CSC high oil content 63# are the dry and hot valley regions in the Yunnan, Guangxi, and Guizhou provinces.

A dwarf and early maturing variety, BT-88 is developed in Malaysia (Tee 2007). The cultivar is a selection from a village in the highlands of Malaysia and produces fruits ready for harvest within 4 months of direct sowing. The tree is dwarf and produces four to six branches close to the ground. Under humid conditions of Malaysia, production is continuous and is being planted in 20 ha in the lowlands of Malaysia.

In Latin American regions, Jatropha is being promoted for the supply of feedstock to oil refineries, but information regarding the genotypes grown is not available. In Brazil, a named variety "Goncalo" is used but its pedigree is not known.

Few private firms have done systematic breeding for the development of elite varieties that are commercially available. The Jatropower (http://www.jatropower.ch/) has developed nontoxic (edible) cultivars (JPNT-1) and toxic cultivars (JP1010, JP47, JP40, JP1003, JP1064) with 2–3 tonnes of dry seeds in the fifth year, and specific characteristics of all these elite varieties are detailed.

18.4.3 Heterosis and Hybrid Breeding

Productivity can be enhanced through the exploitation of heterosis particularly in cross-pollinated crops like *J. curcas*. To achieve this, germplasm needs to be organized to constitute heterotic gene pools, develop an economically viable mechanism for the production of hybrids, identify and stabilize pistillate plants, establish suitable tools for the prediction of heterosis and hybrid performance. Till such time pistillate plants are developed; physiological manipulation of sexuality can also be attempted. Two-year-old plants of the best intraspecific hybrids (IC565735 × IC565739) using accessions collected from India gave a yield of 300 g/plant with 39% oil content (Prakash et al. 2016). Hence, divergent genotypes need to be identified and heterosis is exploited through crosses of elite × elite and elite × local germplasm. The accessions from Mexico and Central American regions should receive the first priority for pre-breeding and breeding efforts as they possess morphological diversity and allelic richness (Montes Osorio et al. 2008a; Basha et al. 2009; Francis et al. 2018).

Intercrossing elite *J. curcas* (Cabo Verde), low-toxic and toxic Guatemalan accessions are being carried out at Plant Research International, Netherlands

(PRI), as a starting point for breeding (Montes Osorio et al. 2008a). It should be a targeted exploration with an emphasis on the trait (gene)-specific germplasm. The direction and speed with which plants are domesticated depend upon the size of the population, the heritability of traits under selection, the mating system, the intensity of selection, and the inherent variability of the traits. The genetics of key traits has to be studied, and the nature of inheritance has to be determined. The heterotic pattern is a key factor for utilizing germplasm to maximize the performance of the population crosses, and derived hybrids and pre-breeding can identify heterotic patterns for breeding programs. High-yielding selections being made in India, China, Africa, and other countries could be further improved through crosses with exotic material from the center of origin.

J. curcas is unisexual and monoecious with female and male flowers borne on cymose inflorescences. The female flowers are borne on the central axes of the main and lateral inflorescence branches, while the male flowers are produced in clusters on the laterals. Under stress conditions, the female flower buds on the lateral inflorescence branches tend to become male. For the production of hybrids, emasculation has to be done on the inbreds/parental lines to be used as females. Interestingly, accessions producing completely pistillate flowers are reported (Alfredo and Quintero 2017; Francis et al. 2018) which can be exploited in the hybrid breeding programs. The number of female flowers in these pistillate plants varied from 5 to 25. In crops like castor, a pistillate plant produces female flowers all along the inflorescence axis while in pistillate plants of *J. curcas*, only the female flowers on the central inflorescence axes are differentiated, and male bud formation is completely suppressed. Studies on floral differentiation in different sex types of both the crops might provide information on the genes governing female flower production. For utilization in the breeding programs, selections among the pistillate lines for a higher number of flowers and the combining ability (GCA and SCA) with diverse parents need to be determined.

Very few studies are being carried out to determine the extent of heterosis, and most of the studies were confined to estimate heterosis for seedling traits which may not be relevant when seed yield and oil content are the key productivity traits. Laviola et al. (2012) reported that estimates of genetic parameters along with predicted gains with high accuracy can be obtained from plants that are 2-year-old enabling early selection. Mathur (https://www.bio.org/sites/default/files/Eric%20Mathur.pdf) predicted 50-fold oilseed yield improvement in Jatropha by maximizing heterosis and further, SNP analysis disclosed 18 heterotic clades in 800 accession families. Tar et al. (2011) attempted crosses between three each of toxic accessions from Myanmar and Thailand with the nontoxic accession from Mexico and heterosis for seed yield in the six F_1 hybrids over better parent ranged from 11.7 to 195.9%. A high correlation among heterosis was found for seed yield with the number of fruits, inflorescences, and fruits per plant and 100 seed weight. Islam et al. (2011) reported heterosis of 42.4–202% over better parent for seed yield per plant. In most of the studies, crosses were made in the half-diallel mating design to produce F_1 hybrids (Martin and Montes 2015).

Currently, hybrid breeding is the main focus of private firms. JatroSolutions developed elite jatropha hybrid cultivars in the edible and non-edible jatropha segments which showed impressive and augmenting seed yields over the years (https://jatrosolutions.com/). In the warm and arid tropics, the seed yields of the new hybrids in the biofuel segment were 4 times higher than those of the best selected first-generation cultivars and 14 times higher than those of wild jatropha accessions. Among edible jatropha plants lacking toxic compounds (phorbol esters), edible hybrid cultivars had 4–5 times higher nut yields than the best selected first-generation cultivars. Jatropower (http://www.jatropower.ch/) has the F_1 hybrid Jatropha seeds/plants (JPH1, JPH2) available for commercial purposes. The advantage of *J. curcas* is its ease of establishment through vegetative stem cuttings which enables rapid propagation of elite cultivars.

18.5 Mutation Breeding

Several physical (UV radiation, gamma, and X-rays) and chemical mutagens [ethane methane sulphonate (EMS), *N*-Nitroso-*N*-methyl urea (NMU), and colchicine] are known to induce variability in economic traits of crop plants. Mutation breeding was widely used for the development of plants with high yield, desired quality, and resistance to biotic/abiotic stresses (Tester and Langridge 2010; Veronese et al. 2001). In vitro mutagenesis technique was also used extensively to develop disease-resistant plants and the creation of genetic variants (Arène et al. 2006). So far, 3200 mutant varieties of different crops were released, of which 63 and 25 mutant varieties were oilseed crops and oilseed rape, respectively (http://www-infocris.iaea.org/MVD/).

Mutation breeding is long drawn and not a directed approach, but it is one of the available options for genetic improvement of *J. curcas* with modest levels of variability. The major breakthrough in castor (*Ricinus communis*—a closely related genus) improvement was through mutation breeding, and there are several reports on induced mutants of breeding value. Using fast neutrons for variety HC-6 in castor, a variety (Aruna) was developed which had reduced plant height, earliness, increased number of spikes per plant, increased oil content, and seed yield when compared to the parent genotype and, thus, converting a semi-wild perennial type to a cultivated annual variety with fourfold yield increase (Kulkarni and Ramanamurthy 1977).

In *J. curcas*, both physical and chemical mutagenesis experiments were conducted by treating dry seeds, vegetative cuttings, or shoot tips to create mutants with gross morphological changes and altered plant architecture. Mutation breeding studies carried out in Thailand by the Agricultural Development Research Center in Northeast Thailand (ADRC) by treating dried seeds with ^{60}CO gamma-ray doses of 0–20 KR resulted in the identification of dwarf, pigmented lines and early flowering mutants in the M_3 generation (Sakaguchi and Somabhi 1987). In the early flowering variants, flowering was observed within 130 days after sowing when compared to 190 days in the control plants. However, the potential productivity of these variants under intensive cultivation conditions was not proved (Sakaguchi and Somabhi

1987). In Indonesia, vegetative shoots were subjected to gamma irradiation for the induction of genetic variability and enhanced oil content (Ita and Ishak 2004). Aranez and Guia (1990) reported that Jatropha seeds treated with different concentrations of gamma radiation showed morphological changes in M_2 population, and plants with tricotyledonous seedlings, early flowering and early branching stems were isolated. Dhakshanamoorthy et al. (2010; 2011) also reported mutants that exhibited early flowering, higher fruit and seed yield with a lower dose of gamma radiation. Conversely, higher doses of gamma radiation lead to a reduction in agro-morphological parameters. Azhar et al. (2010) (inis.iaea.org) determined the LD50 for physical mutagenesis at 320 Gy. Seed mutagenesis resulted in six early flowering mutants, seven dwarf mutants, and 17 high branching plants in the M_1 generation itself.

Maghuly et al. (2013) reported that TILLING (targeting induced local lesions in genomes) can be used for the description of functional genes. Maghuly et al. (2017) described the principles and methods of physical and chemical mutagenesis for the generation of large numbers of induced mutants under both in vivo and in vitro conditions and evaluation of phenotypic and genotypic traits. For the development of mutagenized populations, seeds, in vivo stem cuttings, embryogenic callus, and in vitro grown shoot cultures were used as target tissues. For seeds, an EMS concentration of 1.6% for up to 3 h was found optimal while for in vitro cultures, a lower dose (0.8%) for 3 h is recommended. Similarly for physical mutagenesis, the irradiator doses chosen for mutagenic treatment of stem cuttings (15–35 Gy for 3–12 s) are far lower than that used for seeds (100–500 Gy for 40–212 s).

While the primary focus was to introduce novel genetic variation, recent studies used mutagenized populations as tools of reverse genetics (TILLING) to understand gene function and to unravel the biological function of the candidate gene(s) (Maghuly and Laimer 2013; 2017). The advent of next-generation sequencing (NGS) and genotype by sequencing (GBS) tools not only facilitated studies on gene function but also enabled precise identification of mutations in plant genomes of desired phenotypes. Maghuly et al. (2018) detected SNPs and Indels in 82 EMS-induced mutants along with 14 wild accessions of *J. curcas* using nGBS and ddGBS approaches where a single or two (double digestion) restriction enzymes were used for digestion. The frequency of transitions (G/C to A/T) was high (64%) as compared to the transversions (36%). EMS treatment of 0.8% for 3 h resulted in a higher number of heterozygous SNPs. Metabolic pathway reconstruction showed a total of 16 SNPs in six KEGG pathways of which the ether–lipid metabolism, glycerophospholipid metabolism, starch, and sucrose metabolism were highly represented by nGBS and two pathways (triterpenoid and steroid) by ddGBS.

Despite the efforts at the induction of mutations and the development of tools and techniques for mutagenesis and mutant characterization, stable mutants with agronomically desirable characteristics are not available in *J. curcas* as in the case of castor. Mutation breeding and generating mutants for specific traits require the generation of extensive mutant collections which is a laborious and time-consuming process and is difficult for a cross-pollinated shrub-like *J. curcas*. Further in most of the studies, the population size maintained for the M_1, M_2, and M_3 generations, spacing followed for the crop in each of the generations, and the method adopted for generation advancement are not provided. It is necessary to expand the genome engineering tool kit like site-directed or site-specific nucleases (SDNs/SSNs) such as zinc-finger nucleases (ZFNs), transcription activator-like effector nucleases (TALENs), and CRISPR-Cas9 for the generation of mutations with precision.

18.6 Interspecific Hybridization

One of the possible approaches for genetic enhancement is to exploit the variation existing in the secondary and tertiary gene pools. The genus *Jatropha* is morphologically diverse and geographically widespread encompassing 175 species of herbaceous perennials, shrubs, woody trees, rhizomatous sub-shrubs, succulents, facultative annuals, and geophytes each having a narrow geographical range in the seasonally dry tropics (Dehgan 1984). Jatropha species are naturalized throughout the South American and Meso-American regions in Mexico, Cuba, Peru, Bolivia, Costa Rica, Paraguay, Jamaica, Brazil, El Salvador, Guatemala, Argentina, Dominican Republic, Columbia, Nicaragua, and parts of North America in the states of Arizona and Texas (Heller 1996). *J. villosa* and related species have their origin in India (Ramamurthy 1967). *J. mahafalensis* is a species native of Nicaragua.

Several *Jatropha* species are cultivated for their ornamental leaves and flowers, while some are grown in the tropics for their economic uses. The evolutionary trends in the genus is toward xerophytic adaptation with change in growth habit from arborescent perennial growth habit to facultative annual growth habit; series of reduction in reproductive structures (stamens, style branches, number of flowers, number of locules, seeds), reduced number of vascular bundles, gradual shift from hermaphrodite to monoecy to gynodioecy and dioecy, increase in chromosome numbers from diploidy to tetraploidy (section *Mozinna*; $2n = 44$). These changes exhibit a morphological continuum from south to north with the southern taxa possessing the more primitive and the northern species the most advanced features and increasing aridity (Dehgan and Schutzman 1994). *J. curcas* and its allied taxa grow in tropical-mesic forest regions, whereas the taxa with reduced vascular bundles are most advanced and occur in dry, warm deserts (Dehgan and Webster 1979).

Interspecific hybridization played a vital role in the genetic improvement of several crop plants. The genus *Jatropha* could also be benefited through introgressive breeding and hence, there is a need for collection, assembly, conservation, characterization, evaluation, and utilization of *Jatropha* species in broadening the genetic base of *J. curcas*. Analysis of seed oil fatty acids showed the predominance of linoleic acid with a higher linoleic to oleic acid ratio in all *Jatropha* species except *J. curcas*, which is rich in oleic acid (Banerji et al. 1985; Rao and Lakshminarayana 1987; Sujatha 1996). Cetane number is one of the most important factors for biodiesel which should be 47 as per ASTM D6751 and 51 as per EN 14214. Variation in fatty acid profile significantly influences the cetane number (King et al. 2009), and interspecific derivatives with variable cetane value could be developed. *Jatrophas* are rich sources of hydrocarbons and *J. multifida* with big round seeds possesses higher oil content (50%) as compared to *J. curcas* (23–38%) (Banerji et al. 1985; Sujatha 1996). Barriers between these two species are weak (Basha and Sujatha 2009), and the cross combination can aid in the enhancement of oil content. Seeds of *Jatropha* species (e.g., *J. podagrica*, *J. integerrima*, and *J. gossypifolia*) have thin hull when compared to *J. curcas*. Thin hull types are desirable for the efficient recovery of oil. Nontoxic (edible) genotypes are preferred to enhance the value of the crop for food and feed use, and *J. platyphylla* has been identified as a new nontoxic *Jatropha* species (Makkar et al. 2011).

Determination of the energy values of the oils indicated much higher energy content for *J. gossypifolia* (42.2 MJ/kg), *J. glandulifera* (47.2 MJ/kg), and *J. multifida* (57.1 MJ/kg) than for *J. curcas* (39.8–41.8 MJ/kg) (Banerji et al. 1985; Jones and Miller 1991). *J. mahafalensis* is predicted to have equal energetic promise. *J. gossypifolia*, a facultative annual, has heavy fruit-bearing ability and is adapted to saline regions in Northeast Thailand and India. *J. gossypifolia* is reported to have 18.5% ricinoleic acid in its seed oil (Hosamani and Kotagi 2008), and physico-chemical properties of biodiesel derived from this species are in the acceptable range for use in diesel engines (de Oliviera et al. 2009). *J. nana* and *J. villosa* are found in dry stony places; *J. nana* and *J. heterophylla* are dwarfs of African type. The crop should be of manageable height for mechanization. The availability of species with such diverse plant types and wide adaptability offers immense scope in improving the genetic architecture and agronomic attributes of *J. curcas*. Phylogenetic advancement of the genus *Jatropha* has evolved with adaptations to arid conditions, and Jatropha species adapted toward the Northern hemisphere could be a valuable source for the development of drought-resistant cultivars. Thus, there lies the vast scope for transfer of beneficial traits from wild *Jatropha* species to *J. curcas* such as heavy bearing, photoperiod insensitivity, improved fuel characteristics, high oil content, desired oil quality, plant architecture, earliness, reduced toxicity of endosperm proteins, and wider adaptability.

Interspecific gene transfer is generally limited by crossability barriers, ploidy differences and genetic distance of the taxa. Mc Vaugh (1945), Wilbur (1954) and Dehgan and Webster (1979) regarded *J. curcas* as the most primitive member of the genus because of its ability to interbreed with species from the subgenera, its palmately lobed leaves, arborescent growth habit, and occasional hermaphrodite flowers. Neither geographical isolation nor extensive morphological diversification particularly with respect to growth habit has produced strong barriers to interspecific compatibility; and inter and intra-sectional hybrids were produced with *J. curcas* (Dehgan 1984; Sujatha 1996; Basha and Sujatha 2009). Dehgan (1984) attempted interspecific hybridization of 20 species which showed unilateral compatibility with preferential fertilization, and viable hybrids were obtained in crosses involving *J. curcas* as the ovule parent. Analysis of interspecific crosses that failed to set seed indicated the existence of postfertilization barriers (Dehgan 1984; Reddy et al. 1987; Sujatha 1996). Although several economically important *Jatropha* species are available, the interspecific cross studied extensively was that involving *J. integerrima* (Sujatha and Prabakaran 2003; Basha and Sujatha 2009; Dhillon et al. 2009; One et al. 2014a; Laosatit et al. 2014; Muakrong et al. 2014; Fukuhara et al. 2016; Nagesh et al. 2019). One et al. (2014b) reported high genetic diversity in the F_2 population derived from *J. curcas* and *J. integerrima* hybrid. The same cross combination was used in the development of linkage maps and in mapping studies (Hong 2008; Wang et al. 2011; Wu et al. 2015). The F_2 mapping population of *J. curcas* (susceptible parent) and *J. integerrima* (resistant parent) was used for the identification of QTLs for Jatropha mosaic virus (JMV) resistance (Nagesh et al. 2019). Morphologically, *J. gossypifolia* and *J. glandulifera* have close resemblances with *J. curcas*, and molecular studies showed that *J. stevensii* (Avendano et al. 2015) and *J. cineria* (Nagesh et al. 2019) are the species closest to *J. curcas*. The monoecius *J. curcas* crosses readily with the dioecious species, *J. cinerea* (Dehgan 1984). The intergeneric hybrid of *J. curcas* with *Ricinus communis* ($2n = 2x = 20$) is reported (Laosatit et al. 2017), but such hybrids are of limited value for exploitation in the breeding programs owing to variations in basic chromosome numbers which often lead to severe meiotic irregularities and hybrid failure. Keeping in view the potential genetic wealth in different jatropha species and their easy crossability with *J. curcas*, there is a need to intensify the interspecific hybridization program with several other economically important species.

18.7 Genetic Engineering

J. curcas is reported to display narrow genetic diversity for some of the agronomically desirable traits and seed characteristics warranting optimization of genetic transformation techniques and the development of transgenics. *J. curcas* and related species are reported to respond to in vitro manipulations and shoot regeneration through direct and callus-mediated regeneration from various juvenile and mature plant tissues (reviewed in Sujatha et al. 2013b). In most of the shoot regeneration experiments, caulogenesis was achieved by culturing explants on medium supplemented with the auxin indole-3-butyric acid (IBA) with the cytokinins, benzyl adenine (BA), or thidiazuron (TDZ). Based on these media combinations, protocols for the genetic transformation of *J. curcas* were optimized (reviewed in Kumar et al. 2013). Genetic transformation protocols through both *Agrobacterium tumefaciens*-mediated and particle gun bombardment methods were established at varying frequencies (4.3–62.7%) in different laboratories (Reviewed in Warra et al. 2019). Although transformation through direct and vector-mediated methods is reported, the development of transgenic events through *Agrobacterium tumefaciens*-mediated was the most preferred method due to technical ease of the method. Despite the amenability of various tissues to regenerate in vitro, the target tissues for genetic modification were mostly the seedling tissues such as cotyledons, hypocotyls, primary leaves, roots, and embryos.

The traits that could be manipulated through genetic engineering techniques include seed quality traits, accumulation of high seed oil content, reduced seed toxicity, enhanced resistance to abiotic and biotic stresses, and modification of plant architecture for varied purposes. The trait modification addressed through genetic engineering is mostly the single-gene-controlled traits.

Seed traits: As *J. curcas* is valued as a potential biodiesel crop, the fatty acid composition of the seed oil assumes importance. Oils rich in monounsaturated fatty acids and low in polyunsaturated fatty acids are preferred for biodiesel purpose as it positively impacts the ignition quality, heat of combustion, and oxidative stability with negligible amounts of nitrogen oxides (NO_x) emissions. The seeds of *J. curcas* contain high levels of polyunsaturated fatty acids, which negatively impact the biofuel quality. Hence, to enhance the levels of monounsaturated fatty acids in the seed oil, RNA interference (RNAi) technology was deployed to downregulate the expression of the *JC-FAD2-1* (delta 12 fatty acid desaturase) gene in a seed-specific manner which dramatically increased the oleic acid content to up to 78% with concomitant reduction in polyunsaturated fatty acids (<3%) (Qu et al. 2012).

The oil content of *J. curcas* varies from 32 to 38%, and oil productivity per unit area could be increased by enhancing the seed oil content. Attempts were made to increase either the seed size or seed oil content. Enoki et al. (2017) developed transgenic jatropha by introducing four genes (LOC_Os08g41910 encoding Sua5/YciO/YrdC/YwlC family protein, LOC_Os04g43210 encoding probable inositol transporter 2-like, LOC_Os03g49180 encoding alkaline ceramidase, and LOC_Os10g40934 encoding putative flavonol synthase/flavanone 3-hydroxylase or 2OG-Fe(II) oxygenase containing protein) from rice under CaMV35S promoter to make larger Jatropha seeds. Kim et al. (2014) used RNAi technology with a native JcSDP1 promoter to silence endogenous JcSDP1 expression to generate SDP1-deficient (sugar-dependent 1 gene encoding a patatin-domain triacylglycerol lipase) transgenic Jatropha plants. Seeds of JcSDP1 silenced jatropha plants accumulated up to 30% higher total lipid and had reduced FFA content as compared to the control (CK; 35S:GFP) plants. The *J. curcas* transcription factor (JCMYB1) is reported to bind to the diacylglycerol acyltransferase 1 (DGAT 1) promoter which is a rate-limiting enzyme of the triacylglycerol biosynthesis and activates its expression. Khan et al. (2019) demonstrated that the expression of the R2R3MYB (JCMYB1) enhances seed oil content besides altering the fatty acid composition in Arabidopsis and tobacco. This proof of concept was successfully validated in *J. curcas* by using virus-induced gene silencing (VIGS) technique.

Seeds of *J. curcas* have toxic compounds and antinutritional factors such as phorbol esters (cancer potentiating diterpenes), curcin (a type I ribosome-inactivating protein), phytates, and trypsin inhibitors. Elimination of the major toxic substances in the seed meal is essential to render the seed meal after oil extraction suitable in animal rations. Concerted efforts resulted in the identification of naturally occurring nontoxic accessions devoid of phorbol esters (PE) in the Mexican germplasm which have been adequately characterized for their constituents, feeding use, and diversity (Makkar et al. 1998a, b; Martinez-Herrera et al. 2006; Vandepitte et al. 2019). Attempts were made to develop transgenic jatropha with reduced levels of PE using RNAi technology for downregulating casbene synthase activity (Li et al. 2016). The other major toxic constituent is curcin for which variability in the available germplasm has not been reported. For development of accessions with reduced curcin content, RNAi technology was used for introduction of the curcin precursor gene which reduced curcin transcripts by 98% to undetectable levels (Patade et al. 2014) and the curcin 1 (C1) gene which silenced C1 transcripts in the endosperm (Gu et al. 2015).

Abiotic stresses: *J. curcas* is reported to be a hardy crop but cultivation in different situations showed that the crop is unproductive with poor survivability under conditions of drought, salinity and frost necessitating genetic engineering of the crop for abiotic stresses. This assumes priority as the crop is targeted for production in semi-arid lands unsuitable for food production and also in problematic sites. Jha et al. (2013) developed transgenic lines with improved salt tolerance through the introduction of the *SbNHX1* gene isolated from *Salicornia brachiata.* Such transgenic lines hold promise for the promotion of the crop in saline soils and cultivation with brackish water.

Toward the improvement of *J. curcas* for drought tolerance, Trivedi et al. (2009) transformed cotyledons with *DREB2A* gene and likewise, Zhang et al. (2007; 2008) introduced genes like aquaporin (*JcPIP2*) and betaine aldehyde dehydrogenase (*JcBD1*) for growth suitability of the crop under drought and saline conditions, respectively. Tsuchimoto et al. (2012) developed transgenic lines with three different types of candidate genes which included (1) the phosphopantetheine adenyltransferase (*AtPPAT*) gene encoding an enzyme that catalyzes the penultimate step in the CoA-biosynthetic pathway; (2) the nuclear factor (*NF-YB*) gene that encodes a subunit of NF-Y transcription factor; and (3) overexpressing the *Synechococcus* GSMT (glycine sarcosine methyltransferase) and DMT (sarcosine dimethylglycine methyltransferase) genes which encode enzymes that catalyze glycine betaine production. The lines overexpressing AtPPAT exhibited approximately 1.6-fold higher CoA+ acetyl-CoA levels, enhanced vegetative and reproductive growth, tolerance to salt/osmotic stress in addition to the dry seeds containing 35–50% more fatty acids than the wild type. Putative transgenics harboring the NF-YB gene were not subjected to phenotypic characterization. The glycine betaine content of two plantlets expressing GSMT and DMT (1 and 2) genes was higher (about 0.5 and 0.9 nmol·mg FW^{-1}) as compared to the non-transgenic plantlets (0.2 nmol·mg FW^{-1}).

Biotic stresses: There are no instances of major diseases and pests on wild populations of *J. curcas*. However, under monoculture plantations, *J. curcas* is reported to be vulnerable to the attack of various fungi, viruses, insects, and other pests emphasizing the need for systematic research on biotic stress resistance and development of alternate control measures over chemical control (Anitha and Varaprasad 2012). Efforts at the development of transgenics for conferring resistance to biotic stresses are sporadic as the problem unlike abiotic stresses is location-specific and confined to the respective plantation. Gu et al. (2014) introduced the *Cry1Ab/1Ac gene to confer protection against larvae of Archips micaceanus (tortrix moth) and the selected transgenic lines resulted in feeding cessation and 80–100% larval mortality.* Franco et al. *(*2016*) genetically engineered leaf explants of J. curcas with* chitinase gene (*ech42*) from the *Trichoderma viride* fungus for offering protection against fungal diseases, but the resultant transgenic lines were not subjected to fungal assays.

Some of the genes isolated from *J. curcas* and characterized for their functionality in model crops like Arabidopsis and tobacco serve as valuable candidates for genetic improvement of the crop through transgenic approaches. Agarwal and Agarwal (2016) reported that overexpression of the *J. curcas* pathogenesis-related gene (*JC-PR 10a*) exhibiting both RNase and DNase activity improved shoot regeneration, salinity tolerance, and reduced susceptibility to the fungus *Macrophomina* in transgenic tobacco. Overexpression of salicylic acid-inducible JcWRKY TF in tobacco improved salinity tolerance potential of the transgenics by maintaining reactive oxygen species (ROS) homeostasis and high K^+/Na^+ ratio (Agarwal et al. 2016). Dwarf phenotypes coupled with high productivity are required not only for adapting the crop to mechanization and reducing production costs but also for expanding the suitability of the crop to fit in various cropping systems. Shi et al. (2018) extopically expressed *JcZFP8*, a C2H2 zinc-finger protein gene from *J. curcas in tobacco which resulted in dwarf phenotype paving the way to expedite the* breeding progress in *J. curcas* through engineering of the gibberellic acid (GA) metabolic pathway. Remediation of toxic metals and metalloids through phytoremediators is cost-effective, sustainable, and environmentally friendly as compared to conventional remediation technologies. *J. curcas* is considered as one of the most efficient phytoremediators, and genetic engineering with genes encoding metallothioneins, phytochelatins, and glutathione is envisaged to further improve the efficiency of the phytoremediators for revegetating the toxic metal- and metalloids-contaminated sites (Warra et al. 2019).

18.8 Gene Editing

Maghuly et al. (2019) established molecular tools to modify the biosynthetic pathways related to fatty acid, protein, and toxin biosynthesis, by CRISPR/Cas9 knockout constructs. The authors analyzed the gene structures, identified the number of isoforms, and designed the guide RNAs for potential target sites. Transformation of leaf discs via *A. tumefaciens* with the Cas9 gene, the gRNA expression cassettes, and the *nptII* selectable marker to *J. curcas* resulted in INDEL mutations due to frameshift at the expected positions. The successful gene knockouts were validated by phenomic and genomic analyses. The seed yield in *J. curcas* is low, potentially because of the relatively low number of total flowers and/or the ratio of female to male flowers. It is well established that exogenous cytokinin treatments like benzyl adenine (BA), paclobutrazol, and thidiazuron (TDZ) increase the seed yield of *J. curcas* (Fröschle et al. 2017; Pan and Xu 2011; Chen et al. 2014; Pan et al. 2014; Xu et al. 2016). Based on these findings, Cai et al. (2018) cloned six isopentenyl transferase (IPT) genes, one cytochrome P450 monooxygenase, family 735, subfamily A (*JcCYP735A*) gene, and seven cytokinin oxidase/dehydrogenase (*JcCKX*) genes which showed various expression patterns in different organs of *Jatropha*, while few exhibited tissue-specific expression. The authors analyzed the function of *JcCYP735A* using the CRISPR-Cas9 system and found that the concentrations of *trans*-zeatin (tZ), tZ-riboside *cis*-Zeatin, and cZ-riboside decreased significantly in the *Jccyp735a* mutants and was associated with severely retarded growth. These results are helpful for further studies of the functions of cytokinin metabolic genes and understanding the roles of cytokinins in *Jatropha* growth and development.

18.9 Functional Genomics, Genomic Resources, and Marker-Assisted Selection

The genome of *J. curcas* is relatively small and is estimated to be between 210 and 220 Mb (Hong 2008), 300 Mbp (Graham 2006), 416 Mb (Carvalho et al. 2008), 980 Mb (Kohli et al. 2008) which is equivalent to castor (400 Mb) and rice (430 Mb) but smaller than that of other species of *Euphorbiaceae* (1.3–28.6 pg). The 2C value is 0.85 pg with an average base composition of 38.7% GC. The Jatropha genome being small in size is amenable for genetic improvement through the use of marker-assisted selection (MAS), genome-wide association studies (GWAS), and genomic selection (GS) methods. Additionally, the genomic sequence analysis expressed sequence tag analysis, and transcriptome studies are being done (Hirakawa et al. 2012; Sato et al. 2011; Chen et al. 2011; Eswaran et al. 2012; Natarajan et al. 2010; Natarajan and Parani 2011; Costa et al. 2010; Zhang et al. 2014; Vandepitte et al. 2019). The development of genomic resources, the establishment of high-throughput genotyping platforms, and functional genomics are important for the molecular breeding and functional gene analysis of Jatropha. As on date information on 25,000 genes, 115 gene expression omnibus (GEO) datasets, 102 pop sets, 3 genome assemblies, 57 bio projects, 521 sequence read archives (SRA) which provide sequences of most of the important genes governing fatty acid biosynthesis, seed development, abiotic stress tolerance (drought, waterlogging), and floral sex differentiation under natural and cytokinin-induced conditions is documented (https://www.ncbi.nlm.nih.gov/search/all/?term=jatropha). These genomic resources facilitate the mining of SSRs and SNPs and identification of candidate gene(s) for genetic improvement through genetic engineering tools.

The sequencing data generated by Nagesh et al. (2019) was used to identify genes related to Jatropha geminivirus resistance and characterization of genes specific to drought conditions. Further, differentially expressed genes specific to biotic stress-related pathways such as terpenoid backbone biosynthesis, MAPK signaling pathway, oxidative phosphorylation, carbon fixation in photosynthetic organisms, and putative genes such as LRR receptor-like serine/threonine-protein kinase At3g47570, ABC transporter G family member, and GDSL esterase/lipase 4-like, which are mainly involved in combating against biotic and abiotic stresses in Jatropha, were identified. One of the approaches to increase seed yield is to increase the female to male flower ratio. To enhance female flower production, spraying of plant growth regulators such as gibberellins, cytokinin, and thidiazuron on immature inflorescences prior to sex differentiation is done, but it is labor-intensive and uneconomical with flowering extended for over a period of 3 months. The advancement in genomics and transcriptomics enables identification of candidate genes involved in plant growth regulator biosynthesis pathways and manipulation of sex differentiation toward femaleness. Seesangboon et al. (2018) studied the expression analysis of 23 putative genes following the application of 6-benzyl adenine (BA) on *Jatropha* flower buds which is known to significantly increase female flower numbers and seed yield. The studies showed that *CYTOKININ OXIDASE/DEHYDROGENASE5* (*JcCKX5*) was highly expressed at the transition stage and subsequently enhanced the inflorescence area, increased organogenic capacity, and ovule primordia formation. Application of BA increased the expression of *SUPERMAN* (*JcSUP*) implying its role in female flower formation and arrested stamen formation through the downregulation of *TASSELSEED2* (*JcTS2*) during sex organs differentiated stage.

Jatropha is characterized by a long gestation period and large plant size and hence, elimination of undesirable progenies in breeding populations through marker-assisted selection reduces cost and allows breeders to select population comprised of individuals carrying desirable genes of interest. Unlike in the past, the advent of the molecular marker era has accelerated crop breeding programs. The prerequisites for mapping the genes include identification of germplasm with beneficial characteristics, development of genomic resources, high-throughput genotyping techniques, appropriate mapping populations and tools for mapping the genes. The major concern is the narrow genetic diversity for key traits besides the low productivity of the germplasm being characterized. The narrow genetic base limited the use of germplasm in association mapping studies and hence most of the studies on linkage mapping were confined to traits with Mendelian inheritance. During the past decade, efforts at the development of genomic resources to facilitate genetic improvement through the marker-assisted selection and transgenic breeding resulted in the development of genomic resources, linkage maps, understanding gene function, identification of candidate genes, and loci governing some of the key traits in *J. curcas*. At Temasek Life Sciences Laboratory, a linkage map with 219 microsatellites, 200 SNPs, and 160 AFLP markers has been constructed using backcross populations (Yue 2008). At CNAP, >400 SNPs were detected that could be sufficient for a dense map and in marker-assisted breeding (Graham 2006). A first-generation linkage map (1441 cM) was developed by using two backcross populations with 93 progeny derived from *J. curcas* and *J. integerrima* cross based on 216 microsatellite and 290 SNPs with a marker space of 2.8 cM (Wang et al. 2011). The backcross population from the same cross combination was used to identify 18 QTLs underlying the oil traits from 286 individuals (Liu et al. 2011). Sun et al. (2012) reported identification of 28 QTLs of which 11 were for growth and seed traits, 2 QTLs which control seed yield were conferred by the alleles from *J. curcas*, 5 QTLs which control plant height, branch number, female flower number, and fruit number were conferred by the alleles from *J. integerrima*. Wu et al. (2015) also established a linkage map using 1208 SNP, InDel, and SSR markers on a BC_1 population of 190 individuals from the cross *J. curcas* × *J. integerrima*. Microsatellite and SCAR markers linked to low phorbol ester levels have been identified which could be used to fast-track the breeding programs designed to develop low-toxic genotypes (Padmidiamarri et al. 2009; Basha et al. 2009). Subsequently, linkage maps based on crosses involving toxic and nontoxic genotypes resulted in the identification of the locus governing phorbol ester biosynthesis on linkage group 8 (King et al. 2013; Amkul et al. 2017). Popluechai et al. (2011) reported mapping of SNPs for three oleosin genes. Liu et al. (2011) identified the expression of 3 QTLs, which control oleic acid, oil content, and oleic gene expression. King et al. (2015) identified QTLs markers contributing to plant growth, oil yield, and fatty acid composition in two biparental mapping populations.

Xia et al. (2018) reported the identification of 13 fruit yield QTLs and 2 candidate genes based on an ultrahigh density Jatropha linkage map. Intraspecific populations are more useful than interspecific segregants for identification and exploitation of molecular markers in breeding programs as it overcomes the linkage drag of several undesirable genes in the latter. Ha et al. (2019) reported ~339 Mbp whole genome sequence. Comprehensive transcriptome analysis of *J. curcas* along with nine jatropha species led to the discovery of genes related to the biosynthesis of lipids and the toxic compounds in seeds. Marker-assisted selection can be useful for traits that are difficult to measure, exhibit low heritability, and/or are expressed late in development, but the key for success in MAS lies in the selection of diverse parental lines for generating segregating populations reiterating the need for collection and characterization of germplasm from the center of origin.

18.10 Conclusion

J. curcas has assumed importance as a potential bioenergy crop in several parts of the world and use of the seed oil as biodiesel has been successfully demonstrated. The major challenges lie in the availability of good planting stock and the agro-production techniques and systems with which productive yields could be realized. The first stage of Jatropha cultivation was unsuccessful largely due to the nonavailability of extensive germplasm exhibiting diversity for economically and agronomically desirable attributes, lack of inbred lines with economic breeding values, and sole dependence on populations growing wild. During the past decade, efforts at genetic improvement of *J. curcas* were made by few research groups toward broadening the genetic base, development of elite cultivars, and establishment of genomic resources and tools. However, there are still a large number of challenges to be overcome. "Rapid domestication" of *J. curcas* demands for strong, vibrant, and joint collaborative programs at national and international levels for addressing issues of common interest like collection and exchange of germplasm, development of germplasm resources base, trait discovery, development of appropriate mapping populations and molecular maps, development of genomic resources, and integration of biotechnology into breeding toward selective breeding.

References

Agarwal P, Agarwal PK (2016) *Jatropha curcas* pathogenesis related-10a protein: a jack of many trades via cytokinin signaling. Clon Transgen 5:2

Agarwal P, Dabi M, Sapara KK, Joshi PS, Agarwal PK (2016) Ectopic expression of JcWRKY transcription factor confers salinity tolerance via salicylic acid signaling. Front Plant Sci 7:1541

Akintayo ET (2004) Characteristics and composition of *Parkia biglobbossa* and *Jatropha curcas* oils and cakes. Bioresour Technol 92:307–310

Alfredo ZC, Quintero VP (2017) New clonal varieties of Jatropha. In: Tsuchimoto S (ed) The Jatropha genome, Compendium of plant genomes. Springer, Cham, pp 275–288

Alves AA, Bhering LL, Rosado TB, Laviola BG, Formighieri EF, Cruz CD (2013) Joint analysis of phenotypic and molecular diversity provides new insights on the genetic variability of the Brazilian physic nut germplasm bank. Genet Mol Biol 36:371–381

Amkul K, Laosatit K, Somta P, Shim S, Lee SH, Tanya P, Srinives P (2017) Mapping of QTLs for seed phorbol esters, a toxic chemical in *Jatropha curcas* (L.). Genes 8:E205. https://doi.org/10.3390/genes8080205

Anggraeni TDA, Satyawan D, Kang YJ, Ha J, Kim MY, Chitikineni A, Varshney RK, Lee SH (2018) Genetic diversity of *Jatropha curcas* collections from different islands in Indonesia. Plant Genet Resour 16:334–342

Anitha K, Varaprasad KS (2012) Jatropha pests and diseases: an overview. In: Carels N, Sujatha M, Bahadur B (eds) Jatropha, challenges for a new energy crop. Springer, New York, pp 175–218

Anonymous (2008) Third report on tree borne oilseeds. NOVOD Report, Gurgaon, India

Aranez AT, Guia EE (1990) Effects of gamma radiation on development and structural features of *Jatropha curcas* L. *Euphorbiaceae*. Asia J Plant Sci 2:1–12

Arène L, Bellenot-Kapusta V, Belin J, Cadic A, Clérac M, Decourtye L (2006) Breeding program on woody ornamental plants in Angers, France. The INRA open archive. https://prodinra.inra.fr/?locale=es#!ConsultNotice:22630

Avendano R, Garcia E, Valdez M, Chaves N (2015) Genetic diversity analysis of Jatropha species from Costa Rica using AFLP markers. Am J Plant Sci 6:2426–2438

Azhar M, Mohd Zaim MN, Farim H, Md Razali NM, Sobri H, Abd Rahim H (2010) Mutagenesis of *Jatropha curcas*—exploring new traits in the breeding of a biofuel plant. Inis.iaea.org

Banerji R, Chowdhury AR, Misra G, Sudarsanam G, Verma SC, Srivastava GS (1985) *Jatropha* seed oils for energy. Biomass 8:277–282

Basha SD, Sujatha M (2007) Inter and intra-population variability of *Jatropha curcas* (L.) characterized by RAPD and ISSR markers and development of population specific SCAR markers. Euphytica 156:375–386

Basha SD, Sujatha M (2009) Genetic analysis of *Jatropha* species and interspecific hybrids of *Jatropha curcas* using nuclear and organelle specific markers. Euphytica 168:197–214

Basha SD, Francis G, Makkar HPS, Becker K, Sujatha M (2009) A comparative study of biochemical traits and molecular markers for assessment of genetic relationships between *Jatropha curcas* L. germplasm from different countries. Plant Sci 176:812–823

Becker K, Makkar HPS (1998) Toxic effects of phorbol esters in carp (*Cyprinus carpio* L.). Vet Hum Toxicol 40:82–86

Cai YL, Rakshit KD, Jian XL, Jian ML, Makkar HPS, Becker K (2010) Toxicity of *Jatropha curcas* phorbol esters in mice. Food Chem Toxicol 48:620–625

Cai L, Zhang L, Fu Q, Xu Z (2018) Identification and expression analysis of cytokinin metabolic genes *IPTs*, *CYP735A* and *CKXs* in the biofuel plant *Jatropha curcas*. Peer J 6:e4812

Carvalho CR, Clarindo WR, Praca MM, Araujo FS, Carels N (2008) Genome size, base composition and karyotype of *Jatropha curcas* L., an important biofuel plant. Plant Sci 174:613–617

Chen F (2007) Advances in *Jatropha* industry research and development. International workshop on the development of the JCL industry. 29–31 October 2007, Hainan, China

Chen F, Xu T, Tang L, Whenghua W, Fang Y, Jun W, Yiran G (2008) *Jatropha* biodiesel research and development in China. Jatropha international congress 17–18 December 2008, Singapore

Chen K, Peng R, Chunying Y, Qi J, Xiuqing J (2011) Genetic relationships among *Jatropha curcas* L. clones from Panzhihua as revealed by RAPD and ISSR. Afr J Agric Res 6:2582–2585

Chen MS, Pan BZ, Wang GJ, Ni J, Niu L, Xu ZF (2014) Analysis of the transcriptional responses in inflorescence buds of *Jatropha curcas* exposed to cytokinin treatment. BMC Plant Biol 14:318

Costa GG, Cardoso KC, Del Bem LE, Lima AC, Cunha MA, de Campos-Leite L, Vicentini R, Papes F, Moreira RC, Yunes JA, Campos FAP, Da Silva MJ (2010) Transcriptome analysis of the oil-rich seed of the bioenergy crop *Jatropha curcas* L. BMC Genomics 11:462

Daniel JN, Hegde NG (2007) Tree borne oilseeds in agroforestry. In: Hegde DM (ed) Changing global vegetable oils scenario: issues and challenges before India. Indian Society of Oilseeds Research, Hyderabad, pp 263–276

DBTIndia (2007). http://dbtindia.nic.in/Energy_bioscience/presentations/TS_4_Dr%20Nutan%20Kaushik_TERI_N%20Delhi.pdf.

De Oliviera JS, Leite PM, De Souza LB, Mello VM, Silva EC, Rubim JC, Meneghetti SMP, Suarez PAZ (2009) Characteristics and composition of *Jatropha gossypifolia* and *Jatropha curcas* oils and application for biodiesel production. Biomass Bioenergy 33:449–453

Dehgan B (1984) Phylogenetic significance of interspecific hybridization in *Jatropha* (Euphorbiaceae). Syst Bot 9:467–478

Dehgan B, Schutzman B (1994) Contributions toward a monograph of neotropical *Jatropha:* phenetic and phylogenetic analysis. Ann Missouri Bot Gard 81:349–367

Dehgan B, Webster GL (1979) Morphology and infrageneric relationships of the genus Jatropha (Euphorbiaceae). University of California Publications in Botany, Berkeley vol 74, pp 1–73

Dhakshanamoorthy D, Selvaraj R, Chidambaram A (2010) Physical and chemical mutagenesis in *Jatropha curcas L.* to induce variability in seed germination, growth and yield traits. J Biol Plant Biol 55:113–125

Dhakshanamoorthy D, Selvaraj R, Chidambaram ALA (2011) Induced mutagenesis in *Jatropha curcas* L. using gamma rays and detection of DNA polymorphism through RAPD marker. C R Biol 334:24–30

Dhillon RS, Hooda MS, Jattan M, Chawla V, Bhardwaj M, Goyal SC (2009) Development and molecular characterization of interspecific hybrids of *Jatropha curcas* × *J. integerrima*. Indian J Biotechnol 8:384–390

Enoki H, Funato A, Nabetani Y, Takahashi S, Ichikawa T, Masui M, Motohashi R (2017) *Agrobacterium*-mediated genetic transformation for larger seed size in Jatropha. In: Suguru T (ed) The Jatropha genome. Springer, Cham, pp 191–203

Eswaran N, Parameswaran S, Anantharaman B, Kumar GRK, Sathram B, Johnson TS (2012) Generation of an expressed sequence tag (EST) library from salt-stressed roots of *Jatropha curcas* for identification of abiotic stress-responsive genes. Plant Biol 14:428–437

Francis G, Edinger R, Becker K (2005) A concept for simultaneous wasteland reclamation, fuel production, and socio-economic development in degraded areas in India: need, potential and perspectives of *Jatropha* plantations. Nat Resour Forum 29:12–24

Francis G, Oliver J, Sujatha M (2018) High yielding and trait specific genotypes and genetic associations among yield and yield contributing traits in *Jatropha curcas* L. Agrofor Syst 92:1417–1436

Franco MC, Gomes KA, de Carvalho Filho MM, Harakava R, Carels N, Siqueira WJ, Latado RR, Marques DA (2016) *Agrobacterium*-mediated transformation of *Jatropha curcas* leaf explants with a fungal chitinase gene. Afr J Biotechnol 15:2006–2016

Fröschle M, Horn H, Spring O (2017) Effects of the cytokinins 6-benzyladenine and for chlorfenuron on fruit-, seed-and yield parameters according to developmental stages of flowers of the biofuel plant *Jatropha curcas* L. (Euphorbiaceae). Plant Growth Regul 81:293–303

Fukuhara S, Muakrong N, Kikuchi S, Tanya P, Sassa H, Koba T, Srinivas B (2016) Cytological characterization of an interspecific hybrid in Jatropha and its progeny reveals preferential uniparental chromosome transmission and interspecific translocation. Breed Sci 66:838–844

Ganesh Ram S, Parthiban KT, Kumar RS, Thiruvengadam V, Paramathma M (2008) Genetic diversity among *Jatropha* species as revealed by RAPD markers. Genet Resour Crop Evol 55:803–809

Ginwal HS, Phartyal SS, Rawat PS, Srivastava RL (2005) Seed source variation in morphology, germination and seedling growth of *Jatropha curcas* Linn., in Central India. Silvae Genetica 54:76–80

Gour VK (2006) Production practices, including post-harvest management of *Jatropha curcas*. In: Singh B, Swaminathan R, Ponraj V (eds) Biodiesel conference towards energy independence-focus on *Jatropha*. Rashtrapati Bhawan, New Delhi, pp 223–251

Graham I (2006) Towards the development of new *Jatropha* varieties: molecular and biochemical analysis of toxic and non-toxic lines. At the Queens Anniversary Prizes, The University of York

Grativol C, Lira-Medeiros CF, Hemerly AS, Ferreira PCG (2011) High efficiency and reliability of inter-simple sequence repeats (ISSR) markers for evaluation of genetic diversity in Brazilian cultivated *Jatropha curcas* L. accessions. Mol Biol Rep 38:4245–4256

Gu K, Mao H, Yin Z (2014) Production of marker-free transgenic *Jatropha curcas* expressing hybrid *Bacillus thuringiensis* δ-endotoxin Cry1Ab/1Ac for resistance to larvae of tortrix moth (*Archips micaceanus*). Biotechnol Biofuels 7:68

Gu K, Tian D, Mao H, Wu L, Yin Z (2015) Development of marker-free transgenic *Jatropha curcas* producing curcin-deficient seeds through endosperm-specific RNAi-mediated gene silencing. BMC Plant Biol 15:242

Gupta S, Srivastava M, Mishra GP, Naik PK, Chauhan RS, Tiwari SK, Kumar M, Singh R (2008) Analogy of ISSR and RAPD markers for comparative analysis of genetic diversity among different *Jatropha curcas* genotypes. Afr J Biotechnol 7:4230–4243

Ha J, Shim S, Lee T, Kang YJ, Hwang WJ, Jeong H, Laosatit K, Lee J, Kim SK, Satyawan D, Lestari P, Yoon MY, Kim MY, Chitikineni A, Tanya P, Somta P, Srinives P, Varshney RK, Lee SH (2019) Genome sequence of *Jatropha curcas* L., a non-edible biodiesel plant, provides a resource to improve seed-related traits. Plant Biotechnol J 17:517–530

Hasnam (2007) Improvement of *Jatropha curcas* L. in Indonesia: promise and performance. In: International workshop on the development of the JCL industry. 29–31 October 2007, China, p 22–27

Heller J (1996) Physic nut—*Jatropha curcas* L. promoting the conservation and use of underutilized and neglected crops. International Plant Genetic Resources Institute, Rome, p 66

Henning RK (2006) The Jatropha system, integrated rural development by utilization of *Jatropha curcas* L. (JCL) as raw material and as renewable energy. www.jatropha.org

Hirakawa H, Tsuchimoto S, Sakai H, Nakayama S, Fujishiro T, Kishida Y, Kohara M, Watanabe A, Yamada M, Aizu T, Toyoda A, Fujiyama A, Tabata S, Fukui K, Sato S (2012) Upgraded genomic information of *Jatropha curcas* L. Plant Biotechnol 29:123–130

Hong Y (2008) Genetic markers for Jatropha biodiversity evaluation and breeding. International Consultation on Pro-Poor Jatropha Development, Rome

Hosamani KM, Kotagi KS (2008) Characterization and structural elucidation of 12-hydroxyoctadec-cis-9-enoic acid in *J. gossypifolia* and *Hevea brasiliensis* seed oils: a rich source of hydroxyl fatty acid. Chem Phys Lipids 152:9–12

Ikbal K, Boora S, Dhillon RS (2010) Evaluation of genetic diversity in *Jatropha curcas* L. using RAPD markers. Indian J Biotechnol 9:5057

Islam AKMA, Anuar N, Yaakob Z, Osman M (2011) Heterosis for seed yield and its components in Jatropha (*Jatropha curcas* L.). Int J Plant Breed 5:74–79

Ita D, Ishak (2004) Mutation breeding and biotechnology on Jatropha (*Jatropha curcas* L.) for biodiesel future energy. In: Mutation Breeding in Forum Nuclear Cooperation in Asia. 30 August–3 September 2004, Jogyakarta, Indonesia

Jha B, Mishra A, Jha A, Joshi M (2013) Developing transgenic *Jatropha* using the *SbNHX1* gene from an extreme halophyte for cultivation in saline wasteland. PLoS One 8:e71136

Jones N, Miller JH (1991) *Jatropha curcas*—a multipurpose species for problematic sites. Land Resour Series 1:1–12

Jongschaap REE, Corre WJ, Bindraban PS, Branderburg WA (2007) Claims and facts on *Jatropha curcas* L. global *Jatropha curcas* evaluation, breeding and propagation programme. Plant Research International B.V. Wageningen, the Netherlands, Report 158

Kaushik N, Roy S, Biswas GC (2006) Screening of Indian germplasm of *Jatropha curcas* for selection of high oil yielding plants. Indian J Agroforestry 8:54–57

Kaushik N, Kumar K, Kumar S, Kaushik N, Roy S (2007) Genetic variability and divergence studies in seed traits and oil content of Jatropha (*Jatropha curcas* L.) accessions. Biomass Bioenergy 31:497–502

Khan K, Kumar V, Niranjan A, Shanware A, Sane VA (2019) JcMYB1, a Jatropha R2R3MYB transcription factor gene modulates lipid biosynthesis in transgenic plants. Plant Cell Physiol 60:462–475

Kim MJ, Yang SK, Mao HZ, Veena SP, Yin JL, Chua NH (2014) Gene silencing of sugar-dependent 1 (Jc*SDP*1), encoding a patatin-domain triacylglycerol lipase, enhances seed oil accumulation in *Jatropha curcas*. Biotechnol Biofuels 7:36

King AJ, He W, Cuevas JA, Freudenberger M, Ramiaramanana D, Graham IA (2009) Potential of *Jatropha curcas* as a source of renewable oil and animal feed. J Exp Bot 60:2897–2905

King AJ, Montes LR, Clarke JG, Affleck J, Li Y, Witsenboer H, van der Vossen E, van der Linde P, Tripathi Y, Tavares E, Shukla P, Rajasekaran T, van Loo EN, Graham IA (2013) Linkage mapping in the oilseed crop *Jatropha curcas* L. reveals a locus controlling the biosynthesis of phorbol esters which cause seed toxicity. Plant Biotechnol J 11:986–996

King AJ, Montes LR, Clarke JG, Itzep J, Perez CAA, Jongschaap REE, Visser RGF, van Loo EN, Graham IA (2015) Identification of QTL markers contributing to plant growth, oil yield and fatty acid composition in the oilseed crop *Jatropha curcas* L. Biotech Biofuels 8:160

Kohli A, Popluechai S, Raorane M, Syers KJ, O'Donnell AG (2008) Chapter 14. Jatropha as a novel non-edible oilseed plant for biodiesel. In: Ferry N, Gatehouse AMR (eds) Environmental impact of genetically modified novel crops. CAB International, London, pp 294–322

Kulkarni LG, Ramanamurthy GV (1977) Castor. ICAR, New Delhi

Kumar RV, Tripathi YK, Izhaki I, Yadav VP, Ahlawat SP (2008) Intraspecific variation and interrelationships between morphology, nutritional content and enzymatic activity of *Jatropha curcas* L. Curr Sci 95:239–243

Kumar N, Reddy MP, Sujatha M (2013) Genetic transformation of *Jatropha curcas*: current status and future prospects, Jatropha, challenges for a new energy crop. Science + Business Media, New York, pp 535–546

Laosatit K, Tanya P, Muakrong N, Srinives P (2014) Development of interspecific and intergeneric hybrids among jatropha-related species and verification of the hybrids using EST–SSR markers. Plant Genet Resour 12:58–61

Laosatit K, Mokrong N, Tanya P, Srinivas P (2017) Overcoming crossing barriers between Jatropha (*Jatropha curcas* L.) and castor bean (*Ricinus communis* L.). Crop Breed Appl Biotechnol 17:164–167

Laviola BG, Alves AA, Gurgel FL, Rosado TB, Rocha RB, Albrecht JC (2012) Estimates of genetic parameters for physic nut traits based in the germplasm two years evaluation. Ciência Rural 42:429–435

Li K, Yang WY, Li L, Zhang CH, Cui YZ, Sun YY (2007) Distribution and development strategy for *Jatropha curcas* L. in Yunnan province, Southwest China. Forestry Stud China 9:120–126

Li C, Ng A, Xie L, Mao H, Qiu C, Srinivasan R, Yin Z, Hong Y (2016) Engineering low phorbol ester *Jatropha curcas* seed by intercepting casbene biosynthesis. Plant Cell Rep 35:103–114

Liu P, Wang CM, Li L, Sun F, Yue GH (2011) Mapping QTLs for oil traits and eQTLs for oleosin genes in Jatropha. BMC Plant Biol 11:132

Luis RMO, Andres FTS, Raymond EEJ, Cesar AAP, Julio EBS, Luisa MT, Richard GFV, Eibertus NVL (2014) High level of molecular and phenotypic biodiversity in *Jatropha curcas* from Central America compared to Africa, Asia and South America. BMC Plant Biol 14:77

Maghuly F, Laimer M (2013) *Jatropha curcas,* a biofuel crop: functional genomics for understanding metabolic pathways and genetic improvement. J Biotechnol 8:1172–1182

Maghuly F, Laimer M (2017) Forward and reverse genetics for the improvement of *Jatropha curcas*. In: Tsuchimoto S (ed) The Jatropha genome, Compendium of plant genomes. Springer, Cham, pp 131–148
Maghuly F, Jankowicz J, Till B, Laimer M (2013) The use of EcoTILLING for the genetic improvement of *Jatropha curcas* L. In: Bahadur B, Sujatha M, Carels N (eds) Jatropha, challenges for a new energy crop, vol 2, pp 335–350
Maghuly F, Geslak JJ, Pabinger S, Till BJ, Laimer M (2015) Geographic origin is not supported by the genetic variability found in a large collection of *Jatropha curcas* with accessions from three continents. Biotechnol J 10:536–551
Maghuly F, Bado S, Jankowicz-Cieslak J, Laimer M (2017) Chemical and physical mutagenesis in *Jatropha curcas*. In: Jankowicz-Cieslak J, Tai T, Kumlehn J, Till B (eds) Biotechnologies for plant mutation breeding. Springer, Cham, pp 21–38
Maghuly F, Pabinger S, Krainer J, Laimer M (2018) The pattern and distribution of induced mutations in *J. curcas* using reduced representation sequencing. Front Plant Sci 9:524
Maghuly F, Laimer M, Freudhofmaier M, Parrott W (2019) Establishment of CRISPR/CAS9-mediated gene editing approaches in *Jatropha curcas*. In: FAO/IAEA international symposium on plant mutation breeding and biotechnology. https://conferences.iaea.org/indico/event/145/contributions/4989/contribution.pdf
Makkar HPS, Aderibigbe AO, Becker K (1998a) Comparative evaluation of non-toxic and toxic varieties of *Jatropha curcas* for chemical composition, digestibility, protein degradability and toxic factors. Food Chem 62:207–215
Makkar HPS, Becker K, Schmook B (1998b) Edible provenances of *Jatropha curcas* from Quintana Roo state of Mexico and effect of roasting on antinutrient and toxic factors in seeds. Plant Foods Hum Nutr 52:31–36
Makkar HPS, Kumar V, Oyeleye OO, Akinleye AO, Angulo-Escalante MA, Becker K (2011) *Jatropha platyphylla*, a new non-toxic Jatropha species: physical properties and chemical constituents including toxic and anti-nutritional factors of seeds. Food Chem 125:63–71
Martin M, Montes JM (2015) Quantitative genetic parameters of agronomic and quality traits in a global germplasm collection reveal excellent breeding perspectives for *Jatropha curcas* L. GCB Bioenergy 7:1335–1343
Martinez-Herrera J, Siddhuraju P, Francis G, Davile Ortiz G, Becker K (2006) Chemical composition, toxic/antimetabolite constituents, and effect of different treatments on their levels, in four provenances of *Jatropha curcas* L. from Mexico. Food Chem 96:80–89
Martinez-Herrera J, Evangelista-Lozano S, Martinez-Ayala AL (2007) Perfil nutricional de semillas de Jatropha curcas L. provenientes de Michoacan. CIBIA VI. VI Congreso Iberoamericano de Ingenieria de Alimentos, Ambato, Ecuador 16, p 313–314
Mc Vaugh R (1945) The genus *Jatropha* in America: principal intergeneric groups. Bull Torrey Bot Club 72:271–294
Montes Osorio LR, Van Loo EN, Jongschaap REE, Visser RGF, Azurdia C (2008a) A to Z of *Jatropha curcas* L. genetics, breeding and propagation techniques. Jatropha world 2008, Miami, USA
Montes Osorio LR, Van Loo EN, Jongschaap REE, Visser RGF, Azurdia C (2008b) Global *Jatropha curcas* genetic diversity study and its application in breeding program. Jatropha world 2008, Miami, USA
Montes Osorio LR, Torres Salvador AF, Jongschaap REE, Azurdia Perez CA, Berduo Sandoval JE, Trindade LM, Visser RGF, van Loo EN (2014) High level of molecular and phenotypic biodiversity in *Jatropha curcas* from Central America compared to Africa, Asia and South America. BMC Plant Biol 14:77
Montes JM, Technow F, Martin M, Becker K (2014) Genetic diversity in *Jatropha curcas L.* assessed with SSR and SNP markers. Diversity 6:551–566
Muakrong N, One KT, Tanya P, Srinivas P (2014) Interspecific Jatropha hybrid as a new promising source of woody biomass. Plant Genet Resour 12:517–520

Murty SG, Patel F, Punwar BS, Patel M, Singh AS, Fougat RS (2013) Comparison of RAPD, ISSR and DAMD markers for genetic diversity assessment between accessions of *Jatropha curcas* L. and its related species. J Agr Sci Technol 15:1007–1022

Nagesh K, Saakshi J, Narasimham JV, Vinod N, Vijay Y, Bijal T, Reddy VB, Boney K, Neeta M, Arockiasamy S (2019) *De novo* sequencing and hybrid assembly of the biofuel crop *jatropha curcas* L.: identification of quantitative trait loci for geminivirus resistance. Genes 10:69

Natarajan P, Parani M (2011) *De novo* assembly and transcriptome analysis of five major tissues of *Jatropha curcas* L. using GSFLX titanium platform of 454 pyrosequencing. BMC Genomics 12:191

Natarajan P, Kanagasabapathy D, Gunadayalan G, Panchalingam J, Shree N, Sugantham PA, Singh KK, Madasamy P (2010) Gene discovery from *Jatropha curcas* by sequencing of ESTs from normalized and full-length enriched cDNA library from developing seeds. BMC Genomics 11:606

One KT, Muakrong N, Tanya P, Velette J, Girard P, Srinives P (2014a) Physicochemical properties of seeds and oil from an F_2 population of *Jatropha curcas* × *Jatropha integerrima*. Sci Asia 40:428–435

One KT, Tanya P, Murakrong N, Laosatit K, Srinives P (2014b) Phenotypic and genotypic variability of F_2 plants derived from *Jatropha curcas* × *integerrima* hybrid. Biomass Bioenergy 67:137–144

Padmidiamarri DVNS, Pandya N, Reddy MP, Radhakrishnan T (2009) Comparative study of interspecific genetic divergence and phylogenic analysis of genus *Jatropha* by RAPD and AFLP: genetic divergence and phylogenic analysis of genus *Jatropha*. Mol Biol Rep 36:901–907

Pan BZ, Xu ZF (2011) Benzyladenine treatment significantly increases the seed yield of the biofuel plant Jatropha curcas. Plant Growth Regul 30:166–174

Pan BZ, Chen MS, Ni J, Xu ZF (2014) Transcriptome of the inflorescence meristems of the biofuel plant *Jatropha curcas* treated with cytokinin. BMC Genomics 15:974

Pant KS, Khosla V, Kumar D, Gairola S (2006) Seed oil content variation in *Jatropha curcas* Linn. in different altitudinal ranges and site condition in H.P. India. Lyonia 11:31–34

Patade VY, Khatri D, Kumar K, Grover A, Kumari M, Gupta SM, Kumar D, Nasim M (2014) RNAi mediated curcin precursor gene silencing in Jatropha (*Jatropha curcas* L.). Mol Biol Rep 41:4305–4312

Pazeto MSR, Trevisoli SHU, Correa AAP, Vianna VF, Leite DC, Mauro AOD (2015) Genetic diversity in Jatropha species from different regions of Brazil based on morphological characters and inters-simple sequence repeat (ISSR) molecular markers. Afr J Biotechnol 14:2066–2079

Pecina-Quintero V, Anaya-Lopez JL, Zamarripa-Colmenero A, Montes-Garcia N, Nunez-Colin CA, Solís-Bonilla JL, Aguilar-Rangel MR, Langarica HRG, Bustamante DJM (2011) Molecular characterisation of *Jatropha curcas* L. genetic resources from Chiapas, México through AFLP markers. Biomass Bioenergy 35:1897–1905

Pecina-Quintero V, Anaya-Lopez JL, Zamarripa-Colmenero A, Nunez-Colın CA, Montes-Garcıa N, Solıs-Bonilla JL, Jimenez-Becerril MF (2014) Genetic structure of *Jatropha curcas* L. in Mexico and probable centre of origin. Biomass Bioenergy 60:147–155

Popluechai S, Froissard M, Jolivet P, Breviario D, Gatehouse AM, O'Donnell AG, Chardot T, Kohli A (2011) *Jatropha curcas* oil body proteome and oleosins: L-form JcOle3 as a potential phylogenetic marker. Plant Physiol Biochem 49:352–356

Prakash AR, Singh S, Prakash CR, Ghosh A, Agarwal PK (2016) Development of Jatropha hybrids with enhanced growth, yield and oil attributes suitable for semi-arid wastelands. Agrofor Syst 90:541–553

Qu J, Mao HZ, Chen W, Gao SQ, Bai YN, Sun YW, Geng YF, Ye J (2012) Development of marker-free transgenic *Jatropha* plants with increased levels of seed oleic acid. Biotechnol Biofuels 5:10

Ramamurthy K (1967) A new variety of *Jatropha villosa* from Madras state. Bull Bot Surv India 9:278–279

Ranade SA, Srivastava AP, Rana TS, Srivastava J, Tuli R (2008) Easy assessment of diversity in *Jatropha curcas* L. plants using two single—primer amplification reaction (SPAR) methods. Biomass Bioenergy 32:533–540

Rao KS, Lakshminarayana G (1987) Characteristics and composition of six newer seeds and their oils. Fat Sci Technol 89:324–326

Rao GR, Korwar GR, Shanker AL, Ramakrishna YS (2008) Genetic associations, variability and diversity in seed characters, growth, reproductive phenology and yield in *Jatropha curcas* (L.) accessions. Trees 22:697–709

Reddy KRK, Ramaswamy N, Bahadur B (1987) Cross incompatibility between *Ricinus* and *Jatropha*. Plant Cell Incomp Newslett 19:60–65

Rosado TB, Laviola BG, Faria DA, Pappas M (2010) Molecular markers reveal limited genetic diversity in a large germplasm collection of the biofuel crop *Jatropha curcas* L. in Brazil. Crop Sci 50:2372–2382

Saadaoui E, Martin JJ, Bouazizi R, Romdhane CB, Grira M, Abdelkabir S, Khouja ML, Cervantes E (2015) Phenotypic variability and seed yield of *Jatropha curcas* L. introduced to Tunisia. Acta Bot Mex 110:119–134

Sakaguchi S, Somabhi M (1987) Exploitation of promising crops of Northeast Thailand. Siriphan Offset, Khon Kaen, p 50

Santos DND, Ferreira JL, Setotaw TA, Cancado GMA, Pasqual M, Londe LCN, Saturnino HM, Vendrame WA (2016) Genetic structure from the oldest Jatropha germplasm bank of Brazil and contribution for the genetic improvemnt. Ann Braz Acad Sci 88:2363–2374

Saptadi D, Hartati RS, Setiawan A, Heliyanto B, Sudarsono S (2017) Gentic diversity of Indonesian physic nut (*J. curcas*) based on molecular marker. Agrivita J Agr Sci 39:160–171

Sato S, Hirakawa H, Isobe S, Fukai E, Watanabe A, Kato M, Kawashima K, Minami C, Muraki A, Nakazaki N, Takahashi C, Nakayama S, Kishida Y, Kohara M, Yamada M, Tsuruoka H, Sasamoto S, Tabata S, Aizu T, Toyoda A, Shin-i T, Minakuchi Y, Kohara Y, Fujiyama A, Tsuchimoto S, Kajiyama S, Makigano E, Ohmido N, Shibagaki N, Cartagena JA, Wada N, Kohinata T, Atefeh A, Yuasa S, Matsunaga S, Fukui K (2011) Sequence analysis of the genome of an oil-bearing tree, *Jatropha curcas* L. DNA Res 18:65–76

Schmook B, Seralta-Peraza L (1997) *J. curcas*: distribution and uses in the Yucatan Peninsula of Mexico. Jatropha 97, Managua, Nicaragua

Seesangboon A, Pokawattana T, Eungwanichayapant PD, Tovaranonte J, Popluechai S (2018) Effects of 6-benzyladenine on *Jatropha* gene expression and flower development. Russ J Plant Physiol 65:345–356

Shi X, Wu Y, Dai T, Gu Y, Wang L, Qin X, Xu Y, Chen F (2018) *JcZFP8*, a C2H2 zinc-finger protein gene from *Jatropha curcas*, influences plant development in transgenic tobacco. Electron J Biotechnol 34:76–82

Sujatha M (1996) Genetic and tissue culture studies in castor (*Ricinus communis* L.) and related genera. (PhD Dissertation, Osmania University, Hyderabad)

Sujatha M, Prabakaran A (2003) New ornamental Jatropha hybrids through interspecific hybridization. Genet Resour Crop Evol 50:75–82

Sujatha M, Makkar HPS, Becker K (2005) Shoot bud proliferation from axillary nodes and leaf sections of non-toxic *Jatropha curcas* L. Plant Growth Regul 47:83–90

Sujatha M, Tarakeswari M, Francis G (2013a) Start codon targeted (SCoT) polymorphism in toxic and non-toxic accessions of *Jatropha curcas* L. and development of a codominant SCAR marker. Plant Sci 207:117–127

Sujatha M, Nithiyanantham S, Reddy MP (2013b) Plant regeneration and genetic transformation in Jatropha. In: Jain S, Dutta Gupta S (eds) Biotechnology of neglected and underutilized crops. Springer, Dordrecht

Sun WB (2008) Evaluation of *Jatropha curcas* germplasm in Southwest China. Jatropha International Congress, 7–18 December 2008, Singapore

Sun QB, Li LF, Li Y, Wu GJ, Ge XJ (2008) SSR and AFLP markers reveal low genetic diversity in the biofuel plant *Jatropha curcas* in China. Crop Sci 48:1865–1871

Sun F, Liu P, Ye J, Lo LC, Cao S, Li L, Yue GH, Wang CM (2012) An approach for Jatropha improvement using pleiotropic QTLs regulating plant growth and seed yield. Biotechnol Biofuels 5:42

Sunil N, Sivaraj N, Anitha K, Abraham B, Kumar V, Sudhir E, Vanaja M, Varaprasad KS (2009) Analysis of diversity and distribution of *Jatropha curcas* L. germplasm using geographic information system (DIVA-GIS). Genet Resour Crop Evol 56:115–119

Sunil N, Kumar V, Sujatha M, Rao GR, Varaprasad KSV (2013) Minimal descriptors for characterization and evaluation of *Jatropha curcas* L. germplasm for utilization in crop improvement. Biomass Bioenergy 48:239–249

Tar MM, Tanya P, Srinives P (2011) Heterosis of agronomic characters in Jatropha (*Jatropha curcas* L.). Kasetsart J (Nat Sci) 45:583–593

Tatikonda L, Wani SP, Kannan S, Beerelli N, Sreedevi TK, Hoisington DA, Devi P, Varshney RK (2009) AFLP-based molecular characterization of an elite germplasm collection of *Jatropha curcas* L. a biodiesel plant. Plant Sci 176:505–513

Tee TS (2007) Produce a crop of *Jatropha* within four months from seed. In: International Workshop on the Development of the JCL Industry. 29–31 October 2007, China, p 44

Tester M, Langridge P (2010) Breeding technologies to increase crop production in a changing world. Science 327:818–822

Trebbi D, Papazogloub EG, Saadaoui E, Visch M, Baldini M, Stevanato P, Cettul E, Sanzone AP, Gualdi L, Fabbri A (2015) Assessment of genetic diversity in different accessions of *Jatropha curcas*. Ind Crop Prod 75:35–39

Trivedi S, Gaudani H, Gupta M, Gupta N, Patil P, Gupta G, Vamsi KK, Reddy MP (2009) Establishment of agrobacterium mediated genetic transformation in *Jatropha curcas* L. Int J Agric Sci 1:11–20

Tsuchimoto S, Cartagena J, Khemkladngoen N, Singkaravanit S, Kohinata T, Wada N, Sakai H, Morishita Y, Suzuki H, Shibata D, Fukui K (2012) Development of transgenic plants in Jatropha with drought tolerance. Plant Biotechnol 29:137–143

Van Loo EN, Montes LR, Azurdia C, Jongschaap REE, Barillas E, Visser RGF (2008) Global evaluation of genetic variability in *Jatropha curcas*: comparing South and Meso-American, African, Indian and Asian accessions. In: Jatropha International Congress, 17–18 December 2008, Singapore

Vandepitte K, Valdés-Rodríquez OA, Sánchez-Sánchez O, De Kort H, Martinez-Herrera J, García-Pérez E, De Meyer T, Pérez-Vázquez A, Muys B, Honnay O (2019) High SNP diversity in the non-toxic indigenous *Jatropha curcas* germplasm widens the potential of this upcoming major biofuel crop species. Ann Bot XX:1–8. https://doi.org/10.1093/aob/mcz008

Veronese P, Li X, Niu X, Weller SC, Bressan RA, Hasegawa PM (2001) Bioengineering mint crop improvement. Plant Cell Tissue Organ Cult 64:133–144

Wang CM, Liu P, Yi C, Gu K, Sun F, Li L, Lo LC, Liu X, Feng F, Lin G, Cao S, Hong Y, Yin Z, Yue GH (2011) A first generation microsatellite- and SNP-based linkage map of Jatropha. PLoS One 6:e23632

Warra AA, Prasad MNV, Tarakeswari M, Sujatha M (2019) Approaches for genetic improvement and transformation of *Jatropha curcas* and *Ricinus communis* for efficient remediation of toxic metals and metalloids. In: Transgenic plant technology for remediation of toxic metals and metalloids. Academic, Cambridge, pp 131–154

Wei X, Sujatha M, Liu A (2012) Genetic diversity in the Jatropha genus and its potential application. CAB Rev 7:059

Wilbur RL (1954) A synopsis of *Jatropha*, subsection *Eucurcas*, with the description of two new species from Mexico. J Elisha Mitchell Sci Soc 70:92–101

Wu P, Zhou C, Cheng S, Wu Z, Lu W, Han J, Chen Y, Ni P, Wang Y, Xu X, Huang Y, Song C, Wang Z, Shi N, Zhang X, Fang X, Yang Q, Jiang H, Chen Y, Li M, Wang Y, Chen F, Wang J,

Wu G (2015) Integrated genome sequence and linkage map of physic nut (*Jatropha curcas* L.), a biodiesel plant. Plant J 81:810–821

Xia Z, Zhang S, Wen M, Lu C, Sun Y, Zou M, Wang W (2018) Construction of an ultrahigh-density genetic linkage map for *Jatropha curcas* L. and identification of QTL for fruit yield. Biotechnol Biofuels 11:3

Xu G, Huang J, Yang Y, Yao Y (2016) Transcriptome analysis of flower sex differentiation in *Jatropha curcas* L. using RNA sequencing. PLoS One 11:e0145613

Yi C, Zhang S, Liu X, Bui HTN, Hong Y (2010) Does epigenetic polymorphism contribute to phenotypic variances in *Jatropha curcas* L.? BMC Plant Biol 10:259

Yi C, Reddy C, Varghese K, Bui TNH, Zhang S, Kallath M, Hong Y (2014) A new *Jatropha curcas* variety (JO S2) with improved seed productivity. Sustainability 6:4355–4368

Yue G (2008) Genetics and genomics of *Jatropha* for genetic improvement. In: Jatropha International Congress. 17–18 December 2008, Singapore

Yue GH, Lo LC, Sun F, Cao SY, Yi CX, Hong Y, Bang W (2014) No variation at 29 microsatellites in the genome of *Jatropha curcas*. J Genomics 2:59–63

Zhang Y, Wang Y, Jiang L, Xu Y, Wang Y, Lu D, Chen F (2007) Aquaporin JcPIP2 is involved in drought responses in *Jatropha curcas*. Acta Biochim Biophys Sin 39:787–794

Zhang FL, Niu B, Wang YC, Chen F, Wang SH, Xu Y, Jiang LD, Gao S, Wu J, Tang L, Jia YJ (2008) A novel betaine aldehyde dehydrogenase gene from *Jatropha curcas*, encoding an enzyme implicated in adaptation to environmental stress. Plant Sci 174:510–518

Zhang L, Zhang C, Wu P, Chen Y, Li M, Jiang H, Wu G (2014) Global analysis of gene expression profiles in physic nut (*Jatropha curcas* L.) seedlings exposed to salt stress. PLoS One 9:e97878

Sujatha Mulpuri is a researcher at the Indian Institute of Oilseeds Research, Hyderabad (India), holds a university degree in plant sciences, and a doctoral degree in genetics. Her current focus is on pre-breeding for genetic improvement of mandated oilseed crops of the institute through genetic and biotechnological tools.

Srinivasan Nithiyanantham is pursuing his postdoctoral research at the China Medical University and Hospital, China Medical University, Taichung, Taiwan. His doctoral research is based on the genetic improvement of oilseed crops through modern biotechnological techniques.

Plant Cell Manipulation Technology for Biorefinery

19

Most Tanziman Ara, Nurhidayah Syahira Muhammad Radzi, Misaki Nishibe, and Shinjiro Ogita

Abstract

Plant-based fuels are generated from renewable sources. They produce less greenhouse gas (GHG) emissions and provide us with an alternative to fossil fuels for future energy security. Global warming and GHG emissions are the biggest global threats to mankind. Therefore, sustainable development using safer products is the focus of our research. The biorefinery is one of the key stakeholders of industrialization, which is essential for the sustainable development of any country. The development of different types of biofuels indicates that there is a continuous interest in renewable fuels. This study investigates the present status of biorefinery production, plant cell manipulation technologies, and future perspectives. There are advantages and disadvantages to every generation of biorefinery. Conventional breeding and transgenic breeding with molecular analysis by modern tools for genetic manipulation can potentially increase genetic diversity and develop new cultivars. The present generation of biorefinery focuses on bioengineering of microorganisms to increase the target product. Benefits aside, genetically engineered (GE) or genetically modified organisms (GMOs) have been considered a threat to the environment and human health. Therefore, our research group focused on the combination of metabolic engineering (ME) and plant cell manipulation technology (PCMT) to create alternatives for safer biorefinery production. Satisfactory improvement in metabolic engineering of bamboo and other energy crops was achieved. Therefore, bamboo, as the highest biomass producer, or other energy crops can be the target organisms for PCMT and ME technologies, substituting GE for a safer biorefinery. Perhaps this technology will create a new generation of biorefinery. We hope that, in the future, biofuel will be a safe and economical alternative to fossil fuels.

M. T. Ara · N. S. M. Radzi · M. Nishibe · S. Ogita (✉)
Program in Biological System Science, Graduate School of Comprehensive Scientific Research, Prefectural University of Hiroshima, Hiroshima, Japan
e-mail: ogita@pu-hiroshima.ac.jp

A. Kumar et al. (eds.), *Climate Change, Photosynthesis and Advanced Biofuels*,
https://doi.org/10.1007/978-981-15-5228-1_19

Keywords

Biorefinery · Cell manipulation · Biomass · Biofuel · Cell wall engineering · Metabolic engineering · Bamboo

19.1 Introduction

The IPCC special report (2018) describes the impacts if the global temperature increases by 1.5 °C compared with pre-industrial temperatures. This increase can be related to global greenhouse gas (GHG) emissions. Human activities are estimated to have caused approximately 1.0 °C increase, with a likely range of 0.8–1.2 °C. Global warming is likely to reach 1.5 °C between 2030 and 2052 if it continues to increase at the current rate. Reflecting the long-term warming trend since pre-industrial times, observed global mean surface temperature for the decade 2006–2015 was 0.87 °C (likely between 0.75 and 0.99 °C) higher than the average over the 1850–1900 period (very high confidence). Estimated anthropogenic global warming matches the level of the observed warming to within ±20% (likely range). Estimated anthropogenic global warming is currently increasing at 0.2 °C (likely between 0.1 and 0.3 °C) per decade due to past and ongoing emissions (high confidence) (IPCC 2018).

Accounting for indirect emissions raises the contributions of buildings and industrial sectors (high confidence). When emissions from electricity and heat production are attributed to the sectors that use the final energy (i.e., indirect emissions), the shares of the industry and buildings sectors in global GHG emissions increase to 31% and 19%, respectively. On Earth, human activities are changing the natural greenhouse effect. Over the last century, the burning of fossil fuels, such as coal and oil, has increased the concentration of atmospheric carbon dioxide (CO_2). This happens because the coal or oil burning process combines carbon with oxygen in the air to make CO_2. To a lesser extent, the clearing of land for agriculture, industry, and other human activities has increased concentrations of GHG. The industrial activities that our modern civilization depends upon have raised atmospheric CO_2 levels from 280 to 400 ppm in the last 150 years. The reduction of GHG emissions can benefit global climate change (Yau and Easterling 2018). Co-emitted air pollutants from burning fossil fuels threaten public health by contributing to cardiorespiratory diseases which can cause even death (Rodhe and Muller 2015).

Global demand for energy continues to grow, including from conventional fossil fuels such as oil, coal, and natural gas (Zakir et al. 2016). With the increased world's energy demand and progressive depletion of oil reserves, the search for alternative energy resources is urgent, especially for those derived from renewable materials such as biomass (Saxena et al. 2009). Excessive use of fossil fuels over the last century and the following years has drastically increased the level of GHG in the atmosphere (Ballesteros et al. 2006). Global concern about climate change and its consequences and the consequent need to diminish GHG emissions have encouraged the use of bioethanol as an energy source (Balat et al. 2008).

Biorefineries produce fuels, solvents, plastics, and food for human beings (Ohara 2003). Many hybrid technologies from different fields, such as bioengineering, polymer chemistry, food science, and agriculture, were developed for use in biorefineries. Currently, there is a significant interest in biofuel research due to the increased energy demand by emerging economies and the increased global oil prices (Elshahed 2010). First-generation biofuels were from corn and food-based crops, which are used as a direct substitute for fossil fuel for transportation. However, they are not sufficient to replace petroleum. Second-generation biofuels derive from forest and crop residues, energy crops, and municipal and construction waste. They can reduce net carbon emissions, increase energy efficiency, and reduce energy dependency, potentially overcoming the limitations of the first-generation biofuels. Nevertheless, the implementation of second-generation biofuels requires sustainable energy management or development of local bioenergy systems.

It is well known that biotechnology is the science of applied biological processes. The wide concept of "biotechnology" encompasses a wide range of procedures to modify living organisms according to human purposes, including the domestication of animals, cultivation of plants, and improvements through breeding programs that employ artificial selection and hybridization. Modern usage also includes genetic engineering, cell and tissue culture technologies, and cell manipulation technologies. Plant cell walls represent the most abundant renewable resource on this planet (Pauly and Keegstra 2008). Despite their great abundance, only 2% of this resource is currently used by humans. Hence, research into the feasibility of using plant cell walls in the production of cost-effective biofuels is desirable. Biomass utilization is increasingly considered as a practical way to supply sustainable energy and provide long-term environmental benefits (Xie and Peng 2011). Conversion of lignocellulosic residues from food crops is a potential alternative for that. Because of its recalcitrance, current biomass processes are unacceptably expensive, whereas genetic breeding of energy crops is a promising solution. To meet the required demands by definition, energy crops should have a high yield of food and biofuel material. Plant cell manipulation technology can increase or modify desirable chemicals or substances through our target. Biorefinery production using different cell manipulation technologies is a potential strategy that must be addressed to achieve sustainable development (see Fig. 19.1). This study overviews the present status of biorefinery production, the effects of plant cell manipulation, and future perspectives.

19.2 Cell Manipulation Technology: History and Types

The cell is the structural unit of a plant. It forms the whole plant via mitosis cell division. It performs photosynthesis and respiration, keeping O_2 and CO_2 balanced in the atmosphere. The development of plant cell culture is historically linked with the discovery of the cell itself and the subsequent propounding of the cell theory. Henri-Louis Duhamel du Monceau (1756) pioneered experiments on wound healing in plants. In his experiments, elm plants demonstrated spontaneous callus formation

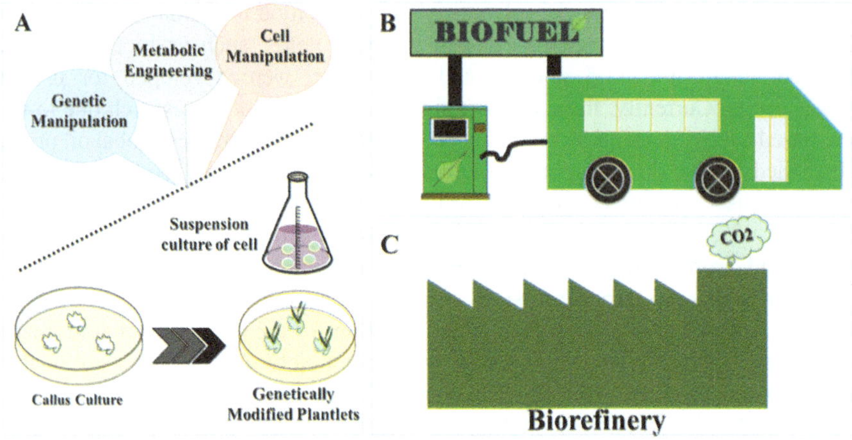

Fig. 19.1 Schematic diagram of production of biorefinery using different cell manipulation technologies

in a decorticated region. His studies, according to the notable biologist Gautheret (1985), can be considered a foreword for the discovery of plant tissue culture. Further contributions to plant tissue culture can be attributed to the cell doctrine, which implicitly admitted that a cell is capable of autonomy and demonstrated the potential for totipotency.

In 1902, the German botanist Gottlieb Haberlandt developed the concept of in vitro cell culture. Later, other scientists were essential for the development of tissue culture (Razdan 2006). After that, a new development for mankind was the genetic improvement through *Agrobacterium* transformation of plants in the twentieth century. Plant biotechnology was founded on the principles of cellular totipotency and genetic transformation, which can be traced back to the cell theory of Matthias Jakob Schleiden and Theodor Schwann, and the discovery of genetic transformation in bacteria by Frederick Griffith, respectively. Plant biotechnology led to the production and commercialization of biotech (transformed or transgenic) plants expressing useful genes, which emphasizes the beneficial effects of plant biotechnology on food security, human health, the environment, and conservation of biodiversity (Vasil 2008).

Modifications to produce desired traits in plants, animals, and microbes used for food began about 10,000 years ago. Advantageous outcomes of these genetic modifications include increased food production, reliability, and yields; enhanced taste and nutritional value; and decreased losses due to various biotic and abiotic stresses, such as fungal and bacterial pathogens. These objectives continue to motivate modern breeders and food scientists, who have designed newer genetic modification methods to identify, select, and analyze individual organisms that possess genetically enhanced features.

For plant species, it can take up to 12 years to develop, evaluate, and release a new variety of crops. However, while advances in modification methods can reduce

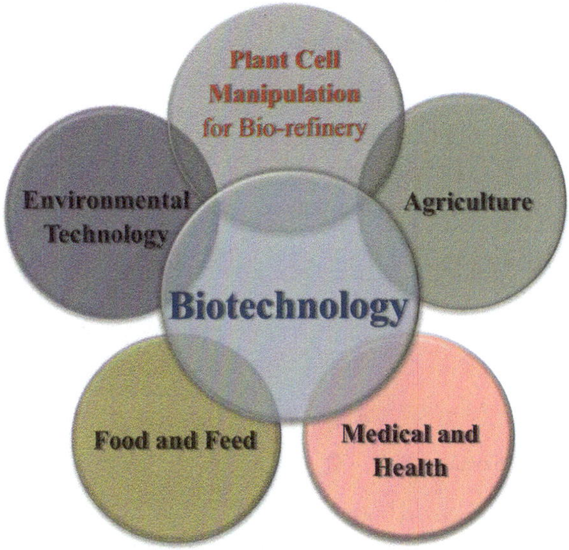

Fig. 19.2 Schematic diagram showing the branches of biotechnology

the time it takes to introduce new foods to the marketplace, an important benefit of a long evaluation is that it provides greater assurance that deleterious features will be identified, and potentially harmful varieties are eliminated before commercial release.

As shown in Fig. 19.2, plant cell manipulation technology is a branch of biotechnology. Different types of genetic modification of plants could be the tools to produce the fourth-generation biorefinery, such as simple and marker-assisted selection; crossing; interspecies crossing (Kozukue et al. 1999; Sanford et al. 1948; Zimnoch-Gozowska et al. 2000); chromosome engineering (Sears 1956, 1981); embryo rescue; somatic hybridization; somaclonal variation (Larkin and Scowcroft 1981; Rowland et al. 2002); mutation breeding, including chemically and X-ray-induced mutagenesis; cell selection (Sebastian and Chaleff 1987; Swanson et al. 1988; Rowland et al. 1989); genetic engineering of cells using microbial vector; microprojectile bombardment (Klein et al. 1992); electroporation; microinjection; transposons and transposable elements; genome editing (Ding et al. 2016); and non-transgenic molecular methods of manipulation.

19.3 Biorefinery: Different Generations and Present Status

19.3.1 Biorefinery System

Biorefinery has become a leading alternative in the bioenergy industry and in the production of bio-based products. Biorefineries sustainably process biomass into a spectrum of marketable products and energy, as stated by the International Energy Agency (IEA Bioenergy Task 42) (Cherubini 2010). The refinery utilizes biomass as

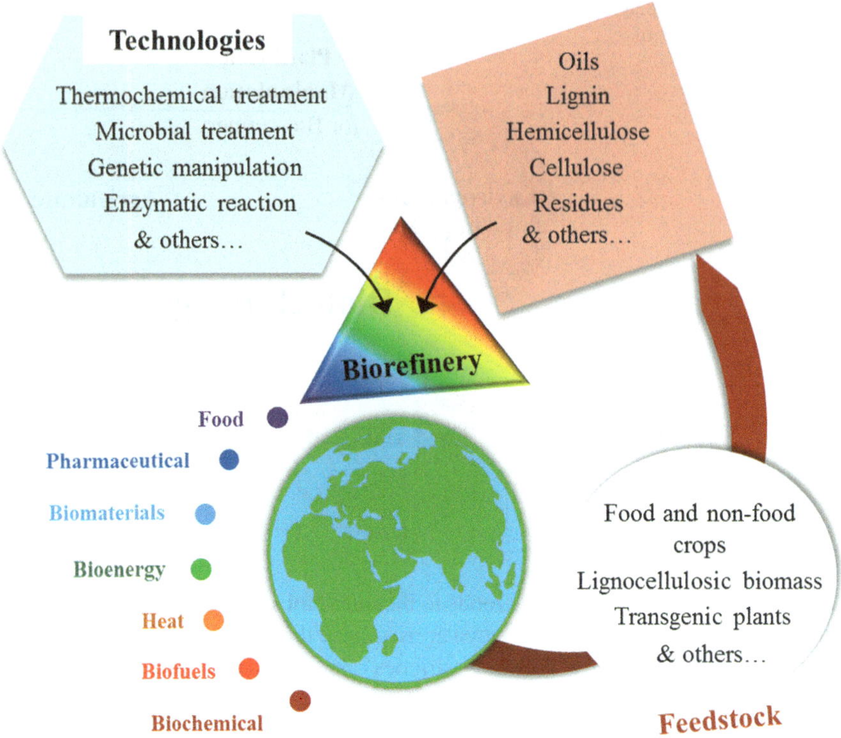

Fig. 19.3 Illustration of the biorefinery overview and its function in biomass transformation

a processing input (feedstock) to produce multiple bio-based products (Fig. 19.3). Biorefinery products range can be applied from petroleum refineries products to other resources such as lignocellulosic feedstock biorefineries, pretreatment and effective separation of lignin, cellulose, and hemicellulose, extensive development of thermal, mechanical, chemical, and biological processes, and combination of substantial conversion (biotechnological and chemical processes) (Kamm and Kamm 2004; Kamm et al. 2006). It is necessary to produce a broad variety of bio-based products in an efficient construction set system.

19.3.2 Biomass as Feedstock

Biomass is a practical and suitable bio-resource that can produce renewable fuels. It can potentially replace a large fraction of fossil resources as feedstock for energy and non-energy sectors (Cherubini 2010). It is defined as a collection of organic matter that includes biological organisms, lipids, and hydrocarbons, such as simple sugar, starch, and lignocellulose, which are the main components for biofuel production

(Hill et al. 2006; Roy and Kumar 2013). Approximately 200 billion tons of biomass per year can be produced by land plants, globally.

19.3.3 Different Generations of Biofuel Production

Biofuels are fuels produced from biological sources. They can be used to produce electricity, heat, and transportation fuels (Yau and Easterling 2018). They are classified into solid, liquid, and gaseous biofuels (Petrou and Pappis 2009). Their raw materials are extracted mainly from sustainable sources, without significant environmental impact, providing an alternative source of energy (Demirbas 2007a, b). The two main liquid biofuels are bioethanol and biodiesel. Approximately 80% of the global liquid biofuel production was in the form of ethanol in 2012–2014. Global bioethanol and biodiesel production have reached 108 and 28 billion liters, respectively (Popp et al. 2016). Currently, several research activities seek to identify prospective renewable energy sources or biomass feedstock and develop their processing systems to produce alternatives to fossil fuels for transport, such as bioethanol, biodiesel, biomethanol, and hydrogen (Antizar-Ladislao and Turrion-Gomez 2008). Biofuels can be divided into primary and secondary biofuels. Primary biofuels utilize unprocessed fuelwoods primarily for heating, cooking, and electricity production. Secondary biofuels refer to bioethanol and biodiesel, which are derived from biomass and can be used in vehicles and numerous industrial practices (Dragone et al. 2010). Currently, there are four generations of secondary biofuels created using variations in feedstock and processing factors (Aro 2016).

First-generation biofuels comprise biodiesel and bioethanol derived from conventional technology from energy-rich food-based feedstock such as sugar, corn, starch, vegetable oils, or animal fats. However, this generation of biofuel energy colludes with food security, food shortage, and price issues (Nadarajah 2018). Examples of first-generation biofuel are starch-based corn-grain ethanol, and sugarcane ethanol/biodiesel. The first process to produce corn-grain ethanol involves enzymatic conversion (i.e., cellulase, hemicellulase, and β-glucosidase) of corn starch into fermentable sugars. Sugarcane provides fibrous residues called bagasse, which are also used as lignocellulosic feedstock for second-generation biofuel production. They represent approximately 25% of the total sugarcane weight with 60–80% carbohydrates (Betancur and Pereira 2010; Rezende et al. 2011). The sugar is then fermented using ethanol-producing microorganisms, such as bacteria, fungi, or yeast. *Saccharomyces cerevisiae* is a superior, widely studied, ethanol-producing microorganism. It is considered one of the pillars of the current biofuel industry (Bujis et al. 2013).

Second-generation biofuels (also known as advanced biofuels) are defined based on the types of feedstock and technology used for their derivatization and the biorefinery process (Nadarajah 2018). The biofuel is sourced from non-food biomass and non-food crops. It is produced from lignocellulosic biomass, which can originate from various sources. Typical feedstock sources include crops, forest, wood process residues, or cultivated energy trees. The products of second-generation lignocellulosic biofuels are cellulosic-ethanol, edible oil, and non-edible oil. These oils have been used in the USA and European countries as sources for biofuel such as biodiesel (Yau and Easterling 2018). Pretreatment and saccharification are important processes for second-generation biofuel production, while fermentation is an important factor affecting the overall production efficiency. After pretreatment, saccharification, and fermentation, the ethanol-producing microorganisms, such as strains of *Saccharomyces cerevisiae* and *Zymomonas mobilis*, can each reach optimum conditions for maximum yield (Yau and Easterling 2018).

Third-generation of biofuels consists of advanced biofuels that utilize unicellular photosynthetic algae and microbes as feedstock (Nadarajah 2018). These photosynthetic algae use solar energy to convert carbon dioxide into carbon-rich lipids. Their high productivity and ability to make and store energy-rich compounds [e.g., triglycerides (TAG)], which can be extracted and converted to biofuels, make them a very attractive alternative for biofuel production (Hu et al. 2008). Energy production from microalgae-derived biofuels is viable because they do not present the main downsides associated with first and second-generation biofuels, such as increased agriculture competition between available arable lands used for food production and energy crops, and high-cost technology involving enzyme-mediated pretreatment. The advantage of microalgae is that they can grow at higher rates, and synthesize and accumulate high amounts of neutral lipids with 20–50% dry weight of biomass. It can produce oil for biodiesel production with a 15–300 times higher yield per area than traditional crops. While the harvest of conventional crops is once or twice a year, microalgae harvesting cycles are approximately 1–10 days, depending on the process and condition. The process of microalgal-biomass conversion into energy can be categorized into biochemical conversion, chemical reaction, thermochemical conversion, and direct combustion. For the production process, there are four types of culture systems that depend on the purpose of the production facility, microalgae strain, and product of interest. Integrated production of biofuels from microalgae includes microalgae cultivation, separation of the cells from the growth medium, and lipid extraction for biodiesel production through transesterification process, or starch hydrolysis and subsequent distillation for bioethanol production through distillation (Dragone et al. 2010).

Fourth-generation biofuels are products of the bioconversion of living organisms such as microorganisms and plants, through the application of biotechnological tools (Rutz and Janseen 2007; FAO 2008) (Gopinathan and Sundhakaran 2009). These biofuels combine the properties of third-generation biofuels with the genetic optimization of their producers (Al-Thani and Potts 2012; Nozzi et al. 2013). Bioenergy technology is integrated into carbon capture and storage (CCS) technology, so the biofuel production is not just carbon neutral but also carbon negative. Carbon negativity is defined as the removal of CO_2 from the atmosphere (Yau and Easterling 2018). Biohydrogen and bioelectricity using photosynthetic mechanisms are included in this biofuel generation (Antizar-Ladislao and Turrion-Gomez 2008). Recent findings show the molecular biological ability to reconstitute plant metabolic pathways in a heterologous host, and that the metabolism of crop plants can be engineered to improve the production of food and biofuels as shown in Table 19.1 (Bhansali and Kumar 2018).

Table 19.1 List of cell and genetic manipulation technology used on microorganism and plant materials for biomass and biorefinery

Manipulation technology	Target/Modification	Plants/Microorganism materials	Outcome	References
Down-regulation by RNAi	COMT gene	*Panicum virgatum* L. (Switchgrass)	Lignin content decrease by 24%	Fu et al. (2011)
Engineering	Furfural resistance	*Escherichia coli*	Resistance improvement towards a sugarcane bagasse hydrolysate	Jönsson and Martín (2016)
Gene deletion	Gene encoding lactate dehydrogenase	*Caldicellulosiruptor bescii*	Higher yield and titer of acetate and hydrogen	Cha et al. (2013)
Gene deletion and expression of enzyme	– Genes encoding glycerol-3-phosphate dehydrogenase – Expression of an acetylating acetaldehyde dehydrogenase	*Saccharomyces cerevisiae*	Conversion of inhibitory acetic acid to ethanol, and eradication of glycerol formation in yeast anaerobic cultures	
Gene silencing by artificial microRNA	COMT and CAD genes	*Brachypodium distachyon* (purple falsebrome)	Higher ethanol yield (by 9–17%)	Li et al. (2014)
Metabolic engineering	Xylose metabolism	*Corynebacterium glutamicum*	Higher lignocellulosic substrate utilization	Kawaguchi et al. (2006)
Mutagenesis	Catabolite repression	*Bacillus pumilus*	High cellulase yield (by four times)	Kotchoni et al. (2003)
Mutagenesis and selection	High-yield-grain crops, cell-wall-altered mutant plants	*Oryza sativa, Miscanthus* spp., *Zea mays* (maize)	Remarkable alteration of cell wall composition and increased rate of biomass degradation	Xie and Peng (2011)
Overexpression of gene	Yap1 transcription factor	*Saccharomyces cerevisiae*	Hyper-resistance to lignocellulose hydrolysates	Jonsson et al. (2013)
	Cellulose, oxidase, xylanase, and hydrolytic enzyme	*Pichia stipitis, Phanerochaete chrysosporium*	Active lignocelluloses and xylose transformation into ethanol	Mood et al. (2013)
	Trans-aldose and alcohol dehydrogenase	*Saccharomyces cerevisiae*	Slight increase in ethanol yield	Jönsson and Martín (2016)

(continued)

Table 19.1 (continued)

Manipulation technology	Target/Modification	Plants/Microorganism materials	Outcome	References
Recombinant DNA	Endoglucanase processing	*Neocallimastrix patriciarum*	Higher endoglucanase activity	Kumar et al. (2008)
	Expression of cellulase gene	*Penicillium chrysogenum, Trichoderma reesei, Pseudomonas fluorescens*	Higher cellulase activity	Kumar et al. (2008)
RNAi silencing	4 CL gene	*Panicum virgatum L. (Switchgrass)*	Plant protein extract 4CL gene activity decrease by 80%, leading to lignin content decline by 32%	Li et al. (2014)
Targeted expression	Fungal xylanase	*Arabidopsis* sp.	Xylanase accumulation in chloroplast and peroxisomes (160% and 240%)	Mood et al. (2013)
Ultraviolet treatment and chemical mutagenesis		*Fusarium oxysporum*	Higher Cellulolytic activity (>80%)	Kuhad et al. (1994)
In planta transformation	Addition of gene encoding enzyme activities from *Cupriavidus necator* (bacteria) that convert an endogenous plant metabolite into polymeric structure that is novel to the plant	Maize, sugarcane, switchgrass, *Brassica napus, Arabidopsis thaliana*	Production of polyhydroxyalkanoates (PHAs), i.e., poly[(R)-3-hydroxybutyrate] (PHB) biopolymer with lignocellulosic as coproduct	Snell et al. (2015)
CRISPR/Cas9	Biallelic mutation of 4 coumarate:CoA ligase (4CL) gene	Stably transformed Populus	Lignin content is reduced by 23%, ratio of S lignin/G lignin decreased by 30%	Zhou et al. (2015)
CRISPR/Cas9	Suppress monolignol biosynthetic-pathway gene expression	*Arabidopsis thaliana*	Decreased lignin content and increased scarification efficiency	Van Acker et al. (2013)

Fourth-generation biofuels also include bioengineering of photosynthetic microorganisms towards direct solar fuel production, i.e., fuel production without a biomass phase. This photosynthesis-based technology demands synthetic biology for efficiency and thorough engineering of novel metabolic pathways for fuel and chemical production in hosts and photosynthetic microorganisms (i.e., algae and cyanobacteria) (Aro 2016). Photosynthetic water splitting (water oxidation) by solar energy is done both through artificial photosynthesis (Inganas and Sundström 2016) and direct solar biofuel production technologies. Synthetic biology can produce various biofuel arrays, potentially becoming a large contributor to fuel production on a global scale (Aro 2016).

19.3.4 Current Status of Biorefinery Production

The conventional process of ethanol or hydrogen production from cellulose feedstock through fermentation involves a complex process of pretreatments such as cellulase production, hydrolysis of cellulose and hemicellulose, and fermentation of hexose sugars (product of cellulose hydrolysis), and pentose sugars (product of hemicellulose hydrolysis). Current technologies include simultaneous saccharification and fermentation (SSF) or simultaneous saccharification and co-fermentation (SSCF) to produce bioethanol from cellulosic feedstock. However, both techniques require an extensive cellulose pretreatment by steam explosion and/or acid treatment, followed by the addition of exogenous cellulolytic enzyme (to hydrolyze cellulose chains and release glucose monomers as fermentation material). Consolidated bioprocessing (CBP) is a system in which cellulase production, substrate hydrolysis, and fermentation processes are accomplished in a single step by cellulolytic microorganisms (Carere et al. 2008).

Current biomass conversion technology generally consists of three steps: thermochemical pretreatment, enzymatic saccharification, and microbial fermentation. The pretreatment aims to achieve a higher yield of sugars and minimize product degradation that might restrict microbial fermentation. This process will set biomass feedstock to be amenable to structural polysaccharide breakdown into fermentable sugar by the action of cellulolytic enzymes. Physical pretreatment involves the reduction of biomass particle size by mechanical milling, steam explosion, and hydrothermolysis. Thermochemical pretreatment utilizes a dilute acid or base for hydrolysis and removal of hemicellulose and lignin (Sun et al. 2013).

First-generation biofuels are produced from natural oils extracted from plants. In contrast, the development of second-generation biofuels is a complex process. Complex catalysis and chemical alteration procedures are required to process wastes and wooden materials into oils (Antizar-Ladislao and Turrion-Gomez 2008). However, studies to maximize the amount of renewable carbon and hydrogen sources that can be converted to fuels from "second generation" biomass are needed. The technology to produce third-generation biofuels is based on algal biomass production. However, the integration of biotechnology and fourth-generation biofuels is a more innovative solution that is expected to create a breakthrough in the biorefinery and biofuel industries (Aro 2016).

In this context, the selection of energy crops from natural germplasm resources and genetic manipulation pools has been used as an initial approach to increase the biomass yield. The genetic breeding strategy includes traditional breeding with related molecular approaches (i.e., genomic mapping and gene markers), which are essential to discovering key genes associated with the plant traits that regulate growth, disease resistance, and biomass production quantity (Xie and Peng 2014).

The acceptance of new breeding practices is essential for food security. New plant varieties can be created to enhance the establishment of bioenergy crops on marginal lands and suitability of the inedible parts of food-crop plants for biorefinery purposes and bioenergy production. An efficient plant breeding motion was clearly given to EU policy in reports by the European Academies Science Advisory Council (EASAC 2013 and 2015). Breeding technologies include transgenesis (GM), cisgenesis, intragenesis, targeted mutagenesis, other transient introduction of recombinant DNA, as well as gene silencing and reverse breeding (Aro 2016).

19.4 Plant Cell Manipulation Technologies for Sustainable Production of Biorefinery

Conventional breeding and transgenic breeding using genetic manipulation with molecular analysis by modern tools can be utilized to increase genetic diversity and develop new cultivars (Table 19.1). These technologies can improve lignocellulosic biomass yield and alter cell-wall quality for better biofuel production efficiency (Yau and Easterling 2018). Genetic breeding and biotechnology have been applied to increase the biomass yield of food crops and energy plants such as switchgrass, *miscanthus*, rice, wheat, maize, and sweet sorghum (Xie and Peng 2011). These perennial and annual crops are the most common sources for bioenergy production worldwide because of their high yield and quality of food supply, easily disrupted cell walls in the stalks or straw for biofuel production, and ability to grow in geographical regions with limited light and water (Xie and Peng 2014).

CRISPR/Cas9 (see Fig. 19.4) is a rapidly developing genome editing technology that has been successfully applied in many organisms, including model and crop plants. As an RNA-guided DNA endonuclease, Cas9 can target specific genomic sequences by constructing a separately encoded guide RNA with which it forms a complex. The CRISPR/Cas9 technology has been successfully applied in model plants such as *Nicotiana benthamiana*, *Nicotiana tabacum*, *Arabidopsis* sp., and crop plants such as wheat, maize, rice, tomato, sorghum, and sweet orange. The CRISPR/Cas9 technique creates a genetically modified plant that carries the gene of interest and allows multiplex gene editing by simultaneous expression of two or more single-guide RNA (sgRNAs) (Belhaj et al. 2015).

Lignin, the largest non-carbohydrate phenolic polymers, is not involved in the fermentation process, which is one of the most significant obstacles to convert lignocellulosic biomass into fermentable sugars. Known lignin-biosynthetic

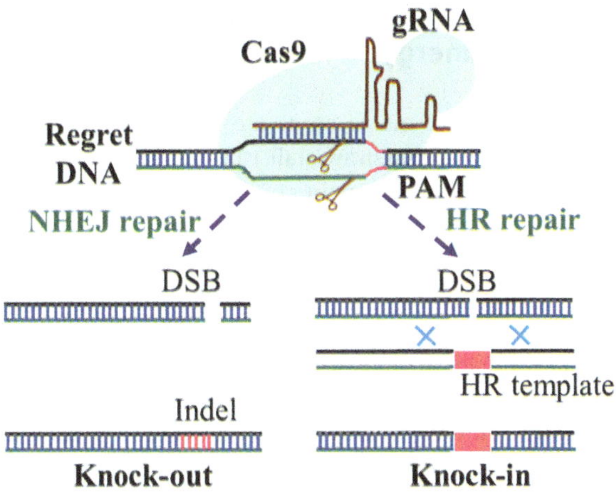

Fig. 19.4 Illustration of Cas9/gRNA genome editing. Created by modifying "Figure-Ding et al. 2016" (©Ling-Ling Chen and Kabin Xie (Licensed under CC BY 4.0))

pathways have been modified to decrease lignin content or to change its composition for a better suited lignocellulosic feedstock through genetic engineering (Van Acker et al. 2014). As for lignin biosynthesis, RNAi technology has been used for many years to suppress lignin-biosynthetic genes. However, significant variations in the efficacy of gene silencing and potential errors in data collection have been reported (McGinnis 2010; Lin et al. 2017). Nevertheless, lignin has attracted a lot of attention as a potential biomass feedstock for fuel because of its large heating value and chemical applications (Ragauskas et al. 2014; Gao et al. 2015). Therefore, to improve the economic feasibility of a biorefinery, biomass, including the lignin stream, must be comprehensively converted into value-added products (Rinaldi et al. 2016).

19.5 Current and Future Challenges

Current issues such as high gasoline prices, environmental concerns, food security, and energy security have drawn public concern (Sun et al. 2013). The global demand for fuel oil for energy is approximately 84 million barrels per day and is projected to rise to about 116 million barrels per day by 2030, with a 60% increased demand by the transportation sector (Cherubini 2010). Transportation is the largest energy-consuming sector. It consumed 28.58 quadrillion kJ in 2011, approximately 28% of the total world's energy consumption (103.08 quadrillion kJ) (Maity 2015). Technical challenges and the main components, such as sustainable feedstock supply, optimized bioconversion technologies, and integrated biorefinery must be resolved in order to establish a sustainable biorefinery system on an industrial scale (Sun et al. 2013). The chemical industry is experiencing a fundamental shift as cost-competitive bio-based chemicals become a commercial challenge.

19.6 Assessment of the Suitability of Bamboo as a Source of Energy

Rice, wheat, and maize crops are major food sources worldwide. However, their enormous biomass (straw/stalk) production has not been well used for biofuels. Carbohydrates are the main components of the wall currently used for biorefining. These carbohydrates are mainly fermented by microbes into monosaccharides, particularly hexoses such as glucose (Loqué et al. 2015). However, the cell walls of crop biomass are mainly made of three components: cellulose (30–45%), a β-1,4-glucan polymer which is crystalline; hemicelluloses (20–30%), branched polymers that are mainly composed by xylose and other five-carbon sugars; and lignins (25–35%), non-carbohydrates that interlink other polymers into a robust cell wall structure and architecture (Pauly and Keegstra 2008, 2010). Cellulose-crystallinity and lignin crosslinking are barriers that critically hinder biomass pretreatment and enzyme digestion (Chen and Dixon 2007; Abramson et al. 2010).

The annual production of cell walls by land plants has been estimated as 150–170 billion tons (Pauly and Keegstra 2008). To improve the biomass quality of these energy crops, it is essential to modify the structure of the plant cell wall. A study by Xie and Peng (2011) considered main grasses (rice, wheat, maize, sorghum, etc.) for high biomass production, focusing on cell wall mutants for energy crop breeding, and cell wall-related genes for genetic manipulation.

19.6.1 Energy Generation and Fuel Characteristics

As with other bioenergy crops, energy can be recovered from bamboo biomass in three main ways: thermal, thermochemical, and biochemical conversion (Boyle 2004). Thermal conversion through direct combustion in the presence of oxygen is the most common way of converting solid biomass to energy (Demirbas 2001). The traditional method commonly uses bamboo as firewood to generate heat for household purposes, such as cooking and boiling water. However, these conventional applications are relatively inefficient, often result in high indoor air pollution, and are a major health concern in the developing world (Fullerton et al. 2008). At the industrial scale, biomass like bamboo can be used in power plants to produce heat and power for electricity and district heating plants (Eisentraut and Brown 2012). The heat produced by direct biomass combustion in a boiler under controlled conditions can be used to generate electricity by running a steam turbine or engine. Direct combustion in power plants is the cheapest and most reliable route to producing power from biomass in standalone applications (IEA 2009). Another thermal conversion method, which is more efficient, is pyrolysis. Pyrolysis is the thermal degradation of biomass at a moderate-to-high temperature in the absence of oxygen. It can be used to convert bamboo biomass to solid fuels (charcoals), liquid fuels, and gas (syngas) (Kerlero and De Bussy 2012). Bamboo charcoal can be used as a fuel the same way as coal, and it is a byproduct of the biomass gasification process. The liquid fuels or pyrolysis fuels can be processed in a biorefinery to produce biofuels. Syngas can be used to produce power or electricity. In biochemical conversion, different strains of microorganisms are used to transform biomass to biogas or biofuels. The basic principle of biochemical conversion is the fermentation of sugar or other substances in the bamboo biomass into bioethanol, methane, and other fuels. Thus, bamboo biomass can be utilized in a variety of forms.

Bamboo biomass energy has great potential to be an alternative for fossil fuel (Truong and Le 2014). Bamboo biomass can be processed in various ways (thermal or biochemical conversion) to produce different energy products (charcoal, syngas, and biofuels), which can be substitutions for existing fossil fuel products. Moreover, bamboo has good fuel characteristics, such as high heat values and volatile contents and lower ash and moisture content, which makes it a suitable crop for bioenergy production (Scurlock et al. 2000; Sritong et al. 2012; Kumar and Chandrashekar 2014). In addition, in comparison to other biomass, bamboo has high cellulose and lignin content (Kuttiraja et al. 2013). These properties may differ according to species, location, maturity stage, and management practices, among others (Kumar and Chandrashekar 2014). However, in general, its overall heating value and composition lie between herbaceous biomass and hardwoods. The fuel characteristics (e.g., heating value and chemical composition) of bamboo are similar to those of other dedicated biomass feedstocks. Table 19.2 shows the fuel characteristics of some bamboo species comparison with other energy crops.

Bamboo biomass has both advantages and drawbacks in comparison to other energy sources. It has better fuel characteristics than most biomass feedstocks and suitable for both thermal and biochemical pathways. Bamboo has a number of desirable fuel characteristics such as low ash content and alkali index (Truong and Le 2014). The high heat value (HHV) of bamboo is higher than most agriculture residue (Table 19.2). It indicated that bamboo biomass is a good candidate for direct combustion (e.g., co-combustion in thermal power plant). The moisture content in bamboo is relatively low (8–23%) (Scurlock et al. 2000) in comparison to other types of plants. The low moisture content reduces the energy input to dry the biomass, hence increases the efficiency of utilization.

Balancing food supply and biofuel production is regarded as a long-term national economic policy in modern countries. To satisfy the above goal, selection of energy crops is a promising solution through a precise cell wall modification of food crops (rice, wheat, maize) and an extensive selection of the biomass-rich perennial plants (bamboo, sweet sorghum, and miscanthus) that are of high sugar level and/or high lignocellulose yield, even if grown on marginal lands. Accordingly, three practicable approaches are recommended for energy crop discovery: natural germplasm collection, cell wall mutant selection, and genetic manipulation. As a result, the energy crops should remain high-yield grain with the reconstructed cell walls in their mature straw/stalk that could be efficiently converted into biofuels. Meanwhile, the genetic model is predicted in order to elucidate the dynamic relationships between plant cell wall remodeling and lignocellulose bioconverting.

Table 19.2 Fuel characteristics of some bamboo species and other energy crops

Type of biomass	Moisture (%)	Ash (%)	Volatile compound (%)	Fixed carbon (%)	Higher heating value (kJ/kg)
Bamboo (BD)	14.3	3.7	63.1	18.9	15.7
Bamboo (DS)	5.8	2.7	71.7	19.8	17.585
Bamboo (PN)	13.62	0.41	72.27	13.7	19.27
Bamboo (PB)	9.54	0.53	75.55	14.38	19.49
Bamboo (PBI)	21.97	0.9	64.99	12.14	19.51
Rice husk	12.05	12.73	56.98	18.88	14.638
Rice straw	10.12	10.42	60.87	18.8	13.275
Bagasse	50.76	1.75	41.99	5.86	9.664
Palm shell	12.12	3.66	68.31	16.3	18.446
Corncob	40.11	0.95	45.55	13.68	11.198
Corn stalk	41.69	3.8	46.98	8.14	11.634
Acacia (AA)	11.2	0.36	65.7	22.7	17.40
Acacia (AM)	10.8	0.25	66.0	23.0	17.5

Here, Bamboo (BD) = *Bambusa deecheyama*, Bamboo (DS) = *Dendrocalamus asper*, Bamboo (PN) = *Phyllostachys nigra*, Bamboo (PB) = *Phylotachys bambosoides*, Bamboo (PBI) = *Phyllostachys bissetii*, Acacia (AA) = *Acacia mangium*, and Acacia (AM) = *Acacia auriculiformis* (Source: Sharma et al. 2018; Truong and Le 2014; Marsoem and Irawati 2016)

19.6.2 Availability and Familiarity

Bamboo is a major non-wood forest product and wood substitute. It belongs to the Gramineae family and has about 90 genera with over 1200 species. It is found in all regions of the world and is economic and culturally important. Bamboo is naturally distributed in the tropical and subtropical belt between approximately 46° north and 47° south latitude and is commonly found in Africa, Asia, and Central and South America. Some species may also grow successfully in mild temperate zones in Europe and North America. Bamboo is an extremely diverse plant, which easily adapts to different climatic and soil conditions. Used for housing, crafts, pulp, paper, panels, boards, veneer, flooring, roofing, fabrics, oil, gas, and charcoal (for fuel and as an excellent natural absorbent), it is also a healthy vegetable (bamboo shoot). Bamboo industries are thriving in Asia and are quickly spreading to Africa and America (FAO 2007). The occupied area might be larger than 36 million hectares worldwide or an average of 3.2% of the total forest area (in the reporting countries) if bamboo outside forest land is included. Sixteen countries in Asia reported a total of approximately 24 million hectares of bamboo forest, constituting 4.4% of the total forest area in the surveyed countries. Although the information gathered from Africa is partial, a total of over 2.7 million hectares of bamboo forest was reported by six countries (Ethiopia, Kenya, Nigeria, Uganda, the United Republic of Tanzania, and Zimbabwe). In Latin America, at least ten countries have significant bamboo resources. Although precise assessments are still to be done, a total of over ten million hectares is considered a realistic estimate for the region. Brazil, Chile, Colombia, Ecuador, and Mexico have the richest bamboo resources (FAO 2007).

Bamboo plantations are easy to establish and can be harvested for bioenergy production after 3–5 years, opening new avenues of income generation and boosting the local economies in a short period. Bamboo also requires less agricultural input compared to other bioenergy crops (Sharma et al. 2018), so its production will be a cost-saving resource for the people. Further, in contrast to the estate plantations that offer casual jobs (Sinaga 2013), bamboo plantations under active management may also offer a high number of long-term jobs for the local people (Xuhe 2003). Indeed, the diversification of the income streams will broaden the livelihood options and reduce the vulnerability of the farmers to crop failure, helping them adapt to the changing climate (Bradshaw et al. 2004). In addition, the electricity generated through bamboo power plants, mainly in regions that lack modern energy sources, could support the local communities to increase their household income by engaging in economic activities, such as running small industries. The bioenergy from the bamboo power plants could thus catalyze rural economic activities and could provide a basis for the alleviation of poverty in the country.

19.6.3 Avoiding the Food–Energy–Environment Trilemma

The production of feedstocks for bioenergy requires land and water. Thus, bioenergy is often debated because of its potential to cause negative impacts on food production and biodiversity due to the direct and indirect land-use change and competition for resources (Immerzeel et al. 2014). Using bamboo as a feedstock for bioenergy can avoid these conflicts, especially when bamboo is grown on degraded and underutilized land. Bamboo is abundantly available, fast-growing, and can grow on degraded and marginal lands or in combination with other crops in forestry or agroforestry systems; thus, there will be no competition for land (Mishra et al. 2014). As a fast-growing species that can develop in degraded lands, it can also establish a habitat for biodiversity. Also, the increased availability of bamboo for bioenergy will help to replace the use of firewood from forests, thereby lowering the pressure on the forests. Bamboo crops are usually ready in 5–12 years and can be systematically harvested without removing the clump every year for the next life cycle of 30–50 years (Sharma et al. 2018), while other bioenergy crops require replanting after harvest. In fact, managing the stand's age and density by annual thinning—using the derived material as feedstock—can increase bamboo productivity (Sharma et al. 2018). Thus, bamboo can provide a sustained source of feedstock for bioenergy production.

19.6.4 Climate Action

The land-use sector is a major contributor to the country's GHG emissions (Sharma et al. 2018). Bamboo bioenergy offers a number of opportunities for emission reduction in the land-use sector. First, bamboo contributes to reducing emissions by replacing the use of fossil fuels for energy generation. Second, as a fast-growing species, it can rapidly sequester and store carbon in its biomass (Lou et al. 2010). The availability of carbon storage and sequestration data for bamboo in different cultivation systems is limited, but several studies around the world have made estimations based on the type of species composition, geographic location, environmental conditions, and management practices. In China, the managed bamboo ecosystems have high carbon storage (102 t ha^{-1} to 288 t ha^{-1}) which is comparable to other forest types (122 t ha^{-1} to 337 t ha^{-1}) (Lou et al. 2010). A managed Moso bamboo (*Phyllostachys edulis*) ecosystem can store up to 106.36 t ha^{-1} (34.3 t ha^{-1} in the aboveground green vegetation and 72.2 t ha ha^{-1} on the forest floor and soil up to 60 cm in depth) (Lou et al. 2010; Zhou and Jiang 2004). On the other hand, the carbon sequestration potential of many bamboo species is comparable to and often higher than that of many fast-growing tree species. For instance, a study in Bangladesh shows that the total carbon sequestration of five-year-old common bamboo (*Bambusa vulgaris*) is higher (15.53 t ha^{-1} $year^{-1}$) than that of other fast-growing hardwood species like Acacia (*Acacia auriculiformis*) (10.21 t ha^{-1} $year^{-1}$ for an 11-year-old crop) (Sohel et al. 2015). In India, the rate of aboveground carbon sequestration in a mixed patch of Bambusa species (*B. vulgaris*, *B. balcooa*, and *B. cacharensis*) ranges between 18.93 and 23.55 t ha^{-1} $year^{-1}$ (Nath and Das 2012). These findings suggest that bamboo can be valuable in sequestering carbon and can be leveraged in bioenergy production to help mitigate climate change.

19.7 Our Laboratory Focus Towards Biorefinery

Significant advances in plant cell, tissue, and organ culture (PCTOC) have been made in the last five decades. PCTOC is now thought to be the underlying technology to understand general or specific biological functions of the plant kingdom. It is one of the most flexible foundations for morphological, physiological, and molecular biology applications of plants (Ogita 2015). Plant cell manipulation laboratories focus on the applications of PCTOC methodologies to all research and development areas of traditional and modern plant biotechnology, e.g., PCTOC, transformation, plant stem cell manipulation, histochemical analysis, and metabolic engineering. We apply these technologies into different aquatic and terrestrial plants, especially woody, medicinal, flowering, aromatic, and staple food species. In the present study, we established an efficient in vitro node culture system (Ogita et al. 2008) and a prominent cell suspension culture system (Ogita 2005) for bamboo.

Bamboo is an ecologically and economically important grass species belonging to the Poaceae family. Worldwide interest in bamboo as a source of biomass in sustainable agriculture and agroforestry system has increased rapidly in recent years (Malini and Anandkumar 2013). Bamboo has been exploited for a range of uses such as food, medicine, charcoal, and housing materials, especially in Asia. Owing to their wide utility and productivity, bamboo species are increasingly regarded as a valuable resource for use in renewable energy in the development of a low-carbon society (Ogita et al. 2018).

Considering that bamboo is one of the main renewable resources of biomass production, our research focuses on the application of different gene or factor delivery to modify bamboo cells. First, we developed a particle bombardment-mediated transformation protocol for *Phyllostachys* bamboo by optimizing the growth efficiency of a target cell culture system (Ogita et al. 2011). Stable transgenic bamboo cells were generated with constructs that express the hygromycin-phosphotransferase gene and enhanced fluorescent protein genes, namely *AcGFP1* and *mCherry*.

Lignocellulosic biomass is the most abundant organic material for biofuel production. Both the biomass yield of a biofuel crop and the cell wall components of the lignocellulosic biomass will determine the efficiency of the overall biofuel production. For that reason, we standardized a novel xylogenic suspension culture model to unveil the process of lignification in living bamboo cells (Ogita et al. 2012). Before starting this research, we believed that there were prominent cultured plant cell lines, such as the tobacco cells (BY-2 line) (*Nicotiana tabacum* cv. Bright Yellow 2) and the Arabidopsis T87 cell line (*Arabidopsis thaliana* (L.) Heynh. ecotype Columbia), which have been used as model plant cells. These suspension cell culture systems are highly applicable to investigate various aspects of plant cell biology. However, no such prominent cultured cell lines exist in bamboo species. Therefore, two types of xylogenic differentiation, fiber-like elements (FLEs) with cell wall thickening and tracheary elements (TEs) with formation of perforations in the cell wall, were observed. The suspension cells rapidly formed secondary cell wall components that were highly lignified, accounting for approximately 25% of the cells on a dry weight basis within 2 weeks. Detailed features of cell growth, differentiation, and death during lignification were characterized by laser scanning microscopic (LSM) imaging. Changes in transcript levels of xylogenesis-related genes were assessed by RT-PCR, which showed that the transcription of essential genes such as PAL1, C4H, CCoAOMT, and CCR was induced at day 4 under lignification conditions. Furthermore, interunit linkage in lignin was compared between mature bamboo culms and the xylogenic suspension cells by heteronuclear single-quantum coherence nuclear magnetic resonance (HSQC NMR) spectroscopy. The most common interunit linkages structures, including β-aryl ether (β-O-4), phenylcoumaran (β-5) and resinol (β-β), were identified in the bamboo cultured cell lignin (BCCL) by HSQC NMR. In addition to these common features of lignin, several differences in lignin substructures were also observed between the BCCL and the bamboo milled wood lignin (BMWL). Our xylogenic suspension culture model was used for a detailed characterization of physiological and molecular biological events in the living bamboo cells. Later, the chemical basis of xylogenesis in bamboo cells was explored (Nomura et al. 2013). The secondary metabolite compositions between xylogenic and non-xylogenic suspension cell cultures of bamboo (*Phyllostachys nigra*), which we developed previously, were compared. Two compounds, one major and one minor, showed a significant increase in the cells cultured under two lignification (xylogenic) conditions, in comparison with cells cultured under proliferation (non-xylogenic) conditions. Based on spectroscopic analyses, the major compound was feruloylputrescine (FP) and the minor compound was *p*-coumaroylputrescine (pCP). We examined the accumulation profiles of the hydroxycinnamic acid amides (HCAAs) during a 16-day culture period. The cells were kept under proliferation conditions for 16 days, the contents of FP and pCP peaked at 2 days [(0.32 and 0.25 nmol·mg^{-1}) fresh weight, respectively] and decreased to trace levels thereafter. In contrast, the FP content increased throughout the 16-day culture period, reaching maximum levels of 4.3 and 6.8 nmol mg^{-}1 fresh weight in the two xylogenic conditions. The pCP content was lower than that of FP under both xylogenic conditions. The pattern of FP accumulation resembled that of lignin accumulation, as monitored by phloroglucinol-HCl staining. It is likely that FP affects xylogenesis in the suspension-cultured bamboo cells.

Our group also focused on the analysis of metabolites of different important plant species such as the biosynthetic activities of primary and secondary metabolites in suspension cultures of *Aquilaria microcarpa* (Ogita et al. 2015a), and the potential of plant cell culture systems to produce a target natural bioactive compound. We proposed a stepwise protocol for β-thujaplicin production (Ogita et al. 2015b). In the suspension-culture system, secondary metabolites in target cells of *Aquilaria microcarpa* were detected using LSM imaging with diphenylboric acid 2-amino ethyl ester (DPBA) and 9-diethylamino-5H-benzo[alpha]phenoxazine-5-one (Nile red) staining. Accumulation of flavonoids and lipids, observed from the scanning of aggregated cells, produced clear fluorescent images (Fig. 19.5).

As shown in Fig. 19.5, the biosynthetic activity of primary and secondary metabolites in suspension cells was visualized through LSM imaging combined with DPBA and Nile red staining. Positive fluorescent images scanned from a portion of the suspension cells clearly showed an accumulation of flavonoids and lipids. By using the prominent emission spectra obtained from the images, we quantitatively evaluated the productivity of several different metabolites in each individual cell mass (lambda scan mode).

An unknown whitish-gray resinous compound sedimented in the medium was also visualized in the extracellular portion of aggregated cells. This research provided an opportunity to investigate the metabolite productivity of aggregated suspension cells. Some prominent extracellular compounds were detected in the used liquid medium, as well as in the resinous residue within the medium. The characteristics of these metabolites were investigated in detail via gas chromatography-mass spectrometry (GC-MS). In another research, Ogita et al. (2015b) demonstrated the potential of plant cell culture systems to produce a target natural bioactive compound and proposed a stepwise protocol for β-thujaplicin production as mentioned before. In the present experiment, different conifer species were used, e.g., *Chamaecyparis* (*C. obtusa* Sieb. et Zucc., and *C. pisifera* Sieb. et Zucc.), *Juniperus* (*J. chinensis* L. "Kaizuka," *J. chinensis* L. var. *sargentii*, and *J. conferta* Parlatore), *Thuja* (*T. occidentalis* L. and *T. standishii* (Gord.) Carr.), *Thujopsis* (*T. dolabrata* Sieb. et Zucc., and *T. dolabrata* Sieb. et Zucc. var. *hondae*), and *Cryptomeria* (*C. japonica* D. Don). We observed that the phenotypes of each callus determine the optimal conditions to induce callus and infer biosynthetic

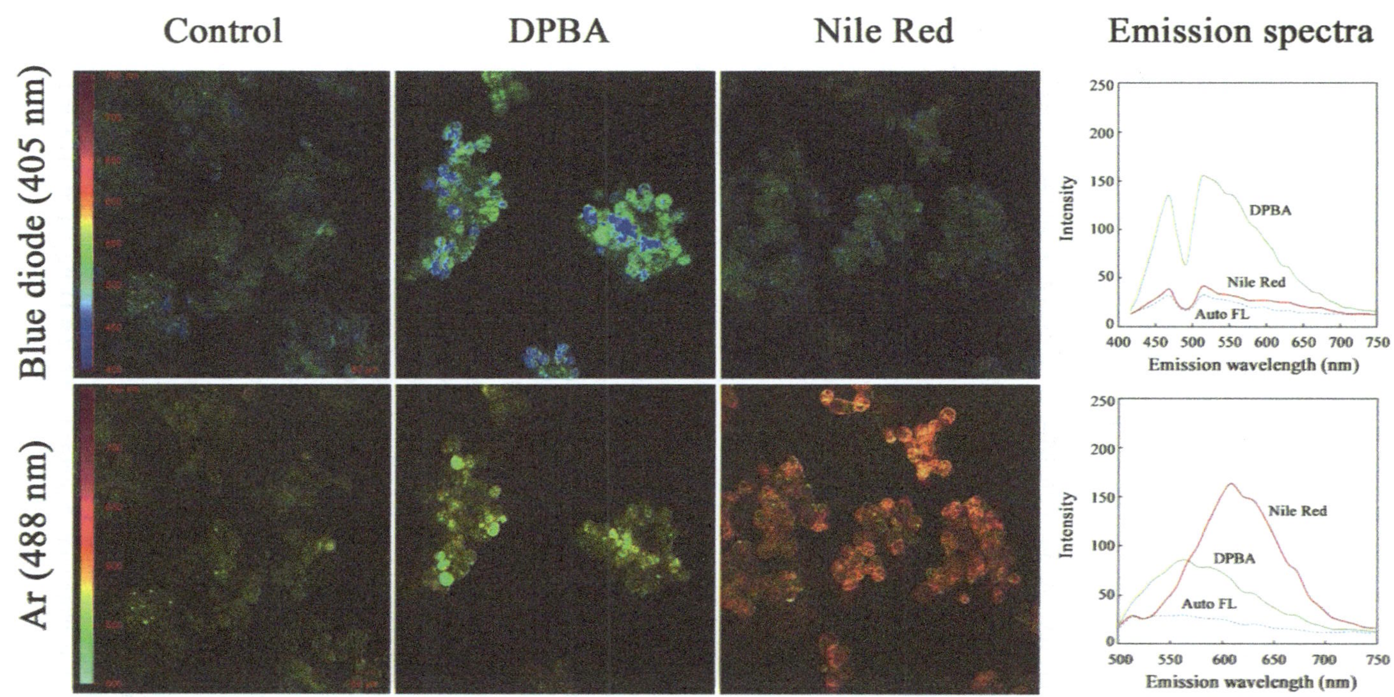

Fig. 19.5 LSM imaging of *Aquilaria* suspension cells stained with florescent reagents (Lambda scan mode). Control: no staining, DPBA: 0.25% w/v, saturated solution in PBS, Nile red: 0.3% w/v, saturated solution in 2-propanol (diluted to 0.005% with PBS) ("Figure was reproduced from Ogita et al. 2015a" by the permission of Natural Product Communications)

activity of the calli over 4–8 weeks. In the habituation phase, each of the cell cultures obtained was transferred to a modified MS (Murashige and Skoog 1962) medium containing 680 mg L^{-1} of KH_2PO_4 and 10 μM of picloram to select the habituated cells with synchronous growth pattern. The growth of each cell culture was highly improved in the habituation medium, except for *J. chinensis* "Kaizuka." In the metabolite-production phase, the habituated cell cultures and the concentration of μ-thujaplicin (known as hinokitiol in Japan) in the shoots of donor trees were analyzed via high-performance liquid chromatography (HPLC). Histochemical characteristics of the cells were also observed using LSM imaging. After the third step, we tested the biosynthetic activity of two habituated calli (*C. obtusa* and *J. conferta*) on a 0.3% w/v yeast extract (YE)-containing medium. We found significant improvement in β-thujaplicin production in *J. conferta* callus (4600 μg g DW^{-1}), which was up to 20-fold higher than in the habituation phase.

To explore the functional differentiation of cells responsible for cell wall and fiber formation, efficient PCTOC models of *Phyllostachys* and *Bambusa* bamboo plants and their application were established by Ogita et al. (2016). Standardization of a novel xylogenic suspension culture model to understand the process of xylogenic cell differentiation during lignification in living *Phyllostachys nigra* (Pn) was the first step. The Pn cells rapidly formed secondary cell wall components that were highly lignified, representing approximately 25% of the dry weight of the cells under the xylogenic differentiation condition (1/2 MS medium supplemented with 10 μM BA).

We systematically maintained in vitro node culture stocks of two prominent *Bambusa* bamboo species: *B. multiplex* (Bm), which has a normal "hollow" culm; and *B. glaucescens* f. *houraikomachi* (Bg), which has a thick-walled "solid" culm. As node portions have apical and intercalary meristems, they could be directly used as explant sources for the establishment of callus and organ cultures without sterilization. When node portions were cultured on an optimized proliferation medium (MSp680 medium supplemented with 10 μM picloram), active callus induction and organ differentiation were observed. Although calli usually proliferate as irregular tissue masses and vary widely in texture, it is still possible to generate different cell lines such as whitish and greenish callus, and bunches of adventitious roots, under the same medium conditions by carefully separating these cultures from the explant.

Highly lignified culms of bamboo show distinctive anatomical and mechanical properties compared with the culms of other grass species. A cell culture system for Pn has enabled investigating the alterations in cellular states associated with secondary cell wall formation during its proliferation and lignification in woody bamboos. To know the transcriptional alteration during proliferation and lignification in Pn cells, we analyzed transcriptome treated with the synthetic auxin 2,4-dichlorophenoxyacetic acid (2,4-D) and the synthetic cytokinin benzylaminopurine (BA) by RNAseq analysis (see Fig. 19.6, Ogita et al. 2018). We observed that some genes putatively involved in cell wall biogenesis and cell division were upregulated in response to the 2,4-D treatment. The induction of lignification by the BA treatment was correlated with the upregulation of genes involved in the shikimate pathway. We also observed that genes encoding MYB

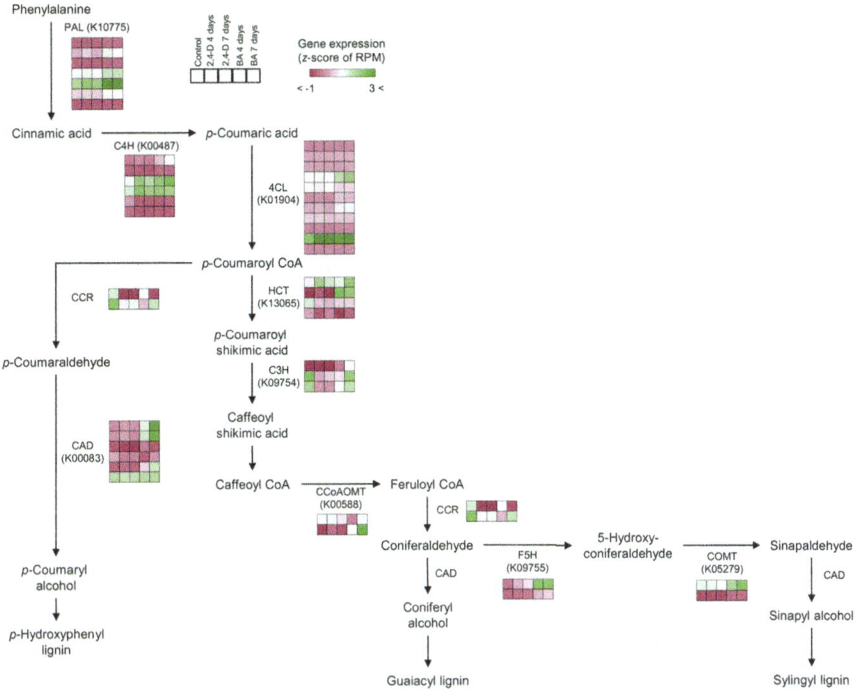

Fig. 19.6 Expression of *P. nigra* genes involved in the monolignol pathway. *P. edulis* genes encoding phenylalanine ammonia-lyase (PAL), cinnamate 4-hydroxylase (C4H), 4-coumarate-CoA ligase (4CL), hydroxycinnamoyl CoA: shikimate transferase (HTC), p-coumarate 3-hydroxylase (C3H), caffeoyl CoA O-methyltransferase (CCoAOMT), cinnamoyl CoA reductase (CRR), ferulate 5-hydroxylase (F5H), caffeic acid O-methyltransferase (COMT), and cinnamyl alcohol dehydrogenase (CAD) were represented in the monolignol pathway. The expression patterns of *P. nigra* genes corresponding to their homologs in *P. edulis* were estimated from the RPM values obtained from the cross-species mapping of *P. nigra* RNA-seq reads to the *P. edulis* genome. The color gradient represents normalized gene expression based on *z*-score of the RPM values ("Figure was reproduced from Ogita et al. 2018", ©Shinjiro Ogita (Licensed under CC BY 4.0))

transcription factors (TFs) show correlated expression patterns with those encoding cinnamyl alcohol dehydrogenase (CAD), suggesting that MYB TFs regulate secondary cell wall formation in the bamboo cells. These findings suggest that cytokinin signaling may regulate lignification in Pn cells through coordinated transcriptional regulation and metabolic alterations.

Our results have also produced a useful resource to better understand secondary cell wall formation in bamboo plants. In another research on Pn, Nomura et al. (2018) focused on metabolic-flow switching to produce exogenous secondary metabolites in bamboo suspension cells. Based on synthetic biology, the production of high-value plant secondary metabolites in microbial hosts has attracted extensive attention despite various challenges, including correct protein expression and supply of starting materials. In contrast, plant cell cultures are rarely used for this purpose owing to their slow proliferation rates and laborious transformation processes. Therefore, we propose a "rational metabolic-flow switching" strategy to efficiently produce exogenous secondary metabolites using suspension-cultured bamboo (Pn) cells as model production hosts. The Pn cells biosynthesize HCAAs of putrescine as major secondary metabolites, which indicates that the phenylpropanoid and polyamine biosynthetic pathways are highly active, and the Pn cells may produce alternative secondary metabolites derived from those pathways. Stable transformants of Pn cells expressing agmatine coumaroyl transferase of barley (*Hordeum vulgare*) were generated with the expectation of metabolic-flow switching from HCAAs of putrescine to those of agmatine. In the recombinant Pn cells, the levels of HCAAs of putrescine decreased and HCAAs of agmatine were produced. The production titer of the major product, *p*-coumaroylagmatine, reached approximately 360 mg/L, providing a proof-of-concept for the usefulness of "rational metabolic-flow switching" in synthetic biology using plant cell hosts.

Plant metabolic engineering is a significant subject because it can create alternatives for safer sustainable biorefinery production, providing alternatives for GM.

19.8 Concluding Remarks

Different generations of biofuels indicate that there is a continuous interest in alternative renewable fuels. From the first generation until today, scientists attempt to use modern technologies for biorefinery with the ultimate goal of attenuating the consequences of global warming. The use of biofuels can reduce GHG and other air pollutant emissions, providing a more eco-friendly environment. Different generations of biofuel depend on different materials, e.g., the first generation uses grain or food-based feedstocks; the second uses lingo-cellulosic biomass; and the third, algal material. We are now in the fourth-generation biorefinery, which focuses on CCS. This generation is not only carbon neutral but also carbon negative.

Every generation has advantages and disadvantages. However, it is known that second generation biofuel plants have achieved encouraging success, e.g., commercial-scale production currently open and running. Biofuels have replaced fossil fuels in some industrial settings, reportedly reducing GHG. Considering the future of biofuels, the greatest development potential is the use of genetic engineering to maximize the potential of feedstock and microbial drivers of production.

Benefits aside, GE or GMOs have been considered a threat to the environment and human health. Therefore, testing the feasibility of GMOs in contained and controlled environments for any potential risks is considered necessary by biosafety regulations of individual countries. Considering these factors, we established a common PCTOC methodology to increase the target component in organisms and maximize biorefinery production. Bamboo, as the highest biomass producer, or other energy crops can be the target organisms of PCTOC methods, which can be used instead of GE technology for a safer biorefinery. Perhaps this technology will create a new generation of biorefineries. Finally, we hope that in the future, biofuels will represent a safe and economical alternative to fossil fuels.

References

Abramson M, Shoseyov O, Shani Z (2010) Plant cell wall reconstruction toward improved lignocellulosic production and processability. Plant Sci 178:61–72

Al-Thani RF, Potts M (2012) Cyanobacteria, oil and cyanofuel? In: Whitton BA (ed) Ecology of cyanobacteria II: their diversity in space and time. Springer, Dordrecht, pp 427–440

Antizar-Ladislao B, Turrion-Gomez JL (2008) Second-generation biofuels and local bioenergy systems. Biofuels Bioprod Biorefin 2:455–469

Aro EM (2016) From first generation biofuels to advanced solar biofuels. Ambio 445(Suppl 1): S24–S31

Balat M, Balat H, Öz C (2008) Progress in bioethanol processing. Prog Energy Combust Sci 34:551–573

Ballesteros I, Negro MJ, Oliva JM, Cabanas A, Manzanares P, Ballesteros M (2006) Ethanol production from steam-explosion pretreated wheat. Appl Biochem Biotechnol 130:496–508

Belhaj K, Chaparro-Garcia A, Kamoun S, Patron NJ, Nekrasov V (2015) Editing plant genomes with CRISPR/Cas9. Curr Opin Biotechnol 32:76–84

Betancur GJV, Pereira N (2010) Sugarcane bagasse as feedstock for second generation ethanol production. Part I: diluted acid pretreatment optimization. Electron J Biotechnol 13:1–9

Bhansali S, Kumar A (2018) Synthetic and semisynthetic metabolic pathways for biofuel production. In: Kumar A et al (eds) Biofuels: greenhouse gas mitigation and global warming. Springer, New Delhi

Boyle G (2004) Renewable energy: power for a sustainable future. Oxford University Press, Oxford, p 464

Bradshaw B, Dolan H, Smit B (2004) Farm-level adaptation to climatic variability and change: crop diversification in the Canadian prairies. Clim Chang 67:119–141

Bujis NA, Siewers V, Nielsen J (2013) Advanced biofuel production by the yeast *Saccharomyces cerevisiae*. Curr Opin Chem Biol 17:480–488

Carere CR, Sparling R, Cicek N, Levin DB (2008) Third generation biofuels via direct cellulose fermentation. Int J Mol Sci 9:1342–1360

Cha M, Chung D, Elkins JG, Guss AM, Westpheling J (2013) Metabolic engineering of *Caldicellulosiruptor bescii* yields increased hydrogen production from lignocellulosic biomass. Biotechnol Biofuels 6:85

Chen F, Dixon RA (2007) Lignin modification improves fermentable sugar yields for biofuel production. Nat Biotechnol 25:759–761

Cherubini F (2010) The biorefinery concept: using biomass instead of oil for producing energy and chemicals. Energy Convers Manag 51(7):1412–1421

Demirbas A (2001) Biomass resource facilities and biomass conversion processing for fuels and chemicals. Energy Convers Manag 42:1357–1378

Demirbas A (2007a) Importance of biodiesel as transportation fuel. Energy Policy 35(9):4661–1670

Demirbas A (2007b) Progress and recent trends in biofuels. Prog Energy Combust Sci 33(1):1–18

Ding Y, Li H, Chen LL, Xie K (2016) Recent advantages in genome editing using CRISPR/Cas 9. Front Plant Sci 7:703

Dragone G, Fernandes B, Vicente AA, Teixeira JA (2010) Third generation biofuels from microalgae. In: Méndez-Vilas A (ed) Current research, technology and education topics in applied microbiology and microbial biotechnology. FORMATEX, Badajoz, pp 1355–1366

Duhamel du Monceau HL (1756) Contenant les experiences et reflexions sur la culture des terres et sur la conservation des grains, faites pendant les années 1755 et 1756. In: Duhamel du Monceau

HL, Tull J (eds) Traite de la culture des terres. Hippolyte-Louis Guerin et Louis-Francois Delatour, Paris
Eisentraut A, Brown A (2012) Technology roadmap: bioenergy for heat and power. Technol Roadmap 2:1–41
Elshahed M (2010) Microbiological aspects of biofuel production: current status and future directions. J Adv Res 1:103–111
FAO (2007) Non wood forest product: World bamboo resources: a thematic study prepared in the framework of the global forest resources assessment. FAO, Rome, p 2005
FAO (2008) Bioenergy, food security and sustainability—towards an international framework. In: Paper prepared for the high-level conference on world food security: the challenges of climate change and bioenergy, 3–5 June 2008
Fu C, Mielenz JR, Xiao X, Ge Y, Hamilton CY, Rodriguez M, Chen F, Foston M, Ragauskas A, Boutona J, Dixon RA, Wang ZY (2011) Genetic manipulation of lignin reduces recalcitrance and improves ethanol production from switchgrass. Proc Natl Acad Sci U S A 108:3803–3808
Fullerton DG, Bruce N, Gordon SB (2008) Indoor air pollution from biomass fuel smoke is a major health concern in the developing world. Trans R Soc Trop Med Hyg 102:843–851
Gao L, Li D, Gao F, Liu Z, Hou Y, Chen S, Zhang D (2015) Hydroxyl radical-aided thermal pretreatment of algal biomass for enhanced biodegradability. Biotechnol Biofuels 8:194
Gautheret RJ (1985) History of plant tissue culture and cell culture: a personal account. In: Vasil IK (ed) Cell culture and somatic cell genetics of plants, Cell growth, nutrition, cytodifferentiation and cryopreservation, vol 2. Academic, Orlando, pp 1–59
Gopinathan MC, Sundhakaran R (2009) Biofuels: opportunities and challenges in India. In Vitro Cell Dev Biol Plant 45(3):350–371
Hill J, Nelson E, Tilman D, Polasky S, Tiffany D (2006) Environmental, economic, and energetic costs and benefits of biodiesel and ethanol biofuels. Proc Natl Acad Sci U S A 103:11206–11210
Hu Q, Sommerfeld M, Jarvis E, Ghirardi M, Posewitz M, Seibert M, Darzins A (2008) Microalgaltriacylglycerols as feedstocks for biofuel production: perspectives and advances. Plant J 54:621–639
IEA (2009) Bioenergy—a sustainable and reliable energy source. IEA, Paris
Immerzeel DJ, Verweij P, Hilst F, Faaij AP (2014) Biodiversity impacts of bioenergy crop production: a state-of-the-art review. GCB Bioenergy 6:183–209
Inganas O, Sundström V (2016) Solar energy for electricity and fuels. V Ambio 45(Suppl 1):15–23
IPCC (2018) Summary for policymakers. In: Masson-Delmotte V, Zhai P, Pörtner HO, Roberts D, Skea J, Shukla PR, Pirani A, Moufouma-Okia W, Péan C, Pidcock R, Connors S, Matthews JBR, Chen Y, Zhou X, Gomis MI, Lonnoy E, Maycock T, Tignor M, Waterfield T (eds) Global warming of 1.5°C. An IPCC special report on the impacts of global warming of 1.5°C above pre-industrial levels and related global greenhouse gas emission pathways, in the context of strengthening the global response to the threat of climate change, sustainable development, and efforts to eradicate poverty. World Meteorological Organization, Geneva, pp 3–24
Jönsson LJ, Martín C (2016) Pretreatment of lignocellulose: formation of inhibitory by-products and strategies for minimizing their effects. Bioresour Technol 199:103–112
Jonsson LJ, Alriksson B, Nilvebrant NO (2013) Bioconversion of lignocellulose: inhibitors and detoxification. Biotechnol Biofuels 6:16
Kamm B, Kamm M (2004) Principle of biorefineries. Appl Microbiol Biotechnol 64(2):137–145
Kamm B, Kamm M, Gruber PR, Kromus S (2006) Biorefinery systems—an overview. In: Kamm B et al (eds) Biorefineries—industrial process and products (status quo and future directions). Wiley-VCH, Weinheim, pp 3–40
Kawaguchi H, Vertes AA, Okino S, Inui M, Yukawa H (2006) Engineering of axylose metabolic pathway in *Corynebacterium glutamicum*. Appl Environ Microbiol 72:3418–3428
Kerlero DR, De Bussy J (2012) Electrical valorization of bamboo in Africa. ENEA Consulting, Paris
Klein RM, Wolf ED, Wu R, Sanford JC (1992) High-velocity microprojectiles for delivering nucleic acids into living cells. 1987. Bioechnology 24:384–386

Kotchoni OS, Shonukan OO, Gachomo WE (2003) *Bacillus pumilus* BpCRI 6, a promising candidate for cellulase production under conditions of catabolite repression. Afr J Biotechnol 2:140–146

Kozukue N, Misoo S, Yamada T, Kamijima O, Friedman M (1999) Inheritance of morphological characters and glycoalkaloids in potatoes of somatic hybrids between dihaploid *Solanum acaule* and tetraploid *Solanum tuberosum*. J Agric Food Chem 47(10):4478–4483

Kuhad RC, Kumar M, Singh A (1994) A hypercellololytic mutant of *Fusarium oxysporum*. Lett Appl Microbiol 19:397–400

Kumar R, Chandrashekar N (2014) Fuel properties and combustion characteristics of some promising bamboo species in India. J For Res 25:471–476

Kumar R, Singh S, Singh OV (2008) Bioconversion of lignocellulosic biomass: biochemical and molecular perspectives. J Ind Microbiol Biotechnol 35:377–391

Kuttiraja M, Sindhu R, Varghese PE, Sandhya SV, Binod P, Vani S, Pandey A, Sukumaran RK (2013) Bioethanol production from bamboo (*Dendrocalamus* sp.) process waste. Biomass Bioenergy 59:142–150

Larkin PJ, Scowcroft WR (1981) Somaclonal variation—a novel source of variability from cell cultures for plant improvement. Theor Appl Genet 60(4):197–214

Li Q, Song J, Peng S, Wang JP, Qu GZ, Sederoff RR, Chiang VL (2014) Plant biotechnology for lignocellulosic biofuel production. Plant Biotechnol J 12:1174–1192

Lin W-H, Giachello CNG, Baines RA (2017) Seizure control through genetic and pharmacological manipulation of Pumilio in Drosophila: a key component of neuronal homeostasis. Dis Model Mech 10:141–150

Loqué D, Scheller HV, Pauly M (2015) Engineering of plant cell walls for enhanced biofuel production. Curr Opin Plant Biol 25:151–161

Lou Y, Li Y, Buckingham K, Henley G, Zhou G (2010) Bamboo and climate change mitigation. Technical report. International Network for Bamboo and Rattan (INBAR), Beijing

Maity SK (2015) Opportunities, recent trends and challenges of integrated biorefinery: part I. Renew Sustain Energ Rev 43:1427–1445

Malini N, Anandakumar CR (2013) Micropropagation of bamboo (*Bambusa vulgaris*) through nodal segment. Int J Forestry Crop Improv 4(1):36–39

Marsoem SN, Irawati D (2016) Basic properties of *Acacia mangium* and *Acacia auriculiformis* as a heating fuel. AIP Publishing, Melville

McGinnis KM (2010) RNAi for functional genomics in plants. Brief Funct Genomics 9(2):111–117

Mishra G, Giri K, Panday S, Kumar R, Bisht N (2014) Bamboo: potential resource for eco-restoration of degraded lands. J Biol Earth Sci 4:130–136

Mood SH, Golfeshan AH, Tabatabaei M, Jouzani GS, Najafi GH, Gholami M, Ardjmand M (2013) Lignocellulosic biomass to bioethanol, a comprehensive review with a focus on pretreatment. Renew Sust Energ Rev 27:77–93

Murashige T, Skoog F (1962) A revised medium for rapid growth and bioassays with tobacco tissue cultures. Physiol Plant 15:473–497

Nadarajah KK (2018) Biofuel sector in Malaysia: challenges and future prospects. In: Kumar A et al (eds) Biofuels: greenhouse gas migitation and global warming. Springer, New Delhi

Nath AJ, Das AK (2012) Carbon pool and sequestration potential of village bamboos in the agroforestry system of northeast India. Trop Ecol 53:287–293

Nomura T, Shiozawa M, Ogita S, Kato Y (2013) Occurrence of hydroxycinnamoylputrescines in xylogenic bamboo suspension cells. Plant Biotechnol 30:447–453

Nomura T, Ogita S, Kato Y (2018) Rational metabolic-flow switching for the production of exogenous secondary metabolites in bamboo suspension cells. Sci Rep 8:13203

Nozzi NE, Oliver JWK, Atsumi S (2013) Cyanobacteria as a platform for biofuel production. Font Bioeng Biotechnol 1:7

Ogita S (2005) Callus and cell suspension culture of bamboo plant, *Phyllostachys nigra*. Plant Biotechnol 22:119–125

Ogita S (2015) Plant cell, tissue and organ culture: the most flexible foundations for plant metabolic engineering applications. Nat Prod Commun 10:815–820

Ogita S, Kashiwagi H, Kato Y (2008) In vitro node culture of seedlings in bamboo plant, *Phyllostachys meyeri* Mcclure. Plant Biotechnol 25:381–385
Ogita S, Kikuchi N, Nomura T, Kato Y (2011) A practical protocol for particle bombardment-mediated transformation of *Phyllostachys* bamboo suspension cells. Plant Biotechnol 28:43–50
Ogita S, Nomura T, Kishimoto T, Kato Y (2012) A novel xylogenic suspension culture model for exploring lignification in *Phyllostachys* bamboo. Plant Biotechnol 28(1):43–50
Ogita O, Leeb J-B, Kurosaki F, Kato Y (2015a) The biosynthetic activities of primary and secondary metabolites in suspension cultures of *Aquilaria microcarpa*. Nat Prod Commun 10 (5):779–782
Ogita S, Shichiken M, Ito C, Yamashita T, Nomura T, Kato Y (2015b) A stepwise protocol for induction and selection of prominent coniferous cell cultures for the production of β-thujaplicin. Nat Prod Commun 10(5):783–787
Ogita S, Kishimoto T, Nomura T, Kato Y (2016) Plant cell, tissue, and organ culture approaches to explore the functional cell differentiation in *Phyllostachys* and *Bambusa* bamboo plants. Fiber Plant:111–126
Ogita S, Nomura T, Kato Y, Uehara-Yamaguchi Y, Inoue K, Yoshida T, Sakurai T, Shinozaki T, Mochida K (2018) Transcriptional alterations during proliferation and lignification in *Phyllostachys nigra* cells. Sci Rep 8:11347
Ohara H (2003) Biorefinery. Appl Microbiol Biotechnol 62:474–477
Pauly M, Keegstra K (2008) Cell-wall carbohydrates and their modification as a resource for biofuels. Plant J 54:559–568
Pauly M, Keegstra K (2010) Plant cell wall polymers as precursors for biofuels. Curr Opin Plant Biol 13:305–312
Petrou EC, Pappis CP (2009) Biofuels: a survey on pros and cons. Energy Fuel 23:1055–1066
Popp J, Harangi-Rákos M, Gabnai Z, Balogh P, Antal G, Bai A (2016) Biofuels and their co-products as livestock feed: global economic and environmental implications. Molecules 21:285
Ragauskas AJ, Beckham GT, Biddy MJ, Chandra R, Chen F, Davis MF, Davison BH, Dixon RA, Gilna P, Keller M, Langan P, Naskar AK, Saddler JN, Tschaplinski TJ, Tuskan GA, Wyman CE (2014) Lignin valorization: improving lignin processing in the biorefinery. Science 344:6185
Razdan MK (2006) Introduction to plant tissue culture, 2nd edn. Oxford and IBH Publishing Co. Pvt. Ltd, New Delhi
Rezende CA, Lima MA, Maziero P, Azevedo ER, Garcia W, Polikarpov I (2011) Chemical and morphological characterization of sugarcane bagasse submitted to a delignification process for enhanced enzymatic digestibility. Biotechnol Biofuels 4:54
Rinaldi R, Jastrzebski R, Clough MT, Ralph J, Kennema M, Bruijnincx CA, Weckhusyen BM (2016) Paving the way for lignin valorisation: recent advances in bioengineering, biorefining and catalysis. Angew Chem Int Ed 55:2–54
Rodhe RA, Muller RA (2015) Air pollution in China: mapping of concentrations and sources. PLoS One 10:e0135749
Rowland GG, McHughen AG, Hormis YA, Rashid KY (2002) CDC Normandy flax. Can J Plant Sci 82:425–426
Rowland GG, MoHughen A, McOnie C (1989) Field performance at saline-affected sites of a somaclonal variant of McGregor flax selected for salt tolerance *in vitro*. Can J Plant Sci 69:49–60
Roy A, Kumar A (2013) Pretreatment methods of lignocellulosic materials for biofuel production: a review. J Emerg Trends Eng Appl Sci (JETEAS) 4(2):181–193
Rutz D, Janseen R (2007) Biofuel technology handbook. WIP Renewable Energies, Munich
Sanford KK, Earle WR, Likely GD (1948) The growth in vitro of single isolated tissue cells. JNCL, Natl Cancer Inst 9:229–246
Saxena RC, Adhikari DK, Goyal HB (2009) Biomass-based energy fuel through biochemical routes: a review. Renew Sust Energ Rev 13:167–178
Scurlock J, Dayton D, Hames B (2000) Bamboo: an overlooked biomass resource? Biomass Bioenergy 19:229–244

Sears EJ (1956) The transfer of leaf-rust resistance from *Aegilops umbellulata* into wheat. Brookhaven Symp Biol 9:1–22

Sears ER (1981) Transfer of alien genetic material to wheat. In: Evans LT, Peacock WJ (eds) Wheat science—today and tomorrow. Cambridge University Press, Cambridge, pp 75–89

Sebastian SA, Chaleff RS (1987) Soybean mutants with increased tolerance for sulfonylurea herbicides. Crop Sci 27:948–952

Sharma R, Wahono J, Baral H (2018) Bamboo as an alternative bioenergy crop and powerful ally for land restoration in Indonesia. Sustainability 10:4367

Sinaga H (2013) Employment and income of workers on Indonesian oil palm plantations: food crisis at the micro level. Future Food J Food Agric Soc 1:64–78

Snell KD, Singh V, Brumbley SM (2015) Production of novel biopolymers in plants: recent technological advances and future prospects. Curr Opin Biotechnol 32:68–75

Sohel MSI, Alamgir M, Akhter S, Rahman M (2015) Carbon storage in a bamboo (*Bambusa vulgaris*) plantation in the degraded tropical forests: implications for policy development. Land Use Policy 49:142–151

Sritong C, Kunavongkrit A, Piumsombun C (2012) Bamboo: an innovative alternative raw material for biomass power plants. Int J Innov Manag Technol 3:759

Sun J, Ding S-Y, Doran-Peterson J (2013) Biomass and its biorefinery: novel approaches from nature-inspired strategies and technology. In: Sun J et al (eds) Biological conversion of biomass for fuels and chemicals: explorations from natural utilization systems. Royal Society of Chemistry, London

Swanson EB, Coumans MP, Brown GL, Patel JD, Beversdorf WD (1988) The characterization of herbicide tolerant plants in Brassica napus L. after in vitro selection of microspores and protoplasts. Plant Cell Rep 7(2):83–87

Truong AH, Le TMA (2014) Overview of bamboo biomass for energy production. HAL Id:halshs-01100209. https://halshs.archives-ouvertes.fr/halshs-01100209/document

Van Acker R, Vanholme R, Storme V, Mortimer JC, Dupree P, Boerjan W (2013) Lignin biosynthesis perturbations affect secondary cell wall composition and saccharification yield in *Arabidopsis thaliana*. Biotechnol Biofuels 6:46

Van Acker R, Leple JC, Aerts D, Storme V, Goeminne G, Ivens B, Legee F, Lapierre C, Piens K, Van Montagu MCE, Santoro N, Foster CE, Ralph J, Soetaert W, Pilate G, Boerjan W (2014) Improved saccharification and ethanol yield from field-grown transgenic poplar deficient in cinnamoyl-CoA reductase. Proc Natl Acad Sci U S A 111:845–850

Vasil IK (2008) A history of plant biotechnology: from the cell theory of Schleiden and Schwann to biotech crops. Plant Cell Rep 27:1423–1440

Xie G, Peng L (2011) Genetic engineering of energy crops: a strategy for biofuel production in China. J Integr Plant Biol 53(2):143–150

Xie G, Peng L (2014) Genetic engineering of bioenergy crops toward high biofuel production. In: Wang L (ed) Sustainable bioenergy production. CRC Press, Boca Raton

Xuhe C (2003) Promotion of bamboo for poverty alleviation and economic development. J Bamboo Rattan 2:345–350

Yau YY, Easterling M (2018) Lignocellulosic feedstock improvement for biofuel production through conventional breeding and biotechnology. In: Kumar A, Ogita S, Yau YY (eds) Biofuels: greenhouse gas mitigation and global warming. Springer, New Delhi

Zakir HM, Hasan M, Shahriar SMS, Ara T, Hossain M (2016) Production of biofuel from agricultural plant wastes: corn stover and sugarcane bagasse. Chem Eng Sci 4(1):5–11

Zhou G, Jiang P (2004) Density, storage and spatial distribution of carbon in *Phyllostachy pubescens* forest. Sci Silvae Sin 40:20–24

Zhou X, Jacobs TB, Xue LJ, Harding SA, Tsai CJ (2015) Exploiting SNPs for biallelic CRISPR mutations in the outcrossing woody perennial *Populus* reveals 4-coumarate:CoA ligase specificity and redundancy. New Phytol 208:298–301

Zimnoch-Gozowska E, Marczewski W, Lebecka R, Flis B, Schäfer-Pregl R, Salamini F, Gebhardt C (2000) QTL analysis of new sources of resistance to *Erwinia carotovora* ssp. *atroseptica* in potato done by AFLP, RFLP, and resistance-gene-like markers. Crop Sci 40(4):1156–1167

Most Tanziman Ara Master of Science and Master of Philosophy in Botany from University of Rajshahi, Bangladesh, is doing PhD on Studies on Versatile Methods for Controlling the Development of Shoot and Root Apical Meristems of Bamboo and Some Model Grass Species through Plant Cell Tissue and Organ Culture Techniques in Prefectural University of Hiroshima, Japan.

Nurhidayah Syahira Muhammad Radzi research student at Prefectural University of Hiroshima, in the Plant and Cell Manipulation laboratory for a year, and MSc in the Program of Biological System Sciences from 2018 to present. Her research interest focuses on the improvement of quality and secondary metabolite production of medicinal plants through cell manipulation biotechnology and transformations.

Misaki Nishibe graduated from Faculty of Life and Environmental Sciences, Department of Life Sciences, Prefectural University of Hiroshima. She became a master student at Prefectural University of Hiroshima, in Program of Biological System Sciences from 2018 to the present. Now she has a keen interest in plant biotechnology and mainly focus on plant tissue culture technology to improve morphological and physiological characteristic of target plants.

Shinjiro Ogita has over two decades of experience in the field of plant biotechnology. Since 1992 he started his research career as a master student at the Graduate School of Agriculture, Tokyo University of Agriculture and Technology (TUAT), Japan. In 1997 he received a Ph.D. in Agriculture (subject; sciences of resources and environment) by the United Graduate School of Agriculture, TUAT, Japan (1997). He is an expert in the field of cell and tissue culture, transformation technologies for higher plants. He presently works as a full professor and the director of field sciences center, Prefectural University of Hiroshima (PUH), Japan.

The manufacturer's authorised representative in the EU is Springer
Nature Customer Service Centre GmbH, Europaplatz 3, 69115 Heidelberg,
Germany. If you have any concerns regarding our products, please
contact ProductSafety@springernature.com

Printed and bound by CPI Group (UK) Ltd, Croydon, CR0 4YY
30/09/2026
02225333-0015